METHODS AND TECHNIQUES OF COSTING

METHODS AND TECHNIQUES OF COSTING

M.E. THUKARAM RAO
Head, Department of Commerce
Sri Sathya Sai University
Brindavan Campus
Bangalore–560066

PUBLISHING FOR ONE WORLD
NEW AGE INTERNATIONAL (P) LIMITED, PUBLISHERS
New Delhi • Bangalore • Chennai • Cochin • Guwahati • Hyderabad
Jalandhar • Kolkata • Lucknow • Mumbai • Ranchi
Visit us at www.newagepublishers.com

Published by New Age International (P) Ltd., Publishers
First Edition: 2011

Branches:

- 37/10, 8th Cross (Near Hanuman Temple), Azad Nagar, Chamrajpet, **Bangalore**-560 018. Tel.: (080) 26756823, Telefax: 26756820, E-mail: bangalore@newagepublishers.com
- 26, Damodaran Street, T. Nagar, **Chennai**-600 017. Tel.: (044) 24353401, Telefax: 24351463 E-mail: chennai@newagepublishers.com
- CE-39/1016, Carrier Station Road, Ernakulam South, **Cochin**-682 016. Tel.: (0484) 2377004, Telefax: 4051303, E-mail: cochin@newagepublishers.com
- Hemsen Complex, Mohd. Shah Road, Paltan Bazar, Near Starline Hotel, **Guwahati**-781 008. Tel.: (0361) 2513881, Telefax: 2543669, E-mail: guwahati@newagepublishers.com
- No. 105, 1st Floor, Madhiray Kaveri Tower, 3-2-19, Azam Jahi Road, Nimboliadda, **Hyderabad**-500 027. Tel.: (040) 24652456, Telefax: 24652457, E-mail: hyderabad@newagepublishers.com
- RDB Chambers (Formerly Lotus Cinema) 106A, 1st Floor, S.N. Banerjee Road, **Kolkata**-700 014. Tel.: (033) 22273773, Telefax: 22275247, E-mail: kolkata@newagepublishers.com
- 16-A, Jopling Road, **Lucknow**-226 001. Tel.: (0522) 2209578, 4045297, Telefax: 2204098, E-mail: lucknow@newagepublishers.com
- 142C, Victor House, Ground Floor, N.M. Joshi Marg, Lower Parel, **Mumbai**-400 013. Tel.: (022) 24927869 Telefax: 24915415, E-mail: mumbai@newagepublishers.com
- 22, Golden House, Daryaganj, **New Delhi**-110 002. Tel.: (011) 23262370, 23262368, Telefax: 43551305, E-mail: sales@newagepublishers.com

---

ISBN: 978-81-224-3040-0

Rs. 395.00

C-10-06-4653

Printed in India at Ram Printograph, Delhi.
Typeset at ABRO Enterprises, Delhi.

PUBLISHING FOR ONE WORLD

NEW AGE INTERNATIONAL (P) LIMITED, PUBLISHERS
4835/24, Ansari Road, Daryaganj, New Delhi - 110002
Visit us at **www.newagepublishers.com**

PREFACE

I have immense pleasure in placing into the hands of my esteemed readers this book on "METHODS AND TECHNIQUES OF COSTING". This is the only book which is available under semester system at present. The need for this book covering all the important aspects of Methods and Techniques of Costing has been felt both by teachers and students for a long time. My long experience in teaching, and the knowledge I have acquired over many years, have made it possible for me to satisfy this need.

This book has been divided into three parts. Part-I deals with Methods of Costing, Part-II deals with Systems of Maintaining Cost Accounting and Part-III deals with Cost Control Techniques. Two new chapters, viz. (*a*) Activity Based Costing and (*b*) Target Costing are also included in this book. The book is written in a simple and lucid style. At the end of each chapter, a large number of standard questions of different varieties are given to test the understanding ability of students. A large number of exercises with answers are also given.

The book serves the needs of students preparing for the examinations of B.Com. (Pass and Hons.) BBM, BBA, BA (Corporate Secretaryship), M.Com., MBA and other diploma courses of management.

I shall appreciate comments and suggestions from readers for the improvement of the book.

—Author

CONTENTS

PART II: SYSTEMS OF MAINTAINING COST ACCOUNTING

PART III: COST CONTROL TECHNIQUES

Part I

METHODS OF COSTING

CHAPTER 1

SINGLE OR OUTPUT OR UNIT COSTING

INTRODUCTION

The methods of costing can be broadly classified into two types, *viz.*, specific order costing and operation costing. Single or output or unit costing is a method of costing that belongs to the group of operation costing method. It is defined by the ICMA terminology as "the basic costing method applicable where goods or services result from a series of continuous or repetitive operations or processes to which costs are charged before being averaged over the units produced during the period". This method is known as "single" method of costing as industries adopting this method manufactures, in most cases a single variety of product. It is termed as unit costing because cost units are identical. Though under this method of costing a single variety of product is manufactured it may vary in respect of size, grade, colour etc. The examples of industries which make use of this method of costing are: quarries, brick, sugar, cloth, coal, cement, fisheries, food canning, plantation industry, etc.

FEATURES AND OBJECTS OF OUTPUT COSTING

The features of industries which adopt single or output method of costing are as follows:

(*a*) The production of goods are undertaken on a continuous basis in anticipation of demand both in the short-run and long-run.

(*b*) The goods are produced on a large scale basis.

(*c*) Excepting in newly started industry in all other cases there always exists opening and closing work-in-progress.

(*d*) Cost of production is ascertained at the end of accounting year which may be a financial year or calender year.

(*e*) The cost is ascertained both per unit of finished product and for all the units manufactured.

(*f*) The end products are always homogeneous or uniform in all respects.

(*g*) Equality of costs is one of the basic feature of unit costing. In other words, identical cost unit will have identical cost.

The specific objects of this method of costing are (*a*) To ascertain total cost and unit cost of production, (*b*) Comparing cost of one period with another period to know efficiency or otherwise of it, and (*c*) To ascertain profit or loss from production.

COST COLLECTION

The various elements of cost is accumulated for the entire accounting period. As under this method, the variety of goods manufactured is restricted to one or to a few varieties, the method does not demand maintaining of elaborate records. The accounting procedure can be summarised under the following stages:

(*a*) The accumulation and tabulating various elements of costs for a given period say, a month, quarter, or year.

(*b*) Measurement of output either in terms of number or quantity.

(*c*) Calculating the cost per unit by dividing the total cost by number or quantity produced.

Collection of Materials Cost

The materials cost is accumulated from materials abstract. Alternatively, cost of raw materials consumed is ascertained by adding purchased raw materials during the period to the opening stock and deducting closing stock of materials therefrom. Any normal loss of materials is adjusted by inflating the rate of good units of raw materials.

Collection of Labour and Overhead

The labour cost is obtained from the pay rolls prepared separately for different sections or departments of the factory.

In majority of the cases overheads are charged at a predetermined rate where quotation prices are to be prepared. Otherwise overheads are collected under the heads of factory, office and administration and selling and distribution for the period for which they relate. When more than one variety of product is manufactured the overheads are apportioned on some suitable basis.

TREATMENT OF WORK-IN-PROGRESS AND SCRAP

The incomplete units of production is termed as work-in-progress. It is valued on the basis of material cost, labour cost and factory overheads. The opening stock of work-in-progress is added to the current cost of production (factory cost) and the closing stock of work-in-progress is deducted from the above total. After this treatment is given, the office and administration overheads are added to arrive at office cost.

Alternatively if the work-in-progress is analysed in terms of materials cost, labour cost and overheads, then the amount of materials, wages and overheads included in the work-in-progress in the beginning will be added and that included in work-in-progress at the end, deducted to ascertain the cost of materials, labour and factory overheads respectively. There afterwards no other treatment need be given to the work-in-progress.

The value of "scrap" realised is to be deducted from works cost or works overhead. This is so because all the elements of costs are incurred on scrapped units also.

Any extra expenses incurred on rectifying "defectives" is added to works cost. However, if the cost of rectifying is high due to abnormal reasons it should be transferred to a separate "Defective account".

COST PRESENTATION-PREPARATION OF COST SHEET

The information derived from various records is presented in the form a statement of cost or a cost sheet. According to Walter W. Biggs, "the expenditure which has been incurred upon production for a period, is extracted from the financial books and the stores' records and set out in a memorandum or statement. If this statement is confined to the disclosure of the cost of the units produced during the period, it is termed as a cost sheet".

Problem 1. The Trading and Profit and Loss Account for the year ended 30th June, 2006 is as follows:

To opening stock (500 units @ Rs. 35 each)	17,500	By sales (10250 units)	7,17,500
To materials consumed	2,60,000	By closing stock (250 units @ Rs. 50 each)	12,500
To wages	1,50,000		
To gross profit	3,02,500		
	7,30,000		7,30,000
To factory overhead	94,700	By Gross profit	3,02,500
To office overhead	1,06,000	By interest	250
To selling expenses	55,000	By rent	10,000
To bad debts	4,000		
To goodwill written off	5,000		
To net profit	48,050		
	3,12,750		3,12,750

The cost sheet shows the following:

(*a*) Cost of materials at Rs. 26 per unit and labour cost Rs. 15 per unit produced.

(*b*) Factory overheads are absorbed at 60% of labour cost.

(*c*) Office overheads are absorbed at 20% factory cost.

(*d*) Selling expenses are charged at Rs. 6 per unit.

(*e*) Opening stock of finished goods is valued at Rs. 45 per unit and closing stock as in financial books. Prepare a cost sheet. *(Acharya Nagarjuna University, B.Com., July 2007)*

Solution: ***Working Note***

(1) Calculation of units produced:

= Sales + Closing stock – Opening stock = 10250 + 250 – 500 = 10000 units

COST SHEET

(for the year ended 30-6-2006)

		Amount	*Cost per unit*
Direct materials		2,60,000	26
Direct labour		1,50,000	15
	Prime cost	4,10,000	41
Factory overheads $\frac{60}{100} \times 1,50,000$		90,000	9
	Factory cost	5,00,000	50
Office & Administration oh (20% of 5,00,000)		1,00,000	10
	Office cost	6,00,000	60
Add: Opening stock of finished goods 500 units × Rs. 45		22,500	–
		6,22,500	60
Less: Closing stock of finished goods 250 units × Rs. 50		12,500	–
Cost of goods sold		6,10,000	60
Selling and distribution oh 10250 units × Rs. 6		61,500	6
	Cost of sales	6,71,500	66
	Profit	46,000	4
	Sales	7,17,500	70

Problem 2. A factory can produce 60,000 units per annum at its 100% capacity. The estimated cost of production are as under:

Direct materials	Rs. 3 per unit
Direct labour	Rs. 2 per unit
Fixed overheads	Rs. 1,50,000 per annum
Variable overheads	Rs. 5 per unit

Semi-variable Rs. 50,000 per annum up to 50% capacity and an extra expenses of Rs. 10,000 per annum for every 25% increase in capacity or part thereof.

The factory produces only against orders. If the production programme of the factory is as indicated below and the management desires to ensure a profit of Rs. 1,00,000 for the year. Work out the average selling price at which unit should be quoted.

First 3 months of the year 50% of capacity, remaining 9 months of the year 80% of capacity.

(Bangalore University, B.Com., Nov. 1991)

Solution:

COST SHEET

Direct materials	43,500 units @ Rs. 3 per unit		1,30,500
Direct labour	43,500 units @ Rs. 2 per unit		87,000
	Prime cost		2,17,500
Indirect expenses:			
Fixed		1,50,000	
Variables 43,500 units @ Rs. 5		2,17,500	
Semi-variable:			
For first 3 months at 50% capacity $50{,}000 \times \frac{3}{12}$		12,500	
For 9 months at 80% capacity $70{,}000 \times \frac{9}{12}$		52,500	4,32,500
	Cost of production		6,50,000
	Estimated profit		1,00,000
	Estimated sales		7,50,000

$$\text{Selling price per unit} = \frac{\text{Estimated sales}}{\text{Total units}}$$

$$= \frac{7{,}50{,}000}{43{,}500} = \text{Rs. } 17.24$$

Working Note: Calculation of units produced

Output for first 3 months at 50% capacity

$$60{,}000 \times \frac{50}{100} \times \frac{3}{12} = 7{,}500 \text{ units}$$

For next 9 months 80% capacity

$$60{,}000 \times \frac{80}{100} \times \frac{9}{12} = 36{,}000 \text{ units}$$

Total units produced 43.500 units

Problem 3. A manufacturing company has an installed capacity of 1,20,000 units per annum. The cost structure of the product manufactured is as under

	Rs.
(*i*) Variable cost per unit:	
Materials	8
Labour (subject to a minimum of Rs. 56,000 per month)	8
Overheads	3

(*ii*) Fixed overhead—Rs. 1,68,750 per annum.

(*iii*) Semi-variable overheads Rs. 48,000 per annum at 60% capacity, which increase by Rs. 6,000 per annum for increase of every 10% of the capacity utilisation or any part thereof, for the year as a whole.

The capacity utilisation for the next year is estimated at 60% for two months: 75% for six months and 80% for the remaining part of the year. If the company is planning to have a profit of 25% on the selling price, calculate the selling price per unit. Assume that there are no opening and closing stocks.

(CA. Inter., November 1997)

Solution: Calculation of capacity utilisation for the next year

$$\frac{60}{100} \times \frac{2}{12} \times 1{,}20{,}000 = 12{,}000 \text{ units}$$

$$\frac{75}{100} \times \frac{6}{12} \times 1{,}20{,}000 = 45{,}000 \text{ units}$$

$$\frac{80}{100} \times \frac{4}{12} \times 1{,}20{,}000 = 32{,}000 \text{ units}$$

89,000 units

Calculation of labour cost (subject to a minimum of Rs. 56,000 PM)

For 2 months	=			1,12,000	(Minimum being Rs. 56,000 PM for 2 months it is 56,000 × 2)
For 6 months	=	45,000 × Rs. 8	=	3,60,000	
For 4 months	=	32,000 × Rs. 8	=	2,56,000	
				7,28,000	

Calculation of semi-variable overhead

At 60% capacity	=	48,000
For 80% capacity at Rs. 6,000 for every 10% capacity utilisation	=	12,000
		60,000

COST SHEET

Materials–89,000 units @ Rs. 8 per unit		7,12,000
Labour cost		7,28,000
	Prime cost	14,40,000
Variable overhead–89,000 units @ Rs. 3 per unit		2,67,000
Semi variable overhead		60,000
Fixed overhead		1,68,750
	Factory cost	19,35,750
	Profit $\frac{1}{3}$ on cost price	6,45,250
	Sales	25,81.000
Selling price per unit		$\frac{25{,}81{,}000}{89{,}000}$
		Rs. = 29.00

Problem 4. Enterprising Engineering works manufacture product "useful' having a very good demand in the market. The annual accounts of the concern for the year ending March 31, 1995 reveal the following details.

	Rs.
Purchase of raw materials	7,75,800
Wages paid to workmen	
Direct	3,12,400
Indirect	34,000
Salaries:	
Works	1,30,900
Administrative	1,20,000
Sales staff	2,40,000
Travelling expenses (for sales staff)	35,000
Income-tax paid	2,00,000
Interest on Bank borrowing	15,000
Advertisements	2,20,000
Power	62,000
Fuel	47,500
Maintenance:	
Plant	51,500
Building	50,000
Freight (outwards)	2,05,000
Rent	40,000
Depreciation	
Plant	38,500
Building	25,000
Rates & Taxes	12,000
Sundry office expenses	56,000
Bad debts	30,000
Opening stock:	
Raw materials	55,000
Work-in-progress	1,56,200
Finished goods (4,000 Pcs)	40,000

During the year the concern produced a total output of 1,42,000 pieces out of which 6,000 pieces remained unsold. The stock of raw materials and work-in-progress as on 31st March, 1995 were Rs. 1,35,000 and Rs. 56,800 respectively. Other data available are as under:

(*i*) Expenses relating to building are apportioned between works and administration in the ratio of 80 : 20.

(*ii*) Rent is shared between plant and administration in the ratio of 75 : 25.

(*iii*) Entire rates and taxes are identified as administrative expenses.

(*iv*) The Sales Manager is to receive a commission calculated at the rate of 6% of the net profit.

(*v*) Total sales turnover for the year was Rs. 30 lakhs and bad debts to the tune of 1.5% thereof are considered normal.

Prepare a statement of cost showing (*a*) cost of production in aggregate and per unit (*b*) cost of goods sold and (*c*) cost of sales and profit earned.

(ICWA, Inter., June 1995)

Solution:

COST SHEET OF ENTERPRISING ENGINEERING WORKS
(for the year ending 31st March 1995)

		Total	*per unit*
Opening stock of raw materials	55,000		
Add: Purchases	7,75,800		
	8,30,800		
Less: Closing stock of raw materials	1,35,000		
Cost of materials consumed		6,95,800	4.90
Direct wages		3,12,400	2.20
Prime cost		10,08,200	7.10
Works overheads:			
Indirect wages	34,000		
Power	62,000		
Fuel	47,500		
Maintenance–plant	51,500		
– Building (80%)	40,000		
Rent (75%)	30,000		
Depreciation–plant	38,500		
– Building	20,000		
Salaries	1,30,900		
		4,54,400	3.20
		14,62,600	10.30
Add: Opening stock of work-in-progress		1,56,200	1.10
		16,18,800	11.40
Less: Closing stock of work-in-progress		56,800	0.40
Work cost		15,62,000	11.00
Administration overheads:			
Rates & Taxes	12,000		
Repairs – Building (20%)	10,000		
Rent (25%)	10,000		
Depreciation – Building (20%)	5,000		
Salaries	1,20,000		
Sundry expenses	56,000		
		2,13,000	1.50
Cost of production		17,75,000	12.50

STATEMENT OF COST OF SALES AND PROFIT

	Rs.	*Units*
Opening stock of finished goods	40,000	4,000
Add: Cost of production	17,75,000	1,42,000
	18,15,000	1,46,000
Less: Closing stock of finished goods @ Rs. 12.50 per unit	75,000	6,000
Cost of goods sold	17,40,000	1,40,000

Selling and Distribution overheads:			
Freight	2,05,000		
Travelling expenses	35,000		
Advertisement	2,20,000		
Salaries	2,40,000		
Sales Manager's Commission @ 6% on Net profit	30,000		
Bad debts	30,000		7,60,000
		Cost of sales	25,00,000
		Profit	5,00,000
		Sales	30,00,000

Working Note:

Calculation of Sales Manager's Commission

Let profit = x

Sales manager's commission 6% of x or $\frac{6x}{100}$

Total profit before charging commission

$$x + \frac{6x}{100}$$

or $$x + \frac{6x}{100} = \text{Sales} - \text{Cost of sales excluding sales manager's commission}$$

$$x + \frac{6x}{100} = 30,00,000 - 24,70,000$$

$$x + \frac{6x}{100} = 5,30,000$$

$$100x + 6x = 5,30,000 \times 100$$

$$106x = 5,30,00,000$$

$$x = \frac{5,30.00.000}{106} = 5,00,000$$

Sales Manager's Commission $= \frac{6}{100} \times 5,00,000 =$ Rs. 30,000

Problem 5. Delta Engineering Ltd. produces a uniform type of product and has a manufacturing capacity of 3,000 units per week of 48 hrs. From the cost records of the company, the following data are available relating to output and cost for three consecutive weeks.

Week no.	*Units manufactured*	*Direct materials Rs.*	*Direct labour Rs.*	*Factory overhead (variable and fixed) Rs.*
1	1,200	9,000	3,600	31,000
2	1,600	12,000	4,800	33,000
3	1,800	13,500	5,400	34,000

Assuming that the company charges a profit of 20% on selling price find out the selling price per unit when the weekly output is 2,000 units. *(ICWA, Inter, June 1990)*

Solution: COST SHEET

		Per unit	Total for 2,000 units
Materials (In the last three weeks material per unit has been Rs. 7.50)		7.50	15,000
Labour (In the last three weeks labour cost per unit is Rs. 3)		3.00	6,000
	Prime cost	10.50	21,000
Overhead:			
Variable overhead 5 per unit *(see note 1)*		5.00	10,000
Fixed overhead *(see note 2)*		12.50	25,000
	Total cost	28.00	56,000
	Profit 25% on cost	7.00	14,000
	Selling price	35.00	70,000

Working Note 1: **Calculation of variable overhead**

Week		*Units manufactured*	*Factory overhead*
1		1.200	Rs. 31,000
2		1.600	Rs. 33,000
	(Difference)	400	Rs. 2,000

For 400 units: Rs. 2,000

For 2,000 units – ? $\frac{2{,}000 \times 2{,}000}{400}$ = Rs. 10,000

Verification of the above figure of Rs. 10,000 considering figures of third week.

Week		*Units manufactured*	*Factory overhead*
2		1,600	33,000
3		1,800	34,000
	Diffference	200	1,000

For 200 : 1.000 : 2.000 : x

i.e., $\frac{1.000 \times 2.000}{200}$ = Rs. 10.000

Working Note 2: **Calculation of fixed cost**

Total factory overheads for 1,200 units = 31,000

Less: Total variable overhead for 1,200 units

For 400 : 2,000

For 1,200 : $\frac{2{,}000 \times 1{,}200}{400}$ – 6,000

(output of Ist week)

Fixed overhead 25,000

Fixed overhead by definition remains fixed throughout.

PROBLEMS ON CALCULATION OF MISSING INFORMATION

Problem 6. *(Where selling price and cost of sales is missing):* A firm purchased a plant to manufacture a new product, the cost data for which is given below:

Estimated annual sales	24,000 units
Estimated costs:	
Materials	Rs. 4 per unit
Direct labour	Re. 0.60 per unit

Overheads	Rs. 24,000 per year
Administrative expenses	Rs. 28,800 per year
Selling expenses	15% on sales

Calculate the selling price if profit per unit is Rs. 1.02. *(ICWA Inter., Dec. 1990)*

Solution: COST SHEET

		Cost per unit	*Amount*
Materials		4.00	96,000
Direct labour		0.60	14,400
	Prime cost	4.60	1,10,400
Overheads		1.00	24,000
	Factory cost	5.60	1,34,400
Administrative expenses		1.20	28,800
	Office cost	6.80	1,63,200
Selling & distribution overheads		1.38	33,120
	Cost of sales	8.18	1,96,320
	Profit	1,0200	24,480
	Sales	9.20	2,20,800

Working Note: **Calculation of selling price**

Let the selling price be $= x$ Rs.

∴ Sales value $= 24{,}000 \times x$ Rs.

Office cost + Selling expense + Profit = Sales

$$1{,}63{,}200 + \left(\frac{15}{100} \times 24{,}000\right) + 24{,}480 = 24{,}000\ x$$

$$1{,}63{,}200 + 3{,}600\ x + 24{,}480 = 24{,}000\ x$$

$$1{,}63{,}200 + 24{,}480 = -\ 3{,}600\ x + 24{,}000\ x$$

$$1{,}87{,}680 = 20{,}400\ x$$

$$x = \frac{1{,}87{,}680}{20{,}400} = 9.2$$

$$\text{Sales} = 24{,}000 \times 92 = 2{,}20{,}800$$

$$\text{Selling and distribution overhead} = \frac{15}{100} \times 2{,}20{,}800 = 33{,}120$$

Problem 7. *(When the quantity of scrap, value of closing stock of raw materials and value of finished goods are missing):* The following particulars relating to the year 2009 have been taken from the books of a chemical works manufacturing and selling a chemical mixture :

	Kg.	*Rs.*
Stock on 1-1-2009:		
Raw materials	2,000	2,000
Finished mixture	500	1,750
Factory stores		7,250
Purchases:		
Raw materials	1,60,000	1,80,000
Factory stores		24,250
Sales		
Finished mixture	1,53,050	9,18,000
Factory scrap		8,170
Factory wages		1,78,650

Power		30,400
Depreciation of machinery		18,000
Salaries:		
Factory		72,220
Office		37, 220
Selling		41,500
Expenses:		
Direct		18,500
Office		18,200
Selling		18,000
Stock on 31-12-86:		
Raw materals	1,200	?
Finished mixture	450	?
Factory stores		5,550

The stock of finished mixture at the end of 2009 is to be valued at the factory cost of the mixture for that year. The purchase price of raw materials remained unchanged throughout 2009.

Prepare a statement giving the maximum possible information about cost and its break up for the year 2009.

Solution:

COST SHEET

(for the year ending 31.12.2009)

		Quantity (Kg.)	*Amount (Rs.)*
Opening stock of raw materials		2,000	2,000
Add:	Purchases	1,60,000	1,80,000
		1,62,000	1,82,000
Less:	Closing stock of raw materials (*see note 1*)	1,200	1,350
	Cost of materials consumed	1,60,800	1,80,650
	Factory wages		1,78,650
	Direct expenses		18,500
	Prime cost		3,77,800
	Factory overhead:		
	Factory stores consumed (*see note 2*)		25,950
	Power		30,400
	Depreciation of machinery		18,000
	Factory salaries		72,220
			5,24,370
Less:	Factory scrap sold (*see note 3*)	7,800	8,170
	Works cost	1,53,000	5,16,200
	Office salaries		37,220
	Office expenses		18,200
	Office cost	1,53,000	5,71,620
Add:	Opening stock of finished mixture	500	1,750
		1,53,500	5,73,370
Less:	Closing stock of finished mixture (*see note 4*)	450	1.518
	Cost of finished mixture sold	1,53,050	5,71,852

Selling & distribution overhead: saaries of selling dept		41,500
Selling expenses		18,000
Cost of sales		6,31,352
Profit		2,86,648
Sales	1,53,050	9,18,000

Working Note 1: **Calculation of value of closing stock of raw materials**

For 1,60,000 kgs. of purchases, the value is 1,80,000

For 1,200 kgs. – ? (Purchases price of raw materials remains same throughout 2009)

$$\frac{1,80,000 \times 1,200}{1,60,000} = \text{Rs. } 1,350$$

Working Note 2: **Calculation of value of factory stores consumed**

Opening stock of factory stores	Rs.	7,250
Add: Purchases of factory stores	Rs.	24,250
		31,500
Less: Closing stock of factory stores		5,550
	Rs.	25,950

Working Note 3: **Calculation of quantity of scrap sold using reverse method**

	Kg.
Quantity of finished mixture sold	1,53,050
Add: Closing stock of finished mixture	450
	1,53,500
Less: Opening stock of finished mixture	500
	1,53,000
Quantity of scrap sold (balancing figure)	7,800
Quantity of materials consumed	1,60,800

Working Note 4: **Calculation of value of closing stock of finished mixture: (at works cost)**

For 1,53,000 kgs. the works cost is Rs. 5,16,200

For 450 kgs – ? $= \frac{5,16,200 \times 450}{1,53,000}$ = Rs. 1,518

Problem 8. *(When purchases is missing):* The books of Adarsh Manufacturing Company present the following data for the month of April 1992:

Direct labour cost Rs. 17,500 being 175% of works overhead

Cost of goods sold excluding administrative expenses Rs. 56,000

Inventory accounts showed the following opening and closing balances.

	April 1 (Rs.)	April 30 (Rs.)
Raw materials	8,000	10,600
Work-in-progress	10,500	14,500
Finished goods	17,600	19,000

Other data are:

Selling expenses	3,500
General & Administration expenses	2,500
Sales for the month	75,000

You are required to
(1) Compute the value of materials purchased.
(2) Prepare a cost statement showing the various elements of cost and also the profit earned.

(CA (Inter). May, 1992)

Solution:

Statement showing value of materials purchased

Cost of goods sold		56,000
Add: Closing stock of finished goods		19,000
		75,000
Less: Opening stock of finished goods		17,600
	Cost of goods manufactured	57,400
Add: Closing stock of work-in-progress		14,500
		71,900
Less: Opening stock of work-in-progress		10,500
	Work cost	61,400
Less: Factory overhead $\frac{100}{175}$ of direct labour cost		10,000
	Prime cost	51,400
Less: Direct labour		17,500
	Raw materials consumed	33,900
Add: Closing stock of raw materials		10,600
		44,500
Less: Opening stock of raw materials		8,000
Value of materials purchased		36,500

COST SHEET

Cost of raw materials consumed		33,900
Direct labour cost		17,500
	Prime cost	51,400
Add: Factory overhead		10,000
		61,400
Add: Opening stock of work-in-progress		10,500
		71,900
Less: Closing stock of work-in-progress		14,500
	Net works cost	57,400
Add: Opening stock of finished goods		17,600
		75,000
Less: Closing stock of finished goods		19,000
	Cost of goods sold	56,000
Add: General & administration expenses		2,500
Add: Selling expenses		3,500
	Cost of sales	62,000
	Profit	13,000
	Sales	75,000

Problem 9. *(When prime cost is missing):* From the following information prepare a cost statement showing maximum possible breakup of cost and total profit:

		Rs.
Sales for January 1990		30,00,000
Cost of goods sold		24,80,000
Administration expenses		1,80,000
Selling expenses		40,000
	1.1.1990 (Rs.)	31.1.1990 (Rs.)
Raw materials stock	3,20,000	4,00,000
Work-in-progress	3,20,000	4,80,000
Finished goods	4,20,000	3,40,000

Direct wages were 30% of prime cost
Raw materials consumed were 50% of prime cost
Direct expenses were 20% of prime cost
Factory overheads were 20% of prime cost

(University of Bombay, B.Com., May 1990)

Solution:

COST SHEET
(for the period ending 31.1.1990)

	Opening stock of raw materials	3,20,000	
Add:	Purchases of raw materials	10,71,667	
		13,91,667	
Less:	Closing stock of raw materials	4,00,000	
	Cost of raw materials consumed		9,91,667
	Wages		5,95,000
	Direct expenses		3,96,667
		Prime cost	19,83,334
	Factory overheads		3,96,666
		Works cost	23,80,000
Add:	Opening stock of work-in-progress		3,20,000
			27,00,000
Less:	Closing stock of work-in-progress		4,80,000
			22,20,000
	Office and administration expenses		1,80,000
		Office cost	24,00,000
Add:	Opening stock of finished goods		4,20,000
			28,20,000
Less:	Closing stock of finished goods		3,40,000
		Cost of goods sold	24,80,000
	Selling expenses		40,000
		Cost of sales	25,20,000
		Profit	4,80,000
		Sales	30,00,000

Note: In this problem prime cost is the basis for calculating various elements of cost Prime cost in this problem is calculated as under:

		Rs.
Let prime cost be		100
Add:	Factory overhead	20
	Works cost	120

Calculation of factory overheads on work cost

$$\frac{20}{120} \times 23,80,000 = 3,96,666$$

Prime cost = Works cost– Factory overheads

= 23,80,000 – 3,96,666

= Rs. 19,83,334

PROBLEMS ON OUTPUT COSTING AS APPLIED TO VARIOUS INDUSTRIES

Problem 10. *(Output costing as applied to plastic industry):* From the following particulars supplied by Pure Plastic Ltd. you are required to prepare a monthly cost sheet showing profit for 1,000 plastic spoons. The opening stock was valued at the same price per 1,000 spoons of the production of the month concerned:

Materials	— Plastic granules	1,400 kgs @ Rs. 5 per kg.
	— Stores	Rs. 5,000
Labour	— Direct	Rs. 16,000
	— Indirect	Rs. 3,000
Overheads	— Works	25% of direct labour
	— Office	10% of works cost

Production for the month of October 2009 was 10,00,000 spoons.
Sales for the month: 9,00,000 spoons @ Rs. 50 per 1,000 spoons.
Stock at the end of the month: 3,00,000 spoons.

Solution:

COST SHEET OF PURE PLASTIC LTD.

	Cost per 1,000 spoons	*Total for 10,00,000 spoons*
Raw materials (*see note 1*)	7,00	7,000
Labour (*see note 2*)	16,00	16,000
Prime cost	23.00	23.000
Stores (*see note 3*)	5.00	5,000
Indirect labour (*see note 4*)	3.00	3,000
Works overhead (*see note 5*)	4.00	4,000
Works cost	35.00	35,000
Office overhead	3.50	3,500
Office cost	38.50	38,500
Add: Opening stock of spoons (*see note 6*)	7.70	7,700
	46.20	46,200
Less: Closing stock of spoons (*see note 7*)	11.55	11,550
Cost of goods sold	34.65	34,650
Profit	10.35	10,350
Sales (*see note 8*)	45.00	45,000

Working Note 1: **Calculation of material cost per 1,000 spoons**

for 10,00,000 spoons: Rs. 7,000

$$\text{for } 1{,}000 \text{ spoons} \rightarrow ? = \frac{7{,}000 \times 1{,}000}{10{,}00{,}000} = \text{Rs. } 7.00$$

Working Note 2: **Calculation of labour cost per 1,000 spoons**

for 10,00,000 spoons: Rs. 16,000

$$\text{for } 1{,}000 \text{ spoons} \rightarrow ? = \frac{16{,}000 \times 1{,}00}{10{,}00{,}000} = \text{Rs. } 16.00$$

Working Note 3: **Calculation of stores cost per 1,000 spoons**

for 10,00,000 spoons: Rs. 5,000

$$\text{for } 1{,}000 \text{ spoons} \rightarrow ? = \frac{5{,}000 \times 1{,}000}{10{,}00{,}000} = \text{Rs. } 5.00$$

Working Note 4: **Calculation of indirect labour per 3,000 spoons**

for 10,00,000 spoons: Rs. 3,000

$$\text{for } 1{,}000 \text{ spoons} \rightarrow ? = \frac{3{,}000 \times 1{,}000}{10{,}00{,}000} = \text{Rs. } 3.00$$

Working Note 5: **Calculation of works overheads per 1,000 spoons**

for 10,00,000 spoons: Rs. 4,000

$$\text{for } 1{,}000 \text{ spoons} \rightarrow ? = \frac{4{,}000 \times 1{,}000}{10{,}00{,}000} = \text{Rs. } 4.00$$

Working Note 6: **Calculation of value of opening stock of spoons**

for 1,000 spoons cost of production is Rs. 38.50

$$\text{for } 2{,}00{,}000 \text{ spoons} \rightarrow ? = \frac{2{,}00{,}000 \times 38.50}{1{,}000} = \text{Rs. } 7{,}700$$

Opening stock = Closing stock + sales – production

= 3,00,000 + 9,00,000 – 10,00,000 = 2,00,000

Working Note 7: **Calculation of value of clolsing stock of spoons**

for 1,000 spoons the cost of production is 38.50

$$\text{for } 3{,}00{,}000 \text{ spoons} \rightarrow ? = \frac{3{,}00{,}000 \times 38.50}{1{,}000} = \text{Rs. } 11{,}550$$

Working Note 8: $\frac{50}{1000} \times 9{,}00{,}000 = 45{,}000$

Working Note 9: **Calculation of selling price**

for 10,00,000 spoons 45,000

$$\text{for } 1000 = \frac{45000 \times 1000}{10{,}00{,}000} = 45$$

Problem 11. *(Output costing as applied to brick manufacturing industries):* From the following information prepare a monthly cost sheet of the Hard Brickworks showing cost per 1,000 bricks. Also find out the profit per 1,000 bricks.

Materials used:

Lime	895 tonnes @ Rs. 30 per tonne
Coal	820 tonnes @ Rs. 25 per tonne

Sand	Rs. 10 per 1,000 bricks made
Others	Rs. 700
Labour:	
Sand digging	Rs. 1,800
Brick making	Rs. 8,000
Factory overhead	25% of direct charges
Office overhead	10% of factory cost
Bricks sold	35,00,000 @ Rs. 20 per 1,000 bricks
Stock of bricks at the beginning of the month	1,00,000
Stock of bricks at the end of the month	6,00,000

Solution:

COST SHEET OF HARD BRICKWORKS

	Cost per 1,000 bricks	*Amount for 40,00,000 bricks*
Material used:		
Lime	6.712	26,850
Coal	5.125	20,500
Sand	10.000	40,000
Others	0.175	700
Total materials cost	22.012	88,050
Labour Sand digging	0.450	1,800
Brick making	2.000	8,000
Prime cost	24.426	97,850
Factory overheads		
25% direct charges *i.e.,* prime cost	6.115	24,462
Factory cost	30.577	1,22,312
Office overheads		
10% of factory cost	3.058	12,231
Office cost or cost of production	33.635	1,34,543

STATEMENT SHOWING PROFIT FOR 1,000 BRICKS

	No. of bricks	*Per 1,000 bricks*	*Total*
Opening stock	1,00,000	33,635	3363.50
Add: Cost of production	40,00,000	–	1,34,540.00
	41,00,000	33,635	1,37,903.50
Less: Closing stock	6,00,000	–	20,181.00
Cost of goods sold	35,00,000	33,635	1,17,222.50
Loss			47,222.50
Sales			70,000.00

Working Note 1: **Calculation of materials cost, lime per 1,000 bricks**

for 40,00,000 bricks, materials cost is 26,850

for 1,000 bricks → ? $= \dfrac{26,850 \times 1,000}{40,00,000}$ = Rs. 6.712

Working Note 2: **Calculation of coal consumed per 1,000 bricks**

for 40,00,000 bricks coal cost is Rs. 20,500

for 1,000 bricks → ? $= \frac{20,500 \times 1,000}{40,00,000}$ = Rs. 5.125

Working Note 3: **Calculation of sand cost for 40,00,000 bricks**

for 1,000 bricks: sand cost is 10

for 40,00,000 bricks → ? $= \frac{40,00,000 \times 10}{1,000}$ = Rs. 40,000

Working Note 4: **Calculation of other cost per 1,000 bricks**

for 40,00,000 bricks, other costs is Rs. 700

for 1,000 bricks → ? $= \frac{700 \times 1,000}{40,00,000}$ = Rs. 0.175

Working Note 5: **Calculation of labour cost per 1,000 bricks**

for 40,00,000 bricks, labour cost is Rs. 1,800

for 1,000 bricks → ? $= \frac{1,800 \times 1,000}{40,00,000}$ = Rs. 0.450

Working Note 6: **Calculation of factory overheads per 1,000 bricks**

for 40,00,000 bricks, factory overhead is Rs. 24,462

for 1,000 bricks → ? $= \frac{24,462 \times 1,000}{40,00,000}$ = Rs. 6.115

Working Note 7: **Calculation of office overheads per 1,000 bricks**

for 40,00,000 bricks, office overheads is Rs. 12,231

for 1,000 bricks → ? $= \frac{12,231 \times 1,000}{40,00,000}$ = Rs. 3.058

Working Note 8: **Calculation of value of opening stock**

for 1,000 bricks: cost of production is Rs. 33,635

for 1,00,000 bricks → ? $= \frac{33,635 \times 1,00,000}{1,000}$ = Rs. 3,363.50

Working Note 9: **Calculation of value of closing stock**

for 1,000 bricks cost of production is Rs. 33,635

for 6,00,000 bricks → ? $= \frac{33,635 \times 6,00,000}{1,000}$ = Rs. 20,181

Working Note 10: **Calculation of sales**

for 1,000 bricks sale price is Rs. 20

for 35,00,000 bricks → ? $= \frac{20 \times 35,00,000}{1,000}$ = Rs. 70,000

Working Note 11: **Number of bricks manufactured**

Number of bricks sold	35,00,000
Add: Closing stock	6,00,000
	41,00,000
Less: Opening stock	1,00,000
Bricks manufactured	40,00,000

Problem 12. *(Output costing as applied to quarries):* The following particulars relate to a manufacturing company:

Slates quarried: 10,000 tons on which a royalty @ Rs. 2 per ton paid

	Rs.
Wages	40,000
Coal	10,000
Internal transport	20,000
Works overhead—20% on direct charges	
Office overhead—10% on works cost	
Cost of removal of overburden @ 0, 01 paise per slate manufactured	
Opening stock	10,000 slates
Manufactured slates purchased	20,000 @ Re. 1 per slate
Slates sold	1,00,000 @ Rs. 2 per slate
Closing stock	30,000 slates
Selling expenses	0.10 paise per slate sold

Prepare a cost sheet showing the cost per unit of each and every element

Solution:

COST SHEET

	Cost per unit	*Total for 1,00,000 slates*
Royalty on slates quarried—10,000 tons @ Rs. 2 per ton	0.20	20,000
Wages	0.40	40,000
Coal	0.10	10,000
Internal transport	0.20	20,000
Direct cost or charge	0.90	90,000
Works overhead—20% on direct charges $\frac{20}{100} \times 90{,}000$	0.18	18.000
Cost of removal of overburden @ 0.01 paise on 1,00,000 slates	0.01	1,000
Works cost	1.09	1,09,000
Office overhead — 10% of works cost	0.10	10,900
Cost of production	1.19	1,19,900

Statement showing profit

	Number of slates quarried	*Amount*
Opening stock	10,000	11,900
Add: Quarried quantity	1,00,000	1,19,900
Add: Purchases of manufactured slates	20,000	20,000
	1,30,000	1,51,890
Less: Closing stock of slates	30,000	35,700
Cost of goods sold	1,00,000	1,16,190
Selling expenses — 1,00,000 slates @ 0.10 paise		10,000

	Cost of sales		1,26,190
	Profit		73,810
	Sales		2,00,000
Note:	Calculation of number of slates quarried		
Sale of slates			1,00,000
Add:	Closing stock		30,000
			1,30,000
Less:	Opening stock	10,000	
Less:	Slates purchased	20,000	30,000
	Number of slates quarried		1,00,000

Problem 13. *(Cost sheet when certain rate of return is expected on share capital):* The government of India has instituted a dual pricing system in the industry in which your organisation operates. You are the head of the costing division of Rajan Taxtiles Co. Ltd. Your company produces a standard type of cloth 50% of which is procured by the Government at a price of Rs. 4 per metre. You are required by the managing director of your company to suggest a suitable price for the cloth to be sold in the open market. Production during 1990-91 has been 20,00,000 meteres of cloth. Relevant information is given below:

	Rs.
Cotton consumed	10,00,000
Direct labour in factory	10,00,000
Carriage inwards	50,000
Indirect labour in factory	4,00,000
Salary of works manager	2,50,000
Water, power and local taxes (factory)	5,00,000
Dyeing, bleaching, etc.	10,00,000
Excise and other taxes	30,00,000
Depreciation of factory	2,00,000
Depreciation of office	1,00,000
Miscellaneous expenses of factory	1,00,000
Miscellaneous expenses of office	1,00,000
Purchase of computer (office)	20,00,000
Purchase of furniture and machines (office)	5,00,000
Expenditure on sales department	4,00,000
Dividends paid	1,22,00,000
Office salaries	10,00,000
Director's fee	2,00,000
Advertisement & publicity	10,00,000
Salary to M.D.	1,00,000
Commission paid on sales	10,00,000
Commission paid to foreign buyers	1,00,000
Packing and forwarding (sales)	2,00,000

The following further information is made available:

(i) The company expects a fair return of 20% on its paid-up capital

(ii) The paid up capital of the company is Rs. 1,00,00,000

(iii) Marketing expenses outstanding are Rs. 1,00,000

Suggest the open market price after preparing a cost sheet

(Sri Sathya Sai University, B.Com. March 1993)

Solution:

COST SHEET OF RAJAN TEXTILES CO. LTD.

Cotton consumed		10,00,000
Direct labour cost		10,00,000
Carriage inwards		50,000
Prime cost		20,50,000
Excise & other duties	30,00,000	
Indirect labour	4,00,000	
Salary of works manager.	2,50,000	
Water, power, etc.	5,00,000	
Dyeing & bleaching	10,00,000	
Depreciation of factory	2,00,000	
Miscellaneous factory expenses	1,00,000	54,50,000
Factory cost		75,00,000
Depreciation of office	1,00,000	
Miscellaneous office expenses	1,00,000	
Salary to M.D.	1,00,000	
Office salaries	10,00,000	
Director's fee	2,00,000	
		15,00,000
Office cost		90,00,000
Expenditure onsales dept.	4,00,000	
advertisement	10,00,000	
Commission paid	10,00,000	
Commission to foreigners	1,00,000	
Packing	2,00,000	
Marketing expenses o/s	1,00,000	28,00,000
Cost of sales		1,18,00,000
Expected rate of return		20,00,000
Sales		1,38,00,000

Working Note:

Cost of production and sale of cloth per meter = $\frac{1,18,00,000}{20,00,000}$	=	5.9
Less: Government procurement price	=	4.0
Loss per meter		1.9
Total loss on 10,00,000 mts. @ Rs. 1.9 per meter		19,00,000
Fair return on capital (20% on 1 crore)		20,00,000
		39,00,000

Amount to be recovered per meter

$$\frac{39,00,000}{10,00,000} = 3.90$$

Thus the price of remaining 10,00,000 mts. should be fixed up as follows:

Cost of sale per meter	5.90
Amount to be recovered per meter	3.90
	9.80

Note: Purchase of computers, furnitures and equipment are assets. They will not come in cost sheet. Dividend paid is a matter of pure finance. It will not affect cost of production and hence excluded from cost sheet.

PREPARATION OF COMPARATIVE COST SHEET

A comparative cost sheet can be prepared for any of the following purposes:

(*a*) To know the comparative profitability of two or more than two lines of existing products.

(*b*) To know the cost of production, profit and sales of two or more than two products over two different period of time.

(*c*) To compare the actual cost of production with predetermined cost to know the efficiency or otherwise.

(*d*) To compare the profit of one line of product with the existing product and thereby to decide whether a new line of product could be introduced.

(*e*) To evaluate the performance of different types of machineries.

Problem 14. *(On comparative cost sheet):* M/s Vidya Pen Company manufactures two types of pens 'Sharada' and 'Viveka'. The particulars for the year ended 31st March, 1999 were as follows:

	Rs.
Direct materials	5,00,000
Direct wages	2,25,000
Direct expenses	75,000
Total sales	10,00,000

There was no work-in-progress at the beginning or at the end of the year. On the study it is ascertained that:

(1) Direct material per unit in 'Sharada pen' consists twice as much as that in type 'Viveka pen'.

(2) The direct wages per unit for 'Viveka pen' were 40% of those for 'Sharada pen'.

(3) Direct expenses were same per unit for Viveka as well as 'Sharada pen'.

(4) Factory overhead were 20% of the prime cost.

(5) Administrative overhead were 50% of direct wages.

(6) 2,500 units of Sharada pen were produced of which 2,000 were sold and 5,000 units of Viveka pen were produced of which 4,000 were sold, during the year.

(7) Selling overheads were Rs. 8 per unit for Sharada pen and Rs. 9 per unit for Viveka pen.

(8) Selling price per unit for Sharada pen was Rs. 250 and Viveka pen was Rs. 125 respectively.

You are required to prepare a statement showing cost and profit in total as well as per unit for Sharada pen and Viveka pen. *(University of Bombay, B.Com., October 1999)*

Solution:

COST SHEET OF VIDYA PEN CO.

(for the year ended 31st March)

	Sharada Pen		*Viveka Pen*	
	Amount	*Cost per pen*	*Amount*	*Cost per pen*
Direct materials	2,50,000	100	2,50,000	50
Direct wages	1,25,000	50	1,00,000	20
Direct expenses	25,000	10	50,000	10
Prime cost	4,00,000	160	4,00,000	80
Factory overheads 20% of Prime cost	80,000	32	80,000	16
Works cost	4,80,000	192	4,80,000	96
Administrative overhead 50% of direct wages	62,500	25	50,000	10
Cost of production	5,42,500	217	5,30,000	106
Less: Closing stock of finished goods	1.08,500	–	1,06,000	–
Cost of goods sold	4,34,000	217	4,24,000	106
Selling overheads	16,000	8	36,000	9
Cost of sales	4,50,000	225	4,60,000	115
Profit	50,000	25	40,000	10
Sales	5,00,000	250	5,00,000	125

Working Notes:

(1) Calculation of direct materials:

Let material used per unit in Viveka be Rs. x

Material used per unit in 'Sharada' = $2x$

$$2x\ (2{,}500) + x\ (5{,}000) = 5{,}00{,}000$$

$$10{,}000x = 5{,}00{,}000$$

$$x = 50$$

per unit in Viveka = Rs. 50

per unit in Sharada = Rs. 100

(2) Calculation of direct wages:

Let direct wages per unit in 'Sharda' be Rs. x

$\therefore$ Direct wages per unit in 'Viveka' = $\frac{40}{100}x$ or $0.4x$

$$x\ (2{,}500) + 0.4x\ (5{,}000) = 2{,}25{,}000$$

$$2{,}500x + 2{,}000x = 2{,}25{,}000$$

$$4{,}500x = 2{,}25{,}000$$
$$x = \text{Rs. } 50$$

Wages per unit in Sharada = Rs. 50

Wages per unit in Viveka $= \frac{40}{100} \times 50$

= Rs. 20

(3) Calculation of direct expenses:

per unit: $\frac{75{,}000}{7{,}500}$ = Rs. 10

(4) Valuation of closing stock of finished goods

	Sharada	*Viveka*
Quantity produced	2,500	5,000
Less: Quantity sold	2,000	4,000
Qty. in closing stock	500	1,000
Value of closing stock		
Sharada 500 × 217	1,08,500	
Viveka 1,000 × 106		1,06,000

Problem 15. A & Co. manufactures two types of products viz. A and B. The following information is available for the year ended 31st March, 1998.

	Rs.
Direct material	6,75,000
Direct wages	9,90,000
Works overheads	1,95,000

(1) Direct material used per unit in product A were 3 times that of product B.

(2) Direct wages per unit in product B where $\frac{2}{3}$ that of product A.

(3) Works overhead per unit were the same for both the products.

(4) Administration overheads were 100% of the prime cost in each of the products.

(5) Selling and distribution cost per unit was Rs. 6 for both A and B.

(6) 35,000 units of product A were produced, out of which 32,000 units were sold @ Rs. 100 per unit.

(7) 30,000 units of product B were produced, out of which 25,000 units were sold @ Rs. 65 per unit.

Prepare cost sheet showing total cost and cost per unit for both the products.

(University of Bombay, B.Com., October 1998)

Solution:

COST SHEET

(for the year ended 31st March, 1998)

	Product A		Product B	
	Per unit	*Total*	*Per unit*	*Total*
Direct materials	15	5,25,000	5	1,50,000
Direct wages	18	6,30,000	12	3,60,000
Prime cost	33	11,55,000	17	5,10,000
Factory overhead	3	1,05,000	3	90,000
Works cost	36	12,60,000	20	6,00,000
Administration overhead	33	11,55,000	17	5,10,000
Cost of production	69	24,15,000	37	11,10,000
Less: Closing stock	–	2,07,000	–	1,85,000
Cost of goods sold	69	22,08,000	37	9,25,000
Selling & Distribution overheads	6	1,92,000	6	1,50,000
Cost of sales	75	24,00,000	43	10,75,000
Profit	25	8,00,000	22	5,50,000
Sales	100	32,00,000	65	16,25,000

Working Notes:

(1) Calculation of direct materials:

Let materials cost per unit of product B = x

$\therefore$ Materials cost per unit of product A = $3x$

$\therefore$ $3x \times 35{,}000 + x \times 30{,}000 = 6{,}75{,}000$

$1{,}05{,}000x + 30{,}000x = 6{,}75{,}000$

$1{,}35{,}000x = 6{,}75{,}000$

$$x = \frac{6{,}75{,}000}{1{,}35{,}000} = 5$$

Product A $3x = 15$

$\therefore$ Materials for A = $35{,}000 \times 15 = 5{,}25{,}000$

Materials for B = $30{,}000 \times 5 = 1{,}50{,}000$

(2) Calculation of direct wages:

Let wages per unit of product A = x

$\therefore$ Wages per unit of product B = $\frac{2}{3}x$

$= x \times 35{,}000 + \frac{2}{3}x \times 30{,}000 = 9{,}90{,}000$

$= 35{,}000x + 20{,}000x = 9{,}90{,}000$

$= 55{,}000x = 9{,}90{,}000$

$$x = \frac{9,90,000}{55,000} = 18$$

$$\therefore \frac{2}{3} x = 12$$

∴ Wages for products = 35,000 × 18 = 6.30,000

Wages for product B = 30,000 × 12 = 3,60,000

9,90,000

(3) Calculation of closing stock valuation:

	Product A	*Product B*
Units produced	35,000	30,000
Units sold	32,000	25,000
Units in stock	3,000	5,000
Value of closing stock		
A — 3,000 units @ Rs. 69	2,07,000	
B — 5,000 units @ Rs. 37		1,85,000

Problem 16. M/s Vishal manufacturing company manufactures two types of products viz., A and B. The information for the year ended on 31st March, 1998 is as under:

Particulars	Products A	Products B
	Rs.	Rs.
Direct material per unit	100	120
Direct labour per unit	60	50
Direct expenses per unit	40	80

Additional Information:

(1) Factory expenses are charged at 20% of prime cost.

(2) Office expenses are charged at 25% of works cost.

(3) 2,000 units of product A were produced of which 1500 units were sold and 5,000 units of product B were produced of which 4,500 units were sold.

(4) Selling expenses are Rs. 15 per unit for product A and Rs. 20 per unit for product B.

(5) Company charges a profit at 20% on sales for both the products.

Prepare a cost sheet showing the cost and profit in total as well as in per unit.

(University of Bombay, B.Com., April 1998)

Solution:

COMPARATIVE COST SHEET

(M/s Vishal Manufacturing Company)

	Product A Total	*Product A Per unit*	*Product B Total*	*Product B Per unit*
Direct materials	2,00,000	100	6,00,000	120
Direct labour	1,20,000	60	2,50,000	50
Direct expenses	80,000	40	4,00,000	80
Prime cost	4,00,000	200	12,50,000	250

Factory overheads 20% of prime cost	80,000	40	2,50,000	50
Works cost	4,80,000	240	15,00,000	300
Office overheads 25% of works cost	1,20,000	60	3,75,000	75
Cost of production	6,00,000	300	18,75,000	375
Less: Closing stock	4,50,000	–	1,87,500	–
Cost of Goods sold	4,50,000	300	16,87,500	375
Selling & Distribution overheads	22,500	15	90,000	20
Cost of sales	4,72,500	315	17,77,500	395
Profit @ 20% on Sales or 25% on cost	1,18,125	78.75	4,44,375	98,75
Sales	590625	393.75	22,21875	493.75

Working Notes:

(1) Calculation of materials:

A = 100 × 2.000 units = 2,00,000

B = 120 × 5,000 units = 6,00,000

(2) Calculation of wages:

A = 60 × 2,000 units = 1,20,000

B = 50 × 5,000 units = 2,50,000

(3) Calculation of selling overheads:

A = 15 × 1,500 units = 22,500

B = 20 × 4,500 units = 90,000

PREPARATION OF TENDER STATEMENT

Very often the purchase of goods and component parts by the buyers are based on the tender price or quotation price quoted by the supplier, *i.e.,* manufacturer. This is to take advantage of the lowest price quoted by the manufacturer. So it is implied that to increase the sales and to attract more customers towards business it is very essential to prepare the quotation price very carefully. At the same time once the tender price is issued it cannot be reversed. Any increased, quotation price leads to rejection of the tender and consequently loss of orders. Any low quotation price results in a loss to the business. Therefore care must be taken to prepare tenders based on actual facts and in precise terms.

The cost accountant plays a significant role in preparing the tender price. First of all, he has to collect all the information relating to cost of materials, wages and other expenses. The exact requirements of the various inputs can be obtained from production planning department. If necessary, advise can also be obtained from industrial engineer. The estimated price of raw materials can be known from market survey. The labour cost for manufacturing so many units can be known from previous years records and providing for additional charges in case of revision of wage rates. Similarly direct expenses and overheads are worked out based on past experience and considering new factors. Wherever it is necessary overheads may be classified into fixed and variable to know the impact of variable overhead. Fixed overheads should be critically examined to ascertain the possible increase for the new work load. The office and administration overheads can be charged on a predetermined basis. Once the estimated cost is prepared the last step is to add a margin of profit to ascertain the quotation price or tender price.

Problem 17. (Tender Statement) *In respect of a factory the following particulars have been extracted for the year 2006.*

	Rs.
Cost of materials	6,00,000
Wages	5,00,000
Factory overheads	3,00,000
Administration overheads	3,36,000
Selling overheads	2,24,000
Distribution overheads	1,40,000
Profit	4,20,000

A work order has to be executed in 2007 and the estimated expenses are materials Rs. 8,000 wages Rs. 5,000.

Assume that in 2007 the rate of factory overheads had gone up by 20%, distribution overheads have gone down by 10% and selling and administration overheads have each gone up by 15%: Factory overheads are based on wages. Administration, selling and distribution overheads are based on factory cost.

At what price should the product be sold so as to earn the same rate of profit on the selling price as in 2006. *(Sri Sathya Sai University, B.Com. (Hons) October 2007)*

Solution:

COST SHEET FOR THE YEAR 2006

Materials cost		6,00,000
Wages		5,00,000
	Prime cost	11,00,000
Factory overheads		3,00,000
	Factory cost	14,00,000
Administration overheads		3,36,000
		17,36,000
Selling overheads		2,24,000
Distribution overheads		1,40,000
	Cost of sales	21,00,000
	Profit	4,20,000
	Sales	25,20,000

Overhead recovery percentages

(1) *Percentage of Factory Overheads on Wages*

$$= \frac{\text{Factory overhead}}{\text{wages}} \times 100$$

$$= \frac{3,00,000}{5,00,000} \times 100 = 60\%$$

(2) *Percentage of Administration Overheads on Factory Cost:*

$$= \frac{\text{Administration overhead}}{\text{Factory cost}} \times 100$$

$$= \frac{3,36,000}{14,00,000} \times 100 = 24\%$$

(3) *Percentage of Selling Overheads on Factory Cost:*

$$= \frac{\text{Selling overhead}}{\text{Factory cost}} \times 100$$

$$= \frac{2,24,000}{14,00,000} \times 100 = 16\%$$

(4) *Percentage of Distribution Overheads on Factory Cost:*

$$= \frac{\text{Distribution overhead}}{\text{Factory cost}} \times 100$$

$$= \frac{1,40,000}{14,00,000} \times 100 = 10\%$$

(5) *Percentage of Profit on Sales:*

$$= \frac{\text{Profit}}{\text{Sales}} \times 100$$

$$= \frac{4,20,000}{25,20,000} \times 100 = 16.67\%$$

TENDER STATEMENT

Materials				8,000
Wages				5,000
	Prime cost			13,000
Factory overhead:				
60% on wages	$\frac{60}{100} \times 50000$	=	3,000	
Add: 20% increase			600	3,600
			Factory cost	16,600
Administration overhead:				
24% on Factory cost	$\frac{24}{100} \times 16,600$	=	3,984	
Add 15% increase		=	598	4,582
			Office cost	21,182
Selling overhead:				
16% on Factory cost	$\frac{16}{100} \times 16,600$	=	2,656	
Add: 15% increase		=	398	3,054
Distribution overhead:				
10% on Factory cost	$\frac{10}{100} \times 16,600$	=	1,660	
Less 10% decrease		=	166	1,494
			Cost of sale	25,730
			Profit 16.67% on sales or 20% on cost price	5,146
			Selling price	30,876

Problem 18. Pleasant cold limited manufactured and sold 1000 refrigerators in the year ending 31st March 2004. The summarised trading and profit and loss account is given below.

To cost of materials	80,000	By Sales	4,00,000
To direct wages	1,20,000		
To manufacturing cost	50,000		
To gross profit c/d	1,50,000		
	4,00,000		4,00,000
To management Salaries	60,000	By Gross profit	1,50,000
To rent and rates of office	10,000		
To other office expenses	20,000		
To selling expenses	30,000		
To Net profit	30,000		
	1,50,000		1,50,000

During 2005, a tender order for the supply of 1200 refrigerators was received. It is estimated that:

(*a*) Price of material will rise by 20% on the previous year's level.

(*b*) Wages will rise by 5%

(*c*) Manufacturing cost will rise in proportion to the combined cost of materials and wages.

(*d*) Selling expenses per unit will remain unchanged.

(*e*) Other expenses will remain unaffected by the rise in output.

You are required to submit a statement showing the quotation price at which refrigerators would be sold so as to shows a profit of 10% on selling price.

(Sri Sathya Sai University, B.Com. (Hons) October 2006)

Solution:

COST SHEET

(For 1000 Refrigerators)

Materials			80,000
Wages			1,20,000
		Prime cost	2,00,000
Manufacturing overhead			50,000
		Works cost	2,50,000
Office and Administration overhead:			
Management salaries	60,000		
Rent & Rates of office	10,000		
Other office expenses	20,000		90,000
		Office cost	3,40,000
Selling overhead			30,000
		Cost of sales	3,70,000
		Profit	30,000
		Sales	4,00,000

Working Note:

(1) Calculation of percentage of manufacturing overhead to prime cost

For a prime cost of 2,00,000 mfg. overhead is 50,000

For 100 → ?

$$= \frac{50,000 \times 100}{2,00,000} = 25\%$$

TENDER STATEMENT FOR 1200 REFRIGERATORS

			Rs.
Material cost	80 per unit		
Add: 20%	16		
	96		1,15,200
Labour	120 per unit		
Add: 5%	6		
	126		1,51,200
		Prime cost	2,66,400
Manufacturing overhead $\frac{25}{100} \times 2,66,400$			66,600
		Works cost	3,33,000
Office & adm. overhead:			
Management salaries	60,000		
Rent & Rates of office	10,000		
Other office expenses	20,000		90,000
		Office cost	4,23,000
Selling & Distribution overhead @ Rs. 30 per unit			36,000
		Cost of sales	4,59,000
		Profit $\frac{1}{9}$ on cost	51,000
		Sales	5,10,000

Selling price per unit = $\frac{5,10,000}{1200}$ = 42.50

Problem 19. From the following figures prepare a cost sheet of popular stoves manufacturing company for the year 2004.

	Rs.
Stock of materials on 1-1-2004	35,000
Stock of materials on 31-12-2004	4,900
Purchase of materials	52,500
Direct wages	95,000
Factory overheads	17,500
Establishment expenses	10,000
Finished goods on hand on 1-1-2004	NIL
Finished goods on hand on 31-12-2004	35,000
Sales	1,89,000

The number of stoves manufactured during the year 2004 was 4000.

The company wants to quote for a contract for the supply of 1000 electric stoves during the year 2005. The stoves to be quoted are of uniform quality and make and similar to those manufactured in the previous year, but the cost of materials has increased by 15% and cost of factory labour by 10%.

Prepare a tender statement showing the price to be quoted to give the same percentage of net profit on turnover as was realised during the year 2004. assuming that the cost per unit of overheads will be the same as in the previous year. *(Sri Sathya Sai University, B.Com. (Hons) February 2006)*

Solution: COST SHEET

			Amount (4000 units)	*Cost (per unit)*
Opening stock of materials		35,000		
Add:	Purchases of materials	52,500		
		87,500		
Less:	Closing stock of materials	4,900		
	Cost of materials consumed		82,600	20.65
Direct wages			95,000	23.75
		Prime cost	1,77,600	44.40
Factory overhead			17,500	4.37
		Factory cost	1,95,100	48.77
Establishment expenses			10,000	2.50
		Office cost	2,05,100	51.27
Less:	Closing stock of finished goods		35,000	
		Cost of goods sold	1,70,100	
		Profit	18,900	
		Sales	1,89,000	

TENDER STATEMENT (1000 Stoves)

Materials cost		20,650		
Add:	15% increase	3,098		23,748
Wages		23,750		
Add:	10% increase	2,375		26,125
			Prime cost	49,873
Factory overheads				4,375
			Factory cost	54,248
Establishment expenses				2,500
			Office cost	56,748
			Profit $\frac{1}{9}$ of cost	6,305
			Sales	63,053

Problem 20. The following expenses were increased for a job during the year ending 31st December 2000.

	Rs.
Direct materials	6,000
Direct wages	8,000
Chargeable expenses	2,000
Factory overheads	4,000
Selling and distribution overhead	4,000
Administration overhead	6,000

Selling price for the above job was Rs. 36,000. You are required to prepare a statement showing the profit earned for the year 2000 from the job and an estimated price of a job which is to be executed in the year 2001. Materials, wages and chargeable expenses will be required of Rs. 10,000, Rs. 1,4000 and Rs. 4,000 respectively for the job.

The various overheads to be recovered on the following basis while calculating the price.

(a) Factory overheads as a percentage on direct wages.

(b) Administration and selling and distribution as percentage of factory cost

(Bangalore University, B.Com., December 2006)

Solution:

COST SHEET

Direct materials		6,000
Direct wages		8,000
Chargeable expenses		2,000
	Prime cost	16,000
Factory overheads		4,000
	Factory cost	20,000
Administration overhead		6,000
	Office cost	26,000
Selling and Distribution overhead		4,000
	Cost of sales	30,000
	Profit	6,000
	Sales	36,000

TENDER STATEMENT

Material cost		10,000
Wages		14,000
Chargeable expenses		4,000
	Prime cost	28,000

Factory overhead:

For direct wages 8,000: Factory overhead is 4,000

For 100 → ?

$$\frac{4{,}000 \times 100}{8{,}000} = 50\%$$

50% of wages

	7000
Factory cost	35000

Administration factory & Selling overhead:

For factory cost of 35,000: Adm & Selling overhead is Rs. 10,000

For 100 → ?

$$\frac{10{,}000 \times 100}{35{,}000} = 28.57\% \text{ of Factory cost}$$

	10,000
Cost of sales	45,000
Profit 20%	9,000
Sales	54,000

Working Note:

(1) Calculation of % of profit on cost of sales:

For cost of 35,000 Profit is Rs. 6,000

For 100 → ?

$$\frac{6{,}000 \times 100}{30{,}000} = 20\%$$

Problem 21. The following figures ralate to the manufacture of electric fans for three months ending 31st Dec., 2000:

	Rs.
Completed stock on 1st Oct. 2000	NIL
Completed stock on 31st dec. 2000	20,250
Stock of raw materials on 1st Oct. 2000	5,000
Stock of raw materials on 31st Dec. 2000	3,500
Factory wages	75,000
Indirect charges	12,500
Materials purchased	32,500
Sales	1,12,500

The number of fans manufactured during these three months was 3,000.

Prepare a statement showing the cost per fan and the price to be quoted for 750 fans to realise the same percentage of profits as was realised during the said period, assuming the same condition.

(Sri Sathya Sai University, B.Com., Feb. 2001)

Solution:

STATEMENT OF COST
(for the three months ending)

		31.12.1982	
		Cost for 3,000 fans	*Cost for fan*
Opening stock of raw materials	5,000		
Add: Purchases	32,500		
	37,500		
Less: Closing stock	3,500		
		34,000	11,33
Factory wages		75,000	25.00
Prime cost		1,09,000	36.33
Indirect charges		12,500	4.17
Cost of production		1,21,500	40.50
Less: Closing stock of fans		20,250	
Cost of goods sold		1,01,250	
Profit		11,250	
Sales		1,12,500	

QUOTATION FOR 750 FANS

			Total	*Per fan*
Material	Rs. $\frac{34,000}{3,000} \times 750$		8,500	11.33
Wages	Rs. $\frac{75,000}{3,000} \times 750$		18,750	25.00
		Prime cost	27,250	36.33
Indirect wages	Rs. $\frac{12,500}{3,000} \times 750$		3,125	4.17
		Cost of production	30,375	40,50
Add: Profit $\frac{1}{9} \times 30,375$			3,375	4,50
		Quotation price	33,750	45.00

Problem 22. The trading profit & loss a/c of Vijaya manufacturing company for the year ending 31.12.1998 was as follows:

To Raw materials purchased	80,000	By sales (2,500 units)	2,50,000
To Direct wages	30,000	By closing stock of Raw materials	5,000
To Direct expenses	25,000		
To Factory expenses	40,000		
To Gross profit c/d	80,000		
	2,55,000		2,55,000

To Office salaries	25,000	By gross profit b/d	80,000
To Office rent	12,000	By dividend received	10,000
To Selling expenses	12,500	By discount received	7,500
To Preliminary expenses written-off	2,500		
To Goodwill written off	5,500		
To Net profit c/d	40,000		
	97,500		97,500

For the year 1999, it is estimated that:

(1) Units produced and sold will rise by 20%

(2) Prices of raw materials per unit will rise by 10%

(3) Direct wages per unit will increase by 25%

(4) Direct expenses will increase by Rs. 5,000 in total

(5) Factory expenses per unit will increase by 25%

(6) The office premises which was on rental basis in 1998 would be purchased by the company, on which depreciation would be Rs. 6000 on 1999

(7) Selling expenses per unit will remain same

You are required to prepare a statement showing estimated cost and profit for the year ended 31.12.1999 considering that company shall charge a profit at 20% on sales.

(University of Bombay, B.Com., April 1999)

Solution:

COST SHEET AND TENDER STATEMENT

		1998 (for 2,500 units)		*1999 (for 3,000 units)*	
		Total	*Per unit*	*Total*	*per unit*
Raw materials		75,000	30	99,000	33
Direct wages		30,000	12	45,000	15
Direct expenses		25,000	10	30,000	10
Prime cost		1,30,000	52	1,74,000	58
Factory expenses		40,000	16	60,000	20
Works cost		1,70,000	68	2,34,000	78
Office & administration overheads		37,000	14.8		
Office salaries	25,000				
Depreciation	6,000			31,000	10.33
		207,000	82.8	2,65,000	88.33
Selling expenses		12,500	5	15,000	5
Cost of sales		2,19,500	87.8	2,80,000	93.33
Profit		30,500	12.2	70,000	23.33
Sales		2,50,000	100	3,50,000	116.66

Working Notes:

(1) Calculation of units produced in 1999

No. of units produced in 1998	=	2,500
Add: increase of 20% in 1999		500
		3,000

(2) Calculation of raw materials in 1999

Raw materials per unit = $\frac{75,000}{2,500}$	=	Rs. 30
In 1998 per unit cost	=	Rs. 30
Add: Increase of 10% in 1999	=	3
		33

(3) Calculation of direct wages

In 1998 wages per unit cost	=	Rs. 12
Add: Increase of 25% in 1999	=	3
		15

(4) Calculation of factory expenses

Factory expenses in 1998	=	40,000
Per unit in 1998 = $\frac{40,000}{2,500}$	=	Rs. 16
Add: Increase by 25%	=	4
		20

Factory expenses in 1999 = 20 × 3,000 = 60,000

(5) It is assumed that salaries will remain same

(6) Rent will not appear in 1999, instead of the rent, depreciation will be Rs. 6,000.

So total expenses = 25,000 + 6,000 = 31,000

(7) Selling expenses for 1999 per unit Rs. 5 as in 1998 will remain same

(8) Profit = 20% on sales or 25% on cost

$$= 2,80,000 \times \frac{25}{100} = 70,000$$

(9) Preliminary expenses written off and goodwill written off are purely financial expenses and dividend received and discount received are purely financial incomes, hence excluded from cost sheet.

Problem 23. Swadeshi Electronics Ltd., furnishes to you the following information for the year ended 31 March, 1996:

Production and sales	15,000 units
Sales	Rs. 12,75,000
Direct wages	Rs. 2,70,000
Direct materials	Rs. 3,30,000
Factory overheads	Rs. 2,25,000
Administrative overheads	Rs. 1,05,000
Sales overheads	Rs. 90,000

On account of intense competition following changes are estimated in the subsequent year.

(1) Production and sales activity will be increased by one-third

(2) Material rate will be lower by 25%. However there will be increase in consumption by 20% due to quality differences

(3) Direct wages cost would be reduced by 20% due to automation

(4) Out of the above factory overhead Rs. 45,000 are of fixed nature. The remaining factory expenses are variable in proportion to the number of units produced

(5) Total administrative overheads will be lower by 40%

(6) Sales overhead per unit would remain the same

(7) Sale price per unit would be lower by 20%

Prepare a statement of cost for both the years ending 31st March, 1996 and 31st March, 1997, showing maximum possible details. *(University of Bombay, B.Com., April 1996)*

Solution:

COST SHEET

(for the Years 1996 and 1997)

	1996		*1997*	
	Total	*Per unit*	*Total*	*Per unit*
Materials	3,30,000	22	3,96,000	19.80
Wages	2,70,000	18	2,88,000	14.40
Prime cost	6,00,000	40	6,84,000	34.20
Factory overhead	2,25,000	15	2,85,000	14.25
Works cost	8,25,000	55	9,69,000	48.45
Administration overhead	1,05,000	7	63,000	3.15
Cost of production	9,30,000	62	10,32,000	51.60
Selling overheads	90,000	6	1,20,000	6.00
Cost of sales	10,20,000	68	11,52,000	57.60
Profit	2,55,000	17	2,08,000	10.40
Sales	12,75,000	85	13,60,000	68

Working Notes:

(1) Calculation of production & sales

$$15{,}000 + \frac{1}{3} \times 15{,}000 = 20{,}000 \text{ units}$$

(2) Calculation of material rate

$1996 = \dfrac{3{,}30{,}000}{15{,}000}$	=	22.00
Less Reduction in price by 25%		5.50
		16.50
Add Rise in consumption by 20		3.30
Rate of materials		19.80

(3) Calculation of wages

Wage rate	18.00
Less. 20% reduction	3.60
Rate of wages	14.40

(4) Calculation of factory overheads

	1996	*1997*
Fixed	45,000	45,000
Variable	1,80,000	2,40,000
	2,25,000	2,85,000

(5) Administration cost in 1996	1,05,000
Less decrease in expenditure 40%	42,000
Cost in 1997	63,000

(6) Sales overheads per unit is Rs. 6
So for 20,000 units it would be 1,20,000

(7) Selling price in 96	85
Less 20% reduction	17
Selling price in 97	68

Problem 24. The following information for the year ending 31st Dec. 1989 is taken from the books of a company manufacturing TV valves:

Materials consumed	6,00,000
Direct wages	4,00,000
Driect expenses	2,00,000
Indirect wages	30,000
Stereo and spares consumed	55,000
Workman's welfare expenses	40,000
Cost of rectifying defective work	15,000
Depreciation on machinery	25,000
Other factory expenses	1,50,000
Sale of factory scrap	15,000
Administrative staff salaries	1,55,000
Other administrative expenses	1,10,000
Professional charges, audit fee. director's fee	35,000
Commission paid to selling agents	1,20,000
Neon sign expenses	25,000
Show room expenses	35,000
Sales	26,40,000

10,000 units has been manufactured and sold for the year ending 31st Dec. 1989.

For the year ending 31st Dec. 1990 the following estimates have been made:

(*i*) Production and sales will be 12,000 units

(*ii*) Materials cost per unit will rise by 50%

(*iii*) Wage rates per unit will rise by 25%

(*iv*) Direct expenses per unit will be in the same proportion to wages as before

(*v*) Factory expenses will be in the same proportion to prime cost. Administration overheads in the same proportion to factory cost and selling overheads in the same proportion to cost of production as before

(*vi*) Profit desired per unit is 25% on selling price

Prepare a cost statement showing total cost, cost per unit, total profit and profit per unit for 1989 and 1990. *(University of Bombay, B.Com., Oct. 1990)*

Solution:

COST SHEET FOR 1989
(output 10,000 units)

Particulars			Cost per unit	Amount
Materials			60.00	6,00,000
Wages			40.00	4,00,000
Direct expenses			20.00	2,00,000
		Prime cost	120.00	12,00,000
Works overheads:				
Indirect wages	30,000			
Stereo and spare parts	55,000			
Workmen's welfare expenses	40,000			
Cost of rectifying defectives	15,000			
Depreciation on machinery	25,000			
Other factory expenses	1,50,000			
	3,15,000			
Less: Sale of factory scrap	15,000		30	3,00,000
		Factory cost	150.00	15,00,000
Administration overheads:				
Administrative staff salaries	1,55,000			
Other administrative expenses	1,10,000			
Profession charges etc.	35,000		30.00	3,00,000
		Office cost	180.00	18,00,000
Selling & distribution overheads:				
Commission paid	1,20,000			
Neon sign expenses	25,000			
Show room expenses	35,000		18.00	1,80,000
		Cost of sales	198.00	19,80,000
		Profit	66.00	6,60,000
		Sales	264.00	26,40,000

QUOTATION PRICE FOR 1990
(output 12,000 units)

Particulars			Cost per unit	Amount
Materials	60		90.00	10,80,000
Add: 50% Increase	30			
Labour	40		50.00	6,00,000
Add: 25% Increase	10			
Direct expenses (*see note 1*)			25.00	3,00,000
		Prime cost	165.00	19,80,000
Works overhead (*see note 2*)			41.25	4,95,000
		Factory cost	206.25	24,75,000
Administrative expenses (*see note 3*)			41.25	4,95,000

	Office cost	247.50	29,70,000
Selling and distribution expenses (*see note 4*)		24.75	2,97,000
	Cost of sales	272.25	32,67,000
	Profit	90.75	10,89,000
	Sales	363.00	43,56,000

Working Notes:

(1) Calculation of direct expenses

In 1989, for Rs. 40 wages: the direct expenses 20

In 1990 for Rs. 50 wages → ? $\frac{50 \times 20}{40}$ = Rs. 25 per unit

(2) Calculation of factory overheads

For 12,00,000 prime cost, factory expenses was 3,00,000

For 100 → ? $\frac{3,00,000 \times 100}{12,00,000} = 25\%$

(3) Calculation of administration overhead

For 15,00,000 of factory cost, the adm. overhead is Rs. 3,00,000

For 100 → ? $\frac{3,00,000 \times 100}{15,00,000} = 20\%$

(4) Calculation of selling overheads

For 18,00,000 of cost of production the selling overhead is Rs. 1,80,000

For 100 → ? $\frac{1,80,000 \times 100}{18,00,000} = 10\%$

Problem 25. A company produces a plant that sells for Rs. 1,800. An increase of 7½% in cost of materials and 6¼% in cost of labour is anticipated. What must be the selling price to produce the same percentage of gross profit as before. Only data available are:

Material cost have been 50% of cost of sale

Wages cost have been 20% of cost of sale

Overhead cost have been 30% of cost of sale

The anticipated increased costs in relation to the present sales price would cause a 25% decrease in the amount of present gross profit.

Prepare a statement of profit or loss per unit, showing the new selling price desired and new cost per unit.

Solution:

Let x be the present cost of sales

Let y be the present gross profit

Then for the present condition $x + y = 1,800$ ...(1)

Rs. x is made up of:

Materials	50% or	=	$0.5x$
Labour	20% or	=	$0.2x$
Overheads	30% or	=	$0.3x$

This would increase:

Materials by	7½% = $0.5x + 0.0375x$	=	$0.5375x$
Labour by	6¼% = $0.2x + 0.0125x$	=	$0.2125x$
Overhead as before			$0.3000x$
			$1.0500x$

This increase in cost would result in a decrease of gross profit by 25% at the same selling price. So the gross profit would be 0.75 *y*

i.e.,

$$1.05\,x + 0.75\,y = 1{,}800 \ldots (2)$$
$$x + y = 1{,}800 \times 1.05$$
$$1.05\,x + 0.75\,y = 1.800$$
$$1.05\,x + 1.05\,y = 1.890$$
$$1.05\,x + 0.75\,y = 1.800$$
$$\text{Subtracting } 0.30\,y = 90$$
$$y = \frac{90}{0.30} = 300$$

Substituting the value of *y* in the 1st equation

$$x + y = 1{,}800$$
$$x + 300 = 1{,}800$$
$$x = 1{,}800 - 300$$
$$= 1{,}500$$

COST SHEET SHOWING PRESENT AND REVISED COST

	Present cost		*Revised cost*
Materials 50%	750,00	Add 7½%	806
Labour 20%	300,00	Add 6¼%	318
Overhead 30%	450,00		450
Cost of sales	1,500		1.575
Profit	300		315
Selling price	1.800		1.890

Problem 26. *(When fixed and variable overheads are given)*: The following information is available from the books of a company manufacturing luxury ceiling fans. Production and sales during the year ending 31.3.1988 was 1,000 units.

	Rs.
Direct materials	2,00,000
Direct wages	1,50,000
Factory expenses	1,37,500
Administration expenses	60,000
Selling expenses	45,000
Sales	7,30,000

The following estimates have been made for 1988-89:

(i) Production and sales will be 1,500 units

(ii) Materials price per unit will increase by 25% but due to economy in consumption the cost per unit will reduce by 12%

(iii) The wages rates per unit will increase by 20%

(iv) Factory expenses Rs. 50,000 are fixed. The remaining factory expenses will be in the same proportion to materials consumed and wages as in the previous year

(v) The total administration expenses will increase by $66\frac{2}{3}\%$

(vi) Selling expenses will be Rs. 90,000

(vii) The profit desired is 20% on sales.

Prepare a cost statement showing possible break-up of cost per unit and total cost for 1987-88 and 1988-99 profit per unit and total profit for 1987-88 and 1989-89.

(University of Bombay, B.Com., Oct. 1988)

Solution

COST SHEET FOR 1987–88
(output 1,000 fans)

		Cost per unit	*Amount*
Direct Materials		200	2,00,000
Direct Wages		150.00	4,00,000
	Prime cost	350.00	3,50,000
Factory overheads: — Fixed		50.00	50,000
— Variable		87.50	87,500
	Factory cost	487.50	4,87,500
Administriation expenses		60.00	60,000
	Office cost	547.50	5,47,500
Selling expenses		45.00	45,000
	Cost of sales	592.50	5,92,500
	Profit	137.50	1,37,500
	Sales	730.00	7,30,000

QUOTATION FOR 1,5000 FANS

			Cost per unit	*Amount*
Materials	300		226.00	3,39,000
Add: 13% Increase	26			
Wages	150			
Add: 20% Increase	30		180	2,27,000
		Prime cost	406	6,09,000
Factory overhead	—Fixed		33.33	50,000
— Variable (*see note*)			101.50	1,52,250
		Factory cost	540.83	8,11,250
Administration expenses			66.67	1,00,000
		Office cost	607.50	9,11,250
Selling expenses			60.00	90,000
		Cost of sales	667.50	10,01,250
		Profit	166.875	2,50,312.50
		Sales	834.375	12,51,562.50

Note : For prime cost of 350 variable overhead is 87.50
For prime cost of 406 variable overhead is 101.50.

ESTIMATED COST SHEET

Business environment is never static. It is dynamic which changes frequently. Hence, the management will decide the quantity of output depending upon the need. It is the endeavour of cost accountant to estimate the cost of production of such output by taking into account the past and future considerations. A cost sheet which is prepared to estimate the cost of a given volume of output is known as estimated cost sheet.

Problem 27. *(Estimated Cost Sheet).* Following information is available from cost records for the year ended 31st December 2004:

Direct materials	:	Rs. 36 per unit
Direct labour	:	Rs. 28 per unit
Chargeable expenses	:	Rs. 11 per unit
Factory overheads	:	Fixed Rs. 15,00,000
	:	Variable Rs. 10 per unit
Office overheads	:	Fixed Rs. 5,00,000
Selling overheads	:	Fixed Rs. 5,00,000
	:	Variable Rs. 25 per unit

Units produced and sold 50,000

Selling price per unit Rs. 210

Following changes are intercipated during the year ended 31st December 2005.

(1) Production and sales will increase by 60%

(2) Direct material cost per unit will increase by 12.5%

(3) Direct labour per unit will decrease by 5%

(4) Chargeable expenses per unit will decrease by 10%

(5) Various factory overheads per unit will increase by 25%

(6) Variable selling overheads will decrease by 25%

(7) All fixed overheads will increase by 20%

(8) 75% of the output will be sold in domestic market at a profit of 20% on sales

(9) Balance 25% output will be sold in export market at a profit of 50% on sales.

You are required to:

(1) Prepare a cost sheet for the year ended 31st December 2004 and estimated cost sheet for the year inded 31st December 2005, showing total and per unit cost.

(2) Calculate total and per unit profit for the year ended 31st December 2004.

(3) Calculate total sales and profit for domestic market and export market.

(University of Bombay, B.Com., October 2005)

Solution: ***Working Notes:***

(1) *Working Note 1:* Materials cost

$$36 \times \frac{12.5}{100} = 4.5 + 36 = 40.50 \text{ per unit}$$

(2) *Working Note 2:* Direct labour cost

$$28 \times \frac{5}{100} = 1.4 - 28 = 26.6 \text{ per unit}$$

(3) *Working Note 3:* Chargeable expenses

$$11 \times \frac{10}{100} = 1.1 - 11 = 9.9 \text{ per unit}$$

(4) *Working Note 4:* Variable Factory overhead:

$$10 \times \frac{25}{100} = 2.5 + 10 = 12.5 \text{ per unit}$$

(5) *Working Note 5:* Fixed Factory overhead:

$$15,00,000 \times \frac{20}{100} = 3,00,000 + 15,00,000 = 18,00,000$$

(6) *Working Note 6:* Office overhead:

$$12,50,000 \times \frac{20}{100} = 2,50,000 + 12,50,000 = 15,00,000$$

(7) *Working Note 7:* Variable S & D overhead

$$25 \times \frac{25}{100} = 6.25 - 25 = 18.75 \text{ per unit}$$

(8) *Working Note 8:* Fixed S & D overhead:

$$5,00,000 \times \frac{20}{100} = 1,00,000 + 5,00,000 = 6,00,000$$

Solution: ESTIMATED COST SHEET

	2004		*2005*	
	Per unit	*Total*	*Per unit*	*Total*
Direct material cost	36	18,00,000	40.50	32,40,000
Direct labour cost	28	14,00,000	26.60	21,28,000
Chargeable expenses	11	5,50,000	9.90	7,92,000
Prime cost	75	37,50,000	77.00	61,60,000
Factory overheads:				
– Variable	10	5,00,000	12.50	10,00,000
– Fixed	30	15,00,000	22.50	18,00,000
Works cost	115	57,50,000	112.00	89,60,000
Office overheads	25	12,50,000	18.75	15,00,000
Office cost	140	70,00,000	130.75	1,04,60,00
Selling & Distribution overhead:				
– Variable	25	12,50,000	18.75	15,00,000
– Fixed	10	5,00,000	7.50	6,00,000
Cost of sales	175	87,50,000	157	1,25,60,000
Profit	35	17,50,000		
Sales	210	1,05,00,000		

Working Note 9

Total Sales of 2005

$$50,000 \times \frac{60}{100} = 30,000 + 50,000 = \text{Rs. } 80,000$$

Domestic sales $= 80{,}000 \times \frac{75}{100} = 60{,}000$

Exports market $= 80{,}000 \times \frac{25}{100} = 20{,}000$

Selling Price of Domestic Market:

Selling price = Cost price + profit

100 = 80 + 20

For a cost price of 80: The selling price is Rs. 100

For cost price of 157 → ?

$$\frac{157 \times 100}{80} = \text{Rs. } 196.25$$

Total Domestic sales = 60000 × 196.25

= 1,17,75,000

Selling Price of Export Market:

For a cost price of Rs. 50: The selling price is Rs. 100

For a cost price of 157 → ?

$$\frac{157 \times 100}{50} = 314$$

Total sales of Export Market

20,000 × 314 = 62,80,000

Problem 28. K.T. manufacturing company gives you the following particulars for the year 2004. Production and sales during the year was 10,000 units.

	Rs.
Materials	2,50,000
Direct wages	1,50,000
Administrative overhead (Fixed)	1,00,000
Sales	12,00,000
Profit	2,50,000
Factory overheads:	
Fixed	1,00,000
Variable	2,00,000
Selling and Distribution overheads:	
Fixed	60,000
Variable	90,000

The company has worked its maximum capacity of 10,000 units during 2004. The management has decided to increase production capacity to 15,000 units for the year 2005 and it is estimated that:

(i) There will be all round rise in all variable expenditure by 10%

(ii) There will be increase of 20% in all fixed overheads

(iii) There will be no need to change the selling price for the year 2005

Prepare a statement showing total as well as unit cost and profit for 2004. Also prepare a statement showing estimated profit for 2005 taking into consideration the changes in 2005.

(University of Bombay, B.Com., Oct. 2003)

Solution: ***Working Notes***

(1)	Material cost per unit		25.00
	Add: 10% increase		2.50
			27.50
(2)	Wages cost per unit		15.00
	Add: 10% increase		1.50
			16.50
(3)	Factory overhead (fixed)		1,00,000
	Add: 20% increase		20,000
			1,20,000
(4)	Factory overhead (variable)		
	Unit cost	$\frac{2,00,000}{10,000}$ =	20
	Add: 10% Increase	=	2
			22
(5)	Administration overhead (fixed)		1,00,000
	Add: 20% increase		20,000
			1,20,000
(6)	Selling overhead (fixed)		60,000
	Add: 20% increase		12,000
			72,000
(7)	Selling overhead (variable)		
	Unit cost	$\frac{90,000}{10,000}$ =	9
	Add: 10% increase		0.90
			9.90

COST SHEET

(for the year ended 31st March 2004 and 2005)

	2004 10000 units		*2005 15,000 units*	
	Total	*Cost per unit*	*Total*	*Cost per unit*
Direct materials	2,50,000	25.00	4,12,500	27.50
Direct wages	1,50,000	15.00	2,47,500	16.50
Prime cost	4,00,000	40.00	6,60,000	44.00
Factory overheads				
– Fixed				
– Variable	1,00,000	10.00	1,20,000	8.00
	2,00,000	20.00	3,30,000	22.00

Factory cost	7,00,000	70.00	11,10,000	74.00
Administration overhead	1,00,000	10.00	1,20,000	8.00
Office cost	8,00,000	80.00	12,30,000	82.00
Selling overheads				
– Fixed				
– Variable	60,000	6.00	72,000	4.80
	90,000	9.00	1,48,500	9.90
Cost of sales	9,50,000	95	14,50,000	96.70
Profit	2,50,000	25	3,49,500	23.70
Sales	12,00,000	120	18,00,000	120

PRODUCTION ACCOUNT

A production account is another form of cost sheet which is prepared to ascertain cost of production, profit and sales. Strictly speaking a cost sheet is prepared to ascertain the cost of production of goods and services. But when the cost sheet is extended further so as to reveal profit and sales, it is termed as production a/c. However, production account is prepared in the form ledger accounts so as to give the exact cost of production, loss in output, sale of scrap, etc.

Problem 29. *(Problem on iron production):* The Sandur Iron and Manganese production for the quarter ending 31st March, 2003 amounted to 32.000 tons. The standard cost of production of one ton of pig iron is Rs. 21.00.

From the particulars given below, prepare pig iron production account showing profit on production and cost per ton of each expenditure.

	Stock on 1-1-03 (Rs.)	*Purchases (Rs.)*	*Stock on 31-1-03 (Rs.)*
Lime stone	3,100	42.300	5,400
Coal and coke	17,400	2,93,500	14,900
Iron ore	14,600	1,77,200	27,800
	Stock on 1-1-03	Sales during the quarter	Stock on 31-3-03
Pig iron	(Rs.) 2,97,000	10,33,000	2,99,000

Other production expenses during the quarter were —Furnace wages Rs. 92,000, carriage inwards Rs. 22,000, furnace repairs Rs. 26,000

Solution:

PRODUCTION ACCOUNT

	Cost per ton	*Total*		*Cost per ton*	*Total*
To materials used:			By trading a/c		
Lime stone	1.25	40,000	(32,000 tons @ Rs. 21)	21	6,72,000
Coal and coke	9.25	2,96,000			
Iron ore	5.12	1,64,000			
To carriage inwards	0.69	22,000			
To furnace wages	2.88	92,000			
To furnace repairs	0.81	26,000			
to profit on production	1.00	32,000			
	21.00	6,72,000		21	6,72,000

Solution:

TRADING ACCOUNT
(for the quarter ended 31.3.91)

To opening stock	2,97,000	By sales	10,33,000
To production a/c	6,72,000	By closing	2,99,000
To gross profit	3,63,000		
	13,32,000		13,32,000

***Working Note:* Calculation of cost of materials consumed**

	Lime stone	*Coal & coke*	*Coal & coke*
Opening stock	3,100	17,400	14,600
Add: Purchases	42,300	2,93,500	1,77,200
	45,400	3,10,900	1,91,800
Less: Closing stock	5,400	14,900	27,800
Materials consumed	40,000	2,96,000	1,64,000

Problem 30. *(Problem on coal production):* The following are the balances of the impersonal ledger of a colliery relating to revenue at the end of the year:

	Rs.
Wages paid for coal production	5,80,000
Coal for colliery consumption	45,000
Timber used in coal production	64,000
Ropes used in coal production	12,000
Stores used in coal production	76,000
Royalties paid for coal production	42,000
General charges for coal production	70,000
Salaries for coal production	36,000
Coal sold (including colliery consumption 1,12,000 tons)	8,84,000
Wages paid for coke making	50,000
Stores used	37,000
Salaries paid	8,000
Coke sold 43,500 tons	5,40,000

The stock of coal at the beginning of the year amounted to 7,000 tons valued at Rs. 5 per ton and at the end of the year 15,000 tons at a similar rate. The stock of coke at the beginning of the year amounted to 2,000 tons valued at Rs. 10 per ton and at the end of the year 500 tons valued at a similar rate.

The total production of the colliery was 1,85,000 tons of coal and 42,000 tons of coke, 65,000 tons of coal being used for coke making.

Prepare separate production accounts for coal and coke showing the cost of each item of expenses per ton of coal and coke separately, taking coal used for coke making at cost price

(Sri Sathya Sai University, B.Com., March 1992)

Solution:

COAL PRODUCTION ACCOUNT

	Per ton	Amount		Per ton	Amount
To wages	3.14	5,80,000	By cost of production	5.00	9,25,000
To coal	0.24	45,000			
To timber	0.34	64,000			
To ropes	0.06	12,000			
To stores	0.42	76,000			
To royalties	0.23	42,000			
To general charges	0.38	70,000			
To salaries	0.19	36,000			
	5.00	9,25,000		5.00	9,25,000
To opening stock		35,000	By sales		8,84,000
To cost of production		9,25,000	By coke production		
To gross profit c/d		3,24,000	65,000 @ Rs. 5		3,25,000
			By closing stock		75,000
		12,84,000			12,84,000

COKE PRODUCTION ACCOUNT

To wages	1.19	50,000	By cost of production a/c	10	4,20,000
To stores	0.88	37,000			
To salaries	0.19	8,000			
To coal used	7.74	3,25,000			
	10.00	4,20,000		10	4,20,000
To opening stock		20,000	By sales		5,40,000
To cost of production		4,20,000	By closing stock (500 × 10) =		5,000
To profit & loss a/c		1,05,000			
		5,45,000			5,45,000

Problem 31. *(Problem on yarn and cloth production a/c):* Given below is the cloth and yarn manufacturing a/c of Uttar Pradesh Cotton Mills for the year ended 31.12.99:

To opening stock		By sale of 2,24,000 kgs	
Yarn 44,800 kgs	36,600	of waste being that portion	
Cloth 88,200 kgs	88,200	of 11,53,600 kgs of cotton	28,000
To cotton consumed			
11,53,600 kgs	9,01,250	By sales:	
To spinning wages	57,050	Yarn — 1,68,000 kgs	1,89,000
To weaving wages	91,700	Cloth — 7,56,000 kgs	8,97,750
To stores consumed in spinning which increased yarn 5,600 kgs in weight	41,300	By closing stock:	
		Yarn — 1,40,000 kgs	1,48,750
To stores and sizing materials consumed in weaving which increased cloth by 1,86,200 kgs	1,48,400	Cloth — 1,90,400 kgs	2,14,200
To spinning fuel	36,050		
To weaving fuel	22,400		
To gross profit	57,750		
	14,77,700		14,77,700

For the satisfaction of income tax authorities, who contend that gross profit is suppressed by the company. You are required to prepare (1) yarn production and (2) cloth production account separately showing the following details:

(1) Gross profit of the department.

(2) Total quantity produced (including the opening balance).

(3) The cost production including the value of opening stock which was at cost.

Solution:

YARN PRODUCTION ACCOUNT

	Qty.	*Amt.*		*Qty*	*Amt.*
To opening stock	44,800	33,600	By sales of waste	2,24,000	28,000
To cotton consumed	11,53,600	9,01,250	By cost of production		
To stores consumed	5,600	41,300	including opening stock		
To spinning wages		57,050	@ Rs. 1.0625	9,88,000	10,41,250
To spinning fuel		36,050			
	12,04,000	10,69,250		12,04,000	10,69,250
To cost of production	9,80,000	10,41,250	By sale of yarn	1,68,000	1,89,000
To gross profit		10,500	By cloth production	6,72,000	7,14,000
			a/c @ 1.0625		
			By closing stock	1,40,000	1,48,750
	9,80,000	10,51,750		9,80,000	10,51,750

CLOTH PRODUCTION ACCOUNT

	Qty.	*Amt.*		*Qty*	*Amt.*
To opening stock	88,200	88,2000	By cost of production	9,46,400	10,64,700
To yarn	6,72,000	7,14,000			
To stores consumed	1,86,200	1,48,400			
To weaving wages		91,700			
To weaving fuel		22,400			
	9,46,400	10,64,700		9,46,400	10,64,700
To cost of production	9,46,400	10,64,700	By sale	7,56,300	8,97,750
To gross profit		47,250	By closing stock	1,90,400	2,14,200
	9,46,400	11,11,950		9,46,400	11,11,950

Problem 32. From the following particulars of Swadeshi Cotton Mills for the half year ended 30.6.2009, prepare two separate production accounts. viz., yarn production a/c and cloth production a/c.

	Rs.
Opening stock of yarn at cost (60,000 kgs)	30,000
Opening stock of cloth at cost (1,20,000 kgs)	90,000
Cotton purchased (10,40,000 kgs)	2,60,000
Weaving wages	60,000
Stores weaving	40,000
Wages of spinning	30,000
Stores	20,000
Fuel	10,000

Yarn sales (2,00,000 kgs)	40,000
Stock at the end at cost:	
Yarn (2,00,000 kgs)	41,000
Cloth (2,05,000 kgs)	1,50,000
Fuel —Weaving	10,000
Sales:	
Waste (2,11,000 kgs)	29,000
Cloth (4,20,000 kgs)	3,00,000

Stores consumed increased the weight of production by 4,000 kgs for spinning stores and 1,20,000 kgs for weaving stores.

Solution:

YARN PRODUCTION ACCOUNT

	Weight	*Amount.*		*Weight*	*Amount*
To cotton	10,40,000	2,60,000	By sales of waste	2,11,000	29,000
To stores	4,000	20,000	By cost of production	8,33,000	2,91,000
To wages		30,000			
To fuel		10,000			
	10,44,000	3,20,000		10,44,000	3,20,000
To opening stock	60,000	30,000	By sales	2,00,000	40,000
To cost of production	8,33,000	2,91,000	By cloth production a/c (*see note 1*)	4,93,000	1,77,215
			By closing stock	2,00,000	41,000
			By P&L a/c		62,785
	8,93,000	3,21,000		8,93,000	3,21,000

CLOTH PRODUCTION ACCOUNT

To yarn production	4,93,000	1,77,215	By loss in weight (*see note 2*)	1,08,000	
To stores	1,20,000	40,000			
To wages		60,000	By cost of production a/c	5,05,000	2,87,215
To fuel		10,000			
	6,13,000	2,87,215		6,13,000	2,87,215
To opening stock	1,20,000	90,000	By sale	4,20,000	30,00,000
To cost of production	5,05,000	2,87,215	By closing stock	2,05,000	1,50,000
To P & L a/c		72,785			
	6,25,000	4,50,000		6,25,000	4,50,000

***Working Note 1:* Calculation of the value of yarn transferred to cloth production a/c:**

For processing a total weight of yarn of 8.93.000 kgs the cost of production is 3.21.000.

For 4,93,000 kgs of yarn output the value is

$$\frac{3{,}21{,}00 \times 4{,}93{,}000}{8{,}93{,}000} = \text{Rs. } 1{,}77{,}215$$

***Working Note 2:* Calculation of weight lost**

Total production	=	Sales + Closing stock – Opening stock
	=	4,20,000 + 2,05,000 – 1,20,000 = 5,05,000

Total weight put in	6,13,000
Actual weight of production	5,05,000
Loss in weight	1,08,000

Problem 33. *(Mica and talcum production account):* The following particulars are extracted from the books of M/s Rao Insulating Company as on 31.12.2009.

Wages paid	6,04,000
Timber used	58,000
Sundry stores	97,000
Rent and taxes $\left(\frac{1}{3}\text{applicable to talcum powder}\right)$	72,000
Wages paid for manufacturing talcum powder	71,000
Royalties paid on mica produced	72,000
Ropes used	12,000
Salaries paid $\left(\frac{1}{4}\text{applicable to talcum powder}\right)$	28,000
Mica sold, 1,94,000 tons	7,02,000
Talcum powder sold 72,000 tons	4,50,000

The total production of the company was 3,04,000 tons of mica and 72,000 tons of talcum powder, 1,10,000 tons of mica were used for making talcum powder.

You are required to prepare mica production and talcum powder production account showing the profit resulting from the sale of Mica and talcum powder, showing the cost per ton of mica and the cost per ton of talcum powder taking mica used for talcum powder at cost price.

Solution:

MICA PRODUCTION ACCOUNT
(output 3.04.000)

To wages paid		6.04,000	By cost of production @ Rs. 3		9,12,000
To timber used		58,000			
To sundry stores		97,000			
To rent & taxes $\frac{2}{3}$, 72,000		48,000			
To royalties		72,000			
To ropes used		12,000			
To salaries $\frac{3}{4}$, 28,000		21,000			
		9,12,000			9.12.000
To cost of production	3,04,000	9,12,000	By sales	1,94,000	7,02.000
To P & L a/c	–	1,20,000	By talcum powder a/c	1,10,000	3,30,000
	3,04,000	10,32,000		3.04.000	10,32,000

Solution:

TALCUM POWDER PRODUCTION ACCOUNT

To Mica production a/c		3,30,000	By cost of production @ Rs,		4,32,000
To rent & taxes $\frac{1}{3} \times 72{,}000$		24,000			
To salaries $\frac{1}{4} \times 28{,}000$		7,000			
To wages		71,000			
		4,32,000			4,32,000
To cost of production	1,10,000	4,32,000	By normal waste	38,000	
To profit	–	18,000	By sales	72,000	4,50,000
	1,10,000	4,50,000		1,10,000	4,50,000

Questions

Short and Long Answer Questions

1. Define "unit costing". What are its characteristics ? What are the types of industries to which it is applicable ? *(Calicut University, B.Com., April 1988)*
2. Describe the objects of cost sheet. Give an example of cost sheet indicating clearly the headings and important items supplying imaginary figures. *(University of Kerala, B.Com., April 1990)*
3. Define unit costing How does it differ from job costing ? *(S.V. University, B.Com., April 1985)*
4. Describe the role of Cost Accountant in determining the quotation price as against a tender. *(ICWA, Inter, December 1989)*
5. Write a short note on unit costing and job costing. *(Andhra University, B.Com., May 1992)*
6. What is a cost sheet ? What are its purposes ? *(University of Kerala, M.Com., May 1992)*
7. What is a cost sheet ? Give a speciman cost sheet for the manufacture of an acid measured in terms of kiloliters and explain the different items included in that cost sheet. *(Osmania University, B.Com., March 1992)*

Exercises

Cost Sheet

1. From the following particulars prepare cost sheet showing various elements of cost.

	Rs.
Opening stock of raw materials	1,10,000
Purchases of raw materials	8,25,000
Carriage outwards	28,500
Direct wages	4,21,400

Direct power	25,840
Technical directors salary	40,590
Factory rent, rates & insurance	10,140
Sale of factory scraps	1460
Depreciation on factory buildings	75,200
Closing work in progress	1,20,260
Factory stationery	12,340
Opening stock of finished goods	45,280
Closing stock of raw materials	36,920
Fees to brand ambassador	2,00,000
Stationary and printing	12,200
Staff salaries	6,30,000
Trade discount	1,20,000
Office rent	60,000
Free sample expenses	20,320
Closing stock of finished goods	50,240

Sales are made to earn profit @ 10% on cost price.

(University of Bombay, B.Com., October 2006)

[**Ans.** Sales Rs. 25,38,723]

2. From the books of accounts of M/s Avdhoot Enterprises, the following details have been extracted for the quarter ending 31st December 2005.

	Rs.
Opening stock of raw materials	2,70,000
Stock of materials-closing	3,00,000
Purchases of materials	12,48,000
Direct wages	3,57,600
Direct expenses	1,20,000
Indirect wages	24,000
Salaries to Administrative staff	60,000
Carriage inwards	48,000
Carriage outwards	37,500
Manager's salary	72,000
General charges	37,200
Legal changes for criminal suit	20,000
Commission on sales	28,000
Fuel	96,000
Electricity charges (Factory)	72,000
Directors fees	36,000
Repairs to plant and machinery	63,000
Rent, rates and taxes – factory	18,000
Rent, rates and taxes – office	9,600
Deprication on plant and machinery	45,000
Depreciation on furniture	3,600
Salesmen's salaries	50,000
Audit fees	18,000

(1) The manager's time is shared between the factory and the office in the ratio of 20 : 80.

(2) Carriage outwards include Rs. 7,500 being carriage inwards on plant and machinery.

(3) Selling price is 120% of the cost price.

From the above details prepare cost sheet for the quarter ending 31-12-2005 and ascertain sales. *(University of Bombay, B. Com., April 2006)*

[**Ans.** Sales Rs. 28,87,200]

3. The following details are available for the year ending 2004

	Rs.
Direct wages	60,000
Purchase of materials	72,000
Indirect materials	3,600
Indirect wages	5.400
Office salaries	7.200
Employer's contribution to employees	
State insurance	600
Printing and stationery	1.200
Power and fuel	5.400
Legal charges	864
Office rent	1.200
Sales (9000 units)	1,80,000
Opening stock:	
Raw materials	12,000
Work in progress	2.880
Finished goods (600 units at the rate of Rs. 16.25 per unit)	
Closing stock:	
Raw materials	13,344
Work in progress	9,600
Finished stock (1200 units)	?

Value the finished stock cost of production

Prepare a cost sheet showing different elements of cost.

(University of Bombay, B.Com., April 2005)

[**Ans.** Profit Rs. 39525]

Tender Statement

4. The following figures have been given from a factory for the year 1996.

	Rs.
Materials	12,00,000
Wages	10,00,000
Factory overhead	6,00,000
Administration expenses	6,72,000
Selling overhead	4,48,000
Distribution overhead	2,80,000
Profit	8,40,000

In 1997, the firm wants to execute a work order which requires Rs. 16,000 materials and Rs. 10,000 for direct wages. Determine the price at the same rate of profit at 1996.

(University of Madras, B. Com. May 2005)

[**Ans.** Sales Rs. 57600]

5. The accounts of the Modern Engineering Co. Ltd. shows the following details for the year ending 31st December 2003:

	Rs.
Materials consumed	17,50,000
Direct labour	13,50,000
Works overhead	3,71,250
Establishment expenses	2,32,500

Show the works cost and total cost, the percentage that the works overhead cost bears to direct labour and percentage that the establishment and general expenses bear to the works cost.

What price should the company quote to manufacture a machine which it is estimated, will require an expenditure of Rs. 7,500 in material and Rs. 6,000 in wages, so that it will yield a profit of 20% on selling price? *(Osmania University, B. Com., October 2004)*

[**Ans.** Quotation price Rs. 20,206]

6. The data pertaining to "XYZ" factory using a job costing system are as follows at the end of 31-03-2002. Direct materials Rs. 9,00,000. Direct wages Rs. 7,50,000: selling and distribution overhead Rs. 5,25,000. Administrative overhead Rs. 4,20,000 and Factory overhead Rs. 4,50,000.

(*i*) Prepare a cost sheet showing all the details.

(*ii*) For 2002-03 the factory has received an order for a number of jobs. It is estimated that the direct materials would be Rs. 12,00,000 and direct labour cost Rs. 7,50,000. What would be the price of job if the factory intends to earn the same rate of profit on sales assuming that the selling and distribution overhead has gone up by 15%. The factory recovers factory overheads as a percentage of direct wages and administrative and selling and distribution overhead as a percentage of works cost, based on the cost rates prevalent in the previous year. *(Osmania University, B. Com., March 2003)*

[**Ans.** Tender price Rs. 42,84,000]

7. From the following particulars, you are required to prepare cost sheet

	Rs.
Stock of finished goods 31-12-98	72,800
Stock of raw materials 31-12-98	33,280
Purchase of raw materials	7,59,200
Productive wages	5,16,880
Sales	15,39,200
Stock of finished goods 31-12-99	78,000
Stock of raw materials 31-12-99	35,360
Works overhead charges	1,29,220
Office overheads	70,000

The company is intending to send a quotation for a large plant. The estimated material cost is Rs. 52,000; wages Rs. 30,000. The quotation is to make a profit of 20% on selling price. Show that the amount of the quotation would be if based on the above percentage.

(Sri Krishnadevaraya University, B. Com., March 2006)

[**Ans.** Quotation price Rs. 1,19,438]

8. In respect of a factory the following figures have been obtained for the year 2003

	Rs.
Cost of materials	12,00,000
Factory overheads	6,00,000
Selling overheads	4,48,000
Distribution overheads	2,80,000
Direct wages	10,00,000

Administrative overheads	6,72,000
Profit	8,40,000

A work order has been executed in 2004 and the following expenses have been incurred: Materials Rs. 16,000 and wages Rs. 10,000. Assuming that in 2004, the rate of factory overheads has increased by 20%, distribution overheads have gone down by 10% and selling and administrative overheads have each give up by $12\frac{1}{2}$%. At what price should be product by sold as to earn the same rate of profit on the selling price as in 2003 ? Factory overheads are based in direct wages while all other overheads are based on the factory cost.

(University of Bombay, B. Com. October 2002)

[**Ans.** Quotation price for 2004 is Rs. 61,354]

9. The Managing Director of a company producing consumer durables seeks your assistance in the matter of fixation of selling price for one of its products called 'X'. The cost structure of product 'X', the unit selling price of which is Rs. 45,000 is as under:

Direct materials	50%
Direct labour	20%
Overheads	30%

An increase of 15% in the cost of materials and 25% in the cost of labour is anticipated. These increased costs in relation to the present selling price would cense a 25% decrease in the amount of present profit per unit of X.

You are required to:

(*a*) Prepare a statement of profit per unit as at present.

(*b*) Find out the revised selling price to produce the same percentage of profit to sales as before.

(University of Delhi, B. Com. (Hons), 2002)

[**Ans.** Selling price Rs. 50,625.]

Estimated Cost Sheet

10. The trading, profit and loss account of Vijaya manufacturing company for the year ending 31-12-2003 was as follows:

TRADING PROFIT & LOSS ACCOUNT
(for the year ending 31-12-2003)

To raw materials purchased	80,000	By sales (2500 units)	2,50,000
To direct wages	30,000	By closing stock of raw materials	5,000
To direct expenses	25,000		
To factory expenses c/d	40,000		
To gross profit c/d	80,000		
	2,55,000		2,55,000
To office salaries	25,000	By gross profit c/d	80,000
To office rent	12,000	By Dividend Received	10,000
To selling expenses	12,500	By Discount received	7,500
To preliminary expenses written off	2,500		
To Goodwill written off	5,500		
To Net profit c/d	40,000		
	97,500		97,500

For the year 2004, it is estimated that:

(1) Units produced and sold will rise by 20%

(2) Prices of raw materials per unit will rise by 10%

(3) Direct wages per unit will increase by 25%

(4) Direct expenses will increase by Rs. 5,000 in total

(5) Factory expenses per unit will increase by 25%

(6) The office premises which was on rental basis in 2003 would be purchased by the company on which depreciation would be Rs. 6000 in 2004.

(7) Selling expenses per unit will remain same.

You are required to prepare a statement showing estimated cost and profit for the year ended 31-12-2004 considering that company shall charge a profit at 20% on sales.

(University of Bombay, B.Com., April 2003)

[**Ans.** Profit for the year 2003 Rs. 30500 Profit for the year 2004 Rs. 70,000.]

11. Super vision company furnishes you with the following information about its 1000 TV sets manufactured and sold during the year.

Materials	18,00,000
Direct wages	10,00,000
Power and stores	2,40,000
Indirect wages	3,00,000
Factory lighting	1,20,000
Cost of rectifying defective work	60,000
Office and administration expenses	6,80,000
Selling and distribution expenses	1,20,000
Sale of scrap	40,000
Sale of 1000 TV sets	62,00,000
Repairs and depreciation of machinery	2,00,000

Prepare the cost sheet for the above year showing the elements of cost per unit. Prepare also the estimated cost sheet for the next year assuming that:

(1) Materials cost and direct wages cost will increase by 10% and 15% respectively.

(2) Factory overheads will be recovered as a percentage of works cost as last year.

(3) Office overheads and selling overheads will be recovered as percentage of works cost as last year.

(4) 1500 TV sets will be produced and sold at Rs. 6,600 each in the next year.

(University of Bombay, B. Com., April 2002)

[**Ans.** Estimated sales Rs. 99,00,000]

12. From thc following profit and loss account presented by sun and company for the year ended 31st March 2005, prepare the estimated cost sheet for the year ended 31st March 2006, bearring in mind that the rate of profit will be @ 25% on cost.

PROFIT AND LOSS ACCOUNT

(for the year ended on 31-3-2005)

To purchase of raw materials	4,30,000	By sales (4000 units)	9,86,800
To Direct wages	2,75,000	By closing stock	
To Direct expenses	72,000	Raw material	24,200
To other chargeable expenses	93,500		
To gross profit c/d	1,39,900		
	10,11,000		10,11,000

To office rent	34,600	By Gross profit c/d	1,39,900
To office salary	72,000	By discount received	21,000
To selling expenses	17,290	By Dividend received	42,000
To preliminary expenses written off	4,720		
To depreciation of office premises	10,000		
To net profit	64,290		
	2,02,900		2,02,900

It is estimated for the year ended 31st March 2006 that:

(*i*) Production and sales will increase by 15%

(*ii*) Price of raw materials will go up by 20%

(*iii*) Direct wages will go up by 20%

(*iv*) Direct expenses will increase to Rs. 86,400

(*v*) Other chargeable expenses will increase by 15% per unit.

(*vi*) Depreciation on office premises for the year 2006 will be Rs. 8,000.

(University of Calcutta, B. Com. (Hons), 2005)

[**Ans.** Estimated sales Rs. 16,07,553]

❑❑❑

CHAPTER 2

JOB AND BATCH COSTING

INTRODUCTION

Job costing as a distinctive method of costing is a form of specific order costing which is adopted to execute the work strictly according to customers specification. The production process depends upon the number of orders received from customers. As such production is not standardised and intermittent in nature. The goods manufactured is not for stocking but for each job as every work order differs from customer to customer. The purpose of job costing is to ascertain the profit or loss made on each job. Further cost of job is compared with the estimated cost to indicate whether estimation was defective or the actual cost incurred is excessive. Such an analysis helps in taking remedial action to improve efficiency and also facilitate revision of estimates.

DEFINITION AND FEATURES OF JOB COSTING

Kohler in his 'Dictionary for Accountants' defines job costing as "a method of cost accounting whereby cost is compiled for a specific quantity of product, equipment, repair or other service that moves through the production process as a continuously identifiable unit, applicable material, direct labour, direct expense and usually a calculated portion of the overhead being charged to a job order".

Industries which employ job costing have the following characteristic features:

(*a*) Each job is dissimilar in nature requiring special treatment. Hence the operations, processes and the sequence may differ from job to job.

(*b*) Production is intermittent and not continuous.

(*c*) Production is undertaken against customer's orders and not for stock.

(*d*) The industries need not incur selling and distribution expenses as the customers themselves come, place orders and collect the goods after production.

(*e*) Each job is treated as a cost unit under this method of costing.

(*f*) Each job is distinctively identified by a production order throughout the production stage.

(*g*) The cost of production of every job is ascertained after the completion of the job.

Examples of industries which employ job costing are: foundries, printers, machine tool making industries, engineering workshops, toy making concerns, furniture making concerns, management consulting concerns, interior decorations, musical instruments, advertising concerns and so on.

Though job costing is applied in manufacturing concerns where production process is not standardised it can still be used in conjunction with standard costing. For example, a company which manufactures machine tool against customers specification, can do so by the use of standard parts and sub-assemblies that can be pre-costed. Subsequently job costing can be applied for accounting non-standard component parts and for final assembly of all parts.

PROCEDURE OF JOB COSTING

The procedure that is commonly applicable to a normal sale transaction equally applies in case of job costing. This is explained under the following steps:

1. Receiving an Enquiry

Before placing an order with the manufacturer, usually the customer will enquire about the price, quality to be maintained, the duration within which the order is to be executed and other specification of the job.

2. Estimation of the Price of the Job

The cost accountant estimates the cost of job after considering the various elements of cost and keeping in mind the specification of customer. This is based on the cost of execution of similar job in the previous years and considering the possible changes in the various elements of cost. The estimated cost of the job is then informed to the prospective customer.

3. Receiving of Order

The customer will then place the order if he is satisfied with the quotation price and other terms of executing the job. The production control department receives the order and it will give a number for every order thus received which is known as job order number. The job is known by this number until it is completed.

4. Preparation of Production Order

A production order is prepared by the production control department and is sent to the concerned persons such as the employees to enable them to carry out the job, to the storekeeper to facilitate him to stock all the required materials, to cost accountant to enable him to prepare job cost sheet in order to ascertain the profit on every job completed. The production order consists of the following particulars:

(*a*) Date on which the order is prepared.
(*b*) Job order number.
(*c*) Description of the article's to be produced.
(*d*) Number of articles to be produced.
(*e*) Date of starting the work.
(*f*) Date of completing the work.
(*g*) Listing of materials to be used.
(*h*) Sequence of production process.
(*i*) Signature of production manager.

5. Preparation of Design

When the job to be executed requires special treatment, a design to meet the customers specification is prepared by the production planning department. This is done by the engineering department in consultation with the production planning department.

6. Execution of Job and its Inspection

The job will be started as per the schedule of production. Necessary materials, employees, tools are used to complete the production. The production process is supervised by the production manager or supervisor from time to time to ensure that the job executed is in accordance with customer's specification and that it is completed as per the production schedule.

7. Despatch of Goods

The finished products are then packed and delivered to the customer as per the delivery schedule. Payment is settled as per the agreed mode of payment.

ACCOUNTING SYSTEM UNDER JOB COSTING

As in any method of costing, the four important elements of cost enter into the cost of production. They are:

1. Materials Cost

Materials are classified into direct and indirect on the basis of traceability of materials to the job. Those materials which are traceable to the job are treated as prime cost element and those not traceable to the job are treated as manufacturing overhead.

The following procedure is followed for recording materials:

(*a*) **When materials are purchased from supplier:** The following entry is passed:

Materials a/c *Dr.*

To cash a/c

Stock of materials is an asset account and appears in the balance sheet. The cost of any unused materials will be included in the stock of materials account at the end of the accounting period.

(*b*) **When materials are issued to job:** A material, requisition note or a bill of materials is prepared to draw materials from stores to start the work on the job. The following entry is passed to record the issue of materials to the job:

Job order no. *Dr.*

To materials

The production manager identifies the nature, quantity and quality of materials required and places the materials requisition note with the storekeeper. Once the materials are issued a copy of the requisition is sent to the cost accounting department to show that such materials have been used and to that extent the stock of materials has been reduced.

However, in large-sized concern, after the issue and pricing of materials are made, the cost of materials so issued are recorded in a material abstract. This facilitates the analysis and collection of cost of direct materials according to job or work orders. The cost of direct materials are then posted to individual job cost sheet. The transactions are posted weekly or fortnightly. However overheads from materials abstract is posted to jobs periodically.

(*c*) Any surplus materials unused on the job is returned back to storeroom supported by a material transfer note. The following entry passed:

Materials a/c *Dr.*

To job order no.

2. Direct Labour Cost

The direct labour cost is debited to the job on which work is directly performed. Clock card is used to record the hours worked by each employee. At the end of each week, the payroll department uses the information on the clock card to calculate each employee's pay. The clock card only records the total time worked by each employee. It is used by the employee to indicate how much time he worked in a period and by the employer to determine how much to pay the employee for that work. A clock card does not record the time spent by each employee on the jobs. This information is recorded by the employees on a time ticket. A time ticket shows the date on which the work was completed, the employee's name, name of the department, the time of starting and finishing the work. If the work is performed directly on a job, the employee would record the job number, indicating the labour as direct labour.

The cost accounting department collects all the labour time tickets, records the pay rate of the employee on the ticket and calculate the labour cost of the operation.

Treatment of overtime wages in Job Costing. When overtime work is done on a job to speed up production with a view to expedite the delivery, the overtime wages is to be debited to that job on which overtime work is performed. Where overtime work is performed as a routine work it forms part of factory overhead.

3. Direct Expenses

The direct expenses under job order industries usually include cost of designs, moulds, hiring of special tools and equipments and maintenance cost of such tools.

4. Factory Overheads

In most cases predetermined overhead rates are used for the absorption of factory overheads. The predetermined overhead rate is calculated at the beginning of the year by estimating the total overhead cost for the year and then dividing either by direct labour cost or direct labour hours. Throughout the year the overhead rate is applied to calculate the overheads and thus charged to job.

PREPARATION OF JOB COST SHEET

A separate job cost sheet is prepared for every job undertaken. This is to facilitate the calculation of cost of the job separately, with a view to know the amount of profit earned on every job. All the elements of cost discussed above are recorded in the job cost sheet. Its contents are as follows:

(*a*) Identity or name of the customer.
(*b*) Date of starting the job.
(*c*) Date on which the job is to be completed.
(*d*) Date on which the job is actually completed.
(*e*) The job number.
(*f*) The name of the product – its description in brief.

The job cost sheet constitutes the heart of job order industries. In the words of Blocker and Weltmer (*Cost Accounting*) "the heart of job order system is the job cost sheet which is used as a daily record of material, direct labour and estimated factory overhead cost for each department or operation for the entire course of production". A specimen form of job cost sheet is given below:

JOB COST SHEET

Name of the customer .... Job no. ....
Quantity on order .... Date promised ....
Name of the product .... Date started. ....
Date completed ....

Materials cost			*Labour cost*			*Overheads*		
Date	*Requsition number*	*Amount*	*Date*	*Hours*	*Amount*	*Hours*	*Rate*	*Amount*

Cost summary	
Materials cost	....
Labour cost	....
Direct expenses	....
Prime cost	....
Overheads	....
Total profit	....
Selling price	....

Fig. 2.1. *Job cost sheet.*

Advantages of Job Costing

1. It records costs more accurately.
2. Comparison of two jobs enables to know profit or loss on jobs. Only profitable jobs alone can be taken up by the enterprise.
3. It facilitates estimation of cost of similar jobs.
4. It facilitates cost control by comparing actuals with estimates.
5. As cost of materials, labour and overhead for every job is available daily, the trend of costs can be known by management.
6. The use of predetermined overhead rates involve application of budgetary control technique.
7. It facilitates identification and control of spoilages and defectives.
8. It is useful in quoting cost plus contract.

Disadvantages

1. It is expensive as cost is accumulated and ascertained for each job separately.
2. The scope of committing mistakes are high as the costs of one job may wrongly be posted to the cost of other job.
3. When drastic economic changes take place cost comparison becomes difficult.

CHAPTER 2

TREATMENT OF SPOILAGES, DEFECTIVES, UNDER JOB COSTING

Spoiled Goods

Spoilages are goods that do not meet production standards and are either sold for their salvage value or discarded. When spoiled goods are discovered, they are taken out of production and no further work is performed on them.

Spoilages can be of two types, *viz.,* (1) Normal spoilage and (2) Abnormal spoilage. Normal spoilage arises from efficient production and are considered as an unavoidable cost of producing good units. It is treated either as factory overheads or the sale proceeds of spoilages are deducted from the cost of production.

Any spoilage of goods which are in excess of normal spoilage is treated as abnormal spoilage. It arises on account of inefficient operations. The cost of abnormal spoilages are kept outside the purview of cost of production. Instead they are charged to costing profit and loss account.

Defective Goods

The goods which do not meet production standards and requires further processing in order to sell them as good units are known as defective goods.

Defectives are again classified into normal and abnormal defectives. The cost of reworking normal defective goods are treated as factory overheads. The total cost of reworking abnormal defective units are to be charged to costing profit and loss account because it is the result of inefficient operation and it should not become part of the production cost.

REPORTS IN JOB COSTING

The following are the important reports prepared under job costing:

1. Monthly Gross Profit Report on the Jobs Completed

This report shows an analysis of gross profit earned on various jobs completed during the month. A proforma of such a report is given below:

ABC CO. LTD.
Gross profit analysis on completed jobs
For the month of ....

Job no.	*Production cost*	*Selling price*	*Gross profit*
101	–	–	–
102	–	–	–
103	–	–	–
104	–	–	–
105	–	–	–
Total			

Fig. 2.2. *Monthly gross profit report.*

2. Completed Job Reports

Such reports are prepared and submitted to the concerned authority to investigate the differences between actual cost and estimated cost. A proforma of such a report is given below:

ABC CO. LTD.

Completed Jobs Report

Job No. ....
Started on ....
Completed on ....

Elements of cost	*Actual cost*	*Estimated cost*	*Defference*
Materials cost	–	–	–
Labour	–	–	–
Overhead	–	–	–
Total			
Gross Profit			
Sales			

Fig. 2.3. *Completed job report.*

3. Defective Work Reports

This type of report enables management to know the extent of defective units obtained and cautions management to take remedial steps to minimise the defectives.

4. Other Reports

Some of the reports which are prepared in addition to those mentioned above are,

(*a*) Purchasing report,
(*b*) Inventory report,
(*c*) Idle time report,
(*d*) Overhead analysis report.

Problem 1. The following direct costs, were incurred on Job no. 678 of standard T.V. company.

Materials				Rs. 4,020
Wages:	Dept. A	—	60 hours @ Rs. 3 per hour	
	Dept. B	—	40 hours @ Rs. 2 per hour	
	Dept. C	—	20 hours @ Rs. 5 per hour	

Overhead expenses were estimated as follows:

Variable overheads:

Dept. A Rs. 5,000 for 5,000 labour hours
Dept. B Rs. 3,000 for 1,500 labour hours
Dept. C Rs. 2,000 for 500 labour hours

Fixed overhead

Estimation at Rs. 20,000 for 10,000 normal working hours

You are required to calculate the cost of job no. 678 and calculate the price to give profit of 25% on selling price. *(S.V. University, B.Com., Oct. 1999)*

Solution: JOB COST SHEET FOR JOB NO. 678

Direct materials			4,020
Direct wages:			
Dept. *A* — 60 × 3	=	180	
Dept. *B* — 40 × 2	=	80	
Dept. *C* — 20 × 5	=	100	360
		Prime cost	4,380
Overheads:			
Variable:			
Dept. *A* $\frac{5{,}000}{5{,}000} \times 60$	=	60	
Dept. *B* $\frac{3{,}000}{1{,}500} \times 40$	=	80	
Dept. *C* $\frac{2{,}000}{500} \times 20$	=	80	220
Fixed $\frac{20{,}000}{10{,}000} \times 120$			240
		Total cost	4,840
		Profit 25% on SP or $\frac{1}{3}$ on cost price	1,613
		Selling price	6,453

Problem 2. A factory has three production departments. The policy of the factory is to recover the production overheads of the entire factory by adopting a single blanket rate based on the percentage of total factory overheads to total factory wages. The relevant data for a month are given below:

Department	*Direct materials Rs.*	*Direct wages Rs.*	*Factory overheads Rs.*	*Direct labour hours*	*Machine hours*
Budget:					
Machining	6,50,000	80,000	3,60,000	20,000	80,000
Assembly	1,70,000	3,50,000	1,40,000	1,00,000	10,000
Packing	1,00,000	70,000	1,25,000	50,000	—
Actuals:					
Machining	7,80,000	96,000	3,90,000	24,000	96,000
Assembly	1,36,000	2,70,000	84,000	90,000	11,000
Packing	1,20,000	90,000	1,35,000	60,000	—

The details of one of the representative jobs produced during the month are as under:

JOB NO. CW 7,083

Department	*Direct materials Rs.*	*Direct wages Rs.*	*Direct labour hours*	*Machine hours*
Machining	1,200	240	60	180
Assembly	6,00	360	120	30
Packing	300	60	40	—

The factory adds 30% on the factory cost to cover administration and selling overheads and profit.

Calculate the overhead absorption rate as per the current policy of the company and determine the selling price of the job no. CW 7083. *(C.A. Inter, Nov. 1999)*

Solution: CALCULATION OF OVERHEAD ABSORPTION RATE

Department	*Budgeted factory overhead*	*Budgeted direct wages*
Machining	3,60,000	80,000
Assembly	1,40,000	3,50,000
Packing	1,25,000`	70,000
	6,25,000	5,00,000

$$\text{Overhead absorption rate} = \frac{\text{Budgeted factory overhead}}{\text{Budgeted direct wages}} \times 100$$

$$= \frac{6,25,000}{5,00,000} \times 100$$

= 125% of direct wages

JOB COST SHEET OF JOB NO. CW 7083

Direct materials (1,200 + 600 + 300)		2,100
Direct wages (240 + 360 + 60)		660
	Prime cost	2,760
Overheads 125% of 660		825
	Factory cost	3,585
	Profit 30% of Rs. 3,585	1,075.50
	Selling price	4,660.50

Problem 3. New product company is contemplating to launch a new product in the market. The estimated cost details are as follows:

(*i*) Materials cost per unit — Rs. 40

(*ii*) Labour cost per unit — Rs. 36

(*iii*) Production overheads are to be calculated from the following data:

Production departments	*Hourly overhead rate (Rs.)*	*Normal monthly hours for overhead rates*	*Fixed overhead included in overhead (Rs.)*	*Time to be taken by new product (Hour)*
A	3.60	30,000	36,000	5
B	4.80	20,000	12,000	2.5
C	6.00	40,000	60,000	4

(*iv*) Annual administration and selling expenses — Rs. 2,50,000 (applicable to the new product)

(*v*) Estimated sales quantity — 50,000 per annum.

Based on the above information prepare a cost-sheet and find out the unit selling price with a profit margin of 40% on total cost. *(ICWA, Inter, Dec. 1992)*

Solution: CALCULATION OF VARIABLE AND FIXED OVERHEAD RATES

	A.	B	C
Total monthly overhead (normal hours × overhead rate)	1,08,000	96,000	2,40,000
Less Fixed overhead	36,000	12,000	60,000
Variable overhead	72,000	84,000	1,80,000
Variable overhead rate per hr.	2.40	4.20	4.50
Fixed overhead rate per hr.	1.20	0.60	1.50

JOB COST SHEET

Direct materials			40.00
Direct labour			36.00
		Prime cost	76.00
Variable overhead:			
Dept. *A* — Rs. 2.40 × 5 hrs	=	12.00	
Dept. *B* — Rs. 4.20 × 2.5 hrs	=	10.50	
Dept. *C* — Rs. 4.50 × 4	=	18.00	
		40.50	
Fixed overhead:			
Dept. *A* — Rs. 1.20 × 5 hrs	=	6.00	
Dept. *B* — Rs. 0.60 × 2.5 hrs	=	1.50	
Dept. *C* — Rs. 1.50 × 4	=	6.00	
			54.00
Factory cost			130.00
Administration and selling expense $\frac{\text{Rs. 2,50,000}}{\text{50,000 pieces}}$			5.00
Cost of sales			135.00
Profit 40% on cost of sales			54.00
Selling price			189.00

Problem 4. The Hind Manufacturing Co., used to absorb overheads by means of a blanket rate based on the direct labour hours employed. From 1st July, 2009, it decides to adopt separate rates for the three main production activities, viz., (a) Store keeping and material handling, (b) Machining and (c) Assembly. The estimates of costs and output upon which overhead absorption rates are predetermined remain unchanged as are the rates of absorption for selling and distribution costs.

Overhead absorption rates are:

Prior to 1st July, 2009:

0.50 paise per direct labour hour

Selling and distribution overhead — 20% of production cost

After 1st July, 2009:

Production overhead:

Storekeeping and material handling — 5% of direct material cost

Machining — 0.75 paise per machine hour

Assembly — 0.30 paise per direct labour hour

Selling and distribution overhead — 20% of production cost

Direct costs of job 120 have been:

Direct materials cost	400
Direct wages:	
Machining — 200 hrs. at 0.60 P	120
Assembly — 100 hrs. at 0.40 P	40
	560

The quotation price of the job is Rs. 896 and it required 184 machine hours to complete. Show the job cost sheet for job 120.

(*a*) As it would appear if the job had been completed entirely in the month of June, and

(*b*) As it would appear if the job were completed entirely in the month of July.

Solution: JOB COST SHEET FOR THE MONTH OF JUNE

Customer: Job no. 120

Particulars:

Started on .... June Finished .... June

Direct materials cost		400
Direct wages:		
Machining — 200 hrs. @ 0.60 P		120
Assembly — 100 hrs. @ 0.40 P		40
	Prime cost	560
Production overhead — 300 hrs. @ 0.50 P		150
	Factory cost	710
Selling and distribution overheads 20% on factory cost		142
	Cost of sales	852
	Profit	44
	Selling price	896

JOB COST SHEET FOR THE MONTH OF JULY

Customer.............. Job no. 120

Particulars ....

Commenced no: 1st July 2009 Finished 31st July 2009

Direct materials	400	
Storekeeping and material handling	20	420
Machining:		
Direct wages: 200 hrs. @ 0.60 P		120
Overhead: 184 hrs. @ 0.75 P (*see note*)		138

Assembly:		
Direct wages: 100 hrs. @ 0.40 P		40
Overhead: 100 hrs. @ 0.30 P		30
	Factory cost	748
Selling and Distribution overhead: 20% on 784		150
	Cost of sales	898
	Loss	2
	Selling price	896

Note: Overhead is absorbed on 184 machine hours, which is actual.

Problem 5. A factory produces goods against customer's order. Its manufacturing section consists of three department *P*, *Q* and *R*. It is the practice of the company to prepare quarterly budgets for the purposes of control and the absorption of overheads. The following information is extracted from the first quarter budget of the company:

	Departments			
	P (*Rs.*)	*Q* (*Rs.*)	*R* (*Rs.*)	*Total* (*Rs.*)
Direct materials	10,000	20,000	30,000	60,000
Direct labour cost	7,000	8,000	8,500	23,500
Variable factory overheads	5,000	4,000	6,000	15,000
Fixed factory overhead	—	—	—	15,000
Administrative overhead	—	—	—	10,000
Machine hours	6,000	4,000	10,000	20,000

The factory overheads are absorbed on the basis of machine hours and the administration overhead is absorbed as a percentage of factory cost.

The factory has received an order from its regular customer for the supply of a specific variety of products that pass through two departments only, *i.e.*, *P* and *Q*. In respect of the order, the following particulars are available:

	P	*Q*
Material cost (Rs.)	1,500	2,500
Direct labour cost (Rs.)	700	800
Machine hours	600	400

Prepare a job order cost sheet and ascertain the price of the order if the margin of 25% on sales is added to the total cost of production:

Solution: JOB ORDER COST SHEET

Materials cost		
Dept. *P*	1,500	
Dept. *Q*	2,500	4,000
Direct wages		
Dept. *P*	700	
Dept. *Q*	800	1,500
Prime cost		5,500

Factory overheads (on the basis of machine hours)

Variable: Dept. P – $\frac{5,000}{6,000} \times 600$	= 500		
Dept. Q – $\frac{4,000}{4,000} \times 400$	= 400		900
Fixed overhead $\frac{15,000}{20,000} \times 1,000$			750
	Facotry cost		7,150
Administration overhead — 8.81% of factory cost			629.91
	Cost of production		7,779.91
	Profit $\frac{1}{3}$ of cost		2,593.30
	Selling price		10,373.21

Problem 6. A jobber used to quote to his customer on the basis of cost plus a certain percentage to cover overheads and profit. He now follows departmental rate of recovery of overheads based on direct labour hours. The following information is available from his books for the period ending 31.3.2009:

Expenses	*Total (Rs.)*	*Suggested base for apportionment to departments*
Consumables stores	500	Direct labour hours
Rent and Rates	3,000	Floor area
Depreciation	2,500	Plant value
Indirect labour	4,000	Direct labour hours
Repairs and renewals	1,000	3: 3: 4
Labour amenities	1,500	No. of employees
E.S. Insurance	400	- Do-
Works manager's salary	3,600	Equally
General administration	6,000	5: 4: 3
Sundry expenses	400	12: 7: 1

Departments	*X*	*Y*	*Z*
Direct labour hours	5,000	3,000	2,000
Wage rate per hour (Re.)	0.60	0.50	0.40
Number of employees	20	20	10
Floor area (in sq. ft.)	3,000	2,000	1,000
Value of plant (Rs.)	15,000	9,000	6,000

You are required to: (*a*) Calculate the hourly overhead rates to be charged for work in each department, (*b*) Prepare a quotation for a job to which the following data relates:

Material		Rs. 300
Direct labour:	*X*	20 hrs.
	Y	15 hrs
	Z	10 hrs.

Profit = 20% of selling price

Solution: STATEMENT SHOWING OVERHEAD APPORTIONMENT

Expenses	*Ratio*	*Dept. X*	*Dept. Y*	*Dept. Z*
Consumable stores	5: 3: 2	250	150	100
Depreciation	5: 3:2	1,250	750	500
Indirect labour	5: 3: 2	2,000	1,200	800
Rent and rates	3: 2: 1	1,500	1,000	500
Repairs and renewals	3: 3: 4	300	300	400
Labour amenities	2: 2: 1	600	600	300
E.S. Insurance	2: 2: 1	160	160	80
Works manager's salary	1: 1: 1	1,200	1,200	1,200
General administration	5: 4: 3	2,500	2,000	1,500
Sundry expenses	12: 7: 1	240	140	20
		10,000	7,500	5,400

	Dept. X	Dept. Y	Dept. Z
Labour (hrs.)	5,000	3,000	2,000
Labour hour rate	$\frac{10,000}{5,000}$ = Rs. 2	$\frac{7,500}{3,000}$ Rs. 2.50	$\frac{5,400}{2,000}$ = Rs. 2.70

Estimate for job

Materials			300
Direct labour cost:			
Dept. *X*	20 hrs. @ Rs. 0.60	= 12.00	
Dept. *Y*	15 hrs. @ Rs. 0.50	= 7.50	
Dept. *Z*	10 hrs. @ Rs. 0.40	= 4.00	23.50
	Prime cost		323.50
Overheads:			
Dept. *X*	20 hrs. @ Rs. 2.00	= 40.00	
Dept. *Y*	15 hrs. @ Rs. 2.50	= 37.50	
Dept. *Z*	10 hrs. @ Rs. 2.70	= 27.00	104.50
	Factory cost		428.00
	Profit 20% on S.P.		107.00
	Selling price		535.00

Problem 7. A job manufacturing company has received an enquiry for the supply of 2,00,000 number of a special type of machine screw. Capacity exists for manufacture of the screws in the company's department number 2 but a fixed investment of Rs. 60,000 and working capital to the extent of 25% of the sale value will be required if the job is undertaken. The costs are estimated as follows:

Raw materials — 20,000 kgs. at Rs. 2.30 per kg.

Labour hours — direct 18,000 of which 2,000 — would be overtime hours payable at double the labour rate.

Labour rate — Re. 1 per hour.

Factory overhead — Re. 1 per direct labour hour.
Selling and distribution on cost — Rs. 23,000.
Material recovered as scrap at the end of the operations is estimated at Rs. 2,000.
The company expects a net return of 25% on the capital employed.
Prepare a job cost sheet indicating the price which should be quoted to the customer.

Solution: JOB COST SHEET

Direct materials: 20,000 kgs. @ Rs. 2.30			46,000
Direct labour:			
Normal wages 16,000 hrs. @ Re. 1		16,000	
Overtime wages 2,000 hrs. @ Rs. 2		4,000	20,000
	Prime cost		66,000
Factory overhead: Re. 1 for 18,000 hrs.			18,000
			84,000
Less: Sale of scrap			2,000
	Factory cost		82,000
Selling and distribution cost			23,000
	Cost of sales		1,05,000
	Profit		23,000
	Selling price		1,28,000

Note: Calculation of selling price

Let sales $= x$

Then S.P. = C.P. + Profit

$$x = 1{,}05{,}000 + \frac{25}{100} \text{ (Fixed capital + Working capital)}$$

$$x = 1{,}05{,}000 + \frac{25}{100} \text{ (Fixed capital + 25\% of sales)}$$

$$x = 1{,}05{,}000 + \frac{1}{4}\left(60{,}000 + \frac{x}{4}\right)$$

$$x = 1{,}05{,}000 + 15{,}000 + \frac{x}{16}$$

$$x = 16{,}80{,}000 + 2{,}40{,}000 + 16x$$

$$19{,}20{,}000 = -x + 16x$$

$$19{,}20{,}000 = 15x$$

$$x = \frac{19{,}20{,}000}{15} = 1{,}28{,}000$$

Problem 8. The process plant division of a group of companies has built a food packaging machine to a customer's requirement. A price of Rs. 49,000 had been quoted with the intention of achieving a profit of 25% on the selling price.

Customer:	Bond Food Ltd.	
Customer's order no.	72, dated 3.2.09	
Job order no.	141	
Date of commencement	5.3.09	
Date of completion	29.4.09	
Date of delivery	2.5.09	
	March	*April*
Material used:		
Machine dept.	Rs. 2,900	700
Assembly dept.	Rs. 1,900	1,400
Direct wage rate per hour:		
Machine dept.	Rs. 4	4.40
Assembly dept.	Rs. 5	5.25
Direct labour hours:		
Machine dept.	Hrs. 200	100
Assembly dept.	Hrs. 50	500
Machine hours in machinery dept:	Hrs. 350	180
Technical drawing (direct cost)	Rs. 2,115	

Production overhead is absorbed at the predetermined rate of Rs. 10 per direct labour hour in the assembly department and Rs. 15 per machine hour in the machining department.

Commissioning cost, i.e., installation and initial running of the machine were Rs. 750 and these are to be treated as a direct production cost.

Selling and general administration cost are charged to jobs at the rate of $33\frac{1}{3}\%$ of production cost.

You are required to prepare a job order cost sheet.

Solution: **Job Order Cost Sheet**

Customer: Bond Food Ltd. Job order no: 141

Customer order: 72 Date of delivery: 2.5.09

Date commenced: 5.3.09

Date completed: 29.4.09

Materials:					
March	—	Machine		2,900	
	—	Assembly		1,9000	4,800
April	—	Machine		700	
	—	Assembly		1,400	2,100
Direct wages:					
March	—	Machine	200 × 4	800	
	—	Assembly	50 × 5	250	1,050

April	—	Machine	100 × 4.40	440	
	—	Assembly	500 × 5.25	2,625	3,065
Direct cost:					
Technical drawing				2,115	
Commission cost				750	2,865
			Prime cost		13,880
Production overhead:					
March	—	Machine	350 × 15		5,250
	—	Assembly	50 × 10		500
April	—	Machine	180 × 15		2,700
	—	Assembly	500 × 10		5,000
			Factory cost		27,330
Selling and administration cost					9,110
			Cost of sales		36,440
			Profit		12,560
			Sales		49,000

Note: Profit is the difference between sales value given in the problem and cost of sales.

Problem 9. The elements comprising the cost of a factory output for the three years 2007, 2008, 2009 were as follows:

	Years		
	2007	*2008*	*2009*
Materials	Rs. 1,50,000	1,12,500	1,87,500
Labour	Rs. 67,000	65,000	69,000
Works on cost	Rs. 30,000	22,500	37,500
Administration on cost	Rs. 18,000	18,000	18,000
Selling and distribution cost	Rs. 9,000	7,000	5,000
Output	Tons 10,000	7,500	12,500

Following an examination of the figures, you are required to state on what basis would you allocate oncost in quoting for a job to supply 17,500 tons of the commodity representing the whole of the anticipated output for the year 2010. What price would you quote assuming there was no change in the organisation and that a profit of 15% on works cost is considered adequate?

Solution: QUOTATION FOR A JOB TO SUPPLY 17,500 TONS

Raw materials	—	It is Rs. 15 for all the 3 years so 17,500 × 15		2,62,500
Labour	—	It has been increasing @ Rs. 2,000 for every 2,500 tons		
Labour cost for 12,500 tons		=	69,000	
Labour cost for 5,000 tones		=	4,000	73,000
		Prime cost		3,35,500
Works oncost	—	It bears 20% relationship with materials		
	∴	20% of 2,62,500		52,500

	Works cost	3,88,000
Administration oncost: It is same for all the years.		
This year also it remains same		18,000
	Office cost	4,06,000
Selling and distribution cost — It shows a reduction by Rs. 2,000 for ever year		
So this year also it goes down by 2,000 (*i.e.,* 5,000 – 2,000)		3,000
	Cost of sales	4,09,000
	Profit	58,200
	Selling price	4,67,200

Problem 10. A company has two manufacturing shops. The shop floor supervisor presented the following cost for job *A* to determine the selling price:

	Rs. per unit
Material	*Rs. per unit*
Direct wages: 14 hrs. @ Rs. 2.50 per hr.	70
(Dept. *X* – 8 hrs. and Dept. *Y* – 6 hrs)	35
Chargeable expenses (stores)	5
	110
Add: $33\frac{1}{3}$% for expenses (overhead)	37
	147

Analysis of the P and L account shows the following:

Materials	1,50,000	Sales less returns	2,50,000
Direct wages:			
Dept. *X*	10,000		
Dept. *Y*	12,000		
Stores expenses	4,000		
Overheads:			
Dept. *X*	5,000		
Dept. *Y*	9,000		
Gross profit	60,000		
	2,50,000		2,50,000

It is noted that average hourly rates for the two department *X* and *Y* are similar. You are required to:

(a) Draw up a job cost sheet.

(b) Calculate the revised cost using overhead figures as shown in the P & L a/c as the basis for charging overhead to Dept. *X* and *Y*.

(c) Add 20% of total cost to determine the selling price. *(C.S. Inter, June 1991)*

Solution:

JOB COST SHEET

Job: A		Date started	
Description		Date completed	
Materials			70.00
Direct wages:			
	Dept. X — 8 × 2.50		20.00
	Dept. Y — 6 × 2.50		15.00
Chargeable expenses			5.00
		Prime cost	110.00
Overheads:			
	Dept. X (50% of direct wages) or, 8 × 1.25		10.00
	Dept. Y (75% of direct wages) or 6 × 1.875		11.25
		Total cost	131.25
		Profit (20% on cost)	26.25
		Selling price	157.50

Working Note:

Direct wages as per P & L a/c	Rs. 10,000.00	12,000.00
Hourly rate	2.50	2.50
Number of hours	$\frac{10{,}000.00}{2.50}$	$\frac{12{,}000.00}{2.5}$
	= 4,000 hrs.	= 4,800 hrs.
Calculation of overhead rate	$\frac{5{,}000}{4{,}000}$	$\frac{9{,}000}{4{,}800}$
	= 1.25 hrs.	= 1.875 hrs.

At the above rates for 8 and 6 hrs., for X and Y it comes to Rs. 21.25 (10 + 11.25)

On the basis of percentage on direct wages

Problem 11. *(Job Cost Sheet):* The estimated material cost of a job is Rs. 5,000 and direct labour cost is likely to be Rs. 1,000. In machine shop the job required machining by machine No. 8 for 20 hours and by machine No. 11. for 6 hours. Machine hour rates for machine No. 8 and machine No. 11 are Rs. 10 and Rs. 15 respectively. Considering only machine shop cost. The direct wages of all jobs last year amounted to Rs. 80,000 as against Rs. 48,000 factory overhead. Last year's factory cost of all jobs amounted to Rs. 2,50,000 as against Rs. 37,500 office expenses.

Prepare a quotation which guaranteed 20% profit on selling price

(Sri Sathya Sai University, B.Com. (Hons) October 2001)

CHAPTER 2

Solution: QUOTATION FOR THE JOB

Materials			5,000
Direct labour cost			1,000
		Prime cost	6,000
Factory Overhead:			
(*a*) Machine shop:			
Machine No. 8	= 20 × 10 = 200		
Machine No. 11	= 6 × 15 = 90		
(*b*) Others: $\frac{48,000 \times 1,000}{80,000}$	= 600		
	890		
		Factory cost	6,890
Office Overheads:			
For 2,50,000 of Factory cost the office overhead is 37500			
For Rs. 6890 → $\frac{37,500 \times 6890}{250,000}$			1,034
		Office cost	7,924
		profit $\frac{1}{4}$ %	
		Cost	1,981
		Selling price	9,905

Problem 12. The equipment division of M/s Centurion Engineering Works Limited have completed the manufacture of a rubber solution mixing equipment as per order received from one of its customers at an agreed price of Rs. 2,55,000. From the following details available, you are required to prepare a job order cost sheet for consideration by the management:

(*A*) Details of the Job:

Customer	:	RSA Enterprises
Order no. of the customer	:	Rs. 1611 dated Dec. 21, 2,000
Job order no allotted	:	30,631
Date of starting the job at works	:	January 27, 2001
Date of completion at works	:	March 15, 2001
Date of delivery at site	:	March 28, 2001
Date of commissioning	:	March 30, 2001

(*B*) Details of Expenditure	Jan.	Feb.	March
Materials:			
Machinery section	Rs. 22350	1875	675
Assembly section	Rs. 4683	517	6100
Direct Labour Hours:			
Machinery Rs. 12 per hr.	160 hrs.	2000 hrs.	1650 hrs.
Assembly Rs. 15 per hr.	40 hrs	800 hrs.	1260 hrs.

Total machine hours spent in machinery section	60 hrs.	1440 hrs.	500 hrs.

Cost of Drawings Rs. 6500

The company had to produce some dies and tools with Rs. 15,000 specifically required for use in executing the above order. However on completion the value of such used dies and tools was estimated at 30% of the original cost and the company envisages to use them up in executing a future order already in the pipeline.

The company has a system of absorbing production overheads at predetermined rates on machine hours utilised for machining section and direct labour hours spent for assembly section. The current rates are Rs. 18 per machine hour and Rs. 6 per labour hour for machining and assembly sections respectively.

The company also incurred an expenditure of Rs. 6150 towards installing and carrying out trial runs at the customer's premises. As per the company's policy these expenses are to be charged as direct production cost.

Selling and distribution costs are to be charged to individual jobs at 16.67% of production cost.

(ICWA, Inter, Stage 1 June 2001)

Solution: JOB COST SHEET

Name of customer	:	RSA Enterprises
Customer order Ref	:	R.S, 1691 date 21-12-2000
Date of commencement	:	27.1.01
Date of completion	:	15.3.01
Job order no.	:	30631
Date of delivery	:	28.3.01
Date of commissioning	:	30.3.01
Price quoted	:	Rs. 2,55,000

Materials:			
January :	Machining	22,350	
	Assembly	4,683	27,033
February :	Machining	1,875	
	Assembly	517	2,392
March :	Machining	675	
	Assembly	6,100	6,775
			36,200
Labour:			
Janaury :	Machining 160 hrs. @ Rs. 12	1,920	
	Assembly 40 hrs. @ 15	600	2,520
February :	Machining 2000 hrs @ Rs. 12	24,000	
	Assembly 800 hrs. @ Rs. 15	12,000	36,000
March :	Machining 1650 hrs. @ Rs. 12	19,800	
	Assembly 1260 hrs. @ Rs. 15	18,900	38,700
Other Direct Costs:			
Cost of Drawing			6,500
Dies and Tools		15,000	
Less: Residual value @ 30%		4,500	10,500

Installation and commissioning cost				6,150
			Prime cost	1,36,570
Production Overhead				
January :	Machining 60 hrs. @ Rs. 18	= 1080		
	Assembly 40 hrs. @ Rs. 6	= 240		1320
February :	Machining 1440 hrs. @ Rs. 18	= 25920		
	Assembly 800 hrs. @ Rs. 6	= 4800		30720
March :	Machining 500 hrs @ Rs. 18	= 9,000		
	Assembly 1260 hrs. @ Rs. 6	= 7,560		16,560
			Works cost	1,85,170
Selling and Distribution overhead 16.67% on 1,85,170				30,868
			Cost of sales	2,16,038
			Profit	38,962
			Sales	2,55,000

Problem 13. A company engaged in the manufacture of air cooler pumps supplies the following information relating to September 1998

		Job 100	Job 101	Job 102	Job 103
Materials issued	(Rs.)	500	250	240	300
Purchases	(Rs.)	600	–	50	150
Wages	(Rs.)	500	150	200	250
Machine hours		50	30	40	15

Overhead is absorbed on the basis of machine hours and the hourly rate is Rs. 10. Overhead incurred during the month was Rs. 2,000. Job no. 100 was finished during the month and invoiced to customers at Rs. 4,000. The other jobs are still in progress.

Prepare a job cost sheet for jobs 100, 101, 102, and 103.

(Sri Sathya Sai University, B.Com. (Hons), October 2000)

Solution: JOB COST SHEET FOR SEPTEMBER 1998

		Job 100	Job 101	Job 102	Job 103
Materials issued		500	250	240	340
Purchases		600	–	50	150
Wages		500	150	200	250
	Prime cost	1600	400	490	700
Overheads		500	300	400	150
	Factory cost	2,100	700	890	850
	Profit	1,900	–	–	–
	Sales	4,000			
	Work-in-progress	–	700	890	850

WORK-IN-PROGRESS CONTROL A/C

	Rs.		Rs.
To stores issued (500 + 250 + 240 + 300)	1,290	By Job No. 100	2,100
To Purchases (600 + 50 + 150)	800	By Balance c/d	2,440
To wages (500 + 150 + 200 + 250)	1,100		
To overheads (500 + 300 + 400 + 150)	1,350		
	4,540		4,540

OVERHEAD CONTROL A/C

To total overhead incurred	2,000	By overhead absorbed 135 hrs @ Rs. 10	1,350
		By Balance c/d overhead under absorbed	650
	2,000		2,000

COSTING P & L A/C

To overhead under absorbed	650	By profit on job 100	1900
To Net profit	1,250		
	1,900		1900

Problem 14. The following information for the year ended 31st December 2002 is obtained from the books and records of a factory.

	Completed Jobs *Rs.*	*WIP* *Rs.*
Raw material supplied from: Stores	88,000	32,000
Wages	1,00,000	40,000
Chargeable expenses	10,000	4,000
Materials returned to stores	1,000	–

Factory overheads 80% of wages. Office overheads are 25% of factory cost and selling and distribution overheads are 10% of cost of production.

The completed jobs realised Rs. 4,10,000. Write up:

(1) Work-in-progress ledger control account

(2) Completed job ledger control account

(3) Cost of sales

(University of Delhi, B. Com. (Hons) 2001)

Solution:

Work in-Progress Ledger Control a/c

To materials	32,000	By Balance c/d	1,35,000
To wages	40,000		
To chargeable expenses	4,000		
To factory overheads (80% of 40,000)	32,000		
Factory cost	1,08,000		
To office overhead $\frac{25}{100} \times 1{,}08{,}000$	27,000		
	1,35,000		1,35,000

COMPLETED JOBS A/C

To materials	88,000		By cost of sales	3,46,250
Less returns	1,000			
		87,000		
To wages		1,00,000		
To chargeable expenses		10,000		
To factory overheads		80,000		
Factory cost		2,77,000		
To office overhead		69,250		
		3,46,250		3,46,250

COST OF SALES A/C

To completed job a/c	3,46,250	By sales	4,10,000
To selling and distribution overheads	34,625		
To profit	29,125		
	4,10,000		4,10,000

Problem 15. A fire occured in a factory and records containing information about the allocation of overhead charges were destroyed. It is however known that works overhead is absorbed at 100% of wages and office overhead is a fixed percentage of factory cost.

The following further information regarding two previous jobs is also available:

Job No.	*Materials*	*Wages*	*Direct expenses*	*Profit (% on total cost)*	*Selling price*
101	20,000	16,000	4,000	10%	80,080
102	32,000	24,000	2,000	15%	1,22,590

Find out the percentage and actual overheads and profit for each of the two jobs.

Solution:

Let the % of factory cost be F and % of office cost be A

JOBS NO. 101

Materials + Direct wages + Direct expenses + Factory overhead + Adm overhead + Profit = Sales

or Materials + Direct wages + Direct expenses + Factory overhead + Adm overhead = Sales – profit

$$\text{Job 101} = 20{,}000 + 16{,}000 + 4{,}000 + 16{,}000\,F + A\,(40{,}000 + 16{,}000F)$$

$$= 80080 - \frac{10}{110} \times 80080$$

$$= 16{,}000\,F + A\,(40{,}000 + 16{,}000\,F) = 32800 \quad \text{...(1)}$$

$$\text{Job 102} = 32{,}000 + 24{,}000 + 2{,}000 + 24{,}000\,F + A\,(58{,}000) + 24{,}000\,F)$$

$$= 1{,}22{,}590 - \frac{15}{115} \times 1{,}22{,}590$$

$$= 24{,}000\,F + A\,(58{,}000 + 24{,}000\,F) = 48600 \quad \text{...(2)}$$

Solving (1) and (2) equations

$$16{,}000\,F + 40{,}000\,A + 16{,}000\,AF = 32{,}800 \quad \times 3$$

$$24{,}000\,F + 58{,}000\,A + 24{,}000\,AF = 48{,}600 \quad \times 2$$

$$48{,}000\,F + 1{,}20{,}000\,A + 48{,}000\,AF = 98{,}400$$

$$48{,}000\,F + 1{,}16{,}000\,A + 48{,}000\,AF = 97{,}200$$

on Subtracting $\quad 4000\,A = 1200$

$$A = \frac{1200}{4000} = 0.30 \text{ or } 30\ \%$$

Substituting A in the equation (1)

$$16000\,F + 12000 + 4800\,F = 32800$$

$$= 20800\,F = 20800$$

$$F = \frac{20800}{20800} = 1 \text{ or } 100\%$$

∴ Factory overhead is 100% on Direct wages

Administration overhead is 30% of factory cost

Job Cost Sheet

		Job 101	Job 102
Materials		20,000	32,000
Direct wages		16,000	24,000
Direct expenses		4,000	2,000
	Prime cost	40,000	58,000
Factory overhead		16,000	24,000
	Factory cost	56,000	82,000
Office & Adm. overhead		16,800	24,600
	Office cost	72,800	106,600
	Profit	7,280	15,990
	Sales	80,080	1,22,590

CHAPTER 2

Problem 16. The Asiatic Engineering Company accept three jobs to which they allocate numbers 34, 35 and 36. The following costs relate to these jobs.

		Job 34	Job 35	Job 36
Materials issued	(Rs.)	70,000	1,30,000	2,50,000
Direct labour	(Rs.)	90,000	1,40,000	1,50,000
Direct expenses	(Rs.)	14,000	14,000	6,000

Overhead are absorbed as follows:

Works overhead	100% of Direct wages
Administration overhead	25% of Factory cost
Selling and Distribution overhead	50% on Factory cost

During the period raw materials were purchased for Rs. 7,00,000 and placed in stores. Actual overhead expenses were incurred as follows:

Works overhead	4,00,000
Administration overhead	2,70,000
Selling and Distribution overhead	70,000

All the jobs were commercial and completed in the period.

The selling prices were:

Job 34—Rs. 4,00,000, Job 35—Rs. 5,00,000, Job 36—Rs. 8,00,000

Prepare necessary ledger accounts.

Solution:

JOB 34 A/C

To materials	70,000	By cost of sales	3,43,000
To Direct Labour	90,000		
To Direct expenses	14,000		
To Works overhead	90,000		
To Administration overhead	66,000		
To Selling and Distribution overhead	13,000		
	3,43,000		3,43,000

JOB 35 A/C

To materials	1,30,000	By cost of sales	5,51,000
To Direct labour	1,40,000		
To Direct expenses	14,000		
To Works overhead	1,40,000		
To Administration overhead	1,06,000		
To Selling and Distribution overhead	21,000		
	5,51,000		5,51,000

JOB ON 36 A/C

To Materials	2,50,000	By cost of sales	7,23,000
To Direct labour	1,50,000		
To Direct expenses	6,000		
To Works overhead	1,50,000		
To Administration overhead	1,39,000		
To Selling & Distribution overhead	28,000		
	7,23,000		7,23,000

WIP CONTROL A/C

To Direct materials	4,50,000	By Balance c/d	16,17,000
To Direct labour	3,80,000		
To Direct expenses	34,000		
To Works overhead	3,80,000		
To Administration overhead	3,11,000		
To Selling & Distribution overhead	62,000		
	16,17,000		16,17,000

WORKS OVERHEAD A/C

To overhead incurred	4,00,000	By overhead estimated	3,80,000
		By under absorption	
		— Costing P & L a/c	20,000
	4,00,000		4,00,000

ADMINISTRATION OVERHEAD A/C

To overhead incurred	2,70,000	By overhead estimated	3,11,000
To over absorption			
overhead of costing P & L a/c	41,000		
	3,11,000		3,11,000

S & D OVERHEAD

To overhead incurred	70,000	By overhead estimated	62,000
		By under absorption	
		— Costing P & L a/c	8,000
	70,000		70,000

COSTING P & L A/C

To Cost sales	16,17,000	By sales	17,00,000
(3,43,000 + 5,51,000		By Administration overhead	
+ 7,23,000)		Over absorbed	41,000
To works overhead—under absorption	20,000		
To S & D overhead—under absorption	8,000		
To Balance c/d	96,000		
	17,41,000		17,41,000

BATCH COSTING

As the name itself suggest, under this method the cost is ascertained in respect of a batch of goods or components manufactured. There are certain products whose cost of production cannot be ascertained in isolation *i.e.,* for each and every unit of article produced. Examples of such products are hardware (such as bolts, nuts, screws, pins etc.), Bakery products, such as breads, biscuits, cakes, etc.) Such industries which manufacture products of this nature make use of batch costing. Other examples of industries which adopt batch costing are readymade garments, drugs and pharmaceuticals, spare parts and components parts as in the case of automobiles, radios, TV's, refrigerators, machineries, etc.

Batch costing does not differ from job costing in respect of accounting procedure. However, there are two points of differences between job costing and batch costing. Firstly, under job costing, each job is treated as a cost unit, whereas in case of batch costing, a batch of articles produced constitute a cost unit. Secondly, a job is undertaken for execution only on receiving an order, whereas, production under batch costing is taken up in anticipation of demand for such products as well as on receiving specific orders.

In order to know the cost of production of a batch of articles, a batch cost sheet is prepared. The production planning department determines the batch of articles to be produced depending upon the orders received or the potential demand for such articles. The other factors considered in determining a batch of articles to be manufactured are the capacity of machines, flow of work, treatment of work-in-progress and such other operation factors. The cost per article is obtained by dividing the total cost of batch by the number of articles manufactured in the batch. The units manufactured in a batch are identical in nature.

Advantages of Batch Costing

When compared to job costing, batch costing has the following advantages:

1. The accounting work is considerably reduced as a group of homogeneous jobs constitute a batch.
2. The variations in the costs arising under job costing is smoothened by means of averaging such costs and spreading over the batch of articles. This gives a consistent cost of production of every article in the batch.
3. It takes the benefit of reduced cost of production arising out of economic batch quantity.
4. Supervision becomes easy and more effective by means of spreading over the supervisor's time over all the units constituting the batch. Thus, the idle time of supervisor's as well as workers are eliminated.
5. The loss of time under job costing arising out of inter-job transfer of materials, labourers and tools is minimised under batch costing.

Disadvantages

1. Determination of a batch from various jobs often pose problem. It is difficult to come across absolute homogeneity of jobs.
2. When quantity of goods to be manufactured differs from customer to customer, again it becomes difficult to determine the batch.
3. If the production of a batch is wrongly undertaken, say, owing to sub-standard materials or defective operations, the whole batch of articles are to be discarded which will become a great loss to the manufacturer. Drugs and pharmaceutical products offer best example to this point of disadvantages.

Determination of Economic Batch Quantity

The concept of economic batch quantity is quite similar to economic order quantity. In batch costing, the determination of economic lot size or economic batch quantity assumes more importance. In fact, determining the size of the batch is a problem by itself under batch costing. This is so because, if the batches are many, the (*i.e.,* quantity of production is less in every job) ECONOMIES arising out of large scale production is not taken advantage of. On the other hand, if a batch is very large, it involves more of inventory and it has its own disadvantages. Therefore, it is always necessary to determine the optimum size of the batch or economic batch quantity before the production is started. The economic batch quantity also helps in eliminating the setting up time involved whenever a batch of articles are produced.

Thus it can be said that the concept of 'Economic batch quantity' is an example of the law of increasing returns and takes advantages of economies of large scale production. By producing the goods in an optimum batch, it reduces the cost of production, thereby maximising the profit. In determining the economic batch quantity, two conflicting forces operate. They are:

(1) Production in a very small batch quantity will involve more production cost which include set up cost and other factors such as loss of time due to change of work, loss of speed of work when work is frequently changed on different jobs. Wastage in materials, increased supervision cost, are also responsible for increased cost of production.

(2) Production in a very large batch quantity will lead to certain disadvantages such as, increased wear and tear of machines, breakdown of machines, problem of inspection, excessive defectives of goods, large inventory, locked up capital and so on. Thus economic batch quantity strikes a balance between these two factors in order to take maximum benefit from production process.

Different formulae are developed to determine the economic batch quantity. One such formula is given below:

$$EBQ = \sqrt{\frac{2US}{C}}$$

where
EBQ = Economic Batch Quantity
U = Units to be produced in a year
S = Set up cost per batch
C = Carrying cost per unit of production.

Where rate of interest and cost of production per unit is given, the following formula is applicable

$$EBQ = \sqrt{\frac{2US}{i \times C}}$$

where
EBQ = Economic Batch Quantity
U = Units to be produced
S = Set up cost per batch
i = Interest rate per year
C = Cost of manufacture per unit.

CHAPTER 2

Problem 17. *(Determination of economic batch quantity):* M/s. Kobo Bearings Ltd., is committed to supply 24,000 bearing per annum to M/s Deluxe Fans on a steady daily basis. It is estimated that it costs 10 paise as inventory holding cost per bearing per month and that the set up per run of bearing manufacture is Rs. 324.

1. What should be the optimum run size for bearing manufacture?
2. What would be due interval between two consecutive optimum runs?
3. Find out the minimum inventory holding cost. *(ICWA, Final, Dec. 1979)*

Solution:

1. Determination of economic batch quantity

$$EBQ = \sqrt{\frac{2US}{C}}$$

$$= \sqrt{\frac{2 \times 24{,}000 \times 324}{12 \times 0.10}}$$

$$= 3{,}600 \text{ units}$$

2. Interval between two consecutive optimum runs expressed in terms of days

$$\frac{24{,}000}{12} = 2{,}000 \text{ bearings per month of 25 working days}$$

$$= \frac{EBQ}{\text{Monthly production}} \times \text{No. of days in the month}$$

$$\frac{3{,}600}{2{,}000} \times 25 = 45 \text{ days interval}$$

3. Average inventory

$$= \frac{3{,}600}{2} = 1{,}800 \text{ bearings}$$

Minimum inventory holding cost per month is:

$$1{,}800 \times 0.10 = \text{Rs. } 180$$

Per annum it is

$$1{,}800 \times 0.10 \times 12 = \text{Rs. } 2{,}160$$

Problem 18. A contractor has to supply 10,000 paper cones per day to a Textile unit. He finds that, when he starts a production run he can produce 25,000 paper cones per day. The cost of holding a paper cone in stock for one year is 2 paise and the set up cost of production run is Rs. 18. How frequently should production run be made. *(ICWA, Final, Dec. 1983)*

Solution:

For 365 days at the rate of 10,000 paper cones it works out to 36,50,000 cones

$$EBQ = \sqrt{\frac{2US}{C}}$$

$$= \sqrt{\frac{2 \times 36{,}50{,}000 \times 18}{0.02}}$$

$$= 81{,}100 \text{ cones}$$

Production run in terms of days

$$= \frac{EBQ}{\text{Daily production}}$$

$$= \frac{81{,}100}{10{,}000} = 8 \text{ calendar days App.}$$

Problem 19. *(Problem involving setting up cost and production cost):* Component 'Pee' is made entirely in cost centre 100. Material cost is 6 paise per component and each component takes 10 minutes to produce. The machine operator is paid 0.72 paise per hour, and the machine hour rate is Rs. 1.50. The setting up of the machine to produce component pee takes 2 hrs. 20 minutes.

On the basis of this information prepare cost sheets showing the production and setting-up costs, both in total and per component, assuming that a batch of (a) 10 components, (b) 100 components and (c) 1,000 components are produced. *(C.A. Inter, Nov. 1992)*

Solution: COST SHEET FOR A BATCH OF 10 COMPONENTS

	Cost of the batch	Cost per unit
Setting up cost		
Wages: 2 hrs., 20 mins., at 0.72 per hr.	1.68	
Overheads: 2 hrs. 20 mins at 1.50 per hr.	3.50	
	5.18	0.518
Production cost		
Materials cost — 10 @ 0.06 P	0.60	0.060
Wages — 1 hr. 40 mins @ 72 paise per hr.	1.20	0.120
Overheads — 1 hr. 40 mins @ 1.50 per hr.	2.50	0.250
Total production and setting up cost	9.48	0.948

COST SHEET FOR A BATCH OF 100 COMPONENTS

	Cost of the batch	Cost per unit
Setting up cost		
Wages: 2 hrs. 20 mins 0.72 per hr.	1.68	
Overheads: 2 hrs. 20 mins 1.50 per hr.	3.50	
	5.18	0.05
Production cost		
Materials cost — 100 @ 0.06 P	6.00	0.06
Wages — 16 hr. 40 mins @ 72 paise per hr.	12.00	0.12
Overheads — 16 hr. 40 mins @ 1.50 per hr.	25.00	0.25
Total production and setting up cost	48.18	0.48

COST SHEET FOR A BATCH OF 1,000 COMPONENTS

	Cost of the batch	Cost per unit
Setting up cost		
Wages: 2 hr., 20 mins @ 72 per hr.	1.68	
Overheads: 2 hr. 20 mins @ 1.50 per hr.	3.50	
	5.18	0.005
Production cost		
Materials cost — 1000 @ 0.06 P	60.00	0.060
Wages — 166 hr. 40 mins @ 72 paise per hr.	120.00	0.120
Overheads — 166 hr. 40 mins @ 1.50 per hr.	250.00	0.250
Total production and setting up cost	435.18	0.435

Problem 20. A printing press is proposing to offer a leaflet advertising service to local trader. The following costs have been estimated for a batch of 10,000 leaflets.

Setting up cost		
Setting up machine	:	6 hrs. @ Rs. 10 per hr.
Art work	:	Rs. 20 per batch
Production cost		
Paper	:	Rs. 1.80 per 100 sheets
Other printing materials	:	Rs. 15
Direct labour cost	:	4 hrs. @ Rs. 6 per hr.

Fixed overheads allocated to this side of business are Rs. 1,000 per annum and recovered on the basis of orders received, which are expected to be 2 per week for 50 week in the year:

The management requires 25% profit on selling price.

You are required to

(*a*) Prepare a price to be quoted for the leaflets for batches of 2,000, 5,000, 10,000 and 20,000 leaflets

(*b*) Show the price per 1,000 leaflets on the basis of price ascertained above.

Solution: BATCH COST SHEET

	Batch sizes			
Setting up cost	*2,000*	*5,000*	*10,000*	*20,000*
Cost of setting up machine	60	60	60	60
Art work	20	20	20	20
Production cost				
Paper	36	90	180	360
Other printing materials	3	7.50	15	30
Labour cost	4.80	12	24	48
Fixed overhead 1,000 ÷ 100	10	10	10	10
Total cost	133.80	199.50	309	528
Profit	44.60	66.50	103	176
Selling price	178.40	266	412	704
Selling price 1,000 leaflets	89.20	53.20	41.20	35.70

Working Note: Calculation per 1,000 leaflets

For 2,000 leaflets the selling price is 178.40

For 1,000 →? $\dfrac{178.40 \times 1{,}000}{2{,}000} = 89.20$

For 5,000 leaflets the selling price is 266

For 1,000 →? $\dfrac{266 \times 1{,}000}{5{,}000} = 53.20$

For 10,000 leaflets the selling price is 412.20

For 1,000 ×? $\dfrac{412 \times 1{,}000}{10{,}000} = 41.20$

For 20,000 leaflets the selling price is 704

For 1,000 → $\dfrac{704 \times 1{,}000}{20{,}000} = 35.70$

Problem 21. K. Ltd., designs and manufactures deluxe containers for various industries. The company has been requested to submit quotations for supplying a specially designed container to which the following data relate:

Esimated design and Development costs

Engineer's time	—	50 hrs. @ Rs. 6 per hour
Draughtman's time	—	30 hrs. @ Rs. 3 per hour
Materials	—	Rs. 30
Engineering machine shop time	—	20 hrs. @ 5 per hour
Project supervision	—	Rs. 200
General overhead allocation	—	Rs. 350

Manufacturing cost including setting up costs

The production machine to be used require 20 hours of an engineer's time @ Rs. 3 per hour and has an output of 500 containers per hour.

Operatives are paid an hourly rate of Rs. 1.50 and production overhead is absorbed at Rs. 2.50 per direct labour hour worked. The direct material cost per container is Rs. 1.25. It is companies practice to provide for a profit of 10% on design and development costs and 30% on production cost.

Calculate the price to be quoted for the supply of (*a*) 5,000 containers, (*b*) 20,000 containers and (*c*) 50,000 containers.

Solution: BATCH COST SHEET

	Batch size		
	5,000	20,000	50,000
Estimated design and development cost:			
Engineer's time — 50 hrs. @ 6 per hr.	300	300	300
Draughtsman's time — 30 hrs. @ Rs. 3 per hr.	90	90	90
Materials	30	30	30
Engineer's machine shop time — 20 hrs. @ Rs. 5 per hr.	100	100	100
Project supervision	200	200	200
General overhead allocation	350	350	350
(A) Total estimated design and development cost	1,070	1,070	1,070
Manufacturing cost including Setting up cost:			
Material cost:	6,250	25,000	62,500
Wages:			
(*a*) Engineer's 20 hrs. @ Rs. 3	60	60	60
(*b*) Operatives (*see working note 1*)	150	600	1,500
Overheads (*see working note 2*)	250	1,000	2,500
(B) Total manufacturing and setting up cost	6,710	26,660	66,560
Profit — 10% on design and development cost	107	107	107
— 30% on manufacturing cost	2,013	7,998	19,968

(C) Total profit	2,120	8,105	20,075
Statement showing selling price:			
(*a*) Total estimated design and development cost	1,070	1,070	1,070
(*b*) Total manufacturing cost	6,710	26,660	66,560
(*c*) Total profit	2,120	8,105	20,075
Selling price	9,900	35,835	87,705

Working Note 1: Calculation of wages of operatives

Calculation of direct labour hours $= \dfrac{\text{Total containers}}{\text{No. of containers per hour}} = \dfrac{5{,}000}{50} = 100$ hrs.

$= \dfrac{20{,}000}{50} = 400$ hrs.

$= \dfrac{50{,}000}{50} = 1{,}000$ hrs.

Wages 100 hrs. @ 1.50 = Rs. 150
400 hrs. @ 1.50 = Rs. 600
1,000 hrs. @ 1.50 = Rs. 1,500

Working Note 2: Calculation of overheads

100 hrs. @ 2.50 = 250
400 hrs. @ 2.50 = 1,000
1,000 hrs. @ 2.50 = 2,500

Problem 22. A Jobbing factory has undertaken to supply 200 pieces of a component per month for the ensuing 6 months. Every month a batch order is opened against which materials and labour hours are booked at actuals. Overheads are leved at a rate per labour hour. The selling price contracted for is Rs. 8 per piece. From the following data present the cost and profit per piece of each batch order and overall position of the order for 1,200 pieces:

Month	*Batch output*	*Material*	*Direct wages*	*Direct labour hrs.*
Jan	210	650	120	240
Feb.	200	640	140	280
Mar.	220	680	150	280
Apr.	180	630	140	270
May	200	700	150	300
June	220	720	160	320

The other details are:

Month	*Chargeable expense*	*Direct labour hrs.*
Jan.	12,000	4,800
Feb.	10,560	4,400
Mar.	12,000	5,000
Apr.	10,580	4,600
May	13,000	5,000
June	12,000	4,800

Solution: BATCH COST SHEET

	Jan.	Feb.	Mar.	Apr.	May	June	Total
Materials	650	640	680	630	700	720	4,020
Wages	120	140	150	140	150	160	860
Charageable expenses (*see working note 1*)	600	672	672	621	780	800	4,145
Total	1,370	1,452	1,502	1,391	1,630	1,680	9,025
Profit	310	148	258	49	– 30	80	815
Sales	1,680	1,600	1,760	1,440	1,600	1,760	9,840
Total cost per unit	$\frac{1,370}{210}$ = 6.52	$\frac{1,452}{200}$ = 7.26	$\frac{1,502}{220}$ = 6.83	$\frac{1,391}{180}$ = 7.73	$\frac{1,600}{200}$ = 8	$\frac{1,680}{220}$ = 7.64	$\frac{9,025}{1,230}$ = 7.33

Overall position of the order for 1,200 units

Sales value of 1,200 units @ Rs. 8 per unit	9,600
Total cost for 1,200 units @ Rs. 7.33	8,796
Profit	804

Working Note:

Jan: For 4,800 labour hrs.: The chargeable expenses expended were 12,000

For 240 → ? $\frac{12,000 \times 240}{4,800} = 600$

Problem 23. The output for April 1987 for ABC Ltd., includes one batch of the Ranji model consisting of 500 units. The following data are taken from the works records:

1. Standard batch data for Ranji:

Material:

	Body and lid	Rs. 136
	Fittings	Rs. 27
Department	I : Labour time	20 hrs.
	II : Machine time	10 hrs.
	III : Machine time	8 hrs.
	IV : Labour time	20 hrs.

2. Department cost for April 1987:

Dept.	*Machine hours*	*Labour hours*	*Wages*	*Variable overheads*	*Fixed expenses*
I	—	6,000	1,800	3,600	0.50 P per labour hour
II	4,000	8,000	3,600	4,000	0.60 P per machine hour
III	6,000	12,000	7,200	9,000	0.80 P per machine hour
IV	—	10,000	3,000	4,500	0.40 P per labour hour

3. Administrative overheads are absorbed at the rate of $12\frac{1}{2}\%$ of works conversion cost. Selling and distribution costs are charged to the products at 10% of the selling value.

Required: Prepare the batch cost sheet of the Ranji model and suggest the minimum unit selling price which will yield a profit of 10% on the selling price.

(Sri Sathya Sai University, B.Com., March, 1989)

Solution: BATCH COST SHEET

Materials			
Body and lid		136	
Fittings		27	
Labour cost (*see working note 1)*			
Dept. I		6.00	
Dept. II		9.00	
Dept. III		9.60	
Dept. IV		6.00	30.60
	Prime cost		193.60
Overhead			
Variable (*see working note 2)*			
Dept I		12.00	
Dept. II		10.00	
Dept. III		12.00	
Dept. IV		9.00	43.00
Fixed			
Dept I	20 × 0.50	10.00	
Dept. II	10 × 0.60	6.00	
Dept. III	8 × 0.80	6.40	
Dept. IV	20 × 0.40	8.00	30.40
	Work cost		267.00
Administration overhead:			
12 ½% of conversion cost (*i.e.*, wages + overheads)			
Wages	30.63		
Overhead —Fixed	43.00		
Variable	30.00		
Conversion cost	104.00		13.00
Cost of production			280.00
Selling and distribution overhead (*see working note 3*)			35.00
	Cost of sales		315.00
	Profit $\frac{1}{9}$ on cost		35.00
	Sales		350.00

Working Note 1: Calculation of labour cost

Dept. I $\frac{\text{Rs. } 1,800}{6,000} = 0.30$

20 hrs. @ 0.30 = Rs. 6

Dept. II $\frac{\text{Rs. } 3,600}{4,000} = 0.90$

10 hrs. @ 0.90 = Rs. 9

Dept. III $\frac{\text{Rs. } 7,200}{6,000} = 1.20$

8 hrs. @ 1.20 = Rs. 9.60

Dept. IV $\frac{\text{Rs. } 3,000}{10,000} = 0.30$

20 hrs. @ 0.30 = Rs. 6.00

Working Note 2: Calculation of variable overheads

Dept. I $\frac{\text{Rs. } 3,600}{6,000} = 0.60$

20 hrs. @ 0.60 = 12.00

Dept. II $\frac{\text{Rs. } 4,000}{4,000} = 1.00$

10 hrs. @ 1.00 = 10.00

Dept. III $\frac{\text{Rs. } 9,000}{6,000} = 1.50$

8 hrs. @ 1.50 = 12.00

Dept. IV $\frac{\text{Rs. } 4,500}{10,000} = 0.45$

20 hrs. @ 0.45 = 9.00

Working Note 3: Calculation of selling and distribution overheads

Selling and distribution overheads is 10% of sales. But sales itself is not given. So

Let selling price	100
Less: profit 10%	10
Cost of sales	90

Cost of sales = Cost of production + Selling and distribution overhead

90 = 80 + 10

$$\frac{10}{80} \times 280 = \text{Rs. } 35$$

On verification of this amount of selling and distribution overhead, *i.e.,* 35 with sales, *i.e.,* 10% on 350 we get the same amount of Rs. 35.

Questions

I. Choose the Correct Answer:

1. Which of the following methods of costing can be used in a factory where the products differ in type of materials and work performed
   (a) Single or output costing (b) Job costing
   (c) Process costing (d) Contract costing
2. Tyre and tubes manufacturing company adopts
   (a) Output costing (b) Job costing
   (c) Contract costing (d) Batch costing
3. Given cost of production of a job Rs. 7,50,000 and profit percentage 20% on sales, the profit would be
   (a) Rs. 1,87,500 (b) Rs. 1,78,500
   (c) Rs. 1,57,800 (d) Rs. 1,85,700
4. Job costing is used in
   (a) Repairs shops (b) Brick making
   (c) Road repairs (d) Cycle manufacturing
5. Job order means
   (a) An order received from a customer for the production of a specific item.
   (b) Instructions to the shops to proceed with the production of the product.
   (c) An order to the storekeeper to issue necessary materials for production
   (d) All of the above
6. Which one of the following industries is not properly matched with its method of costing
   (a) Boiler — job costing
   (b) Hotel — operating costing
   (c) Readymade garments — batch costing
   (d) Oil refinery — process costing
7. Which one of the following method of costing is best suited for interior decoration
   (a) Job costing (b) Contract costing
   (c) Process costing (d) Operating costing
8. Which one of the following method of costing is best suited for garment manufacturing
   (a) Unit costing (b) Job costing
   (c) Batch costing (d) Process costing
9. In shoe manufacturing industries the most suitable method of costing is
   (a) Output costing (b) Batch costing
   (c) Job costing (d) Process costing
10. If the total cost of a product is Rs. 40,000 and 20% of the sale price is the profit to be added to cost, then the profit is
    (a) Rs. 6,000 (b) Rs. 10,000
    (c) Rs. 8,000 (d) Rs. 12,000
11. Pharmaceutical industries adopt
    (a) Output costing (b) Job costing
    (c) Batch costing (d) Process costing

12. At the level of production of the economical lot size

(*a*) Set-up cost is minimum

(*b*) Carrying cost is minimum

(*c*) Total cost of production is minimum

(*d*) Set-up cost is greater than carrying cost

13. Job costing is used in

(*a*) Paper mills (*b*) Printing press

(*c*) Paints and varnishing (*d*) Pharmaceutical products

[***Answers:*** 1. (*b*), 2. (*d*), 3. (*a*) 4. (*a*) 5. (*b*), 6. (*a*), 7. (*a*), 8. (*c*), 9. (*c*), 10. (*c*), 11. (*b*), 12. (*c*), 13. (*b*)]

II. Fill in the blanks:

1. The formula for determining economic batch quantity is ....

2. Under job costing each job is known by its ....

3. In industries where batch costing is employed an important point is the determination of ....

4. An item of cost which is common to prime cost and conversion cost is.............. cost.

5. The backbone of job costing is ....

6. In .... method of costing, the cost of a group of products is ascertained.

7. Under job costing each .... is a cost unit to which all elements of costs are assigned.

[***Answers:*** 1. $\sqrt{\frac{2US}{C}}$ 2. job number, 3. economic batch quantity, 4. labour, 5. job cost sheet, 6. batch, 7. job]

III. Mark true or false in the space provided:

1. In job costing each job should be continuously identifiable from the raw materials stage to the stage or completion. *T/F*

2. In Asian Paints Company job costing method should be applied. *T/F*

3. A job cost sheet is used to collect, classify and summarise costs. *T/F*

4. In job costing method a cost sheet is prepared for each job. *T/F*

5. Job costing cannot be used in industries using standard costing. *T/F*

6. Job costing can be more conveniently used in concerns producing uniformly any specific product. *T/F*

7. Batch costing is a variant of job costing. *T/F*

8. A production order is an order received from a customer for a particular job. *T/F*

9. Compared to process costing, cost control under job costing is more complicated. *T/F*

10. The concept of economic batch quantity is similar to that of economic order quantity. *T/F*

[**Answer:** 1. *F*, 2. *F*, 3. *T*, 4. *T*, 5. *T*, 6. *F*, 7. *T*, 8. *F*, 9. *T*, 10. *T*]

IV. Match the following:

A	*B*
1. Road transport	(*a*) Multiple costing
2. Cement	(*b*) Operating costing
3. Engineering industry	(*c*) Output costing
4. Bicycle industry	(*d*) Process costing
5. Pharmaceuticals	(*e*) Job costing
	(*f*) Batch costing
	(*g*) Contract costing

[***Answer:*** 1. (*b*), 2. (*c*), 3. (*e*), 4. (*a*), 5. (*f*)]

CHAPTER 2

V. Short Answer Questions:

1. What are the salient features of batch costing?

   (Calicut University, B. Com., Oct. 1989 ; Kerala University, B. Com., April 1990)

2. What are the salient features of job costing? Indicate the industries to which it is suitable.

   (Calicut University, B. Com., Oct. 1989)

3. What is the importance of estimation in job costing?

   (Calicut University, B. Com., April 1988)

4. What do you understand by Batch costing? In what industries is it applied?

   (C.A. Inter, May 1987)

Exercises

Exercise 1. 'A' Co. Ltd., had absorbed overhead by means of a blanket rate based on direct labour hours. As from 1.1.09, it decides to adopt separate rates for the three main activities—storekeeping and material handling, machining and assembly. The estimates of costs and absorption rates for selling and distribution cost remain unchanged. Overhead absorption rates are:

Prior to 1st January 2009:

Production overhead — 0.50 paise per direct labour hour. Selling and distribution overhead — 25% of factory cost.

From 1.1.2009:

Storekeeping and material handling — 10% of direct material cost

Machining — 0.75 paise per machine hour

Assembly — 0.30 paise per labour hour

Selling and distribution overhead — 25% of factory cost

The direct cost of job no. 501 have been:	90
Direct material cost	
Direct wages:	
Machine 200 hrs. @ 0.60	120
Assembly 100 hrs. @ 0.40	40
	250

The estimated price of the job is Rs. 525 and it requires 180 machine hours to complete. Show the job cost sheet for job 501

(*a*) As it would appear if the job had been completed prior to 1st Jan. 2009.

(*b*) As it would appear if the job were completed on 1.1.2009.

[***Answer:*** Profit prior to 1.1.09 is Rs. 25, Loss on 1.1.09 is Rs. 5]

Exercise 2. M/s Diamond Tools Co. Ltd., has undertaken a job no. 5. As per cost ledger, the following expenses have been debited to this job. You are required to prepare a job cost sheet and calculate the total amount to be charged to the customer in order to secure 20% profit on sale.

Additional information based on normal capacity is furnished below:

Capacity	Department A	Department B	Total for the factory as a whole
Budgeted production in terms of direct labour hours	6,000	9,000	20,000
Budgeted variable overheads (Rs.)	9,000	10,800	20,000
Budgeted fixed overhead (Rs.)			
Job cost details:			
Direct materials:			
Cast Iron — 10 kgs. @ Rs. 1 per kg.			
Gun metal — 5 kgs. @ Rs. 10 per kg.			
Direct labour:			
Dept. A — 20 hrs. @ Re. 0.50 per hr.			
Dept. B — 15 hrs. @ Re. 0.40 per hr.			
Direct expenses			
Special tool — Rs. 21			

CHAPTER 2

[*Answer:* Selling price Rs. 225]

Exercise 3. The information given below has been taken from the records of an engineering works in respect of job no. 303.

Materials Rs. 4,010

Wages: Dep. A — 60 hrs. at Rs 3 per hr.
Dep. B — 40 hrs. at Rs. 2 per hr.
Dep. C — 20 hrs. at Rs. 5 per hr.

The works overheads are as follows:

Variable: Dep. A — Rs. 5,000 for 5,000 labour hours
Dep. B — Rs. 3,000 for 1,500 labour hours
Dept. C — Rs. 2,000 for 500 labour hours

Fixed expenses: Rs. 20,000 for 10,000 working hours

Calculate the cost of job 303 and price for the job to earn a profit of 25% on the selling price.

(Bangalore University, B. Com., Nov. 1988)

[*Answer:* Selling price Rs. 6,440]

Exercise 4. The following information is extracted from the books of Kohinoor Mossac tiles factory:

	2008 (Rs.)	2009 (Rs.)	2010 (Rs.)
Materials	75,000	50,000	1,00,000
Labour	50,000	40,000	60,000
Works overheads	35,000	30,000	40,000
Administrative overheads	20,000	20,000	20,000
Selling overheads	10,000	9,000	8,000
Number of tiles	15,000	10,000	20,000

On the basis of the data given above prepare an estimate for the supply of 40,000 tiles under job number 55, to fetch a profit of 25% on works cost.

[*Answer:* Selling price for 40,000 tiles is Rs. 4,77,000]

❑❑❑

CHAPTER 3

CONTRACT COSTING

FEATURES OF CONTRACT COSTING

Contract costing is a variation of job costing applied in those undertakings which are engaged in constructional and repair works. There lies a thin line of difference between these two methods of costing—the difference lies in the size of order received. In fact a big order is termed as a contract and a small contract as a job. Contract costing is also known as terminal costing as the preparation of contract account is terminated or closed after the completion of contract. Examples of undertakings which adopt contract costing are builders, civil engineering contractors, road repairing concerns, dams and bridge constructional concerns. The following are the distinctive characteristic features of industries which adopt contract costing:

1. The work is executed in the premises of contractee.
2. Most of the contracts involve longer duration for completion.
3. Expenses chargeable to contracts are direct in nature.
4. The work to be executed depends upon customer's specification. Hence very rarely two contracts resemble one another.
5. Each contract is treated as a cost unit, *i.e.*, the cost of construction is ascertained in respect of each and every contract separately.
6. The payment is received depending on the stage of completion of work.

SIMILARITIES AND DISSIMILARITIES BETWEEN JOB AND CONTRACT COSTING

Points of Similarities

In both the methods of costing:

1. Production starts on received orders from customers.
2. Quotation price is asked before giving orders.
3. Every job and contract is dissimilar in nature.
4. No need to create demand, customers come on their own.
5. Profit is determined in respect of each job and contract separately.
6. Work is to be executed according to the specification of customers.
7. A job and contract is identified by a separate number and is known by that number until it is completed.
8. They belong to the category of specific order costing.

Points of Dissimilarities

1. A job is small in size, contract is big in size.
2. Work under job costing is performed in the premises of manufacturer. Whereas in contract costing, a contract is executed in the premises of customer.

3. A job usually takes less time to complete. Whereas a contract takes more time to complete.
4. The selling price of a job is paid after completing the job in full. Under contract costing, the price is paid in various instalments depending upon the progress of work.
5. Job costing involves heavy investment on assets initially. This is not so under contract costing.
6. Expenses under job costing takes the form of both direct and indirect, whereas under contract costing most of the expenses are direct in nature.
7. Profit earned on a job is entirely taken to P and L a/c. In case of incomplete contract only proportionate profit is taken to P and L a/c.

PROCEDURE OF CONTRACT COSTING

The preparation of contract account is the essense of contract costing. The contract account is prepared by the contractor in his books. In addition to this account the contractor also prepares contractee's account. A separate account is opened for each contract and a serial number is given to it in order to distinguish it from other contracts. The purpose of preparing a contract account is to know the profit or loss on every contract executed. The preparation of contract account involves recording of the following items:

1. Materials

Materials required for the execution of contract in most of the cases are ordered specifically and then used on the contract. The following entries are passed to record the materials cost:

(*a*) When materials are specifically purchased for the contract

Materials a/c	*Dr.*
To cash or supplier's a/c	

Note: Usually the materials bought are sent to the storeroom of the contractor.

(*b*) When materials are sent to site by the contractor

Contract a/c	*Dr.*
To materials a/c	

(*c*) When materials are transferred from one contract to another

Transferee contract a/c	*Dr.*
To transferor contract a/c	

(*d*) When materials are supplied by the contractee, the value of such materials should not be charged to the contract account. Instead, a separate record for such materials should be kept because the unused materials will have to be returned back to the contractee.

(*e*) When any materials are sold at work site

Cash a/c	*Dr.*
To contract a/c	

(*f*) For recording accidental loss of materials

Insurance company a/c	*Dr.* (with the admitted claim)
Profit and loss a/c	*Dr.* (with the residue)
To contract a/c	

CHAPTER 3

(*g*) For recording materials remaining at site at the end of the year

(*i*) At the end of the current year:
Materials at site a/c *Dr.*
To contract a/c

(*ii*) At the beginning of the next year
Contract a/c *Dr.*
To materials on site a/c

(*h*) When materials are sent to stores
Stores control a/c *Dr.*
To contract

2. Labour Cost

All labour engaged at the contract site is regarded as direct irrespective of the nature of the job performed by the employee. Even salary of supervisors are charged as direct labour cost when they spend all their time on a particular contract. However salaries payable to the staff of head office and central stores are treated as indirect wages.

3. Direct Expenses

Some of the direct expenses which are charged to contract account are; Architects' and consultant fees, the hire of plant from outside firms, electricity charges, insurance, telephone, postages and other general expenses.

4. Plant and Machinery

Some of the assets that are to be depreciated while on use on a contract are: Bulldozers, mobile crane, cement mixer, tiles-polishing machines, tractors and lorries. There are two ways of dealing with the plant and machineries used on a contract.

(*a*) Where a plant is specially purchased for a particular contract to be used for longer duration, the contract account is debited with the value of plant. At the end of the accounting period the depreciated value of the plant is credited to the contract account.

(*b*) When the plant is used relatively for a shorter duration on a contract, the contract account is charged with the depreciation of the plant.

5. Indirect Expenses

When a contractor undertakes more than one contract simultaneously, he will set up a common office and engages common supervisory staff. The administration expenses incurred and the supervisor's salary is apportioned among the contracts on some suitable basis.

6. Sub-contract

The contractor may entrust certain types of specialised work such as electrical, plumbing, painting, carpentry, special flooring, etc. to a sub-contractor. The sub-contractor is responsible to the main contractor in terms of performing the work and he will get the payment from the main contractor. The cost of such sub-contract is debited to contract account.

7. Retention Money

Usually the contractee stipulates in the contract deed that he would withhold a part of the contract price to be paid at a later stage after completion of the contract. This is to make sure that the contractor has performed all work relating to contract on the most satisfactory manner and that no repair work arises within a prescribed time limit. The amount so withheld by the contractee is known as retention money. It safeguards the interest of the contractee against the contractor, who may at time perform sub-standard work and gain therefrom.

8. Progress Payment

In a large contract, which takes longer duration to complete, the contractee pays to the contractor a certain amount from time to time depending upon the stage of satisfactory completion of work. The progress of work from time to time will be certified by the architect or engineer of the contractee. Thus every instalment of money paid by the contractee to the contractor depending upon the progress of work is known as progress payment.

9. Escalation Clause

Sometimes, owing to fluctuation in the prices of materials and labour costs the contract price is altered so that neither party suffers the loss arising out of the change in price level. To protect his interest against the rise in prices, the contractor inserts a clause known as the 'escalation clause', under which, the contractee will be obliged to pay the enhanced price of the contract because of increase in the rates of materials, labour and other expenses. Similarly, to protect the interest of the contractee against the fall in the rates of materials, labour and overheads, a 'descalation clause' is inserted. However, it is to be noted that the terms and condition under which the contract price is to be altered is to be specifically mentioned. The reasons which gives rise to change in price level is to be stipulated. There should be no ambiguity in the wordings of this clause. The essentials of the escalation clause are as follows:

(*a*) The elements of costs on the basis of which quotation price is submitted must be specified.

(*b*) The elements in relation to which escalation clause applies should be specified.

(*c*) The escalation clause should apply to only those factors which are beyond the control of contractor.

(*d*) The escalation clause should mention the date from which the rise in price of the contract comes into effect.

(*e*) The records of the contractor is to be made available to the contractee for inspection.

(*f*) The provision for alteration of contract price must relate only to change in design, or any major alteration of the work but not on account of defective work.

10. Cost-plus Contract

It is a modified method of contract costing. This method of costing is resorted to when it is not possible to determine the cost of the contract in advance with a reasonable degree of accuracy. Under such circumstance, the contractee agrees to pay to the contractor, the actual cost incurred together with an agreed amount of profit which the contractor earns in the usual course of business. This type of contract is mostly followed during the period of urgency when certain types of products are to be manufactured and supplied as in the case of defence products, component parts and so on.

Advantages of Cost-plus Contract to the Contractor

(*a*) The contractor will not suffer any risk of loss as he will receive the contract price as is assured by the contractee.

(*b*) There is bargain in the contract price in future under this type of contract.

(*c*) The contractor is relieved from the botheration of preparing quotation price for the sake of submitting it to the contractee.

CHAPTER 3

Advantages to the Contractee

(*a*) Since the contract price is governed by the contract, the contractee will also not suffer from risk of loss.

(*b*) The contractee also stands to benefit in a period of uncertain market condition as he is expected to pay only a reasonable price after satisfying the ruling prices.

11. Work Certified and Work Uncertified

Work certified represents that portion of the contract that has been duly approved by the architect of the contractee. This is denoted in terms of money value in contract account and appears on the credit side of the contract account. Work uncertified refers to that portion of work completed by the contractor but disapproved by the architect on the ground that it has not reached a stipulated stage. The value of work uncertified also appears on the credit side of the contract account.

12. Work-in-Progress

The work-in-progress represents the value of work which is in progress on a contract and requires further completion. The value of work-in-progress appears on the asset side of the balance sheet and is ascertained as under:

Amount of work certified		xxx
Amount of work uncertified		xxx
		xxx
Less: Profit transferred to WIP	xxx	
Less: Cash received	xxx	xxx
Work-in-progress		xxx

13. Profit on Uncompleted Contract

As regards profit on a completed contract, it may be safely taken to profit and loss account. This is so because it is the actual profit earned on a contract which is taken for granted to be the profit earned on a completed contract. But in case of incomplete contract the profit made on it cannot be taken to be the actual profit. Since the completion of contract in the subsequent years may be subjected to risk of loss, a prudent contractor will not consider the profit on incomplete contract to be the actual profit. Instead, it is treated as notional profit and necessary provision is made in anticipation of probable losses. The question is what percentage of notional profit is to be transferred to profit and loss account on an incomplete contract in an accounting year. There are no hard and fast rules laid down to calculate a proportionate amount of profit to be taken into profit and loss account. However, the following rules are followed in practice as laid down by contractors of yester years.

(*a*) **When the work completed is less than 25 per cent of the contract.** Under this situation no profit is transferred to profit and loss account. The entire amount of profit is carried forward in the form of reserve.

(*b*) **When the work completed is more than 25 per cent but less than 50 per cent.** In this case $\frac{1}{3}$ of the notional profit is transferred to profit and loss account and the balance is carried forward in the form of reserve. The following formula is used:

$$\frac{1}{3} \times \text{Notional profit} \times \frac{\text{Cash received}}{\text{Work certified}}$$

(c) **When the amount of work completed is more than 50 per cent but not nearing a stage of completion.** Under such situation $\frac{2}{3}$ of notional profit is transferred to profit and loss account and the balance to work-in-progress account. The following formula is applicable:

$$\frac{2}{3} \times \text{Notional profit} \times \frac{\text{Cash received}}{\text{Work certified}}$$

(d) **When the contract is nearing the stage of completion.** In this case the cost of completing the contract is estimated. Then the estimated profit is calculated by deducting the total estimated cost from the contract price. For calculating the profit to be taken to profit and loss account, the following formula is adopted:

$$\text{Estimated profit} \times \frac{\text{Work certified}}{\text{Contract price}}$$

14. Contractee's Account

The contractee's account is prepared by the contractor in his books. When the various instalments of contract price is received from the contractee, the following entry is passed.

Cash a/c *Dr.*
  To contractee's a/c

When the contract is fully completed the following entry is passed:

Contractee's a/c *Dr.*
  To contract a/c

Thus it is clear that the contractee's account will show a debit balance indicating the amount due from him to the contractor till it is paid fully.

15. Loss on Incomplete Contract

The loss on incomplete contract is taken into the debit side of profit and loss account.

16. Balance Sheet

At the time of preparation of balance sheet, the contractee's account deserves a special mention. The contractee's a/c is not to be shown as a debtor for the full contract price unless the work has been completed. Likewise the sum received from the contractee under various instalments should not be shown as a liability on the balance sheet. On completion of contract, if the contractee still owes the amount to the contractor, his account is shown as a debtor for the amount due from him. When the contractee pays full amount, his account is closed and his account will not appear in the balance sheet.

17. Target Costing

This is a variation of cost-plus contract. Under this method, the contractee agrees to pay the profit as per the agreement on the total contract price. In addition to the profit, sometimes, it is agreed upon by the contractor to complete the contract within a target price. In case, if he completes the contract within the target price, he is entitled to receive a bonus which is in proportion to the savings made, saving being difference between original contract price and target price.

Problem 1. *(Treatment of materials and plant).* Gangaram Construction Limited undertook a number of contracts during 1987. Following particulars are available as regards the construction of a canteen building:

Materials at site on 1.1.87	2,000
Materials purchased	50,000
Materials supplied from stores	10,000

Materials returned to stores	1,000
Materials costing Rs. 2,000 were stolen	
Materials worth Rs. 200 destroyed by fire	
Materials costing Rs. 1,000 were sold for 800	
Materials in hand at the end of the year	10,000
Materials received from hospital contract which was completed	20,000
Materials transferred to cinema contract	500
Plant issued to contract on 1.1.87	50,000
Plant returned to stores	1,000
Plant costing Rs. 2,000 sold for Rs. 2,500	
Plant worth Rs. 500 was stolen	
Plant worth Rs. 300 was destroyed by an accident	
Plant in hand at the site at the end of the year was Rs. 10,000	

Show how these transactions will appear in a contract account.

(Mangalore University, B.Com., April 1988)

Solution:

CONTRACT A/C

To materials at site	2,000	By materials returned to stores	1,000
To materials purchased	50,000	By profit and loss a/c — materials stolen	2,000
To materials supplied from stores	10,000	— materials destroyed by fire	200
To materials received from		By loss on sale of materials (1,000 – 800)	200
hospital contract	20,000	By materials sold	800
		By materials in hand	10,000
		By materials transferred to cinema contract	500
To plant	50,000	By plant returned to stores	1,000
To profit and loss a/c:		By profit and loss a/c	
Profit on sale of plant (2,500 – 2,000)	500	Plant stolen	500
		Plant lost by fire	300
		By plant sold	2,500
		By plant in hand	10,000

Note: Since the rate of depreciation is not given the net value of plant is not shown on the credit side of the contract.

Problem 2. *(Treatment of plant in contract a/c).* Mr. Amar undertook a contract on 1.1.2009. The books are closed on 30th June every year, A second-hand plant of Rs. 1,50,000 was purchased on 1st March 2009 and Rs. 10,000 were spent on reconditioning it. On 30th June a portion of the plant costing Rs. 20,000 was sold for Rs. 14,000. On the same date a component of the plant costing Rs. 5,000 was returned to stores. Plant costing Rs. 6,000 was stolen. Depreciate plant by 15% per annum. Show contract account.

Solution:

CONTRACT A/C

To plant	1,50,000		By plant returned to stores	5,000	
Add: Expenses	10,000	1,60,000	*Less*: Depreciation @ 15%		
			for 4 months	250	4,750
			By plant sold		14,000
			By P and L a/c – cost	20,000	
			Less: Depreciation for		
			4 months @ 15%	1,000	

	19,000	
Less: Sale	14,000	
	5,000	5,000
Plant stolen		6,000
By plant in hand		1,22,550

Working Note: Plant in hand

Cost of plant installed		1,60,000
Less: Cost of plant sold	20,000	
Less: Cost of plant returned	5,000	
Less: Cost of plant stolen	6,000	31,000
		1,29,000
Less: Depreciation @ 15% p.a. for 4 months		6,450
Plant in hand		1,22,550

Problem 3. (*Contract completed in the same year*). The following particulars are in respect of contract no. 20 which commenced on 1.1.1990:

	Rs.
Materials issued	58,000
Direct wages	75,000
Cost of special plant	30,000
Direct expenses	12,000
Establishment charges	8,000

The contract was completed by 31st December, 1990. The contract price was Rs. 1,75,000. The value of materials returned to stores was Rs. 5,000. The special plant was returned to stores subject to depreciation at 20% per annum. Contract price was received in full on 31st December, 1990. Prepare contract account. *(Bangalore University, B.Com., Nov. 1991)*

Solution:

CONTRACT A/C

To materials issued	58,000	By materials returned		5,000
To direct wages	75,000	By plant returned	30,000	
To plant	30,000	*Less:* Depreciation	6,000	24,000
To direct expenses	12,000			
To establishment charges	8,000	By contractee's a/c		1,75,000
To P and L a/c	21,000			
	2,04,000			2,04,000

CONTRACTEE'S A/C

To contract a/c	1,75,000	By bank	1,75,000

Problem 4. *(Incomplete contract).* Mr. Sai undertook a contract for the construction of a building on 1st January 1998. The contract price is Rs. 7,50,000. The following particulars are available prepare contract account and show the amount that should be transferred to profit and loss a/c and prepare contract a/c..

	Rs.
Materials purchased	1,20,000
Materials issued from stores	1,50,000
Labour employed on site	45,000
Plant installed on site	60,000
Direct expenses	30,000
Proportionate establishment charges	7,500
Cash received (80% of work certified)	3,00,000
Work uncertified	75,000
Materials returned to stores	7,500
Materials in hand at the end	3,000
Plant on hand at the end	45,000
Wages outstanding	9,000
Direct expenses accrued	12,000

(S.V. University, B.Com., April 1999)

Solution: CONTRACT A/C FOR THE YEAR ENDED 31.12.1998

To materials purchased	1,20,000	By materials returned	7,500
To materials issued	1,50,000	By materials on hand	3,000
To labour employed	45,000	By plant on hand	45,000
To plant installed	60,000	By work-in-progress	
To direct expenses	30,000	Work certified	3,75,000
To establishment charges	7,500	Work uncertified	75,000
To wages o/s	9,000		
To direct expenses accrued	12,000		
To balance c/d	72,000		
	5,05,500		5,05,500
To profit and loss a/c	38,400	By balance b/d	72,000
$72{,}000 \times \frac{2}{3} \times \frac{3{,}00{,}000}{3{,}75{,}000}$			
To work-in-progress a/c: —reserve	33,600		
	72,000		72,000

Working Note:

CALCULATION OF WORK CERTIFIED

$$\text{Let work certified} = x$$

$$\text{Cash received is } \frac{80}{100}x = 3{,}00{,}000$$

$$80x = 3{,}00{,}000 \times 100$$

$$80x = 3{,}00{,}00{,}000$$

$$x = \frac{3{,}00{,}00{,}000}{80}$$

$$x = 3{,}75{,}000$$

Problem 5. *(When work uncertified is not given).* X Ltd. closes its accounts on 31st December each year. The company commenced work on a contract on 1st January 1998. The following information relates to the contract as on 31st December 1998:

	Rs.
Materials issued	1,25,500
Wages	2,82,800
Salary to foreman	40,650

A machine costing Rs. 1,30,000 had been on the site for 146 days. Its working life is estimated at 7 years and its final scrap value at Rs. 7,500. A supervisor, who is paid Rs. 4,000 per month has devoted half of his time to this contract. Other expenses and administration charges amounted to Rs. 68,250. Materials at site at the end of the year cost Rs. 17,700. The contract price is Rs. 10 lakhs. On 31st December, 1998, two-thirds of the contract was completed. The architect had issued certificate of approval covering 50% of the contract price and the contractor had been paid Rs. 3,75,000 on account.

Prepare the contract account and work-in-progress account. Also show how the work-in-progress will appear in the balance sheet of the contractor as on 31st December 1998. *(C.S., Inter, June 1999)*

Solution:

CONTRACT A/C FOR THE YEAR ENDED 31ST DECEMBER 1998

To materials	1,25,500	By materials at site	17,700
To wages	2,82,800	By work-in-progress a/c	
To salary to foreman	40,650	Work certified	5,00,000
To depreciation on machinery	7,000	Work uncertified	1,32,625
To other expenses and adm. charges	68,250		
To supervisor's salary	24,000		
To balance c/d	1,02,125		
	6,50,325		6,50,325
To profit and loss a/c	51,062.50	By balance b/d	1,02,125
To WIP a/c —(reserve)	51,062.50		
	1,02,125		1,02,125

WORK-IN-PROGRESS A/C

To contract a/c:		By contract a/c	51,062.50
Work certified	5,00,000	(reserve)	
Work uncertified	1,32,625	By balance c/d	5,81,562.50
	6,32,625		6,32625.00

BALANCE SHEET AS ON 31ST DEC. 1998

Work-in-progress:	
Work certified	5,00,000
Work uncertified	1,32,625
	6,32,625
Less: Reserve	51,062.50
	5,81,562.50
Less: Cash received	3,75,000
	2,06,562.50

Working Note:

(1) Calculation of profit taken into P and L a/c

$$1,02,125 \times \frac{2}{3} \times \frac{3,75,000}{5,00,000} = 51062.50$$

(2) Calculation of work uncertified

Work completed $= \frac{2}{3}$ and work certified $= \frac{1}{2}$

Hence work uncertified $= \frac{2}{3} - \frac{1}{2}$

$$= \frac{4-3}{6} = \frac{1}{6}$$

Cost of $\frac{2}{3}$ work completed:

Materials	1,25,500
Wages	2,82,800
Salary to foreman	40,650
Depreciation on machinery	7,000
Other expenses	68,250
Supervisor's salary	24,000
	5,48,200
	17,700
Less: Materials at site	5,30,500

Cost of $\frac{1}{6}$ work completed =

For $\frac{2}{3}$rd completion the cost is 5,30,500

For $\frac{1}{6}$ $\rightarrow \dfrac{5,30,500 \times \frac{1}{6}}{\frac{2}{3}}$

$$= 5,30,500 \times \frac{3}{2} \times \frac{1}{6} = 1,32,625$$

Alternatively, cost of work uncertified can also be found out by the following calculation:

The cost of $\frac{2}{3}$ or 66.67% is 5,30,500

For 100% $\rightarrow \dfrac{5,30,000 \times 100}{66.67}$

$= 7,95,710$

The cost of 50% of the contract which has been certified by the architect is

$$7,95,710 \times \frac{50}{100} = 3,97,855$$

Cost of work uncertified is

5,30,500 – 3,97,855 = 1,32,645

(3) Calculation of Depreciation

1,30,000 – 7,500 = 17,500

$$7 \text{ years} = \frac{17{,}500 \times 146}{365} = 7{,}000$$

Problem 6. A company undertook a contract for construction of a large building complex. The construction work commerced on 1st April 1993 and the following data are available for the year ended 31st March 1994.

	Rs '000.
Contract price	35,000
Work certified	20,000
Progress payment received	15,000
Materials issued to site	7,500
Planning and estimating costs	1,000
Direct wages paid	4,000
Materials returned from site	250
Plant hire charges	1,750
Wage related costs	500
Site office costs	678
Head office expenses apportioned	375
Direct expenses incurred	902
Work not certified	149

The contractor owns a plant which originally cost Rs. 20 lakhs has been continuously in use in this contract throughout the year. The residual value of the plant after 5 years of life is expected to be Rs. 5 lakhs. Straight line method of depreciation is in use.

As on 31st March, 1994 the direct wages due and payable amounted to Rs. 2,70,000 and the materials at site were estimated at Rs. 2,00,000.

Required:

(i) Prepare the contract a/c for the year ended 31st March, 1994.

(ii) Show the calculation of profit to be taken to the profit and loss account of the year.

(iii) Show the relevant balance sheet entries. *(C.A. Inter, November 1994)*

Solution:

CONTRACT A/C FOR THE YEAR ENDED 31ST MARCH 1994 (Rs. '000)

To materials issued	7,500	By materials returned	250
To direct wages paid	4,000	By materials at site	200
To direct wages accrued	270	By work-in-progress:	
To wage related costs	500	Work certified	20,000
To direct expenses	902	Work uncertified	149
To plant hire charges	1,750		
To planning and estimating cost	1,000		
To site office cost	678		
To head office expenses apportioned	375		

To plant depreciation	300		
To balance c/d	3,324		
	20,599		20,599
To P and L a/c $\frac{2}{3}\times 3{,}324\times\frac{15{,}000}{20{,}000}$	1,662	By balance b/d	3,324
To WIP – Reserve	1,662		
	3,324		3,324

BALANCE SHEET

P and L a/c	1,662	Plant at site (2,000 – 300)		1,700
Wages accrued	270	Materials at site		200
		Work-in-progress		
		Work certified	2,000	
		Work uncertified	149	
			20,149	
		Less: reserve	1,662	
			18,487	
		Less: Cash received	15,000	
				3,487

Working Note:

Calculation of Depreciation: ('000)

Original cost of plant	2,000
Less: Residual value	500
	1,500
Life of plant 5 year	
Annual depreciation	$\frac{1{,}500}{5} = 300$

Problem 7. The Maharashtra Construction Company undertook the construction of a building at a contract price of Rs. 12,00,000. The date of commencement of contract was 1st April 1998.

The following cost information is given for the year ended 31st March 1999

	Rs.
Materials sent to the site	3,00,000
Wages	4,40,000
Architect fees	55,500
Office and administration overhead	1,51,000
Uncertified work	55,000
Materials at the site at the end of the year	10,000
Cash received from the contractee (Being 90% of the work certified)	9,45,000
Materials destroyed by fire	5,000
Plant and Machinery at cost	2,00,000

(Date of purchase — 1st July 1998.). The estimated working life of the plant 10 years and its estimated scrap value at the end Rs. 20,000.

Supervisor's salary 60,000

You are required to prepare a contract a/c for the year ended 31st March 1999.

(University of Bombay, B.Com., April 1999)

Solution: CONTRACT A/C FOR THE YEAR ENDED 31.3.1999

To materials sent to site	3,00,000	By P and L a/c: Materials lost	5,000
To wages	4,40,000	By WIP:	
To architect's fees	55,500	Work certified	10,50,000
To office and adm. oh	1,51,000	Work uncertified	55,000
To depreciation	13,500	Materials at site	10,000
To supervisor's salary	60,000		
To balance c/d	1,00,000		
	11,20,000		11,20,000
To P and L a/c $\frac{2}{3}\times 1{,}00{,}000\times\frac{9{,}45{,}000}{10{,}50{,}000}$	60,000	By balance b/d	1,00,000
To WIP – reserve	40,000		
	1,00,000		1,00,000

Working Notes:

(1) *Calculation of depreciation*

$$\text{Depreciation for 1st year} = \frac{\text{Cost} - \text{Scrap value}}{\text{Working life}}$$

$$= \frac{2{,}00{,}000 - 20{,}000}{10}$$

$$= \frac{1{,}80{,}000}{10} = 18{,}000$$

$$\text{For nine months } 18{,}000 \times \frac{9}{12} = 13{,}500$$

(2) *Calculation of work certified*

$$\text{Let work certified} = x$$

$$\text{Cash received is } \frac{90}{100}x = 9{,}45{,}000$$

$$90x = 9{,}45{,}000 \times 100$$

$$90x = 9{,}45{,}00{,}000$$

$$x = \frac{9{,}45{,}00{,}000}{90} = 10{,}50{,}000$$

Problem 8. (Incomplete contract) The Perfect Construction Company Ltd. has undertaken the construction of a bridge for a value of Rs. 45,00,000 subject to a retention of 20% until one year after the certified completion of the contract. The following information is available for the year ended 31st March 2004.

CHAPTER 3

	Rs.
Labour on site	11,55,000
Materials sent to the site	12,30,000
Materials from stores	2,35,500
Plant hire	34,800
Direct expenses	63,000
General overheads allocated to the contract	1,18,200
Material at site (31-3-2004)	22,800
Wages accrued on 31-3-2004	28,800
Direct expenses accrued on 31-3-2004	5,100
Work not yet certified at cost	43,500
Value of work certified	39,00,000
Cash received on account	31,20,000

You are required to prepare:

(1) Contract a/c

(2) Contractee's a/c

(3) Show relevant items in the Balance Sheet *(University of Bombay, B.Com., April 2004)*

Solution:

CONTRACT A/C

To materials	1,23,000	By work-in-progress:	
To labour	11,55,000	Work certified	39,00,000
To materials from stores	2,35,500	Work uncertified	43,500
To plant hire	34,800	Materials at site	22,800
To direct expenses	63,000		
To general overhead	1,18,200		
To wages accured	28,800		
To direct expenses accrued	51,00		
To balance c/d	10,95,000		
	39,66,300		39,66,300
To P and L a/c	5,84,480	By Balance b/d	10,95,000
$10{,}95{,}000 \times \frac{2}{3} \times \frac{80}{100}$			
To WIP – (Reserve)	5,11,420		
	10,95,000		10,95,000

CONTRACTEE'S A/C

To Balance c/d	31,20,000	By Bank	31,20,000

BALANCE SHEET (EXTRACTS)

P and L a/c	5,84,480	**Work in Progress:**	
Accrued wages	28,800	Work certified	39,00,000
Accrued direct expenses	5,100	Work uncertified	43,500
		Materials at site	22,800
			39,66,300
		Less: Cash received	31,20,000
			8,46,300
		Less: Reserve	5,11,420
			3,34,880

Problem 9. Mr. Vivek undertook a contract for the construction of building on 1st January 2004. The contract price being Rs. 15,00,000.

The following details are available for the year 2004:

	Rs.
Materials purchased	2,40,000
Materials issued from stores	3,00,000
Labour employed on site	90,000
Plant installed on site	1,20,000
Direct expenses	60,000
Proportionate establishment charges	15,000
Cash received (80% of work certified)	6,00,000
Work uncertified	1,50,000
Materials returned to stores	15,000
Materials in hand at the end	6,000
Wages outstanding	18,000
Direct expenses accrued	24,000

Prepare the contract account and show the amount that would appear in the Balance Sheet. A part of plant costing Rs. 20,000 was stolen at the beginning of the year and the insurance company paid Rs. 12,000. Plant is depreciated @ 20% p.a. *(University of Bombay, B.Com., October 2002)*

Solution:

CONTRACT A/C

To material purchased		2,40,000	By materials returned to stores	15,000
To materials from stores		3,00,000	By plant stolen	20,000
To labour cost	90,000			
Add: wages o/s	18,000	1,08,000	**By work in progress:**	
			Work certified	7,50,000
To plant		1,20,000	Work uncertified	1,50,000
To direct expenses	60,000		Plant at site	80,000
Add: Accrued	24,000	84,000	Materials in hand	6,000
To establishment charges		15,000		
To balance c/d		1,54,000		
		10,21,000		10,21,000
To P and L a/c		82,133	By balance b/d	1,54,000
To work in progress–Reserve		71,867		
		1,54,000		1,54,000

CHAPTER 3

BALANCE SHEET FOR THE YEAR ENDED 2004 (EXTRACT)

P and L a/c:			**Work in Progress:**	
Profit on contract	82,133		Work certified	7,50,000
Less: Loss on stolen plant	8,000	74,133	Work uncertified	1,50,000
				9,00,000
			Value of plant	80,000
			Materials in hand	6,000
O/s wages		18,000		9,86,000
O/s Direct expenses		24,000	*Less* Reserve	71,867
				9,14,133
			Less: Cash received from contracter	6,00,000
				3,14,133

Working Note:

(1) **Calculation of plant at site:**

Plant installed at site	1,20,000
Less: Stolen at the beginning of the year	20,000
Plant at site at the end of the year	1,00,000
Less: Depreciation @ 20% on 1,00,000	20,000
	80,000

(2) **Profit taken to P and L a/c**

Contract price = Rs. 15,00,000

Work certified $= \frac{100}{80} \times 6{,}00{,}000 = 7{,}50{,}000$

Percentage of work completed 50%

$$\therefore \quad \frac{2}{3} \times \text{Notional profit} \times \frac{\text{Cash received}}{\text{Work certified}}$$

$$= \frac{2}{3} \times 1{,}54{,}000 \times \frac{6{,}00{,}000}{7{,}50{,}000} = \text{Rs. } 82{,}133$$

(3) Insurance claim received is credited to costing. Profit and loss and not the contract a/c.

Problem 10. DS Ltd. started a contract on 1st June 2002 for construction of a Link Road for a contract price of Rs. 5,00,000 to be completed by 31st March, 2004. The budgeted cost of the contract was Rs. 4,50,000. The following particulars are obtained for the year ended 31st March 2003:

Materials issued from stores	48,000
Materials bought direct to site	40,000
Materials from other contract	24,500
Materials at site (closing)	6,000
Materials returned	2,000
Materials lost in accident	8,000
Insurance claim accepted	5,000
Wages paid	74,500
Plant issued to site at WDV	30,000
Hire charges of plant	38,500

Supervision expenses	13,500
Normal loss of materials	2,000
Uncertified work in progress	17,500
Share of other expenses	5,500
Paid to sub-contractors	25,000
Outstanding wages on 31-3-2003	1500
Payment due to sub-contractor on 31-3-2003	2,000
Value of work certified	2,75,000
Advance from contractee (90% of work certified)	
Sale of unused materials (Cost Rs. 10,000)	8,000
Opening balance of a plant lost in accident	6,000
Depreciation on plant @ 200% p.a.	

Draw the Contract a/c and the Contractee a/c. Also compile the value of work-in-progress.

(University of Calcutta, B.Com. (Hons) 2004)

Solution: CONTRACT NO. .... a/c

FOR THE PERIOD 1-6-2002 TO 31-3-2003

To materials issued		48,000	By Materials returned	2,000
To materials bought		40,000	Materials lost	8,000
To materials from other contracts		24,500	By closing stock of materials	6,000
To wages paid	74,500		By materials sold	10,000
Add: Outstanding	1,500	76,000	(Cost)	
To hire charges of plant		38,500	Work certified	2,75,000
To supervision expenses		13,500		
To share of other expenses		5500		
To paid to sub-contractor	25,000			
Add: Due	2,000	27,000		
To Depreciation of plant		5,000		
To Balance c/d		23,000		
		301,000		3,01,000
To P and L a/c		13,800	By Balance b/d	23,000
To WIP a/c		9,200		
		23,000		23,000

CONTRACTEE'S A/C

To Balance c/d	2,47,500	By Bank a/c	2,47,500
		(Advance received)	
		$2{,}75{,}000 \times \frac{90}{100}$	

CHAPTER 3

VALUE OF WORK-IN-PROGRESS

Work certified	2,75,000
Less: Portion of profit transferred to WIP	9,200
Value of work-in-progress	2,65,800

Working Note:

(1) **Depreciation on plant:**

Assuming that the plant is damaged at the end of accounting period depreciation is calculated as under

On Rs. 30,000 plant issued	$30{,}000 \times \frac{20}{100} \times \frac{10}{12}$	= 6,000
Less on plant lost	$6000 \times \frac{20}{100} \times \frac{10}{12}$	= 1,000
	Depreciation on plant	5,000

(2) Normal loss of materials is not an item of cost records. It is shown in costing P and L a/c

(3) Cash received = $2{,}75{,}000 \times \frac{90}{100} = 2{,}47{,}500$

(4) Profit taken to P and L a/c

$$23{,}000 \times \frac{2{,}47{,}500}{2{,}75{,}000} \times \frac{2}{3} = \text{Rs. } 13{,}800$$

Problem 11. Mirik Construction Ltd. entered into a contract for construction of a building. The contract value is Rs. 6,50,000 to be realised in instalments on the basis of the value of work certified by the architect subject to a retention of 10%. The work commenced on 1-4-2001 when the final accounts are to be prepared. The detailed information are

Plant charged to contract at the commencement	32,000
Materials charged to contract	1,80,000
Wages paid	87,000
Expenses incurred on the contract	38,750

Total establishment expenses amounted to Rs. 41,000 out of which 25% is attributable to this contract. Out of the materials issued to the contract, material costing Rs. 4,000 were sold for Rs. 5,000. A part of the plant (cost Rs. 2,000) was damaged on 1-10-2001 and the scrap realised Rs. 300 only. Plant costing Rs. 3,000 was transferred to another contract site on 31-12-2001. Plant is to be depreciated @ 10% p.a. Material on hand on 31-12-2001, Rs. 17,500: cash received from the contractee Rs. 3,06,000: Cost of work yet to be certified is Rs. 30,000

Prepare a contract account *(University of Calcutta, B.Com. (Pass) 2003)*

Solution: CONTRACT NO. .... A/C FOR THE PERIOD 1-4-2001 TO 31-12-2001

To materials charged	1,80,000	By materials sold (at cost)	4,000
To wages	87,000	Materials on hand	17,500
To expenses	38,750	By work certified	3,40,000
To establishement expenses	10,250	By work uncertified	3,000
To depreciation on plant	2,350		
To balance c/d	46,150		
	3,64,500		3,64,500

To P and L a/c	27,690	By balance b/d	4,6150
To WIP a/c	18,460		
	46,150		46,150

Working Note:

(1) **Depreciation on plant:**

on Rs. 32,000 at 10% for 6 months (From 1-4-01 to 1-10-01)	= 1600
on Rs. 30,000 @ 10% for 3 months (From 1-10-01 to 31-12-01)	= 750
	2350

(2) Work certified $= \dfrac{\text{cash received}}{\text{\% of cash received}}$

$$= \frac{\text{work certified + work uncertified}}{\text{contract price}}$$

$$= \frac{3,40,000 + 30,000}{6,50,000} \times 100 = 56.9\%$$

(3) Profit taken to P and L a/c:

$$\text{Notional profit} \times \frac{\text{cash received}}{\text{work certified}} \times \frac{2}{3}$$

$$= 46150 \times \frac{3,06,000}{3,40,000} \times \frac{2}{3} = 27,690$$

Problem 12. The following is the summarised information relating to contract account no. 100 :

Contract price	6,00,000
Wages	1,64,000
General expenses	8,600
Materials	1,20,000
Cash received (80% of certified work)	2,40,000
Materials at site	10,000
Plant	20,000

Included in the above information are wages Rs. 3,500 materials Rs. 4,000 and other expenses Rs. 2500. Which were incurred since certification. Depreciate plant at 10%. Prepare contract a/c No. 100

(University of Delhi, B.Com. (Hons) External 2004)

Solution: CONTRACT A/C NO. 100

To materials	1,20,000	By work in progress:	
To wages	1,64,000	Work certified	3,00,000
To general expenses	8,600	Work uncertified	10,000
To depreciation on plant	2,000	(3500 + 2500 + 4000)	
To balance c/d	25,400	To materials at site	10,000
	3,20,000		3,20,000

CHAPTER 3

To P and L a/c	13,547	By balance b/d	25,400
$25400 \times \frac{2}{3} \times \frac{80}{100}$			
To work-in-progress	11853		
	25,400		25,400

Problem 13. Engineers Ltd. undertook several contracts during the year 2002. The following information relate to contract no. 107.

		Rs.
Direct materials		20,250
Direct wages		15,500
Stores issued		10,500
Loose tools		2,400
Tractor expenses:		
Running material	2,300	
Wages of driver	3,000	5,300
Other direct charges		2,650

The contract took 13 weeks on its completion. The value of loose tools and stores returned at the end of the period were Rs. 2,000 and 3,000 respectively. The plant was also returned at a value of Rs. 16,000 after charging depreciation at 20%. The value of tractor was Rs. 20,000 and the depreciation was to be charged to the tractor @ 15% per annum. The administration and office expenses are to be provided at 10% on works cost. Profit is to be charged at 20% of the total cost.

Prepare the aforesaid contract a/c assuming the balance of the contract was duly received from the contractee. *(University of Delhi, B.Com. (Hons) 2002)*

Solution:

CONTRACT A/C

To direct materials	20,250	By loose tools returned	200
To direct wages	15,500	By stores returned	3,000
To stores	10,500	By contractee's a/c	76,758
To loose tools	2,400	(Balancing figure)	
To tractor expenses (2,300 + 3,000)	5,300		
To depreciation on tractor	750		
To depreciation on plant (20,000 – 16,000)	4,000		
To other direct charges	2,650		
To administrative and office expenses (10% of works cost)	5,815		
To profit and loss a/c (20% of Total cost)	12,793		
	79,958		79,958

Working Note:

(1) Work cost = 20,250 + 15,500 + 10,500 + 2,400 + 5300 + 750 + 4,000 + 2,650 – 3,000 – 200 = Rs. 58,150

(2) Total cost = 58,150 + 5,815 = Rs. 63,965

(3) Depreciation on Tractor

$$20{,}000 \times \frac{15}{100} \times \frac{13}{52} = \text{Rs. } 750$$

Problem 14. *(Comparative contract costing):* Siddesh construction company has undertaken three contracts during three contracts during the year and the following particulars are available as on 31-12-2004.

	Contract A (Rs)	*Contract B (Rs.)*	*Contract C (Rs.)*
Contract price	10,00,000	25,00,000	7,50,000
Material issued to contract	1,65,200	2,24,500	1,89,600
Labour	1,02,800	1,26,500	1,75,500
Sub-contract charges	72,800	65,900	28,500
Supervision charges	12,000	18,000	15,000
Architect fees	10,000	15,000	25,000
Insurance charges	3,000	6,100	7,400
Work certified	4,00,000	5,00,000	5,00,000
Work uncertified	35,000	40,000	25,000
Amount received from contractee	3,20,000	4,50,000	3,75,000
Closing stock of materials	9,000	10,000	20,000

All contracts were commenced during the current year. Total depreciation on plants amounted to Rs. 11,200 and allocate the same to all contracts in the ratio of work certified.

Prepare Contract A/c. Show the calculation of profit transferred to Profit and Loss Account.

(University of Bombay, B.Com October, 2005)

Solution:

CONTRACT 'A' A/C

To materials	1,65,200	By work-in-progress	
To labour	1,02,800	By work certified	4,00,000
To sub-contract	72,800	By work uncertified	35,000
To supervision	12,000	By Materials at site	9,000
To architect fees	10,000		
To insurance	3,000		
To depreciation	3,200		
To balance c/d	75,000		
	4,44,000		4,44,000
To P and L a/c	20,000	By Balance c/d	75,000
To WIP Reserve	55,000		
	75,000		75,000

Working Note:

(1) Calculation of profit transferred to P and L a/c

Contract price	10,00,000
Work certified	4,00,000
% of work complete	40%

∴ Profit to be transferred to P and L a/c

$$\frac{1}{3} \times 75000 \times \frac{3,20,000}{4,00,000} = \text{Rs. } 20,000$$

(2) Depreciation of Rs. 11,200 allocated to contracts ABC in the ratio of 4: 5: 5

CONTRACT 'B' A/C

To materials	2,24,500	By work in progress	
To labour	1,26,500	By work certified	5,00,000
To sub-contract	65,900	By work uncertified	40,000
To supervision	18,000	By Materials at site	10,000
To architect fees	15,000		
To insurance	6,100		
To depreciation	4,000		
To work-in-progress — Reserve	90,000		
	5,50,000		5,50,000

Note:

Contract price	25,00,000
Work certified	5,00,000
% of work completed	20%

As work completed is below 25% no. profit is taken into P and L a/c

CONTRACT 'C' A/C

To materials	1,89,600	By work in progress:	
To labour	1,75,500	By work certified	5,00,000
To sub-contract	28,500	By work uncertified	25,000
To supervision	15,000	To Materials at site	20,000
To architect fees	25,000		
To insurance	7,400		
To depreciation	4,000		
To balance c/d	1,00,000		
	5,45,000		5,45,000
To P and L a/c	50,000	By balance b/d	1,00,000
To WIP Reserve	50,000		
	1,00,000		1,00,000

Working Note:

Calculation of profit taken to P and L a/c

$$\frac{2}{3} \times \text{Notional profit} \times \frac{\text{cash received}}{\text{work certified}}$$

$$= \frac{2}{3} \times 1,00,000 \times \frac{3,75,000}{5,00,000} = 50,000$$

Problem 15. *(Comparative contract a/c):* Mr Behram contractor has undertaken two contracts one at Mumbai and another at Thane. The details of the contracts are given below for the year ended 31st March 1998:

Particulars	*Contract at Mumbai*	*Contract at Thane*
Date of commencement	1st July 97	1st Oct. 97
	Rs.	*Rs.*
Contract price	10,00,000	15,00,000
Direct labour	2,55,000	1,82,000
Materials issued from stores	2,20,000	2,00,000
Materials returned to stores	10,000	15,000
Plant installed at site	2,00,000	3,50,000
Direct expenses	40,000	30,000
Office overheads	15,000	10,000
Materials sold (cost Rs. 8,000)	10,000	–
Materials at site	18,000	16,000
Cash received from contractee (representing 80% of work certified)	4,80,000	2,40,000
Work uncertified	13,000	9,000
Architects fees	7,000	3,000

(*i*) Provide depreciation on plant at 20% p.a.

(*ii*) During the year materials costing Rs. 10,000 were transferred from Thane contract to Mumbai contract.

You are required to prepare contract a/c of Mumbai and Thane contracts.

(University of Bombay, B.Com., Oct. 98)

Solution:

MUMBAI CONTRACT A/C FOR THE YEAR ENDED 31ST MARCH 1998

To materials issued	2,20,000	By work certified	6,00,000
To direct labour	2,55,000	By materials returned	10,000
To plant installed	2,00,000	By materials sold	8,000
To direct expenses	40,000	By materials at site	18,000
To office overheads	15,000	By work uncertified	13,000
To Thane contract	10,000	By plant at site	1,70,000
To architect fees	7,000		
To balance c/d	72,000		
	8,19,000		8,19,000
To P & L a/c $\left(\frac{2}{3} \times 72{,}000 \times \frac{4{,}80{,}000}{6{,}00{,}000}\right)$	38,400	By Balance c/d	72,000
To work-in-progress (Reserve)	33,600		
	72,000		72,000

CHAPTER 3

Working Note:

Calculation of work certified:

Let work certified = x

$$\text{Cash received is } \frac{80}{100}x = 4{,}80{,}000$$

$$\text{or} \quad 80x = 4{,}80{,}000 \times 100$$

$$80x = 4{,}80{,}00{,}000$$

$$x = \frac{4{,}80{,}00{,}000}{80} = 6{,}00{,}000$$

THANE CONTRACT A/C
for the year ended 31st March 1998

To materials issued	2,00,000	By work certified	3,00,000
To direct labour	1,82,000	By materials returned	15,000
To plant installed	3,50,000	By Mumbai contract	10,000
To direct expenses	30,000	By materials at site	16,000
To office overheads	10,000	By work uncertified	9,000
To architect fees	3,000	By plant at site	3,15,000
		By P and L a/c	1,10,000
	7,75,000		7,75,000

Working Note:

CALCULATION OF WORK CERTIFIED

Let work certified = x

$$\text{Cash received is } \frac{80}{100}x = 2{,}40{,}000$$

$$\text{or} \quad 80x = 2{,}40{,}00{,}000$$

$$x = \frac{2{,}40{,}00{,}000}{80} = 3{,}00{,}000$$

Problem 16. Navnirman Ltd. has undertaken three contracts. It furnishes the following information for the year ended 31st March 1996.

(Figures in 00's)

	Goa Contract *Rs.*	*Roha Contract* *Rs.*	*Surat Contract* *Rs.*
1. Balances on 1.4.95:			
Materials at site	100	2,000	–
Uncertified work	2,500	4,000	–
Plant at site	2,200	3,100	–
Work certified	19,500	1,400	–
Provision for contingencies	1,000	600	–
2. Transactions during the year:			
Materials issued	–	6,200	8,000
Sub-contract charges	600	11,800	9,000
3. Balances on 31.3.96:			
Materials at site	–	1,000	800
Uncertified work	–	1,000	3,850

Plant at site	–	2,000	950
Work certified	25,000	30,000	12,000
4. Contract price	25,000	40,000	50,000
5. Amount received	25,000	27,000	10,800

6. Value of plant transferred from Goa contract to Surat contract Rs. 1,550.

The company consistently adopts the policy of taking credit for the contract profit considering the proportion of amounts received to the contract price.

You are required to:

(*i*) Prepare the respective contract accounts for the year ended 31st March 1996.

(*ii*) Find the net profit as per profit and loss account.

(University of Bombay, B.Com. April. 1996)

Solution:

GOA CONTRACT A/C *(Fig. in 00's)*

To opening balance:			
Materials	100	By provision for contingencies	1,000
Uncertified work	2,500	By contractee's a/c	25,000
Certified work	19,500		
To sub-contract charges	600		
To depreciation (*see note*)	650		
To net profit	2,650		
	26,000		26,000

ROHA CONTRACT A/C

To opening balance:			
Materials	2,000	By provision for contingencies	600
Uncertified work	4,000	By work in progress:	
Certified work	1,400	Work certified	30,000
To materials issued	6,200	Work uncertified	1,000
To sub-contract charges	11,800	Materials at site	1,000
To depreciation on plant (see note)	1,100		
To balance c/d	6,100		
	32,600		32,600
To P and L a/c	3,600	By balance b/d	6,100
$6{,}100 \times \frac{2}{3} \times \frac{27{,}000}{30{,}000}$			
To work-in-progress	2,440		
	6,100		6,100

SURAT CONTRACT

To materials issued	8,000	By work-in-progress:	
To sub-contract charges	9,000	Work certified	12,000
To depreciation on plant	600	Work uncertified	3,850
(*see note*)		Materials at site	800
		By P and L a/c	950
	17,600		17,600

CHAPTER 3

PROFIT AND LOSS A/C

To Surat contract	950	By Goa contract	2,650
To net profit	5,360	By Roha contract	3,660
	6,310		6,310

Working Note:

Calculation of depreciation

(1) *Goa contract A/c*

Opening balance	2,200
Less: Value of plant transferred to Surat contract	1,550
Depreciation	650

(2) *Roha contract*

Opening balance	3,100
Less: Closing balance	2,000
Depreciation	1,100

(3) *Surat contract*

Value of plant transferred from Goa contract	1,550
Less: Closing balance	950
Depreciation	600

Problem 17. From the following information, prepare contract a/c in tabular form and show how these figures appear in the balance sheet as on 31st Dec. 1995.

	Contract		
Particulars	*A*	*B*	*C*
Commencement	1.1.95	1.7.95	1.10.95
	Rs.	*Rs.*	*Rs.*
Contract price	80,000	54,000	60,000
Raw materials	14,400	11,600	4,000
Wages	22,000	22,400	2,800
General charges	800	560	200
Plant installed	4,000	3,200	2,400
Materials on hand	800	800	400
Wages accrued	800	800	360
Work certified	40,000	32,000	7,200
Work uncertified	1,200	1,600	420
Cash received in respect of work certified	30,000	24,000	5,400

Depreciation is to be charged on the plant @ 10% which was installed on the opening date of the contract in each case. (*C.S. Inter. June 1997*)

Solution: CONTRACT ACCOUNTS

	A	B	C		A	B	C
To raw materials	14,400	11,600	4,000	By materials-in-hand	800	800	400
To wages paid	22,000	22,400	2,800	By plant in hand	3,600	3,040	2,340
To general charges	800	560	200	By work-in-progress:			
To plant	4,000	3,200	2,400	Work certified	40,000	32,000	7,200
To wages accrued	800	800	360	Work uncertified	1,200	1,600	420
To balance c/d	3,600	–	600	By profit and loss a/c	–	1,120	–
	45,600	38,560	10,360		45,600	38,560	10,360
To P and L a/c	1,800	–	–	By balance b/d	3,600	–	600
To WIP Reserve	1,800	–	600				
	3,600	–	600		3,600	–	600

Working Note:

(1) Depreciation on plant is provided at 10% per annum. So depreciation is deducted in order to arrive at the closing value of plant in hand.

(2) Profit transferred to P and L a/c of contract 'A'

$$3{,}600 \times \frac{2}{3} \times \frac{3}{4} = 1{,}800$$

BALANCE SHEET AS ON 31.12.95

P and L a/c:		**Materials on hand:**	
Profit of contract A	1,800	Contract A	800
Less: loss of contract B	1,120	B	800
	680	C	400
Wages o/s		**Plant in hand**	
Contract A	800	Contract A	3,600
Contract B	800	B	3,040
Contract C	360	C	2,340
		WIP	
		Contract A	9,400
		B	9,600
		C	1,620
			20,620

Work-in-progress:

	A	B	C
Work certified	40,000	32,000	7,200
Work uncertified	1,200	1,600	420
	41,200	33,600	7,620
Less: cash received	30,000	24,000	5,400
	11,200	9,600	2,220
Less: reserve	1,800	–	600
	9,400	9,600	1,620

CHAPTER 3

Problem 18. *(Contract a/c with Trial Balance)*: M/s AB and Associates a partnership firm comprising of partners A and B undertook a contract to build a bridge for Rs. 20,00,000 and commenced the work on 1-4-2006. The following is the trial balance of the firm as on 31-3-2007.

	Dr. (Rs)	*Cr. (Rs.)*
Plant and machinery	2,50,000	—
Office buildings	3,00,000	—
Materials purchased	4,20,000	—
Wages	1,40,000	—
Sub-contract charges	80,000	—
Interest	10,000	—
Office overheads	50,000	—
Capitals:		
A	—	1,20,000
B	—	80,000
Advance from contractee	—	6,00,000
Bank overdraft	—	1,40,000
Outstanding wages	—	10,000
Creditors	—	1,50,000
Loans	—	1,50,000
	12,50,000	12,50,000

Additional information:

(1) Material worth Rs. 4,00,000 were sent to site
(2) Outstanding sub-contracting charges Rs. 20,000 at the year and
(3) Allocate 50% of office overheads and 100% of wages to contract
(4) Plant and machinery were used for the whole year on contract. Provide depreciation on it @ 10% p.a.
(5) Partner 'A' was entitled to a salary of Rs. 20,000 for site supervision for the year. Provide the same in account.
(6) Contractee pays 75% of the work certified
(7) Partners A and B share profits and losses in the ratio of 6: 4 respectively
(8) At the end of the year, work uncertified valued at Rs. 10,000 and materials at site Rs. 20,000

Prepare contract account, profit and loss account for the year ended 31.3.2007 and balance sheet as on that date *(Sri Sathya Sai University, B.Com. (Hons), March 2007)*

Solution:

CONTRACT A/C

To materials		4,00,000	By work in progress:	
To wages		1,40,000	Work certified	8,00,000
To subcontracting			Work uncertified	10,000
charges	80,000		By materials at site	20,000
Add: o/s	20,000			
		1,00,000		
To office overhead (50%)		25,000		
To Dep. on plant		25,000		
To salary to Mr 'A'		20,000		
To balance c/d		1,20,000		
		8,30,000		8,30,000

To profit and loss a/c	30,000	By balance c/d	1,20,000
To WIP—reserve	90,000		
	1,20,000		1,20,000

Note:

As work done is 40% (between 25% and 50%) $\frac{1}{3}$ of notional profit is considered

$$\frac{1}{3}\times 1{,}20{,}000\times\frac{6{,}00{,}000}{8{,}00{,}000}$$
$$= \text{Rs. } 90{,}000$$

PROFIT AND LOSS A/C

		By contract a/c	30,000
To Interest	10,000	By net loss	
To office oh undercharged	25,000	A's capital	3,000
		B's capital	2,000
	35,000		35,000

BALANCE SHEET

Capital a/c:			Plant and machinery		
'A'	1,20,000			2,50,000	
			Less: Dep	25,000	
Add: Salary	20,000				
	1,40,000			2,25,000	
			Office building	3,00,000	
Less: Net loss	3,000		**Work in progress:**		
		1,37,000	Work certified	8,00,000	
			Work uncertified	10,000	
'B'	80,000				
Less: Net loss	2,000		Material at site	20,000	
		78,000		8,30,000	
Bank overdrafts		1,40,000			
Outstanding wages		10,000	*Less:* Cash received	6,00,000	
Creditors		1,50,000			
Loan		1,50,000		2,30,000	
Outstanding subcontract charges		20,000	*Less:* Reserve	90,000	
				1,40,000	
			Stock of materials	20,000	
		6,85,000		6,85,000	

CHAPTER 3

Problem 19. The following is the trial balance of a construction company as on 31-3-2006

	Dr.	*Cr.*
Contractee's account	—	7,20,000
Building	3,20,000	—
Creditors	–	1,44,000
Bank balance	1,90,000	—
Capital	—	10,00,000
Materials	4,00,000	
Wages (only for contract)	3,60,000	
Expenses	94,000	
Plant and machinery	5,00,000	
	18,64,000	18,64,000

Only one contract was taken up during 2005-06 and the work on the same commenced on 1-4-2005.

Materials of Rs. 3,40,000 was sent to the contract out of which materials of Rs. 12,000 was destroyed in an accident at the site.

Plant and Machinery costing Rs. 1,00,000 was used throughout the year. The balance plant and machinery costing Rs. 4,00,000 was used from 1-4-2005 to 31-12-2005 and thereafter returned to the stores. Materials on site on 31-3-2006 were valued at Rs. 8,000.

The contract was for Rs. 12,00,000 and the contractee was required to pay 75% of the work certified. The work certified was 80% of the total contract value and uncertified work was estimated at Rs. 30,000; on 31-3-2006.

Overhead expenses are to be charged to the contract at 25% of wages. Plant and machinery is to be depreciated at 10% for entire year.

Prepare contract account showing the calculation of profit to be taken credit for profit and loss a/c and balance sheet as on 31-3-2006. *(Sri Sathya Sai University, B.Com. (Hons) March 2006)*

Solution:

CONTRACT ACCOUNT

To materials issued	3,40,000	By P and L a/c:		
To wages	3,60,000	Materials destroyed		12,000
To overheads (25% of wages)	90,000	Plant at site:		
To Plant and Machinery	5,00,000	Cost	5,00,000	
To balance c/d	1,80,000	*Less:* Depreciation $\left(\frac{10}{100}\times 10{,}000\right)$	10,000	
			4,90,000	
		$\left(\frac{10}{100}\times 40{,}000\times\frac{3}{12}\right)$	30,000	
				4,60,000
		Materials at site		8,000
		By work-in-progress:		
		Work certified $\frac{80}{100}\times 12{,}00{,}000$		9,60,000
		Work uncertified		30,000
	14,70,000			14,70,000

To P and L a/c	90,000	By balance b/d	1,80,000
$1,80,000 \times \frac{2}{3} \times \frac{75}{100}$			
To WIP reserve	90,000		
	1,80,000		1,80,000

BALANCE SHEET AS ON 31-3-2006

Capital a/c:		10,00,000	Building		3,20,000
P and L a/c:			Plant and machinery		4,60,000
			Materials:		
Profit on contract	90,000		Bought	4,00,000	
Less: Materials destroyed	12,000		*Less:* Issued	3,40,000	
		78,000		60,000	
			Add: at site	8,000	
Creditors		1,44,000			68,000
			Expenses prepaid		
			Incurred	94,000	
			Charged	90,000	
			Prepaid		4,000
			Work-in-progress:		
			Work certified	9,60,000	
			Work uncertified	30,000	
				9,90,000	
			Less: Reserve	90,000	
				9,00,000	
			Less: Cash received	7,20,000	
					1,80,000
			Bank		1,90,000
		12,22,000			12,22,000

Problem 20. Crystal Construction Company Limited engaged in contract work has the following trial balance on 31-12-2004.

	Dr. (*Rs*)	*Cr.* (*Rs.*)
Share capital—shares of Rs. 10 each	—	35,180
Profit and loss a/c as on 1-1-2004	—	2,500
Provision for depreciation on plant	—	6,300
Contractee's a/c—contract no 303	—	1,28,000
Creditors		8,120
Land and Building (*at cost*)	7,400	—
Plant (*at cost*)	5,200	
Bank balance	4,500	
Contract no. 303:		
Materials issued	60,000	
Direct labour	83,000	

Expenses		4,000	
Plant at site (*at cost*)		16,000	
		1,80,100	1,80,100

Contract No. 303 having a contract price of Rs. 2,40,000 was begun on 1-1-2004 and contractee pays 80% of the work completed and certified. The cost of work done since certification is estimated to be Rs. 1,600.

After the trial balance was extracted on 31st December 2004, plant costing Rs. 3,200 was returned to the stores and materials at site on that date were valued at as 3,000.

Provision is to be made for sub-standard cost amounting to Rs. 600 incurred on contract no. 303 and for depreciation of all plant @ 12.5% on cost.

Prepare contract No. 303 a/c showing the computation of profit if any for which credit may be taken in 2004 and prepare balance sheet as on 31-12-2004.

(Sri Sathya Sai University, B. Com. (Hons), October 2005)

Solution:

CONTRACT ACCOUNT

To material	60,000	By plant returned	3,200	
To direct labour	83,000	*Less:* Dep. @ 12.5%	400	2800
To expenses	4,000			
To plant	16,000	By materials site		3,000
To sub-standard cost	600	By plant at site:		
To balance c/d	15,000	Sent to site	16,000	
		Less: Returned to stores	3,200	
			12,800	
		Less: Depreciation	1,600	11,200
		To work-in-process:		
		Work certified		1,60,000
		Work uncertified		1,600
	178,600			1,78,600
To P and L a/c	8,000	By balance b/d		15,000
To WIP–reserve	7,000			
	15,000			15,000

BALANCE SHEET

Share capital		35,180	Land and building		7,400
Creditors		8,120	Plant		
P and L a/c	2,500		At cost	5,200	
Add: Profit	8,000		At site	16,000	
	10,500			21,200	
Less: Depreciation Provision	650	9,850	*Less:* **Depreciation:**		
			Depreciation provision	(–) 6300	
Provision for sub-standard cost		600	Depreciation on plant Sent to stores	(–) 400	
			Depreciation on plant on contract	(–) 1600	

		Depreciation on other plant(–) 650		
				12,250
		Work in Progress:		
		Work certified	1,60,000	
		Work uncertified	1,600	
			1,61,600	
		Less: Reserve	7,000	
			1,54,600	
		Less: Cash received	1,28,000	
				26,600
		Materials at site		3,000
		Bank		4,500
	53,750			53,750

Problem 21. Surya Construction Ltd. with a paid-up share capital of Rs. 50 lakh undertook a contract to construct. MIG apartments. The work commenced on the contract on 1st April 2,000. The contract price was Rs. 60 lakhs, cash received on account of the contract upto 31st March 2001 was Rs. 18 lakhs (being 90% of the work certified). Work completed but not certified was estimated at Rs. 1,00,000. As on 31st March 2001 material at site was estimated at Rs. 30,000, machinery at site costing Rs. 2,00,000 was returned to stores and wages outstanding were Rs. 5,000. Plant and machinery at site is to be depreciated at 50%.

The following were the ledger balances (Dr) as per trial balance as on 31st March, 2003

Land and Building	23,00,000
Plant and Machinery (60% at site)	25,00,000
Furniture	60,000
Materials	14,00,00
Fuel and power	1,25,000
Site expenses	5,000
Office expenses	12,000
Rates and taxes	15,000
Cash at bank	1,33,000
Wages	2,50,000

Prepare the contract account and balance sheet.

(University of Delhi, B.Com. (Hons), 2001)

Solution: CONTRACT A/C

To materials	14,00,000	By plant returned		2,00,000
To wages	2,50,000	*Less:* Depreciation		10,000
To wages o/s	5,000			1,90,000
To plant and machinery	15,00,000	By plant at		
To fuel and power	1,25,000	site	13,00,000	
To office expenses	12,000	*Less:* Depreciation	65,000	
To site expenses	5,000			12,35,000
To Rates and taxes	15,000	By materials at site		30,000
To balance c/d	2,43,000			

		By work in progress:	
		Work certified	20,00,000
		Work uncertified	1,00,000
	35,55,000		35,55,000
To P and L a/c	72,900	By Balance b/d	2,43,000
$2,43,000 \times \frac{1}{3} \times \frac{90}{100}$			
To Reserve–WIP a/c	1,70,100		
	2,43,000		2,43,000

BALANCE SHEET AS ON 31-3-2001

Share capital	50,00,000	Land and Building		23,00,000
Wages o/s	5,000	Plant and machinery		11,90,000
P and L a/c	72,900	(10,00,000 + 1,90,000)		
		Plant at site		12,35,000
		Cash at Bank		1,33,000
		Materials at site		30,000
		Furniture		60,000
		Work in progress:		
		Work certified	20,00,000	
		Work uncertified	1,00,000	
			21,00,000	
		Less: Cash received	18,00,000	
			3,00,000	
		Less: Reserve	1,70,100	1,29,900
	50,77,900			50,77,900

Problem 22. *(Contract a/c with balance sheet):* On 1.1.1999, Archana and Company undertook one contract for a price of Rs. 25,00,000. Of the plant and materials sent to the contract, plant which cost Rs. 25,000 and materials worth Rs. 20,000 were lost in transit. On 31st December 1999, plant which originally cost Rs. 25,000 was returned to stores. The cost of work done but uncertified was Rs. 10,000 and materials costing Rs. 20,000 were at site. Charge depreciation at 10% pa on plant. Prepare contract account and balance sheet, from the following trial balance as on 31.12.99.

	Dr.	*Cr.*
Share capital		6,00,000
Creditors		50,000
Contractee's a/c		10,00,000
(80% of work certified)		
Land and Buildings	2,15,000	—
Bank	1,25,000	—
Contract a/c:		
Materials	4,50,000	
Plant at cost	1,25,000	

Wages	7,00,000	
Expenses	35,000	
	16,50,000	16,50,000

(Sri Sathya Sai University, B.Com., February 2001)

Solution:

CONTRACT A/C

To materials	4,50,000	By P and L			
To plant	1,25,000	Materials lost			20,000
To wages	7,00,000	Plant lost			25,000
To expenses	35,000	By plant returned			
To balance c/d	1,05,000	to stores		25,000	
		Less: Depreciation		2,500	22,500
		By materials at site			20,000
		By work-in-progress			
		Work certified		12,50,000	
		Work uncertified		10,000	
					12,60,000
		By plant at site		1,25,000	
		Less: Lost		25,000	
				1,00,000	
		Less: Returns		25,000	
				75,000	
		Less: Depreciation		7,500	
					67,500
	14,15,000				14,15,000
To P and L a/c	56,000	By balance b/d			1,05,000
To work-in-progress –reserve	49,000				
	1,05,000				1,05,000

BALANCE SHEET

Share capital		6,00,000	Land and Buildings		2,15,000
Creditors		50,000	Bank		1,25,000
P and L A/c			Plant returned		22,500
Profit on contract	56,000		Materials at site		20,000
Less: Plant lost	25,000		**Work-in-progress:**		
	31,000		Work certified	12,50,000	
Less: Loss of materials	20,000	11,000	Work uncertified	10,000	
				12,60,000	
			Less: Cash received	10,00,000	
				2,60,000	

		Less: Reserve	49,000	2,11,000
		Plant at site		67,500
	6,61,000			6,61,000

Problem 23. The following trial balance was extracted on 30th April, 1998 from the books of General Contractors Ltd.

	Debit	*Credit*
Share capital		70,360
Profit and loss a/c (last year)		5,000
Provision for depreciation on plant		12,600
Cash received on contract no. 15		2,56,000
Creditors		16,240
Land and buildings at cost	14,800	
Plant at cost	10,400	
Cash at bank	9,000	
Contract no. 15:		
Materials issued	1,20,000	
Direct labour	1,66,000	
Expenses	8,000	
Plant at cost	32,000	
	3,60,200	3,60,200

Contract no. 15 was begun on May 1, 1997. The contract price is Rs. 4,80,000 and the customer has so far paid Rs. 2,56,000 being 80% of the work certified. The cost of work done since certification is estimated at Rs. 3,200 on April 30, 1998. After the above trial balance was extracted, plant costing Rs. 6,400 was returned to stores and materials then on site were valued at Rs. 5,400. Provision is to be made for labour accured Rs. 1,200 and depreciation on plant at 12½%.

You are required to (*a*) write up contract a/c (*b*) prepare balance sheet as on that date.

(Sri Sathya Sai University, B.Com., October 1998)

Solution:

CONTRACT A/C

To materials used	1,20,000	By materials at site	5,400
To direct labour	1,66,000	**By work-in-progress**	
To expenses	8,000	Work certified	3,20,000
To depreciation on plant	4,000	$2,56,000 \times \frac{100}{80}$	
$\frac{12.5}{100} \times 32,000$		Work uncertified	3,200
To provision for labour accrued	1,200		
To balance c/d	29,400		
	3,28,600		3,28,600
To P and L a/c	15,680	By balance b/d	29,400
$\frac{2}{3} \times 29,400 \times \frac{2,56,000}{3,20,000}$			

To WIP–reserve	13,720		
	29,400		29,400

BALANCE SHEET

Share capital	70,360	Plant	42,400	
P and L a/c (*see note*)	19,380	(32,000 + 10,400)		
Creditors	16,240	*Less:* Provision for dep.	17,900	24,500
Provision for labour accrued	1,200	(*see note*)		
		Land and building		14,800
		Work-in-progress		58,880
		(*see note*)		
		Cash at bank		9,000
	1,07,180			1,07,180

Working Note:

(1) **Calculation of profit transferred to B/s**

P AND L A/C

To depreciation on plant (other than issued to contract)	1,300	By balance b/d	5,000
		By profit from contract no. 15	15,680
To profit transferred to B/s	19,380		
	20,680		20,680

(2) *Calculation of depreciation:*

Provision for depreciation on plant	12,600
Add: Depreciation on plant issued to contract	4,000
Add: Depreciation on general plant	1,300
	17,900

(3) *Calculation of WIP*

Work certified	3,20,000
Work uncertified	3,200
	3,23,200
Less: Cash received	2,56,000
	67,200
Less: Reserve	13,720
	53,480
Add: Materials at site	5,400
	58,880

(4) The value of plant at stores of Rs. 10,400 includes the plant value of 6400 returned to stores.

Problem 24. Modern Construction Company with a paid-up share capital of Rs. 50 lakhs undertook a contract to construct LIC houses. The contract was commenced on 1.1.1994 and the contract price was Rs. 50 lakhs. Cash received on account of contract on 31.12.94 was Rs. 18 lakhs (90% of the work certified) work completed but not certified was estimated at Rs. 1,00,000. As on 31.12.94, materials at site was estimated at Rs. 30,000 and machinery at site costing Rs. 2,00,000 was returned to stores. Plant and machinery at site is to be depreciated at 5%. Wages outstanding on 31.12.94 was Rs. 5,000.

The following trial balance was extracted as on 31.12.94.

	Debit	*Credit*
Paid-up share capital		50,00,000
Cash received (90% of work certified)		18,00,000
Land and Buildings	15,00,000	
Machinery at cost (60% at site)	25,00,000	
Lorries and vehicles	8,00,000	
Furniture	50,000	
Office equipment	10,000	
Materials sent to site	14,00,000	
Site expenses	5,000	
Postage etc.	4,000	
Office expenses	8,000	
Rates and taxes	15,000	
Fuel and power	1,25,000	
Cash at bank	1,33,000	
Wages	2,50,000	
	68,00,000	68,00,000

Prepare the contract a/c to ascertain the profit from the contract and show the work-in-progress in the balance sheet. *(ICWA, Inter, December 1995)*

Solution:

CONTRACT A/C

To materials		14,00,000	By materials at site		30,000
To wages	2,50,000		**By work-in-progress**		
Wages o/s	5,000	255,000	Work certified		20,00,000
To site expenses		5,000	Work uncertified		1,00,000
To postage		4,000			
To power and fuel		1,25,000	Plant returned	2,00,000	
To office expenses		8,000			
To rates and taxes		15,000	*Less:* Depreciation	10,000	1,90,000
To plant		15,00,000			
			Plant at site:		
			Plant value	15,00,000	
			Less: Returned	20,000	
				13,00,000	
			Less: Depreciation	65,000	
				12,35,000	
To balance c/d		2,43,000			
		35,55,000			35,55,000
To P and L a/c		72,900	By balance b/d		2,43,000
$\frac{1}{3} \times \frac{90}{100} \times 2{,}43{,}000$					
To WIP–reserve		1,70,100			
		2,43,000			2,43,000

Working Note:

(1) *Calculation of work-in-progress:*

Work certified 18,00,000 × $\frac{100}{90}$	= 20,00,000
Work uncertified	1,00,000
	21,00,000
Less: Cash received	18,00,000
	3,00,000
Less: Reserve	1,70,100
Work-in-progress	1,29,900

(2) *Value of plant returned to stores:*

Plant returned to stores	2,00,000
Less: Depreciation $\frac{5}{100}$ × 2,00,000	10,000
	1,90,000

BALANCE SHEET

Paid-up capital	50,00,000	Land and Building		15,00,000
Wages o/s	5,000	Lorries and vehicles		8,00,000
P and L. a/c	72,900	Furniture		50,000
		Office equipment		10,000
		Machinery at site	15,00,000	
		Less: Machinery returned	2,00,000	
			13,00,000	
		Less: Depreciation	65,000	
				12,35,000
		Machinery in stores	10,00,000	
		Add: Machinery returned	1,90,000	
				11,90,000
		Materials at site		30,000
		Work-in-progress		1,29,900
		Cash at bank		1,33,000
	50,77,900			50,77,900

Problem 25. *(Contract a/c with trial balance when two contracts are in progress):* The following trial balance was extracted from the books of Gemini Contractor as on 31.12.89:

	Dr.	*Cr.*
Contractees a/c		3,30,000
Buildings	1,00,000	
Creditors		62,000
Bank	35,000	
Capital a/c		3,00,000
Materials	1,00,000	
Wages	70,000	

Expenses	37,000	
Plant	2,50,000	
Work-in-progress (contract no. 83)	1,00,000	
Contract 83 a/c (1.1.89) unadjusted profit		30,000
	6,92,000	6,92,000

Contract no. 83 which was-in-progress on 1.1.89 was completed on 31.3.89. Contract no. 84 commenced on 1.1.89. Rs. 20,000 materials and Rs. 10,000 wages were paid for contract no. 83. Rs. 60,000 materials were sent to contract no. 84 but Rs. 3,000 worth was lost there by accident. Rs. 60,000 wages were paid for contract no. 84. Rs. 50,000 plant was used in contract no. 84 all through but plant costing Rs. 2,00,000 was used on contract no. 84 from 1st April 1989; prior to that, above machinery was used on contract no. 83. Rs. 4,000 materials were at site on contract no. 84 at the end of the year. Provide 10% depreciation of the plant and 2% on buildings.

Contract no. 83 was for Rs. 1,50,000 and certified work up to last year was Rs. 1,00,000. The work has been certified up to the full extent but payment has been received up to 80% of the certified amount. The balance has not been paid yet nor any entry has been passed on completion of the contract.

Expenses are charged to contracts on the basis of 50% of direct wages. The new contract is for Rs. 4,00,000 and 90% is paid on certification. The uncertified work of contract as on 31st Dec. 1989 is estimated at Rs. 15,000.

You are required to prepare: (*a*) Contract 83 a/c, (*b*) Contract no. 84 a/c, (*c*) P and L a/c for 1989, (*d*) Contract no. 83 contractee's a/c (*e*) Contract no. 84 contractee's a/c, (*f*) Balance sheet as on 31.12.89.

(Sri Sathya Sai University, B.Com., March 1993)

Solution:

CONTRACT NO. 83

To work-in-progress a/c (1.1.89)	1,00,000		By contractee's a/c	1,50,000
Less: Unadjusted profit on 1.1.89	30,000	70,000		
To materials		20,000		
To wages		10,000		
To expenses (50% of direct wages)		5,000		
To depreciation on plant		5,000		
To profit and loss a/c		40,000		
		1,50,000		1,50,000

CONTRACT NO. 84

To materials	60,000	By P and L a/c— Abnormal loss	3,000
To wages	60,000	By materials at site	4,000
To expenses 50% of wages	30,000	By work certified	2,00,000
To depreciation on plant		$\frac{100}{90} \times 1{,}80{,}000$	
(5,000 + 15,000)	20,000	By work uncertified	15,000
To P and L a/c	31,200		
To work-in-progress a/c—Reserve	20,800		
	2,22,000		2,22,000

CONTRACTEE'S A/C ON CONTRACT 1983

Particulars	Amount	Particulars		Amount
1988 To Balance c/d	80,000	1988 By Bank		80,000
1989 To Balance c/d	1,20,000	1989 By Balance b/d		80,000
		By cash		
		$\frac{80}{100} \times 1,50,000 =$	1,20,000	
		Less: Cash already received	80,000	40,000
	1,20,000			1,20,000
1990 To Contract a/c	1,50,000	1990 By Balance b/d		1,20,000
		By Balance c/d		30,000
	1,50,000			1,50,000

PROFIT AND LOSS A/C

Particulars	Amount	Particulars	Amount
To expenses not charged to contract	2,000	By contract no. 83	40,000
To contract no. 84:		By contract no. 84	31,200
Abnormal loss	3,000		
To depreciation on building	2,000		
To balance c/d	64,200		
	71,200		71,200

BALANCE SHEET

Liabilities	Amount	Assets		Amount
Creditors	62,000	Bank		35,000
P and L a/c	64,200	Amount due from contractee of no. 83		30,000
Capital	3,00,000	Materials at stores	20,000	
		Add: material at site on contract 84	4,000	24,000
		Work-in-progress a/c:		
		Work certified	2,00,000	
		Work uncertified	15,000	
			2,15,000	
		Less: Reserve	20,800	
			1,94,200	
		Less: Cash received	1,80,000	14,200
		Plant	2,50,000	
		Less: Depreciation	25,000	2,25,000
		Building	1,00,000	
		Less: Depreciation	2,000	
				98,000
	4,26,200			4,26,200

Working Note:

(1) *Calculation of work certified*

It is calculated on the basis of percentage of cash received to work certified

Let work certified = x

Cash received to work certified ratio

$$\frac{90x}{100} \times 1,80,000 = 2,00,000$$

(2) *Calculation of cash received on contract* 84

Contractee's a/c cash received as per trial balance	3,30,000
Less: cash due (see contractee's a/c)	30,000
Total cash received from contractee	3,00,000
Less: Cash received on contract no. 83 upto 1989	1,20,000
∴ Cash received on contract no. 84 (balancing figure)	1,80,000

(3) *Calculation of profit taken to P & L a/c*

Profit made to date 52,000

Since half of the contract is complete. Profit taken to the P & L a/c has been calculated as under

$$52,000 \times \frac{2}{3} \times \frac{90}{100} = 31,200$$

Problem 26. *(Long Term Contract):* The following information relates to a building contract for Rs. 10,00,000 for which 80% of the value of work-in-progress as certified by the architect is being paid by the contractee:

	I Year *(Rs.)*	*II Year* *(Rs.)*	*III Year* *(Rs.)*
Materials issued	1,20,000	1,45,000	84,600
Direct wages	1,10,000	1,55,000	1,10,000
Direct expenses	5,000	17,000	6,000
Indirect expenses	2,000	2,600	500
Work certified	2,35,000	7,50,000	10,00,000
Uncertified work	2,800	8,000	—
Plant issued	14,000	Nil	Nil
Materials at site	2,000	5,000	8,000

The value of the plant at the end of 1st, IInd year and IIIrd years were Rs. 11,200, Rs. 7,000 and Rs. 3,000 respectively. Prepare contract account for these three years, taking into account appropriate profit on incomplete contract. *(Sri Sathya Sai University, B.Com., (Hons), February 2006)*

Solution: CONTRACT ACCOUNT

To materials	1,20,000	By work-in-progress:	
To wages	1,10,000	Work certified	2,35,000
To direct expenses	5,000	Work uncertified	2,800
To indirect expenses	2,000	By plant at site	11,200
To plant	14,000	By materials at site	2,000
	2,51,000		2,51,000
To work-in-progress		By work in progress:	
work certified	2,35,000	– work certified	7,50,000

Particulars	Rs.	Rs.	Particulars	Rs.
work uncertified		11,200	work uncertified	8,000
materials at site		2,000	materials at site	5,000
To materials		1,45,000	plant at site	7,000
To direct wages		1,55,000		
To direct expenses		17,000		
To indirect expenses		2,600		
To balance c/d		1,99,400		
		7,70,000		7,70,000
To P and L a/c		1,06,347	By Balance b/d	1,99,400
$1,99,400 \times \frac{2}{3} \times \frac{80}{100}$				
To WIP- Reserve		93,053		
		1,99,400		1,99,400
To work-in-progress:			By contractee a/c	10,00,000
Work certified	7,50,000		By plant at site	3,000
Work uncertified	8,000		By materials at site	8,000
	7,58,000			
Less: Reserve	93,053	664947		
To Materials		84,600		
To direct wages		1,10,000		
To direct expenses		6,000		
To indirect expenses		500		
To P and L a/c		1,44,953		
		10,11,000		10,11,000

Problem 27. S.V. Construction Ltd. have obtained a contract for construction of a building. The value of the contract is Rs. 45,00,000. The work commenced on 1st July, 2004 and was completed on 31st December 2005. The following information relates to the contract.

	31-12-2005 *(Rs.)*	*31-12-2004* *(Rs.)*
Materials issued	13,50,000	3,75,000
Direct wages	10,35,000	4,70,000
Direct expenses	1,00,000	45,000
Indirect expenses	27,000	6,000
Plant issued	—	63,000
Sub-contract charges	60,000	15,000
Work certified (cumulative)	45,00,000	10,00,000
Work uncertified	—	35,000

The above plant was specially issued for the contract. The residual value of the plant at the end of the project was estimated to be Rs. 3,000.

The contractee has agreed to pay 90% of the work certified. The accounts are closed on 31st December every year. Prepare (1) contract account and (2) contractee account for two years 2004 and 2005. Show the relevant items in balance sheet as on 31-12-2004.

(University of Bombay, B.Com., April 2006)

Solution:

CONTRACT A/C
(for the year ended 31-12.2004)

To materials issued	3,75,000	By work-in-progress:	
To direct wages	4,70,000	Work certified	10,00,000
To direct expenses	45,000	Work uncertified	35,000
To indirect expenses	6,000	By plant at site	
To special plant	63,000	(W.D.V.)	43,000
To sub-contract charges	15,000		
To work-in-progress	1,04,000		
	10,78,000		10,78,000

Note: Since only 22% of work is complete the entire amount of notional profit is taken to work-in-progress, *i.e.,*

$$= \frac{\text{work certified}}{\text{contract}} \times 100$$

$$= \frac{10,00,000}{45,00,000} \times 100 = 22\%$$

CONTRACT A/C
(for the year ended 31-12.2005)

To Work-in-progress b/d		By work-in-progress:	
Work certified	10,00,000	B/D–Reserve	1,04,000
Work uncertified	35,000	By plant scrap	30,000
Plant at site (W.D.V)	43,000	By contractee's a/c	45,00,000
To material issued	13,50,000		
To direct wages	10,35,000		
To direct expenses	1,00,000		
To indirect expenses	27,000		
To sub-contract charges	60,000		
To profit and loss a/c	9,57,000		
	46,07,000		46,07,000

BALANCE SHEET at 31-12-2004 *(Extracts)*

	Work-in-progress:	
	Work certified	10,00,000
	Work uncertified	35,000
	Plant at site	43,000
		10,78,000
	Less: Cash received	9,00,000
		1,78,000
	Less: Reserve	1,04,000
		74,000

CONTRACT A/C *(for 2004)*

To balance c/d	9,00,000	By Bank	9,00,000
To contract a/c	45,00,000	By Balance b/d	9,00,000
		By Bank a/c	36,00,000
	45,00,000		45,00,000

Working Note: Plant value on 31-12-2004

Plant issued	63,000
Less: Depreciation for 2004:	
$63{,}000 - 3{,}000 = 60{,}000 \times \frac{6}{18}$	20,000
W.D.V. on 31.12.2004	43,000

Problem 28. A company took up a contract for Rs. 10 crore and as per the agreement it would receive 75% of the work certified each year. The contract was commenced on 1st April 2000 and completed on 1st October 2003. Further details are as follows:

Particulars	*2000-01*	*2001-02*	*2002-03*	*2003-04*
Machinery purchased	50,00,000	—	—	—
Materials purchased	20,00,000	50,00,000	1,00,00,000	2,00,00,000
Labour	10,00,000	30,00,000	50,00,000	1,40,00,000
Direct expenses	5,00,000	12,18,000	40,00,000	90,00,000
Stock of materials at year end	1,00,000	2,00,000	3,20,000	5,50,000
Work certified (Cumulative)	20,00,000	2,00,00,000	5,00,00,000	10,00,00,000
Work uncertified	8,00,000	10,00,000	60,00,000	—

During 2000-2001 materials costing Rs. 20,000 were returned to stores.

During 2002-03 certain materials costing Rs. 30,000 were found unsuitable and sold at a loss of Rs. 4,000. Materials worth Rs. 8,000 were stolen from site.

During 2003-04 there was an accident at site due to which a worker had to be paid Rs. 50,000 as compensation. This amount is included in wages on completion of contract. The machinery was sold for Rs. 25,00,000.

The company provides depreciation at 20% p.a. on machinery on diminishing balance method. The company closes its accounts on 31st March every year.

Prepare Contract account for each of the above years. Also show contractee's account.

(University of Bombay, B.Com, April 2005)

Solution:

CONTRACT A/C
(for the year ended 31-3-2001)

To machinery	50,00,000	By materials returned to stores	20,000
To materials	20,00,000	**By work-in-progress**	
To labour	10,00,000	Work certified	20,00,000
To other expenses	5,00,000	Work uncertified	8,00,000
		Materials at site	1,00,000

		Machinery at site	40,00,000
		By P and L a/c (*loss*)	15,80,000
	85,00,000		85,00,000

CONTRACT A/C

(for the year ended 31-3-2002)

To WIP b/d:		By WIP c/d:	
Work certified	20,00,000	Work certified	2,00,00,000
Work uncertified	8,00,000	Work uncertified	10,00,000
Materials at site	1,00,000	Materials at site	2,00,000
Machinery at site	40,00,000	Machinery at site	32,00,000
To Materials purchased	50,00,000		
To Labour	30,00,000		
To other expenses	12,18,000		
To WIP c/d	82,82,000		
	2,44,00,000		2,44,00,000

Note: Work complete is 20%. So the entire profit is taken to work in progress a/c.

CONTRACT ACCOUNT

(for the year ended 31-3-2003)

To WIP b/d:		By WIP b/d:	82,82,000
Work certified	2,00,00,000	By materials sold	26,000
		By P and L a/c:	
Work uncertified	10,00,000	Loss on sale	4,000
Materials at site	2,00,000	Stolen	8,000
Machinery at site	32,00,000	By WIP c/d:	
To materials purchased	1,00,00,000	Work certified	5,00,00,000
To labour	50,00,000	Work uncertified	60,00,000
To other expenses	40,00,000	Materials at site	3,20,000
To balance c/d	2,38,00,000	Machinery at site	25,60,000
	6,72,00,000		6,72,00,000
To P and L a/c	1,19,00,000	By balance b/d	2,38,00,000
To WIP–Reserve	1,19,00,000		
	2,38,00,000		2,38,00,000

Working Note:

Calculation of profit transferred to P and L a/c

$$\frac{2}{3} \times 2{,}38{,}00{,}000 \times \frac{75}{100} = 1{,}19{,}00{,}000$$

CONTRACT A/C

(for the year ended 31-3-2004)

To WIP b/d:		By WIP b/d:	1,19,00,000
Work certified	5,00,00,000	By material at site	5,50,000
Work uncertified	60,00,000	By P and L a/c–Worker's	

Materials at site	3,20,000	Compensation	50,000
Machinery at site	25,60,000	To sale of machine	25,00,000
To Material purchased	2,00,00,000	To contractee's a/c	10,00,00,000
To labour	1,40,00,000		
To other expenses	90,00,000		
To P and L a/c profit on sale of plant	1,96,000		
To P and L a/c	1,29,24,000		
	11,50,00,000		11,50,00,000

Working Note:

(1) Profit on sale of machinery

I Year: Purchased			50,00,000
Less:	Depreciation @ 20%		10,00,000
	At site (W.D.V)		40,00,000
II Year:	Depreciation @ 20%		8,00,000
	At site (WDV)		32,00,000
III Year:	Depreciation @ 20%		6,40,000
	At site (WDV)		25,60,000
IV Year:	Depreciation @ 20% for 6 months		2,56,000
			23,04,000
	Sale price of machine		25,00,000
		Profit on sale	1,96,000

(2) Payment by contractee

I Year:	Cumulative work certified		20,00,000
Less:	Opening work certified		—
			20,00,000
	75% of 20,00,000		15,00,000
II Year:	Cumulative work certified	2,00,000	
Less:	Opening work certified	2,00,000	
		18,00,000	
	75% of 18,00,000		1,35,00,000
III Year:	Cumulative	5,00,00,000	
Less:	Opening work certified	2,00,00,000	
		3,00,00,000	
	75% of 3,00,00,000		2,25,00,000
			3,75,00,000
Total contract price due			10,00,00,000
Balance paid fully on completion			6,25,00,000

CONTRACTEE'S A/C

31-3-01		31-3-01	
To balance c/d	15,00,000	By Bank	15,00,000
31-3-02		1-4-02	
To balance c/d	1,50,00,000	By Balance b/d	15,00,000
		By Bank	1,35,00,000
	1,50,00,000		1,50,00,000
31-3-03		1-4-03	
To balance c/d	3,75,00,000	By Balance b/d	1,50,00,000
		By Bank	2,25,00,000
	3,75,00,000		3,75,00,000
31-3-04		1-4-04	
To contract a/c	10,00,00,000	By Balance b/d	3,75,00,000
		By Bank	6,25,00,000
	10,00,00,000		10,00,00,000

Problem 29. *(Long-term contract):* Rajesh contractor obtained a contract to build a house at a contract price of Rs. 15,00,000. The contractee agrees to pay 90% of the value of the work done as certified by the architect immediately on receipt of the certificate and to pay the balance after completion of the contract.

The contractor commenced the work on 1st May 1996. A plant costing Rs. 20,000 was specially bought for the contract. The value of the plant at the end of 1996, 1997 and 1998 was Rs. 16,000, Rs. 10,000 and Rs. 4,000 respectively. The work done and certified by the architect as at the end of 1996 and 1997 was Rs. 3,50,000 and Rs. 11,50,000 respectively. Work costing Rs. 20,000 done as at the end of 1997 was not certified as on that date.

Other details of the contract were as under:

	1996 *(Rs.)*	*1997* *(Rs.)*	*1998* *(Rs.)*
Materials sent to site	1,80,000	2,20,000	1,26,000
Wages paid	1,70,000	2,30,000	1,70,000
Direct expenses	7,000	25,000	9,000
Indirect expenses	3,000	4,000	–

You are required to prepare contract a/c for the year ended 31st Dec. 1996, 1997 and 1998, considering that contract is fully completed in 1998.

(University of Bombay, B.Com., October 1999)

Solution:

CONTRACT A/C
(for the year ended 31-12.1996)

To materials	1,80,000	By work-in-progress	
To wages paid	1,70,000	work certified	3,50,000
To direct expenses	7,000	By P and L	14,000
To indirect expenses	3,000		
To depreciation on plant (20,000 – 16,000)	4,000		
	3,64,000		3,64,000

CONTRACT A/C
(for the year ended 31-12-1997)

To work-in-progress: work certified	3,50,000	By work-in-progress:	
To materials	2,20,000	work certified	11,50,000
To wages	2,30,000	work uncertified	20,000
To direct expenses	25,000		
To indirect expenses	4,000		
To depreciation on plant (16,000 – 10,000)	6,000		
To balance c/d	3,35,000		
	11,70,000		11,70,000
To P and L a/c $\frac{2}{3} \times 3{,}35{,}000 \times \frac{90}{100}$	2,01,000	By balance b/d	3,35,000
To WIP reserve	1,34,000		
	3,35,000		3,35,000

CONTRACT A/C
(for the year ended 31-12-1998)

To work-in-progress a/c			By work-in-progress:	
Work certified	11,50,000		Provision for contingencies	1,34,000
Work uncertified	20,000	11,70,000	By contractee's a/c	15,00,000
To materials		1,26,000		
To wages		1,70,000		
To direct expenses		9,000		
To depreciation on plant (10,000 – 4,000)		6,000		
To P and L		1,53,000		
		16,34,000		16,34,000

Problem 30. Balram contractors undertook a contract for Rs. 15,00,000 on 1st July 1996. The contract was complete on 31st March 1998. The contractor prepares his accounts on 31st March. The details of the contract are

Particulars	Period From 1.7.96 to 31.3.97	From 1.4.97 to 31.3.98
	Rs.	Rs.
Materials issued	1,52,000	3,30,000
Direct wages	1,25,000	4,65,000
Direct expenses	30,000	45,000
Materials returned to stores	22,000	15,000
Materials at site	20,000	8,000
Uncertified work	48,000	–
Office overheads	23,000	66,000
Materials lost by fire	–	5,000
Work certified	3,00,000	15,00,000
Plant issued	3,00,000	1,50,000

CHAPTER 3

Provide depreciation @ 20% p.a. on plant. Prepare contract accounts for the years ended 31.3.1997 and 31.3.1998. *(University of Bombay, B.Com., April 1998)*

Solution:

CONTRACT A/C
(for the period 1.7.96 to 31.3.97)

Particulars	Rs.	Particulars		Rs.
To materials issued	1,52,000	By materials returned to stores		22,000
To direct wages	1,25,000	By work-in-progress:		
To direct expenses	30,000	Work certified	3,00,000	
To office overheads	23,000	Work uncertified	48,000	
To depreciation on plant	45,000	Materials at site	20,000	
$3{,}00{,}000 \times \frac{9}{12} \times \frac{20}{100}$				
To profit reserve for contingency	15,000			
				3,68,000
	3,90,000			3,90,000

CONTRACT A/C
(for the year period 1.4.97 to 31.3.98)

Particulars		Rs.	Particulars	Rs.
To work-in-progress:				
Work certified	3,00,000		By reserve for contingency	15,000
Work uncertified	48,000		By materials returned to stores	15,000
Materials at site	20,000		By materials at site	8,000
		3,68,000	By materials lost by fire	5,000
To materials issued		3,30,000	By contractee's a/c	15,00,000
To direct wages		4,65,000		
To direct expenses		45,000		
To office overheads		66,000		
To depreciation on plant		90,000		
4,50,00 (3,00,000 + 150,000) @ 20%				
To net profit		1,79,000		
		15,43,000		15,43,000

Problem 31. Yuti space contractors and Builder have obtained a contract for constructing alternate accommodation to the slum dwellers. The contract work commenced on 1st July 1994 and was completed on 31st January 1997. The year ending of the company is 31st March. The contract price was Rs. 800 lakhs. The contractee agree to pay 90% of the value of the work done as certified by the architect immediately. The entire contract price was received before 31st March 1997.

A machine costing Rs. 60,00,000 was specially bought and used for the contract. The residual value of the machine as on 31.1.1997 was Rs. 29,00,000 depreciation to be effected on straight line basis.

As a cost consultant, you are provided with the following relevant information:

	Accounting years		
	1994–95	*1995–96*	*1996–97*
	(Rs.)	*(Rs.)*	*(Rs.)*
Materials purchased	27,50,000	86,25,000	19,75,000
Direct labour	77,52,500	89,36,500	1,02,00,000
Supervision charges	1,22,000	1,85,000	2,76,000

Architect fees	2,50,000	4,50,000	5,00,000
Construction overheads	49,50,000	32,16,500	80,15,000
Administration overheads	19,25,500	10,50,000	7,96,000
Uncertified work at the end of the year	2,00,000	4,00,000	
Material on site at the end of the year	50,000	1,25,000	75,000
Money received from the contractee during the year	1,80,00,000	3,60,00,000	2,60,00,000

As per the policy of the company no profit is to be considered unless the certified work completed exceeds 20% of the total contract price. Thereafter profit is to be taken credit for in the same proportion as the amount received bears to the contract price.

Prepare contract accounts for all the three years and show the relevant extracts in the balance sheet as on 31st March of every year. *(University of Bombay, B.Com., October 1997)*

Solution:

CONTRACT A/C
(for the year ended 31-12-1995)

To materials purchased	27,50,000	By Work-in-progress:	
To direct labour	77,52,500	Work certified	2,00,00,000
To supervisor's charges	1,22,000	Work uncertified	2,00,000
To architects fee	2,50,000	Materials at site	50,000
To construction overheads	49,50,000		
To administrative overheads	19,25,500		
To depreciation	9,00,000		
To balance c/d	16,00,000		
	2,02,50,000		2,02,50,000
To P and L a/c $16{,}00{,}000 \times \frac{1{,}80{,}00{,}000(\text{C.R.})}{8{,}00{,}00{,}000(\text{C.P.})}$	3,60,000	By balance b/d	16,00,000
To WIP–reserve	12,40,000		
	16,00,000		16,00,000

BALANCE SHEET
(Extract for the year 1995)

P and L a/c	3,60,000	Plant	60,00,000	
		Less: Depreciation	9,00,000	
				51,00,000
		Work-in-progress:		
		Work certified	2,00,00,000	
		Work uncertified	2,00,000	
		Materials at site	50,000	
			2,02,50,000	
		Less: Cash received	1,80,00,000	
			22,50,000	
		Less: Reserve	12,40,000	
				10,10,000

CONTRACT A/C

(for the year ended 31-3-1996)

Particulars		Amount	Particulars		Amount
To work-in-progress b/d			By reserve for contingency b/d		12,40,000
Work certified	2,00,00,000		**By work-in-progress**		
Work uncertified	2,00,000		Work certified	6,00,00,000	
Materials at site	50,000		Work uncertified	4,00,000	
		2,02,50,000	Materials at site	1,25,000	
To materials purchased		86,25,000			6,05,25,000
To direct labour		89,36,500			
To supervision charges		1,85,000			
To architect fees		4,50,000			
To construction overhead		32,16,500			
To administration overhead		10,50,000			
To depreciation		12,00,000			
To balance c/d		1,78,52,000			
		6,17,65,000			6,17,65,000
To P and L a/c		1,20,50,100	By balance b/d		1,78,52,000
$1{,}78{,}52{,}000 \times \frac{5{,}40{,}00{,}000}{8{,}00{,}00{,}000}$					
To WIP–reserve		58,01,900			
		1,78,52,000			1,78,52,000

BALANCE SHEET

(Extract for the year 31.3.1996)

Liabilities	Amount	Assets		Amount
P and L	1,24,10,100	Plant (WDV)	51,00,000	
		Less: Depreciation	12,00,000	
				39,00,000
		Work-in-progress		
		Work certified	6,00,00,000	
		Work uncertified	4,00,000	
		Materials at site	1,25,000	
			6,05,25,000	
		Less: Cash received	5,40,00,000	
			65,25,000	
		Less: Reserve	58,01,900	
				7,23,100

CONTRACT ACCOUNT

(for the year ended 31-3-1997)

Particulars		Amount	Particulars	Amount
To work-in-progress b/d			By reserve for contingency	58,01,900
Work certified	6,00,00,000		By materials at site	75,000
Work uncertified	4,00,000		By contractee's a/c	8,00,00,000
Materials at site	1,25,000			

	6,05,25,000		
To materials purchased	19,75,000		
To direct labour	1,02,00,000		
To supervision charges	2,76,000		
To architects fees	5,00,000		
To construction overheads	80,15,000		
To administrative overheads	7,96,000		
To depreciation	10,00,000		
To P and L	25,89,900		
	8,58,76,900		8,58,76,900

BALANCE SHEET
(Extract for the year 31.3.1997)

P and L a/c	25,89,900	Plant (WDV)	39,00,000
		Less: Depreciation charged to contract	10,00,000
			29,00,000
		Less: Depreciation charged for 31.1.97 to 31.3.97	2,00,000
		(non-contract period)	27,00,000
		Materials at site	75,000

Working Note:

Calculation of depreciation

Purchased on 1.7.94	60,00,000
Written down value (WDV) on 31.1.1997	29,00,000
	31,00,000

Depreciation for period 1.7.94 to 31.1.97 *i.e.,* 31 months

$$\text{Depreciation per month} = \frac{31,00,000}{31} = 1,00,000 \text{ P.M.}$$

Problem 32. *(Contract nearing the stage of completion):* A contractor commenced a building contract on October 1, 1997. The contract price is Rs. 4,40,000. The following data pertaining to the contract for the year 1998-99 has been compiled from his books and is as under:

		(Rs.)
April. 1st 1998	Work-in-progress not certified	55,000
	Materials at site	2,000
1998-99	Expenses incurred:	
	Materials issued	1,12,000
	Wages paid	1,08,000
	Hire of plant	20,000
	Other expenses	34,000
March 31, 1999	Materials at site	4,000
	Work-in-progress:	
	Not certified	8,000
	Certified	4,05,000

The cash received represents 80% of work certified. It has been estimated that further costs to complete the contract will be Rs. 23,000 including the materials at site as on March 31, 1999.

CHAPTER 3

Required:

Determine the profit on the contract for the year 1998-99 on prudent basis, which has to be credited to P and L a/c. *(C.A. Inter, November 1999)*

Solution:

CONTRACT ACCOUNT
(for the year 1998-1999)

1.4.98			
To work-in-progress (Not certified)	55,000	By materials at site	4,000
To materials at site	2,000	By cost of contract c/d	3,27,000
1998-99			
To materials issued	1,12,000		
To wages paid	1,08,000		
To hire of plant	20,000		
To other expenses	34,000		
	3,31,000		3,31,000
31.3.99			
To cost of contract b/d	3,27,000	By work certified	4,05,000
To P and L a/c	66,273	By work not certified	8,000
To reserve–WIP	19,727		
	4,13,000		4,13,000

Profit for the year 1998-99
4,13,000 – 3,27,000 = Rs. 86,000

ESTIMATED PROFIT (ON THE COMPLETION OF THE CONTRACT)

Cost of the contract	3,27,000
Add: Further cost of completing the contract	23,000
Total cost	3,50,000

Estimated profit = contract price – total estimated cost
= 4,40,000 – 3,50,000
= 90,000

PROFIT TO BE TAKEN TO P AND L A/C

$$\text{Estimated profit} \times \frac{\text{cash received}}{\text{contract price}}$$

$$90{,}000 \times \frac{3{,}24{,}000}{4{,}40{,}000} = 66{,}273$$

Problem 33. Compute a conservative estimate of profit on a contract (which has been 80% complete) from the following particulars. Illustrate four methods of computing the profit.

	Rs
Total expenditure to date	1,70,000
Estimated further expenditure to complete the contract (including the contingencies)	34,000
Contract price	3,06,000
Work certified	2,00,000

Work not certified	17,000
Cash received	1,63,200

(C.A. Inter, May 1998)

Solution:

CALCULATION OF ESTIMATED PROFIT

	Total cost to date	1,70,000
Add:	Estimated further cost to complete the contract (including contingencies)	34,000
	Total estimated cost	2,04,000

$$\begin{aligned}\text{Estimated profit} &= \text{contract price} - \text{total estimated cost}\\ &= 3{,}06{,}000 - 2{,}04{,}000\\ &= 1{,}02{,}000\end{aligned}$$

CALCULATION OF NOTIONAL PROFIT

CONTRACT ACCOUNT

To total cost	1,70,000	By work certified	2,00,000
To notional profit	47,000	By work uncertified	17,000
	2,17,000		2,17,000

FOUR METHODS OF COMPUTING CONSERVATIVE ESTIMATES OF PROFIT

(1) $\text{Estimated profit} \times \dfrac{\text{work certified}}{\text{contract price}}$

$$1{,}02{,}000 \times \frac{2{,}00{,}000}{3{,}06{,}000} = 66{,}666.67$$

(2) $\text{Estimated profit} \times \dfrac{\text{cash received}}{\text{contract price}}$

$$1{,}02{,}000 \times \frac{1{,}63{,}200}{3{,}06{,}000} = 54{,}400$$

(3) $\text{Notional profit} \times \dfrac{\text{work certified}}{\text{contract price}}$

$$47{,}000 \times \frac{2{,}00{,}000}{3{,}06{,}000} = 30718.95$$

(4) $\dfrac{2}{3} \times \text{notional profit} \times \dfrac{\text{cash received}}{\text{work certified}}$

$$\frac{2}{3} \times 47{,}000 \times \frac{1{,}63{,}200}{2{,}00{,}000} = 25.568$$

Problem 34. M/s New Century Builders have entered into a contract to build an office building complex for Rs. 480 lakhs. The work started in April 1997 and it is estimated that the contract will take 15 months to be completed. Work has progressed as per schedule and the actual costs charged till March 1998 are as follows:

CHAPTER 3

	Rs. (in lakhs)
Materials	112.20
Labour	162.00
Hire charges for equipments and other charges	36.00
Establishment charges	32.40
	342.60

The following information are available:

	Rs. (in lakhs)
Materials in hand (March 31, 1998)	6,60
Work certified (of which Rs. 324 lakhs have been paid) at March 31.3.1998	400
Work not yet certified at March 31, 1998 at cost	7.50

As per management estimates, the following further expenditure will be incurred to complete the work

	Rs. (in lakhs)
Materials	10.50
Labour	16.00
Sub-contractors	20.00
Equipments hire and other charges	3.00
Establishment charges	6.90

You are required to prepare contract a/c after considering a reasonable margin of profit. Make a provision for contingencies amounting to 5% of total costs. *(ICWA, Inter, December 1998)*

Solution:

CONTRACT ACCOUNT

To materials	1,12,20,000	By materials in hand	6,60,000
To labour	1,62,00,000	By cost of contract c/d	3,36,00,000
To hire charges	36,00,000		
To establishment charges	32,40,000		
	3,42,60,000		3,42,60,000
To cost of contract b/d	3,36,00,000	By work certified	4,00,00,000
To profit c/d	71,50,000	By work uncertified	7,50,000
	4,07,50,000		4,07,50,000
To P and L a/c	50,00,000	By profit b/d	71,50,000
To WIP–reserve	21,50,000		
	71,50,000		71,50,000

CONTRACTEE'S ACCOUNT

To contract a/c	4,00,00,000	By cash	3,24,00,000
		By balance c/d	76,00,000
	4,00,00,000		4,00,00,000

CALCULATION OF ESTIMATED PROFIT

Expenditure upto 31.3.1998 (3,42,60,000 – 6,60,000)			3,36,00,000
Add:	Estimated expenditure to complete		
	Materials	10,50,000	
	Add: Materials in hand	6,60,000	
		17,10,000	
	Labour	16,00,000	
	Sub-contractors	20,00,000	
	Hire charges	3,00,000	
	Establishment charges	6,90,000	
			63,00,000
			3,99,00,000
Add:	Provision for contingencies		
	$\frac{5}{95} \times 3,99,00,000$		21,00,000
		Estimated cost	4,20,00,000

Estimated profit = Contract price – total estimated cost
= 4,80,00,000 – 4,20,00,000
= 60,00,000

PROPORTIONATE PROFIT TAKEN TO P AND L A/C

$$\frac{\text{work certified}}{\text{contract price}} \times \text{estimated profit}$$

$$\frac{4,00,00,000}{4,80,00,000} \times 60,00,000 = 50,00,000$$

Problem 35. M/s Everfine constructions commenced a contract for the construction of a Bunglow on 1st July 1994. Originally the contract price was Rs. 50,00,000 but finally the same was fixed at Rs. 45,00,000.

Their actual expenditure during the year 1994 and estimated expenditure during 1995, till the completion of the contract is as under:

Particulars	*Actual expenditure upto 31.12.94* Rs.	*Estimated expenditure during 1995* Rs.
Building materials	8,00,000	13,00,000
Labour charges	6,00,000	6,00,000
Plant installed at site (at cost)	4,00,000	—
Materials at site on 31.12.94	50,000	—
General expenses	2,50,000	3,55,000
Plant returned to stores at cost at the end of the year	1,00,000	—
Work certified	20,00,000	contract completed
Work uncertified	75,000	—
Cash received	90% of the work certified	45,00,000

The contract is expected to be completed by 30th September 1995. The plant is subject to depreciation at 20% p.a. on the original cost.

In order to calculate the correct account of profit made on the contract for the year 1994, it was decided to take certain proportion of the estimated profit on completion of the contract to the credit of profit and loss a/c; such proportion being cash received to the total contract price.

Prepare the contract a/c for the year ending 31st December 1994 and work out the estimated profit on the completion of the contract by 30th September 1995.

(University of Bombay, B. Com., October 1995)

Solution:

CONTRACT A/C

(for the year ended 31.12.1994)

To building materials	8,00,000	By WIP a/c:	
To labour charges	6,00,000	Work certified	20,00,000
To depreciation $4,00,000 \times \frac{6}{12} \times \frac{20}{100}$	40,000	Work uncertified	75,000
To general expenses	2,50,000	Materials at site	50,000
To balance c/d	4,35,000		
	21,25,000		21,25,000
To P and L a/c	2,04,000	By balance b/d	4,35,000
To WIP–reserve	2,31,000		
	4,35,000		4,35,000

STATEMENT OF TOTAL CONTRACT

	Actual	*Estimated*	*Total*
Materials	8,00,000	13,00,000	21,00,000
Labour	6,00,000	6,00,000	12,00,000
Depreciation plant	40,000	45,000	85,000
General expenses	2,50,000	3,55,000	6,05,000
Total estimated cost			39,90,000

Estimated profit = contract price – total estimated cost
= 45,00,000 – 39,90,000
= 5,10,000

Profit transferred to P and L a/c

$$\text{Estimated profit} \times \frac{\text{Cash received}}{\text{Contract price}}$$

$$5{,}10{,}000 \times \frac{18{,}00{,}000}{45{,}00{,}000} = 2{,}04{,}000$$

Working Note:

Calculation of depreciation on plant for estimated period of 9 months

	Book value issued	4,00,000
Less:	Plant returned to stores at cost at the end of the year	1,00,000
		3,00,000

$$3{,}00{,}000 \times \frac{20}{100} \times \frac{9}{12} = 45{,}000$$

Problem 36. Prabhu Builders commenced work on 1st April 1993 on a contract of which the agreed price was Rs. 5 lakhs. The following expenditure was incurred during the year upto 31st March 1994.

	Rs.
Wages	1,40,000
Plant	35,000
Materials	1,05,000
Head office expenses	12,500

Materials costing Rs. 10,000 proved unsuitable and were sold for Rs. 11,500 and a part of the plant was scrapped and sold for Rs. 1,700.

Of the contract price Rs. 2,40,000 representing 80% of the work certified had been received by 31st March, 1994 and on that date the value of the plant on the job was Rs. 8,000 and the value of materials was Rs. 3,000. The cost of work done but not certified Rs. 25,000.

It was decided to (*a*) estimate what further expenditure would be incurred in completing the contract (*b*) compute from the estimate and the expenditure already incurred, the total profit that would be made on the contract and (*c*) ascertain the amount of profit to be taken to the credit of P and L a/c for the year ending 31st March 1994. While taking profit to the credit of P and L a/c that portion of the total profit should be taken which the value of work certified bears to the contract price. Details of the estimates were as follows:

(1) That the contract would be completed by 30th September 1994.

(2) The wages to complete would amount to Rs. 84,750.

(3) That materials in addition to those in stock on 31st March 1994 would cost Rs. 50,000.

(4) That further Rs. 15,000 would have to be spent on plant and the residual value of the plant on 30th September 1994 would be Rs. 6,000.

(5) The head office expenses to the contract would be the same annual rate as in 1993-94.

(6) That claims, temporary maintenance and contingencies would require Rs. 9,000.

Prepare contract a/c for the year ended 31st March, 1994 and show your calculations of the sum to be credited to P and L a/c for the year. *(C.S. Inter, June 1995)*

Solution:

CONTRACT ACCOUNT

To wages	1,40,000	By cash – Materials sold	11,500
To materials	1,05,000	By cash – plant sold	1,700
To plant	35,000	By materials	3,000
To H.O. expenses	12,500	By plant	8,000
To P and L a/c–profit on materials sold	1,500	**By work-in-progress:**	
To balance c/d	55,200	Work certified	3,00,000
		Work uncertified	25,000
	3,49,200		3,49,200
To P and L a/c	36,120	By balance b/d	55,200
To work-in-progress—reserve	19,080		
	55,200		55,200

CHAPTER 3

CALCULATION OF ESTIMATED PROFIT

Total net expenditure upto 31.3.1994:

Wages				1,40,000
Plant			35,000	
Less:	Scraped and sold	1,700		
	Plant in hand	8,000		
			9,700	
				25,300
Materials			1,05,000	
Less:	Cost of materials sold	10,000		
	Cost of materials in hand	3,000		
			13,000	
				92,000
H.O. expenses				12,500
				2,69,800
Add: Further estimated expenses:				
Wages			84,750	
Materials		50,000		
Add:	Materials on hand	3,000	53,000	
	Plant	15,000		
Add:	Plant on hand	8,000		
		23,000		
Less:	Residual value	6,000		
			17,000	
H.O. expenses (6 months @ last year $\frac{12,500}{2}$)			6,250	
Contingencies			9,000	
				1,70,000
Total estimated cost				4,39,800

Estimated profit = contract price – estimated cost

= 5,00,000 – 4,39,800

= 60,200

Profit to be credited to P and L a/c

$$\text{Estimated profit} \times \frac{\text{work certified}}{\text{contract price}}$$

$$60,200 \times \frac{3,00,000}{5,00,000} = 36,120$$

Problem 37. Excellent engineers undertook a contract which involves the fabrication of pre-stressed concrete blocks and the erection of the same at the consumer's site. The following information is supplied regarding the contract which is incomplete on 31st December:

	Rs.
Cost incurred:	
Fabrication cost to date:	
Direct materials	8,30,000
Direct labour	2,85,000
Overheads	1,65,000
	12,80,000
Erection costs	1,20,000
	14,00,000
Contract price	25,00,000

Cash received being 80% of work certified 12,80,000

Technical estimates of work completed to-date:

Fabrication : Direct materials 80%

: Direct labour and overheads 75%

: Erection 25%

Prepare a statement showing:

(*a*) Estimated profit on completion of contract

(*b*) Estimated profit-to-date the contract *(Sri Sathya Sai University, B.Com., April 1986)*

Solution:

STATEMENT SHOWING ESTIMATED PROFIT

	Cost to date		*Further cost*		*Total*
	Percentage of completion	*Amount*	*Percentage of completion*	*Amount*	
Fabrication cost:					
Direct materials	80%	8,30,000	20%	2,07,500	10,37,500
Direct labour	75%	2,85,000	25%	95,000	3,80,000
Overhead	75%	1,65,000	25%	55,000	2,20,000
Erection cost	25%	1,20,000	75%	3,60,000	4,80,000
Total		14,00,000		7,17,500	21,17,500

Estimated profit = Contract price – Total estimated cost

= 25,00,000 – 21,17,500 = 3,82,500

$$\text{Profit to be taken} = \text{Estimated profit} \times \frac{\text{Cash received}}{\text{Contract price}}$$

$$= 3{,}82{,}500 \times \frac{12{,}80{,}000}{25{,}00{,}000} = \text{Rs. } 1{,}95{,}840$$

Working Notes: Calculation of further cost of materials

For 80% the cost incurred is 8,30,000

$$\text{For 20\%} \quad ? \rightarrow \quad \frac{8{,}30{,}000 \times 20}{80} = \text{Rs. } 2{,}07{,}500$$

Labour cost to be incurred subsequently

For 75% the labour cost incurred is Rs. 2,85,000

$$\text{For 25\%} \quad ? \rightarrow \quad \frac{2{,}85{,}000 \times 25}{75} = \text{Rs. } 95{,}000$$

CHAPTER 3

Overheads to be incurred subsequently
For 75% the overhead incurred is 1,65,000

For 25% ? → $\dfrac{1,65,000 \times 25}{75}$ = Rs. 55,000

Erection cost to be incurred subsequently
For 25% the erection cost incurred is Rs. 1,20,000

For 75% ? → ? → $\dfrac{1,20,000 \times 75}{25}$ = Rs. 3,60,000

Problem 38. *(Estimated Contract):* RSI Construction Limited commenced a contract on 1st April 2005. The total contract was for Rs. 49,21,875. It was decided to estimate the total profit on the contract and to take to the credit of profit and loss account that proportion of estimated profit on cash basis, which work completed bore to total contract. Actual expenditure for the period April 1, 2005 to March 31, 2006 and estimated expenditure for April 1, 2006 to September 30, 2006 are given below:

	April 1, 2005 To March 31, 2006 (Activals)	*April 1, 2006 To Sept. 30, 2006 (Estimated)*
Materials issued	7,76,250	12,99,375
Labour paid	5,17,500	6,18,750
: prepaid	37,500	–
: outstanding	12,500	5,750
Plant purchased	4,00,000	–
Expenses paid	2,25,000	3,75,000
: outstanding	25,000	10,000
: prepaid	15,000	–
Plant returned to store (historical cost)	1,00,000	3,00,000
	On Sept. 30 2005	*On Sept. 30 2006*
Work certified	22,50,000	Full
Work uncertified	25,000	–
Cash received	18,75,000	–
Materials at site	82,500	42,500

The plant is subject to annual depreciation @ 25% on written down value method. The contract is likely to be completed on 30th September 2006.

Required:

(*i*) Prepare the contract a/c. Determine the profit on the contract for the year 2005-06 on prudent basis, which has to be credited to profit and loss account.

(CA PE-II, Group II, May 2006)

Solution:

CONTRACT A/C

To materials issued		7,76,250	By work-in-progress:	
To Labour	5,17,500		Work certified	22,50,000
Add: Outstanding	12,500		Work uncertified	25,000
	5,30,000			

Less: Prepaid	37,500		By plant returned	87,500
		4,92,500	(1,00,000 – 12,500)	
To Plant		4,00,000	By plant at site	2,25,000
To Expenses	2,25,000		(3,00,000 – 25%)	
Add: Outstanding	25,000		By materials at site	82,500
	2,50,000			
Less: Prepaid	15,000			
		2,35,000		
To balance c/d		7,66,250		
		26,70,000		26,70,000
To P and L a/c $\left(10{,}21{,}125 \times \frac{18{,}75{,}000}{49{,}21{,}875}\right)$		3,89,000	By balance b/d	7,66,250
To WIP–reserve		3,77,250		
		7,66,250		7,66,250

CONTRACT A/C
(for entire life period i.e., 1-4-05 to 30-9-2006)

To materials issued (7,76,250 + 12,99,375)	20,75,625	By contractee a/c		49,21,875
		By materials at site		42,500
To labour (5,17,500 – 37,500 + 12500 + 6,18,750 + 37,500 –12,500 + 5,750)	11,42,000	By plant returned on 30-9-95 (1,00,000 – 12,500)		87,500
To plant	4,00,000			
To expenses (2,25,000 + 25,000 – 15,000 + 3,75,000 – 25,000 + 15,000 + 10,000)	6,10,000	By plant returned on 30 Sept. 2006	3,00,000	
		Less: Depreciation for 05-06	75,000	
			2,25,000	
To estimated profit on contract	10,21,125	*Less:* Depreciation for 06-07)	28,125	1,96,875
	52,48,750			52,48,750

Problem 39. *(Escalation clause in contract):* The escalation clause of a long-term contract stipulates the following quantites and rates of materials of *A*, *B* and *C* and the following number of labour hours of *X*, *Y* and *Z* and their rates of pay. The actuals are shown below:

CHAPTER 3

Materials	*Qty. (Tons)*	*Rate (Rs.)*	*Actual Qty. (Tons)*	*Actual Rule (Rs.)*
A	500	50	500	45
B	1,000	30	1,000	35
C	20	1,000	20	1,010
Labour	Hours	Hourly rate	Hours	Hourly rate
X	4,800	2.00	4,800	2.25
Y	2,400	1.00	2,400	1.50
Z	9,600	1.50	9,600	1.50

Compute the amount of final claim so far as the rate is concerned.

Solution:

STATEMENT SHOWING MATERIAL ESCALATION

Material	*As per contract*		*Actual variation material*		
	Qty	*Rate*	*rate*	*in rate*	*escalation claim*
(1)	(2)	(3)	(4)	(5)	(6) = (2) × (5)
A	500 tons	Rs. 50	45	– Rs. 5	– 2,500
B	1,000 tons	Rs. 30	35	+ Rs. 5	+ 5,000
C	20 tons	Rs. 1,000	1,010	+ Rs. 10	+ 200
			Materials escalation		2,700

STATEMENT SHOWING LABOUR ESCALATION

Labour	*As per contract*		*Actual variation wages*		
	hrs	*rate*	*rate*	*in rate*	*escalation claim*
(1)	(2)	(3)	(4)	(5)	(6) = (2) × (5)
X	4,800	2.00	2.25	+ 0.25	+ 1,200
Y	2,400	1.00	1.50	+ 0.25	+ 1,200
Z	9,600	1.50	1.50	–	–
			Wages escalation		2,400

Final escalation claim						
Materials escalation						2,700
Labour escalation						2,400
Add: Claim as per contract:						
Material	*A*	— 500 tons @	Rs. 50	=	25,000	
	B	— 1,000 tons @	Rs. 30	=	30,000	
	C	— 20 tons @	Rs. 1,000	=	20,000	75,000
Labour:	*X*	— 4,800 hrs. @	Rs. 2	=	9,600	
	Y	— 2,400 hrs. @	Rs. 1	=	2,400	
	Z	— 9,600 hrs. @	Rs. 1.50	=	14,400	1,01,400
					Total claim	1,60,500

Problem 40. A contract for reinforced cement concrete building is governed by an escalation clause in respect of prices of steel, cement, and stone aggregate. The prices ruling on the date of the tender for the building and the actual prices paid by the contractor were as under:

Particulars	*Price on the date of tender (Rs.)*	*Actual Price (Rs.)*
Steel (per ton)	610	675
Cement (per ton)	100	105
Stone aggregate (per ton 100 c.ft.)	40	38

3,00,000 c.ft. reinforced concrete was laid in the building. If 100 lb steel, 2,400 lb cement and 90 c.ft. stone aggregate are the net quantities to cost 100 c.ft. of reinforced cement concrete and the wastage was 5.3 and 10 per cent respectively, calculate the difference in selling price according to the escalation clause.

Solution:

Net quantities required for 3,00,000 c.ft. reinforced cement concrete:

Steel: 100 lb × 3,000	=	3,00,000 lbs
Cement: 2,400 lbs × 3,000	=	72,00,000 lbs
Stone aggregate: 90 c.ft. × 3,000	=	2,70,000 c.ft.

Gross quantities required for 3,00,000 c.ft. of reinforced cement concrete

Steel: 3,00,000 lbs + 5%	=	3,15,000 lbs
		140.65 tons
Cement: 72,00,000 lbs + 3%	=	74,16,000 lbs or
		3,31,071 tons
Stone aggregate: 2,70,000 c.ft. + 10%	=	2,97,000 c.ft.

STATEMENT SHOWING INCREASE IN SELLING PRICE AS PER ESCALATION CLAUSE

	Rs.
Steel @ Rs. 65 per ton on 140.63 tons	9,14,095
Cement @ Rs. 5 per ton on 3,310.71 tons	16,55,355
	25,69,450
Less— Decrease for stone aggregate @ 2 per 100 c.ft. on 2,97,000 c.ft	5,94,000
Increase in selling price	19,75,450

Note: It is assumed that the percentage of wastage are based on the net quantities of materials.

AS 7: ACCOUNTING FOR CONSTRUCTION CONTRACTS*

The following is the text of the Accounting Standard 7 (AS-7) issued by the Institute of Chartered Accountants of India on "Accounting for Construction Contracts". The Standard deals accounting for construction contracts in the financial statements of contractors.

In the initial years, this accounting standard will be recommendatory in character. During this period, this standard is recommended for use by companies listed on a recognised stock exchange and other large commercial, industrial and business enterprises in the public and private sectors.

Introduction

1. This statements deals with accounting for construction contracts in the financial statements of enterprises undertaking such contracts (hereafter referred to as 'contractors'). The statement also applies to enterprises undertaking construction activities of the type

* Issued in December, 1983

dealt with in this statement not as contractors but on their own account as a venture of a commercial nature where the enterprise has entered into agreements for sale.

2. The feature which characterises a construction contract dealt with in this statement is the fact that the date at which the contract is secured and the date when the contract activity is completed fall into different accounting periods. The specific duration of the contract performance is not used as a distinguishing feature of a construction contract. Accounting for such contracts is essentially a process of measuring the results of relatively long-term events and allocating those results to relatively short-term accounting periods.

3. For the purposes of this statement a construction contract is a contract for the construction of an asset or of a combination of assets which together constitute a single project. Examles of activity covered by such contracts include the construction of bridges, dams, ships, buildings and complex pieces of equipment.

4. Contracts for the provision of services come within the scope of this statement to the extent that they are directly related to a contract for the construction of an asset. Examples of such service contracts are contracts for the services of project managers and architects and for technical engineering services related to the construction of an asset.

Explanation

5. The principle problem relating to accounting for construction contracts is the allocation of revenues and related costs to accounting periods over the duration of the contract.

6. Types of Construction Contracts

Construction contracts are formulated in a variety of ways but generally fall into two basic types:

(*i*) Fixed price contracts–the contractor agrees to a fixed contract price or rate, in some cases subject to cost escalation clauses;

(*ii*) Cost plus contracts–the contractor is reimbursed for allowable or otherwise defined costs, and is also allowed a percentage of these costs or a fixed fee.

Both types of contracts are within the scope of this statement.

7. Accounting Treatment of Construction Contract Costs and Revenues

7.1 Two methods of accounting for contracts commonly followed by contractors are the percentage of completion method and the completed contract method.

7.2 Under the percentage of completion method, revenue is recognised as the contract activity progresses based on the stage of completion reached. The costs incurred in reaching the stage of completion are matched with this revenue, resulting in the reporting of results which can be attributed to the proportion of work completed. Although (as per the principle of 'prudence') revenue is recognised only when realised, under this method, the revenue is recognised as the activity progresses even though in certain circumstances it may not be realised.

7.3 Under the completed contract method, revenue is recognised only when the contract is completed or substantially completed; that is, when only minor work is expected other than warranty obligation. Costs and progress payments received are accumulated during the course of the contract but revenue is not recognised until the contract activity is substantially completed.

7.4 Under both methods, provision is made for losses for the stage of completion reached on the contract. In addition, provision is usually made for losses on the remainder of the contract.

7.5 It may be necessary for accounting purposes to combine contracts made with a single customer or to combine contracts made with several customers if the contracts are negotiated as a package or if the contracts are for a single project. Conversely, if a contract covers a number of projects and if the costs and revenues of such individual projects can be identified within the terms of the overall contract, each such project may be treated as equivalent to a separate contract.

AS 8. COSTS TO BE ACCUMULATED FOR CONSTRUCTION CONTRACTS

8.1. Costs attributable to a contract are identified with reference to the period that commences with the securing of the contract and closes when the contract is completed.

8.2 Costs not specially attributable to any contract incurred by the contractor before a contract is secured are usually treated as expenses of the period in which they are incurred. However, if costs attributable to securing the contract can be separately identified and either the contract has been secured or there is a clear indication that the contract will be obtained the costs are sometimes treated as applicable to the contract and are deferred. As a practical measure, costs directly identifiable with a contract are sometimes deferred until it is clear whether the contract has been secured or not.

8.3 Costs attributable to a contract include expected warranty costs. Warranty costs are provided for when such costs can be reasonably estimated.

8.4 Costs incurred by a contractor can be divided into:

(*i*) Costs that relate directly to a specific contract;

(*ii*) Costs that can be attributed to the contract activity in general and can be allocated to specific contracts;

(*iii*) Costs that relate to the activities of the contractor generally, or that relate to contract activity but cannot be related to specific contracts.

8.5 Examples of costs that relate directly to a specific contract include:

(*i*) site labour costs, including supervision;

(*ii*) materials used for project construction;

(*iii*) depreciation of plant and equipment required for a contract;

(*iv*) cost of moving plant and equipment to and from a site.

8.6 Examples of costs that can be attributed to the contract activity in general and can be allocated to specific contracts include:

(*i*) insurance;

(*ii*) design and technical assistance;

(*iii*) construction overheads.

8.7 Examples of costs that relate to the activities of the contractor generally or that relate to contract activity but cannot be related to specific contracts, include:

(*i*) general administration and selling costs;

(*ii*) finance costs;

(*iii*) research and development costs;

(*iv*) depreciation of plant and equipment that cannot be allocated to a particular contract.

8.8 Costs referred to in paragraph 8.7 are usually excluded from the accumulated contract costs because they do not relate to reaching the present stage of completion of a

CHAPTER 3

specific contract. However, in some circumstances general administrative expenses, development costs and finance costs are specifically attributable to a particular contract and are sometimes included as part of accumulated contract costs.

Basis for Recognising Revenue on Construction Contracts

9. Percentage of completion method

9.1 Under the percentage of completion method, the amount of revenue recognised is determined by reference to the stage of completion of the contract activity at the end of each accounting period. The advantage of this method of accounting for contract revenue is that it reflects revenue in the accounting period during which activity is undertaken to earn such revenue.

9.2 The stage of completion used to determine revenue to be recognised in the financial statements is measured in an appropriate manner. For this purpose no special weightage should be given to a single factor; instead, all relevant factors should be taken into consideration; for example, the proportion that costs incurred to date bear to the estimated total costs of the contract, by the surveys which measure work performed and completion of a physical proportion of the contract work.

9.3 Progress payments and advances received from customers may not necessarily reflect the stage of completion and therefore, cannot usually be treated as equivalent to revenue earned.

9.4 If the percentage of completion method is applied by calculating the proportion that costs to date bear to the latest estimated total costs of the contract, adjustments are made to include only those costs that reflect work performed. Examples of items which may need adjustment include:

(*i*) The costs of material that have been purchased for the contract but have not been installed or used during contract performance; and

(*ii*) Payments to subcontractors to the extent they do not reflect the amount of work performed under the subcontract.

9.5 The application of the percentage of completion method is subject to a risk of error in making estimates. For this reason, profit is not recognised in the financial statements unless the outcome of the contract can be reliably estimated. If the outcome cannot be reliably estimated, the percentage of completion method is not used.

9.6 While recognising the profit under this method, an appropriate allowance for future unforseeable factors which may affect the ultimate quantum of profit is generally made on either a specific or a percentage basis.

9.7 In the case of fixed price contracts, the conditions which will usually provide this degree of reliability are:

(*i*) Total contract revenues to be received can be reliably estimated,

(*ii*) Both the costs to complete the contract and the stage of contract performance completed at the reporting date can be reasonably estimated, and

(*iii*) The costs attributable to the contract can be clearly identified so that actual experience can be compared with prior estimates.

9.8 Normally, the profit is not recognised in fixed price contracts unless the work on a contract has progressed to a reasonable extent. Ordinarily this is not considered as having been statisfied unless 20 to 25% of the work is completed.

9.9 In the case of cost *plus* contracts, the conditions which usually provide this degree of reliability are:

(*i*) Costs attributable to the contract can be clearly identified, and

(*ii*) Costs other than those that are specifically reimbursable under the contract can be reliably estimated.

10. Completed Contract Method

10.1 The principal advantage of the completed contract method is that it is based on results as determined when the contract is completed or substantially completed rather than on estimates which may require subsequent adjustment as a result of unforseen costs and possible losses. The risk of recognising profits that may not have been earned is therefore, minimised.

10.2 The principal disadvantage of the completed contract method is that periodic reported income does not reflect the level of activity on contracts during the period. For example, when a few large contracts are completed in one accounting period but no contracts have been completed in the previous period or are to be completed in the subsequent period, the level of reported income can be erratic although the level of activity on contracts may have been relatively constant throughout. Even when numerous contracts are regularly completed in each accounting period and reported income any appear to reflect the level of activity on contracts, there is a continuous lag between the time when work is performed and the related revenue is recognised.

11. Selection of Method

11.1 The selection of a method of accounting for a construction contract depends on considerations discussed earlier. The contractor may use both methods simultaneously for different contracts depending upon circumstances.

11.2 When a contractor uses a particular method of accounting for a contract, then in respect of all other contracts that meet similar criteria, the same method is used.

11.3 The methods of accounting used by the contractor and the criteria adopted in selecting the methods represent an accounting policy.*

12. Change in Accounting Policy

When there is a change in the accounting policy used for construction contracts, there is disclosure of the effect of the change and its amount. If the contractor changes from the percentage of completion method to the completed contract method, it is sometimes not possible to quantify the full effect of the change in the current accounting period. In such cases, there is disclosure of at least the amount of attributable profits reported in prior years, in respect of contracts in progress at the beginning of the accounting period.**

13. Provision for Foreseeable Losses

13.1 When current estimates of total contract costs and revenues indicate a loss, provision is made for the entire loss on the contract irrespective of the amount of work done and the method of accounting followed. In some circumstances, the foreseeable losses may exceed the costs of work done to date. Provision is nevertheless made for the entire loss on the contract.

---

* Refer to AS-1 on "Disclosure of Accounting Policies."

** Refers to AS-5 on "Prior Period and Extraordinary Items and Changes in Accounting Policies".

CHAPTER 3

13.2 When a contract is of such magnitude that it can be expected to absorb a considerable part of the capacity of the enterprise for a substantial period, indirect costs to be incurred during the period of the completion of the contract are sometimes considered to be directly attributable to the contract and included in the calculation of the provision for loss on the contract.

13.3 If a provision for loss is required, the amount of such provision is usually determined irrespective of:

(*i*) whether or not work has commenced on the contract; and

(*ii*) the stage of completion of contract activity; and

(*iii*) the amount of profits expected to arise on other unrelated contracts.

13.4 The determination of a future loss on a contract may be subject to high degree of uncertainty. In some of these cases, it is possible for the future loss and in other cases only the existence of a contingent loss is disclosed.*

14. Claims and Variations Arising Under Construction Contracts

14.1 Amounts due in respect of claims made by the contractor and of variations in contract work approved by the customer are recognised as revenue in the financial statements only in circumstances when, and only to the extent that, the contractor has evidence of the final acceptability of the amount of the claim or variation.**

14.2 Claims or penalties payable by the contractor arising out of delays in completion or from other causes are provided for in full in the financial statements as costs attributable to the contract. Claims in the nature of contingency are treated as indicated in Accounting Standard 4 on "Contingencies and Events Occurring after the Balance Sheet Date".

15. Progress Payments, Advances and Retentions

15.1 Progress payments and advances received from customers in respect of construction contracts in relation to the work performed thereon are disclosed in financial statements either as a liability or shown as a deduction from the amount of contract-work-in-progress.

15.2 In case progress payments and advances received from customers in respect of construction contracts are not in relation to the work performed thereon, these are shown as a liability.

15.3 Amounts retained by customers until the satisfaction of conditions specified in the contract for release of such amounts are either recognised in financial statements are receivables or alternatively indicated way of a note.

ACCOUNTING STANDARD

(The Accounting Standard comprises paragraphs 16-21 of this statement. The Standard Should be read in the context of paragraphs 1-15 of this Statement and of the "Preface to the Statements of Accounting Standards'.)

16. In accounting for construction contracts in financial statements, either the percentage of completion method or the completed contract method may be used. When a contractor uses

* Refers to AS-4 on "Contingencies and Events Occuring After the Balance Sheet Date."

** Refers to AS-1 on "Disclosure of Accounting Policies".

a particular method of accounting for a contract, then the same method should be adopted for all other contracts which meet similar criteria.

17. The percentage of completion method can be used if the outcome of the contract can be reliably estimated.

17.1 In the case of fixed price contracts, this degree of reliability would be provided if the following conditions are satisfied:

(*i*) Total contract revenues to be received can be reliably estimated.

(*ii*) Both the costs to complete the contract and the stage of contract performance completed at the reporting date can be reasonably estimated; and

(*iii*) The costs attributable to the contract can be clearly identified so that actual experience can be compared with prior estimates.

17.2 Profit in the case of fixed price contracts, normally should not be recognised unless the work on a contract has progressed to a reasonable extent.

17.3 In the case of cost plus contracts, this degree of reliability would be provided only if both the following conditions are satisfied:

(*i*) Costs attributable to the contract can be clearly identified; and

(*ii*) Costs other than those that are specifically reimbursable under the contract can be reliably estimated.

17.4 While recognising the profit under percentage of completion method, an appropriate allowance for future unforseeable factors should be made on either a specific or a percentage basis.

18. The costs included in the amount at which construction contract work is stated should comprise those costs that relate directly to a specific contract and those that are attributable to the contract activity in general and can be allocated to specific contracts.

19. A foreseeable loss on the entire contract should be provided for in the financial statements irrespective of the amount of work done and the method of accounting followed.

CHAPTER 3

Disclosure

20. There should be disclosure in the financial statements of:

(*i*) The amount of construction work-in-progress;

(*ii*) Progress payments received and advances and retentions on account of contracts included in construction work-in-progress; and

(*iii*) The amount receivable in respect of income accrued under cost plus contracts not included in construction work-in-progress.

If both the percentage of completion method and the completed contract method are simultaneously used by the contractor, the amount of contract work described in (i) above should be analysed to disclose separately the amounts attributable to contracts accounted for under each method.

21. Disclosure of changes in an accounting policy used for construction contracts should be made in the financial statements giving the effect of the change and its amount. However, if a contractor changes from the percentage of completion method to the completed contract method for contracts in progress at the beginning of the year, it may not be possible to quantify the effect of the change. In such cases, disclosure should be made of the amount of attributable profits reported in prior years in respect of contracts in progress at the beginning of the accounting period.

Problem 41. *(Problem on AS-7):* A limited undertakes civil contracts. For the year ending 31st March 2002, it has undertaken two contracts. From the information given below relating to the two contracts, determine the amount of profit (or loss) to be recognised in respect of each contract for the year ended 31st March 2002 in accordance with AS-7.

	Contract 1 (Rs. 000)	*Contract 2* (Rs. 000)
Total contract price	600	1,000
Cost incurried to date	400	100
Progress payment received and receivable	500	80
Estimated cost for completion	400	8,000

(Sri Sathya Sai University, B.Com. Hons., October 2002)

Solution:

CONTRACT-1

	(Rs. 000)
Contract price	600
Add: Total cost of construction (400 + 400)	800
Total foreseable loss	200

$$\text{Stage of completion } \frac{400}{800} \times 100 = 50\%$$

$$\text{Revenue to be recognised } 600 \times \frac{50}{100} = 300$$

∴ The loss to be recognised is Rs. 3,00,000 – Rs. 4,00,000 = Rs. 1,00,000

Since the foreseeable loss is 2,00,000 an additional provision of Rs. 1,00,000 is to be provided in accordance with para 19 of AS-7.

CONTRACT -2

	(Rs. 000)
Contract price	1,000
Add: Total cost of construction (100 + 800)	900
Total foreseable profit	100

$$\text{Stage of completion } \frac{100}{900} \times 100 = 11.1\%$$

Revenue to be recognised 1,000 × 11.11 =	111
Less: Cost incurred till date	100
∴ Profit to be recognised	11

As the work completed is less than 25%, the profit of Rs. 11,000 must be carried as provision for contingencies.

Problem 42. Ram has undertaken two contracts to construct two primary health centres during the calender year 2000. From the information given below, compute for him, the profit to be considered or loss to be provided for in accordance with AS-7.

	Contract 1 (Rs. 000)	*Contract 2* (Rs. 000)
Total contract price (fixed)	1,500	1,000
Cost incurred to date	160	500

Estimated costs for completion	1,440	300
Progress payment received and receivable	100	600

Ram receives 90% of the work certified in cash.

(Sri Sathya Sai University, B.Com. Hons., October 2001)

Solution:

CONTRACT-1 STATEMENT SHOWING ESTIMATED FORESEEABLE LOSS

(Rs. 000)

Total cost of construction (160 + 1440) = 1,600

Stage of completion $\frac{160}{1600} \times 100$ = 10%

Revenue to be recognised 10% of 1500 = 150

Total forseeable loss:

Total cost of construction	1,600
Total contract price	1,500
Foreseeable loss on the contract	100

Loss to be recognised on the accounting date:

Revenue recognised on percentage of completion basis	150
Cost incurred on accounting date	160
Loss for the accounting period	10
Further loss to be provided as per para 19 of AS-7	90
Net loss including provision	100

Note: As the contract has progressed very little, revenue may not be recognised during the period and loss for the period need not be computed. However, the forseeable loss of 1,00,000 must be provided for in accordance with para 19 of AS-7.

CONTRACT-2 ESTIMATED PROFIT OR LOSS ON THE CONTRACT

(Rs. 000)

Total contract price	1,000
Add: Total estimated costs on the contract (500 + 300)	800
Estimated profit	200

State of completion $\frac{500}{800} \times 100$ = 62.5%

Revenue to be recognised 62.5 % of Rs. 1000 = 625

Profit to be recognised in the Balance Sheet:

Revenue recognised on the accounting date	625
Less: Cost incurred to date	500
	125

Note: Since the contract has progressed only to the extent of 62.5%, it is prudent to set a part $\frac{1}{3}$ of the profit towards contingencies. As 90% of the work certified is received in cash, a sum of Rs. 75,000 is to be considered as the profit for the year and carry forward Rs. 50,000 towards contingencies. The working is shown below

$$1,25,000 \times \frac{2}{3} \times \frac{90}{100} = \text{Rs. } 75,000$$

Questions

Short and Long Answer Type Questions

1. State the features of contract costing.
2. How does job costing differ from contract costing?
3. Explain the following terms
   (*a*) cost plus contract
   (*b*) Sub-contract
   (*c*) Retention money
   (*d*) Escalation clause
4. State with formula the principles relating to transfer of profit in case of incomplete contract.
5. What is cost plus contract? State its merits and demerits.

Exercises

Incomplete Contract

1. Reliable Construction Ltd. entered into a contract to construct a building. The contract value is Rs. 13,00,000 to be realised in instalments on the basis of the value of work certified by the architect subject to a rentention of 10%. The work commenced on 1-4-2003 but it remained incomplete on 31-12-2003 when the final accounts are to be prepared. The facts and figures of the contract are:

	Rs.
Plant charged to contract at the commencement	64,000
Materials charged to contract	3,60,000
Wages paid	1,74,000
Expenses incurred on the contract	77,500

Total establishment expenses amounted to Rs. 82,000 out of which 25% is attributable to this contract. Out of the materials issued to the contract material costing Rs. 8,000 were sold for Rs. 10,000. A part of the plant (costing Rs. 4,000) was damaged on 1-10-2003 and the scrap realised Rs. 600 only. Plant costing Rs. 6,000 was transferred to another contract site on 31-12-2003.

Plant is to be depreciated @ 10% p.a.	Rs.
Materials in hand on 31-12-2003	35,000
Cost of work yet to be certified	60,000
Cost received from contractee	6,12,000

Prepare contract account showing therein the amount of profit or loss to be transferred to profit and loss account *(University of Bombay, B.Com. April 2002)*

[**Ans.** Notional Profit Rs. 1,43,100]

2. M/s Aircraft builders undertook a contract for a contract price of Rs. 60,00,000 and commenced the work on 1st July 2,000. The following particulars are available for 9 months ended 31-3-2001.

	Rs.
Materials issued from stores	4,00,000
Materials bought directly	20,50,000
Wages paid	19,00,000
Direct expenses	3,00,000
Establishment charges	1,50,000
Plant	6,50,000
Sub-contract charges	1,00,000
Scrap sold	30,000
Work certified	50,00,000

The following further information was available.

(*a*) Outstanding wages and direct expenses were Rs. 10,000 and Rs. 20,000 respectively on 31-3-2000.

(*b*) Materials at site at the end of the year is valued at Rs. 1,20,000.

(*c*) Value of work uncertified Rs. 2,00,000 on 31-3-2000.

(*d*) Included in wages is the salary paid to a supervisor @ Rs. 30,000 p.m. who had devoted half of the time on this contract.

(*e*) Working life of the plant is estimated to be 5 years at the end of which it is estimated to realise Rs. 50,000 as scrap value. The plant was purchased exclusively for this contract only.

Prepare contract account for the year ended 31-3-2000.

(University of Bombay, B.Com., April 2001)

[**Ans.** P and L a/c Rs. 3,10,000; Work in progress a/c 1,55,000.]

3. The following information relates to a contract for Rs. 12,00,000 (The contractee paying 90% of the value of work done as certified by the Architect) and rest on completion of the contract)

	Rs.
Materials issued from stores	2,85,000
Materials purchased for the contract	91,000
Normal loss of materials	1,960
Plant installed at the site	1,76,000
Wages paid to the workers	4,62,000
Direct expenses paid for the contract	53,000
Overhead expenses paid	39,000
Direct expenses arrived on 31-3-2004	6,200

of the plant and materials charged to the contract, plant which cost Rs. 9,600 and materials costing Rs. 8,400 were lost. Work certified Rs. 9,70,000. Charge depreciation on plant @ 10% p.a. prepare contract account. *(University of Calcutta, B.Com. (Hons), 2005)*

[**Ans.** Profit and Loss a/c Rs. 14,760 work in progress a/c- Rs. 9,940]

Comparative Contract Account

4. Mr. Raj Kumar and Company has undertaken two contracts, viz, A and B the following particulars are available for the year ended 31st March 2001.

Particulars *Date of commencement*	*Contract* *A* *1-7-2000*	*Contract* *B* *1-12-2000*
Contract price	6,00,000	5,00,000
Materials sent to site	1,60,000	60,000
Materials returned	4,000	2,000
Closing stock of materials at site	22,000	8,000
Direct labour	1,50,000	42,000
Direct expenses	66,000	35,000
Establishment expenses	25,000	7,000
Plant installed at site	80,000	72,000
Work uncertified	23,000	10,000
Work certified	4,20,000	1,35,000
Architect fees	2,000	1,000

During the year materials costing Rs. 9,000 have been transferred from contract A to contract B. The contractor charges depreciation @ 25% p.a. on plant.

You are required to prepare contract accounts and show how the relevant items would appear in the Balance Sheet assuming that contractee had paid 90% of the work certified

(University of Bombay, B.Com., October 2001)

[**Ans.** Contract A - Profit transferred to P and L a/c Rs. 36,000 work in progress Rs. 24,000 Contract B - Loss Rs. 5,000]

Contract Account with Trial Balance

5. The following Trial Balance was extracted on 31st December 2002 from the books of Engineering Contractors Limited which has an authorised capital of Rs. 3,50,000.

	Dr. (Rs.)	*Cr. (Rs.)*
Share capital		3,50,000
Creditors		83,000
Profit and Loss a/c on 1-1-2002		25,000
Provision for Depreciation on plant		63,000
Cash received on account		12,80,000
Building	74,000	
Plant	52,000	
Bank	45,000	
Contract No. 5:		
Materials	6,00,000	
Direct labour	8,30,000	
Expenses	40,000	
Plant on site	1,60,000	
	18,01,000	18,01,000

Contract No. 5, was begun on 1st January 2002.

The contract price is Rs. 2,40,000 and the contractee had so far paid Rs. 12,80,000 being 80% of the work certified. The cost of work done since certification is estimated at Rs. 16,000.

On 31st December 2002, after the above Trial Balance was extracted, plant costing Rs. 32,000 was returned to stores and material, then on site was valued at Rs. 27,000.

Provision has to be made for direct labour accrued Rs. 6,000 and for depreciation of all plant at $12\frac{1}{2}$% on cost.

Prepare (*a*) Contract no. 5 a/c and (*b*) Balance Sheet as on 31-12-2002

(Sri Sathya Sai University, B.Com. (Hons), August 2003)

[**Ans.** Notional Profit Rs. 1,47,000 Total of Balance Sheet Rs. 6,25,400]

6. The following is the Trial Balance of a construction company as on 31-12-2002.

	Dr.	*Cr.*
Contractee a/c	–	6,00,000
Building	3,20,000	–
Creditors	–	1,44,000
Bank balance	70,000	–
Capital	–	10,00,000
Materials	4,00,000	
Wages	3,60,000	
Expenses	94,000	
Plant and Machinery	5,00,000	
	17,44,000	17,44,000

only one contract was taken up during 2002 and work on the same was commerical on 1-1-02. Materials of Rs. 3,40,000 were sent to the contract but materials of Rs. 12,000 were destroyed in an accident at the site.

Plant and machinery costing Rs. 1,00,000 was used throughout the year and the remaining plant and machinery costing Rs. 4,00,000 was used from 1-1-2002 to 30-9-2002 and thereafter returned to the stores. Materials at the site on 31-12-2002 were valued at Rs. 8,000.

The contract was for Rs. 12,00,000 and the contractee was required to pay 75% of the work certified. The work certified was 80% of the total contract value and the uncertified work was estimated at Rs. 30,000 on 31-12-2002.

Expenses are to be charged to contract at 25% of wages and the plant and machinery is to be depreciated at 10% for the entire year.

Prepare the contract account and Balance Sheet as on 31-12-2002.

(Sri Sathya Sai University, B.Com. (Hons) October 2003)

[**Ans.** Notional Profit Rs. 1,80,000, Total of Balance Sheet Rs. 11,93,000]

7. The following is the Trial Balance of Hindustan Construction Company engaged on the execution of contract no. 667 for the year ended 31st December 2003.

	Dr. (Rs)	*Cr. (Rs)*
Contractee's a/c	–	3,00,000
Buildings	1,60,000	–
Creditors	–	72,000
Bank balance	35,000	–
Capital a/c	–	5,00,000
Materials	2,00,000	
Wages	1,80,000	
Expenses	47,000	
Plant	2,50,000	
	8,72,000	8,72,000

CHAPTER 3

The work on contract no. 667 commenced on 1st January 2003. Materials costing Rs. 1,70,000 were sent to the site of the contract but those of Rs. 6,000 were destroyed in an accident. Wages of Rs. 1,80,000 were paid during the year. Plant costing Rs. 50,000 was used on the contract all through the year. Plant with a cost of Rs. 2 lakhs was used from 1st January 2003 to 30th September 2003 and was then returned to the stores. Materials of Rs. 4,000 were at site on 31st December 2003.

The contract was for Rs. 6,00,000 Contractee pays 75% of work certified. Work certified was 80% of the contract price. Uncertified work was estimated at Rs. 15,000, on 31st December 2003. Expenses are charged to the contract at 25% of wages. Plant is depreciated at 10% for the entire year.

Prepare contract no. 667 a/c, costing profit and loss account and make out Balance Sheet as on 31st December 2003 *(University of Bombay, B.Com., October 2004)*

[**Ans.** Notional Profit Rs. 90,000, Total of Balance Sheet Rs. 37,500]

Estimated Contract Cost

8. Brock Construction Ltd. commenced a contract on 1st November 2003. The total contract was for Rs. 39,37,500. It was decided to estimate the total profit on the contract and to take to the credit of P and L a/c that proportion of estimated profit on cash basis, which work completed bore to the total contract. Actual expenditure for the period 1st November 2003 to October 31, 2004 and estimated expenditure for Nov. 1, 2004 to March 31, 2005 are given below:

		1st Nov. 2003 To 31st Oct 2004 (Activals)	*Nov. 1, 2004 To 31st Mar 2005 (Estimated)*
Materials issued		6,75,000	12,37,500
Labour:	paid	4,50,000	5,62,500
	prepaid	25,000	–
	outstanding	–	2,500
Plant purchased		3,75,000	–
Expenses:	paid	2,00,000	3,50,000
	outstanding	50,000	25,000
Plant returned to stores (historical cost)		75,000	3,00,000
		On 31 March 2004	*31 March 2005*
Work certified		20,00,000	Full
Work uncertified		75,000	
Cash received		17,50,000	
Material at site		75,000	37,500

The plant is subject to annual depreciation @ 33% on written down value method. The contract is likely to be completed on 31st March, 2005.

Required:

Prepare the contract a/c. Determine the profit on the contract for the year November 2003 to October 2004 on Prudent basis; which has to be credited to P and L a/c.

(CA PE-II, Group II, November 2004)

[**Ans.** Estimated profit Rs. 2,34,305]

9. Paramount Engineers are engaged in construction and erection of a bridge under a long term contract. The cost incurred upto 31-3-2001 was as under:

	Rs. in lakhs
Fabrication:	
Direct materials	2,80
Direct labour	100
Overheads	60
	440
Erection costs to date	110
	550

The contract price is Rs. 11 crores and the cash received on account till 31-3-2001 was Rs. 6 crores.

A technical estimate of the contract indicates the following degree of completion of work.

Fabrication: Direct materials – 70%

Direct labour and overhead 60% Erection – 40%

You are required to estimate the profit that could be taken to profit and loss a/c against this partly completed contract as at 31-3-2001. *(CA Inter, May 2001)*

[Ans. Total estimated profit Rs. 158.33 (Rs. in lakhs)]

CHAPTER 3

CHAPTER 4

PROCESS COSTING

INTRODUCTION

Process costing refers to a method of accumulating cost of production by process. It represents a method of cost procedure applicable to continuous or mass production industries. Output in such industries consists of like units and every unit of product undergoes similar operation in the process. So it is implied that the same cost of material, labour and overhead is charged to each unit of product processed. Process costing is also known as continuous costing because industries which adopt this method of costing undertake manufacturing of goods on a continuous basis. This method is also known as average costing because the total cost of production is charged to all the units produced irrespective of the fact that on any day the expense incurred is more or less. Under this method, costing an individual unit is impossible. The period of time is adopted as the basis for determining the average cost per unit of production.

There are three types of process costing method. They are: (1) Sequential process cost method, under which all products pass through a series of process in sequence. Under this type, output and cost of production is transferred from process to process until finished products are obtained. This type is used by industries which manufacture a single uniform product or products that are uniformly processed; (2) Parallel process cost method, under which two or more products pass through two or more separate sets of processes. The different sets of processes may be carried on simultaneously or one set may be run for a while and then another started; (3) Selective process cost method, under which cost is accumulated from various processes for various products. They are used where the products pass through only some of the processes but not all the processes of the industry. Thus the process cost vary for different products. Meat packing industry is an example of this type of process costing.

DEFINITION AND FEATURES OF PROCESS COSTING

According to Koheler (Dictionary for Accountants) "process costing is a method of cost accounting whereby costs are charged to processes or operations and averaged over units produced".

According to ICMA terminology "process costing is that form of operation costing which applies where standardised goods are produced".

The distinctive features of process costing can be enumerated under the following points:

1. The factory is divided into various processes or departments which perform a certain limited operation. Each process or department is known as cost centre.
2. The finished products are uniform in all respects such as shape, size, weight, quality, colour, chemical content, etc.

3. The output of one precess goes to constitute input of subsequent process/processes until finished product emerges out.
4. It is not possible to distinguish finished products while they are in the stage of processing.
5. The cost of production of one process is transferred to subsequent process/processes just as output of one process is transferred as input of other process.
6. It is quite common to incur normal loss and wastages. Sometimes, owing to abnormal conditions even abnormal loss is also observed.
7. The production of main product is often accompanied by secondary products which are termed as joint and by-products.
8. The semi-finished products are expressed in terms of complete products. This is technically termed as equivalent production.
9. Cost of production is ascertained at each process and finally after completion of production.
10. Production is accumulated and reported by process.
11. Production process is predetermined and a definite sequence of production is followed. The raw materials and work-in-progress flow from process to process according to the sequence of production.
12. The unit of cost is the "process" under this method of costing.
13. Production is undertaken on a continuous and large scale basis in anticipation of demand.

APPLICATION OF PROCESS COSTING

Gillespie in his book *Cost Accounting and Control* states that a process cost system is used where it is not possible or not desirable to identify the successive jobs or lots of production for cost accumulation purposes. Under the following circumstances process costing method is most suitable:

1. Where the product of one process becomes the materials of another process.
2. Where production undertaken is simultaneous.
3. Where it is not possible to distinguish the products when they are under manufacturing process.
4. Where identical units are produced through an ongoing series of uniform production process.
5. Where production is continuous and undertaken on large-scale basis.
6. Where production is standardised.

In addition to the above, Blocker and Weltmer in their book *Cost Accounting* list the following conditions which are favourable for the use of process costing:

1. Production of single product in a plant.
2. Division of a plant into processes or departments, each responsible for the manufacture of a single product.
3. Processing of a single product for a scheduled time, followed by successive runs of other products, each run being separated as to production and costs.
4. Production of several products of standard design in the same plant, process or department under conditions which permit computation of weighted averages to denote relative importance of each product as to quantities and costs.

5. Division of a factory into separate operation or production centres, each performing standard operations.

Examples of industries to which process costing can be applied are: oil refinery, chemical, electronics, baby food, soaps, breweries, dairy, paints and varnish, ink, rubber, textiles, perfumery, fertilizers, beer, ice-cream and many such other industries.

DIFFERENCES BETWEEN JOB AND PROCESS COSTING

Points of difference	*Job costing*	*Process costing*
1. Commencement of work	On receiving order	In anticipation of demand.
2. Continuity of work	Intermitten	Continuous
3. Standardisation of product	To suit the specific requirement of customers. Products are not identical.	Production is standardised. Finished products are identical.
4. Charging of cost	Specifically to the job	Charged to process concerned.
5. Preparation of accounts	On completion of job	As and when required. Generally once in year.
6. Diversification	Possible	Not possible unless altogether a new set of machineries are installed.
7. Reporting	In respect of a job after completion of job.	Reporting is processwise and in respect of time.
8. Investment of capital	Less	More
9. Work-in-progress	It may or may not exist.	It always exists.
10. Creation of demand	No need to create demand.	Demand creation is necessary for the products.
11. Identification	Each job can be identified in respect of other jobs.	Products cannot be identified when they are under processing.
12. Transfer	Generally no transfer of materials or labour is made unless it is necessary.	Output of one process is transferred to other process.
13. Supervision	More because of difference in jobs.	Less because of standardisation of production.
14. Control of losses and wastages	Losses and wastage can be controlled.	Only normal losses can be controlled but not abnormal losses and wastages.

RECORDING OF ELEMENTS OF COSTS IN PROCESS COSTING

1. Direct Materials Cost

Materials used in process costing industries can be classified in to two types, viz., (1) Basic materials or raw materials as in the case of milk in ice-cream industry, fruit in fruit canning industries. (2) Process materials, which are incurred to convert raw materials into finished product as for example, chemicals added to preserve fruit juice content in the finished product. Materials are usually drawn in bulk quantities by various departmental heads from the store-room through the issue of bill of materials. When solid materials are issued, they are weighed, counted or measured and then recorded in the stores ledger or material consumed report. When liquids or gases are issued, the flow of materials can be recorded by installing meters, readings of which are taken for costing purpose at appropriate intervals. In most of the cases raw materials are fed in the first process and occassionally materials are added in other processes. The cost of raw materials consumed can be ascertained in any of the following ways:

(*a*) By recording the issue prices in stores ledger card,

(*b*) The cost of raw materials purchased is added to the opening stock of raw materials and subtracting the closing stock of raw materials therefrom.

(*c*) Materials of identical nature when issued on a continuous basis can be obtained from material consumed report.

2. Direct Labour Cost

Labour cost constitutes least element of production cost. This is because, production is continuous and standardised, for which reason machines are used on large scale and manual labour employed is less. This results in reduced labour cost per unit with increased overhead cost per unit.

Usually employees are engaged on one particular process. Labour cost of such employees are debited to the process account concerned. Where employees attend to more than one process, the time spent on more than one process is recorded and the wages are allocated to all the processes.

3. Direct Expenses

Hire of special plant, maintenance cost of such plant and other direct expenses are debited to process account.

4. Overheads

One problem encountered in accounting the overhead, under process costing is apportioning the overheads when more than one product is manufactured in the factory or in the same process. A suitable basis is to be decided for apportioning overheads to all products so manufactured. Overheads can be charged under any of the two methods: (*i*) Based on predetermined rate which is expressed mostly as a percentage on direct labour. This method is suitable when production capacity or overheads fluctuate from month to month. The pre-determined rate so used eliminates distortion in the monthly unit cost caused by such fluctuations, (*ii*) Actual overhead incurred is charged to process account concerned. This method is used where volume of production and factory overhead remain constant over a period of time. Common items of overhead such as head office cost, lighting, taxes, rent, etc. are apportioned to all the processes. Other expenses such as power, process steam, compressed air etc. are apportioned to the process concerned.

ADVANTAGES AND DISADVANTAGES OF PROCESS COSTING

The following are the advantages of process costing:

1. When standard costing is in operation, it is possible to find out cost of production periodically by using predetermined rates of material, labour cost, and overheads.
2. The accounting system is simpler and is less expensive compared to job costing.

3. Control can be exercised through standard costing technique and it is possible to evaluate the performance of every process.
4. It is easy to quote tender price because of standardised process.
5. Because cost of production is ascertained periodically, management is in a position to receive various reports and review the progress and efficiency of the production process.

The following are the disadvantages of process costing:

1. When costs are recorded at the end of the period, it is not possible to exercise control over costs.
2. It is difficult to value work-in-progress.
3. It is difficult to value losses, wastes and scraps.
4. It is difficult to apportion total cost among joint products and by-products.
5. Cost of production obtained under process costing is only average cost. Hence cost of production is inaccurate. This makes measurement of efficiency difficult.
6. It involves difficulty in ascertaining closing stock value when output of one process is transferred to another process at transfer price or market price.

COSTING PROCEDURE INVOLVED IN PROCESS COSTING

Costing procedure aims at:

1. Accumulation and apportionment of service department expenses.
2. Computation of cost of production of each process at the end of a given period.
3. Transfer of costs from process to process.
4. Valuation of closing stock of work-in-progress.

The following procedure is involved in a process costing method:

1. Each process is separately identified and named and numbered and a separate account is opened for each process.
2. The various elements of cost are recorded in respect of each and every process.
3. Output of every process is recorded in convenient units such as kilogram, tonne, litre etc. on a periodical basis such as daily, weekly, monthly depending upon processing time.
4. The cost per unit in every process is calculated by dividing the net cost by good units manufactured.
5. Incomplete units at the end of every period in every process is converted in terms of completed units. The degree of completion serves as the basis for such conversion.

The various situations and problems encountered and the method of calculating cost in relation to each of the situation are outlined below:

1. Production involving two or more processes, in which case output of one process is transferred in full to the subsequent process.
2. Output of a process partly transferred to the next process and partly retained in stock.
3. Process consisting of fully completed opening and closing stock.
4. Process costing involving normal loss, abnormal loss, and abnormal gain.

5. Process costing of partially completed closing stock (equivalent production) and may involve the following:
   (*a*) Normal loss, abnormal loss, and abnormal gain,
   (*b*) Partially completed opening stock,
   (*c*) Normal loss, abnormal loss, partially completed opening and closing stock.
6. Process costing involving joint products and by-products. This is discussed separately in the next chapter.

Regarding the costing technique to be employed in process costing, standard costing can be conveniently applied because of production of uniform units. The standard cost can be set in respect of each process and compared with actual cost. This technique gives management an excellent measure of efficiency of production. Application of standard costing technique is widely practised by modern process industries.

Problem 1. *(When the output of a process is fully transferred to subsequent process):* The manufacture of product P.J. requires three distinct processes, numbered 1, 2 and 3. On completion, the product is passed from process 3 to finished stock. During May 2009, the following information was obtained in respect of the product:

	Process 1 *(Rs.)*	*Process 2* *(Rs.)*	*Process 3* *(Rs.)*
Direct materials	5,600	1,250	750
Direct labour	620	860	1,880
Direct expenses	800	200	–

Production overhead incurred was Rs. 5,040 which is to be absorbed by processes on the basis of 150% of direct wages. Production during the period was 100 units. There was no stock of raw materials or work-in-progress either at the beginning or end of the period. The output was sold for Rs. 20,000.

Prepare process cost accounts.

Solution:

PROCESS 1 ACCOUNT

	Amount	*Cost per unit*		*Amount*	*Cost per unit*
To direct materials	5,600	56.00	By process 2 a/c	7,950	79.50
To direct labour	620	6.20			
To direct expenses	800	8.00			
To production overhead	930	9.30			
	7,950	79.50		7,950	79.50

PROCESS 2 ACCOUNT

To process 1 A/c	7,950	79.50	By process 3 a/c	11,550	115.00
To direct materials	1,250	12.50			
To direct labour	860	8.60			
To direct expenses	200	2.00			
To production overhead	1,290	12.90			
	11,550	115.50		11,550	115.50

PROCESS 3 ACCOUNT

To process 2 A/c	11,550	115.50	By finished stock a/c	17,000	170.00
To direct materials	750	7.50			
To direct labour	1,880	18.80			
To production overhead	2,820	28.20			
	17,000	170.00		17,000	170.00

FINISHED STOCK ACCOUNT

To process 3 A/c	17,000	170	By sales	20,000	200
By P & L a/c-profit	3,000	30			
	20,000	200		20,000	200

Problem 2. *(When output of one process is partially transferred to subsequent process and balance retained):* The Andhra Products Ltd., manufacture and sell their chemical produced by consecutive processes. The product of the three processes are dealt with as under:

	Process I	*Process II*	*Process III*
Transferred to next process	$66\frac{2}{3}\%$	60%	
Transferred to warehouse for sales	$33\frac{1}{3}\%$	40%	100%

In each process, 4% of the total weight put in is lost and 6% is scrap, which from process *I* realise Rs. 6 per ton, from process *II* Rs. 10 per ton and from process *III* Rs. 12 per ton.

The following particulars relate to October 1989:

Raw materials used:

Process	*I*	2,800 tons at Rs. 40 per ton
Process	*II*	320 tons at Rs. 64 per ton
Process	*III*	2,520 tons at Rs. 28 per ton

Manufacturing wages and expenses

Process	*I*	Rs. 20,608
Process	*II*	Rs. 12,560
Process	*III*	Rs. 11,580

Prepare process accounts. *(Kakatiya University, B.Com., May 1992)*

Solution:

PROCESS I ACCOUNT

	Qty.	*Amt.*		*Qty.*	*Amt.*
To raw materials	2,800	1,12,000	By loss in wt.	112	–
To wages & expenses		20,608	By sale of scrap	168	1,008
			By process *II* a/c	1,680	87,733
			By warehouse a/c	840	43,867
	2,800	1,32,608		2,800	1,32,608

PROCESS II ACCOUNT

To process *I* A/c	1,680	87,733	By loss in wt.	80	
To materials	320	20,480	By sales of scrap	120	1,200
To wages & expenses		12,560	By process *III* a/c	1,080	71,743
			By warehouse a/c	720	47,830
	2,000	1,20,773		2,000	1,20,733

PROCESS III ACCOUNT

To process *II* A/c	1,080	71,743	By loss in wt.	144	–
To raw materials	2,520	70,560	By sales of scrap	216	2,592
To wages & expenses	–	11,580	By finished stock a/c	3,240	1,51,291
	3,600	1,53,883		3,600	1,53,883

Problem 3. *(When part of output is transferred to next process and part of output is sold)*: JAVA manufacturing company supplies you the following information for the year ended 31st March, 1999.

	Process *A*	*B*	*C*
Raw material introduced in the process (units)	12,000	2,440	2,600
Cost of Raw materials per unit (Rs.)	5	5	5
Direct wages (Rs.)	34,000	24,000	15,000
Production overheads (Rs.)	16,160	16,200	9,600
Normal loss (% on number of units entering into the process having no realisable value)	4%	5%	3%
Wastage (% on number of units entering to the process having scrap value)	6%	5%	4%
Scrap value per unit of wastage (Rs.)	3	4	5
Output transferred to subsequent process	70%	60%	–
Output sold at the end of the process	30%	40%	100%
Selling price per unit of the output sold at the end of the process (Rs.)	12	16	17

You are required to prepare the process A, B and C accounts showing the profit earned at each process. *(University of Bombay, B.Com., October 1999)*

Solution: PROCESS 'A' ACCOUNT

	Qty.	*Amt.*		*Qty.*	*Amt.*
To raw materials introduced	12,000	60,000	By normal loss	480	–
To Direct wages		34,000	$\frac{4}{100} \times 12{,}000$		

To production overhead		16,160	By wastage $\frac{6}{100} \times 12{,}000$	720	2,160
			By output c/d @ Rs. 10 per unit	10,800	1,08,000
	12,000	1,10,160		12,000	1,10,160
To output b/d	10,800	1,08,000	By sales	3,240	38,880
To P & L a/c		6,480	By output transferred to process B	7,560	75,600
	10,800	1,14,480		10,800	1,14,480

$$\text{Cost per unit} = \frac{\text{Total Amount of Debit side} - \text{Scrap value}}{\text{Input Qty.} - \text{Normal loss} - \text{Wastage}}$$

$$= \frac{1{,}10{,}160 - 2{,}160}{12{,}000 - 480 - 720}$$

$$= \frac{1{,}08{,}000}{10{,}800} = \text{Rs. } 10$$

PROCESS 'B' ACCOUNT

	Qty.	*Amt.*		*Qty.*	*Amt.*
To output from process 'A' a/c	7,560	75,600	By normal loss $\frac{5}{100} \times 10{,}000$	500	–
To materials introduced	2,440	12,200			
To direct wages		24,000	By wastage $\frac{5}{100} \times 10{,}000$	500	2,000
To production overhead		16,200			
			By output c/d @ Rs. 14 per unit	9,000	1,26,000
	10,000	1,28,000		10,000	1,28,000
To output b/d	9,000	1,26,000	By sales	3,600	57,600
To P & L a/c		7,200	By output transferred to process 'C' a/c	5,400	75,600
	9,000	1,33,200		9,000	1,33,200

$$\text{Cost per unit} = \frac{\text{Total cost of Debit side} - \text{Scrap value}}{\text{Input Qty.} - \text{Normal loss} - \text{Wastage}}$$

$$= \frac{1{,}28{,}000 - 2{,}000}{10{,}000 - 500 - 500} = \frac{1{,}26{,}000}{9{,}000} = \text{Rs. } 14$$

PROCESS 'C' A/C

	Qty.	Amt.		Qty.	Amt.
To output from process 'B' A/c	5,400	75,600	By normal loss	240	–
To materials introduced	2,600	13,000	$\frac{3}{100} \times 8{,}000$		
To direct wages	–	15,000	By wastage	320	1600
To production overhead	–	9,600	$\frac{4}{100} \times 8{,}000$		
			By output c/d @ Rs. 15 per unit	7,440	1,11,600
	8,000	1,13,200		8,000	1,13,200
To output b/d	7,440	1,11,600	By sales	7,440	1,26,480
To P & L a/c		14,880			
	7,440	1,26,480		7,440	1,26,480

$$\text{Cost per unit} = \frac{\text{Total Amount of Debit side} - \text{Scrap value}}{\text{Input Qty.} - \text{Normal loss} - \text{Wastage}}$$

$$= \frac{1{,}13{,}000 - 1{,}600}{8{,}000 - 240 - 320}$$

$$= \frac{1{,}11{,}600}{7{,}440} = \text{Rs. } 15$$

Problem 4. *(When part of the output is transferred to next process and part of the output is sold):* Nilgiri Tea Ltd. manufactures spice tea which passes through three processes. The following particulars are available for the year ended 30-6-2007:

	Process		
	I	*II*	*III*
Raw materials (Kg)	10,000	4,600	1,500
Cost of raw material per kg	5	6	8
Direct wages (Rs.)	24,000	18,000	12,250
Direct expenses (Rs.)	15,200	10,736	8,590
Factory overheads (Rs.)	20,960	6,000	4,255
Normal loss	4%	8%	5%
Weight loss	6%	2%	Nil
Scrap value per kg of normal loss (Rs.)	1.80	2.50	4.00
output transferred to next process	60%	50%	Nil
Output sold	40%	50%	80%
Selling price of output per kg	14	16	17
Transferred to finished stock	Nil	Nil	20%

The percentage of normal loss and percentage of weight lost are based on total input in the process.

Prepare process accounts and costing profit and loss account.

(Sri Sathya Sai University, B.Com., (Hons) October 2007)

CHAPTER 4

Solution:

PROCESS I ACCOUNT

	Qty.	Amt.		Qty.	Amt.
To raw materials	10,000	50,000	By normal loss (4%)	400	720
To direct wages	–	24,000	By weight loss (6%)	600	–
To direct expenses	–	15,200	By output c/d @ Rs. 12.16	9,000	1,09,440
To factory overhead	–	20,960			
	10,000	1,10,160		10,000	1,10,160
To output b/d	9,000	1,09,440	By sales 40% at Rs. 14	3,600	50,400
To costing P & L	–	6,624	By process II A/c @ Rs. 12.16	5,400	65,664
	9,000	1,16,064		9,000	1,16,064

PROCESS II ACCOUNT

	Qty.	Amt.		Qty.	Amt.
To process I A/c	5,400	65,664	By normal loss (8%)	800	2,000
To Raw material	4,600	27,600	By weight loss (20%)	200	–
To direct wages	–	18,000	By output c/d	9,000	1,26,000
To direct expenses	–	10,736			
To factory overhead	–	6,000			
	10,000	1,28,000		10,000	1,28,000
To output b/d	9,000	1,26,000	By sales @ Rs. 16	4,500	72,000
To costing P & L a/c	–	9,000	By process III A/c	4,500	63,000
	9,000	1,35,000		9,000	1,35,000

PROCESS III ACCOUNT

	Qty.	Amt.		Qty.	Amt.
To process II A/c	4,500	63,000	By normal loss (5%)	300	12,000
To raw materials	1,500	12,000	By output c/d @ Rs. 17.35	5,700	98,895
To direct wages		12,250			
To direct expenses		8,590			
To factory overhead		4,255			
	6,000	1,00,095		6,000	1,00,095
To output b/d	5,700	98,895	By sales @ 80% @ Rs. 17.35	4,560	77,520
			By costing P & L a/c	–	1,596
			By output transferred to finished stock	1,140	19,779
	5,700	98,895		5,700	98,895

COSTING P & L A/C

To process III a/c	1,596	By process I a/c	6,624
To balance c/d–Net profit	14,028	By process II a/c	9,000
	15,624		15,624

Working Note:

$$\text{Cost per unit of Process I} = \frac{\text{Cost of process} - \text{Scrap value of normal loss}}{\text{Input Qty.} - \text{Normal loss Qty.} - \text{Weight loss Qty.}}$$

$$= \frac{1,10,160 - 720}{10,000 - 400 - 600}$$

$$= \frac{1,09,440}{9,000} = 12.16$$

$$\text{Cost per unit of Process II} = \frac{1,28,000 - 2,000}{10,000 - 800 - 200}$$

$$= \frac{1,26,000}{9,000} = 14$$

$$\text{Cost per unit Process III} = \frac{1,00,095 - 1,200}{6,000 - 300}$$

$$= \frac{98,895}{5,700} = 17.35$$

Problem 5. KT & Company have three assembly shops, viz, A, B and C part of the output is transferred to the next process and part is sold directly. The company furnished the following information:

Particulars	*A*	*B*	*C*
Raw materials (in litres)	5,000	19,200	3,576
Material cost Per litre	Rs. 60	Rs. 40	Rs. 80
Labour cost	Rs. 4,28,000	1,06,000	2,10,000
Direct expenses	Rs. 88,000	2,85,200	1,04,800
Wastage as % of total input	4%	5%	10%
(*a*) Output transferred:			
To Lower Assembly	60%	–	–
To Higher Assembly	–	40%	–
(*b*) Output sold in the market	40%	60%	100%
Sale price per litre	Rs. 200	Rs. 205	Rs. 250
Administration overhead Rs. 36,000			
Marketing Overhead Rs. 48,000			

Prepare various assembly accounts and costing profit and loss account.

(University of Bombay, B.Com., October 2003)

CHAPTER · 4

Solution:

PROCESS 'A' ACCOUNT

	Qty.	Amt.		Qty.	Amt.
To materials	5,000	3,00,000	By wastage	200	–
To labour	–	4,28,000	By output c/d	4,800	8,16,000
To direct expenses	–	88,000	@ Rs. 170		
	5,000	8,16,000		5,000	8,16,000
To output b/d	4,800	8,16,000	By sales	1,920	3,84,000
To costing P & L a/c	–	57,600	By process 'B' a/c	2,880	4,89,600
	4,800	8,73,600		4,800	8,73,600

PROCESS B ACCOUNT.

	Qty.	Amt.		Qty.	Amt.
To process 'A' a/c	2,880	4,89,600	By wastage	240	–
To materials	1,920	76,800	By output c/d	4,560	9,57,600
To labour	–	1,06,000	@ Rs. 210		
To direct expenses	–	2,85,200			
	4,800	9,57,600		4,800	9,57,600
To output b/d	4,560	9,57,600	By sales	2,736	5,60,880
			By process 'c' a/c @ 210	1,824	3,83,040
			By costing P & L a/c (loss)	–	13,680
	4,560	9,57,600		4,560	9,57,600

PROCESS 'C' ACCOUNT

	Qty.	Amt.		Qty.	Amt.
To process 'B' a/c	1,824	3,83,040	By wastage	540	–
To materials	3,576	2,86,080	By output c/d (202.45)	4,860	9,83,920
To labour	–	2,10,000			
To direct expenses	–	10,43,800			
	5,400	9,88,920		5,400	9,83,920
To output b/d	4,860	9,83,920	By sales	4,860	12,15,000
To costing P & L a/c	–	2,31,080			
	4,860	12,15,000		4,860	12,15,000

COSTING P & L A/C

To process 'B'	13,680	By process A	57,600
To administrative overhead	36,000	By process C	2,31,080
To marketing overhead	48,000		
To net profit	1,91,000		
	2,88,680		2,88,680

Working Note:

$$\text{Cost per unit} = \frac{\text{Net cost of process}}{\text{Input} - \text{wastage}}$$

$$\text{Process A} = \frac{8,16,000}{4,800} = \text{Rs. } 170$$

$$\text{Process B} = \frac{9,57,600}{4,560} = \text{Rs. } 210$$

$$\text{Process C} = \frac{9,83,920}{4,860} = \text{Rs. } 202.45$$

Problem 6. Product 'X' in a manufacturing unit passes through three processes A, B and C. The expenses incurred in the three processes during the year 2001 were as under:

	Process A	*Process* B	*Process* C
Units of input issued	9,000		
	Rs.	*Rs.*	*Rs.*
Cost per unit	150		
Sundry materials	23,500	25,000	15,000
Direct labour	80,000	2,07,200	26,110
Direct expenses	2,250	7,280	8,100
Selling price per unit of output	200	280	600

The actual outputs obtained vis-a-vis normal process losses from the three processes were

	Output (Units)	*Process loss (%)*
Process A	8,400	5
Process B	5,700	10
Process C	3,660	3

During the year $\frac{3}{4}$th of the output of process 'A' and $\frac{2}{3}$rd of the output of process B were transferred to the next process and the balances were sold outside. The entire output of process 'C' was however sold outside. The losses of the three processes were sold at Rs. 5 per unit for process A, Rs. 10 per unit for process B and Rs. 15 per unit for process C.

Prepare the three process accounts and a statement of income considering a total selling and distribution expenses of Rs. 45,000 which is not allocated to processes. *(ICWA Inter, June 2002)*

CHAPTER 4

Solution:

PROCESS 'A' ACCOUNT

	Qty.	*Amt.*		*Qty.*	*Amt.*
To input units	9,000	13,50,000	By normal loss	450	2,250
To sundry materials	–	23,500	By abnormal loss	150	25,500
To direct labour	–	80,000	By process B a/c	6,300	10,71,000
To direct expenses	–	2,250	By P & L a/c $\frac{1}{4}$	2,100	3,57,000
	9,000	14,55,750		9,000	14,55,750

Working Note:

Calculation of cost of abnormal loss

Input Qty.	9,000	Rs.	14,55,750
Less: Normal loss	450	Rs.	2,250
	8,550		14,53,500

$$\frac{14{,}53{,}500}{8{,}550} = \text{Rs. } 170 \text{ per unit}$$

Cost of abnormal loss = 150 × Rs. 170 = Rs. 25,500

PROCESS 'B' ACCOUNT

	Qty.	*Amt.*		*Qty.*	*Amt.*
To process 'A' a/c	6,300	10,71,000	By normal loss a/c	630	6,300
To sundry materials	–	25,000	By process 'C' a/c $\left(\frac{2}{3}\right)$	3,800	8,74,000
To direct labour	–	2,07,200	By P & L a/c $\left(\frac{1}{3}\right)$	1,900	4,37,000
To direct expenses		7,200			
To abnormal gain a/c	30	6,900			
	6,330	13,17,300		6,330	13,17,300

Working Note:

Calculation of cost of abnormal loss

Input Qty.	6,300	Rs.	13,10,400
Less: Normal loss	630	Rs.	6,300
	5,670	Rs.	13,04,100

$$\frac{13{,}04{,}100}{5{,}670} = \text{Rs. } 230 \text{ per unit}$$

Cost of abnormal gain = 30 × Rs. 230

= Rs. 6,900

PROCESS 'C' ACCOUNT

	Qty.	Amt.		Qty.	Amt.
To process 'B' a/c	3,800	8,74,000	By normal loss	114	1,710
To sundry materials	–	15,000	By abnormal loss	26	6,500
To direct labour	–	26,100	By P & L a/c	3,660	9,15,000
To direct expenses	–	8,100			
	3,800	9,23,210		3,800	9,23,210

Working Note:

Calculation of cost of abnormal loss

Input Qty.	3,800	Rs.	9,23,210
Less: Normal loss	114	Rs.	1,710
	3,686	Rs.	9,21,500

$$\frac{9,21,500}{3,686} = \text{Rs. } 250 \text{ per unit}$$

Cost of abnormal loss = 26 × Rs. 250

= Rs. 6,500

PROFIT & LOSS ACCOUNT

To cost of sales:		By sales:	
Process 'A' 2100 units		Process 'A'	4,20,000
@ Rs. 170 p.a.	3,57,000	Process 'B'	5,32,000
Process 'B' 1900		Process 'C'	21,96,000
units @ Rs. 230 p.u.	4,37,000		
Process 'C' 3660		By abnormal gain:	
units @ Rs. 250 p.u.	9,15,000	Process 'B'	6,600
To abnormal loss:			
Process 'A' 150 units			
@ Rs. 17 p.u.	24,750		
Process 'C' 26 units			
@ Rs. 250 p.u.	6,110		
To selling & distribution			
overhead	45,000		
To net profit	13,69,740		
	31,54,600		31,54,600

TREATMENT OF NORMAL LOSS, REJECTS AND RECTIFICATION, ABNORMAL LOSS AND ABNORMAL GAIN ACCOUNT

1. Normal Loss

The loss which is inherent in the production process and which cannot be controlled is known as normal loss. Because they are inherent in any production process they are capable of estimating in advance of production. The estimation is usually expressed as a certain percentage of input quantity. Normal loss may occur due to evaporation, shrinkage, chemical reaction, moisture, withdrawal for test, or for sampling. Normal loss may take the form of (*a*) normal waste, and (*b*) normal scrap.

Normal waste, according to Theodore Lang *(The Cost Accountants Hand Book)* refers to "that portion of a basic raw material lost in processing, having no recovery value". Scrap has been defined by Theodore Lang as "the incidental residue from certain types of manufacture usually of small amount and low value, recoverable without further processing". There are two points of differences between a waste and scrap. Firstly, waste, is invisible, whereas scrap is visible. Secondly, waste does not yield any income whereas scrap does.

Treatment of normal loss in process cost accounts: The cost of normal loss units will be charged to the good units manufactured. Thus the cost of good units is inflated to the extent of normal loss quantity. For example, if the total cost of processing 1,000 units is Rs. 5,400 the cost per unit would be Rs. 5.40. Suppose if wastage is 100 units, then the cost per good unit is Rs. 6 (5,400 ÷ 900). The increase in the unit cost due to normal waste is 0.60.

If the normal loss is in the form of scrap material such as metal turning or faulty parts, it may have a saleable value. The process account is to be credited with the saleable value, thereby reducing the cost of production to this extent.

2. Rejects and Rectification

The ICMA terminology defines rejects as "output units which fail to reach the required standard of quality or specification before despatch to customers or subsequent users".

The rejects may be reworked so as to meet the required standards. This is subject to the condition that the cost of such rectification does not exceed the additional value obtained from the sale precedes of such rejects. Where cost of rectification is considered to be more, rejects can be sold as scrap and treated in the same manner as discussed under normal loss.

Treatment of rejects in process cost accounts: The cost incurred to the point of rejection is credited to the process account and a separate "rejects account is opened to which is debited with the cost incurred on such rejects". The reject account" will also be debited with the rectification cost and is credited with its sale value.

3. Abnormal Loss Account

Losses which are incurred over and above normal loss is known as abnormal loss. Abnormal loss is incurred in abnormal conditions. So it is not possible to anticipate abnormal units in advance of production. Abnormal loss can be controlled by taking corrective steps. Some of the causes which gives rise to abnormal loss are use of sub-standard materials, carelessness and accidents, bad plant design, bad workmanship and so on.

Treatment of abnormal loss in process cost accounts: The treatment of abnormal loss differs from that of treatment of normal loss. Instead of charging the cost of abnormal loss units to good units manufactured, abnormal units have to bear the respective cost of production. The following formula is used to value abnormal loss units:

$$\frac{\text{Normal cost of normal output}}{\text{Normal output}} \times \text{abnormal loss units}$$

Problem 7. *(Treatment of rejected units):* Ten units were rejected at the end of a process and unit cost in the process was calculated as Rs. 5. The management decided that four units should be rectified and sold at the normal price of Rs. 12 each, two items could be sold as "Seconds" at Rs. 8 each and the remainder would be scrapped at nil value. Rectification costs amounted to Rs. 16.

Prepare rejects account.

Solution:

REJECTS ACCOUNT

	Units	*Amt.*		*Units*	*Amt.*
To process a/c	10	50	By sale	4	48
To rectification cost		16	By sale as seconds	2	16
			By scrap	4	–
			By P & L a/c	–	2
	10	66		10	66

Problem 8. ABNORMAL LOSS

Units of raw materials introduced	–	100 units
Rate per units	–	Rs. 10
Normal loss	–	10% of input
Sale of normal loss units	–	Rs. 3 per unit
Output	–	75 units

Wages incurred Rs. 600. Calculate the abnormal loss units and its value.

Solution:

PROCESS ACCOUNT

	Units	*Amt.*		*Units*	*Amt.*
To raw materials	100	1,000	By normal loss	10	30
To wages		600	By abnormal loss	15	262
			By output of the process	75	1,308
	100	1,600		100	1,600

Working Note:

Normal cost of normal output = 1,600 – 30 = Rs. 1,570

Normal output:

Units introduced	100
Less: Normal loss units	10
	90 units

Abnormal loss units:

Normal output	90
Less: Actual output	75
	15 units

$$\text{Value of abnormal loss units} = \frac{\text{Normal cost of normal output}}{\text{Normal output}} \times \text{abnormal loss units}$$

$$= \frac{1,570}{90} \times 15 = \text{Rs. } 262 \text{ APP}$$

CHAPTER 4

4. Abnormal Gain

It was referred earlier that normal loss units are estimated on the basis of past experience. Sometimes the normal loss actually incurred may be less than anticipated loss. This results in abnormal gain or abnormal effectiveness or abnormal efficiency. Abnormal gain can also be considered as the difference between actual units produced and expected units.

Treatment in process cost accounts: The value of abnormal gain is debited to the process account from which it originated. A separate abnormal gain account is opened and the corresponding credit is given to this account. The sale of abnormal gain is utilised to reduce the cost of normal loss units. This is so because abnormal gain is the difference between actual normal loss and anticipated normal loss. Thus in effect the normal loss is set off against abnormal gain by debiting abnormal gain a/c.

Problem 9. *(Calculation of abnormal gain):* From the following data compute abnormal gain and prepare process account:

Input quantity in process A	– 1,000 kg
Total cost incurred in process A	– Rs. 10,000
Output expected	– 900 kgs
Actual output	– 920 kgs
Sale of scraps	– Rs. 2 per kg.

Solution:

PROCESS 'A' ACCOUNT

	Qty.	*Amt.*		*Qty.*	*Amt.*
To input	1,000	10,000	By normal loss	100	200
To abnormal gain	20	218	By output	920	10,018
	1,020	10,218		1,020	10,218

ABNORMAL GAIN ACCOUNT

To scrap not realised 20 units @ Rs. 2	40	By process a/c	218
To P & L a/c	178		
	218		218

Working Note: Calculation of abnormal gain units and its value.

	Units	*Rs.*
Input	1,000	10,000
Less: Normal loss	100	200
Output expected	900	9,800
Abnormal gain $\frac{9,800}{900}$ = 10.89		
20 units @ Rs. 10.80	20	218
Actual production	920	10,018

Problem 10. The following information is available in respect of process *B* of Production *X.*

Input	5,000 units at a cost of Rs. 25,000
Process materials added	Rs. 12,000
Direct labour	Rs. 8,000

Overheads	Rs. 3,000
Output of process B	4,800 units
Normal wastage	5% of input
Scrap value of wastage	Rs. 2 per unit

Prepare process B account

(Delhi University, B.Com. (Pass), April 1991)

Solution:

PROCESS 'B' ACCOUNT

	Qty.	*Amt.*		*Qty.*	*Amt.*
To input	5,000	25,000	By normal wastage	250	500
To process labour		12,000	By output of process	4,800	48,000
To direct labour		8,000			
To overheads		3,000			
To abnormal gain	50	500			
	5,050	48,500		5,050	48,500

Working Note: Calculation of abnormal gain

Input	5,000	Rs.48,000
Less: Normal loss	250	Rs. 500
Output expected	4,750	47,500

$$\frac{47,500}{4,750} \times 50 = \text{Rs. } 500$$

Problem 11. *(Normal loss, Abnormal loss and Abnormal gain):* A product passes through two process A and B. Prepare the process accounts from the following:

	Process A	*Process B*
10,000 units introduced at cost (Rs.)	20,000	–
Materials consumed (Rs.)	24,000	12,000
Direct labour (Rs.)	28,000	16,000
Manufacturing expenses (Rs.)	8,000	8,566
Normal wastage on input	5%	10%
Scrap value of normal wastage (Rs. per 100 units)	40	50
Output (units)	9,400	8,500

Also prepare the Abnormal wastage/effectives account, as the case may be, with each process account. *(C.S. Inter., Dec. 1999)*

Solution:

PROCESS 'A' ACCOUNT

	Qty.	*Amt.*		*Qty.*	*Amt.*
To cost of input	10,000	20,000	By normal loss a/c	500	200
To materials		24,000	By abnormal wastage a/c	100	840
To direct labour		28,000	By process B a/c	9,400	78,960
To manufacturing expenses		8,000			
	10,000	80,000		10,000	80,000

$$\text{Cost per unit} = \frac{\text{Total cost of debit side} - \text{Scrap value of normal loss units}}{\text{Total input Qty.} - \text{Normal loss units}}$$

$$= \frac{80{,}000 - 200}{10{,}000 - 500} = \frac{79{,}800}{9{,}500} = 8.40$$

Cost of abnormal wastage = $100 \times 8.4 = 840$

PROCESS 'B' ACCOUNT

	Qty.	Amt.		Qty.	Amt.
To process 'A' a/c	9,400	78,960	By normal loss	940	470
To materials		12,000	By finished stock a/c	8,500	1,15,600
To direct labour		16,000			
To manufacturing expenses		8,566			
To abnormal effectives	40	544			
	9,440	1,16,070		9,440	1,16,070

$$\text{Cost per unit (good unit)} = \frac{\text{Cost of production} - \text{Sale value of normal loss units}}{\text{Input Qty.} - \text{Normal loss units}}$$

$$= \frac{1{,}15{,}526 - 470}{9{,}400 - 940}$$

$$= \frac{1{,}15{,}056}{8{,}460} = 13.60$$

Cost of abnormal effectives = 40×13.60 = Rs. 544

NORMAL LOSS ACCOUNT

	Qty.	Amt.		Qty.	Amt.
To process 'A' a/c	500	200	By sale of normal loss:		
To process 'B' a/c	940	470	Process A	500	200
			Process B	900	450
			By abnormal effectives	40	20
	1,440	670		1,440	670

ABNORMAL WASTAGE ACCOUNT

To process 'A' a/c	100	840	By sales	100	40
			By P & L a/c	–	800
	100	840		100	840

ABNORMAL EFFECTIVES ACCOUNT

To process loss a/c	40	20	By process B a/c	40	544
To P & L a/c	–	524			
	40	544		40	544

Problem 12. Product 'A' is obtained after it is processed through process X, Y and Z.

The following cost information is available for the month ended 31st March 1999.

Particulars	Processes X	Y	Z
Number of units introduced in the process	500	–	–
Rate per unit of units introduced (Rs.)	4	–	–
Cost of materials	2,600	2,000	1,025
Direct wages	2,250	3,680	1,400
Production overheads	2,250	3,680	1,400
Normal loss (% on units introduced in each process *i.e.,* input)	10%	20%	25%
Value of scrap per unit	2	4	5
Output in units	450	340	270

There is no stock in any process. You are required to prepare:

1. The process accounts
2. The abnormal loss and abnormal gain a/c. *(University of Bombay, B.Com., April 1999)*

Solution:

PROCESS 'X' ACCOUNT

	Qty.	*Amount*		*Qty.*	*Amount*
To input	500	2,000	By normal loss	50	100
To materials		2,600	By process Y a/c	450	9,000
To direct wages		2,250			
To production overheads		2,250			
	500	9,100		500	9,100

$$\text{Cost per unit} = \frac{\text{Total cost of debit side} - \text{Scrap value of normal loss Qty.}}{\text{Input Qty.} - \text{Normal loss units}}$$

$$= \frac{9{,}100 - 100}{500 - 50} = 20$$

PROCESS 'Y' ACCOUNT

	Qty.	*Amount*		*Qty.*	*Amount*
To process 'X' a/c	450	9,000	By normal loss	90	360
To material		2,000	By abnormal loss	20	1,000
To direct wages		3,680	By process Z a/c		
To production overhead		3,680	@ Rs. 50 per unit	340	17,000
	450	18,360		450	18,360

$$\text{Cost per unit} = \frac{\text{Total cost of debit side} - \text{Scrap value of normal loss Qty.}}{\text{Input Qty.} - \text{Normal loss units}}$$

$$= \frac{18{,}360 - 360}{450 - 90} = \frac{18{,}000}{360} = \text{Rs. } 50$$

CHAPTER 4

PROCESS 'Z' ACCOUNT

	Qty.	*Amount*		*Qty.*	*Amount*
To process 'Y' a/c	340	17,000	By normal loss	85	425
To material		1,025	By finished stock	270	21,600
To direct wages		1,400	@ Rs. 80 per unit		
To production overhead		1,400			
To abnormal gain	15	1,200			
	355	22,025		355	22,025

$$\text{Cost per unit} = \frac{20{,}825 - 425}{340 - 85} = \text{Rs. } 80$$

NORMAL LOSS ACCOUNT

	Qty.	*Amount*		*Qty.*	*Amount*
To process A a/c	50	100	By sale of normal loss units:		
To process B a/c	90	360	Process A	50	100
To process C a/c	85	425	Process B	90	360
			Process C	70	350
			By abnormal Gain a/c	15	75
	225	885		225	885

ABNORMAL LOSS ACCOUNT

	Qty.	*Amount*		*Qty.*	*Amount*
To process 'Y' a/c	20	1,000	By sale	20	80
			To P & L a/c		920
	20	1,000		20	1,000

ABNORMAL GAIN ACCOUNT

To normal loss a/c	15	75	By process Z a/c	15	1,200
To P & L a/c		1,125			
	15	1,200		15	1,200

Problem 13. *(Normal loss, abnormal loss and abnormal gain):* Product 'A' is obtained after it passes through three distinct processes. Prepare process account and other accounts from the following:

	Total		*Process*	
		I	*II*	*III*
Materials	15,084	5,200	3,960	5,924
Wages	18,000	4,000	6,000	8,000
Production overhead	18,000	–	–	–

1,000 units of materials @ Rs. 6 per unit were introduced in process I. Production overhead is to be distributed as 100% on wages.

Processes	Total output units	Normal loss	Value of Scrap per units (Rs)
I	950	5%	4
II	840	10%	8
III	750	15%	10

(University of Madras, B.Com., April 2005)

Solution:

PROCESS I ACCOUNT

	Qty.	Amt.		Qty.	Amt.
To input	1,000	6,000	By normal loss	50	200
To direct materials		5,200	By process II A/c	950	19,000
To direct wages		4,000			
To production overhead		4,000			
	1,000	19,200		1,000	19,200

PROCESS II ACCOUNT

	Qty.	Amt.		Qty.	Amt.
To process I a/c	950	19,000	By normal loss	95	760
To direct materials	–	3,960	By abnormal loss	15	600
To direct wages	–	6,000	By process III a/c	840	33,600
To production overhead	–	6,000			
	950	34,960		950	34,960

Working Note:

Calculation of abnormal loss

$$= \frac{\text{Value of Input} - \text{Normal loss amount}}{\text{Input units} - \text{Normal loss units}} \times \text{Abnormal loss}$$

$$= \frac{34{,}960 - 760}{950 - 95} \times 15 = 600$$

PROCESS III ACCOUNT

	Qty.	Amt.		Qty.	Amt.
To process II a/c	840	33,600	By normal loss	126	1,260
To direct materials	–	5,924	By finished goods	750	57,000
To direct wages	–	8,000			
To production overhead	–	8,000			
To abnormal Gain	36	2,736			
	876	58,260		876	58,260

Working Note:

Calculation of abnormal loss

$$= \frac{\text{Value of Input} - \text{Normal loss amount}}{\text{Input units} - \text{Normal loss units}} \times \text{Abnormal gain}$$

$$= \frac{55{,}524 - 1{,}260}{840 - 126} \times 36 = \text{Rs. } 2{,}736$$

NORMAL LOSS ACCOUNT

	Qty.	Amt.		Qty.	Amt.
To process I a/c	50	200			
To process II a/c	95	760	By cash a/c		
To process III a/c	126	1260	— process A	50	200
			— process B	95	760
			— process C	90	900
			By abnormal gain a/c	36	360
	271	2,220		271	2,220

ABNORMAL GAIN ACCOUNT

	Qty.	Amt.		Qty.	Amt.
To normal loss a/c	36	360	By process III a/c	36	2,736
To costing P & L a/c	–	2,376			
	36	2,736		36	2,736

ABNORMAL LOSS ACCOUNT

	Qty.	Amt.		Qty.	Amt.
To process II a/c	15	600	By cash-sale	15	120
			By costing P & L		480
	15	600		15	600

Problem 14. Y Ltd. manufactures a chemical product which passes through three processes. The cost records show the following particulars for the year ended 30th June 2004.

Input to process I - 20,000 units @ Rs. 28 per unit

Particulars	*Process I* Rs.	*Process II* Rs.	*Process III* Rs.
Materials	48,620	1,08,259	1,03,345
Labour	32,865	84,553	77,180
Expenses	2,515	10,588	16,275
Normal loss	20%	15%	10%
Scrap value per unit	1	2	3
Actual output (units)	18,000	16,000	15,000

Prepare process accounts, abnormal gain or loss account. Also show process cost per unit for each process. *(University of Mumbai, B.Com., October 2005)*

Solution:

PROCESS I ACCOUNT

	Qty.	Amt.		Qty.	Amt.
To input	20,000	5,60,000	By normal loss	4,000	4,000
To material	–	48,620	By process II a/c	18,000	7,20,000
To labour	–	32,865	@ Rs. 40 per unit		

	Qty.	Amt.		Qty.	Amt.
To expenses	–	2,515			
To abnormal gain a/c	2,000	80,000			
	22,000	7,24,000		22,000	7,24,000

Working Note:

Calculation of abnormal gain

Input units	20,000	Rs.	6,44,000
Less: Normal loss	4,000	Rs.	4,000
	16,000		6,40,000

$$\frac{6,40,000}{16,000} \times 2000 = 80,000$$

PROCESS II ACCOUNT

	Qty.	*Amt.*		*Qty.*	*Amt.*
To process I a/c	18,000	7,20,000	By normal loss	2,700	5,400
To material	–	1,08,259	By process III a/c		
To labour	–	84,553	@ Rs. 60 per unit	16,000	9,60,000
To expenses	–	10,588			
To abnormal gain	700	42,000			
	18,700	9,65,400		18,700	9,65,400

Working Note:

Calculation of abnormal gain value

Input	Qty.	18,000	Rs.	9,23,400
Less: Normal loss		2,700		5,400
		15,300		9,18,000

$$\frac{9,18,000}{15,300} \times 700 = \text{Rs. } 42,000$$

PROCESS III ACCOUNT

	Qty.	*Amt.*		*Qty.*	*Amt.*
To process II a/c	16,000	9,60,000	By normal loss	1,600	4,800
To materials	–	1,03,345	By finished stock a/c	15,000	12,00,000
To labour	–	77,180			
To expenses	–	16,275			
To abnormal gain a/c	600	48,000			
	16,600	12,04,800		16,600	12,04,800

Working Note:

Calculation of abnormal gain value

Input	Qty.	16,000	Rs.	9,60,000
Less: Normal loss		1,600	Rs.	4,800
		14,400		11,52,000

$$\frac{11,52,000}{14,400} \times 600 = \text{Rs. } 48,000$$

ABNORMAL GAIN ACCOUNT

	Qty.	*Amt.*		*Qty.*	*Amt.*
To Normal loss:			By process I a/c	2,000	80,000
Process I a/c	2,000	2,000	By process II a/c	700	42,000
Process II a/c	700	1,400	By process III a/c	600	48,000
Process III a/c	600	1,800			
To costing P & L a/c	–	1,64,800			
	3,300	1,70,000		3,300	1,70,000

Problem 15. At the end of process 'A', carried on in a factory during the week ending 31st July 2002 the number of units produced was 850 excluding 50 units damaged at the very end of the process. The damaged units realised Rs. 3 per unit as scrap. A normal wastage of 10% occurs during the process. The wastage realised was Rs. 2 per unit.

A unit of raw materials cost Rs. 4. The other expenses for the week were:

	Rs
Wages	500
Power	200
General expenses	450

Fourty percent of the output is sold so as to show a profit of $16\frac{2}{3}\%$ on the selling price; the rest of the output is transferred to process B.

Prepare process 'A' account. *(University of Delhi, B.Com., (Hons) 2002)*

Solution:

PROCESS 'A' ACCOUNT

	Qty.	*Amt.*		*Qty.*	*Amt.*
To materials	1,000	4,000	By normal loss	100	200
@ Rs. 4 per unit			By units damaged	50	150
To wages	–	500	By sale	340	2,304
To power	–	200	By process 'B' a/c	510	2,880
To general expenses	–	450			
To profit	–	384			
	1,000	5,534		1,000	5,534

Working Notes:

(1) Calculation of material quantity introduced

= units produced + units damaged + normal loss

= 850 units + 50 units + 100 units

= 1,000 units

(2) Total cost of 850 units

Rs. 4,000 + 500 + 200 + 450 – 200 – 150 = Rs. 4,800

(3) Quantity sold

$$\frac{40}{100} \times 850 = 340 \text{ units}$$

(4) Quantity transferred to process 'B'

$$\frac{60}{100} \times 850 = 510 \text{ units}$$

(5) Cost of 340 units after damage

$$\frac{340}{850} \times 4{,}800 = 1{,}920$$

(6) Sale value of 340 units

Let sale price	=	100
Less: Profit margin	=	16.6%
Cost price		83.33
Profit cost price	=	$\frac{16.67}{83.33} \times 1{,}920 = 384$
Sale price	=	cost price + profit
	=	1,920 + 384 = 2,304

Problem 16. A product passes through two distinct processes X and Y before completion. During a certain period, 10,000 units of material were introduced in process X at a cost of Rs. 40,000. After processing in dept. X 9,000 units of processed material were transferred to process Y for finishing. From process Y finally 8,600 units of the finished product were obtained and transferred to finished goods store.

Further data regarding normal waste, cost etc are given belows:

	Process X	*Process Y*
Costs incurred:		
	Rs.	Rs.
Material	10,000	5,000
Labour	20,000	15,000
Overheads	10,000	8,000
Normal waste (% of input)	8%	5%
Realisable value of waste per unit (Rs.)	5	8

There was no opening or closing stock in any process.

Required:

(*a*) Process accounts

(*b*) Normal loss account

(*c*) Abnormal loss/gain accounts

(*d*) Selling price per unit of the finished product, if management wants 25% profit on sales.

(ICWA, Inter Stage I, December 2005)

Solution:

PROCESS 'X' ACCOUNT

	Qty.	*Amt.*		*Qty.*	*Amt.*
To units introduced	10,000	40,000	By normal loss	800	4,000
To material	–	10,000	By abnormal loss	200	1,652
To labour	–	20,000	By process 'Y' a/c	9,000	74,348
To overhead	–	10,000			
	10,000	80,000		10,000	80,000

Working Note:

Calculation of abnormal loss

Input	Qty	10,000	Rs.	80,000
Less: Normal loss		800	Rs.	4,000
		9,200		76,000

$$\frac{76,000}{9,200} \times 9,000 = 1652$$

PROCESS 'Y' ACCOUNT

	Qty.	*Amt.*		*Qty.*	*Amt.*
To process 'X' a/c	9,000	74,348	By normal loss	450	3,600
To materials	–	5,000	By finished stock a/c	8,600	99,325
To labour	–	15,000			
To overheads	–	8,000			
To abnormal gain a/c	50	577			
	9,050	1,02,925		9,050	1,02,925

Working Note:

Calculation of value of abnormal gain

Input	Qty.	9,000	Rs.	1,02,348
Less: Normal loss		450	Rs.	3,600
		8,550	Rs.	98,748

$$\text{Cost per unit} = \frac{98,748}{8,550} = \text{Rs. } 11.55$$

Cost of abnormal gain = 50 units × Rs. 11.55
= Rs. 577

NORMAL LOSS ACCOUNT

	Qty.	*Amt.*		*Qty.*	*Amt.*
To process 'X' a/c	800	4,000	By Cash		
To process 'Y' a/c	450	3,600	— Process X	800	4,000
			— Process Y	400	3,200
			By abnormal a/c	50	400
	1,250	7,600		1,250	7,600

ABNORMAL LOSS ACCOUNT

	Qty.	Amt.		Qty.	Amt.
To process 'X' a/c	200	1,652	By cash	200	1,000
			By costing P & L a/c		652
	200	1,652		200	1,652

ABNORMAL GAIN ACCOUNT

	Qty.	Amt.		Qty.	Amt.
To normal loss a/c	50	400	By process 'Y' a/c	50	577
To costing P & L	–	177			
	50	577		50	577

(4) Computation of selling price per unit of the finished product:

Cost per unit of finished goods	Rs.	11.55
Add: profit 25% on sales or $33\frac{1}{3}$% on cost	Rs.	3.85
Selling price per unit	Rs.	15.40

Problem 17. A company manufactures a product by passing the units of raw material through three distinct process A, B and C. During the month of July 2002, 1,000 units of input raw material costing Rs. 12 per unit were introduced in process 'A'. Further details regarding production and costs are given below:

	Output (Units)	*Normal loss (% of input)*	*Value of Scrapped unit*
Process A	900	10%	2
Process B	800	8%	3
Process C	780	5%	5

Cost incurred for processing

	Materials	*Labour*	*Overhead*
Process A	Rs. 8,000	4,000	3,200
Process B	Rs. 7,416	3,600	3,600
Process C	Rs. 1,800	2,000	2,000

There was no work-in-progress at the beginning or end of the month. Prepare the process accounts and accounts relating to abnormal loss/gain, if any, for the month.

(ICWA, Inter, December 2002)

Solution:

PROCESS 'A' ACCOUNT

	Units	Amt.		Units	Amt.
To raw materials	1,000	12,000	By normal loss a/c	100	200
To materials	–	8,000	By process 'B'	900	27,000
To labour	–	4,000			
To overhead	–	3,200			
	1,000	27,200		1,000	27,200

CHAPTER 4

PROCESS 'B' ACCOUNT

	Units	Amt.		Units	Amt.
To process 'A'	900	27,000	By normal loss a/c	72	216
To materials	–	7,416	By abnormal loss a/c	28	1,400
To labour	–	3,600	By process 'C' a/c	800	40,000
To overhead	–	3,600			
	900	41,616		900	41,616

PROCESS 'C' ACCOUNT

	Units	Amt.		Units	Amt.
To process 'B' a/c	800	40,000	By normal loss a/c	40	200
To materials	–	1,800	By finished stock a/c	780	46,800
To labour	–	2,000			
To overhead	–	2,000			
To abnormal gain a/c	20	1,200			
	820	47,000		820	47,000

Problem 18. *(Calculation of normal loss units):* The following data are available pertaining to a product after passing through two processes A and B.

Output transferred to process C from process B – 9,120 units for Rs. 49,263.

Expenses incurred in process C:

Sundry materials	Rs. 1,480
Direct labour	Rs. 6,500
Direct expenses	Rs. 1,605

The wastage of process C is sold at Re. 1.00 per unit. The overhead charges were 168% of direct labour. The final product was sold at Rs. 10 per unit fetching a profit of 20% on sales.

Find the percentage of wastage in process C and prepare C account.

(University of Delhi, B.Com., (Hons.) April 1999)

Solution:

PROCESS 'C' ACCOUNT

	Qty.	Amt.		Qty.	Amt.
To process B	9,120	49,263	By normal loss	456	456
To sundry materials		1,480	By finished goods a/c	8,664	69,312
To direct labour		6,500			
To direct expenses		1,605			
To overhead		10,920			
	9,120	69,768		9,120	69,768

Working Note:

Calculation of normal loss units

Let normal loss units: x

No. of units of finished product	=	$9{,}120 - x$
Value of normal loss @ Re. 1 per unit	=	$1x$
Cost of finished goods per unit	=	10 – 20%
Rs.	=	8
Cost of finished goods	=	$(9{,}120 - x)$ Rs. 8
i.e., cost per unit × No. of units produced	=	$72{,}960 - 8x$

Total cost of process C $= 69{,}768 - 1x$

or $= 69{,}768 - 1x = 72{,}960 - 8x$

or $8x - 1x = 72{,}960 - 69{,}768$

$7x = 3{,}192$

$$x = \frac{3{,}192}{7} = 456$$

% of normal loss in process C

$$\frac{456}{9{,}120} \times 100 = 5\%$$

Finished goods units $= 9{,}126 - 456$

$= 8{,}664$ units

Cost of finished goods $= 8{,}664 \times$ Rs. 8

$= 69{,}312$

Problem 19. (Process cost accounts with stock)

Particulars	*Process A* *Rs.*	*Process B* *Rs.*	*Process C* *Rs.*
Indirect materials	1,00,000	18,750	16,550
Direct wages	56,250	35,000	44,900
Direct expenses	51,250	6,875	11,500
Value of opening Stock per unit	25	31	40
Scrap value per unit	13.50	11.25	21
	Units	*Units*	*Units*
Output	9,750	9,625	8,000
Stock of process output:			
01-01-2005	1500	1,375	2,000
31-12-2005	1250	2,000	1,000
Percentage of wastage	2%	5%	10%

10000 units of direct materials were introduced in process A at the rate of Rs. 5 per unit. The percentage of wastage is computed on the number of units entering the process concerned. From the above information of "DE" enterprise prepare:

(1) Process accounts
(2) Process stock accounts
(3) Normal loss account
(4) Abnormal loss account
(5) Abnormal gain account. Value closing stock at the respective process cost.

(University of Bomaby, B.Com., April 2006)

Solution:

PROCESS 'A' ACCOUNT

	Qty.	*Amt.*		*Qty.*	*Amt.*
To input	10,000	50,000	By normal loss	200	2,700
To indirect materials	–	1,00,000	By abnormal loss	50	1,300
To direct wages	–	56,250	By process stock 'A' a/c	9,750	2,53,500
To direct expenses	–	51,250			
	10,000	2,57,500		10,000	2,57,500

CHAPTER 4

PROCESS STOCK 'A' A/C

To balance b/d	1,500	37,500	By process 'B' a/c	10,000	2,58,500
To process 'A' a/c	9750	2,53,500	By balance c/d	1,250	32,500
	11,250	2,91,500		11,250	2,57,500

Working Note:
Calculation of abnormal loss value

Input	10,000	Rs.	2,57,500
Less: Normal loss	200	Rs.	2,700
	9,800	Rs.	2,54,800

$$\frac{2,54,800}{9,800} \times 50 = \text{Rs. } 1,300$$

PROCESS 'B' ACCOUNT

To process 'A' stock a/c	10,000	2,58,500	By normal loss a/c	500	5,625
To indirect materials	–	18,750	By process 'B' stock a/c @ Rs. 33	9,625	3,17,625
To direct wages	–	35,000			
To direct expenses	–	6,875			
To abnormal gain a/c	125	4,135			
	10,125	3,23,250		10,125	3,23,250

PROCESS 'B' STOCK ACCOUNT

To balance b/d	1,375	42,625	By process 'c' a/c	9,000	2,94,250
To process 'B' a/c	9,625	3,17,625	By balance c/d	2,000	66,000
	11,000	3,60,250		11,000	3,60,250

Working Note:
Calculation of abnormal gain

Input	10,000	Rs.	3,19,125
Less: Normal loss	500	Rs.	5,625
	9,500	Rs.	3,13,500

$$\frac{3,13,500}{9,500} \times 125 = \text{Rs. } 4,125$$

PROCESS 'C' ACCOUNT

To process 'B' stock a/c	9,000	2,94,250	By normal loss	900	18,900
To indirect materials		16,550	By abnormal loss	100	4,300
To direct wages		44,900	By process 'C' stock a/c	8,000	3,44,000
To direct expenses		11,500			
	9,000	3,67,200		9,000	3,67,200

PROCESS 'C' STOCK ACCOUNT

To balance b/d	2,000	80,000	By finished stock a/c	9,000	3,81,000
To process 'c' a/c	8,000	3,44,000	By balance c/d	1,000	43,000
	10,000	4,24,000		10,000	4,24,000

Working Note:

Calculation of abnormal loss value

Input Units	9,000	Rs.	3,67,200
Less: Normal loss	900	Rs.	18,900
	8,100	Rs.	3,48,300

$$\frac{3,48,300}{8,100} \times 100 = \text{Rs. } 4,300$$

NORMAL LOSS ACCOUNT

To process A	200	2,700	By cash A	200	2,700
To process B	500	5,625	By cash B	375	4,219
To process C	900	18,900	By cash C	900	18,900
			By abnormal gain a/c	125	1,406
	1,600	27,225		1,600	27,225

ABNORMAL LOSS ACCOUNT

To process 'A'	50	1,300	By cash A	50	675
To process 'C'	100	4,300	By cash B	100	2,100
			By costing P & L a/c		2,825
	150	5,600		150	5,600

ABNORMAL GAIN ACCOUNT

To normal loss a/c	125	1,406	By process 'B' a/c	125	4,125
To costing P & L a/c		2,719			
	125	4,125		125	4,125

Problem 20. A product passes through three processes and 40,000 units were introduced in process A at cost of Rs. 30,000. The following further information is available:

Particulars	*Process A*	*Process B*	*Process C*
Sundry materials	Rs. 20,000	4,000	20,000
Direct labour	Rs. 6,000	3,000	1,500
Direct expenses	Rs. 1,920	5,600	4,200
Output (units)	38,000	37,000	34,000
Opening stock (units)	6,000	3,000	4,000
Closing stock (units)	4,000	5,000	9,500
Opening stock valuation (per unit)	Rs. 1.40	Rs. 1.80	Rs. 2.50
% of normal wastage	4%	5%	10%
Scrap sale price (per unit)	Re. 0.20	0.30	0.40

The closing stock in each process is valued at respective process cost.

Prepare process accounts and process stock accounts.

(University of Bombay, B.Com., April 2005)

CHAPTER 4

Solution:

PROCESS 'A' ACCOUNT

	Qty.	Amt.		Qty.	Amt.
To input	40,000	30,000	By normal loss	1,600	320
To sundry materials	–	20,000	By abnormal loss	400	600
To direct labour	–	6,000	By process 'A' stock a/c	38,000	57,000
To direct expenses	–	1,920			
	40,000	57,920		40,000	57,920

Working Note:

Calculation of value of abnormal loss

Input Qty.	40,000	Rs.	57,920
Less: Normal loss	1,600	Rs.	320
	38,400	Rs.	57,600

$$\frac{57,600}{38,400} \times 400 = \text{Rs. } 600$$

PROCESS 'A' STOCK ACCOUNT

	Qty.	Amt.		Qty.	Amt.
To opening stock	6,000	8,400	By process 'B' a/c	40,000	59,400
To process 'A' a/c	38,000	57,000	By closing stock @ Rs. 1.50	4,000	6,000
	44,000	65,400		44,000	65,400

PROCESS 'B' ACCOUNT

	Qty.	Amt.		Qty.	Amt.
To process stock a/c	40,000	59,400	By normal loss	2,000	600
To sundry materials	–	4,000	By abnormal loss	1,000	1,880
To direct labour	–	3,000	By process 'B' stock a/c	37,000	69,520
To direct expenses	–	5,600			
	40,000	72,000		40,000	72,000

Working Note:

Calculation of value of abnormal loss

Input Qty.	40,000	Rs.	72,000
Less: Normal loss	2,000	Rs.	600
	38,000		71,400

$$\frac{71,400}{38,000} \times 1,000 = \text{Rs. } 1,880$$

PROCESS 'B' STOCK ACCOUNT

	Qty.	Amt.		Qty.	Amt.
To opening stock	3,000	5,400	By process 'c' a/c	35,000	65,520
To process 'B' a/c	37,000	69,520	By closing stock	5,000	9,400
	40,000	74,920		40,000	74,920

PROCESS 'C' ACCOUNT

	Qty.	Amt.		Qty.	Amt.
To process 'B' stock	35,000	65,520	By normal loss	3,500	1,400
To sundry materials	–	2,000	By process 'c'		
To direct labour	–	1,500	stock a/c	34,000	77,520
To direct expenses	–	4,200			
To abnormal gain	2,500	5,700			
	37,000	78,920		37,500	78,920

Working Note:

Calculation of value of abnormal gain value

Input Qty.	35,000	Rs.	73,220
Less: Normal loss	3,500	Rs.	1,400
	31,500		71,820

$$\frac{71{,}820}{31{,}500} \times 2{,}500 = \text{Rs. } 5{,}700$$

PROCESS 'C' STOCK ACCOUNT

	Qty.	Amt.		Qty.	Amt.
To opening stock	4,000	10,000	By finished goods a/c	28,500	63,770
To process 'c' a/c	34,000	77,520	By closing stock	9,500	23,750
	38,000	87,520		38,000	87,520

Problem 21. *(Process cost accounts with stock):* The product manufactured by the Standard Chemicals Ltd. passes through three processes I, II and III. The following costs have been incurred for the month of September 96.

Details	Process I	Process II	Process III
	Rs.	Rs.	Rs.
1. Materials consumed	40,000	7,500	5,000
2. Direct wages	22,500	10,000	10,000
3. Direct expenses	20,500	2,250	2,505
Total Rs.	83,000	19,750	17,505
	Units	*Units*	*Units*
4. Output	3,900	3,850	3,200
5. Finished process stock			
(*i*) 1-9-1996	600	550	800
(*ii*) 30-9-1996	500	800	Nil
6. Stock valuation on 1-9-96 (Rs. per unit)	24.50	31	37
7. Percentage of wastage	2	5	10
8. Net realisable value of wastage per unit (Rs.)	13.50	16.25	21

4,000 units of raw materials were introduced in process. No. I at a cost of Rs. 20,000.

Stock are valued and transferred to subsequent processes at weighted average cost. The percentage of wastage is computed on the number of units entering the process concerned.

Prepare (*i*) Process a/cs (*ii*) Process stocks a/cs (*iii*) Normal wastage a/c (*iv*) Abnormal wastage/ Effective a/c. *(University of Delhi, B.Com., (Hons.) April 1997)*

Solution:

PROCESS 'I' ACCOUNT

	Qty.	Amt.		Qty.	Amt.
To units introduced	4,000	20,000	By normal wastage	80	1,080
To total cost		83,000	@ Rs. 13.50		
			By abnormal wastage		
			@ 26 Rs.	20	520
			By process I stock a/c	3,900	1,01,400
			@ Rs. 26 per unit		
	4,000	1,03,000		4,000	1,03,000

PROCESS 'I' STOCK ACCOUNT

To opening balance					
@ Rs. 24.50	600	14,700	By process II A/c @ 25.80	4,000	1,03,200
To process A a/c	3,900	1,01,400	By balance c/d	500	12,900
	4,500	1,16,100		4,500	1,16,100

Cost per unit $\dfrac{1{,}03{,}000 - 1{,}080}{4{,}000 - 80} = 26$

Weighted average rate $= \dfrac{1{,}16{,}100}{4{,}500} = 25.80$

PROCESS 'II' ACCOUNT

	Qty.	Amt.		Qty.	Amt.
To process I stock a/c	4,000	1,03,200	By normal wastage	200	3,250
To total cost	–	19,750	@ 16.25 per unit		
To abnormal effectives	50	1,575	By process II stock a/c		
			@ 31.50 per unit	3,850	1,21,275
	4,050	1,24,525		4,050	1,24,525

PROCESS 'II' STOCK ACCOUNT

To balance b/d			By process III a/c		
@ Rs. 31 per unit	550	17,050	@ Rs. 31.44 per unit	3,600	1,13,175
To process II a/c	3,850	1,21,275	By balance c/d	800	25,150
	4,400	1,38,325		4,400	1,38,325

Cost per unit $\dfrac{1{,}22{,}950 - 3{,}250}{4{,}000 - 200} = \dfrac{1{,}19{,}700}{3{,}800} = 31.50$

Weighted average rate $= \dfrac{1{,}38{,}325}{4{,}400} = 31.44$

PROCESS 'III' ACCOUNT

	Qty.	Amt.		Qty.	Amt.
To process II stock a/c	3,600	1,13,175	By normal wastage @ Rs. 21 per unit	360	7,560
To total cost		17,505	By abnormal wastage @ Rs. 38	40	1,520
			By process III stock a/c	3,200	1,21,600
	3,600	1,30,680		3,600	1,30,680

PROCESS 'III' STOCK ACCOUNT

	Qty.	Amt.		Qty.	Amt.
To balance b/d	800	29,600	To cost of goods sold	4,000	1,51,200
To process III a/c	3,200	1,21,600			
	4,000	1,51,200		4,000	1,51,200

$$\text{Cost per unit} = \frac{1{,}30{,}680 - 7{,}560}{3{,}600 - 360} = \frac{1{,}23{,}120}{3{,}240} = 38$$

ABNORMAL EFFECTIVES ACCOUNT

	Qty.	Amt.		Qty.	Amt.
To normal wastage a/c	50	812.50	By process II A/c	50	1,575
To P & L a/c		762.50			
	50	1,575.00		50	1,575

NORMAL WASTAGE ACCOUNT

	Qty.	Amt.		Qty.	Amt.
To process I a/c	80	1,080	By process I a/c	80	1,080
To process II a/c	200	3,250	By process II a/c	150	2,437.50
To process III a/c	360	7,560	By process III a/c	360	7,560
			By abnormal effectives a/c	50	812.50
	640	11,890		640	11,890.00

ABNORMAL WASTAGE ACCOUNT

	Qty.	Amt.		Qty.	Amt.
To process I a/c	20	520	By sales	60	1,110
To process II a/c	40	1,520	By P & L a/c		930
	60	2,040		60	2,040

Problem 22. *(Calculation of input quantity):* Find out the number of units initially introduced in process I when normal wastage is 10%, 5% and 8% of the input respectively. There are 3,933 units of finished stock. There was no Abnormal loss or Effectives in any of the process.

(University of Delhi, B.Com. (Pass) April 1997)

CHAPTER 4

Solution:

Let output quantity = 100 units

Wastage is the difference between units introduced and units produced wastage is 10%, 5% and 8% in three processes. So input is calculated as follows:

	Let the units introduced =	100
Less:	Wastage 10% (process I)	10
	Final units in process I	90
Less:	Wastage 5% (process II)	4.5
	Final units in process II	85.50
Less:	Wastage 8% (process III)	6.84
	Final units	78.66

For 78.66 Finished stock: The actual output is 3,933

$$\text{For } 100 \rightarrow \frac{3{,}933 \times 100}{78.66} = 5{,}000 \text{ units}$$

So, 5,000 units must have been initially introduced.

Problem 23. *(Process costing as applied to oil manufacturing company):* The following details are extracted from the costing records of Parachut Coconut Oil Mills for the year ended 31st March 2009:

Purchase of 500 kgs. of copra Rs. 2,00,000

The production of oil was undertaken in the following three processes:

	Crushing	*Refining*	*Finishing*
	Rs.	*Rs.*	*Rs.*
Wages	2,500	1,000	1,500
Electric power	600	360	240
Sundry materials	100	2,000	–
Repairs to machinery	280	330	140
Steam	600	450	450
Factory expenses	1,320	660	220

Cost of casks Rs. 7,500.

300 tons of crude oil were produced.

250 tons of oil were produced from the refining process.

248 tons of refined oil were finished for delivery.

Copra sacks sold Rs. 400.

175 tons of copra residue were sold for Rs. 11,000.

Loss in weight in crushing 25 tons.

45 tons of by-products obtained from refining process realised Rs. 6,750. The entire output of finished oil was sold for Rs. 2,50,000.

You are required to show the account in respect of each of the following stages of manufacture for the purpose of arriving at the cost per ton of each process and the total cost per ton of finished oil:

(*a*) Copra crushing process, (*b*) Refining process, (*c*) Finished process including casking.

Solution:

COPRA CRUSHING PROCESS

	Qty.	*Amt.*		*Qty.*	*Amt.*
To copra used	500	2,00,000	By sale of sacks	–	400
To labour		2,500	By sales of copra residue	175	11,000
To electric power		600	By loss in weight	25	–

To sundry materials		100	By refining process a/c	300	1,94,000
To repairs to machinery		280			
To steam		600			
To factory expenses		1,320			
	500	2,05,400		500	2,05,400

REFINING PROCESS

To crushing process a/c	300	1,94,000	By sale of by-product	45	6,750
To labour		1,000	By loss in weight	5	–
To electric power		360	By finished process	250	1,92,050
To sundry materials		2,000			
To repairs to machinery		330			
To steam		450			
To factory expenses		660			
	300	1,98,800		300	1,98,800

FINISHING PROCESS

To refining process	250	1,92,050	By loss in weight	2	–
To labour		1,500	By finished stock a/c	248	1,94,600
To power		240			
To repair to machinery		140			
To steam		450			
To factory expenses		220			
	250	1,95,600		250	1,94,600

FINISHING STOCK ACCOUNT

To finishing process a/c	248	1,94,600	By sales	248	2,50,000
To cost of casks		7,500			
To profit & loss a/c		47,900			
	248	2,50,000		248	2,50,000

INTER-PROCESS PROFIT

In some process costing industries, the management may decide not to transfer the output of a process to other process at cost price but at market price. The market price may relate to current market price prevailing in the market or determined by fixing a margin of profit. The market price is also known as transfer price. The ICMA terminology defines transfer price as 'a price related to goods or other benefits transferred from one process or department to another or from one member of a group to another. The extent to which costs and profit are recovered by the price is a matter of policy. The difference between the transfer price and the cost price is known as inter-process profit.

The output of one process is transferred to a subsequent process at transfer price with two objects. Firstly, to know efficiency of one company's production process with other company's production process. This is possible by comparing the cost of production with the sale price of similar products in the market. Secondly, to enable each process to achieve its own economy and efficiency but not to depend upon the efficiency of preceding process. This

CHAPTER 4

is possible by way of transferring the output of transferor process to the transferee process at market price. The input of every process is charged at current market price. Hence, no process will derive any benefit of efficiency or penalised by the variation in the cost. This facilitates timely action to be taken on those processes where true profit is inadequate.

Advantages of Transferring Output at Market Price

1. Comparison of cost of output with market price at each and every process helps in evaluating efficiency or otherwise of the process.
2. It helps in make or buy decision. When the cost of production is higher as compared to market price, it can well be bought rather than to manufacture.
3. The true profit or loss of each process can be ascertained and appropriate action can be taken if profit of any process is inefficient.

Disadvantages

1. It is necessary to adjust closing stock value so as to get cost price of stock. This is because closing stock is valued in the balance sheet at cost price.
2. If the adjustment is not effected in the closing stock, such valuation is not accepted by auditor and tax authorities.
3. The method involves additional clerical work by way of calculating transfer price and subsequently ascertaining cost price of stock.

Problem 24. A product passes through three process *A*, *B* and *C*. The output of process *A* and process *B* is charged to next process at a price calculated to give a profit of $16\frac{2}{3}$% on transfer price while the output process *C* is charged to the finished stock account at a profit of $13\frac{1}{3}$% on the transfer price.

From the following particulars, prepare the process cost accounts and calculate the amount of reserve that should be made in respect of the stock in hand:

	Process A	*Process B*	*Process C*
	Rs.	*Rs.*	*Rs.*
Material & labour	7,000	2,800	4,800
Closing stock	2,000	2,800	2,000

There was no stock in hand at the beginning of the period and the question of oncost has been ignored. The closing stocks are valued at prime cost in each process.

(University of Kerala, M.Com., May 1992)

Solution:

PROCESS 'A' ACCOUNT

	Total	*Cost*	*Profit*		*Total*	*Cost*	*Profit*
To materials and labour	7,000	7,000	–	By process B a/c	6,000	5,000	1,000
Less: Closing stock	2,000	2,000	–				
Prime cost	5,000	5,000	–				
To profit	1,000	–	1,000				
@ $16\frac{2}{3}$ on transfer price							
	6,000	5,000	1,000		6,000	5,000	1,000

PROCESS 'B' ACCOUNT

To process *A* a/c	6,000	5,000	1,000	By process C a/c	7,200	5,318	1,882
To materials & wages	2,800	2,800	–				
	8,800	7,800	1,000				
Less: Closing stock	2,800	2,482	318				
Prime cost	6,000	5,318	682				
To profit @ $16\frac{2}{3}$ on Transfer price	1,200	–	1,200				
	7,200	5,318	1,882		7,200	5,318	1,882

PROCESS 'C' ACCOUNT

To process *B* a/c	7,200	5,318	1,882	By finished stock a/c	11,538	8432	3,106
To materials & labour	4,800	4,800	–				
	12,000	10,118	1,882				
Less: Closing stock	2,000	1,686	314				
Prime cost	10,000	8,432	1,568				
Profit @ $13\frac{1}{3}$% on transfer price	1,538	–	1,538				
	11,538	8,432	3,106		11,538	8,432	3,106

Working Notes:

1. Calculation of profit included in the closing stock

Process *A* = NIL

Process *B* $\frac{\text{Cost}}{\text{Total}} \times$ Closing stock

$\frac{7,800}{8,800} \times 2,800 = 2,482$

Process *C* $\frac{10,118}{12,000} \times 2,000 = 1,686$

2. Calculation of actual realised profit

Process	*Apparent profit*	–	*Unrealised profit in closing stock*	=	*Actual profit*
A	1,000	–	Nil	=	1,000
B	1,200	–	318	=	882
C	1,538	–	314	=	1,224

Problem 25. From the following information given by a manufacturing company which manufactures a product. You are required to prepare process accounts.

	Process I	*Process II*	*Process III*
	Rs.	*Rs.*	*Rs.*
Direct material	30,000	7,500	7,500
Direct wages	22,500	15,000	15,000
Closing stock	7,500	8,750	21,300

CHAPTER 4

Finished goods sold for Rs. 1,30,000. Value of closing finished stock Rs. 5,612. It is the policy of the company to charge 20% on transfer price or 25% on cost price while transferring the goods from process *I* to process *II* and 20% on cost price from process *II* to process *III* and from process *III* to the finished stock.

(Madras University, M.Com., April 1992)

Solution:

PROCESS I ACCOUNT

	Total	*Cost*	*Profit*		*Total*	*Cost*	*Profit*
To direct materials	30,000	30,000	–	By process *II* a/c	56,250	45,000	11,250
To direct wages	22,500	22,500	–				
	52,500	52,500	–				
Less: Closing stock	7,500	7,500	–				
	45,000	45,000	–				
Profit – 25% on cost price	11,250	–	11,250				
	56,250	45,000	11,250		56,250	45,000	11,250

PROCESS 'II' ACCOUNT

To process *I* a/c	56,250	45,000	11,250	By process *III* a/c	84,000	60,000	24,000
To direct material	7,500	7,500	–				
To direct wages	15,000	15,000	–				
	78,750	67,500	11,250				
Less: Closing stock	8,750	7,500	1,250				
Prime cost	70,000	60,000	10,000				
Profit – 20% on cost price	14,000	–	14,000				
	84,000	60,000	24,000		84,000	60,000	24,000

PROCESS 'III' ACCOUNT

To process *II* a/c	84,000	60,000	24,000	By finished stock a/c	1,02,240	66,018	26,122
To direct materials	7,500	7,500					
To direct wages	15,000	15,000					
	1,06,500	82,500	24,000				
Less: Closing stock	21,300	16,482	4,818				
Prime cost	85,200	66,018	19,182				
Profit – 20% on cost price	17,040	–	17,040				
	1,02,240	66,018	26,222		1,02,240	66,018	26,222

FINISHED STOCK ACCOUNT

To process *III* a/c	1,02,240	66,018	36,222	By sales	1,30,000	62,394	67,606
Less: Closing stock	5,612	3,624	1,988				
Cost of goods sold	96,628	62,394	34,234				
Profit	33,372		33,372				
	1,30,000	62,394	67,606		1,30,000	62,394	67,606

Working notes:

Process *I* $\frac{\text{Cost}}{\text{Total}} \times \text{Closing stock} = \text{NIL}$

Process *II* $\frac{67,500}{78,750} \times 8,750 = 7,500$

Process *III* $\frac{82,500}{1,06,500} \times 21,300 = 16,482$

Process *IV* $\frac{66,018}{1,02,240} \times 5,612 = 3,624$

Problem 26. Product B passes through three processes before it is completed and transferred to finished stock. The following data were available for the month of June 1993:

	Process 1 (Rs.)	*Process 2* (Rs.)	*Process 3* (Rs.)	*Finished stock* (Rs.)
Opening stock	5,000	8,000	10,000	20,000
Direct materials	40,000	12,000	15,000	–
Direct labour	35,000	40,000	35,000	–
Production overhead	20,000	24,000	20,000	–
Closing stock	10,000	4,000	15,000	30,000

Output of process 1 is transferred to process 2 at 25% on transfer price, from process 2 to process 3 at 20% on transfer price and from process 3 to finished stock at 10% on the transfer price. Stock in process have been valued at prime cost. Finished stock has been valued at the price at which it was received from process 3. Sales amounted to Rs. 4,00,000. Provisions for internal process profits as at June 1, 1993 were:

Included in process 2	Rs.	1,395
Included in process 3	Rs.	2,690
Included in finished stock	Rs.	6,534

Prepare process accounts.

Solution:

PROCESS 1 ACCOUNT

	Total	*Cost*	*Profit*		*Total*	*Cost*	*Profit*
To opening stock	5,000	5,000	–	By process 2 A/c	1,20,000	90,000	30,000
To direct materials	40,000	40,000	–				
To direct labour	35,000	35,000	–				
	80,000	80,000	–				
Less: Closing stock	10,000	10,000	–				
Prime cost	70,000	70,000	–				
To production overhead	20,000	20,000	–				
Process cost	90,000	90,000	–				
Profit 25% on transfer price	30,000	–	30,000				
	1,20,000	90,000	30,000		1,20,000	90,000	30,000

CHAPTER · 4

PROCESS 2 ACCOUNT

To opening stock	8,000	6,605	1,395	By process 3 a/c	2,50,000	1,69,303	80,697
To process I a/c	1,20,000	90,000	30,000				
To direct material	12,000	12,000	–				
To direct labour	40,000	40,000	–				
	1,80,000	1,48,605	31,395				
Less: Closing stock	4,000	3,302	698				
Prime cost	1,76,000	1,45,303	30,697				
Production overhead	24,000	24,000	–				
Process cost	2,00,000	1,69,303	30,697				
Profit @ 20% on transfer price	50,000	–	50,000				
	2,50,000	1,69,303	80,697		2,50,000	1,69,303	80,697

PROCESS 3 ACCOUNT

To opening stock	10,000	7,310	2,690	By finished stock a/c	3,50,000	2,35,648	1,14,352
To process 2 a/c	2,50,000	1,69,303	80,697				
To direct material	15,000	15,000	–				
To direct labour	35,000	35,000	–				
	3,10,000	2,26,613	83,387				
Less: Closing stock	15,000	10,965	4,035				
Prime cost	2,95,000	2,15,648	79,352				
Production overhead	20,000	20,000	–				
Process cost	3,15,000	2,35,648	79,352				
Profit 10% on transfer price	35,000	–	35,000				
	3,50,000	2,35,648	1,14,352		3,50,000	2,35,648	1,14,352

FINISHED STOCK ACCOUNT

To opening stock	20,000	13,466	6,534	By sales	4,00,000	2,28,916	1,71,084
To process 3 a/c	3,50,000	2,35,648	1,14,352				
	3,70,000	2,49,114	1,20,886				
Less: Closing stock	30,000	20,198	9,802				
Cost of goods sold	3,40,000	2,28,916	1,11,084				
Gross profit	60,000	–	60,000				
	4,00,000	2,28,916	1,71,084		4,00,000	2,28,916	1,71,084

	Apparent profit	+	*Unrealised profit in opening stock*	–	*Unrealised profit in closing stock*	=	*Actual profit*
Process 1	30,000	+	NIL	–	NIL	=	30,000
Process 2	50,000	+	1,395	–	698	=	50,697
Process 3	35,000	+	2,690	–	4,035	=	33,655
Finished stock	60,000	+	6,534	–	9,802	=	56,732

Problem 27. A Ltd., produces a product which passes through two processes it is completed and transferred to finished stock. The following data relate to Dec. 1992:

	Process I	*Process II*	*Process III*
	Rs.	*Rs.*	*Rs.*
Opening stock	7,500	9,000	22,500
Direct material	15,000	15,750	
Direct wages	11,200	11,250	
Factory overhead	10,500	4,500	
Closing stock	3,700	4,500	11,250
Inter-process profit included in opening stock	–	1,500	8,250

Output of process *I* is transferred to process *II* at 25% profit on the transfer price and from process *II* to finished stock at 25% on cost price. Stock in process are valued at prime cost. Finished stock is valued at the price at which it is received from process *II*. Sales during the period are Rs. 1,40,000. Prepare process account and finished stock account showing the profit element at each stage.

Solution:

PROCESS I ACCOUNT

	Total	*Cost*	*Profit*		*Total*	*Cost*	*Profit*
To opening stock	7,500	7,500	–	By process *II* A/c	54,000	40,500	13,500
To materials	15,000	15,000	–				
To labour	11,200	11,200	–				
	33,700	33.700	–				
Less: Closing stock	3,700	3,700	–				
Prime cost	30,000	30,000	–				
Overhead Process cost	10,500	10,500	–				
	40,500	40,500	–				
Profit 25% on transfer price	13,500	–	13,500				
	54,000	40,500	13,500		54,000	40,500	13,500

PROCESS II ACCOUNT

	Total	*Cost*	*Profit*		*Total*	*Cost*	*Profit*
To opening stock	9,000	7,500	1,500	By finished stock	1,12,500	75,750	36,750
To process *I* a/c	54,000	40,500	13,500				
To material	15,750	15,750	–				
To labour	11,250	11,250	–				
	90,000	75,000	15,000				
Less: Closing stock	4,500	3,750	750				
Prime cost	85,500	71,250	14,250				
Overhead	4,500	4,500	–				
	90,000	75,750	14,250				
Profit 25% on 90,000	22,500	–	22,500				
	1,12,500	75,750	36,750		1,12,500	75,750	36,750

FINISHED STOCK ACCOUNT

To opening stock	22,500	14,250	8,250	By sales	1,40,000	82,500	57,500
To process *II* a/c	1,12,500	75,750	36,750				
	1,35,000	90,000	45,000				
Less: Closing stock	11,250	7,500	3,750				
	1,23,750	82,500	41,250				
Profit	16,250	–	16,250				
	1,40,000	82,500	57,500		1,40,000	82,500	57,500

Working Note: Computation of actual profit

	Apparent profit	+	*Unrealised profit in opening stock*	–	*Unrealised profit in closing stock*	=	*Actual profit*
Process *I*	13,500	+	*NIL*	–	*NIL*	=	13,500
Process *II*	22,500	+	1,500	–	750	=	23,250
Finished stock	16,250	+	8,250	–	3,750	=	20,750
							57,500

EQUIVALENT PRODUCTION

The continuous nature of the production process implies that in many cases there always exists opening and closing work-in-progress. This poses a problem in accounting of work-in-progress. The process costing method is based on the 'homogeneity approach'. In other words each unit completed is supposed to have received the same treatment compared to all other units. This means each completed unit contains the same amount of materials, wages and overheads. All units at the same stage of completion are supposed to contain the same amount of each element of cost. In addition to the above assumption, it is also further assumed that the elements of costs are evenly incurred. These two notions provide key to the treatment of the work-in-progress which invariably exists in process costing.

To illustrate from cost point of view, 100 units of opening work-in-progress which are 10% complete are the exact equivalent of 10 units of opening work-in-progress which are 100% complete. In either case, 100 units, 10% complete or 10 units, 100% complete, the number of 'equivalent units' in opening work-in-progress is 10. Similarly, in the absence of opening work-in-progress, suppose 200 units are started during a period and all 200 units are 60% completed at the end of the period, then the equivalent production for the period is 120 units. That is to say, the expected cost for the period equals the cost that would have been expected to incur if 120 units had been started and completed in the period.

It should be emphasised that a separate equivalent production figure must be calculated for each element of cost which varies in its own way. Cost that is applied in an identical manner in the production process can be grouped and a single equivalent production unit figure can be calculated. For instance, labour cost and factory overheads are often incurred at the same rate and therefore are grouped together. This is not true for materials as they are added at discrete point in the production process. But in some processes they may be added at more than one point. For instance, material *X* may be added at the beginning of the process and material *Y* may be added at the midpoint. Thus, any unit of production that is started will be complete as to material *X* but any unit that is less than 50% complete will not have material Y and any unit that is more than 50% complete will be complete as to material Y. So, the number of equivalent production units will be different for material X and Y.

Methods of Computing Equivalent Production

There are two methods of computing equivalent production. They are (1) FIFO method and (2) Average method.

(1) **FIFO method:** Under this method, costs are shown separately for (*a*) Opening work-in-progress carried over and remaining to be completed and (*b*) Units introduced and completed during the current period. Under this method, the units transferred to next process or finished stock are issued on FIFO principle. In other words, the units transferred contains the units from opening inventory as well as the units introduced and completed during the current period. Equivalent units and costs of opening inventory is distinguished from units introduced and completed as if they are different batch of units. Cost per unit under each process is obtained by applying the following formula:

$$\frac{\text{Cost added in each process during the current period}}{\text{Equivalent production units}}$$

The cost added in each process during the current period is then apportioned between opening inventory, units completed and transferred to next process (or finished stock) and the closing inventory. The main object of FIFO method is to value the inventory at current cost.

During inflationary period, cost of production will be less, profit will be more and closing stock value will reflect current market price. Reverse is the trend under deflationary period.

(2) **Average method:** Under this method, no distinction is made between opening work-in-progress and current production units. The cost of opening work-in-progress is added to the cost incurred during current period and the total process cost is divided by the total equivalent units to arrive at an average cost of equivalent units.

Procedure Involved in Calculating Equivalent Units

1. Express the opening stock of work-in-progress in terms of equivalent units. This is done by applying the percentage of work yet to be done. For example, if the opening work-in-progress is 100 units, which is 60% complete, then the equivalent of this will be $100 \times \dfrac{40}{100} = 40$ units.
2. Find out the number of units started and finished during the year. This is found out by deducting the units in closing work-in-progress from the number of new units put into the process. Add these units to the opening stock of work-in-progress as was calculated in the first step.
3. Add to the above, the equivalent completed units of closing work-in-progress. This is calculated by applying the percentage of work already done on unfinished units at the end of the period. For example, if 100 units of closing work-in-progress is 60% then the equivalent completed units of closing work-in-progress is $100 \times \dfrac{60}{100} = 60$ units.
4. Find out the net process cost after considering additional cost incurred, normal losses, etc. in respect of each and every element of cost.

CHAPTER 4

5. Obtain the cost per unit of equivalent production. This is obtained by dividing cost of each element by the number of equivalent units.
6. Ascertain the value of every form of output, such as abnormal loss/gain, finished product and work-in-progress.

In applying the above steps we will prepare three statements, they are (1) Statement of equivalent production, step number 1 to 3 fall under this statement (2) Statement of cost, step number 4 and 5 fall under this statement (3) Statement of evaluation, step number 6 fall under this statement.

Problem 28. *(Equivalent production without losses):* Shakthi Pvt. Ltd. is a manufacturer of processed goods and the results in process 2 for April 2009 were as follows:

Opening stock	NIL
Materials input from process I	4,000 units
Cost of input:	
Materials from process I	6,000
Labour cost	1,080
Overheads	1,720

Output is transferred into the next process, *i.e.*, process 3

Closing work-in-progress amounted to 800 units complete as to:

Process 1 material	100%
Labour cost	50%
Overheads	30%

Prepare the account for process 2 and a statement of equivalent unit, a statement of cost and statement of evaluation.

Solution:

STATEMENT OF EQUIVALENT UNIT

Input qty.	*Particulars of output*	*Output qty.*	*Equivalent units*					
			Process I materials		*Labour cost*		*Overhead*	
			%	*Units*	*%*	*Units*	*%*	*Units*
4,000	Finished stock	3,200	100%	3,200	100%	3,200	100%	3,200
	Closing work-in-progress	800	100%	800	50%	400	30%	240
4,000		4,000		4,000		36,00		3,440

STATEMENT OF COST PER EQUIVALENT UNIT

Materials cost	6,000	÷	4,000	=	1.50
Labour cost	1,080	÷	3,600	=	0.30
Overhead	1,720	÷	3,440	=	0.50

STATEMENT OF EVALUATION

Finished stock:

Materials cost	3,200 × Rs. 1.50	=	Rs. 4,800
Labour cost	3,200 × Rs. 0.30	=	Rs. 960
Overhead	3,200 × Rs. 0.50	=	Rs. 1,600
		=	Rs. 7,360

Closing work-in-progress:

Materials	800 × Rs. 1.50	=	Rs.	1,200
Labour cost	400 × Re. 0.30	=	Rs.	120
Overhead	240 × Re. 0.50	=	Rs.	120
		=	Rs.	1,440

PROCESS ACCOUNT

	Qty.	Amt.		Qty.	Amt.
To process 1 material	4,000	6,000	By process 3 a/c		
To labour		1,080	Finished output	3,200	7,360
To overhead		1,720	By closing stock	800	1,440
	4,000	8,800		4,000	8,800

Problem 29. *(Equivalent production with normal loss):* The following particulars are extracted from the books of East-West Co. Ltd. as on 31.3.2009.

	Units	Cost (Rs.)
Materials introduced in the process I	20,000	1,99,800
Wages		1,37,600
Overheads 100% of wages		
Units completed and transferred to process 2	16,000	
Closing work-in-progress	3,000	

Degree of completion:

60% as to materials

40% as to labour and overhead

Normal loss 5% of materials introduced which realises Rs. 4 per unit

Prepare (1) Statement of equivalent units (2) Statement of cost per unit.

Solution:

STATEMENT OF EQUIVALENT UNIT

Input	Particulars of output	Units	Equivalent units			
			Materials		Labour and overheads	
			%	Units	%	Units
20,000	Normal loss	1,000	–	–	–	–
	Finished goods	16,000	100%	16,000	100%	16,000
	Closing WIP	3,000	60%	1,800	40%	1,200
20,000		20,000		17,800		17,200

STATEMENT OF COST PER UNIT

Materials	1,99,800				
Less: Scrap value of normal loss	4,000				
	1,95,800	÷	17,800	=	11.00
Wages	1,37,600	÷	17,200	=	8,00
Overheads	1,37,600	=	17,200	=	8.00

STATEMENT OF EVALUATION

Finished goods:

Materials	16,000 × 11	=	1,76,000	
Wages	16,000 × 8	=	1,28,000	
Overhead	16,000 × 8	=	1,28,000	
				4,32,000

Closing work-in-progress:

Materials	1,800 × 11	=	19,800	
Wages	1,200 × 8	=	9,600	
Overhead	1,200 × 8	=	9,600	39,000
				4,71,000

PROCESS I ACCOUNT

	Qty.	Amt.		Qty.	Amt.
To materials	20,000	1,99,800	By normal loss	1,000	4,000
To wages		1,37,600	By process 2 a/c	16,000	4,32,000
To overheads		1,37,600	By closing WIP	3,000	39,000
	20,000	4,75,000		20,000	4,75,000

Problem 30. During January, 1,000 units costing Rs. 6,000 were introduced in process *I*. There was no work-in-progress at the beginning. At the end of January, 600 units were transferred to process *II*; 250 units were incomplete and 150 units had been scrapped. The normal process loss was 10% of input. It was estimated that incompleted units had reached the following stages:

Material 80%, Labour 60%, Overhead 60%, Direct materials introduced Rs. 2,700, Direct wages Rs. 3,200, Production overhead Rs. 1,600.

Value of scrap is Rs. 2 each. The units scrapped had passed through the process and is 100% complete as regards material, labour and overhead.

You are required to

(a) Prepare a statement of equivalent production.
(b) Statement of cost for each element.
(c) Statement of apportionment of process cost.
(d) Process *I* account.
(e) Abnormal loss account. *(Madras University, B.Com., March 1992)*

Solution:

STATEMENT OF EQUIVALENT PRODUCTION

Input	*Particulars of output*	*Output*	*Equivalent units*					
			Materials		*Labour*		*Overhead*	
			%	*Units*	*%*	*Units*	*%*	*Units*
1,000	Normal loss	100	–	–	–	–	–	–
	Abnormal loss	50	100%	50	100%	50	100%	50
	Finished goods	600	100%	600	100%	600	100%	600
	Closing WIP	250	80%	200	60%	150	60%	150
1,000		1,000		850		800		800

STATEMENT OF COST PER UNIT

Materials		6,000				
Add:	Further materials	2,700				
		8,700				
Less:	Normal loss	200	8,500 ÷ 850	=	Rs.	10
	Labour		32,00 ÷ 800	=	Rs.	4
	Overhead		1,600 ÷ 800	=	Rs.	2

STATEMENT OF EVALUATION

Abnormal loss:				
Materials	50 × 10	=	500	
Labour	50 × 4	=	200	
Overhead	50 × 2	=	100	800
Finished goods:				
Materials	600 × 10	=	6,000	
Labour	600 × 4	=	2,400	
Overhead	600 × 2	=	1,200	9,600
Closing-work-in-progress:				
Materials	200 × 10	=	2,000	
Labour	150 × 4	=	600	
Overhead	150 × 2	=	300	2,900
				13,200

PROCESS 'I' ACCOUNT

	Qty.	*Amt.*		*Qty.*	*Amt.*
To units introduced	1,000	6,000	By normal loss	100	200
To materials		2,700	By output transferred to process '2' a/c	600	9,600
To labour		3,200	By abnormal loss	50	800
To overhead		1,600	By closing WIP	250	2,900
	1,000	13,500		1,000	13,500

ABNORMAL WASTAGE ACCOUNT

To process 'I' a/c	50	800	By sale of abnormal wastage	50	100
			By P & L a/c		700
	50	800		50	800

Problem 31. *(Equivalent production relating to two process):* Attractive and Company operates a department producing a component which passes through two processes. During November 1990, materials for 40,000 components were put into process. There was no opening process stock 30,000 units were finished and passed to next process. Those not passed forward were calculated to be one-half finished as regards wages and overheads. The costs incurred were as follows:

	Rs.
Direct material	10,000
Factory overhead	12,000
Direct wages	8,000

Of those passed to the second process, 28,000 units were completed and passed to finished stores. 200 units were scrapped, which was not abnormal. 1,800 units remained unfinished in process, one-quarter finished as regards wages and overheads. No further process material costs occur after introduction at the first process until the end of the second process, when protective packing is applied to the completed components. The process and packing costs incurred at the end of the second process were:

	Rs.
Direct material (packing)	4,000
Factory overhead	4,500
Direct wages	3,500

Prepare a cost analysis statement for November 1990 accounting for total costs incurred analysed into elements of costs for process, covering finished and part-finished items and process accounts.

(C.S. Inter, June 1991)

Solution:

STATEMENT OF EQUIVALENT PRODUCTION (COST ANALYSIS)

Input	*Particulars of output*	*Output*	*Equivalent units*					
			Materials		*Labour*		*Overhead*	
			Units	*%*	*Units*	*%*	*Units*	*%*
40,000	Finished goods	30,000	30,000	100%	30,000	100%	30,000	100%
	Closing WIP	10,000	10,000	100%	5,000	50%	5,000	50%
40,000		40,000	40,000		35,000		35,000	

STATEMENT OF COST PER UNIT OF EQUIVALENT PRODUCTION

Materials	10,000 ÷ 40,000	=	0.25
Labour	8,000 ÷ 35,000	=	0.2286
Overhead	12,000 ÷ 35,000	=	0.3429

STATEMENT OF EVALUATION

Finished goods:				
Materials	30,000 × 0.25	=	7,500	
Labour	30,000 × 0.2286	=	6,857	
Overhead	30,000 × 0.3429	=	10,286	24,643
Closing stock WIP				
Material	10,000 × 0.25	=	2,500	
Labour	5,000 × 0.2286	=	1,143	
Overhead	5,000 × 0.3429	=	1,714	5,357
				30,000

PROCESS 'I' ACCOUNT

	Qty.	Amt.		Qty.	Amt.
To direct materials	40,000	10,000	By process *II* A/c	30,000	24,643
To direct wages		8,000	By closing WIP	10,000	5,357
To factory overhead		12,000			
	40,000	30,000		40,000	30,000

PROCESS II ACCOUNT

STATEMENT OF EQUIVALENT PRODUCTION

Input	*Particulars of output*	*Output*	*Equivalent units*					
			Materials		*Labour*		*Overhead*	
			Qty.	*%*	*Qty.*	*%*	*Qty.*	*%*
30,000	Finished goods	28,000	28,000	100%	28,000	100%	28,000	100%
	Normal loss	200	–	–	–	–	–	–
	Closing WIP	1,800	1,800	100%	450	25%	450	25%
30,000		30,000	29,800		28,450		28,450	

STATEMENT OF COST PER UNIT OF EQUIVALENT PRODUCTION

Materials					
Cost of materials	4,000				
From process I	24,643				
	28,643	÷	29,800	=	0.9611
Wages	3,500	÷	28,450	=	0.123
Overhead	4,500	÷	28,450	=	0.158

STATEMENT OF EVALUATION

Finished goods:

Materials	28,000 × 0.9611	=	26,912	
Labour	28,000 × 0.123	=	3,449	
Overhead	28,000 × 0.158	=	4,424	
			34,780	34,785

Closing WIP

Material			1,800 × 0.9611	=	1,730
Labour	450 × 0.123	=	56		
Overhead	450 × 0.158	=	72		
		=	1858		

CHAPTER 4

PROCESS II ACCOUNT

	Qty.	Amt.		Qty.	Amt.
To process I a/c	30,000	24,643	By normal loss	200	–
To direct materials (primary packing materials)		4,000	By finished goods	28,000	34,785
			By closing WIP	1,800	1,858
To direct wages		3,500			
To factory overhead		4,500			
	30,000	36,643		30,000	36,643

Problem 32. *(Equivalent production with abnormal gain):* The books of Sunflower Oil Co. Pvt. Ltd. revealed the following:

Units introduced into process *A*	6000 units
Cost incurred in process *A* was as follows:	
Direct materials	Rs. 55,200
Direct labour cost	Rs. 30,600
Factory overhead	Rs. 14,400
Normal loss in process *A*	10% of the input
Output obtained and transferred to process *B*	4,000 units
Units scrapped in process *A*	500 units
Closing work-in-progress in process *A*	1,500 units
Degree of completion:	
Materials	100%
Direct labour	80%
Factory overhead	60%
Realisable value of scrapped units	Rs. 2 per unit

Compute (*a*) Equivalent production, (*b*) Cost per unit of equivalent production, (*c*) Process a/c, (*d*) Normal loss a/c, (*e*) Abnormal gain a/c. *(Sri Sathya Sai University, B.Com., February 2001)*

Solution:

STATEMENT OF EQUIVALENT PRODUCTION (COST ANALYSIS)

Input	Particulars of output	Output	Equivalent units					
			Materials		Labour		Overhead	
			%	Units	%	Units	%	Units
6,000	Normal loss	600	–	–	–	–	–	–
	Finished goods	4,000	100%	4,000	100%	4,000	100%	4,000
	Closing WIP	1,500	100%	1,500	80%	1,200	60%	900
6,000		6,100		5,500		5,200		4,900
	Less: Abnormal gain	100	100%	100	100%	100	100%	100
6,000		6,000		5,400		5,100		4,800

STATEMENT OF COST PER UNIT OF EQUIVALENT PRODUCTION

Materials	55,200				
Less: Sale of normal units	1,200				
	54,000	÷	5,400	=	Rs. 10

Labour	30,600	÷	5,100	= Rs. 6
Overhead	14,400	÷	4,800	= Rs. 3

STATEMENT OF EVALUATION

Finished goods:				
Materials	4,000 × Rs. 10	=	Rs. 40,000	
Labour	4,000 × Rs. 6	=	Rs. 24,000	
Overhead	4,000 × Rs. 3	=	Rs. 12,000	76,000
Abnormal gain:				
Materials	100 × Rs. 10	=	Rs. 1,000	
Labour	100 × Rs. 6	=	Rs. 600	
Overhead	100 × Rs. 3	=	Rs. 300	1,900
Closing work-in-progress:				
Materials	1,500 × Rs. 10	=	Rs. 15,000	
Labour	1,200 × Rs. 6	=	Rs. 7,200	
Overhead	900 × Rs. 3	=	Rs. 2,700	24,900

PROCESS A ACCOUNT

	Qty.	*Amt.*		*Qty.*	*Amt.*
To materials	6,000	55,200	By normal loss	600	1,200
To labour		30,600	By finished goods	4,000	76,000
To overhead		14,400	By closing WIP	1,500	24,900
To abnormal gain	100	1,900			
	6,100	1,02,100		6,100	1,02,100

NORMAL LOSS ACCOUNT

To process *A* a/c	600	1,200	By sales	500	1,000
			By abnormal gain	100	200
	600	1,200		600	1,200

ABNORMAL GAIN ACCOUNT

To normal loss a/c	100	200	By process A a/c	100	1,900
To P & L a/c		1,700			
	100	1,900		100	1,900

Problem 33. AB Ltd. is engaged in the process engineering industry. During the month of April 1999; 2000 units were introduced in process '*x*'. The normal loss is estimated at 5% of input. At the end of the month 1,400 units had been produced and transferred to process Y. 460 were incomplete units and 140 units had to be scrapped at the end of the process. The incomplete units reached the following degree of completion:

Materials	:	75%
Labour	:	50%
Overhead	:	50%

CHAPTER 4

Following are the further details regarding process x

Cost of 2,000 units introduced	Rs. 58,000
Additional materials consumed	Rs. 14,400
Direct labour	Rs. 33,400
Allocated overheads	Rs. 16,700

Note: The scrapped units fetched Rs. 10 each, Required:

(1) Statement of equivalent production
(2) Statement of cost
(3) Statement of evaluation
(4) Process 'X' a/c

(ICWA, Inter, June 1999)

Solution:

STATEMENT OF EQUIVALENT PRODUCTION

Input	*Output*	*Units*	*Equivalent Production*					
			Materials		*Labour*		*Overhead*	
			% of completion	*Units*	*% of completion*	*Units*	*% of completion*	*Units*
2,000	Normal loss	1,000	–	–	–	–	–	–
	Abnormal loss	40	100%	40	100%	40	100%	40
	Finished goods	1,400	100%	1,400	100%	1,400	100%	1,400
	Closing WIP	460	75%	345	50%	230	50%	230
2,000		2,000		1,785		1,670		1,670

STATEMENT OF COST

Cost of materials introduced		58,000	
Additional materials		14,400	
		72,400	
Less: Sale of scrapped units		1,000	
		71,400	
Cost per unit	=	71,400 / 1.785	= 40
Labour cost per unit	=	33,400 / 1,670	= 20
Overhead per unit	=	16,700 / 1,670	= 10
			70

STATEMENT OF EVALUATION

Finished goods:	1,400 × 70	= 98,000
Abnormal loss	40 × 70	= 2,800
Closing WIP		
Materials	345 × 40	= 13,800
Labour	230 × 20	= 4,600
Overhead	230 × 10	= 2,300
		20,700

PROCESS 'X' ACCOUNT

	Qty.	Amt.		Qty.	Amt.
To units introduced	2,000	58,000	By normal loss	100	1,000
To materials		14,400	By abnormal loss	40	2,800
To labour		33,400	By process Y a/c	1,400	98,000
To overheads		16,700	By closing WIP	460	20,700
	2,000	1,22,500		2,000	1,22,500

Problem 34. *(When opening work-in-progress is given):* Khomainee & Company uses process cost system. Production and inventory data for February 1991 are given below:

Units put in process : 60,000
Units completed in February : 70,000

February 1, work-in-progress was composed of 40,000 units with 30% materials content and 40% of the conversion cost complete.

February 28, work-in-progress was composed of 30,000 units with 60% of the materials content and 70% of the conversion cost complete.

Compute the equivalent units. *(Sri Sathya Sai University, B.Com., March 1991)*

Solution:

STATEMENT OF EQUIVALENT PRODUCTION

Input	Particulars of output	Output	Equivalent units					
			Materials		Labour		Overhead	
			%	Units	%	Units	%	Units
40,000	Opening WIP	40,000	70%	12,000	60%	24,000	60%	24,000
60,000	Finished goods	30,000	100%	30,000	100%	30,000	100%	30,000
	Closing WIP	30,000	60%	18,000	70%	21,000	70%	21,000
1,00,000		1,00,000		60,000		75,000		75,000

Problem 35. From the following particulars extracted from the books of Unique Ltd. for the month of March 1998. You are required to *(i)* Prepare the statement of equivalent units of production, *(ii)* Compute the cost per equivalent unit for each cost element, *(iii)* Compute the cost of finished output and closing work-in-progress and *(iv)* Prepare process account. Assume that FIFO method is used by the company:

		Units
(i)	Opening stock as on 1st March 1998 @ Rs. 4 per unit	200
	Degree of completion	
	— Materials 100% and	
	— Labour and overhead 40%	
(ii)	Introduced during March 1998	1,050
(iii)	Transferred to next process	1,100
(iv)	Closing stock as on 31st March 1998	150
	Degree of completion	
	- Materials 100% and	
	- Labour and overhead 70%	

Other relevant information regarding the process account is as under:

		Rs.
Materials cost		3,150
Labour cost		4,500
Production overheads		2,250
	Total	9,900

(C.S. Inter, Dec. 1998)

Solution:

STATEMENT OF EQUIVALENT PRODUCTION

Input	*Output*	*Units*	*Equivalent units*					
			Materials		*Labour*		*Overhead*	
			%	*Units*	*%*	*Units*	*%*	*Units*
200	Opening stock of WIP	200	--	–	60%	120	60%	120
1050	Finished goods	900	100%	900	100%	900	100%	900
	Closing stock of WIP	150	100%	150	70%	105	70%	105
1250		1,250		1,050		1,125		1,125

STATEMENT SHOWING COST PER UNIT

Materials cost per unit $= \dfrac{3,150}{1,050} = 3$

Wages per unit $= \dfrac{4,500}{1,125} = 4$

Overheads per unit $= \dfrac{2,250}{1,125} = 2$

STATEMENT OF EVALUATION

(1) Opening stock of WIP:		
Materials	Nil	Nil
Labour	120 × 4	480
Overheads	120 × 2	240
		720
(2) Finished goods:		
Materials	900 × 3	2,700
Labour	900 × 4	3,600
Overheads	900 × 2	1,800
		8,100
(3) Closing WIP:		
Materials	150 × 3	450
Labour	105 × 4	420
Overheads	105 × 2	210
		1,080

PROCESS ACCOUNT

	Qty.	Amt.		Qty.	Amt.
To opening stock of WIP	200	800	By output transferred	1,100	9,620
To materials	1,050	3,150	to next process		
To labour		4,500	(800 + 720 + 8100)		
To overheads		2,250	By closing WIP	150	1,080
	1,250	10,700		1,250	10,700

Problem 36. Roy & Johnson (P) Ltd., gives the following particulars relating to process A in its plant for the month of December 1997:

Work-in-progress (opening balance) on 1.12.97				500 units
Cost:	Materials	Rs.	4,800	
	Labour		3,200	
	Overhead		6,400	
			14,400	
Units introduced during the month				19,500
Processing cost incurred during the month				
	Materials	Rs.	1,86,200	
	Labour		72,000	
	Overheads		1,06,400	
			3,64,600	
Output:	Units transferred to process *B*			18,200
	Units scrapped (completely processed)			1,400
	Work-in-process (closing balance)			400
	Degree of completion:			
	Materials	100%		
	Labour and overheads	50%		

Normal loss in processing is 5% of total input and normal scrapped units fetch Re. 1 each. Prepare the following statements for process A for December 1997:

(*a*) Statement of equivalent production
(*b*) Statement of cost
(*c*) Statement of evaluation
(*d*) Process 'A' a/c *(ICWA, Inter, June 1998)*

Solution: STATEMENT OF EQUIVALENT PRODUCTION

Input	*Output*	*Units*	*Equivalent units*			
			Materials		*Labour of Overhead*	
			%	*Units*	*%*	*Units*
500	Opening WIP					
19,500	Units introduced					
	Finished goods	18,200	100%	18,200	100%	18,200
	Normal loss 5% on 20,000	1,000	–	–	–	–
	Abnormal loss	400	100%	400	100%	400
	Closing WIP	400	100%	400	50%	200
20,000		20,000		19,000		18,800

CHAPTER 4

STATEMENT OF COST PER UNIT

Materials:	Opening balance		4,800	
	Materials consumed		1,86,200	
			1,91,000	
Less:	Sale of normal loss units 1,000 units @ 1		1,000	
			1,90,000	
Materials cost per unit		=	$\frac{1,90,000}{19,000}$	= 10
Labour:	Opening balance		3,200	
	Labour cost during the year		72,000	
			75,200	
Labour cost per unit			$\frac{75,200}{18,800}$	= 4
Overhead:				
	Opening balance		6,400	
	Cost during the month		1,06,400	
			1,12,800	
Overhead per unit		=	$\frac{1,12,800}{18,800}$	= 6
			Total	20

STATEMENT OF EVALUATION

(1)	Abnormal loss units- 400 units @ 20	=	8,000
(2)	Output or finished goods - 18,200 units @ 20	=	3,64,000
(3)	Closing work-in-progress:		
	Materials 400 units @ Rs. 10	=	4,000
	Labour & overheads 200 units @ Rs. 10	=	2,000
			6,000

PROCESS 'A' ACCOUNT

	Qty.	*Amt.*		*Qty.*	*Amt.*
To opening WIP	500	14,400	By normal loss	1,000	1,000
To units introduced	19,500	1,86,200	By abnormal loss	400	8,000
To labour		72,000	By process *B* a/c	18,200	3,64,000
To overhead		1,06,400	By closing WIP	400	6,000
	20,000	3,79,000		20,000	3,79,000

Problem 37. The following data relate to process Q

(i)	Opening work-in-process			4,000 units
	Degree of completion:			
	Materials	100%	Rs.	24,000
	Labour	60%	Rs.	14,400
	Overheads	60%	Rs.	7,200
(ii)	Received during the month of April 1998 from process P			
	40,000 units		Rs.	1,71,000
(iii)	Expenses incurred in process Q during the month			
	Materials		Rs.	79,000
	Labour		Rs.	1,38,230
	Overhead		Rs.	69,120
(iv)	Closing work-in-process			3,000 units
	Degree of completion:			
	Materials			100%
	Labour & overheads			50%
(v)	Units scrapped			4,000 units
	Degree of completion:			
	Materials			100%
	Labour & overheads			80%

(vi) Normal loss: 5% of current input

(vii) Spoiled goods realised Rs. 1.50 each on sale.

(viii) Completion units are transferred to warehouse.

Required:

(i) Equivalent units statement.

(ii) Statement of cost per equivalent unit and total costs.

(iii) Process Q account.

(iv) Any other account necessary. *(C.A. Inter, May 1998)*

Solution:

STATEMENT OF EQUIVALENT PRODUCTION

Input	*Output*	*Units*	*Equivalent units*					
			Materials		*Labour*		*Overhead*	
			%	*Units*	*%*	*Units*	*%*	*Units*
4,000	Opening WIP	4,000	–	–	40%	1,600	40%	1,600
40,000	Normal loss	2,000	–	–	–	–	–	–
	Abnormal loss	2,000	100%	2,000	80%	1,600	80%	1,600
	Finished goods	33,000	100%	33,000	100%	33,000	100%	33,000
	Closing WIP	3,000	100%	3,000	50%	1,500	50%	1,500
44,000		44,000		38,000		37,700		37,700

CHAPTER 4

STATEMENT OF COST

Materials from process 'P'		1,71,000		
Materials cost incurred in process Q	79,000			
Less: Sale of scrapped units	3,000	76,000		
		2,47,000		
Materials cost per unit		$\frac{2,47,000}{38,000}$	=	6.50
Labour cost per unit		$\frac{1,38,230}{37,700}$	=	3.66
Overhead per unit		$\frac{69,120}{37,700}$	=	1.84
				12.00

STATEMENT OF EVALUATION

OP WIP:		
Material		Nil
Labour	1,600 × 3.66	5,856
Overhead	1,600 × 1.84	2,944
		8,800
Finished goods:		
	33,000 × 12.00	3,96,000
Abnormal loss		
Materials	2,000 × 6.50	13,000
Labour	1,600 × 3.66	5,856
Overhead	1,600 × 1.84	2,944
		21,800
Closing WIP		
Materials	3,000 × 6.50	19,500
Labour	1,500 × 3.66	5,490
Overhead	1,500 × 1.84	2,760
		27,750

PROCESS 'Q' ACCOUNT

	Qty.	*Amt.*		*Qty*	*Amt.*
To op. WIP	4,000	45,600	By normal loss	2,000	3,000
To input qty.	40,000	1,71,000	By abnormal loss	2,000	21,800
To materials		79,000	By finished stock a/c	37,000	4,50,400
To labour		1,38,230	(45,600 + 8,800 + 3,96,000)		
To overhead		69,120	By cl WIP	3,000	27,750
	44,000	5,02,950		44,000	5,02,950

ABNORMAL LOSS A/C

To process Q a/c	2,000	21,800	By sales	2,000	3,000
			By P & L a/c (Balancing figure)	–	18,800
	2,000	21,800		2,000	21,800

Problem 38. A manufacturing company furnishes the following data in respect of process 3 for the month of April 1993:

Direct material added in process	776
Direct labour	386
Production overhead	768
Transfer from process 2	4,200 units valued at Rs. 1,560
Transfer to process 4	3,650 units
Stock at 1st April 1993	600 units valued at Rs. 390
Degree of completion:	
Material added in process	60%
Labour	50%
Overhead	40%
Stock at 30th April 1993	800 units
Degree of completion:	
Materials added in process	80%
Labour	70%
Overhead	60%
Units scrapped	350 units
Degree of completion:	
Materials added to process	100%
Labour	80%
Overhead	80%

Normal loss is 10% of output (opening stock + input of the process – closing stock). All units scrapped can be sold for Re. 0.10 per unit. You are required to prepare:

(a) A statement showing the cost per unit of production and the value of the output.

(b) Process '3' account.

(c) Abnormal loss or abnormal gain account.

Solution:

STATEMENT OF EQUIVALENT PRODUCTION

Input	*Particulars of output*	*Units*	*Material 1*		*Material 2*		*Labour*		*Overhead*	
			%	*Units*	*%*	*Units*	*%*	*Units*	*%*	*Units*
600	Op. stock	600	–	–	40%	240	50%	300	60%	360
4,200	Normal loss	400	–	–	–	–	–	–	–	–
	Finished goods	3,050	100%	3,050	100%	3,050	100%	3,050	100%	3,050
	Closing stock	800	100%	800	80%	640	70%	560	60%	480
4,800		4,850		3,850		3,690		3,610		3,530
	Less: Ab. gain	50	100%	50	100%	50	100%	50	100%	50
4,800		4,800		3,800		3,640		3,560		3,480

CHAPTER 4

STATEMENT OF COST PER EQUIVALENT UNIT

	Process 1 material	Process 2 material	Labour	Overhead
Cost incurred	1,560	776	386	768
Less: Sale proceeds of normal loss	40	–	–	–
Column (1)	1,520	776	386	768
Equivalent units (Column 2)	3,800	3880	3,860	3,840
Cost per equivalent unit (1 ÷ 2)	0.40	0.20	0.10	0.20

STATEMENT OF EVALUATION

	Process 1 material	Process 2 material	Labour	Overhead
Opening stock	–	48	30	72
Finished goods	1,220	610	305	610
Abnormal gain	(20)	(10)	(5)	(10)
Closing stock	320	128	56	96
	1,520	776	396	768

PROCESS 3 ACCOUNT

	Qty.	Amt.		Qty.	Amt.
To opening stock	600	390	By normal loss	400	40
To process '2' a/c	4,200	1,560	By process '4' a/c	3,650	3,285
To direct materials		776	(390 + 150 + 2,745)		
To direct labour		386	By closing WIP	800	600
To overhead		768			
To abnormal gain	50	45			
	4,850	3,925		4,850	3,925

ABNORMAL GAIN ACCOUNT

To normal loss a/c 50 units @ 10 P	50	5	By process '3' a/c	50	45
To P & L a/c	–	40			
	50	45		50	45

Problem 39. *(Equivalent production under weighted average cost method)*: XYZ Company produces an item which is manufactured in two consecutive processes. Information relating to process '2' during September 2009 is as follows:

Opening stock		800 units
Degree of completion		
Process '1' materials	100%	Cost Rs. 4,700
Added materials	40%	600
Conversion cost	30%	1,000
		6,300

During September 2009, 3,000 units were transferred from process '1' at a valuation of 18,100. Added materials cost Rs. 9,600 and conversion costs were, 11,800.

Closing stock on 30th September 2009 amounted to 1,000 units which were 100% complete with respect to process '1' materials and 60% completed with respect to added materials and conversion cost work was 40% complete.

XYZ Company uses a weighted average cost system for the valuation of output and closing stock. Prepare the process '2' a/c for September 2009.

Solution: Opening stock units count as a full equivalent unit of production when the weighted average cost system is applied. Closing stock equivalent units are assessed in the usualy way.

STATEMENT OF EQUIVALENT PRODUCTION

Input	*Particulars of output*	*Output*	*Equivalent units*					
			Material 1		*Material 2*		*Conversion cost*	
			%	*Units*	*%*	*Units*	*%*	*Units*
800	Opening stock	800	100%	800	100%	800	100%	800
3,000	Finished goods	2,000*	100%	2,000	100%	2,000	100%	2,000
	Closing stock	1,000	100%	1,000	60%	600	40%	400
3,800		3,800		3,800		3,400		3,200

* 3,000 units from process '1' minus closing stock of 1,000 units.

STATEMENT OF COST PER EQUIVALENT UNIT

The cost of opening stock is added to costs incurred in September 2009 and a cost per equivalent unit is then calculated.

	Material 1	*Material 2*	*Conversion cost*
Opening stock	4,700	600	1,000
Added in September 2009	18,100	9,600	11,800
Total cost	22,800	10,200	12,800
Equivalent units	3800 units	3400 units	3200 units

STATEMENT OF EVALUATION

	Material 1	*Material 2*	*Conversion Cost*	*Total*
Finished goods (2800 units)	16,800	8,400	11,200	36,400
Closing stock	6,000	1,800	1,600	9,640
				45,800

PROCESS 2 ACCOUNT

	Qty.	*Amt.*		*Qty.*	*Amt.*
To opening stock	800	6,300	By finished goods	2,800	36,400
To process I A/c	3,000	18,100	By closing stock	1,000	9,400
To added material	–	9,600			
To conversion cost	–	11,800			
	3,800	45,800		3,800	45,800

Problem 40. *(Equivalent Production):* During a month 40,000 units were introduced into process 'A'. The process costs were direct materials Rs. 3,02,000, direct wages Rs. 2,07,000, factory overheads 50% of direct wages.

The normal loss was estimated at 10% on input. At the end of the month 32,000 units have been produced and transferred to process B. 5,000 units had been scrapped (scrapped units had been completely processed and realised Rs. 5 per unit) and 3,000 units were incomplete and the stage of completion in respect of three units was estimated to be: Materials 75%, labour 50%: overheads 50%

Prepare: (*a*) Statement of equivalent product, (*b*) Cost per unit, (*c*) Statement of evaluation (*d*) Process 'A' a/c and (*e*) Abnormal loss account.

(Sri Sathya Sai University, B.Com. (Hons) October 2006)

Solution:

STATEMENT SHOWING EQUIVALENT PRODUCTION

Input	*Output*	*Units*	*Equivalent production*			
			Materials		*Labour and overhead*	
			% of completion	*Units*	*% of completion*	*Units*
40,000	Normal Loss	4,000	–	–	–	–
	Abnormal loss (5,000 – 4,000)	1,000	100%	1,000	100%	1,000
	Finished goods	32,000	100%	32,000	100%	32,000
	WIP	3,000	75%	2,250	50%	1,500
40,000		40,000		35,250		34,500

STATEMENT OF COST

Materials		3,02,000				
Less:	Scrap value 4,000 × 5	20,000	÷	35,250	=	8
		2,82,000				
Labour cost		20,7000	÷	34,500	=	6
Factory overhead		1,03,500	÷	34,500	=	3

STATEMENT OF EVALUATION

Abnormal loss:

Materials	1,000 × 8	=	8,000	
Labour	1,000 × 6	=	6,000	
Overhead	1,000 × 3	=	3,000	17,000

Finished goods:

Materials	32,000 × 8	=	2,56,000
Labour	32,000 × 6	=	1,92,000
Overhead	32,000 × 3	=	96,000
			5,44,000

WIP:

Materials	2,250 × 8	=	18,000
Labour	1,500 × 6	=	9,000
Overhead	1,500 × 3	=	4,500
			31,500

PROCESS 'A' ACCOUNT

	Units	*Amt.*		*Units*	*Amt.*
To units introduced	40,000	3,02,000	By normal loss	4,000	20,000
To direct wages	–	2,07,000	By abnormal loss	1,000	17,000
To overheads	–	1,03,500	By process 'B' a/c	32,000	5,44,000
			By WIP a/c	3,000	31,500
	40,000	6,12,500		40,000	6,12,500

ABNORMAL LOSS ACCOUNT

To process 'A' a/c	1,000	17,000	By sales @ Rs. 5	1,000	5,000
			By P & L a/c	–	12,000
	1,000	17,000		1,000	17,000

Problem 41. The following data pertains to process I a/c for March 2005 of Alpha Ltd:

Opening work-in-progress	– 1500 units amounting to Rs. 15,000
Degree of completion	– Materials 100%
	– Labour and overhead $33\frac{1}{3}\%$
Input of materials	– 18,500 units amounting to Rs. 52,000
Direct labour	– Rs. 14,000
Overheads	– Rs. 28,000
Closing work in progress	– 5,000 units
Degree of completion	– Materials 90%
	– Labour and overheads 30%
Normal process loss	– 10% of total input
Scrap value	– Rs. 2 per unit
Transferred to next process	– 15,000 units

You are required to:

(*a*) Prepare a statement of equivalent production assuming that the FIFO method is followed.

(*b*) Calculate cost per equivalent unit for each element.

(*c*) Statement of evaluation.

(*d*) Process a/c.

(Sri Sathya Sai University, B.Com. (Hons.), October 2005)

CHAPTER 4

Solution:

STATEMENT OF EQUIVALENT PRODUCTION

Input	*Output forms*	*Output unit*	*Equivalent Production*			
			Materials		*Labour and overheads*	
			%	*Units*	%	*Units*
1,500	Opening WIP	1,500	–	–	$66\frac{2}{3}$ %	1,000
18,500	Unit introduced and completed	13,500	100%	13,500	100%	13,500
	Normal loss	2,000	–	–	–	–
	Closing WIP	5,000	90%	4,500	30%	1,500
	Total	22,000		18,000		16,000
	Less: Abnormal gain	2,000	100%	2,000	100%	2,000
20,000		20,000		16,000		14,000

STATEMENT OF COST PER EQUIVALENT UNIT

Materials	52,000			
Less: Scrap	4,000			
	48,000	÷	16,000	= 3
Labour	14,000	÷	14,000	= 1
Overhead	28,000	÷	14,000	= 2

STATEMENT OF EVALUATION

Opening WIP			
Materials	Nil	=	Nil
Labour	1,000 × 1	=	1,000
Overhead	1,000 × 2	=	2,000
			3,000
Units completed			
Materials	13,500 × 3	=	40,500
Labour	13,500 × 1	=	13,500
Overheads	13,500 × 2	=	27,000
			81,000

Closing WIP

Materials	4,500 × 3	=	13,500
Labour	1,500 × 1	=	1,500
Overhead	1,500 × 2	=	3,000
			18,000

Abnormal gain

Materials	2,000 × 3	=	6,000
Labour	2,000 × 1	=	2,000
Overhead	2,000 × 2	=	4,000
			12,000

PROCESS 'I' ACCOUNT

To opening WIP	1,500	15,000	By normal loss	2,000	4,000
To units introduced	18,500	52,000	By process II a/c	15,000	99,000
To direct labour	–	14,000	By closing WIP	5,000	18,000
To overhead	–	28,000			
To abnormal gain	2,000	12,000			
	22,000	1,21,000		22,000	1,21,000

Problem 42. The Vega Manufacturing Co. uses FIFO method of inventory valuation in process costing. The following data relate to process I for the month of April 2004:

(*i*) Beginning work in process:

Quantity	:	1500 units
Value	:	Rs. 4500

(*ii*) Introduced during the month : 5,000 units

(*iii*) Transferred to process II : 5,500 units

(*iv*) Ending work-in-process : 1,000 units

(*v*) Degree of completion:

	Beginning WIP	*Ending WIP*
Materials	100%	100%
Conversion costs	80%	60%

(*vi*) Costs added during the month:

Materials	:	Rs. 10,000
Labour	:	Rs. 9,800
Overheads:	:	Rs. 4,900

You are required to:

(*i*) Prepare a statement of equivalent production.

(*ii*) Prepare process I account. (*University of Delhi, B.Com. (Hons), 2004*)

CHAPTER 4

Solution:

STATEMENT OF EQUIVALENT PRODUCTION

Input	*Forms of output*	*Output Qty*	*Equivalent Units*			
			Materials		*Labour and overheads*	
			%	*Units*	*%*	*Units*
1,500	Opening WIP	1,500	–	–	20%	300
5,000	Introduced and completed	4,000	100%	4,000	100%	4,000
	Closing WIP	1,000	100%	1,000	60%	6,00
6,500		6,500		5,000		4,900

STATEMENT SHOWING COST PER UNIT OF EQUIVALENT PRODUCTION

Materials	=	10,000	÷	5,000	=	Rs. 2
Labour	=	9,800	÷	4,900	=	Rs. 2
Overhead	=	4,900	÷	4,900	=	Re. 1
				Total		5

STATEMENT OF EVALUATION

Finished stock:

Materials	4,000 × 2	=	8,000
Labour	4,000 × 2	=	8,000
Overhead	4,000 × 1	=	4,000
			20,000

Opening WIP

Materials		=	Nil
Labour	300 × 2	=	600
Overheads	300 × 1	=	300
			900

Closing WIP

Materials	1,000 × 2	=	2,000
Labour	600 × 2	=	1,200
Overhead	600 × 1	=	600
			3,800

PROCESS 'I' ACCOUNT

	Qty.	*Amount*		*Qty.*	*Amount*
To opening WIP	1,500	4,500	By process II A/c (4500 + 900 + 20,000)	5,500	25,400
To materials	5,000	10,000			
To labour		9,800	By closing WIP	1,000	3,800
To overhead		4,900			
	6,500	29,200		6,500	29,200

Problem 43. SBL Ltd. furnishes you the following information relating to process B for the month of April 2006:

(*i*) Opening work-in-progress Nil
(*ii*) Units introduced – 10,000 units @ Rs. 5 per unit
(*iii*) Expenses debited to process B Rs. 5,000
(*iv*) 8,000 units of finished output were transferred to the next process during the month
(*v*) Normal loss in process – 10% of input
(*vi*) Closing work-in-progress – 800 units
Degree of completion as to
Materials 100%
Labour & overheads 50%
(*vii*) Degree of completion of abnormal loss as to:
Materials 100%
Labour & overheads 80%
(*viii*) Scrap realisation: Normal loss @ Rs. 2 per unit:
Abnormal loss @ Rs. 4 per unit

You are required to prepare
(1) Statement of equivalent production
(2) Statement of cost of each element
(3) Statement of evaluation
(4) Process B account
(5) Abnormal loss account

(ICWA, Inter, June 2006)

Solution:

STATEMENT OF EQUIVALENT PRODUCTION

Input	*Forms of output*	*Output Qty*	*Equivalent production*			
			Materials		*Labour and overheads*	
			Units	*%*	*Units*	*%*
	Opening WIP	–	–	–	–	–
10,000	Introduced input	–	–	–	–	–
	Finished goods	8,000	8,000	100%	8,000	100%
	Normal loss	1,000	–	–	–	–
	Abnormal loss	200	200	100%	160	80%
	Closing WIP	800	800	100%	400	50%
10,000		10,000	9,000		8,560	

STATEMENT OF COST OF EACH ELEMENT

Materials cost	50,000				
Add: Processing materials	24,600				
	74,600				
Less: Scrap realised	2,000				
Net cost	72,600	÷	9,000	=	8.067
Labour	10,400	÷	8,560	=	1.215
Overhead	5,000	÷	8,560	=	0.584

CHAPTER 4

STATEMENT OF EVALUATION

Finished goods	8,000 × 9.866	=	78,926
Abnormal loss:			
Materials	200 × 8.067	=	1,613
Labour	160 × 1.215	=	194
Overhead	160 × 0.584	=	93
			1,900
Closing WIP:			
Materials	800 × 8.067	=	6,454
Labour	400 × 1.215	=	486
Overhead	400 × 0.584	=	234
			7,174

PROCESS 'B' ACCOUNT

	Qty.	*Amt.*		*Qty.*	*Amt.*
To units introduced	10,000	50,000	By normal loss	1,000	2,000
To processing materials		24,600	By abnormal loss	200	1,900
To direct labour		10,400	By finished goods	8,000	78,926
To overheads		5,000	By closing WIP	800	7,174
	10,000	90,000		10,000	90,000

Problem 44. From the following information relating to process I of a factory for the month of April 2003. Prepare the statement of equivalent production, statement of cost, statement of evaluation and process account, using average cost method:

(*i*) Opening work-in-progress: 500 units

Materials	Rs.	27,000
Labour	Rs.	8,000
Overheads	Rs.	12,500
		47,500

(*ii*) Cost incurred during April 2003:

Input of materials: 14000 units	Rs.	5,74,750
Labour	Rs.	1,19,300
Overheads	Rs.	1,78,450

(*iii*) Process loss:
Normal loss: 10%
Value of scrapped unit: Rs. 10 each
Actual loss during April 2003: 1500 units
Degree of completion:
Materials 100%,
Labour and overheads 60%

(*iv*) Closing work in progress: 1,000 units
Degree of completion:
Materials 100%, Labour and overheads 70%

(*v*) Processed units transferred to process II:
12,000 units during April 2003

(ICWA, Inter, June 2003)

Solution:

STATEMENT OF EQUIVALENT PRODUCTION
(AVERAGE COST METHOD)

Input Qty.	*Forms of output*	*Output Qty.*	*Equivalent production*			
			Materials		*Labour & Overheads*	
			Units	%	*Units*	%
500	Opening WIP		–	–	–	–
14,000	Introduced units					
	Finished goods	12,000	12,000	100%	12,000	100%
	Normal loss	1,450	–	–	–	–
	Abnormal loss	50	50	100%	30	60%
	Closing WIP	1,000	1,000	100%	700	70%
14,500		14,500	13,050		12,730	

STATEMENT OF COST

Materials	27,000		
Add: Further materials added	5,74,750		
	6,01,750		
Less: Normal loss realised	14,500		
	5,87,250	÷ 13,050	= Rs. 45
Labour cost - Opening WIP	8,000		
Add: Additional labour cost	1,19,300		
	1,27,300	÷ 12,730	= Rs. 10
Overhead - Opening WIP	12,500		
Add: Additional overhead	1,78,450		
	1,90,950	÷ 12,730	= Rs. 15

STATEMENT OF EVALUATION

Finished goods:			
	Materials	12,000 × 45	= 5,40,000
	Labour	12,000 × 10	= 1,20,000
	Overheads	12,000 × 15	= 1,80,000
			8,40,000
Abnormal loss:	Materials	50 × 45	= 2,250
	Labour	30 × 10	= 300
	Overheads	30 × 15	= 450
			3,000

Closing WIP:	Materials	1,000 × 45	=	45,000
	Labour	700 × 10	=	7,000
	Overhead	700 × 15	=	10,500
				62,500

PROCESS 'I' ACCOUNT

	Qty.	*Amt.*		*Qty.*	*Amt.*
To opening WIP	500	47,500	By normal loss	1,450	14,500
To materials	14,000	5,74,750	By abnormal loss	50	3,000
To labour	–	1,19,300	By process II A/c	12,000	8,40,000
To overhead	–	1,78,450	By closing WIP	1,000	62,500
	14,500	9,20,000		14,500	9,20,000

Questions

Very Short Answer Questions

1. State any four features of process costing.
2. Define process costing.
3. What do you mean by normal loss? How is it treated in process cost accounting?
4. What do you mean by abnormal loss? How is it treated in process accounting?
5. Distinguish between normal loss and abnormal loss.
6. What do you mean by abnormal effective? How is treated in process cost accounts?
7. What do you mean by inter-process profit? What purpose does it serve?
8. What do you mean by equivalent production?

Short Answer Questions

1. Describe the main features of process costing.
(Madurai Kamraj University, B.Com., April 1992)
2. How do you treat abnormal gain? *(University of Kerala, B.Com., 1991)*
3. Distinguish between job costing and process costing. *(C.A. Inter, June 1990)*
4. Define normal loss and abnormal loss. Explaining their possible causes.
(University of Madras, M.Com., April 1992)
5. What motivational criteria are applicable to transfer pricing?
(University of Madras, M.Com., April 1992)

Exercises

1. (When part of the output as transferred to next process and part of output is sold) KT Ltd. provides you the following information for the year ended 31st March 2004.

Particulars	*Processes*		
	A	*B*	*C*
Raw materials (units)	12,000	2,440	2,600
Cost of raw materials per unit (Rs)	5	5	5
Direct wages (Rs)	34,000	24,000	15,000
Production overhead (Rs)	16,160	16,200	9,600
Normal loss (% of total No. of units entering to the process)	4%	5%	3%
Wastage (% of total no. of units entering to the process)	6%	5%	4%
Scrap per unit of wastage (Rs.)	3	4	5
Output transferred to subsequent process	70%	60%	–
Output sold at the end of the process	30%	40%	100%
Selling price per unit (Rs)	12	16	17

Prepare process A, B and C account. *(University of Bombay, B.Com., April 2003)*

[**Ans.** Profit from process A, Rs. 6,480,
Profit from process B, Rs. 7,200,
Profit from process C, Rs. 14,880]

2. Mr. Kale manufactures a product in two grade. Grade I and Grade II from common raw material. Raw material is introduced in "Basic process" the produce of which is dealt with as follows:

25% sold in open market.

25% transferred to Grade I process and the balance 50% transferred to Grade II process

The details of processes are follows:

Particulars	*Basic Process*	*Grade I Process*	*Grade II Process*
Raw materials	1,000 Units	–	–
Cost per unit	Rs. 200	–	–
Other materials	Rs. 25,000	Rs. 30,000	Rs. 30,000
Labour	Rs. 60,000	Rs. 50,000	Rs. 50,000
Manufacturing overhead	Rs. 75,000	Rs. 60,000	Rs. 60,000
Sale price per unit	Rs. 400	Rs. 1,400	Rs. 900

Prepare process accounts and determine total profit earned by Mr. Kale assuming that there is no stock in any process. *(University of Bombay, B. Com., October 2002)*

[**Ans.** Profit from basic process Rs. 10,000
Profit from Grade I process Rs. 1,20,000
Profit from Grade II process Rs. 1,30,000]

CHAPTER 4

3. (*Normal loss, Abnormal loss and Abnormal gain*): A product passes through three processes, viz, process I, process II and process III. The normal wastage of each process is:

Process I-3% process II-6%: process III -10%. The percentage of normal wastage in each case is computed on the basis of the number of units entering the process concerned. The wastage of process I is sold @ 25 paise per unit, that of process II is sold at 0.50 paise per unit and that of process III is sold @ Re 1 per unit. 15,000 units of raw materials were introduced in process I at Re 1 per unit. The other expenses are:

	Process I	*Process II*	*Process III*
	Rs.	*Rs.*	*Rs.*
Materials	1,500	2,250	750
Direct labour	7,500	12,000	9,750
Manufacturing expenses	1,575	1,425	3,015

The output of each process has been as follows:

Process I	Rs. 14,250
Process II	Rs. 13,650
Process III	Rs. 12,012

Prepare process accounts, abnormal wastage, normal wastage and abnormal gain a/c.

(Sri Sathya Sai University, B.Com (Hons.), February 2006)

[**Ans.** Profit transferred from abnormal Gain a/c to costing P & L a/c Rs. 63,750 and profit transferred from abnormal wastage a/c to costing P & L a/c Rs. 1,357]

4. Product 'A' is obtained after it passes three distinct processes. The following information is obtained from the accounts for the week ending 31st May 2003

Items	*Total*	*Process* I	II	III
Direct materials	7,542	2,600	1,980	2,962
Direct wages	9,000	2,000	3,000	4,000
Production overhead	9,000			

1,000 units at Rs. 3 each were introduced to process I. The output of each process passed direct to the next process and finally to finished stock. Production overhead is recovered as 100% of direct wages. The following additional data are obtained.

Process	*Output during the week*	*Percentage of normal loss to output*	*Value of scrap per unit (Rs)*
Process I	950	5%	2
Process II	840	10%	4
Process III	750	15%	5

Prepare process cost accounts and abnormal gain or loss account.

(Sri Sathya Sai University, B.Com. (Hons.), August 2003)

[**Ans.** Profit transferred to P & L a/c from Abnormal loss a/c is Rs. 240 and from Abnormal gain a/c is Rs. 1,188]

5. A product passes through 3 process A, B and C. From the following information, prepare process accounts and also give abnormal wastage account

	Processes		
	A	*B*	*C*
	Rs.	*Rs.*	*Rs.*
Materials	2,000	3,000	1,000
Labour	10,000	16,000	13,000
Overheads	2,100	2,376	4,018
Actual output (Units)	19,000	18,200	16,200
Normal loss	3%	5%	8%

The wastage of process A was sold at 25 paise per unit, that of process B at 50 paise per unit and that of process C at Re 1 per unit. 20,000 units were issued to process A in the beginning at a cost of Re 1 per unit. *(Andhra University, B.Com, June 2006)*

[**Ans.** Total of process 'A' a/c 20,000 units – Rs. 34,100
Total of process 'B' A/c 19,150 units – Rs. 55,076
Total of process 'C' a/c 18,200 units – Rs. 72,619]

6. A product is obtained after passing through three processes from the following particulars relating to the year ending December 2000. Prepare process accounts and also prepare Abnormal gain or loss accounts. 1,000 units are introduced at the rate of Rs. 3 per unit in the 1st process:

		Process I	*Process II*	*Process III*
Direct material	Rs.	2,600	1,980	2,962
Direct wages	Rs.	2,000	3,000	4,000
Overheads on wages		100%	100%	100%

Other information is as under:

Process	*Production (Units)*	*Normal loss (Rs.)*	*Scrap value per unit (Rs.)*
I	950	5%	2
II	840	10%	4
III	750	15%	5

(Kakatiya University, B.Com., July 2004)

[**Ans.** Total of process A—Units 1,000 Rs. 9,600
Total of process B—Units 950 Rs. 17,480
Total of process C—Units 876 Rs. 29,130]

7. A product passes through three processes A, B & C. 10,000 units at a cost of Rs. 1.10 per unit were issued to process A. The other direct expenses were as follows:

Particulars	*A*	*B*	*C*
Sundry materials	1,500	1,500	1,500
Direct labour	4,500	8,000	6,500
Direct expenses	1,000	1,000	1,503

The wastage of process 'A' was 5% and in process 'B' at 4% of input. The wastage of process 'A' was sold at Re 0.25 per unit and that of process 'B' at Re 0.50 per unit and that of process 'C' at Re. 1.00 per unit. The overhead charges was 160% of direct labour. The final product was sold at Rs. 10 per unit fetching a profit of 20% on sales.

Prepare all process accounts. *(University of Bombay, B.Com., October 2004)*

[**Ans.** Profit transferred to P & L a/c Rs. 16,848]

8. A product passes through three processes in January 2002 the costs of production were as given below.

Particulars	*Process I* Rs.	*Process II* Rs.	*Process III* Rs.
Direct material	12,000	18,120	20,772
Wages	21,000	25,356	30,000
Production overhead	9,000	12,000	15,000
6,000 units were issued to process I @ Rs. 15 each			
Normal loss	10%	5%	10%
Wastages realises	Rs. 6 per unit	Rs. 15 per unit	Rs. 18 per unit
Actual production (in units)	5,520	5,220	4,800

Prepare:

(1) Process accounts I, II & III

(2) Normal loss a/c

(3) Abnormal loss a/c *(University of Bombay, B.Com., April 2002)*

[**Ans.** Total of process I 6,120 units—Rs. 1,34,853

Total of process II 5,520 units—Rs. 1,86,729,

Total of process III 5,322 units—Rs. 2,52,695]

9. (*Process Cost Accounts with Stock*) The product of a company passes through three different processes A, B and C. The company gives you the following information for the month of January 2001

	Process A (Rs)	*Process B (Rs)*	*Process C (Rs)*
Raw materials	15,200	–	–
Other materials consumed	8,000	3,000	2,000
Direct labour	12,000	8,000	6,000
Factory indirect expenses	2,000	1,000	3,000
Output for the month (in units)	1950 units	1925 units	1590 units
Stock of output:			
Jan 1, 2001 (in units)	200 units	300 units	500 units
Jan 31, 2001 (in units)	150 units	400 units	–
Stock valuation on			
Jan 1, 2001 (per unit)	Rs. 19	Rs. 27	Rs. 36.50

Stock on 31st January 2001 are to be valued at cost as shown by months production account. Prepare process accounts. *(Osmania University, M. Com. (Final), June 2002)*

10. The product of a manufacturing unit passes through two distinct processes. From past experience, the incidence of wastage is ascertained as under

Process A	2%
Process B	10%

In each case, the percentage of wastage is computed on the number of units entering the process concerned. The realisation of wastage in process A and B are Rs. 25 per 100 units and Rs. 50 per 100 units respectively.

The following information is obtained for the month of April 2004. 40,000 units of crude material were introduced in process A at a cost of Rs. 16,000.

	Process A (*Rs*)	*Process B* (*Rs*)
Other materials	16,000	5,000
Direct labour	9,000	8,000
Direct expenses	8,200	1,500
Output	39,000	36,500
	Units	*Units*
Finished product stock:		
April 1	6,000	5,000
April 30	5,000	8,000
Value of stock per unit on April 1st	Rs. 1.20	Rs. 1.60

Stocks are valued and transferred to subsequent process at weighted average costs. Prepare respective process accounts and stock accounts. *(University of Andhra, B.Com., March 2007)*

[**Ans.** Total of process 'A' stock a/c 45,000 units amounting to Rs. 55,950
Total of process 'B' stock a/c 41,500 units amounting to Rs. 71,115.]

11. The product of M/s Naveen Traders Ltd. passes through two processes A and B. From the past experience the percentage of loss is calculated on the input at 2% in process A and 5% in process B. The loss of each process possesses a scrap value of Rs. 10 per 100 units in process A and Rs. 20 per 100 units in process B. The following information is available for the year ended 31st March 2002. 40,000 units of raw materials were introduced in process A at a cost of Rs. 16 per unit

Item	*Process A* Rs.	*Process B* Rs.
Materials consumed	8,000	2,800
Direct labour	12,200	14,000
Manufacturing expenses	3,080	1,000
	Units	*Units*
Finished products	39,000	38,500
Finished stock on 1-4-2001	4,000	6,000
Finished stock on 1-4-2002	3,000	8,000
Stock valuation on 1-4-2002 per unit	Re 0.90	Rs. 1.47

Stock on 1-4-2002 is to be valued at cost as shown by the year's relevant process accounts. Prepare process accounts. *(Andhra University, B. Com., June 2007)*

[**Ans.** Total of process 'A' stock–units 43,000 amounting to Rs. 42,600 and Total of process 'B' stock–units 44,500 amounting to Rs. 66,570]

12. The product of a company passes through three distinct processes A, B and C. The company gives you the following information for the month of January 2004.

	Process A *Rs.*	*Process* B *Rs.*	*Process* C *Rs* .
Raw materials (2,000 units introduced in process A)	15,200	–	
Other materials consumed	8,000	3,000	2,000
Direct labour	12,000	8,000	6,000
Factory indirect expenses	2,000	1,000	3,000
Output for the month of January (units)	1950	1925	1590
Stock of output:			
1-1-2004 (units)	200	300	500
31-1-2004 (units)	150	400	–
Stock valuation on 1-1-2004 (per unit)	Rs 19	Rs 27	Rs 36.50

Stock on 31-1-2004 are to be valued at cost as shown by month's production accounts.

Prepare process accounts. *(Andhra University, B. Com., April 2006)*

[**Ans.** Total of process 'A' stock a/c 2,150 units amounting to Rs. 41,000
Total of process 'B' stock a/c 2,225 units amounting to Rs 58,238
Total of process 'C' stock a/c 2,090 units amounting to Rs. 77,068.]

13. (*Applied Process Costing*) In an oil refinery, the product passes through three different processes, viz crushing, refining and finishing. The following information is available for the month of March 2003.

Particulars	*Crushing Process*	*Refining Process*	*Finishing Process*
Raw materials (500 tons copra)	Rs. 9,00,000	–	–
Wages	32,000	23,600	23,500
Power	4,800	4,000	6,000
Sundry materials	2,000	7,600	–
Factory expenses	2,400	4,000	3,800

200 tons of oil cake was sold for Rs. 60,000 and 275 tons of crade oil was obtained from crushing process.

25 tons of by-product of the crushing process fetched Rs. 3,600.

25 tons of by product of the refining process was sold for Rs. 3,600 and 250 tons of refined oil was obtained.

10 tons of finished oil were sold for Rs. 4,800 and 240 tons of finished oil was stored in drums.

The establishment expenses for the month amounted to Rs. 14,000 which is to be charged to the three processes in proportion of 3: 2: 2.

The cost of drums for storing finished oil was Rs. 84,100.

Prepare process accounts for all the three processes.

(University of Bomaby, B.Com., April 2004)

[**Ans.** Cost of production of oil in Finishing process A/c is 240 tons amounting to Rs. 10,39,800]

14. (*Inter Process Profit*) The following are the details in respect of process X and process Y of a processing factory:

	Process X	*Process Y*
Materials	10,000	–
Labour	10,000	14,000
Overhead	4,000	10,000

The output of process X is transferred to process Y at a price calculated to give a profit of 20% on the transfer price and the output of process Y is charged to finished stock at a profit of 25% on the transfer price. The finished goods department realised Rs. 1,00,000 for the finished goods received from process Y. You are required to show the process accounts and total profit assuming there was no opening and closing work-in-progress.(*University of Madras, B.Com., April 2003*)

[**Ans.** Profit on sale of finished goods Rs. 28,000]

15. A certain product passes through three processes before it is transferred to finished stock. The following information is obtained for the month of December 2000.

Process	*I* *Rs.*	*II* *Rs.*	*III* *Rs*	*Finished* *Stock Rs.*
Opening stock	5,000	8,000	10,000	20,000
Direct materials	40,000	12,000	15,000	–
Direct wages	35,000	40,000	35,000	–
Production overheads	20,000	24,000	20,000	–
Closing stock	10,000	4,000	15,000	30,000
Profit % on transfer price (to next process)	25%	20%	10%	–
Inter-process profit for opening stock	–	1,395	2,690	6,534

Closing stock in processes are valued at prime cost and finished stock has been valued at the price at which it is received from process III. Sales during the period were 4,00,000. Prepare process cost accounts showing profit element at each stage.

(*Osmania University, M. Com., (Final) June 2003*)

[**Ans.** Profit in finished stock a/c is Rs. 60,350]

16. (*Equivalent Production*) During the month of April 2004, 4,000 units were introduced into process *A* at a cost of Rs. 23,200. At the end of the month 3,000 units had been produced and transferred to process B, 720 units were still in process and 280 units were scrapped. Normal wastage of 5% on input is allowed. It was estimated that incomplete units had reached a stage of production as follows:

Materials	75% complete
Labour	50% complete
Production overhead	50% complete

Additional cost incurred on process 'A' were:

	Rs.
Direct materials	6,154
Direct wages	13,760
Production overhead	6,880

Unit scrapped realised Rs. 2 each and were 100% complete as regards materials, labour, and overhead.

Prepare process 'A' account, abnormal wastage a/c, statement of equivalent production, statement of cost for each element and statement of evaluation.

(Sri Sathya Sai University, B.Com. (Hons.), October 2004)

[**Ans.** Total of process 'A' a/c 4,000 units Rs. 49,995]

17. The following data are available in respect of process I for January 2002:

Opening work in progress 100 units amounting to Rs. 4500		
Degree of completion:	Materials	100%
	Labour	60%
	Overheads	60%
Input of materials 9,100 units amounting to Rs. 27,300		
Direct wages		8,200
Production overhead		16,400
Units scrapped		1,200 units
Degree of completion:	Materials	100%
	Labour	70%
	Overhead	70%
Closing work-in-progress		1,000 units
Degree of completion:	Materials	100%
	Labour	80%
	Overhead	80%

Units transferred to next process 7,800 units. Normal process loss is 10% of total output (Opening stock plus units put in) scrap value is Rs. 3 per unit. You are required to follow FIFO method and compute:

(*a*) Equivalent production

(*b*) Cost per equivalent unit for each element and cost of abnormal loss.

(*c*) Process I A/c. *(Sri Sathya Sai University, B.Com. (Hons.) October 2003)*

[**Ans.** Total of process I a/c 10,000 units amounting to Rs. 56,400]

18. During the month of April 2003, 4,000 units were introduced into process 'A' at the cost of Rs. 23,200. At the end of the month 3,000 units were completed and transferred to process 'B' a/c. 720 units were still in process and 280 units were scrapped. A normal wastage of 5% was expected. It was estimated that the incomplete units have reached a stage in production as follows:

Materials	75%
Labour and overhead	50%

The additional costs incurred were:

	Rs.
Materials	6,160
Wages	13,760
Overheads	6,880

Unit scrapped realised at Rs. 2 per unit

Prepare:

(1) A statement showing equivalent production.

(2) Statement showing cost per unit. *(University of Delhi, B.Com. (Hons.) E 2004)*

[**Ans.** Material cost per unit Rs. 8, Labour cost per unit Rs. 4, Overhead cost per unit Rs. 2]

19. Data relating to work done in process 'A' of a company during the month of April 2,000 is given below:

Opening work-in-progress (1,000 Units)

	Rs.
Materials	40,000
Labour	7,500
Overheads	22,500
Material introduced in process 'A' (19,000 units) Rs.	7,40,000
Direct labour	1,79,500
Overheads	5,38,500
Units scrapped :	1,500 units

Degree of completion:
Materials 100%
Labour and overheads 80%

Closing work-in-progress : 1,000 units

Degree of completion:
Materials 100%
Labour and overheads 80%

Units finished and transferred to process B = 17,500 units.

Normal loss: 5% of total input including opening WIP

Scrapped units fetch Rs. 20 per piece

Required:

(*a*) Statement of equivalent production.

(*b*) Statement of cost

(*c*) Statement of distribution of cost

(*d*) Process 'A' account and other accounts *(ICWA Inter, Stage 1, December 2,000)*

[**Ans.** Total of process 'A' A/c 20,000 Units Rs. 15,28,000]

❏❏❏

CHAPTER 5

JOINT PRODUCTS AND BY-PRODUCTS

DEFINITION AND FEATURE OF JOINT PRODUCTS

Sam R. Goodman and James S. Reece in their book *Controller's Handbook* define joint product as "when a group of individual products is simultaneously produced, with each product having a significant relative sales value, the outputs are usually called joint products". According to Matz-Curry-Frank, joint products are produced simultaneously by a common process or series of processes, with each product possessing more than a nominal value in the form in which it is produced. This definition emphasises the point that the manufacturing process creates the products in a somewhat definite quantitative relationship. An increase in the output of one product will bring about an increase in the quantity of others or *vice versa,* but not necessarily in the same proportion.

According to Theodore Lang, joint products represent two or more products separated in the course of the same processing operation, usually requiring further processing, each product being in such proportion that no single product can be designated as a major product.

According to this definition when two or more products of equal importance are simultaneously produced from the same basic raw materials from a common process they are known as joint products.

Features of Joint Products

1. Joint products are produced simultaneously with respect to common raw material and common manufacturing process.
2. Each product has a significantly relative sales value.
3. The "Jointness" may be associated with the complete transformation from raw materials into finished goods or it may be characteristic of only part of the complete transformation process.
4. There exist a direct quantitative relationship among various joint products such that an increase in the output of one increases the output of others, although not necessarily in the same ratio.

Some examples of joint products are: In oil refinery, the joint products obtained are, petrol, diesel, gasoline, kerosene, nephtha, lubricant oil and paraffin. In dairy industry, the joint products are, butter, ghee, cheese, etc. In gold mining, the joint products are gold, silver and copper.

Definition of Joint Cost

All costs associated with the manufacture of a joint product up to the point in the production process, where they are clearly differentiable as distinct products are called joint cost. Kohler in his *Dictionary for Accountants* defines joint cost as "the common cost of facilities or of services employed in the output of two or more simultaneously produced or otherwise related operations, commodities or services".

According to NAA Research Series No. 31–Costing Joint products—States "costs are joint when two or more distinguishably different products are producted together from a single input cost factor. The characteristic condition of a joint cost is that the cost of several different products is incurred as a lump-sum for the combination and not separately for the individual products".

Definition of Split-off-point

The exact point in a manufacturing process where two or more products can be distinguished is denoted as the split-off-point. This split-off-point can be either at the stage when the manufacturing process is complete, *i.e.,* when the raw materials have been converted into finished goods or alternatively at some intermediate point in the manufacturing process.

Difference between Joint Cost and Separable Cost

The expenses incurred up to the point of separation is known as joint cost and after the point of split-off is known as separate cost. Joint costs incurred prior to the point of split-off cannot be traced to the particular products involved.

Definition of Minor Products

All jointly produced products of lesser economic significance are termed as "minor products".

Problems Associated with Joint Products

Joint products are obtained incidentally from the same manufacturing process. Some joint products are controllable while others are not. In most of the cases, the management has no alternative course of action. While some are useful, others are not and sometimes even unwanted. To the extent possible, the products that could be made separately are linked together to secure the economies of large scale operation. Further markets are developed for joint products previously discarded as waste. Joint products have assumed more importance today because increasing competition has compelled many industries to consider products which were formerly regarded as waste. Necessary steps are taken to make joint products more useful either at the point of derivation or after post split-off-point.

Some of the problems associated with joint products can be summarised as follows:

1. Classification of joint products into important and incidental.
2. Deciding the method of allocating the joint cost among joint products.
3. The method of treating the revenue earned by the sale of incidental or minor products.
4. Furnishing of information periodically to management for the purpose of deciding what products should be manufactured in what quantities ?
5. Not all joint products are sold at the same time. To that extent it becomes difficult to account joint product.

OBJECTS OF TREATING JOINT COSTS

The main objectives of treating joint cost are as follows:

1. To know unit cost of every joint product.
2. For the purpose of stock valuation.

CHAPTER 5

3. For the determination of profitability of different products.
4. For determining and appraising efficiency of various joint products.
5. For exercising internal control over various joint products obtained.
6. For fixing up the selling price of joint products.
7. For setting up standard output in advance of production.

CO-PRODUCTS

According to the NAA Research Study, "when two or more major products appear in a jointly produced combination, these products are called co-products to indicate that they are products of like importance". Co-products are the products which are manufactured in different varieties by the same industry. They may not arise from the same operation or raw material but are produced in different quantities without any co-relation to the other. Examples of co-products in a furniture-making industries are tables, chairs, benches, desks, almirahs. Similarly in fan manufacturing industry, the various co-products are ceiling fans, table fans, and exhaust fans. In automobile industry, the various co-products are buses, lorries, tractors and so on.

Difference between Joint Product and a Co-product

The important difference between a joint product and co-product is, there exists a linear relationship in terms of quantities among various joint products. Whereas no such relationship exists under co-products. They may be produced in different quantities depending upon the demand for them.

TREATMENT OF JOINT COST OF JOINT PRODUCTS

The purpose of accounting of joint product is to know individual product cost. This is done by means of allocating joint cost to all joint products which are obtained at the point of split-off. The following methods are adopted for allocating the joint expenses to all the joint products.

1. Physical Unit Method

This method of allocating costs uses a physical measure such as weight, volume, length and so on as the criterion for allocating joint costs. The joint products are measured in terms of a measuring rod such as a kg, litre, meter and so on. Process loss if any is borne by joint products in proportion to the volume of output. This method cannot be adopted where it is not possible to measure joint products with a common measuring rod. This situation arises when the joint products obtained are in different forms such as solid, liquid and gases.

Problem 1. One tonne of raw materials put into a common process yields four joint products *P*, *Q*, *R*, and *S*, their weights being 63 kgs, 117 kgs, 180 kgs and 540 kgs respectively. The balance in weight is considered as normal wastage.

Based on the total processing cost of Rs. 20,000 per tonne of raw material input, you are required to apportion the joint cost to products *P, Q, R* and *S*. *(C.S. (Final) Dec. 1986)*

Solution:

STATEMENT SHOWING APPORTIONMENT OF JOINT COST

Products	*Yield (kg.)*	*Basis* = $\frac{\text{Yield}}{\text{Output}}$ × *Joint cost*	*Share of joint cost*
P	63	$\frac{63}{900} \times 20{,}000$	1,400
Q	117	$\frac{117}{900} \times 20{,}000$	2,600
R	180	$\frac{180}{900} \times 20{,}000$	4,000
S	540	$\frac{540}{900} \times 20{,}000$	12,000
Output	900		20,000
Normal loss	100		—
Input	1,000		20,000

Problem 2. In a process line of *X Y* Company three joint products are produced. For the month of May 1990 the following data were available:

	L	M	N
Sales price per kg	Rs. 5	Rs. 10	Rs. 20
Post-separation joint costs	10,000	5,000	15,000
Output in kg	2,500	1,000	1,500

Pre-separation joint costs amounted to Rs. 20,000.

The joint products are manufactured in one common process, after which they are separated and may undergo further individual processing. The pre-separation joint costs are apportioned to joint products according to weight. You are required to prepare a statement showing the estimated profit or loss for each product and in total. *(ICWA, Inter., June 1990)*

Solution:

STATEMENT SHOWING APPORTIONMENT OF JOINT COST

	L (2,500)		*M (1,000)*		*N (1,500)*	
	Total	*Per unit*	*Total*	*Per unit*	*Total*	*Per unit*
Pre-separation cost (5 : 2 : 3)	10,000	4	4,000	4	6,000	4
Post-separation cost	10,000	4	5,000	5	15,000	10
Total cost	20,000	8	9,000	9	21,000	14
Profit/Loss (balancing-figure)	(7,500)	(– 3)	1,000	1	9,000	6
Sales price	12,500	5	10,000	10	30,000	20

CHAPTER 5

Problem 3. The following data have been extracted from the books of Bhaskar Coke Works:

Joint products	*Yield in kgs of recovered products per ton of coal*
Coke	1,420
Coal tar	120
Benzol	22
Ammonium sulphate	26
Gas	412
Total	2,000

The price of coal is Rs. 80 per ton. Direct labour and overhead costs to point of split-off are Rs. 40 and 60 respectively per ton of coal.

Calculate material cost, labour, overhead and total cost of each product on the basis of weight.

Solution:

STATEMENT SHOWING APPORTIONMENT OF JOINT COST PER TON OF COAL

Products	*Yield in kg of product per ton of coal*	*% of weight of products to coal*	*Cost of coal*	*Direct labour*	*Overhead*	*Total*
Coke	1,420	71.0%	56.80	28.40	42.60	127.8
Coal tar	120	6.0%	4.80	2.40	3.60	10.8
Benzol	22	1.1%	0.88	0.44	0.66	1.9
Ammonium sulphate	26	1.3%	1.04	0.52	0.8	2.4
Gas	412	20.6%	16.48	8.24	12.38	37.1
	2,000	100	80	40	60	180

2. Sales Value Method

Under this method, the criterion for allocating joint costs is the relative sales value associated with each individual product. Thus joint costs are allocated to the joint products on the basis of their ability to earn revenue. The assumption is that if a higher selling price exists for one product as compared with another product, then the production costs associated with that product must also be higher. The disadvantages of this method are:

(*a*) Market prices of various products at the point of separation are not always available,

(*b*) The joint product of one year may be sold in the next year. To this extent accounts are distorted. The joint products can be sold at any of the following three stages. So the treatment also differs between these stages of sale.

(*a*) Market value at the point of separation: The market prices of the joint products at the point of separation is ascertained and joint cost is apportioned in the ratio of these prices. This method is followed when further processing involves disproportionate costs.

Problem 4. From the information below, find the profit made by each product, apportioning joint costs on the sale-value basis:

Joint costs	*Rs.*
Direct materials	1,26,000
Power	25,000

Petrol, oil and lubricants		5,000
Labour		7,500
Other charges		4,100

	Product X (Rs.)	Product Y (Rs.)
Selling costs	20,000	80,000
Sales	1,52,000	1,68,000

(Andhra University, B.Com., May 1992)

Solution: STATEMENT SHOWING APPORTIONMENT OF JOINT COST

	X	Y	Total
Direct materials	59,850	66,150	1,26,000
Power	11,875	13,125	25,000
Petrol, oil, lubricants	2,375	2,625	5,000
Labour	3,563	3,937	7,500
Other charges	1,947	2,153	4,100
Cost of production	79,610	87,990	1,67,600
Selling cost	20,000	80,000	1,00,000
Cost of sales	99,610	1,67,990	2,67,600
Profit	52,390	10	52,400
Sales	1,52,000	1,68,000	3,20,000

(*b*) **Market value at finished stage:** Usually it is difficult to obtain the selling price of joint products at the point of separation and it is comparatively easy to know selling price of joint products after further processing. Therefore this method is used more frequently compared to the above method.

Problem 5. In a manufacturing concern a certain product A yield by products B and C. The joint expenses of manufacture are, materials Rs. 10,200, labour Rs. 11,400 and oncost Rs. 7,600 while subsequent expenses are:

	A	B	C
Materials	2,500	1,200	1,400
Labour	1,900	1,600	2,000
Oncost	1,500	900	1,050
	5,900	3,700	4,450

The selling prices are: *A* – 30,000 *B* – 20,000, *C* – 15,000. The estimated profits on turnover are: *A* – 40%, *B* – 30%, *C* – 25%.

Show how would you apportion the joint expenses of manufacture and prepare process accounts.

(Bangalore University, B.Com., May 1992)

Solution: STATEMENT SHOWING APPORTIONMENT OF OVERHEAD

	A	B	C	Total
Sales	30,000	20,000	15,000	65,000
Less: Profit	12,000	6,000	3,750	21,750
Cost of production	18,000	14,000	11,250	43,250
Less: Subsequent expenses	5,900	3,700	4,450	14,050
Share in the joint cost	12,100	10,300	6,800	29,200

MAIN PRODUCT 'A' ACCOUNT

To materials			By by-product B/ac	10,300
Joint	10,200		By by-product C a/c	6,800
Subsequent	2,500	12,700	By sales	30,000
To labour:				
Joint	11,400			
Subsequent	1,900	13,300		
To oncost				
Joint	7,600			
Subsequent	1,500	9,100		
To P & L a/c – profit		12,000		
		47,100		47,100

BY-PRODUCT 'B' ACCOUNT

To main product A a/c	10,300	By sales	20,000
To material	1,200		
To labour	1,600		
To oncost	900		
To P & L a/c – profit	6,000		
	20,000		20,000

BY-PRODUCT 'C' ACCOUNT

To main product A a/c	6,800	By sales	15,000
To material	1,400		
To labour	2,000		
To oncost	1,050		
To P & L a/c – profit	3,750		
	15,000		15,000

Problem 6. A concern while producing product *A*, also produces a by-product *B*. The by-product can be sold after a separate process.

From the under mentioned information prepare accounts to show the cost per kg. of *A* and *B*

	Joint expense	*Separate expense*	
		A	*B*
Material	Rs. 60,000	22,000	2,500
Labour	Rs. 35,000	24,000	8,000
Overhead	Rs. 10,000	4,000	1,500
Output	–	450 kgs	150 kgs

The selling price of B is Rs. 300 per kg. It is estimated that profit is 50% of selling price.

(Sri Sathya Sai University, B.Com., March 1992)

Solution:

STATEMENT SHOWING SHARE OF B IN THE JOINT COST

Sale value product B – 150 × Rs. 300			45,000
Less: Profit – 50% of sales value			22,500
Cost of production			22,500
Less: Separate expenses –	Materials	2,500	
	Labour	8,000	
	Overhead	1,500	12,000
Share of B in the joint cost			10,500

PRODUCT 'A' ACCOUNT

To materials			By by-product B a/c	10,500
Joint	60,000		By cost of production	1,44,500
Separate	22,000	82,000	@ Rs. 321.11	
To labour				
Joint	35,000			
Separate	24,000	59,000		
To overhead				
Joint	10,000			
Separate	4,000	14,000		
		1,55,000		1,55,000

BY- PRODUCT 'B' ACCOUNT

To main product A a/c		10,500	By sales	45,000
To separate expense	– Material	2,500		
	– Labour	8,000		
	– Overhead	1,500		
To profit		22,500		
		45,000		45,000

Problem 7. *(When profit of main product is not given): X, Y, Z* Ltd., manufactures product *A* which yields two by-products *B* and *C*. In a month, the joint cost was Rs. 20,500. Subsequent expenses were:

	A (Rs.)	*B (Rs.)*	*C (Rs.)*
Material	300	200	150
Direct wages	400	300	200
Overhead	300	270	280
	1,000	770	630

Gross sales value of products *A*, *B* and *C* was Rs. 15,000, Rs. 10,000, and Rs. 5,000 respectively. It was estimated that the net profit as percentage of sales in the case of *B* and *C* would be 25% and 20% respectively.

Ascertain the profit earned by *A* and prepare process cost accounts.

(Sri Sathya Sai University, B.Com., March 1989; Kerala University, B.Com., May 1991)

CHAPTER 5

Solution:

Note: First of all calculate the share of joint cost of product B and C and prepare product A a/c. The balancing figure in product A will be the profit. Subsequently the share of joint cost of product A can be known.

STATEMENT SHOWING SHARE OF JOINT COST

	A	*B*	*C*	*Total*
Sales	15,000	10,000	5,000	30,000
Less: Profit	3,500	2,500	1,000	7,000
Cost of production	11,500	7,500	4,000	23,000
Less: Subsequent expenses	1,000	770	630	2,400
Share in the joint cost	10,500	6,730	3,370	20,600

PRODUCT 'A' ACCOUNT

To share in the joint cost	20,600	By product *B* a/c	6,730
To materials	300	By product *C* a/c	3,370
To direct wages	400	By sales	15,000
To overhead	300		
To profit (balancing figure)	3,500		
	25,100		25,100

PRODUCT 'B' ACCOUNT

To product *A* a/c	6,730	By sales	10,000
To materials	200		
To labour	300		
To overhead	270		
To profit	2,500		
	10,000		10,000

PRODUCT 'C' ACCOUNT

To product *A* A/c	3,370	By sales	5,000
To materials	150		
To labour	200		
To overhead	280		
To profit	1,000		
	5,000		5,000

% of profit of product A on sales

$$\frac{3{,}500 \times 100}{15{,}000} = 23.33\%$$

Problem 8. *(Physical unit and sale value method):* In a process line of *XY* company three joint products for the month of May 2004, following data were available

Products	*A*	*B*	*C*
Sale price per kg	5	10	20
Post separation cost	10,000	5,000	15,000
Output in kgs	2,500	1,000	1,500

Pre-separation costs amounted to Rs. 20,000. The Joint products are manufactured in one common process after which they were separated and may undergo further individual processing.

You are required to prepare a statement showing estimated profit or loss for each product in total according to:

(1) Physical unit method

(2) Sales value method *(Sri Sathya Sai University, B.Com. (Hons), October 2004)*

Solution:

STATEMENT SHOWING APPORTIONMENT OF JOINT COSTS

	Physical unit method		*Sales value method*	
A	$\frac{2500}{5,000} \times 20,000$ =	10,000	$\frac{12,500}{52,500} \times 20,000$ =	4,762
B	$\frac{1,000}{5,000} \times 20,000$ =	4,000	$\frac{10,000}{52,500} \times 20,000$ =	3,810
C	$\frac{1,500}{5,000} \times 20,000$ =	6,000	$\frac{30,000}{52,500} \times 20,000$ =	11,428
		20,000		20,000

STATEMENT OF PROFIT OR LOSS UNDER PHYSICAL UNIT METHOD

		A	*B*	*C*	*Total*
(A)	Sales value weight × SP	12,500	10,000	30,000	52,500
(B)	*Less:* Total cost				
	– Pre-separation	10,000	4,000	6,000	20,000
	– Post-separation	10,000	5,000	15,000	30,000
		20,000	9,000	21,000	50,000
Profit/Loss (A) – (B)		– 7,500	1,000	9,000	2,500

STATEMENT OF PROFIT OR LOSS UNDER SALES VALUE METHOD

		A	*B*	*C*	*Total*
(A)	Sales	12,500	10,000	30,000	52,500
Less: Total cost					
	Pre-separation	4,762	3,810	11,428	20,000
	Post-separation	10,000	5,000	15,000	30,000
		14,762	8,810	26,428	50,000
Profit/Loss (A) – (B)		– 2,262	+ 1,190	+ 3,572	2,500

CHAPTER 5

Problem 9. *(Physical unit and sales value method):* Three joint products are produced by passing chemicals through two consecutive processes. Output from process 1 is transferred to process 2 from which three joint products are produced and immediately sold. The data regarding the process for April 1990 is given below:

	Process 1	*Process 2*
Direct materials 2,500 kilo at Rs. 4 per kilo	Rs. 10,000	–
Direct labour	Rs. 6,250	Rs. 6,900
Overheads	Rs. 4,500	Rs. 6,900
Normal loss	10% of input	Nil
Sale value of loss	Rs. 2 per kilo	
Output	2,300 kilo	Joint products
		A – 900 kilo
		B – 800 kilo
		C – 800 kilo

There were no opening or closing stocks in either process and the selling price of the output from process '2' were:

Joint product *A*	–	Rs. 24 per kilo
Joint product *B*	–	Rs. 18 per kilo
Joint product *C*	–	Rs. 12 per kilo

Required:

(*a*) Prepare an account for process 1 together with any loss or gain a/c you consider necessary to record the month's activities.

(*b*) Calculate the profit attributable to each of the joint product by apportioning the total costs from process 2.

(*i*) According to weight of output

(*ii*) By market value of production. *(C.A., Inter., May 1991)*

Solution:

PROCESS '1' ACCOUNT

	Qty.	*Amt.*		*Qty.*	*Amt.*
To direct materials	2,500	10,000	By normal loss	250	500
To direct labour		6,250	By process '2' a/c	2,300	20,700
To overhead		4,500			
To abnormal gain	50	450			
	2,550	21,200		2,550	21,200

NORMAL LOSS ACCOUNT

To process '1' a/c	250	500	By abnormal gain	50	100
			By sales	200	400
	250	500		250	500

ABNORMAL GAIN ACCOUNT

To normal loss a/c	50	100	By process 1 a/c	50	450
To costing P & L a/c		350			
	50	450		50	450

JOINT COST OF THREE PRODUCTS UNDER PROCESS 2

From process '1' – cost of production		20,700
Direct labour		6,900
Overheads		6,900
	Total	34,500

STATEMENT SHOWING APPORTIONMENT OF JOINT COST UNDER WEIGHT OF OUTPUT

A – For 2,300 kilo; joint cost is Rs. 34,500

For 900 kilo = $\frac{900 \times 34{,}500}{2{,}300}$ = Rs. 13,500

B – For 2,300 kilo; joint cost is Rs. 34,500

For 800 kilo = $\frac{800 \times 34{,}500}{2{,}300}$ = Rs. 12,000

C – For 2,300 kilo; joint cost is Rs. 34,500

For 600 kilo = $\frac{600 \times 34{,}500}{2{,}300}$ = Rs. 9,000

STATEMENT OF PROFIT UNDER WEIGHT OF OUTPUT

Joint product	*Output*	*S.P. per unit*	*Sales value*	*Share of joint cost*	*Profit loss*
A	900	24	21,600	13,500	8,100
B	800	18	14,400	12,000	2,400
C	600	12	7,200	9,000	1,800

STATEMENT SHOWING SHARE OF JOINT EXPENSE UNDER SALES VALUE

Joint product	Output	S.P. per unit	Sales revenue
A	900	24	21,600
B	800	18	14,400
C	600	12	7,200
			43,200

A – For total sale of 43,200 total joint expense is 34,500

For 21,600 = $\frac{21{,}600 \times 34{,}500}{43{,}200}$ = Rs. 17,250

B – For total sale of 43,200, total joint expense is 34,500

For 14,400 = $\frac{14{,}400 \times 34{,}500}{43{,}200}$ = Rs. 11,500

C – For total sale of 43,200 total joint expense is 34,500

For 7,200 = $\frac{7{,}200 \times 34{,}500}{43{,}200}$ = Rs. 5,750

CHAPTER 5

STATEMENT OF PROFIT UNDER SALES VALUE METHOD

Joint product	*Output*	*S.P. per unit*	*Sales value*	*Share of joint cost*	*Profit*
A	900	24	21,600	17,250	4,350
B	800	18	14,400	11,500	2,900
C	600	12	7,200	5,750	1,450
			43,200	34,500	8,700

Net Realisable Value

Under this method, in addition to deducting the profit and 'subsequent expenses from the sales value, the non-manufacturing cost such as selling and distribution overheads are also dedicated. This gives rise to net saleable value of the various joint product.

Problem 10. Calculate the estimated cost of production of by-product *X* and *Y* at the point of separation from the main product.

	By products *X*	By products *Y*
Selling price per unit	Rs. 12	Rs. 24
Cost per unit after separation from main product	Rs. 3	Rs. 5
Unit produced	500	200

Selling expenses amount to 25% of the total works cost—*i.e.*, including both pre-separation and post-separation works cost.

Selling prices are arrived at by adding 20% on total cost—*i.e.*, the sum of works cost and selling expense. *(Sri Sathya Sai University, B.Com., March 1991)*

Solution:

STATEMENT SHOWING SHARE OF JOINT EXPENSE

	X	*Y*
Sales	6,000	4,800
Less: Profit $\frac{20}{120} \times 6,000$	1,000	
$\frac{20}{120} \times 4,800$		800
Cost of production	5,000	4,000
Less: Subsequent expense	1,500	1,000
	3,500	3,000
Less: Selling expenses 25% on total cost of production $\left(\frac{25}{100} \times 5,000 \text{ and } \frac{25}{100} \times 4,000\right)$	1,250	1,000
Share in the joint cost	2,250	2,000

PRODUCT 'X' ACCOUNT

To share in joint cost	2,250	By sales	6,000
To subsequent expense	1,500		
To selling expense	1,250		
To profit	1,000		
	6,000		6,000

PRODUCT 'Y' ACCOUNT

	Rs.		Rs.
To share in joint cost	2,000	By sales	4,800
To subsequent expense	1,000		
To selling expense	1,000		
To profit	800		
	4,800		4,800

Problem 11. In a manufacturing concern production of *A* yields by-products *B* and *C*. The joint expenses of manufacture are:

Materials Rs. 8,500, Labour Rs. 9,000, Overheads Rs. 7,500. Subsequent expenses are as follows:

	Material	*Labour*	*Overhead*
A	Rs. 2,500	1,900	1,500
B	Rs. 1,200	1,600	900
C	Rs. 1,400	2,000	1,050

Selling price: A Rs. 30,000, B Rs. 20,000, C Rs. 15,000

Profit on S.P.: A 40%, B 30%, C 25%

Show how you would apportion the joint expenses and ascertain profit of each product.

(Osmania University, B.Com., March 1992)

Solution:

STATEMENT SHOWING APPORTIONMENT OF JOINT COST

	A	*B*	*C*	*Total*
Sales	30,000	20,000	15,000	65,000
Less: Profit	12,000	6,000	3,750	21,750
Cost of production	18,000	14,000	11,250	43,250
Less: Subsequent expenses	5,900	3,700	4,450	14,050
	12,100	10,300	6,800	29,200
Less: Selling & distribution overhead	1,938	1,291	971	4,200
Share in the joint cost	10,162	9,009	5,829	25,000

Note: The total joint cost given in the problem is Rs. 25,000. Whereas the total as per the above statement is Rs. 29,200. The difference of Rs. 4,200 (29,200 – 25,000) is selling and distribution overheads apportioned on the basis of sales value.

MAIN PRODUCT 'A' ACCOUNT

To material	8,500		By by-product B a/c	9,009
Add: Addition	2,500	11,000	By by-product C a/c	5,829
To labour	9,000		By sales	30,000
Add: Addition	1,900	10,900		
To overhead	7,500			
Add: Addition	1,500	9,000		
By selling & distribution overhead		1,938		
To profit		12,000		
		44,838		44,838

BY-PRODUCT 'B' ACCOUNT

To main product *A* a/c	9,009	By sales	20,000
To materials	1,200		
To labour	1,600		
To overhead	900		
To selling & distribution overhead	1,291		
To profit	6,000		
	20,000		20,000

BY-PRODUCT 'C' ACCOUNT

To main product *A* a/c	5,829	By sales	15,000
To materials	1,400		
To labour	2,000		
To overhead	1,050		
To selling & distribution overhead	971		
To profit	3,750		
	15,000		15,000

Problem 12. *(Net realisable value method):*A factory producing article 'A' also yields 'B' and 'C' as by-products. The joint cost of manufacture is:

Materials	Rs.	20,000
Labour	Rs.	4,000
Overheads	Rs.	16,000
	Rs.	40,000

Subsequent costs excluding selling and distribution overheads are as under:

	A	*B*	*C*
	Rs	*Rs*	*Rs*
Materials	3,000	2,600	2,000
Labour	400	300	200
Overhead	1,600	1,100	800
	5,000	4,000	3,000
Sales value	Rs. 60,000	48,000	40,000
Estimated profit on sales value	30%	25%	20%

Prepare a statement showing share in the joint cost and prepare the necessary accounts in respect of A, B and C. *(Sri Sathya Sai University, B.Com., (Hons) October 2007)*

Solution:

STATEMENT OF SHOWING SHARE IN THE JOINT COST

	A	*B*	*C*	*Total*
Sales	60,000	48,000	40,000	1,48,000
Less: Profit	18,000	12,000	8,000	38,000
Total Cost	42,000	36,000	32,000	1,10,000

Less: Subsequent expenses	5,000	4,000	3,000	12,000
	37,000	32,000	29,000	98,000
Less: Selling & distribution overhead	23,514	18,810	15,676	58,000
60 : 48 : 40				
Share in the joint cost	13,486	13,190	13,324	40,000

MAIN PRODUCT 'A' A/C

To Joint cost:		By-product B a/c	13,190
Materials	20,000	By-product C a/c	13,324
Labour	4,000	By-sales	60,000
Overheads	16,000		
To subsequent cost:			
Materials	3,000		
Labour	400		
Overheads	1,600		
To selling and distribution overhead	23,514		
To P & L a/c profit	18,000		
	86,514		86,514

BY-PRODUCT 'B' ACCOUNT

To main product A a/c	13,190	By sales	48,000
To subsequent cost			
– Materials	2,600		
– Labour	300		
– Overheads	1,100		
By selling & distribution overhead	18,810		
By P & L a/c – profit	12,000		
	48,000		48,000

BY-PRODUCT 'C' ACCOUNT

To main product 'A' a/c	13,324	By Sales	40,000
To subsequent costs:			
– Materials	2,000		
– Labour	200		
– Overhead	800		
To selling & distribution overhead	15,676		
To profit & loss a/c–profit	8,000		
	40,000		40,000

Problem 13. From a natural material, four *A, B, C* and *D* were obtained. The first three are joint products and the last product a "by-product" the joint products are further refined at Rs. 2, Rs. 1.60, and Re 1 per kg of output respectively. The selling prices are fixed at Rs. 6, Rs. 4.50, Rs. 3.50 and Re 0.50 per kg respectively to yield a profit margin of 20% on selling price. During the year the production of these four products is 8,000 units, 5,000 units and 6,000 units of the refined products and 2,000 units of by-product. Apportion the joint cost of Rs. 50,000 among these products and prepare process accounts of joint products. *(Sri Sathya Sai University, B.Com., (Hons) October 2006)*

CHAPTER 5

Solution:

STATEMENT SHOWING APPORTIONMENT OF JOINT COST

		A	*B*	*C*	*D*	*Total*
Sales		48,000	22,500	21,000	1,000	92,500
Less:	Profit (20%)	9,600	4,500	4,200	200	18,500
	Cost of production	38,400	18,000	16,800	800	74,000
Less:	Subsequent cost	16,000	8,000	6,000	–	30,000
		22,400	10,000	10,800	800	44,000
Less:	Selling & distribution overhead (224 : 100 : 108 : 8)	3,460	1,545	1,700	125	6,800
Share in the Joint cost		18,940	8,455	9,100	675	50,800

Note: Selling and Distribution overheads can also be apportioned on the basis of sales. The sales ratios are as follows

$$\frac{96}{185} : \frac{45}{185} : \frac{42}{185} \; \frac{2}{185}$$

on apportionment the selling and distribution overheads work out to Rs.

A	–	3528
B	–	1654
C	–	1543
D	–	73

PRODUCT 'A' A/C

To joint cost	18,940	By sales	48,000
To subsequent cost	16,000		
To selling & distribution overhead	3,460		
By P & L a/c–Profit	9,600		
	48,000		48,000

PRODUCT 'B' A/C

To joint cost	8,455	By sales	22,500
To subsequent cost	8,000		
To selling & distribution overhead	1,545		
By P & L a/c – profit	4,500		
	22,500		22,500

PRODUCT 'C' A/C

To joint cost	9,100	By sales	21,000
To subsequent cost	6,000		
To selling & distribution overhead	1,700		
To P & L a/c – profit	4,200		
	21,000		21,000

PRODUCT 'D' ACCOUNT

To joint cost	675	By sales	1,000
To selling & distribution overhead	125		
To P & L a/c – Profit	200		
	1,000		1,000

Problem 14. Bright Chemicals Ltd. electrolyses common salt to obtain three joint products caustic soda, chlorine and hydrogen. During 2004 the expenditure relating to the inputs for the common process amounted to Rs. 3,21,500. Subsequent expenses amounting to Rs. 1,60,000, Rs. 75,000 and Rs. 10,000 were incurred for caustic soda, chlorine and hydrogen respectively. The entire production was sold and Rs. 3,75,000, Rs. 2,50,000 and Rs. 60,000 were realised for caustic soda, chlorine and hydrogen respectively. The selling expenses were estimated at 5% of realisation from sales. The management expected profit @ 15%, 10% and 5%, respectively of realisation from sale of caustic soda, chlorine and hydrogen.

Prepare a statement showing the apportionment of Joint cost and prepare the process accounts of these three products. *(Sri Sathya Sai University, B.Com. (Hons.) October 2005)*

Solution:

STATEMENT SHOWING APPORTIONMENT OF JOINT COST

	Caustic Soda	*Chlorine*	*Hydrogen*
Sales	3,75,000	2,50,000	60,000
Less: Expected profit	56,250	25,000	3,000
	3,18,750	2,25,000	57,000
Less: Selling overheads	18,750	12,500	3,000
Cost of production	3,00,000	2,12,500	54,000
Less: Subsequent cost	1,60,000	75,000	10,000
Share in joint cost	1,40,000	1,37,500	44,000

CAUSTIC SODA ACCOUNT

To joint cost	1,40,000	By sales	3,75,000
To subsequent cost	1,60,000		
To selling expenses	18,750		
To P & L a/c–Profit	56,250		
	3,75,000		3,75,000

CHLORINE ACCOUNT

To joint cost	1,37,500	By sales	2,50,000
To subsequent cost	75,000		
To selling overhead	12,500		
To P & L a/c–profit	25,000		
	2,50,000		2,50,000

CHAPTER 5

HYDROGEN ACCOUNT

To joint cost	44,000	By sales	60,000
To subsequent cost	10,000		
To selling overhead	3,000		
To P & L a/c–Profit	3,000		
	60,000		60,000

3. Points Value Method or Survey Method or Weighted Average Cost Method

Where it is not possible to apply the physical unit method and sales value method because of their disadvantages, survey method is adopted to apportion the joint cost to joint products. Each joint products is assigned weights or points based upon quantity, size, difficulty of manufacture, time taken to manufacture, types of labour engaged, the cost of material used, etc. The output is multiplied by the weights or points thus assigned to get the product of each joint product. The total weight of all joint products is then obtained by adding them together. The product is divided by the total weight of all the joint products to get a quotient. The quotient multiplied by the joint cost gives the share of joint cost for every joint product. Expressed in the form of formula:

$$\text{Joint cost} = \frac{\text{Total weight assigned to each joint product}}{\text{Total weight of all products}} \times \text{Joint cost}$$

where;

Total weight assigned to each joint product

= No. of units produced × Weights or points assigned

Problem 15. From the following particulars calculate the share of every joint product

Units produced:	X	–	30,000 units
	Y	–	32,000 units
	Z	–	20,000 units
Joint cost incurred		–	Rs. 1,32,000
Weights assigned to each joint product	X	–	3 points
	Y	–	2.5 points
	Z	–	4 points

Solution:

STATEMENT SHOWING SHARE OF JOINT COST

Joint product	*Unit*	*Weight*	*Total weight*
X	30,000	3	90,000
Y	32,000	2.5	80,000
Z	20,000	4	80,000
			2,50,000

$$A = \frac{90{,}000}{2{,}50{,}000} \times 1{,}32{,}000 = \text{Rs. } 47{,}520$$

$$B = \frac{80{,}000}{2{,}50{,}000} \times 1{,}32{,}000 = \text{Rs. } 42{,}240$$

$$C = \frac{80,000}{2,50,000} \times 1,32,000 = \text{Rs. } 42,240$$

Problem 16. A and B Chemicals Ltd., manufacture three varieties of mannures by mixing chemical fertilizer in different proportions. The three varieties X, Y, and Z are manufactured in two operations, (a) mixing and (b) packing and had the following relative cost values on a point value basis per cwt of output:

Variety	*Mixing*	*Packing*
X	2	1
Y	3	1.5
Z	5	2.5

The total cost for each process for January is as follows:

Mixing	Rs. 1,45,000
Packing	Rs. 2,900
	1,47,900

Production for the month was

Mixture X	2,500 cwts
Mixture Y	1,500 cwts
Mixture Z	1,000 cwts
	5,000 cwts

Prepare a statement showing the cost per cwt of each variety per operation.

Solution:

STATEMENT SHOWING POINT VALUE OF OUTPUT

Mixture	*Cwt*	*Points value (Total)*	
		Mixing	*Packing*
X	2,500	5,000	2,500
Y	1,500	4,500	2,250
Z	1,000	5,000	2,500
	5,000	14,500	7,250

STATEMENT SHOWING APPORTIONMENT OF OVERHEAD

Mixture	*Mixing*		*Packing*		*Total*	
	Total	*Per cwt*	*Total*	*Per cwt*	*Amt.*	*Per cwt*
X	50,000	20	1,000	0.40	51,000	20.40
Y	45,000	30	900	0.60	45,900	30.00
Z	50,000	50	1,000	1.00	51,000	51.00
	1,45,000		2,900		1,47,900	

Working Note: The costs are apportioned on the basis of their point values. For *X* it is calculated thus:

$$\frac{5,000}{14,500} \times 1,45,000 = \text{Rs. } 50,000$$

and similarly for products *Y* and *Z*.

CHAPTER 5

Problem 17. Growmore chemicals Ltd. manufacture three varieties of mannures X, Y and Z by mixing chemical fertilizers in different proportions. Two operations (i) mixing and (ii) packing are involved. There is no costing system. On technical estimation, relative cost values are fixed on points, basis per weight of output. However, the financial accounts data are segregated into two operations mentioned above. The relevant data for a particular month are as follows:

Varieties	*Point-Values per unit weight*		*Production*
	Mixing	*Packing*	*(Tons)*
X	2	5	250
Y	3	3	150
Z	5	2	100
Total cost (Rs.)	1,45,000	19,000	

Compute cost per ton of each variety per operation. *(C.S. Inter., June 1997)*

Solution:

CALCULATION OF COST PER POINT

Variety	*Production (Tons)*	*Mixing*		*Packing*	
		Points	*Total*	*Points*	*Total*
(1)	(2)	(3)	4(2 × 3)	5	6 (2 × 5)
X	250	2	500	5	1,250
Y	150	3	450	3	450
Z	100	5	500	2	200
			1,450		1,900

Cost per point of mixing $\frac{1,45,000}{1,450} = 100$

Cost per point of packing $\frac{19,000}{1,900} = 10$

STATEMENT SHOWING COST PER TON

Variety	*Mixing*		*Packing*		*Total*
	Points	*Rs.*	*Points*	*Rs.*	
X	2	200	5	50	250
Y	3	300	3	30	330
Z	5	500	2	20	520

4. Average Unit Cost Method

Under this method, the total joint cost of manufacturing is divided by the total number of physical units produced. This results in average cost associated with the production of each unit of product regardless of its physical characteristics. Joint cost is apportioned to all joint products at this average cost.

Problem 18. From the following particulars calculate the share of joint cost of each joint product under average unit cost method:

Joint cost –Rs. 1,00,000

Joint products	X	–	2,500 units
	Y	–	3,500 units
	Z	–	4,000 units

Solution:

$$\text{Average cost per unit} = \frac{\text{Total joint cost}}{\text{Total units of joint product}}$$

$$= \frac{1,00,000}{10,000} = \text{Rs. } 10$$

STATEMENT SHOWING APPORTIONMENT OF JOINT COST

Joint products	*Units*	*Cost per unit*	*Amount*
X	2,500	10	25,000
Y	3,500	10	35,000
Z	4,000	10	40,000
			1,00,000

5. Marginal Cost Method

Under this method marginal cost is allocated to joint products on the basis of weight or quantity of each product. Fixed cost is apportioned on the basis of contribution made by each product. This method has the advantage of maximising profit by proper planning and product mix.

Problem 19. JBS Company operates a chemical process which jointly produces two joint products. The total joint cost is Rs. 2,750. It is the policy of the company to allocate the joint cost to products at different basis as mentioned below:

	Joint cost (Rs.)	*Basis of allocation*
Direct material	1,400	Sales value
Variable conversion	600	Weight
Fixed conversion	750	Contribution
	2,750	

	A	*B*
Sales	kg. 50	100
Selling price per kg.	Rs. 30	20

Calculate cost A and B.

Solution:

STATEMENT SHOWING ALLOCATION OF JOINT COST

	Total	*A*	*B*
Materials 15 : 20	1,400	600	800
Variable conversion 5 : 10	600	200	400
Total variable cost	2,000	800	1,200
Sales	3,500	1,500	2,000
Contribution	1,500	700	800
Fixed cost 7 : 8	750	350	400

STATEMENT OF COST OF A AND B

	Total	*A*	*B*
Total variable cost	2,000	800	1,200
Fixed cost	750	350	400
Total cost	2,750	1,150	1,600

DEFINITION AND FEATURES OF BY-PRODUCT

According to Sam R. Goodman and James S. Reece by-product is a term which is applied to products produced simultaneously that have a very minor sales value as compared with that of main product. According to Theodore Lang, by-products refer to any saleable or usable value incidentally produced in addition to the main product. In the words of Neuner, by-products are produced from the same process as joint products but that the by-products are essentially the secondary results of the operations. According to him the rule that is suggested for interpreting the meaning of "secondary" is that, if the value of a product is less than 10% of the total value of all products, it could be considered a by-product; not a co-product or joint product. According to Schlatter (cost accounting), by-product split-off from the original material in process and thus do not become part of the main product.

The characteristic features of a by-product may be listed out as follows:

(*a*) By-products are obtained simultaneously along with main product.

(*b*) The quantity of by-product obtained is less than the main product.

(*c*) By-products yield only minor sales value when compared to main product.

(*d*) By-products cannot be used as such and therefore require further processing to make it more acceptable by customer.

(*e*) Only limited control can be exercised over by-products.

Some examples of by-products are as follows:

1. *Meat packing*—Meat is main product, whereas bones, fats, hides, hairs are by-products.
2. *Sugar industry*—Sugar is main product—Baggasse and molasses are by-products.
3. *Cotton ginning*—Cotton fibre is main product, cotton seeds are by-product.
4. *Coke making*—Coke is main product, ammonia, coal tar, benzol and gas are by-products.
5. *Soap making*—Soap is main product—glycerin is by-product.

DIFFERENCE BETWEEN JOINT PRODUCTS AND BY-PRODUCTS

1. Obtaining the joint products is the main objective of an industry, whereas obtaining by-product is secondary objective.
2. There exists a direct quantitative relationship among joint products. It may not be true in case of by-products.
3. The value of joint product is significant whereas it is neglegible in case of by-product.
4. Joint product differs from by-products in respect of accounting treatment.
5. Greater degree of control can be exercised on joint products, whereas it is not possible to exercise utmost control over by-products.

TREATMENT OF BY-PRODUCTS

By products need not bear any portion of the joint costs incurred prior to the point of separation. This is so because production is undertaken not with the intention of obtaining by-product but with the intention of obtaining main product. However separate expenses or additional processing costs and selling and distribution expenses are to be charged to by-products obtained.

The only treatment that is to be given to by-products is the proper disposition of the net revenue received from by-product. There are two ways of measuring revenue obtained from by products. First is in terms of net sales revenue obtained from by product actually sold. Secondly, in terms of the net sales value of the by-product recovered, whether sold or not. Under the first method there exists no closing stock on hand at the end of the period. Whereas the second method carries a stock of by-products.

The accounting treatment of by-products depend upon two main consideration, *i.e.,* (*i*) the by-product is of little value, (*ii*) the by-product is of considerable value.

1. Where the By-product is of Little Value

This consideration of treating by-product is based on two difficulties, *viz.,* (*a*) It is uneconomical to incur expenses in determining the price of each unit of by-product, (*b*) It is also equally difficult to ascertain the cost of producing a by-product up to the point of separation. The amount realised on the sales of by-product is treated as a profit and recorded in the profit and loss account. If the amount realised is too less, the amount so realised may be credited to the main product, thereby reducing the cost of the main product.

2. Where By-product is of Considerable Value

Here, the cost of by-product is ascertained accurately on account of increased costs incurred. A separate account is opened for each by-product to which is debited the estimated cost up to the point of split-off. The main product account is credited with the amount realised on sale of by-product.

Accounting of By-products

Two methods are followed in accounting of by-products. This is shown in the following chart:

ACCOUNTING OF BY-PRODUCTS

Non-cost method or sales value method

(1) Other income method.
(2) Sales value of by-product included in the main product sale
(3) Sale value of by-product deducted from total cost
(4) Credit of by-product at sales value less selling and distribution cost
(5) Credit of by-product at sales value less cost incurred after split-off-point
(6) Credit of by-product at sale value less an estimated profit, (or reverse cost method) post split-off-cost, and selling and distribution cost

Cost method

(1) Opportunity or replacement cost method
(2) Standard cost method
(3) Joint cost method

I. Non-cost or Sales Value Method

1. Other income method: Under this method the sale value of by-product is credited to profit and loss account as "other income". The circumstances under which this method is used are as follows:

(*a*) Where the value of by-product is very less;

(*b*) Where the use of a more detailed method entails too much expense in comparison to benefits obtained;

(*c*) Where the carrying of by-product with the main product does not entail any appreciable difference in the cost of the main product.

The chief limitations of this method are as follows:

(*a*) Sometimes by-product may arise in one accounting period but it may be sold in another accounting year. This time gap distorts profits in different year;

(*b*) No value is given to by-product stock. This results in the over-valuation of main product stock. This again distorts the balance sheet;

(*c*) No attempt is made to control the by-product and it may result in losses due to fraud.

Problem 20. From the following data, prepare a profit and loss account to show the total profit earned by the sale of main product and by-product:

Total cost of production – 12,000 units @ Rs. 1.50 per unit
Closing stock – 2,000 units @ Rs. 1.50
Administration & selling expenses – Rs. 2,000
Sales of main product – 10,000 units @ Rs. 2
Sales by-product – Rs. 1,500

Solution:

STATEMENT SHOWING PROFIT EARNED BY MAIN PRODUCT

Cost of production	12,000 × 1.50	18,000		
Less: Closing stock	2,000 × 1.50	3,000		15,000
Administration and selling expenses				2,000
			Cost of sales	17,000
			Profit	3,000
			Sales	20,000

P & L A/C

By main product a/c	3,000
By by-product a/c	1,500
	4,500

2. Sale of by-product included in the main product sales: This method is resorted to (*a*) when by product could be sold without altering the condition in which they arise, (*b*) when by-product is not significant in relation to main product. The costs incurred on main product and by-product are ascertained and the total cost is deducted from the combined sales value to know the profit.

Problem 21. Prepare process account of main product for problem 16 under sale of by-product included in the main product sales.

Solution:

PROCESS ACCOUNT OF MAIN PRODUCT

To cost of production	18,000	By sales a/c	
To administration & selling expenses	2,000	Main product	20,000
To profit & loss a/c	4,500	By product	1,500
		By closing stock	3,000
	24,500		24,500

Note: The total profit credited under this method is Rs. 4,500 whereas in the first method it is only Rs. 3,000.

3. Sales value of by-product deducted from total cost: Under this method the sales value of by-product is deducted from total cost of by-product and main product. Its disadvantages are:

(*a*) Any inefficiency in the cost of manufacture of main product cannot be ascertained correctly if the credit from by-product is fluctuating.

(*b*) Closing stock for balance sheet purpose is valued at selling price of by-product.

4. Credit of by-product at sales less selling and distribution cost: The selling and distribution costs incurred on by-product are deducted from sale value of by-product and the net income thus obtained is deducted from the total cost of main product and by-product together or credited to the process account. The closing stock is valued at selling price less an estimated cost of selling and distribution cost of by-product.

5. Credit of by-product at sales value less cost incurred after split-off-point and selling and distribution cost: If by-product requires further processing after split-off-point then the post-split-off cost and selling and distribution cost incurred are deducted from the sale value of by-product and the net amount thus obtained is deducted from total cost or credited to process account. The closing stock is valued at selling price less an estimated selling and distribution cost and further processing costs after split-off-point.

6. Credit of by-product at sale value less an estimated profit, post-split-off-cost and selling and distribution overhead: Under this method, the credit is given to process account after deducting an estimated profit, post-split-off-cost and selling and distribution cost from the sale of by-products.

Problem 22. During November 2009, Splatter Ltd., recorded the following results:

Opening stock of main product	Nil
Opening stock of by-product	Nil
Cost of production	Rs. 1,20,000

Sale of main product amounted to 90% of output during the period and 10% of production was held as closing stock at 30th November 2009. Sales revenue from the main product during November 2009 was 1,50,000.

A by-product X is produced and output had a net sales value of Rs. 1,000. Of this output value of Rs. 700 was sold during the month and Rs. 300 was still in stock at 30th November 2009.

Calculate the profit under:

1. Other income method.
2. By-product income added to sales of main product.

CHAPTER 5

3. Sale income of the by-product deducted from the cost of production in the period.
4. Saleable value of the by-product deducted from the cost of product in the period.

Solution:

1. Other income method

Opening stock of main product	NIL
Add: Cost of production	1,20,000
	1,20,000
Less: Closing stock 10% of cost of production	12,000
Cost of goods sold	1,08,000
Profit	42,000
Sales	1,50,000
Profit from sale of main product	Rs. 42,000
Profit from sale of by-product	Rs. 700
Total profit	Rs. 42,700

2. By-product income added to sales of main product

Opening stock of main product	NIL
Add: Cost of production	1,20,000
	1,20,000
Less: Closing stock 10% of cost of production	12,000
Cost of goods sold	1,08,000
Profit	42,700
Sales (1,50,000 + 700)	1,50,700

3. Sales income of the by-product deducted from the cost of production in the period

Opening stock	NIL
Add: Cost of production (1,20,000 – 700)	1,19,300
Less: Closing stock 10% of cost of production	11,930
Cost of goods sold	1,07,370
Profit	42,630
Sales	1,50,000

4. Saleable value of the by-product deducted from the cost of product in the period

Opening stock	NIL
Add: Cost of production (1,20,000 – 1,000)	1,19,000
	1,19,000
Less: Closing stock – 10% of cost of production	11,900
Cost of goods sold	1,07,100
Profit	42,900
Sales	1,50,000

Problem 23. From the following particulars calculate the amount of profit under:

1. Other income method.
2. When sales value of by-product is added to sale of main product.
3. When sales value of by-product deducting selling and distribution expenses and expenses incurred after split-off-point is credited to main product sales a/c.

Total production cost	Rs. 31,500
Quantity produced	18,000 units
Sale of main product	15,000 units @ Rs. 2.50 per unit
By-product revenue received	Rs. 2,275
Selling & distribution expenses of main product	Rs. 3,250
Closing stock of main product	3,000 units @ Rs. 1.75
Selling & distribution expenses of by-product	Rs. 500
Additional process cost of packing by-product	Rs. 100

Solution:

1. Other income method—Statement showing profit

Cost of production	18,000 × 1.75 =	31,500	
Less: Closing stock	3,000 × 1.75 =	5,250	26,250
Add: Selling & distribution overhead			3,250
	Cost of sales		29,500
	Profit		8,000
	Sales		37,500

Total income:

Profit from main product		Rs, 8,000
Sale of by-product		Rs. 2,275
	Net income	10,275

2. Statement showing profit when sales value of by-product is added to sale of main product:

Cost of production	18,000 × 1.75 =	31,500	
Less: Closing stock	=	5,250	26,250
Add: Selling & distribution overhead			3,250
	Cost of sales		29,500
	Profit		10,275
	Sales (37,500 + 2,275)		39,775

3. Statement showing profit by crediting sales value less selling and distribution expenses and expenses incurred after split-off-point:

Cost of production	31,500	
Less: Closing stock	5,250	
Cost of goods sold		26,250
Add: Selling & distribution expenses of main product		3,250
		29,500
Profit		9,675
Sales (main product sales *i.e.,* 37,500 + 2,275 − 500 − 100 = 39,175 from sale of by-product less S & D expenses and additional cost of by-product		39,175

Problem 24. *(Reverse cost method):* From the following data obtained from the books of Grace Co. (Pvt.) Ltd., calculate (*a*) unit cost of main product and (*b*) unit cost of by-product:

	By-product Rs.	*Main product* Rs.
Sales	12,000	1,50,000
Process cost before separation	—	75,000
Process cost after separation	2,200	23,000
Selling & distribution expenses	1,500	12,000
Units produced	9,000	20,000

There is no opening and closing stock of inventories. The company allows a 20% gross profit for by-product.

Solution:

STATEMENT SHOWING SHARE IN JOINT COST

	By-product
Sales	12,000
Less: Gross profit 20%	2,400
	9,600
Less: Selling & distribution overhead	1,500
	8,100
Less: Subsequent cost	2,200
Share in the joint cost	5,900

STATEMENT OF COST

	Main product	*By product*
Sales value	75,000	—
Less: Share in joint cost of by-product	5,900	—
Joint cost	69,100	5,900
Add: Subsequent cost	23,000	2,200
Total cost	92,100	8,100
Unit cost	$= \frac{19,100}{20,000}$	$\frac{8,100}{9,000}$
	= 4.605	= 0.9

II. Cost Methods

1. Opportunity cost method or replacement cost method: Under this method, the by-products obtained in the production process is utilised either as indirect material or as raw material (direct material). The by-products are valued at the cost of replacing them. Hence the name replacement cost method. The term opportunity cost is used because it represents the cost which would have been incurred had the by-products been bought from external source. Examples of by-products which are treated under this method is, use of cotton waste arising from the processing of fine yarn may be used either as indirect material, *i.e.*, for cleaning the machines or as raw materials used in producing low quality yarn. Similarly, coke oven which is a by-product of coke may be used for heating the open hearth furnace. The equivalent heat units produced by its consumption may be ascertained and the coke account will be credited with the value thereof.

Problem 25. *(Opportunity cost method):* From problem No. 19 calculate the net income from the sale of main product under opportunity cost method taking the replacement cost of by-product used in production as Rs. 2,275.

Solution:

STATEMENT SHOWING NET INCOME OF MAIN PRODUCT UNDER OPPORTUNITY COST METHOD

Cost of production	18,000 × 1.75 =	31,500	
Less: Replacement cost of by-product used in production		2,275	
		29,225	
Less: Closing stock—3,000 units @ 1.6236		4,871	
i.e., $\frac{29,225}{18,000} = 1.6236$			24,354
Add: Selling & Distribution expenses			3,250
Cost of sales			27,604
Profit			9,896
Sales			37,500

2. Standard cost method: Under this method a standard cost is determined for each type of by-product based upon past experience. The main product account is credited with this standard cost of by-products. This enables to judge efficiency of main product because the standard cost of by-products is a constant figure.

3. Joint cost method: This method is used when the value of by-products is comparatively of a higher value. So the cost incurred up to the point of split-off (joint cost) is apportioned among the main product and all by-products using any of the method discussed under the topic treatment of joint products.

Problem 26. A Ltd., manufactures one main product and two by-products. During January 2009 the following data is available:

	Main product	*By-products*	
		A	B
Sales	Rs. 4,00,000	32,000	48,000
Joint cost	Rs. 1,55,200		
Subsequent expenses	Rs. 40,000	6,400	7,200
Percentage of profit to sales		10%	15%
Selling expenses as % of sales value	10%	5%	6%

No Stock exists. Prepare a statement showing profit under joint cost method.

Solution:

STATEMENT SHOWING APPORTIONMENT OF JOINT COST

	A	B
Sales	32,000	48,000
Less: Profit	3,200	7,200
	28,800	40,800
Less: Separation cost	6,400	7,200
	22,400	33,600
Less: Selling expenses	1,600	2,800
Shares in the joint cost	20,800	30,800

CHAPTER 5

Joint cost to be borne by main product	1,55,200
Less: Share of by product *A* & *B* (20,800 + 30,800)	51,600
	1,03,600

STATEMENT SHOWING PROFIT

	Main product	*A*	*B*
Share of joint expense	1,03,600	20,800	30,800
Post separation cost	40,000	6,400	7,200
Selling expenses	40,000	1,600	2,800
	1,83,600	28,800	40,800
Profit	2,16,400	3,200	7,200
Sales	4,00,000	32,000	48,000

Questions

I. Mark 'T' or 'F' in the space provided:

1. In theory, joint products can be distinguished from by-products. (*T*/*F*)
2. Under the relative sales value method of allocating costs to joint products, each of the joint products will show the same gross margin percentage if sold at the split-off-point. (*T*/*F*)
3. A common processing operations may yield several joint products and a number of by-products. (*T*/*F*)
4. If Rs. 1,00,000 of common processing costs are incurred to produce two products A and B with market value at split-off-point of A—Rs. 1,16,000 and B—Rs. 84,000; Rs. 58,000 of costs should be allocated to product A. (*T*/*F*)
5. Cost per unit of each product is the same if joint costs are apportioned on the basis of the physical measure of product. (*T*/*F*)
6. The sales value method apportions the joint costs on the basis of respective capacities of individual products to bear joint costs. (*T*/*F*)
7. The allocation of joint cost to joint and by-products does not affect the total profit or loss. (*T*/*F*)
8. Joint costs can be obtained separately and are thus divisible. Common costs cannot be obtained separately and are thus indivisible. (*T*/*F*)
9. Standard cost method of evaluating by-products can be used only in standard costing technique. (*T*/*F*)
10. It is possible for a main product of one industry to be the by-product of another. (*T*/*F*)

[***Answers:*** True 1, 2, 3, 4, 6, 7, 10, False 5, 8, 9]

II. Choose the correct answer:

1. Which one of the following statement relative to joint products and by-products is false
   (*a*) The preferred accounting treatment of by-products is to set them up in inventory at their net realisable-value and credit this amount to accounts in which the costs of the main product are recorded.
   (*b*) The common costs incurred in producing joint products may, under certain circumstances, be allocated to the products on a physical unit basis.

(*c*) For product costing purposes, the common costs incurred in producing joint products should be allocated to the products on a physical units basis.

(*d*) Costs allocated to joint products are useful to management for a variety of decisions.[ ]

2. Trimmings in timber industry should be treated as a

(*a*) Waste (*b*) Scrap

(*c*) Joint proudct (*d*) By-product [ ]

3. The primary purpose of apportioning joint costs of a processing centre to various products produced is to:

(*a*) Develop accurate processing cost variance product-wise

(*b*) Report more correct standard product costs for comparative analysis

(*c*) Establish inventory cost assigned to unsold units

(*d*) Record accurate cost of sales by product lines [ ]

4. The method of accounting for joint product costs that will produce the same gross profit rate for all the product is

(*a*) Sales value method (*b*) Other income method

(*c*) Opportunity cost method (*d*) Reverse cost method [ ]

5. Individual products, each of a significant value, produced simultaneously from the same raw material is known as:

(*a*) By-product (*b*) Main product

(*c*) Joint product (*d*) Co-product [ ]

6. Which one of the following method of accounting of by-product should be applied, when the by-product is to be utilised in the undertaking itself as material for some other process:

(*a*) Sales value method (*b*) Joint income method

(*c*) Opportunity cost method (*d*) Standard cost method [ ]

7. A bakery producing cakes, biscuits, and breads should be treated as:

(*a*) Main product (*b*) Joint product

(*c*) By-product (*d*) Co-product [ ]

8. A method of apportionment of joint costs to various joint products on the basis of their respective capacities to bear joint cost is

(*a*) Sales value method (*b*) Physical unit method

(*c*) Average cost method (*d*) Survey method [ ]

9. Credit is given to the process account at a predetermined value of the by-product under:

(*a*) Standard cost method (*b*) Sales value method

(*c*) Opportunity cost method (*d*) Points value method [ ]

10. Which one of the following are virtually always subsequent costs?

(*a*) Direct materials (*b*) Direct labour

(*c*) Administration overhead (*d*) Selling & distribution overhead [ ]

[***Answers:*** 1. (*d*), 2. (*b*), 3. (*d*), 4. (*a*), 5. (*c*), 6. (*c*), 7. (*b*), 8. (*a*), 9. (*a*), 10. (*d*)]

III. Simple Questions

1. Define joint product.
2. Define by-product.
3. Define a co-product.
4. Distinguish between a joint product and a by-product.
5. Distinguish between a by-product and scrap.

6. What is split-off-point ?
7. What is a subsequent expense ?
8. List out four methods of apportioning joint expense.
9. List out four methods of accounting of by-product.
10. List out four examples of industries producing joint products.

IV. Short Answer Questions

1. Explain the methods of treating joint cost in cost accounts.
(Sri Sathya Sai University, B.Com., March 1992)
2. Distinguish between joint product and by-product. Explain as to how cost share allocated between main product and by-products in cost account.
3. Enumerate the methods which may be employed in costing joint products.
(University of Kerala, B.Com., April 1989)
4. Explain briefly the methods of accounting for "by-product".
(Calicut University, B.Com., October 1989)
5. What are three most common methods of allocating joint cost of production to joint product ? Write a brief explantory note on each. *(Delhi University, B.Com., (Hons.) 1990)*
6. How would you deal with by-products in costing ?
   (*i*) Where they are of small total value ?
   (*ii*) Where they are of considerable total value ?
   (*iii*) Where they require further processing ? *(C.A. (Inter.), May 1988)*
7. Distinguish between by-product and joint product. *(Madras University, M.Com., April 1992)*

Exercises

1. (*Physical unit method*): A coke manufacturing company produces the following products by putting 5,000 tonnes of coal @ Rs. 25 per ton into a common process:

Coke	3,500 tonnes
Tar	1,200 tonnes
Ammonium sulphate	52 tonnes
Benzol	48 tonnes

Apportion the joint cost amongst the products on the basis of the physical unit method.

[**Ans.** Share of joint expenses are coke Rs. 91,150, Tar 31,250, Ammonium sulphate 1,350, Benzol 1,250]

2. (*Sales value method*): From the following information find out the cost of product *X* and *Y*. Product *Y* is the by-product of *X* and on sale of *Y* a profit of 20% on selling price is obtained usually:

	Joint cost	*Separate expense*	
		X	*Y*
Materials	15,000	3,000	2,000
Wages	8,000	1,200	800
Other expenses	6,000	1,500	700

Total amount realised by sale of *Y* was Rs. 8,000.

(Bangalore University, B.Com., Nov. 1991)

[**Ans.** Cost of production of product *A* is Rs. 16,100 and total of product *B* a/c is 8,000]

3. (*When profit of a product is not given*): Product X yields by-products Y and Z. The joint expenses of production and separate expenses are as follows:

	Joint expense	*Separate expenses*		
		X	Y	Z
Material	Rs. 4,000	800	640	720
Labour	Rs. 3,200	960	560	680
Overheads	Rs. 3,600	1,040	400	600
Selling price	—	16,800	8,000	7,200
Profit on sale	—	—	50%	$33\frac{1}{3}\%$

Show how would you apportion the joint expenses. Prepare the necessary accounts.

(Bangalore University, B.Com., November 1990)

[**Ans.** Profit of product is Rs. 8,400. Total of product Y and Z accounts are Rs. 8,000 and Rs. 7,000 respectively]

4. A factory producing article P also produces a by-product Q which is further processed into finished product. The joint cost of manufacture is given below:

Material	Rs. 5,000
Labour	Rs. 3,000
Overheads	Rs. 2,000
	Rs. 10,000

Subsequent costs are given below:

	P	*Q*
Material	3,000	1,500
Labour	1,400	1,000
Overheads	600	500
Selling prices	16,000	8,000
Estimated profit on selling price	25%	20%

Assume that selling and distribution expenses are in proportion of sales price. Show how you would apportion joint costs of manufacture and prepare process accounts relating to products *P* and *Q*. *(Sri Sathya Sai University, B.Com., July 1990)*

[**Ans.** Share in the joint costs are P—Rs. 6,733; Q—3,267; Total of product P a/c is Rs. 19,267 and Q a/c is Rs. 8,000]

5. In an oil mill, four products emerge from a refining process. The total cost of input during the quarter ending March 1983 is Rs. 1,48,000. The output, sales and additional processing costs are as under:

Products	*Output in litres*	*Additional processing cost after split-off-point*	*Sales value*
AOXE	8,000	Rs. 43,000	Rs. 1,72,500
BOXE	4,000	Rs. 9,000	Rs. 15,000
COXE	2,000	—	—
DOXE	4,000	Rs. 1,500	Rs. 45,000

CHAPTER 5

In case these products were disposed off at the split-off-point, that is before further processing, the selling price would have been

AOXE	BOXE	COXE	DOXE
Rs. 15	Rs. 6	Rs. 3	Rs. 7.50

Prepare a statement of profitability based on

(*i*) If the products are sold after further processing is carried out in the mill

(*ii*) If they are sold at the split-off-point (*CA Inter., May 1983*)

[**Ans.** Profit under case (i) – AOXE Rs. 30,833, BOXE – loss Rs. 13,733, COXE – profit Rs. 1,067, DOXE – profit Rs. 18,833. Under case (ii) – AOXE Rs. 21,333, BOXE Rs. 4,267, COXE Rs. 1,067, DOXE Rs. 5,333]

6. (*Points value method*): It is the policy of a company to allocate joint cost on points basis. The data are as follows:

Joint cost incurred – Rs. 1,08,000

Product	*No. of units produced*	*Relative evaluation of points*
P	100	40
Q	150	30
R	200	20
S	100	20

Calculate how joint cost will be allocated.

[**Ans.** P's share Rs. 28,800, Q's share – Rs. 32,400, R's share Rs. 28,800, S's Rs. 68,000]

7. (*On by-product*): The yield of a certain process is 80% as to the main product, 15% as to the by-product and 5% as to the process loss. The material put in process (5,000 units) cost Rs. 23,75 per unit and all other charges are Rs. 14,250, of which power cost a accounted for $33\frac{1}{3}\%$. It is ascertained that power is chargeable as to the main product and by-product in the ratio of 10 : 9.

Draw up a statement showing the cost of the by-product. (*CA, Inter., May 1988*)

[**Ans.** Cost of by-products is Rs. 22,500]

8. (*Sale value method*): Product 'A' yields by-products 'B' and 'C', the joint expenses of manufacture are:

Materials Rs. 5,000, Labour Rs. 4,000 on cost Rs. 4,500

The subsequent expenses are as follows:

	A *Rs.*	*B* *Rs.*	*C* *Rs.*
Materials	1,000	800	900
Labour	1,200	700	850
Oncost	1,300	500	750
	3,500	2,000	2,500
Selling price	21,000	10,000	9,000
Profit on sales	50%	50%	$33\frac{1}{3}\%$

Show how would you apportion the joint cost of manufacture and prepare main product *A* account and by-products *B* and *C* account.

(*Sri Sathya Sai University, B.Com., (Hons.) August 2003*)

[**Ans.** Total of main product 'A' a/c Rs. 27,500,
Total of by-product 'B' a/c Rs. 10,000
Total of by-product 'C' a/c Rs. 9,000]

9. A factory producing article 'A' also yields 'B' and 'C' as by products. The joint cost of manufacture as:

Materials	Rs.	10,000
Labour	Rs.	2,000
Overheads	Rs.	8,000
		20,000

Subsequent costs are as under:

	A	*B*	*C*
	Rs.	*Rs.*	*Rs.*
Material	1,500	1,300	1,000
Labour	200	150	100
Overhead	800	550	400
	2,500	2,000	1,500
Selling price	30,000	24,000	20,000
Estimated profit on Selling price	30%	25%	20%

Show how would you propose to apportion the joint cost of manufacture and prepare the necessary statement in respect of *A*, *B* and *C*.

(Sri Sathya Sai University, B.Com. (Hons.) October 2003)

[**Ans.** Total of product '*A*' A/c Rs. 43,257, Total of product '*B*' a/c Rs. 24,000, Total of product '*C*' a/c Rs. 20,000.]

10. Product '*A*' yield by-products *B* and *C*. The Joint expenses of manufacture are:

Materials Rs. 8,500, Labour Rs. 9,000, Overheads Rs. 7,500

Subsequent expenses are as follows:

	Materials	*Labour*	*Overhead*
	Rs.	*Rs.*	*Rs.*
A	2,500	1,900	1,500
B	1,200	1,600	900
C	1,400	2,000	1,050

Selling price: A	Rs. 30,000
B	Rs. 20,000
C	Rs. 15,000
Profit on selling price: A	40%
B	30%
C	25%

Show (*i*) statement showing apportionment of Joint cost and (*ii*) product *A*, *B* and *C* account.

(Sri Sathya Sai University, B.Com. (Hons.) February 2003)

[**Ans.** Total of main product '*A*' a/c Rs. 44,838
Total of by-product '*B*' a/c Rs. 20,000
Total of by-product '*C*' a/c Rs. 15,000]

CHAPTER 5

11. A company manufactures product 'A' which yields two by-products 'B' and 'C'. In a month the joint cost was Rs. 20,600. Subsequent expenses were:

	A	*B*	*C*
	Rs.	*Rs.*	*Rs.*
Materials	300	200	150
Direct wages	400	300	200
Overhead	300	270	280
	1,000	770	630

Gross sales value of product 'A' 'B' and 'C' was Rs. 15,000, Rs. 10,000 and Rs. 5,000 respectively. It was estimated that the net profit as percentage of sales in the case of 'B' and 'C' would be 25% and 20% respectively. Ascertain the profit earned by 'A' and prepare product accounts.

(Sri Sathya Sai University, B. Com. (Hons.), October 2002)

[**Ans.** Total of product 'A' a/c Rs. 25,100
Total of product 'B' a/c Rs. 10,000
Total of product 'C' a/c Rs. 5,000]

12. Product 'A' yields by-products *B* and *C*. The joint expenses of manufacture are material Rs. 5,000, Labour Rs. 4,000: Overheads Rs. 4,500.

The subsequent expenses are as follows:

	A	*B*	*C*
	Rs.	*Rs.*	*Rs.*
Materials	1,000	800	900
Labour	1,200	700	850
Overhead	1,300	500	750
	3,500	2,000	2,500
Selling price	21,000	10,000	9,000
Profit percentage on sale	50%	50%	$33\frac{1}{3}$

Show how would you apportion the joint expenses of manufacture and prepare main product 'A' a/c and by-products 'B' and 'C' accounts.

(Sri Sathya Sai University, B. Com. (Hons.), February 2001)

[**Ans.** Total of product 'A' a/c Rs. 27,500, Total of product 'B' a/c Rs. 10,000, Total of product 'C' a/c Rs. 9,000.]

13. Two products *A* and *B* are obtained in a crude form and require further processing at a cost of Rs. 5 for *A* and Rs. 4 for *B* per unit before sales. Assuming a net margin of 25% on cost, their sale prices are fixed at Rs. 13.75 and Rs. 8.75 per unit respectively. During the period the joint cost was Rs. 88,000 and the output were

A–8,000 Units, B–6,000 units

Ascertain the joint cost per unit. *(University of Delhi, B. Com. (Hons.) 2002)*

[**Ans.** Cost per unit of A–Rs. 8, Cost per unit of B–Rs. 4]

14. In the course of manufacture of the main product 'P', by-products 'A' and 'B' also emerge. The joint expenses of manufacture amount to Rs. 1,19,550. All the products are processed further after separation and sold as per details given below.

	Main products	By-products	
	P	A	B
	(Rs.)	(Rs.)	(Rs.)
Sales	90,000	60,000	40,000
Cost beyond split-off Stage	6,000	5,000	4,000
Profit as percentage of sales	25%	20%	15%

Selling and distribution overheads are absorbed as percentage of cost of sales. Prepare a statement showing the apportionment of joint cost to the main product and by-products. Also prepare main product 'P' account. *(University of Delhi, B.Com. (Hons.), 2001)*

[**Ans.** Total of main-product 'P' a/c Rs. 1,54,800]

❑❑❑

CHAPTER 6

OPERATING OR SERVICE COSTING

DEFINITION, CHARACTERISTICS AND APPLICATIONS OF OPERATING COSTING

The ICMA terminology defines operating cost as "the cost of providing a service" and operating costing as "that form of operating costing which applies where standardised services are provided either by an undertaking or by a service cost-centre within an undertaking". The method may be used where the service is not completely standardised but it is convenient to regard it as such, and to calculate average costs per period in relation to the standardised unit of measurement. From these definitions it is clear that the cost of providing a service is known as operating cost and the method employed to ascertain the cost of such service is known as operating costing.

Characteristic Features of Operating Costing

The characteristic features of operating costing are as follows:

1. The industries which adopt this method of costing do not produce any tangible goods. Instead they are concerned with rendering the services to public at large. They render uniform service to all those who apply for such services.
2. The cost unit used by such undertaking is known as composite unit as compared to the use of simple cost unit by rest of the undertakings. Some of the examples of composite units used by the undertakings are as follows:

Railways	–	Passenger-kilometer or ton-kilometer
Hospitals	–	Patient-day
Hotel (Lodging)	–	Room-day
Hotel (Boarding)	–	Plate-meals
Gas distribution	–	Cubic-meter

3. Classification of expenses into fixed and variable plays an important role. The cost of rendering additional service is affected by variable cost.
4. In most of the cases, it involves many stages and processes in converting basic materials to the ultimate service.
5. Costs are usually computed period-wise. However, under special circumstances costs are computed order-wise as in the case of utilisation of vehicles, use of road-roller, etc.
6. The demand for the services of industries adopting this method of costing fluctuates. Hence it is difficult to estimate the cost accurately.
7. There is no difficulty faced in respect of valuation of work-in-progress or closing stock when compared to other industries as closing stock does not exist in such organisation.

8. Only medium and large sized undertakings deal in manufacturing and rendering services. Such industries assume the position of monopolistic undertakings or public utility undertakings.

Applications: Operating costing is applied to those organisations which render service externally (*i.e.,* to public at large) or internally (*i.e.,* to various departments of the same organisation). Some examples of services rendered to outsiders are electricity supply, water supply, gas supply, boarding and lodging and so on. While services such as repairs and maintenance, purchasing and storage, internal transport, etc. constitute services internally. There is a point of difference in respect of accounting of services rendered. Whereas the objective of accounting external service is to know the total cost of manufacturing and profit on provision of such service, the object of internal service is to facilitate apportionment of service departments cost to various other departments.

Operating costing is applied to the following undertakings:

1. Transport undertakings such as Roadways, Railways, Tramways and Airways.
2. Municipal services such as supply of water, street lights, etc.
3. Steam and electricity undertakings.
4. Hotels covering boarding and lodging.
5. Hospitals.
6. Educational institutions.
7. Public libraries.
8. Service departments in big factories.
9. Cinemas.
10. Distribution of gas, air compressor, air-conditioning.
11. Sports and recreational clubs.
12. Services such as supply cranes, road roller, water pumping, fire extinguishers, etc.

COMPONENTS OF TOTAL COST IN OPERATING COSTING

As compared to other methods of costing, where the total comprises of prime cost, factory cost, and office cost, the total cost under operating costing is classified under the following headings:

1. Capital expenditure: This includes:

(*a*) Buildings

(*b*) Plant

(*c*) Vehicles

(*d*) Equipments from the use of which service is provided.

2. Direct or variable cost: This includes:

(*a*) Expenses on medicines (in hospitals)

(*b*) Fuel (in railways, buses, trucks, taxies)

(*c*) Pilots salaries (in Airways)

(*d*) Provisions (in hotels).

3. Fixed cost: This includes:

(*a*) Depreciation

(*b*) Rent and Rates

(*c*) Taxes and insurance

(*d*) Licence fee

TRANSPORT COSTING

The main objects of transport costing are (*a*) to ascertain the cost of operating a vehicle and (*b*) to fix up a fair for carrying passengers or goods for a certain distance.

Advantages of Transport Costing

The advantages of transport costing are as follows:

1. To decide whether to own a vehicle or to hire a vehicle.
2. To provide a basis for apportioning service departments expenses to production departments.
3. To enable the owner of a vehicle to charge fare to the clients, including concessional and special rates.
4. To know efficiency by comparing the cost of maintaining and running one vehicle with similar type of other vehicle; *e.g.,* car vs. matador van.
5. To choose between two modes of transport, as for example between lorry or railway in order to deliver goods to customers.

Components of Transport Costing

The total cost of operating service by a transport company is as follows:

1. Standing charges or fixed cost: These include:

(*a*) Administrative expenses such as salaries of manager, accountant, mechanic, etc.

(*b*) Garage rent

(*c*) Taxes and insurance of vehicles

(*d*) Wages of drivers, conductors, and cleaners

(*e*) Depreciation of vehicles.

2. Running charges or variable cost: These include:

(*a*) Petrol

(*b*) Diesel, oil

(*c*) Lubricant

3. Maintenance cost:

(*a*) Repairs and maintenance charges

(*b*) Tyres, tubes, batteries, etc.

Recording of Cost Under Transport Costing

Two important documents are used under transport costing. They are as follows:

1. Daily report sheet or log sheet: This document is used when a owner owns a fleet of vehicles. This report is prepared either by a driver, foreman or a clerk at the garage or depot. The purpose of this document is to enable the accumulation and control of cost. This document contains particulars about each journey made by the vehicle. These particulars enable the owner to utilise the vehicle without allowing for idle running capacity. Besides this, the data in this report also enables to apportion expenses to other vehicles. A proforma of this document is shown below:

DAILY REPORT SHEET OR LOG SHEET

Vehicle No .... No. of sheet ....
Route .... Date ....
Driver .... Starting time ....
Registration No. ....

Particulars of trip

Trip No.	From	To	Passenger or Goods		Distance	Time	
			Out	Enroute		Out	In

Employees name:
Driver ....
Cleaner ....
Conductor ....
Machinic ....

Delays:
Loading ....
Traffic ....
Accident ....
Breakdown ....

Supplier:
Petrol ....
Oil ....
Grease ....
Tyre ....

2. Operating cost sheet or cost summary and performance statement: This document helps in calculating cost per carrying passenger or goods for one kilometer. The cost of operating the vehicles help the owner in the following ways:

(*a*) To avoid waste of oil, fuel, etc.
(*b*) To know the efficiency by comparing with other vehicles.
(*c*) Facilitates in quoting hire charges.
(*d*) Cost of idle vehicle and lost running time can be known.
(*e*) Helps in apportioning overheads to other departments.
(*f*) Helps in controlling cost by comparing operating cost of one year with previous year.

The operating cost sheet is prepared on periodical basis. The information for this purpose is collected from log sheet.

A specimen of this document is shown below:

COST SUMMARY AND PERFORMANCE STATEMENT

Registration No. .... Month ended ....
Capacity ....

Monthly Charges

A-Operating charges
1. Petrol ....
2. Oil ....

B-Maintenance charges
1. Tyres ....
2. Repairs ....

C-Fixed charges
1. Insurance ....
2. Interest ....

3. Grease	3. Overhead	3. Depreciation ....
4. Driver	4. Garage	4. Tax, licence ....
5. Assistant		5. Others ....
6. Mechanics		
Total		

Monthly Cost Sheet

1. Total capital cost

Performance record

2. Days operated
3. Days idle
4. Days maintained
5. Total hours operated
6. Total kilometers covered
7. Total trips made

Performance averages

8. Average kilometers per day maintained	6 ÷ 4
9. Average kilometers per day operated	6 ÷ 2
10. Average kilometers per trip	6 ÷ 7

Costs for the month

11. Total expenses for the month (A + B + C)	
12. Cost per day operated	11 ÷ 2
13. Cost per day maintained	11 ÷ 2
14. Cost per kilometer operated	11 ÷ 6
15. Cost per hour	11 ÷ 5

Units of Costs in Transport Costing

To measure the efficiency of running vehicles different cost units are used. Important among those are:

1. The tonne-kilometer: These are of two types:

(*a*) *The absolute tonne-kilometer:* It refers to unit cost of carrying one tonne of goods over a distance of 1 kilometer. Expressed in the form of formula:

Absolute tonne km = Weight carried × Kilometer run for each section of the trip

Problem 1. *(Calculation of absolute tonne-kilometer):* A truck carries 4 tonnes of goods. It unloads 1 tonne after 7 km and the balance after travelling 5 km. It runs empty on the return trip. Calculate absolute tonne-kilometer.

Solution:

(1 ton × 7) + (3 tons × 12) + (0 ton × 12)

7 + 36 + 0 = 43 Absolute ton-km.

(*b*) *Commercial tonne-kilometer:* This unit is used to overcome the limitation of calculating the unloaded capacity at various distance covered. This method is based on the assumption of an average, wherein it is assumed that the vehicle runs with half load throughout the distance covered (after having unloaded part of the goods at different places). Expressed in the form of formula:

$$\text{Commercial tonne-kilometer} = \frac{\text{Tonnage loaded on lorry + Distance travelled}}{2}$$

Problem 2. A truck starts with a load of 10 tonnes of goods from station P. It unloads 4 tonnes at station Q and rest of the goods at station R. It reaches back directly to station P after getting reloaded with 8 tonnes of goods at station R. The distances between P to Q, Q to R and then from R to P are 40 kms., 60 kms., and 80 kms., respectively. Compute absolute tonne-kms.

(University of Delhi, B.Com., (Hons) April 1998)

Solution:

From	P to Q	10 tonnes × 40 kms	=	400
From	Q to R	6 tonnes × 60 kms	=	360
From	R to P	8 tonnes × 80 kms	=	640
	Total Tonne-kms.			1,400

Problem 3. *(Calculation of commercial ton-kilometer):* A vehicle Ka-03-2,758 runs with 6 tonnes on a trip. 2 tonnes were unloaded after travelling a distance of 10 km; 1 ton after a further travels of 8 km; yet another ton after travelling a further distance of 4 kms and the remaining 2 tonnes after a further 5 km travel. It then returns empty, covering a distance of 13 kms.

Calculate (*a*) The absolute tonne-kilometer

(*b*) Commercial tonne-kilometer.

Solution:

(*a*) Absolute tonne-kilometer

(6 × 10) + (4 × 8) + (3 × 4) + (2 × 5) + (0 × 13)

= 60 + 32 + 12 + 10 + 0

= 114 tonne-kilometer

(*b*) Commercial tonne-kilometer

$$\frac{6 \times 40}{2} = 120 \text{ tonne-kilometer.}$$

1. **The kilometer run:** This unit is used to know the performance of vehicles, where carrying of goods is not involved. The cost of transporting the staff of a company can be related to this unit of cost.

2. **Cost per hour:** This unit is used when vehicles are hired out by customers at a certain rate per hour. This unit covers all the costs of operating the vehicle along with a profit margin. The hour for this purpose is taken as "hired-out hour".

Steps to ensure optimum utilisation of vehicles: In order to increase the efficiency of operating the services, the following steps are suggested:

1. To select the shortest routes for the journey.
2. To utilise the full loading capacity. As far as possible the return journey is to be so planned that it does not return empty.
3. Delay in transit is to be totally avoided. Repairs are to be constantly undertaken even if there is minor defect in the vehicle so as to avoid frequent breakdown of vehicle.
4. To ensure economic consumption of petrol diesel.
5. The size of the vehicle should suit the specific purpose. The owner has to buy different sizes of vehicles or with different capacities for the right useful of vehicle or the right purpose.

Problem 4. A lorry starts with a load of 20 tonnes of goods from station *A*. It unloads 8 tonnes at station *B* and rest of goods at station *C*. It reaches back directly to station *A* after getting reloaded with 16 tonnes of goods at station *C*. The distance between *A* to *B*, *B* to *C* and then from *C* to *A* are 80 kms, 120 kms and 160 kms respectively. Compute "Absolute tonnes-kms" and "Commercial tonnes-kms". *(CA Inter., November 1999)*

Solution:

Absolute ton-kms = 20 tonnes × 80 kms + 12 tonnes × 120 kms + 16 tonnes × 160 kms = 5600 ton-kms.

Commercial ton-kms = Average load × Total kms travelled

$$\left(\frac{20+12+16}{3}\right) \times 360 \text{ kms.}$$

= 5,760 tons-kms.

Problem 5. A truck starts with a load of 10 tonnes of goods from station *P*. It unloads 4 tonnes at a station *Q* and rest of the goods at station *R*. It reaches back directly to station *P* after getting reloaded with 8 tonnes of goods at station *R*. The distances between *P* to *Q*, *Q* to *R* and then from *R* to *P* are 40 kms, 60 kms and 80 kms respectively. Compute "Absolute tonne-kms" and commercial tonne km. *(CA, Inter., May 1995)*

Solution:

Absolute ton-km = 10 tons × 40 kms + 6 tons × 60 kms + 8 tons × 80 kms = 1,400 Ton-kms

Commercial ton-km = Average load × Total kilometer travelled

$$= \left(\frac{10+6+8}{3}\right) \times 180 \text{ ton-kms}$$

8 × 180 = 1,440 ton-kms

Problem 6. A 40 seater tourist bus operates in a month as follows:

	Capacity	*From*	*To*	*Distance*
First 10 days	Full	Hyderabad	Khammam	160 kms
Next 15 days	75%	Hyderabad	Khazipet	250 kms
Next 5 days	60%	Hyderabad	Koppidi	100 kms

The total expenses are Rs. 3,77,000. Ascertain cost of one passenger km.

(S.V. University, B.Com., April 1999)

Solution:

No. of passenger kilometer = No. of Days × Capacity × Capacity used × Distance × To and Fro

$$\text{First 10 days} = 10 \times 40 \times \frac{100}{100} \times 160 \times 2 = 1{,}28{,}000$$

$$\text{Next 15 days} = 15 \times 40 \times \frac{75}{100} \times 250 \times 2 = 2{,}25{,}000$$

$$\text{Next 5 days} = 5 \times 40 \times \frac{60}{100} \times 100 \times 2 = 24{,}000$$

Total passenger-kms 3,77,000

$$\text{Cost per passenger km} = \frac{\text{Total expenses}}{\text{Total passenger kms}}$$

$$= \frac{3{,}77{,}000}{3{,}77{,}000} = 1 \text{ per passenger kms}$$

Problem 7. Sri Venkata Rao runs a tempo service at Anatapur. He furnishes you, the following data and wants you to compute the operating cost per running kilometer:

Cost of vehicle	Rs. 1,50,000
Road licence	Rs. 1,500 p.a.
Supervisors salary	Rs. 26,500 p.a.
Driver's wages per hour	Rs. 10
Cost of fuel per litre	Rs. 8
Repairs and Maintenance per km	Rs. 5
Tyre cost per km	Rs. 2
Garage rent p.a.	Rs. 7,200
Insurance	Rs. 2,000 p.a.
Kilometers running per litre	5
Kilometers run during the year	6,000
Estimated life of the vehicle	75,000 kms
Charge interest at 10% p.a. on the cost of vehicle	
The vehicle runs 20 kms per hour on an average	

(S.V. University, B.Com., October 1999)

Solution:

OPERATING COST SHEET

	Per annum	*Per running km*
Standing charges:		
Road licence	1,500	
Supervisor salary	26,500	
Garage rent	7,200	
Insurance	2,000	
Interest $\left(1{,}50{,}000 \times \frac{10}{100}\right)$	15,000	
	52,200	

Per running km $= \frac{52,200}{6,000}$ 8.70

Operating charges:

Driver's wages $\frac{10}{20} \times 6,000$	= 3,000	
Fuel $\frac{8}{5} \times 6,000$	= 9,600	
Repairs and maintenance 5 × 6,000	= 30,000	
Tyre cost 2 × 6,000	= 12,000	
Depreciation $\frac{1,50,000}{75,000} \times 6,000$	= 12,000	
	66,600	
Per running km = $\frac{66,600}{6,000}$		11.10
Operating cost per running km		19.80

Problem 8. A transport service company is running five buses between two towns which are 50 kms apart. Seating capacity of each bus is 50 passengers. The following particulars were obtained from their books for April 1998:

	Rs.
Wages of drivers, conductors and cleaners	24,000
Salaries of office staff	10,000
Diesel oil and other oil	35,000
Repairs and maintenance	8,000
Taxation, insurance etc.	16,000
Depreciation	26,000
Interest and other expenses	20,000
	1,39,000

Actually passengers carried were 75% of seating capacity. All buses ran on all days of the month. Each bus made one round trip per day.

Find out the cost per passenger kilometer *(CA, Inter., Nov. 1998)*

Solution:

CALCULATION OF PASSENGER KILOMETER

No. of Buses × Distance × "To" and "Fro" × Seating capacity × Percentage of seating capacity × No. of days in a month

$$= 5 \times 50 \times 2 \times 50 \times \frac{75}{100} \times 30 = 5,62,500 \text{ kms}$$

OPERATING COST SHEET

Standing charges:

Wages of drivers, conductors and cleaners	24,000
Salaries of office staff	10,000

Taxation, Insurance	16,000	
Interest and other expenses	20,000	70,000
Running and Maintenance cost:		
Repairs and maintenance	8,000	
Diesel and other oil	35,000	
Depreciation	26,000	69,000
		1,39,000

$$\text{Cost per passenger km} = \frac{1,39,000}{5,62,500} = 0.2471$$

Problem 9. Anami Transport company has been given a route 40 km long to run a bus. The bus costs the company a sum of Rs. 1,00,000. It has been insured at 3% p.a. and the annual tax will amount to Rs. 2,000. Garage rent is Rs. 200 p.m. Annual repairs will be Rs. 2,000 and the bus is likely to last for 5 years.

The driver's salary will be Rs. 300 p.m and the conductor's salary will be Rs. 200 p.m. in addition to 10% of takings as commission (to be shared by the driver and the conductor equally)

Cost of stationery will be Rs. 100 p.m. Manager-cum Accountant's salary is Rs. 700 p.m.

Petrol and oil will be Rs. 50 per 100 km. The bus will make 3 up and down trips carrying, on an average, 40 passengers on each trip. Assuming 15% profit on takings, calculate the bus fare to be charged from each passenger. The bus will run on an average 25 days in a month.

(ICWA, Inter., December 1998)

Solution:

CALCULATION OF EFFECTIVE PASSENGER-KM PER MONTH

40 km × 2 (up and down) × 3 trips × 25 days × 40 passengers

= 2,40,000 passenger-kms per month

CALCULATION OF TOTAL KM RUN PER MONTH

40 kms × 2 (up and down) × 3 trips × 25 days

= 6,000 kms

OPERATING COST SHEET

		P.A. Rs.	P.M Rs.
Standing charges:			
Insurance @ 3% on Rs. 1,00,000	=	3,000	
Tax	=	2,000	
Garage rent @ Rs. 200 p.m.	=	2,400	
Driver's salary @ Rs. 300 p.m.	=	3,600	
Conductor's salary @ Rs. 200 p.m.	=	2,400	
Stationery @ Rs. 100 p.m.	=	1,200	
Manager-com-Accountant's salary @ Rs. 700 p.m.	=	8,400	
Total		23,000	1,916,67

Variable charges:

Depreciation	$\frac{1,00,000}{5}$	20,000	1,666.67
Repairs		2,000	166.66
Petrol and oil	$\left(\frac{5}{100} \times 40 \text{ km} \times 2 \times 3 \times 25\right)$		3,000.00
	Total cost without commission		6,750.00
Commission (*See note 1*)			900.00
Profit (*See note 2*)			1,350.00
	Total takings		9,000.00

Fare per passenger km $\frac{9,000}{2,40,000} = 0.0375$

Or Fare per passenger $\frac{9,000}{6,000} = 1.50$

Working Note:

Commission to driver and conductor @ 10% of total takings and profit @ 15% of the same.

Let total taking be x

Commission @ 10% $= \frac{x}{10}$

Profit @ 15% of takings $= \frac{15x}{100} = \frac{3x}{20}$

Total cost without commission $= 6,750$

$$\therefore \quad x = \frac{6,750}{1} + \frac{x}{10} + \frac{3x}{10}$$

$$x = \frac{1,35,000 + 2x + 3x}{20}$$

$$\text{or} \quad 20 \times x = 1,35,000 + 2x + 3x$$

$$20x - 2x - 3x = 1,35,000$$

$$15x = 1,35,000$$

$$x = \frac{1,35,000}{15} = 9,000$$

1. Commission @ 10% on 9,000 = 900
2. Profit @ 15% on 9,000 = 1,350

Problem 10. Mr. X owns a bus which runs according to the following schedule:

(*i*) Delhi to Chandigarh and back on the same day.

Distance covered: 150 kms, one way, number of days run each month 8, seating capacity occupied 90%.

(*ii*) Delhi to Agra and back, the same day:

Distance covered: 120 kms, one way, number of days run each month 10, seating capacity occupied 85%.

(*iii*) Delhi to Jaipur and back the same day distance covered: 270 kms, one way, Number of days run each month: 6 Seating capacity occupied 100%.

(*iv*) Following are the other details:

Cost of the bus	Rs. 6,00,000
Salary of the driver	Rs. 2,800 p.m.
Salary of the conductor	Rs. 2,200 p.m.
Salary of the part time accountant	Rs. 200 p.m.
Insurance of the bus	Rs. 4,800 p.a.
Diesel consumption 4 kms per litre	@ Rs. 6 per litre
Road tax	Rs. 1,500 p.a.
Lubricant oil	Rs. 10 per 100 kms.
Permit fee	Rs. 315 p.m
Repairs and maintenance	Rs. 1,000 p.m.
Depreciation of the bus	@ 20% p.a.
Seating capacity of the bus	50 persons

Passenger tax is 20% of the total takings. Calculate the bus fare to be charged from each passenger to earn a profit of 30% on total takings. The fares are to be indicated per passenger for the journeys:

(*i*) Delhi to Chandigarh

(*ii*) Delhi to Agra

(*iii*) Delhi to Jaipur

(C.A. Inter, May 1994)

Solution:

CALCULATION OF TOTAL RUN PER MONTH

	Km. per trip	×	Trips per day	×	Days per month	=	Kms per month
Delhi to Chandigarh	150	×	2	×	8	=	2,400
Delhi to Agra	120	×	2	×	10	=	2,400
Delhi to Jaipur	270	×	2	×	6	=	3,240
							8,040

CALCULATION OF PASSANGER-KMS PER MONTH

	Total seats available per month			*Capacity utilised % seats*	*Kms. per trip*		*Passenger kms. per month*
Delhi to Chandigarh and back	50 seats × 2 trips × 8 days	=	800	90	720 × 150	=	1,08,000
Delhi to Agra and back	50 seats × 2 trips × 10 days	=	1,000	85	850 × 120	=	1,02,000
Delhi to Jaipur and back	50 seats × 2 trips × 6 days	=	600	100	600 × 270	=	1,62,000
					Total		3,72,000

OPERATING COST SHEET

	P.M
Fixed cost:	
Salary of driver	2,800
Salary of conductor	2,200
Salary of part-time accountant	200
Depreciation $6,00,000 \times \frac{20}{100} \times \frac{1}{2}$	10,000

Insurance $4{,}800 \times \frac{1}{12}$		400
Road tax, $1{,}500 \times \frac{1}{12}$		125
Repairs and maintenance		1,000
Permit fee		315
	Total fixed cost	17,040
Variable cost:		
Diesel $8{,}040 \times \frac{6}{4}$		12,060
Lubricant oil $8{,}040 \times \frac{10}{100}$		804
	Total cost per month	29,904
Profit and passenger tax together account for 50% of total takings or 100% of cost		29,904
	Total takings	59,808
Passenger tax	20% of takings	11,96,1.60
Profit	30% of takings	17,942.60

Rate per passenger kilometer = $\frac{59{,}808}{3{,}72{,}000}$ = 0.16

Fare to be charged

Delhi to Chandigarh, per passenger	=	150 × 0.16	=	Rs. 24.00
Delhi to Agra, per passenger	=	120 × 0.16	=	Rs. 19.20
Delhi to Jaipur, per passenger	=	270 × 0.16	=	Rs. 43.20

Problem 11. The following information is available for estimating the cost for the year 1997. Full operating cost per truck hour applicable to the trucks operated for a service department of a telephone company. The 1996 actual costs are used to calculate the 1997 rate:

(*a*) Cost of rent, insurance etc. of the garage was Rs. 15,576 of which one-third was allocated to the service department.

(*b*) The foreman of the garage was paid a salary of Rs. 4,800. All his time was devoted to the trucks.

(*c*) The trucks belonging to the service department represent one-quarter of the trucks handled by the garage.

(*d*) The heating, lighting cost of the garage amounted to Rs. 3,540.

(*e*) Other running, expenses, which fluctuated with the truck operating hours, amounted to Rs. 4,794 for the six trucks in the service department.

(*f*) The trucks operate 40 hours a week for 52 weeks each year. During the working hours each of the trucks in down for approximately 60 hours per year for repairs and maintenance. As a result of a strike in July, the garage was closed for 2 weeks.

(*g*) Cost of each truck was Rs. 3,500 and its scrap value is estimated at Rs. 400. The life of a truck was estimated at 12,000 operating hours.

(*h*) The estimated repairs over the life of a truck are expected to amount Rs. 1,400.

(*i*) Driver's wages paid amounted to Rs. 18,720 for the year.

You are required to calculate the applicable rate per hour to be used to reflect the full operating cost for 1997 for truck operating in the service department of the telephone company.

(Sri Sathya Sai University, B.Com., March 1998)

Solution:

CALCULATION OF HOURS OPERATED

40 hrs. × 50 weeks	=	2,000	hrs.
Less: Repairs and maintenance	=	60	hrs.
Per truck		1,940	hrs. for the year
For 6 trucks 1,940 × 6 trucks	=	11,640	hrs.

OPERATING COST SHEET OF SERVICE DEPARTMENT OF TELEPHONE COMPANY

Standing charges:		
Depreciation:		
Cost	3,500 per truck	
Less: Scrap value	400	
	3,100	
$\frac{3,100}{12,000}$ × 1,940 hrs. × 6 trucks		3,007
Foreman's salary 4,800 × $\frac{1}{4}$		1,200
Rent and Insurance 15,576 × $\frac{1}{3}$		5,192
Maintenance charges:		
Repairs $\frac{1,400}{12,000 \text{ hrs}}$ × 1,940 × 6 trucks		1,358
Heating, Lighting of garage 3,540 × $\frac{1}{3}$		1,180
Running charges:		
Running expenses for 6 trucks		4,794
Driver's wages		18,720
		35,451
Cost per hour = $\frac{35,451}{11,640}$		3.04

Problem 12. A private transport company has been given a route which is 20 kms. long to run a Mini bus. The cost of the Mini bus is Rs. 4,50,000. It is insured at 3% p.a. and the annual tax will amount to Rs. 5,000. Garage rent is Rs. 500 pm. Repairs expenses will be Rs. 2,500 p.a. and the bus is likely to last for 5 years.

The driver's salary will be Rs. 1,500 pm and the conductor's salary will be Rs. 1,200 pm., in addition to 10% of the takings as commission to be shared equally by the driver and the conductor.

Cost of stationery Rs. 150 pm. Accountant's salary Rs. 1,350 pm, petrol and oil Rs. 50 per 100 km. The bus will make 3 round trips carrying on an average 40 passengers on each trip. The bus will run on an average 25 days in a month.

Assuming 15% profit on takings, calculate the bus fare to be charged from each passenger per kilometer. *(Sri Sathya Sai University, B.Com. (Hons.), October 2007)*

Solution:

Working Notes:

(1) Calculation of passenger kilometres

20 kms × 2 × 3 trips × 25 days × 40 passengers

= 1,20,000 passenger kms

(2) Calculation of running kilometers

20 kms × 2 × 3 trips × 25 days = 3,000 kms

(3) Commission and profit

	Let bus fare be		=	Rs. 100
Less:	Commission	10% of 100	= 10	
	Profit	15% of 100	= 15	25
		Opertaing cost		75

Rate of commission on cost $= \frac{10}{75}$

Rate of profit on cost $= \frac{15}{75}$

OPERATING COST SHEET

Fixed charges per month:

Driver's salary	1,500
Conductor's salary	1,200
Insurance $4,50,000 \times \frac{3}{100} \times \frac{1}{12}$	1,125
Tax $\frac{5,000}{12}$	417
Garage rent	500
Cost of stationery	150
Accountant's salary	1,350
	6,242

Fixed cost per passenger km

$$\frac{6,242}{1,20,000} = 0.0520$$

Variable cost:

Depreciation $\frac{4,50,000}{5} \times \frac{1}{12} \times \frac{1}{1,20,000}$	=	0.0625
Repairs $\frac{2,500}{12} \times \frac{1}{1,20,000}$	=	0.0017
Petrol and oil $\frac{50}{100\text{ km}} \times 3,000\text{ km} \times \frac{1}{1,20,000}$	=	0.0125
		0.1287
Commission		0.0172

$0.1287 \times \frac{10}{75}$	
Profit	0.0258
$0.1287 \times \frac{15}{25}$	
Cost per passenger K.M.	0.1717

Problem 13. *(Comparative operating cost sheet):* Mr. Dinakar runs in tempo service in the town and has two vehicles. He furnishes you the following data and wants you to compute the cost per running mile:

	Vehicle 'A' *(Rs.)*	*Vehicle 'B'* *(Rs.)*
Cost of vehicle	2,50,000	1,50,000
Road licence fee per year	750	750
Supervisor's salary (yearly)	1,800	1,200
Driver's wages per hour	8	8
Cost of fuel per litre	48	48
Repairs and maintenance per mile	1.50	2.00
Tyre cost per mile	1.00	0.80
Garage rent per year	1,600	550
Insurance premium yearly	850	500
Miles run per litre	6 miles	5 miles
Mileage run during the year	15,000 miles	6,000 miles
Estimated life of vehicle	1,00,000 miles	75,000 miles

Charge interest at 1% p.a. on the cost of vehicle. The vehicles run 20 miles per hour on an average. *(Sri Sathya Sai University, B.Com., (Hons) February 2006)*

Solution:

OPERATING COST SHEET

	Vehicle 'A'	Vehicle 'B'
Standing charges per annum:		
Road licence	750	750
Supervision and salary	1,800	1,200
Garage rent	1,600	550
Insurance	8,50	500
Interest @ 10% on cost	2,500	1,500
	7,500	4,500
Fixed expenses per mile	$\frac{7,500}{15,000}$ = 0.50	$\frac{4,500}{6,000}$ 0.75

Running expenses:			
Drivers wages	$\frac{8}{20}$	0.40	0.40
Fuel cost $\frac{48}{6}$	$\frac{48}{5}$	8.00	9.60
Repairs and maintenance		1.50	2.00
Depreciation $\frac{2,50,000}{1,00,000}$ = 2.50	$\frac{1,50,000}{75,000}$ = 2.00	2.50	2.00
Tyre cost		1.00	0.80
		13.90	15.55

Problem 14. Delhi Transport Company has been given a 20 km route to run a bus. The bus costs the company a sum of Rs. 5,40,000 and its scrap value is estimated at Rs. 1,80,000 after its life period of 10 years.

The driver's salary will be Rs. 2,000 per month and conductor's salary will be Rs. 1,500 per month. Cost of stationery will be Rs. 300 per month. Manager's salary is Rs. 3,500 per month.

The bus has been insured at 2% per annum on the original cost and annual tax will amount to Rs. 3,000. Garage rent is Rs. 500 per month. Actual repairs amount to Rs. 6,000 per annum.

Petrol and oil will be Rs. 50 per 100 kms. The bus will make 3 round trips a day carrying on an average 40 passengers on each trip. The bus will run on an average 25 days in a month.

You are required to prepare operating cost sheet and to calculate the cost per passenger kilometer.

(Sri Sathya Sai University, B.Com. (Hons), October 2006)

Solution:

OPERATING COST SHEET

		P.M	Per Pass-km
Fixed charges:			
Driver's salary		2,000	
Conductor's salary		1,500	
Manager's salary		3,500	
Stationery		300	
Insurance $\frac{2}{100} \times 5,40,000$	= 10,800		
per month	$\frac{10,800}{12}$	900	
Tax	$\frac{3,000}{12}$	250	
Garage rent		500	
Repairs	$\frac{6,000}{12}$	500	
		9,450	0.07875

Variable charges:

Depreciation

$$\frac{5,40,000-1,80,000}{10} \times \frac{1}{12} = 3,000 \qquad 0.025$$

Petrol and oil

$$\text{Rs. } \frac{50}{100 \text{ km}} \times 3,000 \text{ kms} = \frac{1,500}{1,20,000} \qquad 0.0125$$

0.11,625

Problem 15. Shankar has been promised a contract to run a tourist car on a 20 km. land route for the chief executive of a multinational firm. He buys a car for Rs. 1,50,000. The annual cost of insurance and taxes are Rs. 4,500 and Rs. 900 respectively. He has to pay Rs. 500 per month for a garage where he keeps the car when it is not in use. The annual repair costs are estimated at Rs. 4,000. The car is estimated to have a life of 10 years at the end of which the scrap value is likely to be Rs. 50,000.

He hires a driver who is to be paid Rs. 300 per month plus 10% of takings as commission. Other incidental expenses are estimated at Rs. 200 per month.

Petrol and oil will cost Rs. 100 per 100 km. The car will make 4 round trips each day. Assuming that a profit of 15% on takings is desired and that the car will be on the road for 25 days on an average per month, what should be the charge per round trip?

(Sri Sathya Sai University, B.Com. (Hons.), October 2005)

Solution:

OPERATING COST SHEET

		PA	*PM*
Depreciation		10,000	
Insurance		4,500	
Taxes		900	
Garage	500 × 12	6,000	
Annual repairs		4,000	
Driver's salary	(300 × 12)	3,600	
Incidental expenses		2,400	
	200 × 12		
		31,400	26,16.67
Petrol $\left(4,000 \text{ km} \times \frac{1}{100 \text{ km}} \times \text{Rs. } 100\right)$			4,000
	Total cost without commission		66,16.67

Calculation of Commission

Let x be the total takings per month

Driver's commission 10% of x or $\frac{10x}{100}$ or $\frac{x}{10}$

Profit 15% of x = $\frac{15x}{100}$ or $\frac{3}{20}$

Total taking per month = Total cost + Driver's commission + Profit

$$x = 6616.67 + \frac{x}{10} + \frac{3x}{20}$$

$$x = \frac{1.32,320 + 2x + 3x}{20}$$

$$20x = 1,32,320 + 5x$$

$$20x - 5x = 1,32,320$$

$$15x = 1,32,320$$

$$x = 8,822$$

Total number of round trips per month = 25 days × 4 round trips per day
= 100

Charge per round trip $= \frac{8,822}{100} = 88.22$

Problem 16. Mr. Veerabhadra owns a bus which runs between Bangalore and Mysore and back for 10 days in a month. The distance from Mysore and Bangalore is 150 kms. The bus completes the trip from Bangalore to Mysore and back on the same day. The bus goes another 10 days in a month towards Ooty. The distance between Mysore and Ooty is 120 kms.

The trip is also completed on the same day. For the rest 4 days of its operation in a month it runs in the local city. Daily distance covered in the local city is 40 kms. Calculate the rate. Mr. Veerabhadra should charge for passengers when he wants to earn a profit of $33\frac{1}{3}\%$ of his taking. The other information are:

	Rs.
Cost of bus	3,00000
Depreciation 20% p.a	
Salary of driver per month	1,750
Salary of conductor per month	1,750
Salary of clerk per month	800
Insurance per year	8,400
Diesel 8 km per litre costing Rs. 20 per litre	
Token tax per year	3,000
Lubricating oil Rs. 50 per 100 km	
Repairs and maintenance per year	3,000
Permit fees per year	1,420

Normal capacity 50 passengers. The bus is generally occupied 90% of the capacity when it goes to Mysore and 80% when it goes to Ooty and 100% in the local trip.

(Bangalore University, B.Com., April 2003)

Solution:

Working Notes:

(1) Calculation of passenger kilometers

(1) Bangalore to Mysore	$150 \times 2 \times 10 \times 50 \times \frac{90}{100}$	=	1,35,000
(2) Mysore to Ooty	$120 \times 2 \times 10 \times 50 \times \frac{80}{100}$	=	96,000
(3) Local trip	$40 \times 4 \times 50 \times \frac{100}{100}$	=	8,000
	Passenger kilometers		2,39,000

(2) Calculation of running kilometers

Bangalore to Mysore	150 × 2 × 10	=	3,000
Mysore to Ooty	120 × 2 × 10	=	2,400
Local	40 × 4	=	160
			5,560

OPERATING COST SHEET

Fixed charges per month:		Cost per pass. km.
Salary of driver		1,750
Salary of conductor		1,750
Salary of clerk		800
Insurance	$\frac{8400}{12}$	700
Permit fees	$\frac{1,420}{12}$	118.33
Token tax	$\frac{3,000}{12}$	2,50.00
		5,368.33
		16,800

Fixed cost per passenger-km

$$\frac{5368.33}{2,39,000} = 0.022$$

Running and operating charges:

Depreciation $3,00,000 \times \frac{20}{100} \times \frac{1}{12} \times \frac{1}{2,39,000} = 0.020$

Diesel

For 8 km = Rs. 20

For 5,560 km = $\frac{20}{8} \times 5560 = \frac{13,000}{2,39,000} = 0.058$

Lubricant $\frac{50}{100} \times 5,560 = \frac{2,780}{2,39,000} = 0.011$

Repairs and maintenance $\frac{3,000}{12 \text{ months}} = \frac{250}{2,39,000} = 0.001$

Operating cost per passanger km 0.112

Add: Profit $33\frac{1}{3}$% on taking or 50% on cost 0.056

Charge per passenger km 0.168

Problem 17. West Road Company runs 10 buses between two places covering a distance of 25 kms. Seating capacity of each bus is 30 passengers. The expenses for the month of April 2003 were as under

Salaries of drivers and conductors	90,000
Salaries of mechanical staff	9,000
Diesel, oil and lubricants	60,000
Taxes, insurance etc	7,800
Repairs and maintenance	12,000
Depreciation	48,000

60% of seating capacity was utilised in each case. All buses ran 25 days a month each making four round trips daily.

(*a*) Find out cost per passenger-kilometer

(*b*) What would have been the cost per round trip per passenger, if seating capacity utilisation were to go up to 80% *(Calcutta University, B.Com. (Hons), 2003)*

Solution:

OPERATING COST SHEET

		Cost per month	*Cost per pass km.*
Fixed expenses:			
Salaries of mechanical staff		9,000	
Taxes, Insurance		7,800	
		16,800	
Fixed cost per passanger-km		$\frac{16,800}{9,00,000}$	0.018,67
Running expenses:			
Salaries of drivers and conductors		90,000	
Diesel, oil and lubricants		60,000	
Repairs and maintenance		12,000	
Depreciation		48,000	
	Total	2,10,000	
Running expenses per passenger-km	$\frac{2,10,000}{9,00,000}$		0.23333
	Cost per passenger km		0.25200

Working Note:

(*a*) Calculation of passenger km for the month

$$10 \times 25 \times 4 \times 2 \times 25 \times 30 \times \frac{60}{100} = 9,00,000$$

(*b*) Total cost for the month:

Fixed expenses	16,800
Running expenses	2,10,000
	2,26,800

Total passenger-km at 80% utilisation of seating capacity

$$10 \times 25 \times 4 \times 2 \times 25 \times 30 \times \frac{80}{100} = 12,00,000$$

So, Cost per passenger per round trip

$$= \frac{2,26,800}{12,00,000} \times 25 \times 2 = \text{Rs. } 9.45$$

Problem 18. Raju the owner of a taxi supplies the following information

Cost of Taxi Rs. 1,50,000 with total life of 2,50,000 kms. Driver's salary–Rs. 700 p.m. Cleaner's salary–Rs. 300 pm., Repairs–Rs. 3,600 p.a. Garage rent–Rs. 2,400 p.a., Tax–Rs. 1,200 p.a. Diesel consumption 8 kms per litre. Diesel cost Rs. 8 per litre. Oil and other sundries Rs. 20 per 125 kms.

The taxi runs 100 kms per day for an average of 25 days in a month. 20% of the distance has been run without any passenger.

Calculate the cost of running the taxi per-km. *(Calcutta University, B.Com (Hons.), 2000)*

Solution:

OPERATING COST SHEET

		PM (Rs.)	*Per km (Rs.)*
Standing charges:			
Driver's salary		700	
Cleaner's salary		300	
Repairs 3600 ÷ 12		300	
Garage Rent 2400 ÷ 12		200	
Tax 1200 ÷ 12		100	
	Total	1,500	
Fixed cost per km.	$\frac{1,500}{2,000}$		0.75
Variable expenses:			
Depreciation	$\frac{1,50,000}{2,00,000}$		0.75
Diesel $\frac{8}{6}$			1.33
Oil and other sundries	$\frac{20}{100}$		0.20
Cost of running the taxi per km			3.03

Problem 19. Sohail has been promised a contract to run a tourist car on a 20 km long route for the CEO of a multinational firm. He buys a car costing Rs. 1,50,000. The annual cost of Insurance and taxes are Rs. 4,500 and Rs. 900 respectively. He has to pay Rs. 500 per month for a garage where he keeps the car when it is not in use. The annual repair costs are estimated at Rs. 4,000. The car is estimated to have a life of 10 years at the end of which the scrap value is likely to be Rs. 50,000.

He hires a driver who is to be paid Rs. 300 per month plus 10% of the takings as commission. Other incidental expenses are estimated at Rs. 200 per month.

Petrol and oil will cost Rs. 100 per 100 km. The car will make 4 round trips each day. Assuming that a profit of 15% on takings is desired and that car will be on the road for 25 days on an average per month, what should be charge per round trip? *(University of Delhi, B. Com. (Hons.), 2005)*

Solution:

OPERATING COST SHEET

		Per annum	Per month
Standing charges:			
Depreciation		10,000	
Insurance		4,500	
Taxes		900	
Garage rent	500 × 12	6,000	
Annual repairs		4,000	
Driver's salary	300 × 12	3,600	
Incidental expenses	200 × 12	24,00	
		31,400	2616.17
Variable charges:			
Petrol and oil			
4,000 kms × $\frac{1}{100}$ × 100			4000.00
	Total cost without commission		6616.17

Working Note:

Calculation of running km

20 km × 2 × 4 round trips × 25 days = 4,000 km

Calculation of charges of per round trip

Let 'x' be the final takings for the month

Driver's commission = 10% of x or $\frac{x}{10}$

Profit = 15% of x or $\frac{3x}{20}$

Takings = Total cost + Driver's commission + Profit

$$x = TC + \frac{x}{10} + \frac{3x}{20}$$

or, $$x - \frac{x}{10} - \frac{3x}{20} = 6616.67$$

or, $$\frac{20x - 2x - 3x}{20} = 6616.67$$

or, $$15x = 6616.17 \times 20$$

or, $$x = \frac{6616.17 \times 20}{15}$$

$$x = \text{Rs. } 8822.22$$

Total number of round trips per month

$$25 \times 4 = 100$$

$$\text{Charge per trip} = \frac{8822.22}{100} = \text{Rs. } 88.22$$

Problem 20. Hira Singh owns a taxi, a bus, and a truck. The bus is 50 seater. The maximum capacity of the truck is 10 tonnes. The taxi runs on an average 3,000 kms; in a month out of which 20% is normal running without fare. Variable cost of running the taxi is Rs. 8 per kilometre.

The bus and the truck run between Delhi and Jaipur, one way distance being 300 kms. The bus makes 25 round trips in a month and is generally 90% occupied. Variable cost of running a truck is Rs. 9.50 per kilometre.

You are required to calculate.

(*i*) Total variable cost per month and variable cost per effective kilometer for the taxi.

(*ii*) Total variable cost per month and variable cost per effective passenger-kilometer for the bus and

(*iii*) Total variable cost per month and variable cost per effective ton-kilometer for the truck.

(University of Delhi, B.Com. (Hons.), 2005)

Solution:

(*i*) Taxi

Total variable cost = 3000 km × Rs. 8 = Rs. 24,000

$$\text{Cost per effective} = \frac{\text{Rs. } 24{,}000}{3{,}000 \text{ km} \times 80\%}$$

= Rs. 10

(*ii*) Bus

Total variable cost = 300 km × 2 × 25 days × Rs. 13.50
= Rs. 2,02,500

$$\text{Cost per effective} = \frac{2{,}02{,}500}{300 \text{ km} \times 2 \times 25 \text{ days} \times 50 \text{ pass} \times 90\%}$$

= 0.30

(*iii*) Truck

Total varriable cost = 300 kms × 2 × 20 trips × Rs. 9.50
= Rs. 1,14,000

Calculation of Ton- km.

$$(300 \times 20 \text{ trips} \times 10 \text{ tons} + 300 \text{ km} \times 20 \text{ trips} \times 10 \text{ ton}) \times \frac{90}{100}$$

= 1,14,000

$$\text{Cost per effective Ton-km.} = \frac{\text{Rs. } 1{,}14{,}000}{1{,}14{,}000 \text{ Ton} - \text{km}}$$

= Re. 1

Problem 21. A transport company is running four buses between Delhi and Alwar, covering a distance of 100 kms. The seating capacity of each bus is 40 passengers. The following particulars are obtained from its books for the month of October 2003:

	Rs.
Wages of drivers, conductors	9,600
Salaries of office staff	3,000
Honorarium of accountant	1,000
Diesel, oil etc.	16,000
Repairs and maintenance	3,200
Road tax and insurance	6,400
Depreciation	10,400
Interest	8,000

Actual passengers carried were 75% of the seating capacity. All the buses ran for 30 days. Each bus made one round trip per day. Find out the fare the company would charge per passenger-km if it wants a profit of 20% on the takings. *(University of Delhi, B.Com. (Hons.), 2004)*

Solution:

Working Note:

Calculation of passenger mile

$$4 \times 40 \times \frac{75}{100} \times 30 \times 100 \times 2 = 7{,}20{,}000$$

OPERATING COST SHEET

Wages of drivers and conductors	9,600
Salaries of office staff	3,000
Honorarium of accountant	1,000
Diesel, oil etc	16,000
Repairs and maintenance	3,200
Road tax and Insurance	6,400
Depreciation	10,400
Interest and other charges	8,000
	57,600

Cost per passenger mile $= \frac{57{,}600}{7{,}20{,}000} = 8$ paise

Fare to be charged:

Cost per passenger mile	= 8	paise
Profit 20% of taking or 25% on Cost	= 2	paise
	10	paise

Problem 22. Mr. Singh started transport business with a fleet of 10 taxis. Expenses of operating the fleet are given below:

	Rs.
(*i*) Cost of each taxi	3,80,000
(*ii*) Salary of office and garage staff	38,000 pm
(*iii*) Rent of garage	12,000 pm
(*iv*) Driver's salary per taxi	4,000 pm
(*v*) Insurance, tax and sundry expenses per taxi	55,200 per yr.

The life of a taxi is 3,00,000 kms at the end of which it is estimated to be sold at Rs. 20,000. A taxi is expected to run on an average 4,000 kms per month. Petrol consumption is 12 kms per litre of petrol costing Rs. 30 per litre. You are required to:

(*i*) Calculate the cost of running a taxi per km by preparing a statement of operating cost and

(*ii*) Find out the profit of Mr. Singh may expect to earn during the first month of operation if the hire charges is Rs. 10 per km. Assume that during the month each taxi run on an average 4,000 kms of which 800 kms it runs empty.

(University of Delhi, B.Com. (Hons.), 2004)

Solution:

OPERATING COST SHEET

		Per Taxi per month (Rs.)	Per km (Rs.)
Fixed Expenses:			
Rent	$\frac{12,000}{10}$	1,200	
Salary of drivers		4,000	
Salary of office and garage staff	$\frac{38,000}{10}$	3,800	
Insurance	$\frac{55,200}{12}$	4,600	
		13,600	
Fixed expenses per taxi per km	$\frac{13,600}{4,000}$		3.40
Variable expenses:			
Depreciation	$\frac{3,80,000 - 20,000}{3,00,000}$		1.20
Petrol			2.50
	Total		7.10

CALCULATION OF PROFIT

Takings 3,200 km × 10 taxis × Rs. 10	=	3,20,000
Less: Cost of 4,000 km × 10 taxis × Rs. 7.10	=	2,84,000
Profit		36,000

Problem 23. A bus started from Delhi for Mussoorie with 50 passengers on board. 20 passengers got off at Dehradun and the bus proceeded with the remaining passengers. In the evening the same bus left Mussoorie with 50 passengers, 10 passengers got off at Dehradun and the bus resumed its journey with remaining passengers for Delhi. The distance between Delhi and Dehradun is 280 kms and between Dehradun to Mussoorie it is 20 km.

Compute the cost per passenger km, if the total cost of running the bus comes out to be Rs. 5,000.

(University of Delhi, B.Com. (Hons.), 2003)

Solution:

Working Note:

Calculation of passenger kms:

From Delhi to Dehradun	(50 × 280)	14,000
From Dehradun to Mussoorie	(30 × 20)	600
From Mussoorie to Dehradun	(50 × 20)	1,000
From Dehradun to Delhi	(40 × 280)	11,200
	Total passenger kms	26,800

Cost per passenger km $= \frac{\text{Rs. } 5,000}{26.800} = 0.19$

Problem 24. From the following information calculate the bus fare to be charged from each passenger for the journeys.

(1) Delhi to Agra
(2) Delhi to Bhiwani
(3) Delhi to Chandigarh

(*i*)	Delhi to Agra	200 km
	Delhi to Bhiwani	120 km
	Delhi to Chandigarh	250 km
(*ii*)	Effective passanger km	3,72,000

(*iii*) Total operating cost (excluding conductor's commission @ 15% and passenger tax @ 5% of total takings) Rs. 148,800.

(*iv*) Desired profit–30% on total taking *(University of Delhi, B.Com. (Hons.), 2002)*

Solution:

Working Note:

Calculation of takings

Total operating cost = Rs. 1,48,800
Let total takings = x

$$x = 1{,}48{,}800 + \frac{15x}{100} + \frac{5x}{100} + \frac{30x}{100}$$

$$x = 1{,}48{,}800 + 0.50x$$

$$x - 0.50x = 1{,}48{,}800$$

$$x = \text{Rs. } 2{,}97{,}600$$

OPERATING COST SHEET

Total cost		1,48,800
Commission	$\frac{15}{100} \times 2{,}97{,}600$	44,640
Passenger tax	$\frac{5}{100} \times 2{,}97{,}600$	14,880
Profit	$\frac{30}{100} \times 2{,}97{,}600$	89,290
	Total takings	2,97,600

Fare per passenger km $= \frac{2{,}97{,}600}{3{,}72{,}000} = 0.80$

Bus fare per passenger

(1) Delhi to Agra = 200 km × 0.80 = Rs. 160
(2) Delhi to Bhiwani = 120 km × 0.80 = Rs. 96
(3) Delhi to Chandigarh = 250 kms × 0.80 = Rs. 200

Problem 25. A company is considering three alternative proposals for conveyance facilities for its sales personnel who have to do considerable travelling, approximately 20,000 kilometres every year. The proposals are as follows:

(*i*) Purchase and maintain its own sheet of cars. The average cost of a car is Rs. 1,00,000
(*ii*) Allow the Executive use his own car and reimburse expenses at the rate of Rs. 1.60 per kilometer and also bear insurance costs.
(*iii*) Hire cars from an agency at Rs. 20,000 per year per car. The company will have to bear costs of petrol, taxes and tyres.

The following further details are available
Petrol Re. 0.60 per kilometer
Repairs and maintenance Re 0.20 per kilometer
Tyre Re 0.12 per kilometer
Insurance Rs. 1,200 per car per annum
Taxes Rs. 800 per car per annum
Life of the car: 5 years with annual mileage of 20,000 kilometers
Resale value: Rs. 20,000 at the end of the fifth year.
You are required to work out the relative costs of the three proposals and rank them.

(University of Delhi, B.Com., (Hons) 2001)

Solution: COMPARATIVE OPERATING COST SHEET

	Company car per year	Executive car per km	Hired car per km
Fixed cost :			
Taxes	800	–	0.04
Depreciation	16,000	–	–
Insurance	1,200	0.06	–
Total	18,000	0.06	0.04
Fixed cost per km	0.90	0.06	0.04
Reimbursement	–	1.60	1.00
Variable costs:			
Petrol	0.60	–	0.60
Repairs	0.20	–	–
Tyre	0.12	–	0.12
Total	1.82	1.66	1.76
Ranking	III	I	II

Problem 26. A transport company maintain a fleet of lorries for carrying goods from Delhi to Panipat, 100 kms off. Each lorry, which operates 25 days on an average in a month, starts every day from Delhi with a load of 4 tonnes and returns from Panipat with a load of 2 tonnes. Calculate the total commercial tonne-kms and cost per commercial tonne-km when the total monthly charges for a lorry are Rs. 27,000. What rate per tonne should the company charge if it plans to earn a gross profit of 20% on the freightage?

(University of Delhi, B.Com. (Hons.), 2000)

Solution: CALCULATION OF TON-KM

		Ton kms
(1) From Delhi to Panipat		
100 kms × 25 days × 4 tons	=	10,000
(2) Return from Panipat to Delhi		
100 kms × 25 days × 2 tons	=	5,000
Total ton kms		15,000

$$\text{Cost per ton-km} = \frac{\text{Rs. } 27{,}000}{15{,}000} = \text{Rs. } 1.80$$

Add: Profit 20% on freightage or 25% on cost = Rs. 0.45

Freight rate per tonne 2.25

Problem 27. *(Problem involving decision-making):* A practising chartered accountant now spends Re. 0.90 per km on taxi for his clients work. He is considering two other alternative, the purchase of a new small car or an old bigger car. The estimated cost figures are:

	New small car (Rs.)	*Old bigger car (Rs.)*
Purchase price	35,000	20,000
Sale price, after 5 years	19,000	12,000
Repairs and servicing per annum	1,000	1,200
Taxes and insurance per annum	1,700	700
Petrol consumption per litre	10 km.	7 km.
Petrol price per litre	3.50	3.50

He estimates that he does 10,000 km annually. Which of the three alternatives will be the cheapest. If his practice expands and he has to do 19,000 kms per annum, what should be his decision? At how many kms per annum, will the cost of the two breakeven and why? Assume petrol only as variable cost. Ignore interest and income tax. *(Delhi University, B.Com. (Hons.), April 88)*

Solution:

STATEMENT OF COMPARATIVE COST

Ist alternative		*New small car*	*Old big car*	*Taxi*
(*a*) Fixed cost:				
Depreciation	$\frac{35,000-19,000}{5}$; $\frac{20,000-12,000}{5}$	3,200	1,600	–
Repairs and servicing		1,000	1,200	–
Taxes and Insurance		1,700	700	–
	Total fixed cost	5,900	3,500	–
(*b*) Variable cost				
Petrol: $\frac{3.50}{10} \times 10,000$, $\frac{3.50}{7} \times 10,000$		3,500	5,000	–
Hire charges (10,000 × 0.90)		–	–	9,000
	Total cost	9,400	8,500	9,000
2nd alternative				
Petrol for 19,000 kms		6,650	9,500	–
Fixed cost		5,900	3,500	–
Hire charges (19,000 × 0.90)		–	–	17,100
	Total cost	12,550	13,000	17,000

It is clear from the above statement that out of the three alternatives, old bigger car is the cheapest for his present practice. But if his practice expands new small car will be the cheapest.

Break-even point:

$$= \frac{\text{Difference in fixed cost}}{\text{Difference in variable cost per km}}$$

$$= \frac{5,900 - 3,500}{50\text{p} - 35\text{p}} = \frac{2,400}{15\text{p}} = 16,000 \text{ kms}$$

At 16,000 km, the cost of operating new small car and old bigger car will be the same.

Alternate Method

Let 'x' be the km per annum where the cost of two cars will be identical

Then total cost for x km under each car will be:

For new car: $0.35x + 5{,}900$

For old car: $0.50x + 3{,}500$

$$= 0.35x + 5{,}900 = 0.50x + 3{,}500$$
$$= 0.35x - 0.50x = -5{,}900 + 3{,}500$$
$$= 0.15x = 2{,}400$$
$$x = \frac{2{,}400}{0.15} = 16{,}000 \text{ kms}$$

POWER HOUSE COSTING OR BOILER HOUSE COSTING

The power house or boiler house is concerned with generation of electricity or steam as a form of power or energy. The purpose of power house costing is to ascertain the cost of electricity or steam produced. The components of total cost in such undertakings include cost of boiler service, depreciation on capital assets, fuel cost, salaries and wages of power house employees, salaries of supervisors and managers, and other overheads. For the sake of exercising control on various elements of cost, the total cost of generating the service is classified into the following types:

(a) *Cost of fuel*. This includes cost of coal or fuel oil, handling and storage cost of fuel.

(b) *Water*. This includes the cost of purification, softening and supply.

(c) *Labour*. This includes coal handlers and ash removers.

(d) *Supervision*. This includes salary of foreman, and supervisors.

(e) *Maintenance*. This includes repairs and maintenance, cost of furnace and boiler house.

(f) *Fixed overhead*. This includes rent, rates, depreciation, insurance and interest on capital.

A cost sheet is prepared to record all the above-mentioned costs. A specimen of such cost sheet is given below:

Bolier house cost sheet for the month of ....

Items of expenses	*Total cost*		*Percentage*
	Current year	*Previous year*	*Increase or decrease*
Fuel			
Fuel handling			
Ash removal and disposal			
Water purchased			
Water softened			
Sundry indirect materials			
Gas and electric light			
Repairs and maintenance			
Wages of boiler house			
Wages of generating house			

Salaries of supervisors Rent and rates Depreciation on plant Depreication on building Share of administration charges Interest on capital			
Total cost			

Problem 28. *(Cost of generating electricity):* From the following data, find out in an appropriate cost sheet form, the cost of generating per unit of electricity in an iron and steel works, during the month of April 1989.

1. **Fuel:** Coal at the beginning of the month — 500 tonnes
   Supply during the month — 1,100 tonnes
   Balance at the end of the month — 400 tonnes
   Annual contract for supply of coal F.O.R. Colliery at Rs. 10 per tonne.
   Add 10% to cover freight and handling charges.
2. **Oil:** 10 tonnes at Rs. 250 per tonne.
3. **Water:** 50,000 litres. Pumping charges at 30 paise per 100 litres.
4. **Depreciation of steam boiler:** Capital value Rs. 24,000 and the rate of depreciation $12\frac{1}{2}$ per cent per annum.
5. **Salaries and wages of the boiler house:**
   10 Men at Rs. 100 each per month
   40 Coolies at Rs. 20 each for month
6. **Recovery on account of sale of ash:**
   100 tonnes at Re. 1 per tonne
7. **Salaries and wages of the generating station:**
   50 Men at Rs. 100 per month each
   20 Collies at Rs. 20 per month each
8. Repairs and maintenance of the generating equipment Rs. 2,600.
9. Depreciation of generating equipment: Capital value Rs. 1,20,000 and the rate of depreciation $12\frac{1}{2}$ per cent per annum.
10. Share of administration charges Rs. 1,750
11. Number of units generated; 1,46,000.
12. Loss in process: 2,000 units generated.

(Sri Sathya Sai University, B.Com., March 1990)

Solution:

OPERATING COST SHEET

Coal at the beginning	500 tons
Add: Supply during the month	1,100 tons
	1,600 tons
Less: Balance of coal at the end	400 tons

		1,2000
Freight and handling charges 10% on coal consumed		1,200
Oil: 10 tons at Rs. 250 per ton		2,500
Water: $\frac{0.30}{100} \times 50{,}000$		150
Depreciation: $24{,}000 \times \frac{12.5}{100} \times \frac{1}{12}$		250
Salaries and wages of boiler house:		
10 Men @ Rs. 100 each p.m.		1,000
40 Coolies @ Rs. 20 each p.m.		800
Salaries and wages of generating equipment		
50 Men @ Rs. 100 each p.m.		5,000
20 Coolies @ Rs. 20 each p.m.		400
Repairs and maintenance of generation equipment		2,600
Depreciation of generating equipment		
$1{,}20{,}000 \times \frac{12.5}{100} \times \frac{1}{12}$		1,250
		27,150
Less: Recovery on account of sale of ash		100
	Works cost	27,050
Share of administration charges		1,750
	Total cost	28,800

Net units generated:	
Total units generated	1,46,000 units
Less: Loss in process	2,000 units
Net units	1,44,000

$$\text{Cost per unit of electricity generated} = \frac{24{,}300}{1{,}44{,}000} = 0.16$$

Problem 29. A manufacturing firm facing shortage of electric power supply from the State Electricity Board has set up its own power generation plant for efficient running of its production units in the factory. The following information has been taken from the records in connection with the generation of power for a month:

1. Number of units generated was 10,00,000 for the month of which 10% was utilised by the generation department.
2. Consumption data of materials etc. for the month:
   (*a*) Coal consumed 300 MTs @ Rs. 3,600 per MT
   (*b*) Oil consumed 4.5 MTs @ Rs. 40,000 per MT
   (*c*) Cost of water extraction and treatment for 6 lakhs litres @ Rs. 1.25 per litre.
3. Steam boiler costs Rs. 20 lakhs with a residual value of Rs. 2 lakhs after a life of 10 years.
4. Salaries and wages per month
   (*a*) For staff of generating plan
      (*i*) 100 skilled workers @ Rs. 3,000 p.m.
      (*ii*) 150 helpers @ Rs. 1,500 p.m.

(*b*) For staff of boiler house:

(*i*) 60 category A workers @ Rs. 1500 p.m.

(*ii*) 100 category B workers @ Rs. 1,000 p.m.

5. Cost of generating plant: Rs. 36 lakhs with no residual value. Depreciation @ 10% on straight line basis is to be charged.
6. Repairs and maintenance of generating plant and boiler Rs. 50,000 p.m.
7. Share of administrative charges Rs. 40,000 p.m.
8. Sales value of ash disposed of Rs. 15,000 p.m.

Calculate the per unit cost of electricity generated using a cost sheet format.

(*ICWA, Inter., June 1996*)

Solution: OPERATING COST SHEET FOR THE MONTH

			Total Rs. in lakhs	*Cost per unit*
1.	Coal 300 MT @ Rs. 3,600 MT		10.80	1.200
2.	Oil 4.5 MT @ Rs. 4,000 MT		1.80	0.200
3.	Water extraction and treatment charges (6 lakhs litres @ 1.25)		7.50	0.833
4.	Depreciation on steam boiler $\frac{20-2}{10}$ = 1.8 per annum or 0.15 p.m.		0.15	0.017
			20.25	2.250
5.	*Less:* Sale of ash		0.15	0.017
			20.10	2.233
6.	Salaries and wages:			
	Generating plant	5.25		
	Boiler house	1.90		
		7.15 p.m.	7.15	0.794
7.	Repairs and maintenance plan		0.50	0.056
8.	Depreciation on generating plant 10% of 36 lakhs or 3.6 lakhs per annum or 0.30 p.m.		0.30	0.033
9.	Share of administrative expenses		0.40	0.045
			28.45	3.161
10.	*Add:* Cost of electricity used by generation dept. 10% of 10 lakhs units or 1 lakh unit and 10% of cost is		2.84	0.315
	Total cost		31.29	3.476

Note: The cost of generation should bear the cost of electricity used by generation department. As such total cost is calculated on 9 lakhs units and not on 10 lakh units.

CANTEEN OR HOTEL COSTING

A canteen is established in most of the factories as a welfare facility. Canteens provide whole-some food and snacks at a subsidised rates. The real problem in canteen costing is use of perishable provisions. This involves careful planning and controlling of various provisions to be bought from time to time. The real problem is determination of menu which again depends upon the canteen manager. Menu is sold against the purchase of coupons and therefore the number of coupons sold in terms of value constitutes the sale value.

For enabling proper accounting three important records are maintained. They are: 1. Stock register, which is maintained by canteen store keeper. As and when issues are made they are recorded to ascertain the cost of materials consumed in the canteen 2. Wage analysis sheet is maintained by the personal department to record labour cost and supervision of canteen 3. Overhead analysis sheet, to calculate proportionate overhead charge.

The purpose of canteen costing is to determine a price at which menu is to be sold. For this purpose it involves classifying the expenses according to the following types:

(*a*) *Provisions:* Such as rice, wheat flour, oil, sugar, vegetables, fruits, milk, coffee and tea powder.

(*b*) *Labour:* Cook's wages, salary of supervisor, wages of kitchen assistants, porter's wages.

(*c*) *Service:* Such as water, gas, electricity, steam, light.

(*d*) *Consumable stores:* Such as table cloths, crockery, glassware, cleaning materials, dustbins, brushes, etc.

(*e*) *Overheads:* Such as rent and rates, insurance, depreciation, etc.

(*f*) *Sales:* This includes revenue from meals, tea and coffee, sale of snacks, etc.

A cost sheet is prepared to record all the above items, a specimen of which is given below:

Cost sheet of a canteen for the month.....................

	This month	*This month last year*
Provisions:		
Rice		
Wheat flour		
Oil		
Sugar		
Milk		
Coffee		
Tea		
Vegetables and fruits		
Soft drinks		
Wages and salaries:		
Cooks		
Supervisor		
Counter helpers		
Miscellaneous cost		
Crockery and glassware		
Maintenance and repairs		
Table clothes		
Consumable stores		

Rent
Gas
Electricity and lighting
Profit
Income from sales

As compared to canteen costing which is concerned with the supply of food and refreshments, big hotels are concerned with the provision of boarding and lodging facilities. The main object of accounting in such hotels is to ascertain rental rates and charges to be fixed for supplying food.

Problem 30. *(Hotel costing):* Following are the information given by an owner of a hotel. You are required to advise him that what rent should he charge from his customers per day so that he is able to earn 25% on cost other than interest.

1. Staff salaries: Rs. 80,000 per annum.
2. Room attendants salary Rs. 2 per day. The salary is paid on daily basis and services of room attendant are needed only when the room is occupied. There is one room attendant for one room.
3. Lighting, heating and power:

The normal lighting expenses for a room if it is occupied for the whole month is Rs. 50. Power is used only in winter and normal charge per month if occupied for a room is Rs. 20.

4. Repairs to building — Rs. 10,000 p.a.
5. Linen etc. — Rs. 4,800 p.a.
6. Sundries — Rs. 6,600 p.a.
7. Interior decoration — Rs. 10,000 p.a.
8. Cost of building — Rs. 4,00,000 rate of depreciation 5%
9. Other equipments — Rs. 1,00,000 rate of depreciation 10%
10. Interest @ 5% is to be charged on its investment of Rs. 5,00,000 on the building and equipment.

There are 100 rooms in the hotel and 80% of the rooms are normally occupied in summer and 30% of the rooms are busy in winter. You may assume that period of the summer and winter is 6 months each. Normal days in a month may be assumed to be 30.

(University of Madras, B.Com., March 1996)

Solution: OPERATING COST SHEET

1. Salaries of staff		80,000
2. Room attendants salary:		
Summer – $2 \times 100 \times \frac{80}{100} \times 30 \times 6$	= 28,000	
Winter – $2 \times 100 \times \frac{30}{100} \times 30 \times 6$	= 10,800	39,600
3. Lighting:		
Summer – $50 \times 6 \times 100 \times \frac{80}{100}$	= 24,000	
Winter – $50 \times 6 \times 100 \times \frac{30}{100}$	= 9,000	33,000
Power – $20 \times 6 \times 100 \times \frac{30}{100}$		3,600

4. **Repairs**		10,000
5. **Linen**		4,800
6. **Sundries**		6,600
7. **Interior decoration**		10,000
8. **Depreciation:**		
On building $\frac{5}{100} \times 4,00,000$	= 20,000	
Other equipment $\frac{10}{100} \times 1,00,000$	= 10,000	30,000
9. **Interest on investment** $\frac{5}{100} \times 5,00,000$		25,000
		2,42,600
10. *Add:* **Profit 25% on cost other than interest** (2,42,600 – 25,000) 2,17,000 × $\frac{25}{100}$		54,400
		2,97,000

$$\text{Rent per room for one day} = \frac{2,97,000}{19,800} = \text{Rs. } 15$$

Calculation of total room days:

Summer $= 100 \times \frac{80}{100} \times 30 \times 6$	= 14,400
Winter $= 100 \times \frac{30}{100} \times 30 \times 6$	= 5,400
Total rooms days	19,800

HOSPITAL COSTING

Hospital costing may relate to ascertaining the cost of medical service rendered by a dispensary belonging to a factory or an hospital or a nursing home owned by government or private doctors. The main object of hospital costing is to ascertain the cost of rendering service per patient per day. For this purpose a cost sheet is prepared to record the various expenses incurred in the hospital. Such costs are classified under the following headings:

1. Medicines
2. Cost of operation theater
3. Cost of out-patient department
4. Cost of in-patient wards
5. Salary of physicians, surgeons and nurses
6. Cost of equipments and instruments
7. Cost of blood bank
8. Cost of X-ray and pathological services.

Problem 31. *(On hospital costing):* A public health centre runs an intensive medical care unit. For this purpose, it has hired a building at a rent of Rs. 5,000 per month with the understanding that it would bear the repairs and maintenance charges also.

The units consists of 25 beds and 5 more beds can be comfortably accommodated when the occassion demands. The permanent staff attached to the unit are as follows:

2 Supervisors, each at a salary of Rs. 500 p.m.

4 Nurses, each at a salary of Rs. 300 p.m.

2 Wardways, each at a salary of Rs. 150 p.m.

Though the unit is open for patients all the 365 days in a year. Scrutiny of accounts in 1990 reveals that only for 120 days in the year the unit had the full capacity of 25 patients per day and for another 80 days it had on an average 20 beds occupied per day. But there were occasions when the beds were full, extra beds were hired at a charge of Rs. 5 per bed per day and this did not come to more than 5 beds extra above the normal capacity on any one day. The total hire charges for extra beds incurred for the whole year amounted to Rs. 2,000.

The unit engaged doctors from outside to attend on the patients and the fees were paid on the basis of the number of patients attended and time spent by them which on an average worked out to Rs. 10,000 p.m. in 1990.

The other expenses for the year were as under:

Repairs and maintenance	3,600
Food supplied to patients	44,000
Janitor services	12,500
Laundry charges for bed linen	28,000
Medicines supplied	35,000
Cost of oxygen, X-ray, etc.	54,000
General administration charges allocated to the unit	49,550

If the unit recovered an overall amount of Rs. 100 per day on an average from each patient, what is the profit per day made by the unit in 1990.

Solution:

OPERATING COST SHEET

Variable cost:		
Food	44,000	
Janitor services	12,500	
Laundry	28,000	
Medicines	3,500	
Doctors fee 10,000 × 12	1,20,000	
Hire charges for extra beds	2,000	2,41,500
Fixed cost:		
Salaries (2 × 500 + 4 × 300 + 2 × 150) 12		30,000
Rent 5,000 × 12		60,000
Repairs and maintenance		3,600
General administration		49,550
Cost of oxygen		54,000
	Total cost	4,38,650
	Profit	61,350
	Income received	5,00,000

$$\text{Profit per patient day} = \frac{61,350}{5,000} = 12.27$$

Calculation of patient days:

25 beds × 120 days	3,000
20 beds × 80 days	1,600
Extra bed days (*See note below*)	400
Total patient days	5,000

$$\text{Extra bed days} = \frac{\text{Total hire charges of extra beds}}{\text{Charges}}$$

$$= \frac{2,000}{5} = 400 \text{ patient days}$$

Questions

I. (A) Choose the Correct Answer:

1. Classification and accumulation of costs by fixed and variable costs is a distinctive feature of

(*a*) Process costing (*b*) Unit costing
(*c*) Operating costing (*d*) Batch costing [ ]

2. Composite unit is distinctive feature of

(*a*) Single or output costing (*b*) Process costing
(*c*) Job costing (*d*) Operating costing [ ]

3. Electricity generating company should employ

(*a*) Unit costing (*b*) Process costing
(*c*) Operating costing (*d*) Multiple costing [ ]

4. Cinema house must adopt

(*a*) Operating costing (*b*) Job costing
(*c*) Batch costing (*d*) Contract costing [ ]

5. For a library the best method of costing suitable is

(*a*) Output costing (*b*) Operating costing
(*c*) Process costing (*d*) Multiple costing [ ]

6. For an educational institutes the right method of costing is

(*a*) Output costing (*b*) Job costing
(*c*) Operating costing (*d*) Process costing [ ]

7. Hospitals must make use of

(*a*) Operating costing (*b*) Batch costing
(*c*) Process costing (*d*) Multiple costing [ ]

8. For hotels the best method of costing is

(*a*) Single or output (*b*) Contract costing
(*c*) Process costing (*d*) Operating costing [ ]

9. Air India Company must make use of

(*a*) Job costing (*b*) Operating costing
(*c*) Batch costing (*d*) Process costing [ ]

10. Indian Railways must adopt
   (a) Operating costing (b) Unit costing
   (c) Batch costing (d) Multiple costing [ ]
11. Public utility undertakings must invariably adopt
   (a) Operating costing (b) Output costing
   (c) Contract costing (d) Multiple costing [ ]
12. Karnataka Electricity Board must make use of
   (a) Single or output costing (b) Job costing
   (c) Process costing (d) Operating costing [ ]
13. The Karnataka State Road Transport Corporation must use
   (a) Operating costing (b) Contract costing
   (c) Process costing (d) Multiple costing [ ]

[***Answers:*** 1. (c), 2. (d), 3. (c), 4. (a), 5. (b), 6. (b), 7. (a), 8. (d), 9. (b), 10. (a), 11. (d), 12. (a), 13. (a)]

(B) Fill in the Blanks

1. The book in which expenses incurred for different trips by a transport company is recorded is known as ....
2. In hospitals, the cost unit is ....
3. The method of costing to be used by a gas distributing company is ....
4. The cost unit in an educational institute is ....
5. The cost unit in an electricity generating concern is ....
6. The cost unit used in a canteen is ....
7. The industries which render services must make use of .... method of costing.
8. Operating costing is a variation of .... category of costing method.

[***Answers:*** 1. log book, 2, bed-day, 3. operating costing, 4. student, 5. kilo-watt, 6. plate-meals, 7. operating, 8. operation]

(C) Simple Questions

1. Define operating cost.
2. Define operating costing.
3. Distinguish between operating cost and operating costing.

Exercises

1. The following figures are extracted from the books of Naveen Limited for the year 2001-2002
   Passenger buses: 5 buses costing Rs. 5,00,000, Rs. 12,00,000, Rs. 4,50,000, Rs. 5,50,000 and Rs. 8,00,000 respectively. Yearly depreciation of vehicles: 20% of the cost. Annual repair and maintenance 80% of depreciation
   Wages of 10 drivers @ Rs. 5,000 each p.m.
   Wages of 20 clearners @ Rs. 2,500 each p.m.
   Yearly rate of interest 8% on capital
   Rent of six garages @ Rs. 2,500 per month
   Directors fees @ Rs. 8,000 p.m.
   Office establishment: Rs. 5,000 p.m.

Licence and taxes @ Rs. 4,000 every 6 months

Realisation of sales proceeds of old tyres and tubes: @ Rs. 6,600 every 6 months

900 passengers were carried over 1,600 kms during the year. Work out the cost per passenger kilometre. *(Sri Sathya Sai University, B.Com. (Hons.), August 2003)*

[**Ans.** Cost per passenger k.m. = Rs. 2,14]

2. A city municipality arranges for the removal of its garbage by means of motor vehicle transport. The following vehicles are manufactured:

No. of Vehicles	*Specification*
20	5 Ton Lorries
30	4 Ton Lorries
40	3 Ton Lorries
50	2 Ton Lorries

On an average each lorry makes six trips a day and in each trip covers an average distance of 5 kilometres. Each lorry carries garbage weighing only 60% of its capacity. Taking on an annual average 20% of the lorries are lovid up for repairs every day, the conservancy work is carried on daily.

The following are the monthly charges incurrred on the conservancy transport:

	Rs.
Salary of superintendent	10,000
Salary of foreman	1,500
Wages of Rs. 1,500 each for 140 drivers	
Wages of cleaners Rs. 7.50 each for 280 cleaners	
Consumable stores	16,000
Diesel	60,000
Lubricants	15,000
Replacement of tyres and tubes	5,000
Garage rent	3,000
Coal and electric charges	1,000
Miscellaneous expenses	12,000
Repairs and maintenance	10,000

Assuming that a month consists of 30 days. Calculate the cost per ton-km for removal of garbage. *(Sri Sathya Sai University, B.Com. (Hons.), October 2003)*

[**Ans.** Cost per ton-km. is Rs. 2.91]

3. Arun runs a tempo service at Anantapur. He furnishes you the following data and wants you to complete the cost of running per km.

	Rs.
Cost of vehicle	1,50,000
Road licence p.a.	7,500
Supervisor's salary p.a.	26,500
Driver's wages per hour	8
Repairs and maintenance per km.	25
Tyre cost per km	4
Garage rent per annum	36,000
Insurance per annum	5,000
Kilometres running per litre	10
Kilometres run during the year	6,000
Estimated life of the vehicle	75,000 kms

Ton per km (average)	5

Charge interest @ 10% p.a. on cost of vehicle

The vehicle runs 20 kms per hour on an average.

(Sri Sathya Sai University, B.Com. (Hons.), February 2003)

[**Ans.** Cost of running per ton-km Rs. 5]

4. A transport company supplies the following details in respect of a truck of 5 ton capacity:

Cost of truck	Rs. 9,00,000
Estimated life	10 years
Diesel oil @ Rs. 150 per trip each way	
Repairs	Rs. 5,000 per month
Driver's wages	Rs. 2,500 per month
Clearner's wages	Rs. 2,500 per month
Insurance	Rs. 48,000 per year
Taxes	Rs. 24,000 per year
General supervision	Rs. 48,000 per year

The truck carries goods to and from a city covering a distance of 50 kms each way. While going to the city, freight is available to the extent of full capacity and on return 20% of the capacity. Assuming that the truck runs on an average 25 days a month, work out operating cost per kilo meter. *(Sri Sathya Sai University, B.Com. (Hons.), October 2002)*

[**Ans.** Operating cost per kilometer Rs. 5]

5. From the following data relating to vehicle 'A', complete the cost per running tonne-km

	Vehicle 'A'
Kilometers run (annual)	15,000
Tonne per km (average)	6
Cost of vehicle	Rs. 250,000
Road licence (annual)	Rs. 800
Insurance (annual)	Rs. 700
Garage rent (annual)	Rs. 1,300
Supervision and salaries	Rs. 2,700
Driver's wages per hour	Rs. 4
Cost of fuel per litre	Rs. 6
Kms run per litre	20
Repair and maintenance per km	Rs. 2
Tyre allocation per km	Re 1
Estimated life of the vehicle	1,00,000 kms

Charge interest at 5% p.a. on cost of vehicle. The vehicle runs 20 kms per hour on an average.

(University of Madras, B.Com., May 2004)

[**Ans.** Cost per ton km = Rs. 1.20]

6. A truck carries goods covering a distance of 60 km each way on upward journey, freight is available for its full capacity but on downward journey only 25% of its capacity is filled up. The truck run an average 20 days a month. Compute ton-km per month for the truck if the capacity of the truck is 10 tons. *(Calcutta University, B.Com. (Hons.), 2004)*

[**Ans.** Ton-kms 15,000]

7. A truck starts with a load of 10 tonnes of goods from station D. It unloads 3 tonnes at station E and rest of goods at station F. It reaches back directly to station D after getting reloaded with 8 tonnes of goods at station F. The distance from D to E, E to F and then from F to D are 80 kms, 120 kms and 160 kms respectively. Compute tonne-kms for the truck service.

(Calcutta University, B.Com. (Hons.), 2001)

[**Ans.** Tonne-kms for the truck service 2,920]

❑❑❑

Part II

SYSTEMS OF MAINTAINING COST ACCOUNTING

Chapter 7

NON-INTEGRAL OR COST LEDGER ACCOUNTING SYSTEM

INTRODUCTION

Just as financial accounting system is maintained with certain objectives in view, cost accounting system is often distinctively maintained with a view to achieve its objectives. All transactions are collected from the same invoices, vouchers or receipts which are also common for financial accounts. Costs are then classified according to functions, departments or products. Though real accounts and nominal accounts are of direct relevance in ascertaining the cost of products, personal accounts and cash or bank account are not directly related to cost ascertainment. When cost accounting system is maintained it involves maintenance of certain books, for recording day-to-day transactions. It is not necessary to maintain cost accounting under double-entry system of book-keeping. However, in order to ensure arithmetical accuracy of data often the principles of double entry system of book-keeping is followed. Under double entry system cost accounts are maintained in the main ledger which is termed as cost ledger. In addition to this, many subsidiary ledgers are also maintained. In the cost ledger, control accounts are maintained pertaining to each subsidiary ledger. In addition to control accounts, two other accounts, viz, cost of sales account and costing profit and loss account are also maintained in the cost ledger, in order to match cost with revenue. Apart from these accounts, a general ledger adjustment account is opened in cost ledger to accommodate entries relating to transactions adjustable against cash, bank, debtors, creditors etc. Entries in the accounts are made once in each accounting period on the basis of periodical totals of transactions contained in subsidiary ledgers.

INTERLOCKING SYSTEM

There two systems of maintaining cost records, viz, interlocking system and integral accounting system. Under interlocking system, cost records are maintained in a separate set of books independent of financial accounting. The ICMA terminology defines interlocking system of accounting as "a system in which the cost accounting are distinct from the financial accounting. The two sets of accounts being kept continuously in agreement or readily recognizable".

The following are some of the advantages of interlocking accounting system:

1. When separate set of costing books are maintained it facilitates ready accomplishment of its objectives.
2. It avoids the complications of recording the entries if it is integrated with financial accounting.
3. It can be maintained according to convenience as it need not be statutorily maintained.

The following are some of the limitations of this accounting system:

1. When cost accounting is independently maintained, it amounts to duplication of expenses along with financial accounting.
2. The profit shown by cost books may vary with that shown by financial accounting. This requires reconciliation which involves time and effort.

The integral accounting system is discussed in a separate chapter.

ENTRIES TO RECORD TRANSACTIONS UNDER INTERLOCKING SYSTEM

1. Materials

(*a*) Purchase of materials for stock (cash or credit basis):

Stores ledger control a/c *Dr.*
To General Ledger adjustment a/c

(*b*) Returns to suppliers:

General ledger adjustment a/c *Dr.*
To stores ledger control a/c

(*c*) Materials purchased specifically for a job (*i.e.,* direct issue)

Work-in-progress control a/c *Dr.*
To General ledger adjustment a/c

(*d*) Direct material issued from stores to Job:

Work-in-progress control a/c *Dr.*
To stores ledger control a/c

(*e*) Materials returned from jobs to stores:

Stores ledger control a/c *Dr.*
To work-in-progress a/c

(*f*) Issue of indirect materials:

Factory overhead control a/c *Dr.*
To stores ledger control a/c

(*g*) Transfer of materials from one job to another:

Receiving job a/c *Dr.*
To giving job

(*h*) Normal wastage of materials and stores:

Factory overhead control a/c *Dr.*
To stores ledger control a/c

(*i*) Abnormal wastage of materials:

Costing P & L a/c *Dr.*
To stores ledger control a/c

(*j*) Abnormal gain of materials: *Dr.*

Stores ledger control a/c
To costing P & L a/c

2. Labour

(*a*) Payment of direct wages:

Wages control a/c *Dr.*
To General Ledger adjustment a/c

(*b*) Allocation of direct labour:

Work-in-progress a/c Dr.

To wage control a/c

(*c*) Payment of indirect labour cost:

Wage control a/c Dr.

To General ledger adjustment a/c

(*d*) Allocation of indirect labour cost:

Overhead control a/c Dr.

To wage control a/c

(*e*) Normal idle time cost:

Factory overhead control a/c Dr.

To wage control a/c

(*f*) Abnormal idle time cost:

Costing P & L a/c Dr.

To wage control a/c

3. Direct Expenses

Work-in-progress control a/c Dr.

To General ledger adjustment a/c

4. Overheads

(*a*) For recording overhead incurred and accrued:

Factory control a/c Dr.

Administration control a/c Dr.

S & D control a/c Dr.

To General ledger adjustment a/c

(*b*) Allocation of factory overheads:

Work-in-progress control a/c Dr.

To factory overhead control a/c

(*c*) Absorption of administration overhead

Finished stock ledger control a/c Dr.

To administration overhead control a/c

(*d*) Absorption of selling and distribution overhead:

Cost of sales a/c Dr.

To S & D overhead control a/c

(*e*) If under/over absorbed amounts are carried forward to subsequent year, the balance of each overhead a/c will have to be transferred to respective overhead suspense (or reserve) account as follows

(*i*) Production overhead a/c Dr.

To production overhead suspense a/c

(For over recovery)

(*ii*) Administration overhead suspense a/c Dr.

To Administration overhead a/c

(For under recovery)

(*iii*) Selling and distribution overhead suspense a/c *Dr.*
To S & D overhead a/c
(For under recovery)

(*f*) In case of under/over absorbed overheads are transferred to costing P & L a/c then the relevant entries will be as follows:

(*i*) For over recovery:
Overhead control a/c *Dr.*
To costing P & L a/c

(*ii*) For under recovery:
Costing P & L a/c *Dr.*
To overhead control a/c

5. Finished Goods or Completed Jobs

(*a*) Transfer of completed jobs or finished goods produced to finished goods ledger:
Finished stock ledger control a/c *Dr.*
To work-in-progress control a/c

(*b*) Transfer of finished goods sold:
Cost of sales a/c *Dr.*
To finished stock ledger control a/c

(*c*) Transfer of cost of sales a/c to P & L a/c:
Costing P & L a/c *Dr.*
To cost of sales a/c

(*d*) To record sales:
General ledger adjustment a/c *Dr.*
To costing P & L a/c

6. Transfer of Profit or Loss

(*a*) In case of profit:
Costing P & L a/c *Dr.*
To General ledger adjustment a/c

(*b*) In case of loss:
General ledger adjustment a/c *Dr.*
To costing P & L a/c

LEDGERS MAINTAINED UNDER COST ACCOUNTING SYSTEM

Under cost accounting system the following ledgers are maintained:

1. Cost Ledger

It is the main ledger maintained in the cost department. It contains two accounts, viz (*a*) control account for each of the subsidiary ledgers. Some of the control accounts maintained in this ledger are stores ledger control account, work-in-progress ledger control account, etc. (*b*) cost ledger control account to make the cost ledger self balancing.

2. Stores Ledger

All transactions relating to materials are found in this ledger. It contains a separate account for each item of stores such as raw materials, component parts, indirect materials. The concerned material account is debited with materials received and credited with materials issued. The entries in each account is made from the invoice, materials received note, material requisition note, etc. The balance in this account represent the cost of unused materials.

3. Work-in-Progress Ledger

This is also known as job ledger. It contains a separate account for each job or work-in-progress. The elements of cost is debited to this account and is credited with the amount of finished goods completed and transferred. The balance in this account represent cost of incomplete job.

4. Finished Goods Ledger

This ledger contains a separate account for each item of finished product or completed job. This account is debited with the cost of finished product and the amount of administration overhead absorbed and credited with the cost of goods sold. The balance in this account shows the closing stock of finished goods in terms of value.

Control Accounts

Under interlocking system, control accounts are maintained in the cost ledger to complete double entry in cost books. These control accounts are nothing but total accounts or adjustment accounts summarising mass of information contained in the subsidiary ledgers, *i.e.,* stores ledger, job ledger and finished stock ledger.

A control account is maintained in the cost ledger so that double entry in the cost ledger may be completed and make it self-balancing. These control accounts are posted with the totals of items which have been debited or credited in detail to the accounts in the ledgers to which they relate. The balance in control accounts represents the total of balances in a number of accounts of similar nature maintained in that subsidiary ledger to which the control account relates. For example, the balance in stores ledger control account represents in aggregate the detailed balances of stores accounts.

In addition to these control accounts for each of the subsidiary ledger, a cost ledger control account is also kept in cost ledger. This is operated to make the cost ledger self-balancing.

Advantages

1. It provides a check for ensuring that all expenditure is accounted for in cost accounts with the help of control account.
2. It provides a basis for reconciliation with the financial accounts.
3. It provides a ready means of preparing monthly or periodical balance sheet, profit and loss account and statistics relating to cost.

IMPORTANT CONTROL ACCOUNTS MAINTAINED UNDER INTERLOCKING SYSTEM

The various control accounts under interlocking system are as follows:

1. Stores Ledger Control Account

This account is maintained in cost ledger. This records a summary of the value of stores received, issued and balances on hand. Receipts are posted from materials received notes to the debit side of this account. Similarly, issue of materials from material requisition or material abstract is posted to the credit side of the account. Thus for each entry in stores ledger, there is a corresponding debit or credit, (though in total) in this account. The balance of this account represents the total balance of stock which should agree with the aggregate of the balances of individual accounts in the stores ledger.

2. Wages Control Account

This account records labour transactions in aggregate *i.e.,* direct and indirect wages. This account is debited with gross wages shown in wages analysis sheet. It is closed by transfer of direct labour to work-in-progress and indirect labour to overhead, *i.e.,* production administration or selling and distribution overhead account; as the case may be. Wages paid for abnormal idle time are transferred to costing profit and loss account.

3. Factory Overhead Control Account

This account records factory overhead expenses in aggregate. It is debited with the amount of indirect materials, indirect labour and indirect expenses as available from indirect materials analysis sheet, wages analysis sheet etc. This account is credited with the amount of overheads recovered. The balance in the control account represents under or over-absorption which is transferred to overhead adjustment.

4. Administration Overheads Control Account

This account is debited with the administrative overheads incurred and credited with the amount of administrative overhead absorbed by finished goods. Any balance in this account represent under or over-absorption of administrative overhead which is transferred to overhead adjustment account.

5. Selling and Distribution Overhead Control Account

This account is debited with the amount of selling and distribution overhead incurred and credited by the amount of such overheads absorbed by the cost of sales. Balance in this account represents under or over-absorption of selling and distribution overhead which is transferred to overhead adjustment.

6. Overhead Adjustment Account

This account is debited with under-absorbed overheads and credited with over-absorbed overhead amount. The net balance in this account is transferred to costing profit and loss account.

7. Work-in-Progress Control Account

This account represents the total work-in-progress at any time. This account is debited with the totals of materials, wages and overheads as transferred from the respective control accounts. This account is credited when a job is completed. Thus, this account shows the total value of unfinished jobs.

8. Stock Ledger Control Account

This account contains the summary of all finished goods transactions in total. It is debited with the cost of finished goods transferred from work-in-progress control account and the amount of administration overhead absorbed which is transferred from Administration overhead control account. This account is credited with the total cost of goods sold which is transferred to the cost of sales account.

9. Cost of Sales Account

This account is debited with the cost of goods sold by transfer from finished goods ledger control account and also by the selling and distribution overhead absorbed. It is closed by transferring its balance to costing profit and loss account.

10. Costing Profit and Loss Account

This account reveals the result of the business *i.e.,* profit or loss of the business. This account is debited with the cost of sales, abnormal losses and under-absorbed overhead and credited with the sales value, abnormal gain and over-absorbed overhead. The balance in this account represents profit and loss which is transferred to cost ledger control account.

11. Cost Ledger Control Account or General Ledger Adjustment Account

This account is also known as financial ledger control account. This account is maintained to make the cost ledger self-balancing. Cost ledger contains only impersonal accounts. As no personal accounts are kept and in order to complete double entry, it becomes necessary to debit or credit all the transactions which arise in financial accounts to cost ledger control account. In fact, the account represents the personal accounts shown in the financial ledger. For example, wages are paid to the extent of Rs. 5,000, as no cash or bank account is maintained in cost ledger, therefore, in order to complete double entry, wages account will be debited and in place of Bank or cash account. General Ledger Adjustment account in the cost ledger will be credited. Thus, all the financial transactions on account of material purchases, wages, salaries and miscellaneous expenses are credited to cost ledger control account by contra debit to various control accounts. In a similar way all the financial receipts are debited to this account. Any transfer from cost books to financial books, *e.g.,* cost of capital, work done in the factory, will also be entered in this account.

The main object of this account is to complete double entry in cost accounting. Therefore, purely cost accounting transactions say transfer entries with no relations to the finances are not passed through this account as double entry is already complete. The balance in this account represents the total of the balances of all personal accounts in the financial ledger.

CHAPTER 7

Problem 1. The following figures have been ascertained from the costing records. You are required to pass the necessary entries in the cost journal. Assume that a system of maintaining control accounts prevails in the organisation.

		Rs.
(1)	Purchases	3,90,000
(2)	Carriage inwards	5,850
(3)	Stores issued	3,58,800
(4)	Productive wages	3,46,320
(5)	Unproductive wages	1,21,680
(6)	Works on cost	3,48,400
(7)	Materials used in repairs	3,120
(8)	Cost of completed jobs	12,80,630

Solution:

COST JOURNAL

(1)	Stores ledger control a/c	*Dr.*	3,90,000	
	To general ledger adj. a/c			3,90,000
	(Being the entry for purchase of materials)			

		Dr.	Cr.
(2)	Stores ledger control a/c *Dr.*	5,850	
	To general ledger adj. a/c		5,850
	(Being carriage inward treated as part of the cost of materials purchased)		
(3)	Work-in-progress ledger control a/c *Dr.*	3,58,800	
	To stores ledger control a/c		3,58,800
	(Being stores issued to production)		
(4)	Wages control a/c *Dr.*	3,46,320	
	To general ledger adj. a/c		3,46,320
	(Being payment of wages)		
(5)	Factory overhead control a/c *Dr.*	1,21,680	
	To cost ledger control a/c		1,21,680
	(Being indirect wages incurred)		
(6)	Factory overhead control a/c *Dr.*	3,48,400	
	To cost ledger control a/c		3,48,400
	(Being works overhead other than indirect wages)		
(7)	Factory overhead control a/c *Dr.*	3,120	
	To stores ledger control a/c		3,120
	(Being materials used in repairs)		
(8)	Finished stock ledger control a/c *Dr.*	12,80,630	
	To work-in-progress ledger control a/c		12,80,630
	(Being completed production transferred to finished stock)		

Problem 2. The following transactions pertaining to materials took place during March 2001 in ABC Company Ltd. Enter the transactions in the cost books.

(1)	Materials purchased	
	Credit purchases	10,000
	Cash purchases	8,000
	Credit purchases for job no. 20	1,000
(2)	Return to suppliers	500
(3)	Direct materials issued to jobs	4,000
(4)	Indirect materials issued to jobs	400
(5)	Materials returned from jobs to stores	200
(6)	Materials transferred from job no. 8 to job no. 12	300

Solution:

COST JOURNAL

		Dr.	Cr.
(1)	(*a*) Stores ledger control a/c *Dr.*	10,000	
	To general ledger adj. a/c		10,000
	(Being the amount of credit purchases)		
	(*b*) Stores ledger control a/c *Dr.*	8,000	
	To general ledger adj. a/c		8,000
	(Being cash purchases)		
	(*c*) Work-in-progress ledger control a/c *Dr.*	1,000	
	To general ledger adj. a/c		1,000
	(Being purchases for a special job no. 20)		
(2)	General ledger adj. a/c *Dr.*	500	
	To stores ledger control a/c		500
	(Being the return to supplier)		

(3) Work-in-progress ledger control a/c	*Dr.*	4,000	
To stores ledger control a/c			4,000
(Being the direct materials issued to jobs)			
(4) Factory overhead control a/c	*Dr.*	400	
To stores overhead control a/c			400
(Being issue of indirect materials)			
(5) Stores ledger control a/c	*Dr.*	200	
To work-in-progress control a/c			200
(Being the materials returned from jobs to stores)			
(6) Job no. 12 a/c	*Dr.*	300	
To job no. 8 a/c			300
(Being the transfer of materials from job no. 8 to job no. 12)			

Problem 3. Pass Journal entries in the cost books (non-integrated system) for the following transactions:

(1) Materials worth Rs. 25,000 returned to stores from job.

(2) Gross total wages paid Rs. 48,000. Employer's contribution to PF and state insurance amount to Rs. 2,000. Wages analysis book detailed Rs. 20,000 direct labour, Rs. 12,000 towards indirect factory labour, Rs. 10,000 towards salaries to office staff and Rs. 8,000 for salaries to selling and distribution staff. *(University of Delhi, B.Com. (Hons.), April 1999)*

Solution:

JOURNAL ENTRIES

(*i*) Stores ledger control a/c	*Dr.*	25,000	
To work-in-progress a/c			25,000
(Being raw materials returned to stores)			
(*ii*) Wages control a/c	*Dr.*	50,000	
To general ledger control a/c			50,000
(Being payment of wages)			
Work-in-progress control a/c	*Dr.*	20,000	
Factory overhead a/c	*Dr.*	12,000	
Office overhead a/c	*Dr.*	10,000	
Selling overhead a/c	*Dr.*	8,000	
To wages control a/c			50,000
(Allocation of wages to direct and indirect cost)			

Problem 4. As at 31st March 2001, the following balances existed in a company's cost ledger

	Dr.	*Cr.*
Stores ledger control a/c	6,02,870	
Work-in-progress control a/c	2,44,730	
Finished stock ledger control a/c	5,03,890	
Manufacturing overhead control a/c		21,050
Cost ledger control a/c		13,30,440
	13,51,490	13,51,490

During the next three months the following items arose

	Rs.
(1) Raw materials purchased	2,46,000
(2) Materials returned to suppliers	5,800
(3) Materials issued to production	2,54,630

CHAPTER 7

(4)	Factory wages	1,01,060
(5)	Manufacturing overhead incurred	1,83,020
(6)	Indirect labour	43,330
(7)	Manufacturing overhead charged to production	1,54,400
(8)	Cost of sales	3,71,780
(9)	Sales returns at cost	10,760
(10)	Finished product at cost	4,21,670

Pass the necessary entries, open ledger accounts and prepare trial balance

Solution:

JOURNAL ENTRIES

(1)	Stores ledger control a/c	Dr.	2,46,000	
	To general ledger adj. a/c			2,46,000
	(Being materials purchased)			
(2)	General ledger adj. a/c	Dr.	5,800	
	To stores ledger control a/c			5,800
	(Entry for materials returned to suppliers)			
(3)	Work-in-progress control a/c	Dr.	2,54,630	
	To stores ledger control a/c			2,54,630
	(Entry for issue of materials to production)			
(4)	Wages control a/c	Dr.	1,01,060	
	To general ledger adj. a/c			1,01,060
	(Entry for direct wages incurred)			
(5)	Work-in-progress control a/c	Dr.	1,01,060	
	To wages control a/c			1,01,060
	(Entry for direct wages charged to production)			
(6)	Works overhead control a/c	Dr.	1,83,020	
	To general ledger adj. a/c			1,83,020
	(Entry for works overhead incurred)			
(7)	Works overhead control a/c	Dr.	43,330	
	To general ledger adj. a/c			43,330
	(Entry for indirect wages incurred)			
(8)	Work-in-progress control a/c	Dr.	1,54,400	
	To works overhead control a/c			1,54,400
	(Entry for overhead charged to production)			
(9)	General ledger adj. a/c	Dr.	3,71,780	
	To finished stock ledger control a/c			3,71,780
	(Entry for cost of sales)			
(10)	Finished stock ledger control a/c	Dr.	10,760	
	To general ledger adj. a/c			10,760
	(Entry for sales return)			
(11)	Finished stock ledger control a/c	Dr.	4,21,670	
	To work-in-progress control a/c			4,21,670
	(Entry for finished goods transferred)			

GENERAL LEDGER ADJUSTMENT ACCOUNT

To stores ledger control a/c	5,800	By balance b/d	13,30,440
To finished stock ledger control a/c	3,71,780	By stores ledger control a/c	2,46,000
To balance c/d	15,37,030	By wages control a/c	1,01,060
		By works overhead control a/c	1,83,020
		By works overhead control a/c	43,330
		By finished stock ledger control a/c	10,760
	19,14,610		19,14,610

STORES LEDGER CONTROL ACCOUNT

To balance b/d	6,02,870	By general ledger control	5,800
To general ledger adj. a/c	2,46,000	By work-in-progress control a/c	2,54,630
		By balance c/d	5,88,440
	8,48,870		8,48,870

MANUFACTURING OVERHEAD CONTROL ACCOUNT

To general ledger control a/c	1,80,020	By balance b/d	21,050
To general ledger control a/c	43,300	By work-in-progress control a/c	1,54,400
		By balance c/d	50,900
	2,26,350		2,26,350

WORK-IN-PROGRESS CONTROL ACCOUNT

To balance b/d	2,44,730	By finished stock ledger control a/c	4,21,670
To stores ledger control a/c	2,54,630	By balance c/d	3,33,150
To wages control a/c	1,01,060		
To Mfg. overhead control a/c	1,54,400		
	7,54,820		7,54,820

FINISHED STOCK LEDGER CONTROL ACCOUNT

To balance b/d	5,03,890	By cost ledger control a/c	3,71,780
To work-in-progress control a/c	4,21,670	By balance c/d	5,64,540
To general ledger adj a/c	10,760		
	9,36,320		9,36,320

TRIAL BALANCE

	Dr.	*Cr.*
Cost ledger control a/c		15,37,030
Stores ledger control a/c	5,88,440	
Manufacturing overhead control a/c	50,900	
WIP control a/c	3,33,150	
Finished stock ledger control a/c	5,64,540	
	15,37,030	15,37,030

CHAPTER 7

Problem 5. A fire destroyed some accounting records of a company. You have been able to collect the following from the spoilt papers/records and as a result of consultation with accounting staff in respect of January 1997.

***(i)* Incomplete ledger entries**

RAW MATERIALS ACCOUNT

Beginning Inventory	32,000		

WORK-IN-PROGRESS ACCOUNT

Beginning Inventory	9,200	Finished stock	1,51,000

CREDITORS ACCOUNT

Closing balance	19,200	Opening balance	16,400

MANUFACTURING OVERHEAD ACCOUNT

Amount spent	29,600		

FINISHED GOODS ACCOUNT

Opening Inventory	2,400		
		Closing Inventory	30,000

***(ii)* Additional information:**

(1) Cash book showed that Rs. 89,200 have been paid to creditors for raw materials.

(2) Ending inventory of work-in-progress included materials Rs. 5,000 on which 300 direct labour hours have been booked against wages and overheads.

(3) The job card showed that workers have worked for 7,000 hours. The wage rate is Rs. 10 per labour hour.

(4) Overhead recovery rate was Rs. 4 per direct labour hour.

You are required to complete the above accounts in the cost ledger of the company.

(C.A. Inter, May 1997)

Solution:

CREDITORS ACCOUNT

To cash & bank (1)	89,200	By balance b/d	16,400
To balance c/d	19,200	By purchases (Balancing figure)	92,000
	1,08,400		1,08,400

WORK-IN-PROGRESS ACCOUNT

To balance b/d	9,200	By finished goods	1,51,000
To raw materials (Balancing figure)	53,000	By balance c/d materials (2)	5,000
To wages (3) 7,000 hrs. × Rs. 10	70,000	Labour (2) (300 hrs. × 10)	3,000
To overheads (4) 7,000 hrs. × Rs. 4	28,000	Overhead (2) 300 hrs. × Rs. 4	1,200
	1,60,200		1,60,200

RAW MATERIALS ACCOUNT

To balance b/d	32,000	By work-in-progress a/c	53,000
To purchases	92,000	By balance c/d	71,000
	1,24,000		1,24,000

FINISHED GOODS ACCOUNT

To balance b/d	24,000	By cost of sales (Balancing figure)	1,45,000
To work-in-progress a/c	1,51,000	By balance c/d	30,000
	1,75,000		1,75,000

MANUFACTURING OVERHEAD ACCOUNT

To sundries (Amount spent)	29,600	By work-in-progress a/c (7,000 × 4)	28,000
		By under-absorbed overhead	1,600
	29,600		29,600

Problem 6. The following balances were extracted from a company's ledger as on 31st December 1997.

	Rs.	Rs.
Raw materials control a/c	48,836	
Work-in-progress control a/c	14,745	
Finished stock control a/c	21,980	
Nominal ledger control a/c		85,561
	85,561	85,561

Further transactions took place during the following quarter as follows:

	Rs.
Factory overhead— allocation to WIP	11,786
Goods finished — at cost	36,834
Raw materials purchased	22,422
Direct wages — allocated to WIP	18,370
Cost of goods sold	42,000
Raw materials — issued to production	17,000
Raw materials — credited by suppliers	1,000
Inventory audit — raw materials losses	1,300
WIP rejected (with no scrap value)	1,800
Customer's returns (at cost) of finished goods	3,000

Prepare all the ledger accounts in cost ledger.

(*C.A. Inter, November 1998*)

Solution:

RAW MATERIALS CONTROL ACCOUNT

To balance b/d	48,836	By WIP control a/c	17,000
To nominal ledger control a/c	22,422	By nominal ledger control a/c	1,000
		By nominal ledger control a/c	1,300
		By balance c/d	51,958
	71,258		71,258

WORK-IN-PROGRESS CONTROL ACCOUNT

To balance b/d	14,745	By finished stock control a/c	36,834
To nominal ledger control a/c	11,786	By nominal ledger control a/c	1,800
To raw material control a/c	17,000	By balance c/d	23,267
To nominal ledger control a/c	18,370		
	61,901		61,901

FINISHED STOCK CONTROL ACCOUNT

To balance b/d	21,980	By nominal ledger control a/c	42,000
To WIP control a/c	36,834	By balance c/d	19,814
To nominal ledger control a/c	3,000		
	61,814		61,814

NOMINAL LEDGER CONTROL ACCOUNT

To raw material control a/c	1,000	By balance b/d	85,561
To raw material control a/c	1,300	By raw materials control a/c	22,422
To finished stock control a/c	42,000	By WIP control a/c	11,786
To WIP control a/c	1,800	By WIP control a/c	18,370
To balance c/d	95,039	By finished stock control a/c	3,000
	1,41,139		1,41,139

Problem 7. From the following balances and transactions extracted from the books of East-West Company Ltd., journalise and write up the accounts in the cost ledger and prepare a trial balance as at 31st December 2010. Also show the profit or loss for the month:

	Dr.	*Cr.*
Balances as on 1.12.2010:		
Work-in-progress a/c	5,200	
Finished goods a/c	2,300	
Factory overhead suspense a/c	50	
Office overhead suspense a/c	30	
Stores ledger control a/c	1,150	
General ledger adjustment a/c		8,730
	8,730	8,730

Transactions for the month were:	*Rs.*
Direct wages	7,500
Indirect wages	500
Works overhead absorbed in production	2,200
Office overhead absorbed in production	1,200
Stores issued to production	4,900
Goods finished during the month	18,000
Finished goods sold	21,000
Stores purchased	5,000
Stores issued to factory repair orders	200
Carriage inwards on stores issued for production	80
Factory expenses	1,450
Office expenses	1,170

Solution: JOURNAL ENTRIES

Work-in-progress ledger control a/c	*Dr.*	5,200	
Finished goods ledger control a/c	*Dr.*	2,300	
Factory overhead suspense a/c	*Dr.*	50	
Office overhead suspense a/c	*Dr.*	30	
Stores ledger control a/c	*Dr.*	1,150	
To general ledger adjustment a/c			8,730
(Being the opening entries for the balances)			
Stores ledger control a/c	*Dr.*	5,000	
To general ledger adjustment a/c			5,000
(Being stores purchased)			
Work-in-progress ledger control a/c	*Dr.*	4,980	
To stores ledger control a/c			4,980
(Being the stores issued to production Rs. 4,900 and carriage inward on stores issued Rs. 80)			
Factory overhead control a/c	*Dr.*	200	
To stores ledger control a/c			200
(Being stores issued to factory repairs)			
Work-in-progress ledger control a/c	*Dr.*	7,500	
To wages control a/c			7,500
(Being direct wages charged to production)			
Factory overhead control a/c	*Dr.*	500	
To wages control a/c			500
(Being indirect wages charged to factory overhead)			
Wages control a/c	*Dr.*	8,000	
To general ledger adjustment a/c			8,000
(Being the total wages brought into costing book from financial books)			
Factory overhead control a/c	*Dr.*	50	
To factory overhead suspense a/c			50
(Being the latter transferred to former a/c)			
Factory overhead control a/c	*Dr.*	1,450	
To general ledger adjustment a/c			1,450
(Being the actual factory expenses brought into costing books)			
Work-in-progress ledger control a/c	*Dr.*	2,200	
To factory overhead control a/c			2,200
(Being the overheads charged to production)			
Office overhead control a/c	*Dr.*	30	
To office overhead suspense a/c			30
(Being suspense a/c transferred to former a/c			

CHAPTER 7

Particulars		Dr. (Rs.)	Cr. (Rs.)
Office overhead control a/c	Dr.	1,170	
To general ledger adjustment a/c			1,170
(Being the actual office overheads brought into costing books)			
Work-in-progress ledger control a/c	Dr.	1,200	
To office overhead control a/c			1,200
(Being the office overheads charged to production)			
Finished goods control a/c	Dr.	18,000	
To work-in-progress ledger control a/c			18,000
(Being the work-in-progress transferred to former a/c			
Cost of sales a/c	Dr.	20,300	
To finished goods control a/c			20,300
(Being the finished stock transferred to former a/c)			
Costing profit & loss a/c	Dr.	20,300	
To cost of sales a/c			20,300
(Being cost of sales transferred to profit & loss a/c)			
General ledger adjustment a/c	Dr.	21,000	
To costing profit & loss a/c			21,000
(Being the amount of sales brought into costing profit & loss a/c)			
Costing profit & loss a/c	Dr.	700	
To general ledger adjustment a/c			700
(Being the amount of profit)			

COST LEDGER

GENERAL LEDGER ADJUSTMENT A/C

Particulars	Rs.	Particulars	Rs.
To costing P & L a/c	21,000	By balance b/d	8,730
To balance c/d	4,050	By stores ledger control a/c	5,000
		By wages control a/c	8,000
		By factory overhead control a/c	1,450
		By office overhead control a/c	1,170
		By costing P & L a/c	700
	25,050		25,050

STORES LEDGER CONTROL ACCOUNT

Particulars	Rs.	Particulars	Rs.
To balance b/d	1,150	By WIP ledger control a/c	4,980
To general ledger adjustment a/c	5,000	By factory overhead control a/c	200
		By balance c/d	970
	6,150		6,150

WAGES CONTROL ACCOUNT

To general ledger adjustment a/c	8,000	By WIP ledger control a/c	7,500
		By factory overhead control a/c	500
	8,000		8,000

FACTORY OVERHEAD CONTROL ACCOUNT

To stores ledger control a/c	200	By WIP ledger control a/c	2,200
To wages control a/c	500		
To factory overhead suspense a/c	50		
To general ledger adjustment a/c	1,450		
	2,200		2,200

OFFICE OVERHEAD CONTROL ACCOUNT

To office overhead suspense a/c	30	By WIP ledger control a/c	1,200
To general ledger adjustment a/c	1,170		
	1,200		1,200

WORK-IN-PROGRESS LEDGER CONTROL ACCOUNT

To balance b/d	5,200	By finished goods control a/c	18,000
To stores ledger control a/c	4,900	By balance c/d	3,080
To wages control a/c	7,500		
To factory overhead control a/c	2,200		
To office overhead control a/c	1,200		
	21,080		21,080

FINISHED GOODS CONTROL ACCOUNT

To balance b/d	2,300	By cost of sales a/c	20,300
To WIP ledger control a/c	18,000		
	20,300		20,300

COST OF SALES ACCOUNT

To finished goods control a/c	20,300	By costing P & L a/c	20,300

COSTING PROFIT & LOSS ACCOUNT

To cost of sales a/c	20,300	By general ledger adjustment a/c (sales)	21,000
To general ledger adjustment a/c (profit)	700		
	21,000		21,000

TRIAL BALANCE AS ON 31.12.1992

	Dr.	Cr.
General ledger control a/c		4,050
Stores ledger control a/c	970	
WIP ledger control a/c	3,080	
	4,050	4,050

CHAPTER 7

Problem 8. On 31.3.1989, the following balances were extracted from the books of the Supreme Manufacturing Company:

	Dr.	Cr.
Stores ledger control a/c	35,000	
WIP control a/c	38,000	
Finished goods control a/c	25,000	
Cost ledger control a/c		98,000
	98,000	98,000

The following transactions took place in April 1989:

	Rs.
Raw materials:	
Purchased	95,000
Returned to suppliers	3,000
Issued to production	98,000
Returned to stores	3,000
Productive wages	40,000
Indirect labour	25,000
Factory overhead expenses incurred	50,000
Selling and administration expenses	40,000
Cost of finished goods transferred to warehouse	2,13,000
Cost of goods sold	2,10,000
Sales	3,00,000

Factory overheads are applied to production at 150% of direct wages and under or over absorbed overhead being carried forward for adjustment in the subsequent months. All administrative and selling expenses are treated as period cost and charged off to the profit and loss a/c of the month in which they are incurred:

Show the following accounts.

(a) Cost ledger control a/c
(b) Stores ledger control a/c
(c) WIP control a/c
(d) Finished goods stock control a/c
(e) Factory overhead control a/c
(f) Costing P & L a/c
(g) Trial balance as at 30.4.1989 *(C.A., Inter, May 1989)*

Solution:

COST LEDGER CONTROL ACCOUNT

To costing P & L a/c (sales)	3,00,000	By balance b/d	98,000
To stores ledger control a/c	3,000	By stores ledger control a/c	95,000
To balance c/d	95,000	By wage control a/c (production and indirect)	65,000
		By factory overhead control a/c	50,000
		By selling & adm. overhead	40,000
		By costing P & L a/c	50,000
	3,98,000		3,98,000

STORES LEDGER CONTROL ACCOUNT

To Balance b/d	35,000	By cost ledger control a/c	3,000
To cost ledger control a/c	95,000	By work-in-progress control a/c	98,000
To WIP control a/c	3,000	By balance c/d	32,000
	1,33,000		1,33,000

WORK-IN-PROGRESS CONTROL ACCOUNT

To Balance b/d	38,000	By stores ledger control a/c	3,000
To stores ledger control a/c	98,000	By finished goods a/c	2,31,000
To wages control a/c	40,000	By balance c/d	20,000
To factory overhead control a/c	60,000		
	2,36,000		2,36,000

FINISHED GOODS CONTROL ACCOUNT

To balance b/d	25,000	By cost of goods sold a/c	2,10,000
To work-in-progress control a/c	2,13,000	By Balance c/d	28,000
	2,38,000		2,38,000

FACTORY OVERHEAD CONTROL ACCOUNT

To wage control a/c (indirect labour)	25,000	By WIP control a/c	60,000
To cost ledger control a/c	50,000	By balance c/d	15,000
	75,000		75,000

COSTING P & L ACCOUNT

To cost of goods sold a/c	2,10,000	By cost ledger control a/c	3,00,000
To selling & adm. overhead	40,000		
To cost ledger control a/c (costing profit)	50,000		
	3,00,000		3,00,000

TRIAL BALANCE AS ON 30.4.1989

	Dr.	*Cr.*
Stores ledger control a/c	32,000	
Work-in-progress control a/c	20,000	
Finished goods control a/c	28,000	
Factory overhead control a/c	15,000	
General ledger control a/c		95,000
	95,000	95,000

Working Notes:

WAGES CONTROL ACCOUNT

To cost ledger control a/c	65,000	By WIP control a/c	40,000
		By factory overhead control a/c	25,000
	65,000		65,000

COST OF GOODS SOLD

To finished goods control a/c	2,10,000	By costing P & L a/c	2,10,000
	2,10,000		2,10,000

SELLING AND DISTRIBUTION OVERHEAD

To cost ledger control a/c	40,000	By costing P & L a/c	40,000

Problem 9. The following balances are extracted from Gujarat Chemical Company's cost ledger as on 31st March 1993:

Figures in 000

	Rs. (Dr.)	Rs. (Cr.)
Control accounts		
Raw materials	500	
Work-in-progress	230	
Finished stock	130	
General ledger adjustment a/c		860
	860	860

Further transactions took place during the following quarter as follows:

	Fig. in 000's
Factory overhead — allocated to WIP	120
Goods finished at cost	400
Raw materials purchased	225
Direct wages allocated to work-in-progress	80
Raw materials issued to production	165
Cost of goods sold	515
Raw materials — returned to suppliers	10
Sales returns (at cost)	50
Inventory audit — raw material shortage	12
Work-in-progress rejected (scrap value—nil)	18

You are required to:

(*a*) Write up the four accounts in the cost ledger

(*b*) Prepare a trial balances as on 31.3.1993

Solution:

RAW MATERIALS CONTROL ACCOUNT (RS. '000)

To Balance b/d	500	By WIP (materials issued)	165
To General ledger adjustment a/c	225	By General ledger adjustment (purchase returns)	10
		By General ledger adjustment (stock shortage)	12
		By Balance c/d	538
	725		725

WORK-IN-PROGRESS CONTROL ACCOUNT

To Balance b/d	230	By finished stock control a/c	400
To General ledger adjustment a/c (factory overhead)	120	By general ledger adjustment a/c (rejections)	18
To Raw materials control a/c (raw materials)	165	By balance c/d	177
To General ledger adjustment a/c (direct wages)	80		
	595		595

FINISHED STOCK CONTROL ACCOUNT

To Balance b/d	130	By General ledger adjustment a/c (sales at cost)	515
To WIP (completed production)	400	By Balance c/d	45
To General ledger adjustment (sales returns)	30		
	560		560

GENERAL LEDGER ADJUSTMENT ACCOUNT

To Finished stock control a/c (sales of cost)	515	By balance b/d	860
To raw material control a/c (purchase returns)	10	By WIP (factory overhead)	120
To raw materials control a/c (stock shortage)	12	By WIP (direct wages)	80
To WIP (rejections)	18	By Raw material control a/c	225
To balance c/d	760	By Finished stock control a/c (sales returns)	30
	1,315		1,315

TRIAL BALANCE AS ON 31.3.1993 (FIGURES IN '000)

	Dr.	*Cr.*
Raw materials ledger control a/c	538	
Work-in-progress control a/c	177	
Finished stock control a/c	45	
General ledger adjustment a/c		760
	760	760

Problem 10. The following balances appear in the books of M.K. Company Ltd. on 1.1.2002:

General ledger adjustment a/c		15,200
Stores ledger control a/c	8,750	
WIP ledger control a/c	4,280	
Finished goods ledger control a/c	2,170	
	15,200	15,200

On 31.12.2002, the following information was supplied:

Purchase of stores		60,640
Purchase for special job		1,950
Direct wages	38,627	

Indirect factory wages	9,543	
Administrative salaries	6,731	
Selling and distribution salaries	4,252	59,153
Production expenses		10,432
Administration expenses		9,546
Selling and distribution expenses		6,430
Stores issued to production		56,501
Stores issued to maintenance a/c		2,556
Returns to suppliers		312
Production overhead absorbed by production		23,410
Administration overhead absorbed by finished goods		15,150
Selling overhead recovered on sales		9,515
Products finished during the year		1,18,517
Finished goods sold at cost		1,33,382
Sales		1,55,000

You are required to record the entries in cost ledger for the year 2002 and prepare a trial balance.

Solution: COST LEDGER

GENERAL LEDGER ADJUSTMENT ACCOUNT

To stores ledger control a/c (returns)	312	By balance b/d	15,200
To P & L a/c (sales)	1,55,000	By stores ledger control a/c purchases	60,640
To balance c/d	18,697	By WIP control a/c (special purchases)	1,950
		By wages control a/c	59,153
		By production overhead a/c	10,432
		By administration overhead a/c	9,546
		By selling overhead a/c	6,430
		By costing P & L a/c	10,658
	1,74,009		1,74,009

STORES LEDGER CONTROL ACCOUNT

To Balance b/d	8,750	By work-in-progress control a/c	56,501
To General ledger adjustment a/c	60,640	By production overhead a/c	2,586
		By General ledger adjustment a/c	312
		By Balance c/d	9,991
	69,390		69,390

WAGES CONTROL ACCOUNT

To General ledger adjustment a/c	59,513	By WIP control a/c	38,627
		By production overhead a/c	9,543
		By Administration overhead a/c	6,731
		By selling & distribution overhead a/c	4,252
	59,513		59,513

PRODUCTION OVERHEAD ACCOUNT

To general ledger adjustment a/c	10,432	By WIP control a/c	23,410
To stores ledger control a/c	2,586		
To wages control a/c	9,543		
To overhead adjustment a/c	849		
	23,410		23,410

WIP LEDGER CONTROL ACCOUNT

To balance b/d	4,280	By Finished goods ledger control a/c	1,18,517
To General ledger adjustment a/c	1,950	By Balance c/d	6,251
To stores ledger control a/c	56,501		
To wages control a/c	38,627		
To production overhead a/c	23,410		
	1,24,768		1,24,768

ADMINISTRATION OVERHEAD ACCOUNT

To General ledger adjustment a/c	9,546	By Finished goods ledger control a/c	15,150
To wages control a/c	6,731	To overhead adjustment a/c	1,127
	16,277		16,277

SELLING AND DISTRIBUTION OVERHEAD ACCOUNT

To General ledger adjustment a/c	6,430	By cost of sales a/c	9,515
To wages control a/c	4,252	By overhead adjustment a/c	1,167
	10,682		10,682

FINISHED GOODS LEDGER CONTROL ACCOUNT

To Balance b/d	2,170	By cost of sales a/c	1,33,382
To administration overhead a/c	15,150	By Balance c/d	2,455
To WIP control a/c	1,18,517		
To wages control a/c	6,731		
	1,35,837		1,35,837

COST OF SALES ACCOUNT

To selling and distribution overhead a/c	9,515	By P & L a/c	1,42,897
To finished goods ledger control a/c	1,33,382		
	1,42,897		1,42,897

OVERHEAD ADJUSTMENT ACCOUNT

To administration overhead a/c	1,127	By production overhead a/c	849
To selling and distribution overhead a/c	1,167	By costing P & L a/c	1,445
	2,294		2,294

COSTING P & L ACCOUNT

To cost of sales a/c	1,42,897	By sales	1,55,000
To overhead adjustment a/c	1,445		
To General ledger adjustment a/c– net profit	10,658		
	1,55,000		1,55,000

TRIAL BALANCE

	Dr.	*Cr.*
Stores ledger control a/c	9,991	
Work-in-progress control a/c	6,251	
Finished goods ledger control a/c	2,455	
General ledger adjustment a/c		18,697
	18,697	18,697

Problem 11. A firm maintains its books under non-integral accounting system. Enter the following transactions in the cost books:

		Rs.
(i)	Credit purchase for a special job	30,000
(ii)	Returned to suppliers	5,000
(iii)	Materials returned from jobs to stores	500
(iv)	Direct materials issued to jobs	10,000

(Osmania University, M.Com., Final, June 2003)

Solution:

COST JOURNAL

(i) Work-in-progress control a/c To general ledger control a/c (Being credit purchase for a special job)	*Dr.*	30,000	30,000
(ii) General ledger control a/c To work-in-progress control (Being materials returned to suppliers)	*Dr.*	5,000	5,000
(iii) Stores ledger control a/c To work-in-progress control a/c (Being materials returned from jobs to stores)	*Dr.*	500	500
(iv) Work-in-progress control a/c To stores ledger control a/c (Being direct material issued to jobs)	*Dr.*	10,000	10,000

Problem 12. Give journal entries for entering the following transactions in the non-integrated accounting system in the cost records.

(a) Purchase of raw materials on credit Rs. 10,000

(b) Depreciation on machinery used for production Rs. 2,000

(c) Absorption of production overhead Rs. 6,000

(d) Cash period for indirect production wages Rs. 1,000

(Osmania University, M.Com., Final, May 1997)

Solution:

COST JOURNAL

(1) Stores ledger control a/c To general ledger adjustment a/c (Being raw materials purchased)	*Dr.*	10,000	10,000
(2) Factory overhead control a/c To general ledger adjustment a/c (Being depreciation on machinery)	*Dr.*	2,000	2,000
(3) Work-in-progress control a/c To factory overhead control a/c (Being absorption of production overhead)	*Dr.*	6,000	6,000
(4) Wages control a/c To General ledger adjustment a/c (Being cash paid to indirect production wages)	*Dr.*	1,000	1,000

Problem 13. R.K. Ltd., operates separate cost accounting and financial accounting systems. The following information has been extracted from the cost records of the company for the month of January 2005.

(A) **Control account balance**	*1-1-05* *Rs.*	*31-1-05* *Rs.*
Raw material	49,500	50,300
Work-in-progress	60,100	56,900
Finished goods	1,15,400	1,37,400

(B) **Additional information for the month:**

Raw materials purchased	1,08,000
Production overhead incurred	91,600
Production overhead absorbed (185% of direct wages)	74,000
Factory cost of goods produced	2,22,000
Cost of goods sold (Excluding selling and Administration overhead)	2,00,000
Selling and Administration overhead incurred and absorbed	30,000
Sales	3,00,000
Loss of materials damaged by flood	2,400

You are required to:

(*i*) Prepare the following control accounts in the cost ledger:

(*a*) Raw material (*b*) Work-in-progress

(*c*) Finished goods (*d*) Production overhead

(*ii*) Ascertain profit as per cost accounts for the month of January 2005 assuming that under or over absorbed overhead is written off to costing profit and loss account.

(University of Delhi, B.Com. (Hons) 2005)

Solution:

RAW MATERIAL CONTROL ACCOUNT

To balance b/d	49,500	By costing P & L a/c	2,400
To cost ledger control a/c	1,08,000	By work-in-progress control a/c	1,04,800
		By Balance c/d	50,300
	1,57,500		1,57,500

WORK-IN-PROGRESS CONTROL ACCOUNT

To Balance b/d	60,100	By finished goods control a/c	2,22,000
To stores ledger control a/c	1,04,800	By Balance c/d	56,900
To wage control a/c (W.N-1)	40,000		
To production overhead control a/c	74,000		
	2,78,900		2,78,900

Working Note:

Wage control $= 74{,}000 \times \frac{100}{185} =$ Rs. 40,000

FINISHED GOODS CONTROL ACCOUNT

To balance b/d	1,15,400	By cost of goods sold	2,00,000
To work-in-progress control a/c	2,22,000	By balance c/d	1,37,400
	3,37,400		3,37,400

PRODUCTION OVERHEAD CONTROL ACCOUNT

To cost ledger adj. a/c	91,600	By work-in-progress control a/c	74,000
		By profit & loss a/c	17,600
	91,600		91,600

COSTING PROFIT & LOSS ACCOUNT

To cost of goods sold	2,00,000	By sales	3,00,000
To Abnormal loss	2,400		
To production overhead under absorbed	17,600		
To selling & administration overhead	30,000		
To profit	50,000		
	3,00,000		3,00,000

Problem 14. The following figures have been extracted from the cost records of a manufacturing unit:

		Rs.
Stores:	Opening balance	32,000
	Purchase of material	1,58,000
Transfer from work-in-progress		80,000
Issues to work-in-progress		1,60,000
Issues to repair and maintenance		20,000
Deficiencies found in stock taking		6,000
Work-in-progress: opening balance		60,000

Direct wages applied	65,000
Overhead applied	2,40,000
Closing balance of WIP	45,000

Finished product; Entire output is sold at a profit of 10% on actual cost from work-in-progress. Wages incurred Rs. 70,000, overhead incurred Rs. 2,50,000.

Items not included in cost records: Income from investment Rs. 10,000. Loss on sale of capital assets Rs. 20,000.

Draw up stores control account, work-in-progress control account, costing profit and loss account and reconciliation statement. *(C.A. PE-II, Group II, May 2005)*

Solution:

IN COSTING BOOKS

STORES CONTROL ACCOUNT

To balance b/d	32,000	By WIP control a/c	1,60,000
To General ledger adjustment a/c	1,58,000	By works overhead control a/c	20,000
To work-in-progress control a/c	80,000	By costing P & L a/c	6,000
		By balance c/d	84,000
	2,70,000		2,70,000

WIP CONTROL ACCOUNT

To Balance b/d	60,000	By stores control a/c	80,000
To stores control a/c	1,60,000	By costing P&L a/c (cost of sales)	4,00,000
To Direct wages control a/c	65,000		
To works overhead control a/c	2,40,000	To balance c/d	45,000
	5,25,000		5,25,000

WORKS OVERHEAD CONTROL ACCOUNT

To general ledger adjustment a/c	2,50,000	By WIP control a/c	2,40,000
To stores ledger control a/c	20,000	By costing P & L a/c (under recovery)	30,000
	2,70,000		2,70,000

COSTING P & L ACCOUNT

To WIP control a/c (cost of sales)	4,00,000	By general ledger adjustment a/c:		
To works overhead control a/c	30,000	Cost of sales	4,00,000	
To stores control a/c (shortage)	6,000	Profit 10%	40,000	
To profit	4,000			4,40,000
	4,40,000			4,40,000

IN FINANCIAL BOOKS

PROFIT & LOSS ACCOUNT

To opening stock:			By sales		4,40,000
– Stores	32,000		By closing stock:		
– WIP	60,000	92,000	– Stores	84,000	
To purchases		1,58,000	– WIP	45,000	1,29,000
To wages		70,000			
To overhead		2,50,000	By income from investment		10,000
To loss on sale of capital assets		20,000	By loss		11,000
		5,90,000			5,90,000

RECONCILIATION STATEMENT

Profit as per cost accounts		4,000
Add: Income from investment recorded in financial accounts		10,000
		14,000
Less: Under absorption of wages in cost accounts	5,000	
Loss on sale of capital asset included in financial at accounts only	20,000	
		25,000
Loss as per financial accounts		11,000

Problem 15. A company operates separate cost accounting and financial accounting systems. The following is the list of opening balances as on 1-4-2001 in the cost ledger

	Debit Rs.	*Credit* Rs.
Stores ledger control a/c	53,375	—
WIP control account	104,595	—
Finished goods control a/c	30,780	—
General ledger adjustment a/c	—	1,88,750

Transactions for the quarter ended 30-6-2001 are as under:

	Rs.
Materials purchased	26,700
Materials issued to production	40,000
Materials issued for factory repairs	900
Factory wages paid (including indirect wages Rs. 23,000)	77,500
Production overhead incurred	95,200
Production overheads under-absorbed and written off	3,200
Sales	2,56,000

The company's gross profit is 25% on factory cost. At the end of the quarter, work-in-progress stocks increased by Rs. 7,500.

Prepare the relevant control accounts, costing profit & loss a/c, and General ledger adjustment account to record the above transactions for the quarter ended 30-6-2001.

(C.A. Inter, November 2001)

Solution:

GENERAL LEDGER ADJUSTMENT ACCOUNT

To sales	2,56,000	By Balance b/d	1,88,750
To balance c/d	1,80,150	By stores ledger control a/c	26,700
		By wages control a/c	77,500
		By overhead control a/c	95,200
		By costing profit & loss a/c	48,000
	4,36,150		4,36,150

STORES LEDGER CONTROL ACCOUNT

To Balance b/d	53,375	By WIP control a/c	40,000
To General ledger adjustment a/c	26,700	By Factory overhead control a/c	900
		By Balance c/d	39,175
	80,075		80,075

WIP CONTROL ACCOUNT

To Balance b/d	1,04,595	By Finished goods control a/c	2,02,900
To stores ledger control a/c	40,000	By Balance c/d	1,12,095
To wages control a/c	54,500		
To Factory overhead control	1,15,900		
	3,14,995		3,14,995

FINISHED GOODS CONTROL ACCOUNT

To Balance b/d	30,780	By cost sales a/c	2,04,800
To WIP control a/c	2,02,900	By balance c/d	28,880
	2,33,680		2,33,680

Note:

Gross profit is 25% on factory cost or 20% on sales

Hence cost of sales = Rs. 2,56,000 – 20% of 2,56,000 = Rs. 2,04,800

FACTORY OVERHEAD CONTROL ACCOUNT

To stores ledger control a/c	900	By costing P & L a/c	3,200
To wages control a/c	23,000	By WIP control a/c	1,15,900
To general ledger adjustment a/c	95,200		
	1,19,100		1,19,100

COST OF SALES ACCOUNT

To finished goods control a/c	2,04,800	By costing P & L a/c	2,04,800

SALES ACCOUNT

To costing P & L a/c	2,56,000	By General ledger adjustment a/c	2,56,000

CHAPTER 7

WAGES CONTROL ACCOUNT

To general ledger adjustment a/c	77,500	By factory overhead control a/c	23,000
		By WIP control a/c	54,500
	77,500		77,500

COSTING PROFIT AND LOSS ACCOUNT

To factory overhead control a/c	3,200	By sales a/c	2,56,000
To cost of sales a/c	2,04,800		
To General ledger adjustment a/c profit	48,000		
	2,56,000		2,56,000

TRIAL BALANCE AS ON 30-6-2001

	Dr.	Cr.
Stores ledger control a/c	39,175	–
WIP control a/c	1,12,095	
Finished goods control a/c	28,880	
General ledger adjustment a/c	–	1,80,150
	1,80,150	1,80,150

Questions

1. The following are the balances in the cost ledger of a manufacturing company on Ist January 1981:

	Dr.	Cr.
Stores ledger	4,500	
Work-in-progress ledger	10,200	
Finished goods ledger	6,800	
Financial ledger		21,500

Summary of transactions during the year 1981:

Materials purchased	20,000
Materials issued to jobs	18,000
Materials issued for repairs in a factory	1,000
Direct wages paid	15,000
Indirect wages paid	5,000
Factory expenses paid	6,000
Administration expenses paid	15,000
Selling expenses paid	7,000
Cost of finished goods produced	56,000
Cost of finished goods sold	70,000
Sales	1,00,000

Prepare control accounts and costing profit and loss account in the cost ledger assuming that the overheads recovered and incurred are the same and that administration overheads are charged to finished goods. *(Madurai University, M.Com., Nov. 1988)*

[**Ans.** Costing profit is Rs. 23,800]

2. Hewlett and Packard company maintains cost control accounts. The following balances have been extracted from the books:

	Dr.	*Cr.*
	Rs.	*Rs.*
Cost ledger control		60,000
Stores ledger control	20,000	
Work-in-progress ledger control	10,000	
Finished goods control	30,000	
	60,000	60,000

Following transactions are recorded for the period:

Materials supplied to stores	1,00,000
Materials supplied directly to production	30,000
Materials returned from stores to supplier	50,000
Materials issued to production	70,000
Indirect materials	10,000
Wages and salaries (Direct wages Rs. 80,000, factory wages Rs. 10,000, administrative salary Rs. 40,000, marketing salary Rs. 20,000)	1,50,000
Production overhead incurred	60,000
Factory overhead recovered at 90% of direct wages	–
Administrative expenses	40,000
Office overhead absorbed	85,000
Marketing expenses	28,000
Marketing expenses recovered at Re. 1 per unit sold	–
Cost of production	25,000
Cost of goods sold	3,20,000
Sales at Rs. 10 per unit	4,00,000

Prepare necessary ledger accounts in cost ledger and extract trial balance as on the last date of period. *(Bangalore University, M. Com., April 1999)*

[**Ans.** Balances: Cost ledger control Rs. 92,000, Stores ledger control Rs. 35,000, WIP ledger control Rs. 12,000, Finished goods control Rs. 45,000, Net profit Rs. 29,000]

❑❑❑

CHAPTER 8

RECONCILIATION OF COST AND FINANCIAL ACCOUNTS

INTRODUCTION

While discussing the relationship between cost and financial accounts under chapter 1, it was stated that cost accounts resemble to financial accounts in one important respect, viz., that the basis for recording transactions is common for both the accounting methods. As for example, invoices constitutes the basis for recording transactions relating to materials, payroll for recording wages and salaries and vouchers for recording indirect expenses such as rent, taxes, insurance, etc. From this point of view, the result shown by cost accounting must be the same as shown by financial accounts. But owing to the subjective recording of transactions (*i.e.,* according to the nature of expense) in financial accounting and objective recording (*i.e.,* according to purpose for which costs are incurred) in cost accounting often gives rise to a difference in profit as revealed by these two sets of books. Hence, a need arises to reconcile the profit as shown by these two accounting systems.

NEED FOR AND OBJECTIVES OF RECONCILIATION

The profit reconciliation is necessary to fulfil the following objectives:

(*a*) **To ensure reliability of cost accounts:** The correct ascertainment of cost of production in cost accounts depends upon proper analysis, allocation, apportionment and recording of costs. If these steps are incorrect, naturally the cost of production will be incorrect. Reconciliation with financial accounts help in checking the reliability of cost accounts.

(*b*) **To check the arithmetical accuracy of cost accounts:** Sometimes some transactions in cost accounts are ignored while recording. At other times some items would have been wrongly recorded. This again is responsible for inaccurate results. So to check the arithmetical accuracy of cost accounts, a reconciliation is necessary.

However the need for reconciliation does not arise when cost and financial accounts are integrated.

REASONS FOR DIFFERENCES IN PROFIT

The various reasons or causes for the differences in the profit as shown by cost and financial accounts may be outlined as under:

I. Items Included in the Financial Accounts but not in Cost Accounts

(A) Purely Financial Charges

(*a*) Expenses incurred on share transfer,
(*b*) Interest paid on bank overdraft,
(*c*) Damages payable at law,
(*d*) Penalties payable at law,
(*e*) Losses on scrapping machinery,
(*f*) Losses arising out of natural calamities such as fire, flood, etc.,
(*g*) Donations and charities,
(*h*) Loss on sale of investments.

(B) Purely Financial Incomes

(*a*) Interest on bank deposit received,
(*b*) Dividends on investments received,
(*c*) Rent receivable,
(*d*) Profit made on sale of assets,
(*e*) Amount received on sale of old magazines and newspapers,
(*f*) Transfer fee received.

(C) Appropriation of Profits

(*a*) Transfer to reserve fund, general reserves, etc.,
(*b*) Transfer to income tax,
(*c*) Dividend paid,
(*d*) Creation of depreciation fund, sinking fund, dividend equalisation fund, workmen's compensation fund, investment fluctuation fund, reserve for bad and doubtful debts,
(*e*) Goodwill, preliminary expenses, underwriting commission, etc. written off.

II. Items Included in Cost Accounts Only

The items of transactions recorded only in cost accounts are in the form of notional costs, such as notional rent, notional interest and notional salary.

III. Under or Over-absorption of Overheads

In financial accounts manufacturing and office expenses are written off as and when they are incurred. However, in cost accounts such expenses are not included until production is completed. Till then, such costs are accumulated in suspense account in the cost ledger. This difference in the treatment gives rise to difference in the amount of profit shown by two sets of books.

One important point that is to be remembered in this connection is, overheads are charged in the cost accounts on the basis of estimates. Manufacturing overheads are recovered as a percentage of wages, office and administration overheads on cost of production and selling and distribution overheads on the basis of sales. The difference between actual and estimated overheads (either over or under absorption) is written off either to overhead adjustment account or to costing profit and loss account. The effect of such transfer is the actual amount shown in the financial accounts will now agree with the amount shown in cost accounts. This means, when reconciling cost and financial accounts no further adjustment is necessary in respect of such overheads.

However, it must also be emphasised that when under or over absorbed overheads are not written off to costing profit and loss, but has been carried forward to next year, the amount of overhead (under or over absorbed) must be considered for reconciliation purpose. This method of treating over or under absorption is not very popular. But when decided by cost accountant, the adjustment of overhead is to be duly considered while preparing reconciliation statement.

IV. Adoption of Different Bases for Stock Valuation

In cost accounts closing stock for balance sheet purpose is valued at cost, while in financial accounts the basis is cost or market price whichever is less. This difference in the basis of valuation of closing stock is responsible for the disagreement of profit in these sets of books.

V. Adoption of Different Bases for Depreciation

In cost accounts the assets may be depreciated on the straightline method, whereas in financial accounts a different method of depreciation such as reducing balance method or sinking policy method or a different method is followed. The difference in the method of depreciation followed in these systems of accounts results in a difference of profit.

VI. Adoption of Different Bases for Valuing Work-in-progress

Work-in-progress is valued either at the stage of prime cost, works cost or cost of production. In cost accounts, the basis followed may be quite different than that followed in financial accounts. This difference in the method of valuing work-in-progress gives rise to preparation of reconciliation statement.

VII. Expenses of Abnormal Nature are Excluded in Cost Accounts

VIII. Recording of Costs by Approximation in Cost Accounts in Some Cases

PROCEDURE OF RECONCILIATION

The following steps are involved in reconciling the profits as revealed by cost and financial accounts:

1. Find out the reasons responsible for the disagreement in the profits. The various reasons are already discussed in the previous sub-topic. However notional charges are to be ignored in reconciliation, as they do not affect costing profit and loss account. If a notional cost is included in cost accounts the entry made is

   Production overhead control a/c Dr.

   To costing P & L a/c

2. Adjust the stock or raw materials, work-in-progress and closing stock depending upon their impact on profit. The impact of stock values are shown below:

   (*a*) More opening stock — Profit is less
   (*b*) Less opening stock — Profit is more
   (*c*) More closing stock — Profit is more
   (*d*) Less closing stock — Profit is less

3. If profit figure is not given prepare a cost sheet to know profit as per cost accounts and a profit and loss account to know the profit as per financial accounts.

The following proforma can be followed to make necessary adjustments while reconciling the profit :

Start with profit as per cost accounts xx

Add:	(1) Items of incomes included in financial accounts but not in cost accounts	xx	
	(2) Items of expenditure included in cost accounts but not in financial accounts	xx	
	(3) Excess income shown in financial accounts over corresponding incomes in cost accounts	xx	
	(4) Excess expenditure shown in cost accounts over corresponding items in financial accounts	xx	
	(5) Over-absorption of overheads in cost accounts	xx	
	(6) Over-valued closing stock in financial accounts	xx	
	(7) Over-valued opening stock in cost accounts	xx	xx
Less:	(1) Items of income included in cost accounts but not in financial accounts	xx	xx
	(2) Items of expenditure included in financial accounts but not in cost accounts	xx	
	(3) Excess income shown in cost accounts over corresponding items in financial accounts	xx	
	(4) Excess expenditure shown in financial accounts over corresponding entries in cost accounts	xx	
	(5) Under-absorption of overheads in cost accounts	xx	
	(6) Under-valuation of closing stock in financial accounts	xx	
	(7) Under-valuation of opening stock in cost accounts	xx	xx
	Profit as per financial accounts		xx

ADVANTAGES OF RECONCILIATION

1. The arithmetical accuracy can be checked in both the sets of books.
2. It enables to set right under or over-absorption of overheads in cost accounts by making use of supplementary rates.
3. Reconciliation facilitates location of areas of inefficiencies. For example, if predetermined overheads are fixed fairly and actual expenditure incurred are more it is a case of inefficiency requiring remedial measure.
4. It helps in detecting frauds as for example, any wrong entry of stock of material in stores ledger owing to theft can be brought to light by comparing with the stock of financial accounts.
5. Separate maintenance of cost accounts has the advantage of exemption from statutory audit as the purpose of cost accounts is to ascertain and control cost rather than ascertainment of profit.

PRESENTATION OF RECONCILIATION STATEMENT

The reconciliation of profit can be presented under two methods, viz., (1) Adjustment method and (2) Memorandum reconciliation account method. Under the first method, the various adjustments are made in a statement form. Under the second method, the adjustments are made in a memorandum reconciliation account. Whichever method is followed is immaterial, because reconciliation is only a memorandum account. For balance sheet purpose, the profit shown in financial accounts holds good. Reconciliation facilitates only internal control by highlighting various causes responsible for increase or decrease in profit. Memorandum reconciliation is only an account to show adjustment of transactions and hence does not form part of double entry books.

A proforma memorandum reconciliation account is shown below:

In the cost ledger

Memorandum reconciliation account

To expenses recorded in financial accounts:	By profit as per cost accounts:
Discount	By Income from financial accounts:
Fines & penalties	Rent
Bank interest on O.D.	Interest
Underwriter's commission	Dividends
Donation	Profit on sale of assets
Goodwill written off	
To under absorption of overheads	By over-absorption of overheads in cost a/cs
To closing stock under-valued in financial accounts	By closing stock overvalued in financial accounts
To opening stock under-valued in cost accounts	By opening stock over-valued in cost accounts
To profit as per cost a/c	

Problem 1. In a factory, works overhead is absorbed at 60% of labour and office expenses at 20% of the works cost. The total expenditure is as follows:

	Rs.
Materials	2,00,000
Labour	1,50,000
Factory expenses	98,000
Office expenses	85,000
Total cost	5,33,000

10% of the total cost is seen as closing stock and sales total up to Rs. 5,10,000. Prepare:

(*i*) Profit & loss account as per financial accounts.

(*ii*) Cost sheet

(*iii*) Reconciliation *(Sri Sathya Sai University, B.Com. (Hons.), October 2008)*

Solution:

P & L ACCOUNT AS PER FINANCIAL ACCOUNTS

To materials	2,00,000	By sales	5,10,000
To labour	1,50,000	By closing stock	53,300
To factory expenses	98,000	(10% of total cost)	
To office expenses	85,000		
To profit c/d	30,300		
	5,63,300		5,63,300

COST SHEET

Materials		2,00,000
Labour		1,50,000
	Prime cost	3,50,000
Factory expenses – 60% of wages		90,000
	Factory cost	4,40,000
Office overhead – 20% of factory cost		88,000
	Office cost	5,28,000
Less: closing stock $\frac{10}{100} \times 5,28,000$		52,800
	Cost of goods sold	4,75,200
	Profit	34,800
	Sales	5,10,000

RECONCILIATION STATEMENT

Profit as per cost accounts		34,800
Add: Over absorption of office Overhead (88,000 – 85,000)	= 3,000	
Add: Under charge of closing Stock in cost accounts (53,300 – 52,800)	= 500	3,500
		38,300
Less: Under absorption of factory Overhead (98,000 – 90,000)		8,000
	Profit as per financial a/c	30,300

Problem 2. The following is a summary of the trading and profit and loss account of Alpha manufacturing company limited for the year ended 31-12-2006.

To materials	27,40,000	By sales	60,00,000
To wages	15,10,000	(1,20,000 units)	
To factory expenses	8,30,000	By finished stock	1,60,000
To selling expenses	4,50,000	(4,000 units)	
To administration expenses	3,81,400	By work-in-progress:	
To preliminary expenses written off	40,000	By materials	64,000
To goodwill written off	20,000	By wages	36,000
To net profit	3,26,600	By factory expenses	20,000
		By dividend	18,000
	62,98,000		62,98,000

The company manufactures a standard unit. The following details pertain to the cost accounts:

(*a*) Factory expenses have been charged to production at 20% on prime cost

(*b*) Administration expenses have been charged at Rs. 3 per unit produced

(*c*) Selling expenses have been charged at Rs. 4 per unit sold.

Prepare a cost sheet and reconcile the profit disclosed as per the cost sheet with that shown in the financial accounts. *(Sri Sathya Sai University, B.Com. (Hons.), October 2007)*

CHAPTER 8

Solution:

COST SHEET

			Total
Materials		27,40,000	
Less:	Materials in closing WIP	64,000	26,76,000
Labour cost		15,00,000	
Less:	Wages in closing WIP	36,000	14,74,000
		Prime cost	41,50,000
Factory expenses $\frac{20}{100} \times 41,50,000$			8,30,000
		Factory cost	49,80,000
Administration overhead: 1,24,000 × 3			3,72,000
		Office cost	53,52,000
Less:	Closing stock $\frac{53,52,000 \times 4,000}{1,24,000}$		1,72,645
		Cost of goods sold	51,79,355
Selling overhead 1,20,000 × Rs. 4			4,80,000
		Cost of sales	56,59,355
		Profit	3,40,645
		Sales	60,00,000

RECONCILIATION STATEMENT

Profit as per cost accounts			3,40,645
Add:	Over-absorption of factory overheads in cost accounts 8,30,000 – 8,10,000	= 20,000	
Add:	Over-absorption of selling overhead in cost accounts (4,80,000 – 4,50,000)	= 30,000	
Add:	Income credited in financial accounting but not in cost accounting	= 18,000	68,000
			4,08,645
Less:	Under absorption of administrative overhead in cost a/c (3,81,400 – 3,72,000)	= 9,400	
Less:	Over valuation of closing stock in cost a/c (1,72,645 – 1,60,000)	= 12,645	

Less:	Expenses included in FA but not in CA:		
	Preliminary expenses	= 40,000	
	Goodwill written off	= 20,000	82,045
	Profit as per financial accounts		3,26,600

Note:

Factory expenses recorded in Financial Account:

Factory expenses debited to P & L a/c		8,30,000
Less:	Factory expenses included in closing WIP	20,000
	Net factory expenses debited to P & L a/c	8,10,000

Problem 3. From the following figures prepare a reconciliation statement:

	Rs.
Net profit as per financial accounts	1,28,755
Net profit as per cost records	1,72,400
Works overhead under recovered in cost records	3,120
Administrative overhead recovered in excess	1,700
Depreciation charged in financial records	11,200
Depreciation recovered in cost records	12,500
Interest received but not included in cost records	8,000
Obsolescence loss charged in financial records	5,700
Income tax provided in financial books	40,300
Bank interest credited in financial books	750
Stores adjustment (credited in financial books)	475
Depreciation of stock charged in financial books	6,750

(Sri Sathya Sai University, B.Com. (Hons.), February 2006)

Solution: RECONCILIATION STATEMENT

Net profit as per costing records		1,72,400
Add: Excess depreciation charged (12,500 – 11,200)	= 1,300	
Add: Over absorption of administrative overhead	= 1,700	
Add: Interest received but not included	= 8,000	
Add: Bank interest not included	= 750	
Add: Stores adjustment	= 475	12,225
		1,84,625
Less: Under absorption of works overhead	= 3,120	
Less: Loss on absolescence charged	= 5,700	
Less: Income tax not provided for	= 40,300	
Less: Depreciation of stock not charged	= 6,750	55,870
Profit as per financial a/c		1,28,755

CHAPTER 8

Problem 4. The profit disclosed by a company's cost accounts for the year was Rs. 30,114, while the net profit shown by the financial accounts amounted to Rs. 19,760. On reconciling the figures the following differences were brought to light:

(*i*) Overheads in the cost accounts were estimated at Rs. 7,500. The charge for the year shown in the financial accounts was Rs. 6,932

(*ii*) Director's fees not charged in the cost accounts amounted to Rs. 750

(*iii*) The company has allocated Rs. 600 to general provision for bad debts

(*iv*) Depreciation of Rs. 600 omitted from cost accounts

(*v*) Transfer fees received amounted to Rs. 28 omitted in cost accounts

(*vi*) The amount set inside for income tax Rs. 9,000

Prepare a reconciliation statement. *(Sri Sathya Sai University, B.Com. (Hons.), October 2006)*

Solution: RECONCILIATION STATEMENT

Profit as per cost accounts			30,114
Add: Over-absorption of overheads in cost accounts (7,500 – 6,932)	=	568	
Transfer fee not recorded in cost accounts	=	28	596
			30,710
Less: Directors fees charged in financial accounts	=	750	
Provision for bad debts made in financial accounts	=	600	
Depreciation omitted from cost accounts	=	600	
Income tax not charged in cost accounts	=	9,000	10,950
Profit as per financial accounts			19,760

Problem 5. The following particulars are extracted from financial and cost records:

	Cost records (Rs.)	Financial records (Rs.)
Material cost	5,70,000	5,20,000
Direct wages	2,40,000	2,55,000
Factory overheads	1,62,000	3,60,000
Administrative overheads	5,40,000	4,00,000
Sales overheads	8,00,000	9,60,000
Preliminary expenses written off	–	30,000
Sales	20,80,000	20,80,000
Rent received		1,20,000
Profit	80,000	–
Loss	–	13,000

Prepare a reconciliation statement *(Sri Sathya Sai University, B.Com. (Hons.), October 2003)*

Solution: RECONCILIATION STATEMENT

Net profit as per cost books		80,000
Add: Materials cost over absorbed in cost books	50,000	
Administrative overhead over absorbed in cost books	1,40,000	
Rent received recorded in FA but not in CA	1,20,000	3,10,000
		3,90,000

Less: Wages under recovered in cost books	15,000	
Less: Factory overhead underabsorbed in cost books	1,98,000	
Less: Selling overhead underabsorbed in cost books	1,60,000	
Less: Preliminary expenses written off in financial books	30,000	
		4,03,000
Loss as per financial books		13,000

Problem 6. The net profit shown by financial accounts of a company amounted to Rs. 18,550, while profit disclosed by cost books for the same period were Rs. 28,660. After completing the books the following points were noted:

(*a*) Director's fees not charged in cost accounts	650
(*b*) Provision for bad debts	570
(*c*) Bank interest (Cr)	30
(*d*) Income tax	8,300
(*e*) Overheads overcharged in cost accounts	180
(*f*) Depreciation charged on machine costing Rs. 16,000 @ 5% only in financial accounts	

Prepare a statement reconciling the profit in both the books.

(Sri Sathya Sai University, B.Com. (Hons.), February 2003)

Solution:

RECONCILIATION STATEMENT

Profit as per cost accounts		28,660
Add: Bank interest credited in financial accounts but not in cost accounts	30	
Add: Over absorption of overhead in cost accounts	180	210
		28,870
Less: Not charged in cost accounts:		
(*a*) Director's fees	650	
(*b*) Provision for bad debts	570	
(c) Income tax	8,300	
(*d*) Depreciation 5% on 16,000	800	10,320
Profit as per financial accounts		18,550

Problem 7. In a factory, works overheads are absorbed at 60% of labour and office expenses at 20% of works cost. The total expenditure is as follows:

	Rs.
Materials	2,00,552
Wages	1,49,448
Factory expenses	98,000
Office expenses	85,000

10% of total cost incurred is the closing stock and the sales total up to Rs. 5,10,000. Prepare:

(*a*) Profit & loss a/c as per financial accounts

(*b*) Cost sheet

(c) Reconciliation statement

(Sri Sathya Sai University, B.Com. (Hons.), October 2002)

CHAPTER 8

Solution:

PROFIT AND LOSS ACCOUNT

To materials	2,00,552	By sales	5,10,000
To wages	1,49,448	By closing stock	53,300
To factory expenses	98,000	(10% of 5,33,000)	
To office expenses	85,000		
Total costs	5,33,000		
To profit	30,300		
	5,63,300		5,63,300

COST SHEET

Materials		2,00,552
Wages		1,49,448
	Prime cost	3,50,000
Works overhead 60% of wages		89,669
	Works cost	4,39,669
Office overhead 20% of works cost		87,934
	Office cost	5,27,603
Less: Closing stock $\frac{10}{100} \times 5{,}27{,}603$		52,760
	Cost of goods sold	4,74,843
	Profit	35,157
	Sales	5,10,000

RECONCILIATION STATEMENT

Profit as per cost accounts		35,157
Add: Closing stock of finished goods undervalued in cost accounts (53,300 – 52,760)	= 540	
Add: Over absorption of office overheads in cost accounts (87,934 – 85,000)	= 2,934	3,474
		38,631
Less: Under absorption of works overhead in cost accounts (98,000 – 89,669)		8,331
Profit as per financial accounts		30,300

Problem 8. The cost accounts of a company show a net profit of Rs. 3,44,800 for the year ending 31st December 2002. Prepare a reconciliation statement showing the net profit as per financial records.

	Rs.
Under recovery of factory overheads in cost accounts	6,240
Over-recovery of office overheads in cost accounts	3,400
Depreciation in financial accounts	22,400
Depreciation in cost accounts	25,000
Interest on investment not included in cost accounts	16,000

Loss on absolescence charged in financial accounts	11,400
Income tax debited in financial accounts	80,600
Bank interest credited to financial accounts	2,450
Loss in stock not charged in cost accounts	13,500

(University of Madras, B.Com., April 2004)

Solution:

RECONCILIATION STATEMENT

Net profit as per cost accounts		3,44,800
Add: Over absorption of expenses in cost accounts		
(*a*) Office overhead over recovered in cost books	3,400	
(*b*) Depreciation shown in excess in cost books	2,600	
(25,000 – 22,400)	6,000	
Add: Income recorded only in financial books:		
(*a*) Interest on investment	16,000	
(*b*) Bank interest credited in financial accounts	2,450	
		24,450
		3,69,250
Less: Under absorption of factory expenses in cost accounts	6,240	
Less: Loss on absolescence charged in financial accounts	11,400	
Less: Income tax debited in financial accounts	80,600	
Less: Loss in stock not charged in cost accounts	13,500	1,11,740
Profit as per financial accounts		2,57,510

Problem 9. From the following, prepare reconciliation statement of M/s XYZ and company as on 30-6-2004

(1) Net profit as per financial accounts Rs. 40,340

(2) Income tax provision made Rs. 30,000

(3) Materials purchases of Rs. 5,000 units were recorded in cost at standard cost Rs. 24 per unit whereas in financial accounts it was recorded at actual cost Rs. 22 per unit.

(4) Old bad debts received Rs. 20,500

(5) Loss on sale of furniture was Rs. 4,120 *(University of Bombay, B.Com., October 2006)*

Solution:

RECONCILIATION STATEMENT

Profit as per financial accounts		40,340
Add: Income tax provision not recorded in cost books	30,000	
Add: Loss on sale of furniture not included in cost accounts	4,120	
		34,120
		74,460
Less: Old bad debt recovered recorded in financial books only	20,500	
Less: Material purchased overcharged in cost accounts	10,000	
		30,500
Profit as per cost accounts		43,960

CHAPTER 8

Problem 10. From the following particulars, prepare reconciliation statement and ascertain costing profit or loss. Net profit as per financial P & L a/c is Rs. 50,000 opening stock was over valued by Rs. 2,000 in cost accounts as compared to financial accounts.

Administrative overheads charged in financial book is Rs. 20,000 but recovered in cost Rs. 40,000

Income tax provision Rs. 1,200

Notional salary of proprietor in cost Rs. 20,000

Interest received Rs. 12,000

Closing stock as per financial books Rs. 16,200 whereas in cost books it was Rs. 19,000

(University of Bombay, B.Com., April 2005)

Solution:

RECONCILIATION STATEMENT

Profit as per P & L a/c		50,000
Add: Income tax provision provided only in financial accounts	1,200	
Add: Difference in closing stock	2,800	4,000
		54,000
Less: Opening stock overvalued in cost accounts	2,000	
Less: Administration expenses over absorbed in cost accounts	20,000	
Less: Notional salary only in CA	20,000	
Less: Inerest received only in FA	12,000	54,000
Profit as per cost accounts		Nil

Problem 11. The following figures have been extracted from the financial accounts of Bawa manufacturing company for the first year of its operations:

	Rs.
Direct material consumption	50,00,000
Direct wages	30,00,000
Factory overheads	16,00,000
Administration overheads	7,00,000
Selling and distribution overheads	9,60,000
Provision for bad debts	80,000
Preliminary expenses written off	40,000
Dividend received	1,00,000
Interest received on deposits	20,000
Sales (1,20,000 units)	1,20,00,000
Closing stock:	
Finished goods (4,000 units)	3,20,000
Work-in-progress	2,40,000

The cost accounts for the same period reveal that the direct material consumption was Rs. 56,00,000. Factory overheads are recovered at 20% on prime cost. Administrative overheads are recovered at Rs. 6 per unit of production. Selling and distribution overheads are recovered at Rs. 8 per unit sold. Prepare profit and loss account as per financial records and cost sheet as per cost records. Reconcile the profit as per the two records. The cost accounts value closing stock of finished goods at cost of production. *(University of Bombay, B.Com., October 2004)*

Solution:

COST SHEET

Direct materials		56,00,000
Direct wages		30,00,000
Prime cost		86,00,000
Works overheads @ 20% of Prime cost	17,20,000	
Less: Closing stock of work-in-progress	2,40,000	14,80,000
Works cost		1,00,80,000
Office & Administrative overhead @ Rs. 6 each (1,24,000 × 6)		7,44,000
Office cost		1,08,24,000
Less: Closing stock of finished goods (4,000 units)		3,49,161
Cost of goods sold		1,04,74,839
Selling & distribution overhead @ Rs. 8 each (1,20,000 × Rs. 8)		9,60,000
Cost of sales		1,14,34,839
Profit		5,65,161
Sales		1,20,00,000

***Note:* Valuation of closing stock of finished goods**

$$\frac{1,08,24,000}{1,24,000} \times 4,000 = 3,49,161$$

PROFIT AND LOSS ACCOUNT

To materials	50,00,000	By sales	1,20,00,000
To wages	30,00,000	By closing stock:	
To factory expenses	16,00,000	– finished goods	3,20,000
To administrative expenses	7,00,000	– work-in-progress	2,40,000
To selling expenses	9,60,000	By interest on deposits	20,000
To bad debts	80,000		
To preliminary expenses written off	40,000	By dividends	1,00,000
To net profit	13,00,000		
	1,26,80,000		1,26,80,000

RECONCILIATION STATEMENT

Profit as per cost accounts		5,65,161
Add: Income credited in FA only:		
– Interest received	20,000	
– Dividends	1,00,000	
Add: Cost over recovered in cost accounts:		
– Materials	6,00,000	
– Factory overheads	1,20,000	
– Administrative overheads	44,000	8,64,000

Less: Bad debts recorded only in financial accounts	80,000	
Less: Preliminary expenses w/off recorded in financial accounts	40,000	
Less: Overvaluation of closing stock in cost accounts	29,161	1,49,161
Profit as per financial accounts		13,00,000

Problem 12. From the following details of KT and Company compute profit as per profit and loss account as well as as per cost sheet and reconcile profit between cost sheet and profit and loss account showing clearly the reasons for the variation of the two profit figures.

	Rs.
Sales	20,000
Purchase of material	3,000
Closing stock of materials	500
Direct wages	1,000
Indirect wages	500
Indirect factory expenses	2,000
Bad debts	100
Interest on overdraft	100
Profit on sale of assets	50
Selling expenses	2,000
Distribution expenses	1,000

In cost sheet manufacturing overheads are recovered at 300% of direct wages. Selling overheads at Rs. 1,500 and distribution overheads at Rs. 700. *(University of Bombay, B.Com., October 2003)*

Solution:

COST SHEET

Purchase of raw materials	3,000	
Less: Closing stock	500	2,500
Direct wages		1,000
Prime cost		3,500
Works overhead – 300 % of direct wages		3,000
Works cost		6,500
Administration overhead		
Office cost		6,500
Selling & distribution overheads (1500 + 700)		2,200
Cost of sales		8,700
Profit		11,300
Sales		20,000

PROFIT AND LOSS ACCOUNT

To materials	3,000	By sales	20,000
To wages	1,000	By profit on sale of asset	1,000
To indirect wages	500	By closing stock of raw materials	500
To factory expenses	2,000		
To selling expenses	2,000		
To distribution expenses	1,000		
To interest	50		
To bad debts	100		
To profit	11,850		
	21,500		21,500

RECONCILIATION STATEMENT

Profit as per cost accounts		11,300
Add: Income credited in FA		
– Profit on sale of assets	1,000	
Add: Over absorption of factory overhead in cost accounts	500	1,500
(3,000 – 500 – 2,000)		12,800
Less: Items debited in FA only		
– Interest	50	
– Bad debts	100	
Less: Under absorption of selling and distribution overhead in cost accounts (2,000 + 1,000 – 2,000)	800	950
Profit as per financial accounts		11,850

Problem 13. From the following, prepare a statement of reconciliation and find out profit or loss as per financial records

	Rs.
Net loss as per cost records	1,72,400
Works overhead under-recovered in costing	3,120
Administrative overheads over-recovered in costing	1,700
Depreciation in financial a/c	11,200
Depreciation in cost a/c	12,500
Interest received	8,750
Obsolescence loss in financial a/c	5,700
Provision for income tax	40,300
Opening stock:	
– Financial records	52,600
– Cost records	54,000
Closing stock:	
– Financial records	52,000
– Cost records	49,600
Interest charges in cost accounts only	6,000
Preliminary expenses written off	950

(University of Bombay, B.Com., October 2001)

CHAPTER 8

Solution:

RECONCILIATION STATEMENT

Loss as per cost records		1,72,400
Add: Obsolescence loss debited in financial accounts	5,700	
Add: Provision for income tax recorded in FA only	40,300	
Add: Preliminary expenses written off in FA only	950	
Add: Works overhead under absorbed in cost accounts	3,120	50,070
		2,20,470
Less: Interest received credited in FA only	8,750	
Less: Closing stock overhead in FA	2,400	
Less: Opening stock under valued in FA	1,400	
Less: Depreciation under charged in FA	1,300	
Less: Over absorption of administrative overhead in CA	1,700	
Less: Interest charges debited in CA only	6,000	21,550
Loss as per financial accounts		2,00,920

Problem 14. The following is the Trading and Profit and Loss account of a manufacturer for the year ended 31st December 2003

To opening stock:		By closing stock:	
To raw materials	59,000	raw materials	64,000
To purchase of raw materials	3,73,000	work-in-progress	25,600
To wages	5,96,000	cost of goods manufactured	13,19,900
To factory expenses	3,81,500		
	14,09,500		14,09,500
To cost of goods manufactured	13,19,900	By sales (15,200 units)	18,24,000
To administration expenses	2,45,000	By finished stock (2,800 units)	2,35,200
To selling & distribution overhead	3,28,000	By interest on investment	2,600
To preliminary expenses written-off	20,000	By Dividend received	11,000
By loss on sale of fixed assets	15,000		
By net profit	1,44,900		
	20,72,800		20,72,800

In cost accounts, factory overheads are allocated to production at 60% of direct labour cost administration overheads are applied at Rs. 12 per unit over the units produced and selling and distribution overheads are charged so as to work out 20% of selling price.

You are required to:

(*i*) Prepare a statement of cost and ascertain the profit as per cost accounts:

(*ii*) Prepare a statement reconciling the profit or loss as per the two accounts.

(University of Delhi, B.Com., 2004)

Solution:

COST SHEET

Opening stock of raw materials	59,000	
Add: Purchases	3,73,000	
	4,32,000	
Less: Closing stock	64,000	
Cost of materials consumed		3,68,000
Direct wages		5,96,000
Prime cost		9,64,000
Factory overheads		3,57,600
		13,21,600
Less: Closing work-in-progress		25,600
Factory cost		12,96,000
Administration overheads (15,200 + 2,800 = 18,000 units)		2,16,000
Office cost		15,12,000
Less: Closing stock of finished goods $\frac{15,12,000}{18,000} \times 2,800$ units		2,35,200
Cost of goods sold		12,76,800
Selling & distribution overheads		3,64,800
Cost of sales		16,41,600
Profit		1,82,400
Sales		18,24,000

RECONCILIATION STATEMENT

Profit as per costing books		1,82,400
Add: Interest on investment	2,600	
Add: Dividend received	11,000	
Add: Over absorption of S & D overhead (3,64,800 – 3,28,000)	36,800	50,400
		2,32,800
Less: Factory overhead underabsorbed in costing books (3,81,500 – 3,57,600)	23,900	
Less: Administration overheads underabsorbed in costing books (2,45,000 – 2,16,000)	29,000	
Less: Preliminary expenses w/off	20,000	
Less: Loss on sale of fixed assets	15,000	87,900
Profit as per financial accounts		1,44,900

Problem 15. The following information is made available to you from the financial books of SV Ltd. for the year ended December 31, 2002

To direct materials	3,00,000	By sales (2,00,000 units)	7,50,000
To direct wages	2,00,000		
To factory expenses	1,20,000		
To office expenses	40,000		
To selling & distribution expenses	80,000		
To net profit	10,000		
	7,50,000		7,50,000

Normal output of the factory is 2,50,000 units. Factory overhead are fixed upto Rs. 60,000 and office expenses are fixed for all practical purposes. Selling and distribution expenses are fixed to the extent of Rs. 50,000: The rest are variable.

Prepare a statement reconciling the profits as per cost and financial accounts assuming that indirect expenses are absorbed on the basis of normal production capacity in cost accounts.

(University of Delhi, B.Com. (Hons))

Solution: RECONCILIATION STATEMENT

Profit as per cost accounts		40,000
Less: Under absorption of overheads in cost accounts		
– Factory overheads	12,000	
– Office overheads	8,000	
– Selling and distribution overheads	10,000	30,000
Profit as per financial accounts		10,000

Problem 16. The profit and loss account of Oil India (Pvt.) Ltd. for the year ended 31st March 2001 is as follows

To materials	4,80,000	By sales	9,60,000
To wages	3,60,000	By work-in-progress:	
To direct expenses	2,40,000	By materials	30,000
To gross profit	1,20,000	By wages	18,000
		By direct expenses	12,000
		By closing stock	1,80,000
	12,00,000		12,00,000
To administrative expenses	60,000	By Gross profit	1,20,000
To net profit	66,000	By dividends received	6,000
	1,26,000		1,26,000

As per the cost records the direct expenses have been estimated at a cost of Rs. 30 per kg and administration expenses at Rs. 15 per kg. During the year production was 6,000 kg and sales were 4,800 kg.

Prepare a statement of costing profit and loss account and reconcile the costing profit with financial profit.

(University of Delhi, B.Com. (Hons.), 2001)

Solution:

COST SHEET

Materials (4,80,000 – 30,000)	4,50,000
Wages (3,60,000 – 18,000)	3,42,000
Direct expenses (6,000 × Rs. 30)	1,80,000
Prime cost	9,72,000

Administrative overhead (6,000 × Rs. 15)		90,000
	Office cost	10,62,000
Less: Closing stock $\frac{10,62,000 \times 1,200}{6,000}$		2,12,400
	Cost of goods sold	8,49,600
	Profit	1,10,400
	Sales	9,60,000

RECONCILIATION STATEMENT

Profit as per costing records			1,10,400
Add: Over absorption of administration overheads in costing records		30,000	
Add: Dividend received		6,000	36,000
			1,46,400
Less: Over-valuation of closing stock in costing records (2,12,400 – 1,80,000)	=	32,400	
Less: Over charged direct expenses in financial a/c (2,40,000 – 12,000 – 1,80,000)	=	48,000	80,400
Profit as per financial accounts			66,000

Problem 17. A manufacturing company disclosed a net loss of Rs. 3,47,000 as per there cost accounts for the year ended March 31, 2003. The financial accounts however disclosed a net loss of Rs. 5,10,000 for the same period. The following information was revealed as a result of scrutiny of the figures of both the sets of accounts.

		Rs.
(1)	Factory overheads under-absorbed	40,000
(2)	Administration overheads over-absorbed	60,000
(3)	Depreciation charged in financial accounts	3,25,000
(4)	Depreciation charged in cost accounts	2,75,000
(5)	Interest on investment not included in cost accounts	96,000
(6)	Income-tax provided	54,000
(7)	Interest on loan funds in financial accounts	2,45,000
(8)	Transfer fees (credit in financial books)	24,000
(9)	Stores adjustment (credit in financial books)	14,000
(10)	Dividend received	32,000

Prepare a memorandum reconciliation account. *(CA PE-II, Group II, May 2003)*

Solution:

MEMORANDUM RECONCILIATION ACCOUNT

To net loss as per costing books	3,47,000	By Administration overheads over recovered in cost a/cs	60,000
To factory overheads under absorbed in cost accounts	40,000	By interest on investment not included in cost a/cs	96,000
To depreciation under charged in cost accounts	50,000	By Transfer fees in financial books	24,000
To income tax not provided in cost accounts	54,000	By stores adjustment (credit in financial books)	14,000

CHAPTER 8

To interest on loan funds in financial accounts	2,45,000	By dividend received in financial books	32,000
		By net loss as per financial books	5,10,000
	7,36,000		7,36,000

Problem 18. The summarised P & L account of a company for the year ended 31-3-2002 are given below:

Materials	44,00,000	Sales (2,00,000 units)	10,00,0000
Wages	24,00,000	Finished goods	
Factory overheads	14,00,000	Stock (12,000 units)	5,00,000
Administrative overheads	5,20,000	Work in progress:	
Selling & distribution overheads	4,80,000	– Materials	1,20,000
Bad debts written off	40,000	– Labour	80,000
Preliminary expenses written off	60,000	– Factory overhead	40,000
Net profit	16,00,000	Agricultural income	32,000
		Miscellaneous receipts	1,28,000
	1,09,00,000		1,09,00,000

The following additional information is also furnished:

(*i*) In cost accounts, factory overheads have been absorbed at 22% of prime cost.

(*ii*) In cost accounts administration overheads have been absorbed at a flat rate of Rs. 3 per unit.

(*iii*) In cost accounts, selling and distribution overheads have been absorbed at Rs. 2.50 per unit.

(*iv*) Closing WIP valued by the cost department has been incorporated in financial accounts.

(*v*) Valuation of finished goods has been independently made by the financial accounts branch.

You are required to prepare the costing profit and loss account and reconcile the profit as per costing and profit and loss account with the profit as per financial account

(ICWA, Inter, June 2002)

Solution:

COST SHEET

Materials		44,00,000
Wages		24,00,000
	Prime cost	68,00,000
Factory overheads 22% of prime cost		14,96,000
		82,96,000
Less: Closing stock of WIP		2,40,000
	Works cost	80,56,000
Administration overhead 2,12,000 × Rs. 3		6,36,000
	Cost of production	86,92,000
Less: Closing stock of finished goods		4,92,000

$$\frac{86,92,000}{2,00,000 + 12,000} = \text{Rs. } 41$$

12,000 units × Rs. 41	Cost of goods sold	82,00,000
Selling & distribution overheads		
2,00,0000 units × 2.50		5,00,000
	Cost of sales	87,00,000
	Profit	13,00,000
	Sales	1,00,00,000

RECONCILIATION STATEMENT

Profit as per financial account		16,00,000
Less: Items not included in cost accounts:		
Agricultural income	32,000	
Miscellaneous receipts	1,28,000	1,60,000
		14,40,000
Less: Overvaluation of finished goods in financial accounts (5,00,000 – 4,92,000)	8,000	
Less: Over absorption of overheads in cost accounts		
Factory overheads	96,000	
Administration overheads	1,16,000	
Selling & distribution overheads	20,000	2,40,000
		12,00,000
Add: Expenses not included in cost accounts		
Bad debts written off	40,000	
Preliminary expenses written off	60,000	1,00,000
Profit as per cost accounts		13,00,000

Problem 19. From the following particulars prepare a statement of reconciliation

Opening stock of materials	1,44,000
Opening stock of finished goods	2,88,000
Purchase of materials	8,64,000
Closing stock of materials	2,16,000
Closing stock of finished goods	72,000
Wages	3,60,000
Actual works expenses	2,27,150
Actual office expenses	1,85,950
Profit as per costing books	3,56,544

Works expenses are recovered at 20% on prime cost and office expenses at 80% on works oncost in costing books. *(S.V. University, B.Com., April 1999)*

Solution:

COST SHEET

Opening stock of materials	1,44,000
Add: Purchases of materials	8,64,000
	10,08,000
Less: Closing stock of raw materials	2,16,000

CHAPTER 8

Cost of materials consumed		7,92,000
Direct wages		3,60,000
	Prime cost	11,52,000
Factory on cost $\frac{20}{100} \times 11{,}52{,}000$		2,30,400
	Factory cost	13,82,400
Office overheads $\frac{80}{100} \times 2{,}30{,}400$		1,84,320
	Office cost	15,66,720

RECONCILIATION STATEMENT

Profit as per cost books	3,56,544
Add: Over-absorption of works oncost (2,30,400 – 2,27,150)	3,250
	3,59,794
Less: Under-absorption of office oncost (1,85,950 – 1,84,320)	1,630
Profit as per financial books	3,58,164

Problem 20. From the following figures, prepare a reconciliation statement and find out profit as per financial accounts:

	Rs.
Net profit as per costing records	17,24,000
Works overhead under-recovered in cost-books	31,200
Administration overhead over recovered in excess	17,000
Depreciation charged in financial accounts	1,12,000
Depreciation recovered in cost books	1,25,000
Interest received but not included in costing	80,000
Obsolescence loss charged in financial records	57,000
Income tax provided in financial books	4,03,000
Stores adjustment (credited in financial books)	4,750
Bank interest credited in financial books	7,500
Depreciation of stock charged in financial books	67,500

(S.V. University, B.Com., October 1999)

Solution:

RECONCILIATION STATEMENT

Net profit as per costing records		17,24,000
Add: Administrative overhead over recovered	17,000	
Depreciation over charged in cost books	13,000	
Interest received but not recovered in cost books	80,000	
Bank interest credited only in financial books	7,500	
Stores adjustment (credited in financial books only)	4,750	1,22,250
		18,46,250
Less: Works overhead under recovered	31,200	
Obsolescence loss charged in financial records only	57,000	
Income-tax provided in financial books only	4,03,000	

Depreciation on stock charged only in financial books	67,500	
		5,58,700
Net profit as per financial records		12,87,550

Problem 21. From the following data prepare a reconciliation statement

Profits as per cost accounts	1,45,500
Works overhead under-recovered	9,500
Administrative overhead under-recovered	22,750
Selling overhead over-recovered	19,500
Over-valuation of opening stock in cost accounts	15,000
Over-valuation of closing stock in cost accounts	7,500
Interest earned during the year	3,750
Rent received during the year	27,000
Bad debts written off during the year	9,000
Preliminary expenses written off during the year	18,000

(ICWA, Inter, June 1998)

Solution:

RECONCILIATION STATEMENT

Profit as per cost books			1,45,500
Add: Over-recovery of selling overheads		19,500	
Over-valuation of opening stock		15,000	
Income excluded from cost accounts:			
Interest earned		3,750	
Rent received		27,000	65,250
			2,10,750
Less: Under-recovery of works overhead		9,500	
Under-recovery of administrative overhead		22,750	
Over-valuation of closing stock		7,500	
Expenses excluded from cost accounts:			
Bad debts	9,000		
Preliminary expenses	18,000	27,000	66,750
Profit as per financial accounts			1,44,000

Problem 22. From the following figures prepare a reconciliation statement:

	Rs.
Net loss as per costing records	1,72,400
Works overhead under recovered in costing	3,120
Administrative overhead recovered in excess	1,700
Depreciation charged in financial records	11,200
Depreciation recovered in costing	12,500
Interest received not included in costing	8,000
Obsolescence charged (loss) in financial records	5,700
Income-tax provided in financial records	40,300
Bank interest credited in financial books	750
Stores adjustment (credit) in financial books	475
Value of opening stock in:	
Cost accounts	52,600
Financial accounts	54,000

CHAPTER 8

Value of closing stock in:		
Cost accounts		52,000
Financial accounts		49,600
Interest charged in cost accounts but not in financial accounts		6,000
Preliminary expenses written off in financial accounts		800
Provision for dubtful debts in financial accounts		150

(C.S. Inter, December 1997)

Solution:

RECONCILIATION STATEMENT

Loss as per cost accounting		1,72,400
Add: Works overhead under recovered in costing	3,120	
Obsolescence loss charged in financial accounts	5,700	
Income-tax provided in financial accounts	40,300	
Opening stock (under-valued) in costing (54,000 – 52,600)	1,400	
Over valued closing stock in costing (52,000 – 49,600)	2,400	
Preliminary expenses written off in financial accounts	800	
Provision for doubtful debts in financial accounts	150	53,870
		2,26,270
Less: Administrative overhead recovered in excess	1,700	
Depreciation overcharged in cost accounts (12,500 – 11,200)	1,300	
Interest received not included in costing	8,000	
Bank interest credited in financial books	750	
Stores adjustment credited in financial books	475	
Interest charged in cost accounts but not in financial accounts	6,000	
		18,225
Loss as per financial account		2,08,045

Problem 23. Enthusiasts Ltd. commenced business on 1st April 1996, cost and financial records are maintained for the year ended 31st March 1997. From the following information prepare statements:

(1) Showing the results as per costing records
(2) Showing result as per financial records
(3) Reconciling these results

Particular	*As per Costing Records*	*As per Financial Records*
Materials consumed (20,000 kgs.)	Rs. 28.50 per kg.	Rs. 26 per kg.
Direct wages (3,000 man-days)	Rs. 80 per man-day	Rs. 85 per man-day
Factory overhead	20% of prime cost	Rs. 3,60,000
Administrative overhead	Rs. 30 per kg. of output produced	Rs. 4,00,000
Sales overhead	Rs. 50 per kg. of output sold	Rs. 9,60,000
Stock (of output produced) as on 31-3-97, 2000 kgs.	At cost of production	Rs. 1,50,000
Work-in-progress as on 31-3-97	Rs. 1,62,000	Rs. 1,62,000
Sales (16,000 kgs.)	Rs. 130 per kg.	Rs. 129.50 per kg.

Rent income	–	Rs. 1,20,000
Preliminary expenses written off	–	Rs. 30,000

(University of Bombay, B.Com., October 1997)

Solution:

COST SHEET FOR THE YEAR ENDED 1997

		Total Cost	*Per kg. (18,000 kgs.)*
Materials consumed		5,70,000	31.67
Direct wages		2,40,000	13.33
	Prime cost	8,10,000	45.00
Factory overheads		1,62,000	9.00
		9,72,000	54.00
Less: Closing stock of work-in-progress		1,62,000	9.00
	Factory cost	8,10,000	45.00
Administration overheads		5,40,000	30.00
		13,50,000	75.00
Less: Closing stock of finished goods		1,50,000	–
Cost of goods sold		12,00,000	75.00
Selling overheads		8,00,000	50.00
	Total cost	20,00,000	125.00
	Profit	80,000	5.00
	Sales (16,000 × 130)	20,80,000	130.00

Working Note:

Quantity produced = Sales + closing stock – op stock
= 16,000 + 2,000 – Nil
= 18,000 kgs.

PROFIT AND LOSS ACCOUNT

To materials consumed	5,20,000	By sales	20,72,000
To direct wages	2,55,000	By rent	1,20,000
To factory overheads	3,60,000	By closing stock:	
To administration overheads	4,00,000	Finished goods	1,50,000
To sales overheads	9,60,000	WIP	1,62,000
To preliminary expenses written off	30,000	By net loss	21,000
	25,25,000		25,25,000

RECONCILIATION STATEMENT

Net profit as per cost books		80,000
Add: Materials consumed over charged in cost books	50,000	
Administrative overheads over absorbed in cost books	1,40,000	
Rent recorded in finance book only	1,20,000	3,10,000
		3,90,000

CHAPTER 8

Less: Wages under recovered in cost books	15,000	
Factory overhead under recovered in cost books	1,98,000	
Selling overhead under recovered in cost books	1,60,000	
Preliminary expenses written off in finance book only	30,000	
Sales per unit differ @ Rs. 0.5 in cost and finance books	8,000	
	(–)	(4,11,000)
Net lost as per financial books	(–)	21,000

Problem 24. A company's profit as per the costing system was Rs. 46,126 whereas the audited financial accounts showed a profit of Rs. 33,248. From the following additional information you are required to prepare a reconciliation statement, showing clearly the reasons for the difference between the two figures:

PROFIT AND LOSS ACCOUNT

(for the year ended 31-12-1993)

Opening stock	4,94,358		Sales	6,93,000
Purchases	1,64,308			
	6,58,666			
Closing stock	1,50,242			
		5,08,424		
Direct wages		46,266		
Factory overheads		41,652		
Gross profit c/d		96,658		
		6,93,000		6,93,000
Administrative expenses		19,690	Gross profit	96,658
Selling expenses		44,352	Sundry income	632
Net profit		33,248		
		97,290		97,290

The cost records show:

(1) Closing stock balance of Rs. 1,56,394

(2) Direct wages absorbed during the year Rs. 49,734

(3) Factory overhead absorbed Rs. 39,428

(4) Administration expenses charged @ 3% on sales

(5) Selling expenses charged @ 5% of value of sales

(6) No mention of sundry income. *(C.S. Inter, June 1995)*

Solution:

RECONCILIATION STATEMENT

Profit as per cost accounts		46,126
Add: Sundry income shown in financial accounts	632	
Over recovery of direct wages	3,468	
Over absorption of administration		

expenses in cost accounts	1,100	
		5,200
		51,326
Less: Over-valuation of closing stock in cost accounts	6,152	
Under-absorption of factory expenses in cost accounts	2,224	
Under absorption of selling expenses in cost accounts	9,702	
		18,078
Profit as per financial accounts		33,248

Problem 25. From the following information prepare:

(a) P & L account

(b) Cost sheet

(c) Reconciliation statement

	Rs.	Units
Sales	2,50,000	20,000
Materials	1,00,000	
Wages	50,000	
Factory overheads	45,000	
Office overheads	26,000	
Selling and distribution overheads	18,000	
Closing stock of finished goods	15,000	1,230
Work-in-progress:		
Materials	3,000	
Wages	2,000	
Factory overheads	2,000	
Good will written off	20,000	
Interest on capital	2,000	

In costing books, factory overheads is charged at 100% on wages, administration overheads at 10% of factory cost and selling distribution overheads at the rate of Re. 1 per unit sold.

(University of Delhi, B.Com., (Pass), April 1996)

Solution:

PROFIT AND LOSS ACCOUNT

To Materials	1,00,000	By sales	2,50,000
To wages	50,000	By closing stock	15,000
To Factory overheads	45,000	By work-in-progress	7,000
To office overheads	26,000	(3,000 + 2,000 + 2,000)	
To selling & distribution overhead	18,000		
To Goodwill	20,000		
To interest	2,000		
To profit	11,000		
	2,72,000		2,72,000

CHAPTER 8

COST SHEET

Materials		1,00,000
Wages		50,000
	Prime cost	1,50,000
Factory overhead		50,000
		2,00,000
Less: Work-in-progress (closing) (3,000 + 2,000 + 2,000)		7,000
	Factory cost	1,93,000
Administration overhead		19,300
	Cost of production	2,12,300
Less: Closing stock of finished goods $\frac{2,12,300}{21,230} \times 1,230 = 12,300$		12,300
	Cost of goods sold	2,00,000
Selling & distribution overhead @ Rs. 1 for 20,000 units		20,000
	Cost of sales	2,20,000
	Profit	30,000
	Sales	2,50,000

RECONCILIATION STATEMENT

Profit as per cost accounts		30,000
Add: Over charged factory overhead in cost accounts		5,000
Overcharged selling and distribution overhead in cost accounts		2,000
Closing stock under valued in cost accounts		2,700
Less: Goodwill written off in financial accounts	20,000	39,700
Interest on capital cost not included in cost accounts	2,000	
Under charged office overhead in cost accounts	6,700	28,700
Profit as per financial accounts		11,000

Problem 26. The following information is available from the financial books of a company having a normal production capacity of 60,000 units for the year ended 31-3-1995:

(i) Sales Rs. 10,00,000 (50,000 units)
(ii) There was no opening and closing stock of finished units
(iii) Direct materials and direct wages cost were Rs. 5,00,000 and Rs. 2,50,000 respectively
(iv) Actual factory expenses were Rs. 1,50,000 of which 60% are fixed
(v) Actual administrative expenses were Rs. 45,000 which are completely fixed
(vi) Actual selling and distribution expenses were Rs. 30,000 of which 40% are fixed
(vii) Interest and dividends received Rs. 15,000. You are required to:

(*a*) Find out profit as per financial books for the year ended 31st March, 1995
(*b*) Prepare the cost sheet and ascertain the profit as per cost accounts for the year ended 31-3-1995 assuming that the indirect expenses are absorbed on the basis of normal production capacity and
(*c*) Prepare a statement reconciling profits shown by financial and cost books.

(C.A. Inter, May 1995)

Solution:

PROFIT AND LOSS ACCOUNT

(for the year ended 31-3-1995)

To direct material	5,00,000	By sales	10,00,000
To direct wages	2,50,000	(50,000 units)	
To factory expenses	1,50,000	By interest and dividends	15,000
To administrative expenses	45,000		
To S & D expenses	30,000		
To profit	40,000		
	10,15,000		10,15,000

COST SHEET

(for the year ended 31-3-1995)

Direct materials			5,00,000
Direct wages			2,50,000
Prime cost			7,50,000
Factory expenses:			
Variable		6,000	
Fixed 90,000 × $\frac{5}{6}$		75,000	1,35,000
Works cost			8,85,000
Administrative expenses 45,000 × $\frac{5}{6}$			37,500
Cost of production			9,22,500
Selling & Distribution expenses			
Variable	= 18,000		
Fixed 12,000 × $\frac{5}{6}$	= 10,000		28,000
Cost of sales			9,50,500
Profit			49,500
Sales			10,00,000

RECONCILIATION STATEMENT

Profit as per cost accounts		49,500
Add: Income from interest & dividends		15,000
		64,500
Less: Factory expenses undercharged in cost accounts (1,50,000 – 1,35,000)	15,000	
Less: Administrative expenses undercharged in cost accounts (45,000 – 37,500)	7,500	
Less: S & D expenses under charged in cost accounts (30,000 – 28,000)	2,000	
		24,500
Profit as per financial accounts		40,000

CHAPTER 8

Problem 27. Following the profit and loss account of M/s Anubhav Manufacturing Company for the year ended 31st December, 1994

To opening stock of:		By sales	9,20,000
Raw materials	60,000	By closing stock of:	
Work-in-process	35,000	Raw materials	60,000
Finished goods	80,000	Work-in-progress	41,000
	1,75,000	Finished goods	30,000
To purchases	2,40,000		
To factory wages	60,000		
To electricity charges	66,000		
To factory overheads	90,000		
To gross profit c/d	4,20,000		
	10,51,000		10,51,000
To administrative expenses	25,000	By gross profit b/d	4,20,000
To selling & distribution expenses	1,15,000	By miscellaneous income	20,000
To bad debts	30,000		
To net profit	2,70,000		
	4,40,000		4,40,000

Their cost accounts showed a profit of Rs. 2,81,750, on scruting of their costing profit and profit and loss account it was found that

(1) Their opening stock and closing stock were valued as under:

Opening stock of raw materials	Rs. 80,000
Opening stock of work-in-progress	Rs. 40,000
Opening stock of finished goods	Rs. 60,000
Closing stock of raw materials	Rs. 70,000
Closing stock of work-in-progress	Rs. 44,000
Closing stock of finished goods	Rs. 20,000

(2) They charged administrative expenses at Rs. 18,000 and selling and distribution expenses at Rs. 1,27,000

(3) They had charged depreciation @ 25% on written down value method on its plant which was purchased on 1st July 1991 for Rs. 80,000. In financial accounts, however, the depreciation was provided on straight line method and the same was included in the factory overheads of Rs. 90,000.

Prepare a statement reconciling the difference in the profits as disclosed by the two records.

(University of Bombay, B.Com., April 1995)

Solution:

RECONCILIATION STATEMENT

Profit as per cost accounts		2,81,750
Add: Under valuation of openting stock of raw materials in financial accounts	20,000	
Under valuation of opening stock of WIP in financial account	5,000	
Over absorption of selling and distribution expenses in cost accounts	12,000	
Over valuation of stock of financial accounts	10,000	
Miscellaneous income taken in financial accounts only	20,000	
		67,000
		3,48,750

Less: Over-valuation of closing stock of finished goods in financial accounts	20,000	
Under absorption of Adm. expenses in cost accounts	7,000	
Under valuation of closing stock of raw materials in financial accounts	10,000	
Under valuation of closing stock of WIP in financial accounts	3,000	
Bad debts written off in financial accounts	30,000	
Under provision for depreciation in cost accounts (20,000 – 9,844)	10,156	
		2,68,594
Add undetected difference		1,406
Profit as per financial accounts		2,70,000

Working Note:

Depreciation as per financial account $80{,}000 \times \frac{25}{100} = 20{,}000$

Depreciation as per cost account

Machine purchased on 1.7.91	80,000
Less: Depreciation upto 31-12-91 @ 25%	10,000
Written down value on 1.1.92	70,000
Less: Depreciation @ 25%	17,500
Written down value on 1.1.93	52,500
Less: Depreciation @ 25%	13,125
Written down value on 1.1.94	39,375
Less: Depreciation @ 25%	9,844
	29,531

Problem 28. From the following figures prepare a reconciliation statement:

	Cost Books (*Rs.*)	*Financial Books* (*Rs.*)
Profit	50,000	?
Marketing overheads	8,000	8,000
Provision for bad debts	–	5,000
Factory overheads	8,500	7,000
Directors fees	–	2,000
Income tax paid	–	15,000
Rent of owned premises	6,000	–
Depreciation	11,250	12,000
Share transfer fee (cr.)	–	1,000
Administrative overheads	5,000	8,000

(University of Delhi, B.Com., April 1995)

CHAPTER 8

Solution:

RECONCILIATION STATEMENT

Profit as per cost accounts		50,000
Add: Over absorption of factory overheads	1,500	
Rent of own building	6,000	
Share transfer fee	1,000	
		8,500
		58,500
Less: Provision for bad debts	5,000	
Directors fees	2,000	
Income tax	15,000	
Depreciation under charged in cost account	750	
Administration overhead under absorbed in cost accounts	3,000	
		25,750
Profit as per finaucial accounts		32,750

Problem 29. A factory is producing two products, Anusha and Bhanusha. Their particulars are as follows:

	Anusha	*Bhanusha*
Materials (per unit)	Rs. 12.50	7.50
Wages (per unit)	Rs. 10.00	6.00

Factory expenses are to be calculated as 100% of wages and office expenses as 25% of works cost. Product, Anusha was produced 200 units and Bhanusha 500 units. Price of product Anusha was Rs. 50 per unit and Bhanusha at Rs. 30. Actual factory expenses were Rs. 4,800 and office expenses Rs. 4,200. Reconcile both the profit. *(Andhra University, B.Com., May 1992)*

Solution:

COST SHEET

	Anusha		*Bhanusha*		*Total*
	Per unit	*Total*	*Per unit*	*Total*	
Materials	12.50	2,500	7.50	3,750	6,250
Wages	10.00	2,000	6.00	3,000	5,000
Prime cost	22.50	4,500	13.50	6,750	11,250
Factory expenses	10.00	2,000	6.00	3,000	5,000
Factory cost	32.50	6,500	19.50	9,750	16,250
Office expenses	8.125	1,625	4.875	2,438	4,063
Office cost	40.625	8,125	24.375	12,188	20,313
Profit	9.375	1,875	5.625	2,812	4,687
Sales	50.00	10,000	30.000	15,000	25,000

PROFIT AND LOSS ACCOUNT

To Materials:		By sales:	
Anusha	2,500	Anusha	10,000
Bhanusha	3,750	Bhanusha	15,000
To Wages:			
Anusha	2,000		
Bhanusha	3,000		
To factory expenses	4,800		
To office expenses	4,200		
To P & L a/c profit	4,750		
	25,000		25,000

RECONCILIATION STATEMENT

Profit as per cost accounts	4,687
Add: Over absorption of factory expenses in cost accounts (5,000 – 4,800)	200
	4,887
Less: Under absorption of office expenses in cost accounts (4,200 – 4,063)	137
Profit as per financial accounts	4,750

CHAPTER 8

Questions

Simple Questions

1. When does the need for reconciling the profit of cost accounts with financial accounts arise?
2. Mention two forms of reconciling the profits of cost and financial accounts. Which do you prefer?
3. State four reasons for profit as per cost books being more than that shown in financial books.
4. State four reasons for profit as per cost books being less than that shown in financial books.
5. List out any four items (transactions) included in cost accounts but not in financial accounts.
6. List out any four transactions included in financial accounts but not in cost accounts.
7. Mention different bases of valuing closing stock in cost and financial accounts.
8. Mention any two methods of depreciation followed in cost and financial accounts.

Short Answer Questions

1. What is the purpose of reconciling cost and financial accounts when two accounts are separately maintained ? State the causes which are responsible for the disagreement in the profit.
2. "When separate sets of books are maintained for financial and cost accounts the profits computed in one set may differ from profit computed in the other set". Indicate the possible reasons for such variation.
3. Compile a proforma reconciliation statement (without figures) to explain profit variation between cost and financial accounts.
4. "For computing the profits, the methods adopted in financial accounts differ from those used in cost accounts". Elucidate.
5. "The reconciliation between cost and financial accounts help not only explaining profit variations in the past but also enables evolving better estimates, for future guidance". Comment.

6. Why is reconciliation of cost and financial accounts necessary ? Give any eight items which appear in the financial accounts but do not appear in the cost accounts.
(Kakatiya University, B.Com., November 1991 and May 1992)

7. Explain the procedure of reconciling the profit as shown by cost and financial accounts.
(Sri Sathya Sai University, B.Com., July 1990)

8. What do you understand by reconciliation of cost and financial accounts ? Indicate the possible reasons (causes) for difference between them. *(Kakatiya University, B.Com., October 1990)*

9. Write a note on profit reconciliation statement. *(Osmania University, B.Com., March 1992)*

10. Explain the causes of difference between profit shown by financial accounts and profit shown by cost accounts. *(Delhi University, B.Com. (Hons.), April 1989)*

Exercises

1. Find out the profit as per financial records from the following data:

	Rs.
(*i*) Profit as per cost records	70,500
(*ii*) Under valuation of closing stock in cost records	10,500
(*iii*) Administration overheads under recovered in cost records	5,200
(*iv*) Bad debts written off in financial accounts only	7,345
(*v*) Depreciation over charged in cost records	3,445

(Sri Sathya Sai University, B.Com., (Hons) October 2004)

[**Ans.** Profit as per financial accounts Rs. 71,900]

2. The following information has been obtained from the records of M/s New Electronics, a manufacturer of T.V. sets.

	Rs.
(*a*) Material cost per T.V. set	450
(*b*) Wages per T.V. set	250
(*c*) Charge 60% of wages as works-overhead	
(*d*) Charge 20% of prime cost as office overhead	
(*e*) Number of units manufactured and sold	4,000
(*f*) Sales price per T.V. set	2,000

Prepare a cost sheet showing profit as per costing records and a statement showing profit as per financial records. Actual factory expenses being Rs. 6,20,000 and office expenses Rs. 6,40,000 Prepare a reconciliation statement of profit as per costing record and profit as per financial records. *(Sri Sathya Sai University, B.Com. (Hons.), August 2003)*

[**Ans.** Profit as per cost accounting Rs. 40,40,000, profit as per financial accounting Rs. 39,40,000]

3. Ascertain the profit as per financial book from the following information

	Rs.
Profit as per cost accounts	25,000
Closing stock over valued in cost books	12,500
Preliminary expenses written off	3,000
Profit on sale of building	30,000

Administrative expenses over recovered in cost books	50,375
Works overhead under recovered in financial books	30,375
Bank interest and transfer fee in financial books	5,000
Interest on investment recorded in financial books	10,000
Depreciation shown excess in cost books	4,000
Provision made for income tax	40,000

(University of Madras, B.Com., April 2002)

[**Ans.** Profit as per financial accounts Rs. 38,500]

4. From the following data prepare a reconciliation statement to find profit as per financial statement

	Rs.
Profit as per cost accounts	2,50,000
Works overhead over-absorbed	20,000
Administration overheads under-absorbed	45,000
Under valuation of opening stock in cost accounts	15,000
Bad debts written off during the year	14,000
Preliminary expenses written off during the year	10,000

(University of Calcutta, B.Com. (Hons), 2005)

[**Ans.** Profit as per financial accounts Rs. 1,86,000]

5. A manufacturing company disclosed a net loss of Rs. 3,47,000 as per their cost accounts for the year ended 31st March 2003. The financial accounts however, disclosed a net loss of Rs. 5,10,000 for the same period. The following information was revealed as a result of scrutiny of the figures of both the sets of accounts.

	Rs.
(i) Factory overheads under-absorbed	40,000
(ii) Administration overheads over-absorbed	60,000
(iii) Depreciation charged in financial accounts	3,25,000
(iv) Depreciation charged in cost accounts	2,75,000
(v) Interest on investments not included in cost accounts	96,000
(vi) Income-tax provided	54,000
(vii) Interest on loan funds in financial accounts	2,45,000
(viii) Transfer fee (credit in financial accounts)	24,000
(ix) Stores adjustment (credit in financial books)	14,000
(x) Dividend received	32,000

Prepare a memorandum reconciliation account. *(University of Delhi, B.Com., 2005)*

[**Ans.** Net loss as per financial accounts Rs. 5,10,000]

6. Prepare a reconciliation account from the following details:

Profit as per cost accounts were Rs. 60,000, while the profit as per financial accounts were Rs. 59,700. The values of opening and closing stock as shown in cost accounts and financial accounts were as under:

	Financial a/c Rs.	*Cost a/c Rs.*
Raw materials:		
Opening	25,300	25,000
Closing	30,000	29,600

CHAPTER 8

WIP:		
Opening	16,000	15,500
Closing	20,000	19,900

(University of Delhi, B.Com. (Hons.), 2003)

[Ans. Profit as per financial accounts Rs. 59,700]

7. A manufacturing company disclosed a net loss of Rs. 3,47,000 as per their cost accounts for the year ended March 31, 2004. The financial accounts however, disclosed a net loss of Rs. 5,10,000 for the same period. The following information was revealed as a result of scrutiny of the figures of both the sets of accounts.

	Rs.
Factory overheads under-absorbed	40,000
Administration overheads over-absorbed	60,000
Depreciation charged in financial accounts	3,25,000
Depreciation recovered in cost accounts	2,75,000
Interest on investments not included in cost accounts	96,000
Income tax provided	54,000
Interest on loan funds in financial accounts	2,45,000

(University of Delhi, B.Com. (Pass), 2005)

[Ans. Loss as per financial accounts Rs. 5,10,000]

8. From the following information reconcile the profit as per cost accounts with financial accounts:

	Cost Account *Rs.*	*Financial Account* *Rs.*
Profit	86,250	?
Opening Stock:		
Material	10,500	10,300
Work-in-progress	8,500	8,000
Closing Stock:		
Material	14,200	15,000
Work-in-progress	6,000	5,600

Dividend and interest received Rs. 600. Loss on sale of investment Rs. 1,000. Interest charged by the bank not considered in financial accounts and cost accounts Rs. 1,500. Goodwill written off during the year Rs. 2,500. Preliminary expenses written off Rs. 3,000 overhead incurred Rs. 40,000. Overhead absorbed in cost accounts Rs. 38,500. Find out profit as per financial accounts.

(University of Delhi, B.Com. (Pass), 2005)

[Ans. Profit as per financial accounts Rs.79,950]

Note: Interest charged by the bank not considered in financial accounts and cost accounts Rs. 1,500 will not come in reconciliation statement.

9. The following is the trading and profit and loss account of Deep Industries Ltd. for the year ended 31st December 2001

To materials	45,000	By sales	96,000
To wages	33,000	(4,800 units)	
To works expenses	24,000	By closing stock	20,400
To administrative expenses	6,000	(1,200 units)	
To net profit	8,400		
	1,16,400		1,16,400

The company's cost records show that:

(*i*) Works overheads have been absorbed at Rs. 3 per unit produced and

(*ii*) Administrative overheads have been absorbed at Rs. 1.50 per unit produced.

Assuming there is nothing by way of work-in-progress either at the beginning or at the end and there is no opening stock of finished goods prepare:

(*i*) A statement of cost indicating the net profit

(*ii*) A statement reconciling the profit as disclosed by cost accounts and that shown in financial accounts. *(University of Delhi, B.Com., April 2003)*

[**Ans.** Profit as per cost accounting Rs. 12,000. Profit as per financial accounting Rs. 8,400]

10. From the following figures prepare a reconciliation statement:

	Rs.
Profit as per costing records	5,000
Factory overheads under-recovered in costing	3,000
Selling and administration overheads over recovered in costing	2,000
Bank interest credited in financial books	500
Preliminary expenses written off in financial books	6,500
Opening stock value:	
in cost books	5,000
in financial books	4,000
Closing stock value:	
in cost books	12,000
in financial books	10,000

(University of Delhi, B.Com. (Pass), 2002)

[**Ans.** Loss as per financial accounts Rs. 3,000]

11. The net profit as per financial accounts of a company amounted to Rs. 18,550, while the profit as per cost accounts were Rs. 28,660. On reconciling the figure, the following were noted:

	Rs.
(1) Director's fees not charged in cost accounts	1,050
(2) A provision for bad and doubtful debts	970
(3) Bank interest credited	30
(4) Provision for income-tax	8,300
(5) Over-recovery of overhead in cost accounts	180

Prepare reconciliation statement *(University of Delhi, B.Com. (Pass), 2001)*

[**Ans.** Profit as per financial accounts Rs. 18,550]

12. The following profit and loss account for the year ending 31st March, 1999 has been extracted from the books of 'A' Ltd.

To Direct material	10,000	By sales	50,000
To Direct labour	20,000	By work-in-progress	
To Factory expenses	9,500	– Direct material	400
To Administrative expenses	5,200	– Direct labour	600
To Selling & distribution expenses	3,800	– Factory expenses	300
To Interest on capital	1,000	By finished stock	2,760
To Goodwill written off	1,500		
To Net profit	3,000		
	54,000		54,000

Cost accounts manual states that the factory overheads are to be recovered at 50% of direct wages, administration overheads at 10% of works cost and selling and distribution overheads @ Re. 1 per unit sold.

The units of product sold and in-hand were 4,000 and 257 respectively. Prepare:

(*i*) Statement of cost and profit as per cost accounts.

(*ii*) Reconciliation statement

(University of Delhi, B.Com. (Pass), 2000)

[**Ans.** Profit as per cost accounting Rs. 6,000 and profit as per financial accounting Rs. 3,000]

13. In a factory, works overhead is absorbed @ 60% of labour and office expenses @ 20% of works cost. The total expenditure is as follows:

	Rs.
Materials	2,00,000
Labour	1,50,000
Factory expenses	98,000
Office expenses	85,000
	5,33,000

10% of the output is in stock and sales Rs. 5,10,000. Prepare a cost sheet and a reconciliation statement. *(Bharathidasan University, B.Com., April 1998)*

[**Ans.** Profit as per cost a/c Rs. 26,000 and as per financial accounts Rs. 21,800]

14. The following information has been obtained from the records of a machine manufacturer.

(*a*) Materials per machine Rs. 1,750

(*b*) Wages per machine Rs. 1,000

(*c*) Number of machines manufactured and sold 80

(*d*) Works overhead to be charged at 60% of wages

(*e*) Office overhead to be charged at 20% of works costs

(*f*) There was no opening or closing stock of machines

(*g*) Selling price per machine Rs. 4,750

Prepare a statement showing the profit per machine sold. Also prepare a statement showing the actual profit. The works expenses were Rs. 48,000 and office expenses were Rs. 47,500 as per financial books.

Reconcite the profit as shown by costing and financial records.

(Bangalore University, B.Com., April 1997)

[**Ans.** Profit as per cost account Rs. 58,400 and as per financial account Rs. 64,500]

15. From the following particulars, prepare (*a*) statement of cost of manufacture for the year (*b*) a statement of profit as per cost accounts (*c*) profit and loss account in the financial books and (*d*) reconciliation of the difference in the profit as shown by (*b*) and (*c*) above.

	Rs.
Opening stock of raw materials	1,00,000
Closing stock of raw materials	1,50,000
Opening stock of finished product	2,00,000
Closing stock of finished product	50,000
Purchase of raw materials	2,50,000
Wages	2,50,000

Calculate factory overhead at 25% of prime cost. Office overhead will be levied at 75% on factory overhead. Actual works expenditure amounted to Rs. 1,93,750 and actual office expenses amounted to Rs. 1,52,500. The selling price was fixed at 25% above cost price.

(Calicut University, B.Com., April 1997)

[**Ans.** Profit as per cost accounts Rs. 3,25,000. As per financial accounts Rs. 3,28,750]

16. Papanna and Company Ltd., maintains separate cost accounting system which revealed a profit of Rs. 59,900 on 31st March 1999. The financial accounting revealed a different profit on verification. The following information was obtained:

(*a*) Preliminary expenses written off	Rs. 12,200
(*b*) Factory overhead incurred Rs. 8,900 but absorbed	Rs. 10,000
(*c*) Provision for doubtful debts	Rs. 800
(*d*) Discount on issue of shares	Rs. 6,200
(*e*) Transfer fees received	Rs. 1,700
(*f*) Underwriting commission paid	Rs. 5,000
(*g*) Dividend received	Rs. 7,300
(*h*) Depreciation charged Rs. 10,800 but recovered only	Rs. 10,000
(*i*) Closing stock valued in financial accounts Rs. 7,800 but in cost accounts	Rs. 9,000
(*j*) Interest on own capital recorded in cost accounts	Rs. 1,300

Prepare a reconciliation statement as on 21-3-1999 and show profit as per financial accounts.

(Bangalore University, B.B.M., April 1999)

[**Ans.** Profit as per financial books Rs. 42,800]

17. From the following data, prepare a reconciliation statement:

	Rs.
Profit as per cost accounts	1,45,500
Works cverhead under recovered	9,500
Administrative overhead under-recovered	22,750
Selling overheads over-recovered	19,500
Over valuation of opening stock in cost accounts	15,000
Over valuation of closing stock in cost accounts	7,500
Interest earned during the year	3,750
Rent received during the year	27,000
Bad debts written off during the year	9,000
Preliminary expenses written off during the year	18,000

[**Ans.** Profit as per financial accounts Rs. 1,44,000].

❑❑❑

CHAPTER 8

CHAPTER 9

INTEGRAL ACCOUNTING

INTRODUCTION

Cost and financial accounts are said to be integrated when only one set of account is operated to serve the purpose of both the accounts. Under this method a general ledger is maintained which contains both cost and financial accounts. In general ledger, purchases, wages and expenses undergo functional analysis instead of grouping under various account heads. Entries in general ledger are made in summarised form supported by detailed day-to-day posting in related subsidiary ledgers. The main objective of integrated account is to avoid duplication of work. It also obviates the necessity of reconciliation between cost and financial accounts. It facilitates short-term closing of accounts without going through the tedious process of stock-taking and valuation.

DEFINITION AND FEATURES OF INTEGRATED ACCOUNTING

The ICMA terminology defines integrated accounts as "a single set of accounts which provides both financial and management accounting information".

The following are the features of integrated accounting system:

(*a*) Only one set of book is maintained for both financial and cost accounts.

(*b*) The general ledger adjustment account is dispensed with under this system.

(*c*) Transactions which affect cost accounts are recorded in the general ledger just as they are recorded in cost accounts.

(*d*) Transactions which affect financial accounts are recorded in the general ledger just as they are recorded in financial accounts.

(*e*) Instead of a double entry for every transaction in the financial books and further double entry in cost books, this system requires only one debit and one credit.

Advantages and Limitations

The following are the advantages of integral accounting system:

1. The preparation of reconciliation statement does not arise as only one profit and loss account is prepared.
2. It avoids duplication of work which is involved in non-integral accounts.
3. It results in economy in maintaining accounts.
4. It facilitates control of accounting procedure and coordination of accounting function in different departments.
5. Because of centralisation principle the accounting procedure is simplified.
6. It facilitates quick reporting as all accounting information is readily available.

Integrated accounting suffers from one limitation. Financial accounts have to be prepared in a proforma prescribed by the Comapanies Act to satisfy statutory requirements. This would necessitate classification of income and expenses according to their natural headings. Similarly, total purchases figures will have to be shown in the accounts as per law, whereas the integrated accounts do not have a purchase account maintained. A common flexible code of accounts to suit both statutory accounting and integrated accounting would be necessary for the purpose. Purchases, wages and expenses would then undergo natural analysis followed by functional analysis.

Books and Accounts Under Integral Accounting

The following are the subsidiary ledgers and the accounts that are maintained under integral accounting system:

(1) **Sales ledger:** This ledger consists of accounts of customers. The account that is prepared in this ledger is known as sales ledger control account.

(2) **Bought ledger:** This ledger is maintained to record the accounts of suppliers. The account that is maintained in this ledger is known as bought ledger control account.

(3) **Stores ledger:** This ledger is maintained for recording accounts relating to individual items of materials and stores kept in stock. The account maintained in this ledger is known as stores ledger control account.

(4) **Stock ledger:** This ledger is maintained for recording accounts relating to individual items of finished goods in stock. The account maintained in this ledger is known as stock ledger control account.

(5) **Job ledger:** This ledger contains accounts of individual jobs on hand. The account maintained in this ledger is known as work-in-progress account.

(6) **Expenses ledger:** This ledger contains accounts of expenses kept separately for factory, office and selling and distribution expenses. The accounts maintained in this ledger are: (*a*) works overhead account, (*b*) office overhead account, and (*c*) selling and distribution overhead account.

JOURNAL ENTRIES UNDER INTEGRAL ACCOUNTS

(1) Materials

(*a*) Materials purchased on credit for stock

Stores control a/c *Dr.*

To creditors a/c

(*b*) Cash purchase of materials for stock

Stores control a/c *Dr.*

To cash

(*c*) Material purchased for special jobs

Work-in-progress a/c *Dr.*

To creditors/cash

(*d*) Purchase of materials for immediate repair work

Factory overhead control a/c *Dr.*

To creditors/cash

(e) Materials returned to suppliers from stock

Sundry creditors a/c *Dr.*

To stores control a/c

(f) Payment to creditors for purchases made

Creditors a/c *Dr.*

To cash

(2) Payment of wages

Wages control a/c *Dr.*

To cash

(3) Payment for expenses

Overhead control a/c *Dr.*

To cash

(4) Recording of goods completed

Finished goods control a/c *Dr.*

To WIP

(5) Recording of cost of goods sold

Cost of sales *Dr.*

To P & L a/c

To finished goods control a/c

Problem 1. Journalise the following transactions assuming that cost and financial transactions are integrated:

Raw materials purchased	2,00,000
Direct materials issued to production	1,50,000
Wages paid (30% indirect)	1,20,000
Wages charged to production	84,000
Manufacturing expenses incurred	84,000
Manufacturing overhead charged to production	92,000
Selling and Distribution costs	20,000
Finished products (at cost)	2,00,000
Sales	2,90,000
Receipts from debtors	69,000
Payments to creditors	1,10,000

Solution:

JOURNAL ENTRIES

Stores ledger control a/c To cash/creditors a/c (Being materials purchased)	Dr.	2,00,000	 2,00,000
Work-in-progress control a/c To stores ledger control a/c (Materials issued to production)	Dr.	1,50,000	 1,50,000

Wages control a/c To cash (Being wages paid)	Dr.	1,20,000	1,20,000
Work-in-progress control a/c To wages control a/c (Direct wages charged to production)	Dr.	84,000	84,000
Factory overhead control a/c To wages control a/c (30% of wages paid being indirect charged to overhead)	Dr.	36,000	36,000
Factory overhead control a/c To Bank a/c (Manufactured overhead incurred)	Dr.	84,000	84,000
Work-in-progress control a/c To factory overhead control a/c (Being manufactured overhead charged to production)	Dr.	92,000	92,000
Selling & Distribution overhead control a/c To Bank (Being selling & distribution overhead incurred)	Dr.	20,000	20,000
Finished goods ledger control a/c To WIP control a/c (Being cost of goods finished)	Dr.	2,00,000	2,00,000
Cost of sales a/c To Finished stock ledger control a/c To S & D overhead a/c (Being cost of goods sold)	Dr.	2,20,000	 2,00,000 20,000
Sundry debtors a/c To sales (Being finished stock sold)	Dr.	2,90,000	2,90,000
Bank a/c To sundry debtors (Receipts from Debtors)	Dr.	69,000	69,000
Sundry creditors a/c To Bank (Being payment made to creditors)	Dr.	1,10,000	1,10,000

Problem 2. Shivalika Enterprises operates an integrated system of accounting. You are required to pass journal entries for the following transactions that took place for the year ended 31st March 1992 (Naration is not required)

(1) Raw materials purchased (50% on credit)	Rs.	6,00,000
(2) Materials issued to production	Rs.	4,00,000
(3) Factory overhead incurred	Rs.	80,000
(4) Sales (50% credit)	Rs.	7,50,000
(5) Receipt from debtors	Rs.	2,00,000
(6) Payment to creditors	Rs.	2,00,000

(University of Delhi, B.Com. (Hons.), April 98)

CHAPTER 9

Solution:

JOURNAL ENTRIES

(1)	Stores ledger control a/c To cash a/c To credit a/c	Dr.	6,00,000	 3,00,000 3,00,000
(2)	Work-in-progress control a/c To stores ledger control a/c	Dr.	4,00,000	 4,00,000
(3)	Factory overhead control a/c To cash	Dr.	80,000	 80,000
(4)	Cash a/c Debtors a/c To sales	Dr. Dr.	3,75,000 3,75,000	 7,50,000
(5)	Cash a/c To debtors a/c	Dr.	2,00,000	 2,00,000
(6)	Creditors a/c To cash	Dr.	2,00,000	 2,00,000

Problem 3. Journalise the following transactions under the integral accounting system

	Rs.
Direct wages paid in cash	60,000
Indirect wages paid in cash	30,000
Purchases made in cash	15,000
Purchases (credit)	2,90,000
Stores issued against production order	2,75,000
Works expenses incurred and paid in cash	55,000
Works expenses allocated to jobs	80,000
Administration expenses paid in cash	40,000
Administration expenses allocated to jobs	48,000
Finished goods transferred to warehouse	4,50,000

(ICWA, Inter., June 1995)

Solution:

JOURNAL ENTRIES

(1)	Wages control a/c To cash (Cash paid towards direct wages Rs. 60,000 + Indirect wages Rs. 30,000)	Dr.	90,000	 90,000
(2)	WIP control a/c To wage control a/c (Direct wages transferred to WIP)	Dr.	60,000	 60,000
(3)	Works expenses control a/c To wages control a/c (Indirect wages transferred to works expenses control a/c)	Dr.	30,000	 30,000

(4)	Stores ledger control a/c To creditors To cash (To record cash and credit purchases)	*Dr.*	3,05,000	 2,90,000 15,000
(5)	WIP control a/c To stores ledger control a/c (To record transfer of stores issued to production)	*Dr.*	2,75,000	 2,75,000
(6)	Works expenses control a/c To cash a/c (To record work expenses disbursed in cash)	*Dr.*	55,000	 55,000
(7)	WIP control a/c To works expenses control a/c (Works expenses allocated to jobs)	*Dr.*	80,000	 80,000
(8)	Administration expenses control a/c To cash (Administration expenses paid in cash)	*Dr.*	40,000	 40,000
(9)	WIP control a/c To Administrative expenses (Administration expenses charged to production)	*Dr.*	48,000	 48,000
(10)	Finished goods control a/c To WIP control a/c (To record finished goods transferred)	*Dr.*	4,50,000	 4,50,000

Problem 4. Journalise the following transactions assuming that the cost and financial records are integrated and prepare necessary ledger accounts.

Raw materials purchased on credit	14,800
Direct materials issued to production	12,800
Wages paid	16,800
Wages productive	14,800
Wages unproductive	2,000
Factory expenses	6,500
Factory expenses charged to production	8,600
Administration expenses	4,400
Administration expenses charged to production	4,350
Selling expenses charged to sales	4,500
Finished goods at cost	30,000
Cash sales	39,000

Solution:

JOURNAL ENTRIES

(1)	Stores control a/c To creditors a/c (Being materials purchased on credit)	*Dr.*	14,800	 14,800
(2)	Work-in-progress control a/c To stores control a/c (Being materials issued to jobs)	*Dr.*	12,800	 12,800

CHAPTER 9

			Dr. (₹)	Cr. (₹)
(3)	Wages control a/c To Bank (Being wages paid)	Dr.	16,800	16,800
(4)	Work-in-progress control a/c To wages control a/c (Being the wages charged to production)	Dr.	14,800	14,800
(5)	Factory overhead control a/c To wages control a/c (Being the indirect wages charged to factory overheads)	Dr.	2,000	2,000
(6)	Factory overhead control a/c To Bank (Being the factory expenses incurred)	Dr.	6,500	6,500
(7)	WIP control a/c To factory overhead control a/c (Being factory overhead charged to production)	Dr.	8,600	8,600
(8)	Administration overhead control a/c To Bank (Being Administration expenses incurred)	Dr.	4,400	4,400
(9)	WIP control a/c To administration overhead control a/c (Being administration overhead charged to production)	Dr.	4,350	4,350
(10)	S & D overhead control a/c To Bank (Being selling expenses incurred)	Dr.	4,500	4,500
(11)	Finished stock control a/c To WIP control a/c (Being finished stock transferred at cost)	Dr.	30,000	30,000
(12)	Cost of sales a/c To finished stock control a/c To S & D overhead control a/c (Being the cost of production of goods sold)	Dr.	34,500	30,000 4,500
(13)	Bank a/c To sales control a/c (Being cash sales)	Dr.	39,000	39,000

STORES CONTROL ACCOUNT

To creditors	14,800	By WIP control a/c	12,800
		By balance c/d	2,000
	14,800		14,800

WAGES CONTROL ACCOUNT

To Bank	16,800	By WIP control a/c	14,800
		By factory overhead control a/c	2,000
	16,800		16,800

WORK-IN-PROGRESS CONTROL ACCOUNT

To wages control a/c	14,800	By finished stock control a/c	30,000
To stores control a/c	12,800	By Balance c/d	10,550
To factory overhead control a/c	8,600		
To Adm. overhead control a/c	4,350		
	40,550		40,550

FACTORY OVERHEAD CONTROL ACCOUNT

To wages control	2,000	By WIP control a/c	8,600
To Bank	6,500		
To overhead suspense a/c (Balancing figure)	100		
	8,600		8,600

ADMINISTRATION OVERHEAD CONTROL ACCOUNT

To Bank	4,400	By WIP control a/c	4,350
		By overhead suspense a/c (Balancing figure)	50
	4,400		4,400

S & D OVERHEAD CONTROL ACCOUNT

To bank	4,500	By cost of sales	4,500

FINISHED STOCK CONTROL ACCOUNT

To WIP control a/c	30,000	By cost of sales	30,000

COST OF SALES ACCOUNT

To finished stock control a/c	30,000	By sales control a/c	39,000
To S & D overhead control a/c	4,500		
To P & L a/c (Balancing figure)	4,500		
	39,000		39,000

BANK ACCOUNT

To sales control a/c	39,000	By wages control a/c	16,800
		By Factory overhead control a/c	6,500
		By administration overhead control a/c	4,400
		By S & D overhead control a/c	4,500
		By Balance c/d	6,800
	39,000		39,000

CREDITORS ACCOUNT

To Balance c/d	14,800	By stores control a/c	14,800

SALES CONTROL ACCOUNT

To cost of sales a/c	39,000	By bank a/c	39,000

CHAPTER 9

OVERHEAD SUSPENSE ACCOUNT

To Adm. overhead control a/c	50	By factory overhead control a/c	100
To balance c/d	50		
	100		100

TRIAL BALANCE

	Dr.	Cr.
Stores control a/c	2,000	
WIP control a/c	10,550	
Bank	6,800	
Creditor's control a/c		14,800
Overhead suspense a/c		50
P & L a/c		4,500
	19,350	19,350

THIRD ENTRY METHOD

Integral accounting is sometimes called third entry accounting. In fact third entry accounting is a modified form of self-balancing system. In a self-balancing ledger, four entries are passed in respect of each transaction affecting accounts maintained in two different ledgers. For example, general ledger and stores ledger are kept under self-balancing system, then in general ledger a total account under the heading 'stores ledger control account' is opened. In stores ledger, in turn, a total account is opened under the heading 'General ledger control account'. When raw materials are purchased for stock the following four transactions are passed.

In general ledger

Stores ledger control a/c *Dr.*

To sundry creditors a/c

In stores ledger

Stores a/c *Dr.*

To general ledger control a/c

But in practice, the general ledger control account in stores ledger is not maintained, in which case the transaction involves only three entries, *i.e.,* two entries in the general ledger and a third entry in stores ledger card. This process is extended to any or all subsidiary ledgers which are "controlled" by a total account in general ledger.

Problem 5. Dutta Enterprises operates an integral system of accounting. You are required to pass the journal entries for the following transactions that took place for the year ended 30.6.2010.

	Rs.
Raw materials purchased (50% on credit)	6,00,000
Materials issued to production	4,00,000
Wages paid (50% direct)	2,00,000
Wages charged to production	1,00,000
Factory overheads incurred	80,000
Factory overheads charged to production	1,00,000
Selling and distribution overhead incurred	40,000
Finished goods at cost	5,00,000
Sales (50% credit)	7,50,000

Closing stock	NIL
Receipts from debtors	2,00,000
Payments to creditors	2,00,000

(C.A., Inter., November 1990)

Solution:

JOURNAL ENTRIES

Particulars		Dr. (Rs.)	Cr. (Rs.)
Stores ledger control a/c	Dr.	6,00,000	
To sundry creditors a/c			3,00,000
To cash			3,00,000
Work-in-progress control a/c	Dr.	4,00,000	
To stores ledger control a/c			4,00,000
Wages control a/c	Dr.	1,00,000	
To cash a/c			1,00,000
Factory overhead control a/c	Dr.	1,00,000	
To wages control a/c			1,00,000
Factory overhead control a/c	Dr.	80,000	
To cash			80,000
Work-in-progress control a/c	Dr.	1,00,000	
To factory overhead control a/c			1,00,000
Selling and distribution overhead control a/c	Dr.	40,000	
To cash			40,000
Finished stock ledger control a/c	Dr.	5,00,000	
To WIP control a/c			5,00,000
Cost of sales a/c	Dr.	5,40,000	
To finished stock ledger control a/c			5,00,000
To selling & distribution overhead control a/c			40,000
Sundry debtors	Dr.	3,75,000	
Cash a/c	Dr.	3,75,000	
To sales a/c			7,50,000
Cash	Dr.	2,00,000	
To sundry debtors			2,00,000
Sundry creditors a/c	Dr.	2,00,000	
To cash			2,00,000

Problem 6. In the absence of the chief accountant, you have been asked to prepare a month's cost accounts for a company which operates a batch costing system fully integrated with the financial accounts.

The following relevant information is provided to you:

	Rs.
Balances at the beginning of the month:	
Stores ledger control a/c	25,000
Work-in-progress control a/c	20,000
Finished goods control a/c	35,000
Prepaid production overhead brought forward from the previous month	3,000

CHAPTER 9

Transactions during the month:		
Materials purchased		75,000
Materials issued		
To production	30,000	
To factory maintenance	4,000	34,000
Materials transferred between batches		5,000
Total wages paid:		
To direct workers	25,000	
To indirect workers	5,000	30,000
Direct wages charged to batches		20,000
Recorded non-productive time of direct workers		5,000
Selling and distribution overheads incurred		6,000
Other production overheads incurred		12,000
Sales		1,00,000
Cost of finished goods sold		80,000
Cost of completed and transferred finished goods during the month		65,000
Physical value of work-in-progress at the end of the month		40,000

The production overhead absorption rate is 15% of direct wages charged to work-in progress:

Prepare the following accounts for the month:

(*a*) Stores ledger control a/c
(*b*) Work-in-progress control a/c
(*c*) Finished goods control a/c
(*d*) Production overhead control a/c
(*e*) Profit & loss a/c

(C.A. Inter., May 1988)

Solution:

STORES LEDGER CONTROL ACCOUNT

To balance b/d	25,000	By WIP control a/c	30,000
To creditors (or bank)	75,000	By production overhead control a/c	4,000
		By balance c/d	66,000
	1,00,000		1,00,000

WIP CONTROL ACCOUNT

To balance b/d	20,000	By finished goods control a/c	65,000
To stores ledger control a/c	30,000	By balance c/d (physical value)	40,000
To wages control a/c	20,000		
To production overhead control a/c (150% of direct wages)	30,000		
To P & L a/c (stock gain)	5,000		
	1,05,000		1,05,000

FINISHED GOODS CONTROL ACCOUNT

To balance b/d	35,000	By cost of goods a/c or P & L a/c	80,000
To work-in-progress control a/c	65,000	By balance c/d	20,000
	1,00,000		1,00,000

PRODUCTION OVERHEAD CONTROL ACCOUNT

To balance b/d (prepaid amount)	3,000	By WIP control a/c	30,000
To stores ledger control a/c	4,000	(150% of direct wages)	
To wages control a/c			
Direct workers	5,000		
Indirect workers	5,000		
To bank	12,000		
To P & L a/c			
(over absorption-balancing figure)	1,000		
	30,000		30,000

PROFIT & LOSS ACCOUNT

To finished goods control a/c or cost of goods sold a/c	80,000	By sales a/c	1,00,000
To selling & distribution overhead a/c	6,000	By production overhead control a/c	1,000
To balance c/d	20,000	By work-in-progress control a/c (stock gain)	5,000
	1,06,000		1,06,000

Notes:

(1) Materials transferred between batches will not affect the control accounts.

(2) Non-production time of direct workers is a production overhead and therefore will not be charged to WIP control a/c.

(3) Production overheads absorbed in WIP control a/c will then equal Rs. 30,000 (150% of Rs. 20,000).

(4) In the WIP control a/c the excess physical value of stock is taken resulting in stock gain. Stock gain is transferred to P & L a/c.

Problem 7. Journalise the following transactions assuming that cost and financial accounts are integrated:

Wages paid (30% indirect)	1,20,000
Wages charged to production	84,000
Manufacturing expenses incurred	48,000
Manufacturing overhead charged to production	82,000

(Osmania University, M.Com., May 2001)

Solution:

JOURNAL ENTRIES

Wage control a/c	*Dr.*	84,000	
Factory overhead control a/c	*Dr.*	36,000	
To bank a/c			1,20,000
(Being wages paid 30% being indirect)			
Work-in-progress control a/c	*Dr.*	84,000	
To wage control a/c			84,000
(Being wages charged to production)			
Factory overhead control a/c	*Dr.*	48,000	
To Bank a/c			48,000
(Being manufacturing expenses incurred)			

CHAPTER 9

Work-in-progress control a/c To factory overhead control a/c (Being factory overhead charged to production)	*Dr.*	69,000	69,000

Problem 8. Journalise the following transactions in the integrated books.

(*i*) Credit purchases Rs. 80,000

(*ii*) Works expenses charged to production Rs. 22,200

(*iii*) Finished goods transferred from production order Rs. 1,40,000

(*iv*) Stores issued to production order Rs. 70,000. *(Osmania University, M.Com., May 2000)*

Solution:

JOURNAL ENTRIES

(1)	Stores ledger control a/c To sundry creditor a/c (Being materials purchased)	*Dr.*	80,000	80,000
(2)	Work-in-progress control a/c To factory overhead control a/c (Being works overheads charged to production)	*Dr.*	22,200	22,200
(3)	Finished stock ledger control a/c To work-in-progress control a/c (Being finished goods transferred)	*Dr.*	1,40,000	1,40,000
(4)	Work-in-progress control a/c To stores ledger control a/c (Being stores issued to production)	*Dr.*	70,000	70,000

Problem 9. Journalise the following transactions in the integrated books

(*i*) Unproductive wages Rs. 50,000

(*ii*) Stores issued to production order Rs. 70,000

(*iii*) Administrative expenses charged to production Rs. 10,200

(*iv*) Works expenses outstanding Rs. 200

(Osmania University, M.Com., May 1999)

Solution:

JOURNAL ENTRIES

(1)	Factory overhead control a/c To wage control a/c (Being unproductive wages recorded)	*Dr.*	50,000	50,000
(2)	Work-in-progress control a/c To stores ledger control a/c (Being stores issued to production order)	*Dr.*	70,000	70,000
(3)	Work-in-progress a/c To Administration overhead control a/c (Being administrative overhead charged to production)	*Dr.*	10,200	10,200
(4)	Wages control a/c To wages payable a/c (Being wages outstanding)	*Dr.*	200	200

Problem 10. Pass journal entries for the following transactions for posting into control accounts under integral accounting

(*i*) Direct material purchased specially for jobs Rs. 900

(*ii*) Actual amount of wages paid Rs. 1,500

(*iii*) Selling and distribution expenses allocated to sales Rs. 1,000

(*iv*) Completed jobs and products passed to finished stock Rs. 2,500

(Osmania University, M.Com., May 1998)

Solution:

JOURNAL ENTRIES

(1)	Work-in-progress control a/c To sundry creditors/Bank (Direct materials purchased)	*Dr.*	900	900
(2)	Wages control a/c To Bank a/c (Being wages paid)	*Dr.*	1,500	1,500
(3)	Cost of sales a/c To S & D overhead control a/c (Being selling & distribution overhead allocated)	*Dr.*	1,000	1,000
(4)	Finished stock ledger control a/c To work-in-progress control a/c (Being completed goods passes to finished stock)	*Dr.*	2,500	2,500

Problem 11. Journalise the following transactions assuming that cost and financial records are integrated.

	Rs.
Raw materials purchased on credit	2,00,000
Direct material issued for production	1,50,000
Wages paid (30% indirect)	1,20,000
Manufacturing expenses incurred	84,000
Manufacturing overhead charged to production	92,000
Selling & Distribution overhead	20,000
Finished product (at cost)	2,00,000
Sales	2,90,000
Closing stock	NIL
Receipts from debtors	69,000
Payments to creditors	1,10,000

(Osmania University, M.Com., May 1998)

Solution:

JOURNAL ENTRIES

(1)	Stores ledger control a/c To sundry creditors/Bank a/c (Being raw materials purchased)	*Dr.*	2,00,000	2,00,000
(2)	WIP control a/c To stores ledger control a/c (Being material issued to production)	*Dr.*	1,50,000	1,50,000

(3)	Wage control a/c	Dr.	84,000	
	Factory overhead control a/c	Dr.	36,000	
	To Bank a/c			1,20,000
	(Being direct and indirect wages paid)			
(4)	Factory overhead control a/c	Dr.	84,000	
	To Bank a/c			84,000
	(Being manufacturing overhead paid)			
(5)	WIP control a/c	Dr.	92,000	
	To Factory overhead control a/c			92,000
	(Being factory overhead charged to production)			
(6)	Selling & distribution overhead control a/c	Dr.	20,000	
	To Bank			20,000
	(Being selling & distribution overhead incurred)			
(7)	Finished stock control a/c	Dr.	2,00,000	
	To WIP control a/c			2,00,000
	(Being cost of production of finished goods)			
(8)	Sundry debtors a/c	Dr.	2,90,000	
	To sales a/c			2,90,000
	(Being credit sales made)			
(9)	Bank a/c	Dr.	69,000	
	To Sundry debtors			69,000
	(Being amount received from debtors)			
(10)	Sundry creditors a/c	Dr.	1,10,000	
	To Bank			1,10,000
	(Being amount paid to creditors)			

Problem 12. Show the journal entries for the following transactions in the integrated books of accounts

		Rs.
(*i*)	Cash purchase	18,000
(*ii*)	Credit purchase	2,45,000
(*iii*)	Materials issued to production	3,25,000
(*iv*)	Wages paid to workers	1,39,612
(*v*)	Finished goods transferred from production	6,29,775
(*vi*)	Administrative overhead allocable to production	78,900
(*vii*)	Works expenses outstanding	2,25,000
(*viii*)	Goods sold during the month	7,65,000

(ICWA, Inter, June 2000)

Solution:

JOURNAL ENTRIES

(1)	Stores ledger control a/c To cash a/c (Being materials purchased on cash)	Dr.	18,000	18,000
(2)	Stores ledger control a/c To sundry creditors a/c (Being materials purchased on credit)	Dr.	2,45,000	2,45,000
(3)	Work-in-progress control a/c To stores ledger control a/c (Being materials issued to production)	Dr.	3,25,000	3,25,000
(4)	Wages control a/c To cash (Being wages paid to workers)	Dr.	1,39,612	1,39,612
(5)	Finished stock ledger control a/c To work-in-progress control a/c (Being finished good transferred to finished stock ledger control a/c)	Dr.	6,29,775	6,29,775
(6)	Work-in-progress control a/c To Administration overhead control a/c (Being administrative overhead charged to production)	Dr.	78,900	78,900
(7)	Works expenses control a/c To works expenses a/c (Being works expenses outstanding)	Dr.	2,25,000	2,25,000
(8)	Sundry debtors a/c To sales a/c (Being goods sold on credit)	Dr.	7,65,000	7,65,000

Problem 13. Rex Enterprises operates an integral system of accounting. You are required to pass the journal entries for the following transactions that took place for the year ended 30th June 2002.

	Rs.
Raw materials purchased (50% on credit)	6,00,000
Materials issued to production	4,00,000
Wages paid to workers	2,00,000
Factory overheads incurred	80,000
Factory overheads charged to production	1,00,000
Selling and distribution overhead incurred	40,000
Finished goods at cost	5,00,000
Sales (50% credit)	7,50,000

(University of Delhi, B.Com. (Hons.), 2002)

Solution:

JOURNAL ENTRIES

(1)	Stores ledger control a/c To bank a/c To sundry creditors (Being raw materials purchased)	Dr.	6,00,000	 3,00,000 3,00,000
(2)	WIP control a/c To stores ledger control a/c (Being materials issued)	Dr.	4,00,000	 4,00,000
(3)	Wages control a/c To Bank a/c (Being payment of wages)	Dr.	2,00,000	 2,00,000
(4)	Factory overhead control a/c To bank (Being factory overhead incurred)	Dr.	80,000	 80,000
(5)	WIP control a/c To factory overhead control a/c (Being overhead charged to production)	Dr.	1,00,000	 1,00,000
(6)	S & D overhead control a/c To bank (Being S & D overhead incurred)	Dr.	40,000	 40,000
(7)	Finished stock control a/c To WIP control a/c (Being finished goods at cost)	Dr.	5,00,000	 5,00,000
(8)	Bank a/c Sundry debtors a/c To sales a/c (Being sale of goods)	Dr.	3,75,000 3,75,000	 7,50,000

Problem 14. BPR Limited keeps books on integrated accounting system. The following balances appear in the books as on April 1, 2002

	Dr. (Rs.)	Cr. (Rs.)
Stores control a/c	40,950	–
Work-in-progress a/c	38,675	–
Finished goods a/c	52,325	–
Bank a/c	–	22,750
Creditors a/c	–	18,200
Fixed assets a/c	1,47,875	–
Debtors a/c	27,300	–
Share capital a/c	–	1,82,000
Provision for depreciation a/c	–	11,375
Provision for doubtful debts a/c	–	3,725
Factory overheads o/s a/c	–	6,250
Prepaid administration overheads a/c	9,975	–
Profit & loss a/c	–	72,800
	3,17,100	3,17,100

The transactions for the year ended March 31, 2003, were as given below:

Direct wages	1,97,925
Indirect wages	11,375
Purchase of materials (on credit)	2,27,500
Materials issued to production	2,50,250
Materials issued for repairs	4,550
Goods finished during the year (at cost)	4,89,125
Credit sales	6,82,500
Cost of goods sold	5,00,500
Production overhead absorbed	1,09,200
Production overheads o/s at the end of year	7,775
Administration overhead paid during the year	27,300
Selling overheads incurred	31,850
Payments to creditors	2,29,775
Payment received from debtors	6,59,750
Depreciation of machinery	14,789
Administration overheads o/s at the end of the year	2,225
Provision for doubtful debts at the end of the year	4,590

Write up accounts in the integrated ledger of BPR Limited and prepare a trial balance.

(CA PE II, Group II, November 2003)

Solution:

STORES CONTROL ACCOUNT

To balance b/d	40,950	By WIP A/c	2,50,250
To creditors a/c	2,27,500	By production overhead a/c	4,550
		By Balance c/d	13,650
	2,68,450		2,68,450

ADMINISTRATION OVERHEAD ACCOUNT

To pre-paid administration overhead a/c	9,975	By Finished goods a/c	39,500
To bank	27,300		
To Administration o/s	2,225		
	39,500		39,500

COST OF SALES

To finished goods a/c	5,00,500	By sales a/c	5,32,350
To selling overhead	31,850		
	5,32,350		5,32,350

SALES ACCOUNT

To cost of sales	5,32,350	By Debtors a/c	6,82,500
To profit & loss a/c	1,50,150		
	6,82,500		6,82,500

CHAPTER 9

FACTORY OVERHEAD O/S ACCOUNT

To Bank a/c	6,250	By Balance b/d	6,250
To balance c/d	7,775	By production overhead	7,775
	14,025		14,025

PREPAID ADMINISTRATION OVERHEAD ACCOUNT

To Balance b/d	9,975	By Administration overhead account	9,975

PROVISION FOR DEPRECIATION ACCOUNT

To Balance c/d	26,164	By Balance b/d	11,375
		By production overhead account	14,789
	26,164		26,164

PROVISION FOR DOUBTFUL DEBTS ACCOUNT

To Balance c/d	4,590	By Balance b/d	3,725
		By P & L a/c	865
	4,590		4,590

WAGES CONTROL ACCOUNT

To Bank a/c	1,97,925	By work-in-progress	1,97,925
To Bank a/c	11,375	By production overhead a/c	11,375
	2,09,300		2,09,300

WORK-IN-PROGRESS ACCOUNT

To Balance b/d	38,675	By Finished goods a/c	4,89,125
To wages control a/c	1,97,925	By Balance c/d	1,06,925
To stores control a/c	2,50,250		
To production overheads a/c	1,09,200		
	5,96,050		5,96,050

PRODUCTION OVERHEADS ACCOUNT

To wages control a/c	11,375	By WIP a/c	1,09,200
To stores control a/c	4,550	By profit & loss a/c	
To bank	84,750	(under-absorbed overheads	
(91,000 – 6,250)		written-off)	14,039
To production overhead o/s	7,775		
To provision for depreciation	14,789		
	1,23,239		1,23,239

FINISHED GOODS ACCOUNT

To Balance b/d	52,325	By cost of sales a/c	5,00,500
To work-in-progress a/c	4,89,125	By balance c/d	80,450
To Adm. overhead a/c	39,500		
	5,80,950		5,80,950

PROFIT AND LOSS ACCOUNT

To Provision for doubtful debts	865	By balance b/d	72,800
To production overhead	14,039	By sales a/c	1,50,150
To Balance c/d	2,08,046		
	2,22,950		2,22,950

DEBTORS ACCOUNT

To Balance b/d	27,300	By Bank a/c	6,59,750
To sales a/c	6,82,500	By balance c/d	50,050
	7,09,800		7,09,800

CREDITORS ACCOUNT

To Bank a/c	2,29,775	By balance b/d	18,200
To Balance c/d	15,925	By stores control a/c	2,27,500
	2,45,700		2,45,700

FIXED ASSETS ACCOUNT

To Balance b/d	1,47,875	By Balance c/d	1,47,875

BANK ACCOUNT

To debtors	6,59,750	By Balance b/d	22,750
		By Direct wages	1,97,925
		By Indirect wages	11,375
		By Production overhead (84,750 + 6,250)	91,000
		By Administrative overhead account	27,300
		By Selling overhead account	31,850
		By creditors account	2,29,775
		By Balance c/d	47,775
	6,59,750		6,59,750

TRIAL BALANCE AS ON 31-3-2003

	Dr.	*Cr.*
Stores control a/c	13,650	
Work-in-progress a/c	1,06,925	
Finished goods a/c	80,450	
Bank a/c	47,775	
Creditors a/c		15,925
Fixed assets a/c	1,47,875	
Debtors a/c	50,050	
Share capital a/c	—	1,82,000
Provision for depreciation a/c	—	26,164
Profit & loss a/c	—	2,08,046
Production overhead o/s a/c	—	7,775
O/S administrative overhead a/c	—	2,225
Provision for doubtful debts a/c	—	4,590
	4,46,725	4,46,725

CHAPTER 9

COST RECORD RULES

INTRODUCTION

The Government of India is empowered under Section 209 of the Companies Act 1956 to prescribe that certain classes of companies, engaged in production, processing, manufacturing or mining activities should keep records showing particulars of utilisation of materials or labour or other costs. The Central government is also empowered to frame rules in respect of such industries to maintain cost accounting records. These rules are framed to ascertain the cost of production in the various units of the selected industries. Accordingly, rules have been framed in the case of 28 products/industries which are consumer oriented, earners of high margin of profit etc., viz, cement, cycles, rubber tyres and tubes, soda, room air conditioners, refrigerators, automobiles, batteries, electric lamps, electric fans, electric motors, motor vehicles, tractors, aluminium, bulk drugs. Sugar, infant milk food, industries alcohol, jute goods, paper, rayon, dyes, soda ash, polyester, nylon, cotton textiles, dry cell batteries and sulphuric acid. Accordingly to these rules proper accounts should be maintained in respect of the various items of cost. If a company defaults in the maintenance of cost records, the company and every officer of the company who is in default, is punishable with fine which may extend to Rs. 500 and if the default is a continuing one, with a further fine upto Rs. 50 for each day of default. It may be mentioned that separate rules have been framed for each industry. However, there are some similar requirements for all the industries in respect of each element of cost. They are briefly stated below:

(1) Raw materials and purchased components:

(*a*) Adequate records should be kept showing the receipts, issues and balances, both in quantities and values of raw materials and components required for the manufacture of the product.

(*b*) The basis on which the value of purchases and issues have been calculated should be indicated in the records and the basis adopted should be followed consistently throughout the relevant period.

(*c*) The value should include all direct charges such as freight, insurance etc. and also the wastage, whether in storage, transmit or for other reasons.

(*d*) If the consumption value is determined on any basis other than actual, the method adopted should be explained and a reconciliation with actuals and the treatment of variations should be dealt with in the records.

(2) Manufactured components: Detailed records showing the cost of manufactured components should be maintained in the form prescribed in the rules. The records should also indicate quantities, values and wastages as in the case of purchased components. Quantities manufactured, rejected and accepted together with the losses on account of rejection should also be mentioned in the records.

(*a*) Adequate records should be kept showing the receipts, issues and balances, both in quantities and value of various items of stores and spares required for repairs, maintenance or manufacture. The value should include all direct charges upto the works such as freight, insurance etc.

(*b*) The value of issues and balances may be recorded monthly or at such shorter intervals as may be decided by the company or may be kept in the form of control accounts. The balances being reconciled at least once a year with the values of the quantity shown by the quantity accounts for each items of stores.

(c) The value of stores consumed should be charged to relevant heads like manufacturing, repairs to plant and machinery, repairs to buildings, maintenance of vehicles, capital items etc.

(d) The value of stores charged to manufacture should further be allocated to the different departments or manufacturing units or cost centres and products, if any, manufactured therein.

(e) Stores issued for capital works like additions to plant and machinery and other assets should be shown separately under the capital heads.

(f) Any wastage, whether in stores or transit or for other causes should be recorded separately and their treatment explained.

(g) Stores should be valued consistently during the relevant period on a reasonable basis and the method adopted should be recorded.

(3) Labour: With regard to labour, proper and systematic account should be maintained in respect of:

(a) Attendance of workers and other staff.

(b) All payments made for overtime work.

(c) Cost of wages and salaries relating to various manufacturing departments or units in the factory being the amount payable and allocated to the different departments or units.

(d) Idle time, giving reasons for such idle time and the method of its treatment in calculating cost of products.

(e) The extent of wages and salaries capitalised as additions to plant and machinery and other assets.

(f) Where piece rate payment system is applicable, the records should be so maintained as to enable ascertainment of wages payable to such workers.

(4) Overheads:

(a) The overhead expenses should be classified as works overhead, administration overhead and selling and distribution overhead.

(b) Adequate records should be kept showing the accounts comprising the overhead expenses and the details of appointment of overhead expenses to the various departments, manufacturing units and cost centres. The details of such expenses should be collected from the financial accounts.

(c) Certain jobs carried out during the year should bear their appropriate share of overhead expenses. The basis of allocation of such expenses should be indicated in relevant records.

(d) The works overhead should include, among other items, indirect wages and salaries relating to works, indirect materials consumed, staff and labour welfare expenses and expenditure on repairs and maintenance.

(e) The method of apporting overheads to various departments or manufacturing units or cost centres should be followed consistently and explained in the records.

(f) It overheads are recovered on a basis other than actuals, the method of reconciliation with actuals and treatment of variances should be disclosed in the cost records.

(*g*) The shares of selling and distribution expenses, if any, applicable to products sold should be shown separately in the cost records stating clearly the basis of apportionment.

(5) Service:

(*a*) Proper records should be maintained so that the cost of services such as power, water, electricity and steam produced and utilised in manufacturing operations can be ascertained.

(*b*) The power cost should include cost of units consumed, fixed charges and duties.

(*c*) If power, water or steam is generated by the company, the records maintained should show the different items making up the cost of such power, water or steam produced and consumed.

(*d*) The allocation of the consumption of services by different departments or manufacturing units should be shown in the records. If actual allocation is not possible, the basis of allocation should be indicated in the cost records.

(6) Depreciation:

(*a*) Adequate records should be maintained in respect of fixed assets showing (*i*) the cost of each item of asset (*ii*) date of acquisition (*iii*) rate of depreciation and (*iv*) the amount of depreciation provided.

(*b*) Depreciation chargeable to the different departments or cost centres should not be less than, the amount chargeable under Section 205(2) of the companies Act, 1956 and should relate to the plant and machinery and other assets utilised in the department or cost centre. The method of depreciation, once adopted should be applied consistently.

(7) Production and Sale:

(*a*) Adequate records should be maintained in respect of production, showing the receipts, issues and balances of the goods produced by the company. The value should be based on the cost of production of the items concerned. The value of the issues and balances may be recorded monthly or at such shorter intervals as may be decided.

(8) Inventories:

(*a*) Inventories may be classified as (*i*) raw materials, (*ii*) stores and spares parts, (*iii*) work-in-progress and (*iv*) finished goods. The records required for raw materials, stores and spares and finished goods production have already been dealt with.

(*b*) Proper records should be maintained to show the quantities and value of work-in-progress at the end of period for which the costs are made up. It should represent the cost incurred upto the relevant stage of manufacture.

(*c*) Stock verification records should be maintained in respect of all raw materials, components, stores and spares including tools, etc. The method of dealing with the discrepancies arising out of such verification should be indicated in the records.

Questions

1. From the following transaction pass journal entries under integral accounting system.

	Rs.	
Purchases	2,00,000	(30% in cash)
Wages	1,00,000	(25% indirect)
Stores issued to production	3,00,000	
Factory overhead paid	60,000	
Factory overhead allocated	80,000	
Administration expenses incurred	15,000	
Outstanding administration expenses	2,000	
Administration expenses allocated	20,000	
Finished goods transferred	4,55,000	

2. From the following information write up the control accounts in the general ledger of Sujatha Enterprises where cost and financial accounts are integrated and prepare a trial balance as on 31st March 2010:

Ledger balances as on 1 April 2009:

Share capital		1,00,000
Reserves		25,000
Sundry debtors		20,000
Plant and machinery		1,25,000
Sundry creditors		30,000
Bank overdraft		40,000
Raw materials		50,000

Transactions during the year:

Raw materials purchases		80,000
Raw materials issued to production		1,00,000
Raw materials on hand 31.3.1993		26,000
Direct wages	– Incurred	95,000
	– Charged to production	93,000
Manufacturing expenses	– Incurred	87,500
	– Charged to production	93,000
Selling & distribution expenses		10,000
Finished stock	– Production (at cost)	1,91,000
	– Sales (at selling price)	86,000
	– Closing stock	NIL
Payments to creditors		85,000
Receipts from debtors		3,00,000

[**Ans.** Total of trial balance Rs. 4,59,000]

3. Give journal entries for the following transactions under integrated accounting

(*i*) Materials purchased on credit:

(*ii*) Payment to creditors

(*iii*) Issue of direct materials for production

CHAPTER 9

(*iv*) Payment of wages

(*v*) Charging wages for production *(University of Delhi, B.Com. (Hons.), 2003)*

4. Shivalika Enterprises operates an integral system of accounting. You are required to pass journal entries for any four of the following transactions that took place for the year ended 31st March 2003:

(*i*) Raw materials purchased (50% on credit)	Rs.	6,00,000
(*ii*) Materials issued to production	Rs.	4,00,000
(*iii*) Factory overheads incurred	Rs.	80,000
(*iv*) Sales (50% credit)	Rs.	7,50,000
(*v*) Receipts from debtors	Rs.	2,00,000
(*vi*) Payments to creditors	Rs.	2,00,000

(University of Delhi, B.Com. (Hons.), 1998)

❑❑❑

Part III

COST CONTROL TECHNIQUES

CHAPTER 10

MARGINAL COSTING

ABSORPTION COSTING

Absorption costing is a conventional technique of ascertaining the cost of production. It is, therefore, sometimes referred to as 'orthodox costing'. It is a technique of ascertaining cost of goods or services manufactured under which both variable and fixed cost are taken into consideration. As under this technique all costs-fixed and variable-are taken into account, it is also known as 'full costing' technique. The total or 'full' cost is classified on the functional basis into production cost, administration cost, selling and distribution cost. Under this technique, the total cost per unit remains constant only when the level of output remains same from time to time. But in today's dynamic world the level of activity differs from time to time and so does the cost. The cost of production may be Rs. 18 today and Rs. 20 next week. This change in the cost of production is on account of change in the volume of output and the way in which fixed cost tend to behave. The differences in the cost of production from time to time often poses a problem to the management in decision-making process. Hence, the marginal costing technique is used to enable management in carrying out its day-to-day functions of planning, decision-making and controlling. However, the proponents of absorption costing technique argue that both fixed and variable costs must be charged to the cost of production in order to meet the requirements of generally accepted accounting principles and income tax reporting.

The total cost under this technique is classified into two types, viz., (*i*) production cost and (*ii*) period cost. These costs are further classified into various types of costs. These costs are shown in the following chart:

COST CLASSIFICATION UNDER ABSORPTION COSTING

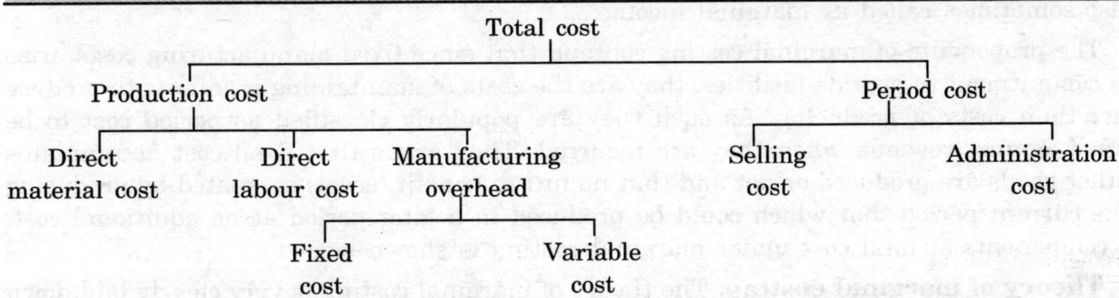

Fig. 10.1 *Chart showing total components of total cost under absorption costing.*

MARGINAL COSTING

Before dealing with the marginal costing technique, it would be appropriate at this stage to know the meaning and definition of 'marginal cost'.

Marginal Cost

From the Economist's point of view, the cost incurred in producing an additional unit of product is known as marginal cost. However, from the cost accounting point of view marginal cost applies to the total cost obtained by adding prime cost and variable cost. In other words all costs other than fixed costs are the marginal cost.

The ICMA Terminology defines marginal cost as 'the amount at any given volume of output by which aggregate costs are changed if the volume of output is increased or decreased by one unit. In practice, this is measured by the total variable costs attributable to one unit. In this context a unit may be a single article, a batch of articles, an order, a stage of production capacity, a man hour, a process or a department. It relates to change in output in the particular circumstance under consideration.

Marginal Costing

This is one of the technique of ascertaining cost of production of goods or services manufactured. It is not a method of costing but it could be used in conjunction with any method of costing such as job or process costing. This technique can also be used along with other techniques of costing such as standard costing and budgetary control. This technique is also known by other names such as direct costing, variable costing, attributable costing, out-of-pocket costing and so on. Whereas in USA it is termed as 'direct costing', in UK it is referred to as 'marginal costing'.

Definition of marginal costing: According to NAA Bulletin on Direct Costing Research Series 23, "direct costing should be defined as seggregation of manufacturing cost between those which are fixed and those which vary directly with volume. Only the prime cost plus variable factory overheads are used to value inventory and cost of sales. The remaining factory expenses are charged off currently to profit and loss account".

In simple words, marginal costing may be defined as "the ascertainment of marginal costs and the effect on profit of changes in volume by differentiating between fixed cost and variable cost".

It is a technique whereby marginal costs of cost units are ascertained. Only variable costs are charged to cost units. The fixed costs attributable to a relevant period being written off in full against the contribution for that period.

The term contribution represents excess of sales over total variable costs. Contribution is also sometimes called as marginal income.

The proponents of marginal costing contend that since fixed manufacturing costs arise from commitment to provide facilities, they are the costs of maintaining readiness to produce rather than costs of producing. As such they are popularly classified as period cost to be charged against revenue when they are incurred. They argue that fixed cost accumulates whether goods are produced or not and that no future benefit (asset) is created by producing in the current period that which could be produced in a later period at no additional cost. The components of total cost under marginal costing is shown below:

Theory of marginal costing: The theory of marginal costing is very clearly laid down by the ICMA, London, in the booklet on *A Report on Marginal Costing*. It can be stated thus; in relation to a given volume of output, additional output can normally be obtained at less than proportionate cost because within limits the aggregate of certain items of cost will tend to remain fixed and only the aggregate of the remainder will tend to rise proportionately with the increase in output. Conversely, a decrease in the volume of output will normally be accompanied by a less than proportionate fall in the aggregate cost. In simple words, when the volume of output increases the cost per unit will decrease. Similarly, if the volume of output is reduced, the cost per unit will increase. For example, if a company produces 100 units at a cost of Rs. 200, and by increasing the output by 2 units, the cost goes up to Rs. 204. The marginal cost of the increased output is Rs. 4.

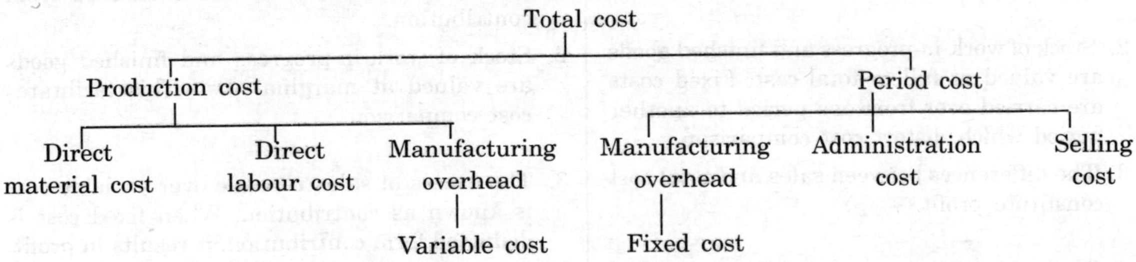

Fig. 10.2 *Chart showing components of total cost under marginal costing.*

Features of marginal costing: The features of marginal costing may be summarised under the following points:

(*a*) It is a technique of costing used to ascertain the marginal cost and to know impact of variable cost on the volume of output.

(*b*) All costs are classified on the basis of variability into fixed and variable cost. Semi-variable costs are segregated into fixed and variable costs.

(*c*) Variable costs alone are charged to production. Fixed costs are recovered from contribution.

(*d*) Stock of work-in-progress and finished goods are valued on the basis of marginal cost.

(*e*) Selling price is based on marginal cost plus the contribution.

(*f*) Profit is calculated not in the usual manner. When fixed cost is deducted from sales it gives rise to contribution. When fixed cost is deducted from contribution it results in profit.

(*g*) Break-even analysis and cost-volume-profit analysis (discussed in the next chapter) are integral part of this technique.

(*h*) The profitability of product or department is based on contribution made available by each department of product.

SIMILARITIES AND DISSIMILARITIES BETWEEN ABSORPTION AND MARGINAL COSTING

Similarities

1. Both the techniques agree that fixed and variable administration and selling expenses are period cost.
2. Both agree that the variable manufacturing costs are product cost.
3. Both agree that marginal costing presents the data for internal use.

Dissimilarities

Absorption Costing	*Marginal Costing*
1. All costs are charged to the cost of production.	1. Only variable cost is charged to cost of production. Fixed costs are recovered from contribution.
2. Stock of work-in-progress and finished goods are valued at full or total cost. Fixed costs are carried over from one period to another period which distort cost comparison.	2. Stock of work-in-progress and finished goods are valued at marginal cost. This facilitates cost comparison.
3. The differences between sales and total cost constitute profit.	3. The excess of sales revenue over variable cost is known as contribution. When fixed cost is deducted form contribution, it results in profit.
4. The apportionment of fixed costs on an arbitrary basis gives rise to under or over absorption of overheads.	4. As only variable costs are charged to products, it does not give rise to over or under absorption of overheads.
5. Costs are classified according to functional basis such as production cost, administration cost, selling and distribution cost.	5. Costs are classified according to variability.

Cost Presentation under Absorption and Marginal Costing

The difference between absorption and marginal costing in respect of cost ascertainment and cost presentation can be better understood by considering the following problems:

Problem 1. *(Without stock balance):* Given the following data:

	Rs.
Direct material cost per unit	5.00
Direct labour cost per unit	9.00
Variable manufacturing overhead per unit	0.60
Total fixed manufacturing overhead per year	92,000
Number of units produced per year	10,000

Calculate the cost of production per unit under (1) absorption costing and (2) marginal costing.

Solution:

STATEMENT OF COST UNDER ABSORPTION AND MARGINAL COSTING

	Absorption costing	*Marginal costing*
Direct material cost	5.00	5.00
Direct labour cost	9.00	9.00
Variable manufacturing cost	0.60	0.60
Fixed mfg. cost $\frac{92,000}{10,000}$	9.20	–
Cost of production per unit	23.80	14.60

Note: Under marginal costing fixed manufacturing cost of Rs. 92,000 is treated as period cost.

Problem 2. *(When closing stock balance is given):* Considering the date given in the above problem and the following additional information. Calculate the profit under (*i*) absorption costing and (*ii*) marginal costing.

Opening stock of finished goods	NIL
Units produced	10,000
Units sold	8,000
Sale price per unit	Rs. 35
Variable selling and administration expenses	Rs. 1.20 per unit sold
Fixed administration and selling expenses	Rs. 58,000

Solution:

STATEMENT OF PROFIT UNDER ABSORPTION AND MARGINAL COSTING

Absorption costing			*Marginal costing*		
Sales 8,000 × 35		2,80,000	Sales 8,000 × 35		2,80,000
Opening stock	NIL		Opening stock	NIL	
Add: Cost of goods produced			*Add:* Cost of goods produced		
10,000 × 23.80	2,38,000		(10,000 × 14.60)	1,46,000	
(as per the 1st problem)				1,46,000	
	2,38,000		*Less:* Closing stock		
Less: Closing stock			2,000 × 14.60		
2,000 × 23.80	47,600	1,90,400		29,200	1,16,800
		8,960			1,63,200
Less: Selling and distribution expenses:			*Less:* Variable selling and		
Fixed	58,000		administration expenses		9,600
Variable	9,600		contribution		1,53,600
8,000 × 1.20		67,600	*Less:* Fixed cost:		
Profit		22,000	Mfg. Overhead	92,000	
			Fixed selling &		
			Admn. exp.	58,000	1,50,000
			Profit		3,600

The difference in the profit of Rs. 18,400 (22,000 – 3,600) is because of the different accounting treatment of fixed manufacturing costs under marginal costing and absorption costing. Under the marginal costing, the fixed manufacturing costs are accounted for as a period cost and the entire fixed manufacturing cost of Rs. 92,000 is deducted in calculating profit. Under absorption costing, the fixed manufacturing costs are treated as a product cost and charged to 10,000 units produced during the year. This means the closing stock of finished goods of 2,000 units will 'absorb' a part of the fixed manufacturing cost. The valuation of closing stock of finished goods under absorption costing is shown below:

Variable manufacturing cost	2,000 × 14.60	=	29,200
Fixed manufacturing cost	2,000 × 9.20	=	18,400
Total manufacturing cost	2,000 units @ 23.80	=	47,600

The fixed manufacturing cost of Rs. 18,400 assigned to closing finished goods is carried forward in the balance sheet until these, 2,000 units are sold in the next accounting year. In fact, the difference in the profit under the two methods shown earlier is this amount of

CHAPTER 10

Rs. 18,400. When this 2,000 units are sold, the fixed cost of Rs. 18,400 is deducted in the profit statement. In other words, under absorption costing, the total fixed manufacturing cost of Rs. 92,000 is divided between the unsold finished goods (18,400) and finished goods that are sold (73,600).

Thus, in calculating net profit under absorption costing only Rs. 73,600 of fixed manufacturing costs are included in cost of goods sold, whereas, under marginal costing the entire fixed manufacturing cost of Rs. 92,000 is included as part of the fixed cost in the profit statement.

Problem 3. *(Preparation of cost statement under marginal costing):* Novelty company submitted the following particulars for the quarter ended 30th September 2008. Find out the profit by using marginal cost technique.

Opening stock	400 units
Closing stock	500 units
Sales 3,000 units @ Rs. 65 each	
Fixed overheads	Rs. 20,000
Direct materials	Rs. 40,000
Direct labour	Rs. 26,000
Variable overheads	Rs. 27,000

Solution:

STATEMENT OF COST UNDER MARGINAL COSTING

Direct materials		40,000
Direct labour		26,000
Variable overheads		27,000
	Marginal cost	93,000
Add: Opening stock		12,000
		1,05,000
Less: Closing stock		15,000
	Marginal cost of goods sold	90,000

STATEMENT OF PROFIT UNDER MARGINAL COSTING

Sales		1,95,000
Less: Marginal cost of goods sold		90,000
		1,05,000
Less: Fixed cost		20,000
	Profit	85,000

Working Notes

(1) Finding out the units produced

= Sales + Closing stock – Opening stock

= 3,000 + 500 – 400

= 3100 units

(2) Valuation of stock

(*a*) In marginal costing stocks are valued by taking variable costs only

(*b*) The variable cost per unit remains the same. Therefore, the opening and closing stocks are valued at the same rate.

(3) Finding out variable cost per unit

$$= \frac{\text{Marginal cost}}{\text{Total output}}$$

$$= \frac{93{,}000}{3{,}100} = \text{Rs. } 30 \text{ per unit}$$

(4) Valuation of opening stock

= No. of units × variable cost per unit

= 400 × 30 = Rs. 12,000

(5) Valuation of closing stock

= 500 × 30 = Rs. 15,000

Problem 4. From the following particulars, prepare cost statement under: (*a*) Absorption costing and (*b*) marginal costing.

Number of units produced :	20,000
Cost per unit	*Rs.*
Material	10
Productive labour	6
Factory overheads	4 (50% fixed)
Administration overhead	3 (60% fixed)
Selling overhead	2 (100% variable on units sold)
Number of units sold	19,000
Selling price	Rs. 40 per unit

Solution:

COST STATEMENT UNDER ABSORPTION COSTING METHOD

		Total 20,000 units	*Cost per unit*
Materials		2,00,000	10
Productive labour		1,20,000	6
	Prime cost	3,20,000	16
Factory overhead		80,000	4
	Works cost	4,00,000	20
Administrative overhead		60,000	3
	Office cost	4,60,000	23
Less: Closing stock of finished goods = 20,000 units – 19,000 units = 1,000 units × Rs. 23		23,000	–
	Cost of goods sold	4,37,000	23
Selling overhead 19,000 × 2		38,000	2
	Cost of sales	4,75,000	25
	Profit	2,85,000	15
	Sales	7,60,000	40

CHAPTER 10

COST STATEMENT UNDER MARGINAL COSTING METHOD

	Total	Per unit
Materials cost	2,00,000	10
Productive labour	1,20,000	6
Prime cost	3,20,000	16
Factory overheads $4 \times \frac{50}{100} = \text{Rs. } 2$	40,000	2
Administrative overheads $3 \times \frac{40}{100} = \text{Rs. } 1.20$	24,000	1.20
	3,84,000	19.20
Less: Closing stock of finished goods = (20,000 units – 19,000 units) = 1,000 × Rs. 19.20	19,200	–
	3,64,800	19.20
Selling overhead 19,000 × 2	38,000	2.00
Total variable cost	4,02,800	21.20
Sales	7,60,000	40.00
Contribution (S – V)	3,57,200	18.80
Less: Total fixed cost (38,000 + 34,200)	72,200	3.80
Profit	2,85,000	15.00

COST BEHAVIOUR AND ITS IMPACT OVER MARGINAL COSTING

The concept of marginal costing is based on the behaviour of the costs with volume of output. The assumption is that some components of costs change directly in proportion to the change in volume of activity and some remain unchanged with the change in the activity level. The pattern of cost behaviour is the result of an interplay of many forces which cause some costs to fluctuate and others to remain constant. The most important of these forces are the following:

(*a*) Volume
(*b*) Inherent nature of the cost
(*c*) Capacity
(*d*) Managerial policies
(*e*) Effective control
(*f*) Prices of input factors
(*g*) Strikes and lockouts
(*h*) Weather and economic conditions

Accordingly, the costs can be classified on the basis of their behaviour into three types.

They are as follows:

1. Fixed Cost

A fixed cost is defined as "a cost which accrues in relation to the passage of time and which within certain output and turnover limits tend to be unaffected by fluctuations in the level of activity, *i.e.,* output or turnover.

The key points in this definition are:

1. Fixed costs are time related.
2. It remains fixed within the limits of output or turnover.
3. It is unaffected by changes in the level of activity.
4. Though it is called fixed cost in the short run, in the long it may vary as for example when the policy of management is to expand after 10 years, fixed cost will increase no matter what volume of output is produced. Therefore, fixed cost is sometimes called as 'policy cost'.

Fixed costs are of two types. They are as follows:

(*a*) **Committed cost:** These costs are related to the provision of a capacity to do business. The amount of committed costs is fixed by decisions which were made in the past and is not subject to management control in the present on a short run basis. Since there is no direct relationship between committed costs and either the planned or actual utilisation of existing facilities, the amount will remain constant over the whole range of operating activity.

(*b*) **Programmed or managed cost:** These costs are related to the utilisation of the capacity provided:

Fixed cost can be shown graphically as under:

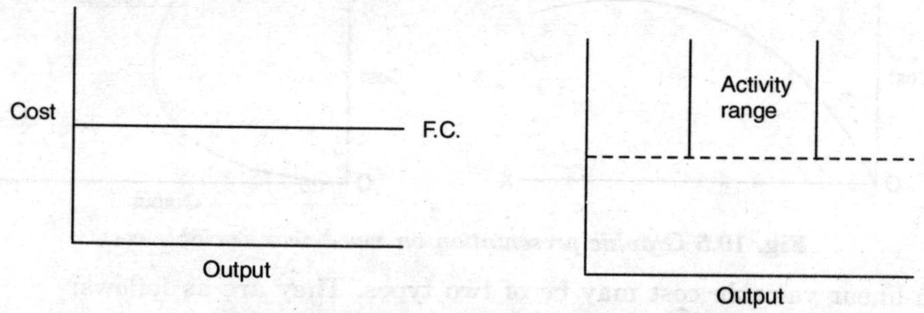

Fig. 10.3 *Graphic presentation of fixed cost.*

Fixed cost can be expressed algebraically thus:

$$\text{Cost} = a$$

where '*a*' is constant, '*V*' the volume of output does not appear in this equation to indicate the change in activity is deemed not to affect the fixed cost.

2. Variable Cost

It is defined as a cost which, in the aggregate, tends to vary in direct proportion to changes in the volume of output or turnover.

Variable costs are of two types. They are as follows:

(*a*) **Linear variable cost or Engineered cost:** When the relationship between variable cost and output can be shown as a straight line on a graph, they termed as linear variable cost. This is shown graphically below:

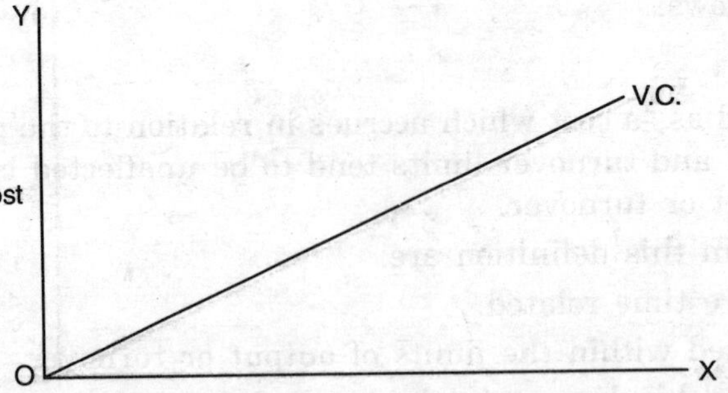

Fig. 10.4 *Graph showing linear variable cost.*

A variable cost is called as engineered cost because an optimum relationship can be carefully determined by work measurement technique between input and output. Direct material cost and direct labour cost are good examples of engineered cost.

For calculation and analysis it is more convenient to express the linear relationship algebraically, thus

$$\text{Cost} = bx$$

where x is the volume of output in units, b is a constant representing the variable cost per unit.

(*b*) **Non-linear or Curvilinear variable cost:** When the relationship between variable cost and output can be shown as a curved line on a graph, it is said to be curvilinear. This is shown graphically below:

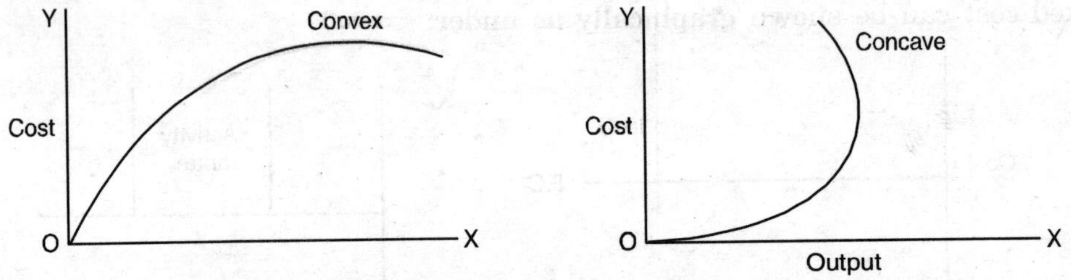

Fig. 10.5 *Graphic presentation on non-linear variable cost.*

The non-linear variable cost may be of two types. They are as follows:

(a) Convex-linear variable cost: It is a cost where each extra unit of output causes less than proportionate increase in cost.

(b) Concave-linear variable cost: It is a cost where each extra unit of output causes a more than proportionate increase in cost. Differential piece rate system of wage payment is a good example for this type of cost.

Curvilinear variable cost — the parabola: When the slope of the cost function changes uniformly with changes in output, the curve is known as a parabola and is algebraically expressed thus:

Cost: $bx + cx^2 + dx^3 ....px^4$

where x is as defined previously and $b, c, d,$ ...., p are constant.

Problem 5. *(Calculation of variable cost):* Using the equation cost $bx + cx^2 + dx^3$, where $b = 8$, $c = 0.5$ and $d = 0.03$, calculate

(*i*) variable cost when production is 10 units.

(*ii*) variable cost when production is 15 units.

Is the function convex or concave?

Solution:

Cost $= bx + cx^2 + dx^3$

$= 8 \times 10 + 0.5 \times 10^2 + 0.03 \times 10^3$

$=$ Rs. 160

$= 8 \times 15 + 0.5 \times 15^2 + 0.03 \times 15^3$

$=$ Rs. 333.75

From the above it is seen that the increase in activity from 10 to 15 units results in more than a doubling of variable cost. This shows that there is a more than proportionate increase in the unit cost of extra production. Hence the function is concave.

3. Semi-variable Cost

It is defined as a cost containing both fixed and variable elements, which is, therefore, partly affected by fluctuation in the volume of output or turnover. It is also sometimes called as a stepped cost or hybrid cost.

Semi-variable cost be shown graphically thus:

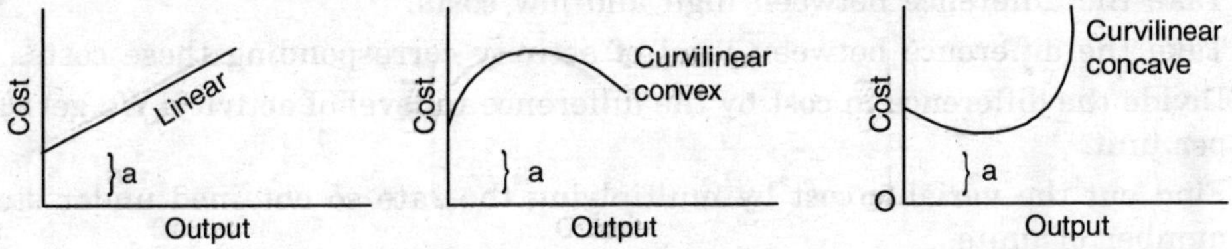

Fig. 10.6 *Graphic presentation of semi-variable cost.*

In the above figures *'a'* represents the fixed element of the cost.

Semi-variable cost can be algebraically expressed by combining the previous equations of fixed and variable cost.

Linear semi-variable cost

Cost $= a + bx$

Curvilinear semi-variable cost

Cost $= a + bx + cx^2 + dx^3 \text{ .... } + px^4$

Problem 6. *(Calculation of total cost):* The analysis of repairs department of Jain & Co. shows that there is a fixed cost of Rs. 500 per month and a variable cost related to machine hours amounting to Rs. 2.25 per machine hour.

What is the expected cost for a month when the budgeted activity level is (*i*) 1,500 machine hours and (*ii*) 1,800 machine hours?

Solution:

Total cost $= a + bx$

(*i*) $500 \times 2.25 \; (1{,}500)$

$=$ Rs. 3,875

(*ii*) $500 \times 2.25 \; (1{,}800)$

$=$ Rs. 4,550

As, under marginal costing technique, only variable cost alone is charged to the product it becomes necessary to segregate semi-variable cost into fixed and variable cost. Besides this, the segregation of semi-variable cost into fixed and variable has the following uses:

1. It facilitates budgeting the expenses for various levels of production.
2. It helps in exercising control over variable cost once semi-variable cost is segregated into fixed and variable.

3. It helps in fixing prices especially during the period of depression.
4. It helps in making useful decisions.

Method of Segregating Semi-variable Cost

1. High-low method: This method is based on the analysis of past records of expenses. This method takes into account only the highest and lowest values contained in the data in order to determine the rate of cost change and thereafter variable cost. The variable cost is then deducted from total cost to get fixed cost.

Limitations:

(*i*) It is not based on all the items of the given data.

(*ii*) It assumes that the variable portion of semi-variable cost has linear relationship. In other words, the variable portion of cost per unit is constant which is not always true.

Steps involved:

1. Take the difference between high and low costs.
2. Take the difference between level of activity corresponding these costs.
3. Divide the difference in cost by the difference in level of activity. We get variable cost per unit.
4. Find out the variable cost by multiplying the rate so obtained under step 3 by the number of units.
5. Deduct this variable cost from semi-variable cost to get fixed cost.

Problem 7. *(Segregation of semi-variable cost under high-low method):* From the following data segregate the semi-variable cost into fixed and variable cost under High-Low method.

Level of activity (Units)	*Cost (Rs.)*	*Level of activity (Units)*	*Cost (Rs.)*
36	800	72	950
48	700	75	1,170
49	970	76	1,020
55	790	82	1,200
57	1,050	86	1,060
59	900	92	1,100
65	880	94	1,310
66	1,020	97	1,250
67	1,180		

Solution:

High point	94	units at a cost of Rs.	1,310
Low point	48	units at a cost of Rs.	700
Difference	46		610

Variable cost per unit $= \frac{610}{46} = 13.26$ unit.

Semi-variable cost at 48 units	=	Rs.	700
Less: Variable cost 48 × 13.26	=	Rs.	636
Fixed cost	=	Rs.	64

In this process for each level of activity, fixed cost and variable cost can be found out.

2. Scatter graph technique: This is a graphic method of segregating semi-variable cost into fixed and variable cost. This methods is also known as regression line method. This method is also based on past data, but it takes into account all the data when compared to the high-low method. The following steps are involved under this method:

1. X-axis is taken to represent the output sales.
2. On the Y-axis is the cost represented.
3. Plot the points on the graph.
4. Draw a line through the points plotted, with the same number of points on each side of the line. In the process of drawing the line ignore any abnormal cost if there is any. The line so drawn is known as the line of best fit.
5. Extend the line so drawn above the Y-axis. The point where the line of best fit intersects the Y-axis is taken to be the amount of fixed cost.

One main limitation of this method is that there will be a change in fixed cost even if there is a slight bias in drawing the straight line.

Problem 8. *(Scatter graph method of segregating semi-variable cost into fixed and variable cost):* From the following data extracted from the books of Arun Food Products Ltd. Calculate the fixed cost using scatter graph method:

Output (units)	*Semi-variable cost (Rs.)*	*Output (units)*	*Semi-variable cost (Rs.)*
500	2,500	900	3,700
600	2,800	840	3,520
700	3,100	800	3,400
940	3,820	860	3,580
740	3,220	760	3,280
880	3,640	540	2,620

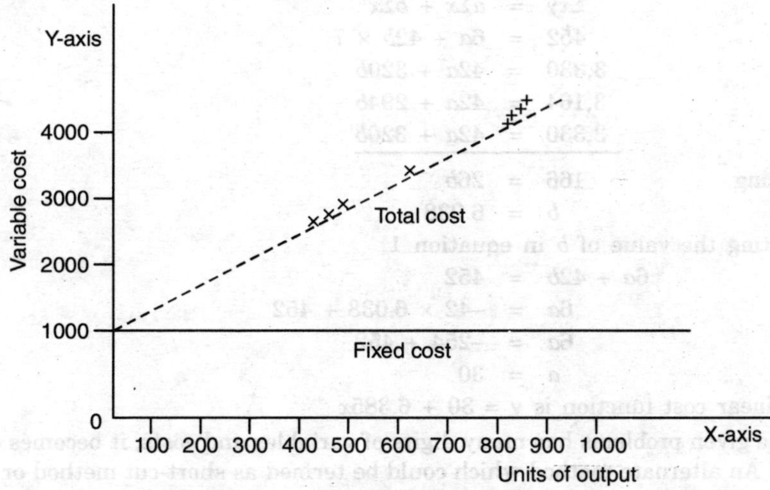

Fig. 10.7 *Graph I.*

Solution:

From the graph it is clear that fixed cost is Rs. 1,000.

3. Method of least squares: This is a statistical method used for establishing a line of best fit to data. The linear cost function can be represented by the following equation:

$$y = a + bx$$

where y = cost. Other symbols are as defined previously.

To find out the values of 'a' and 'b' which are constants, we have to solve two simultaneous equations:

$$\Sigma y = ax + b\Sigma x$$
$$\Sigma xy = a\Sigma x + b\Sigma x^2$$

where x = number of pairs of cost and activity figures.

Problem 9. *(Simultaneous equation method of segregating semi-variable cost):* From the following data obtain an equation to indicate the linear cost function under the method of least squares:

Unit x	*Cost y*
4	56
5	62
7	80
7	72
9	88
10	94

Solution:

x	y	xy	x^2
4	56	224	16
5	62	310	25
7	80	560	49
7	72	504	49
9	88	792	81
10	94	940	100
$\Sigma x = 42$	$\Sigma y = 452$	$\Sigma xy = 3,330$	$\Sigma x^2 = 320$

$$\Sigma y = ax + b\Sigma x$$
$$\Sigma xy = a\Sigma x + b\Sigma x^2$$
$$452 = 6a + 42b \times 7$$
$$3,330 = 42a + 320b$$
$$3,164 = 42a + 294b$$
$$3,330 = 42a + 320b$$

Subtracting
$$166 = 26b$$
$$b = 6.038$$

Substituting the value of b in equation 1

$$6a + 42b = 452$$
$$6a = -42 \times 6.038 + 452$$
$$6a = -254 + 452$$
$$a = 30$$

∴ The linear cost function is $y = 30 + 6.385x$

When the given problems has many digits of variables and costs, it becomes difficult to adopt the above method. An alternate method which could be termed as short-cut method or indirect method can be followed in such situations. The steps involved under this method are as follows:

(*a*) Calculate the average of 'x' series. This is found out by the formula $x = \frac{\Sigma x}{N}$

(*b*) Calculate the average of 'y' series. This is found out by the formula $y = \frac{\Sigma y}{N}$

(*c*) Take the deviations from the average of 'x' series for each and every item of 'x' series.

(*d*) Take the deviations from the average of 'y' series for each and every item of 'y' series.

(*e*) Square the deviations of 'x' series to get x^2.

(*f*) Obtain the products of the deviations of 'x' and 'y' series. This is denoted under the column 'xy'.

(*g*) Add up the products so obtained under step (*f*) and denote it with Σxy.

(*h*) Apply the following formula $b = \dfrac{\Sigma xy}{\Sigma x^2}$.

Problem 10. *(Short-cut or indirect method of segregating semi-variable cost):* The following semi-variable costs are taken from Ajanta Metal Works for the period January to June, 2003:

Month	*Machine hours*	*Semi-variable cost*
Jan.	2,000	300
Feb.	2,200	320
Mar.	1,700	270
Apr.	2,400	340
May	1,800	280
June	1,900	290

Find out the amount of variable and fixed cost.

Solution:

Month	*Machine hours (x)*	*Semi-variable cost (y)*	*Deviations from x (2,000)*	*Deviations from y (300)*	x^2	*xy*
Jan.	2,000	300	0	0	0	0
Feb.	2,200	320	+ 200	+ 20	40,000	+ 4,000
Mar.	1,700	270	– 300	– 30	90,000	+ 9,000
Apr.	2,400	340	+ 400	+ 40	1,60,000	+ 16,000
May	1,800	280	– 200	– 20	40,000	+ 4,000
June	1,900	290	– 100	– 10	10,000	+ 1,000
	Σx = 12,000	Σxy = 1,800			Σxy = 3,40,000	Σxy = 34,000

$$x = \frac{\text{Total machine hrs}}{\text{No. of months}} = \frac{12,000}{6} = 2,000$$

$$y = \frac{\text{Total semi-variable cost}}{\text{No. of months}} = \frac{18,00}{6} = 300$$

$$b = \frac{\Sigma xy}{\Sigma x^2} = \frac{34,000}{3,40,000} = 0.10$$

The variable element of semi-variable cost is 0.10 per unit.

The fixed cost can be calculated thus:

$$y = a + bx$$

Jan.

$$300 = a + 2,000 \times 0.10$$
$$300 = a + 200$$
$$a = 100$$

Feb.

$$320 = a + 2,200\ (0.10)$$
$$320 = a + 220$$
$$a = 100$$

Fixed cost of Rs. 100 is thus same whatever may be the level of activity.

ARGUMENTS IN FAVOUR OF MARGINAL COSTING

The exclusion of fixed costs in the ascertainment of cost of production (marginal costing) is justified on the following grounds:

1. Fixed costs relate to a particular period of time and therefore, it should be charged for that period. Thus, it is treated as a period cost not as a product cost.
2. There is no accurate basis to apportion fixed cost. So it is charged to profit and loss account instead of charging to cost of production.
3. Since fixed cost remain constant irrespective of change in the level of activity, it does not affect production capacity. It is only variable cost that affects various levels of productive activity. Hence fixed cost is excluded from cost of production.
4. To faciliate cost comparison and to maintain accuracy of trading results, fixed costs are excluded from cost of production.
5. Fixed costs under absorption costing is carried over from one period to another as part of inventory value. Such carrying forward is not justified as it does not facilitate matching of cost with revenue.
6. For the current operation of any level of activity only variable cost alone is relevant but not the fixed cost. Hence it is excluded from cost of production.

MARGINAL COST EQUATION

The marginal cost equation is stated below:

Sales – Variable Cost = Contribution

or $S - V = C$

Contribution = Fixed cost + Profit

or $C = FC + P$

or $S - V = FC + P$

If there is loss

$S - V = FC - L$

At the point of no-profit no-loss (break-even point)

$S - V = F + \text{Zero}$

The usefulness of marginal cost equation is, given any of the three particulars, we can find out the fourth unknown particular. From the above equation it is clear that contribution is used to meet fixed cost and any balance left out is equal to profit.

Problem 11. From the following particulars extracted from the books of Sunshine Enterprises for the period ending 31st December 2010. Find out the amount of profit.

Number of units produced	=	500 units
Variable cost for the period	=	Rs. 2 per unit
Fixed cost for the period	=	Rs. 800
Selling price per unit	=	Rs. 4

Solution:

Sales – Variable cost = Fixed cost + Profit

or Sales – Variable cost – Fixed cost = Profit

$500 (4 - 2) - 800 = P$

$1{,}000 - 800 = P$

$P = \text{Rs. } 200$

MARGINAL COSTING AND VALUATION OF FINISHED STOCK

Under marginal costing technique closing stock is valued at marginal cost. However, under absorption costing, fixed cost is included in the value of closing stock. Because of this treatment the amount of fixed overhead charged under absorption costing for the current year will differ from the actual fixed overhead pertaining to that year. The reason is the amount of fixed cost carried down from the previous year, will differ from the amount of fixed cost carried forward to the next year. This can better be understood by the following examples:

Example:

On 1.1.2009, Opening stock of finished goods of 2,000 units is valued at Rs. 4 of which fixed cost is Rs. 1.

On 31.12.2009, closing stock of finished goods of 1,600 units valued at Rs. 5 per unit of which fixed cost is Rs. 1.50.

Value of fixed cost brought in is 2,000 (2,000 × Re. 1). Whereas the value of fixed cost carried forward is Rs. 2,400 (1.600 × 1.50).

If the actual fixed cost for the year 2009 is Rs. 50,000, then the fixed cost charged for the year will be Rs. 49,600 (50,000 + 2,000 – 2,400).

When compared to absorption costing, considering the output, sales and closing stock the following conclusions are arrived at:

1. When production volume is equal to sales volume, net profit is same under both marginal costing and absorption costing.
2. When sales are less than production, profit under marginal costing are less than that of absorption costing.
3. When sales exceed production, profit under marginal costing are more than that of absorption costing.

UTILITY, ADVANTAGES AND LIMITATIONS OF MARGINAL COSTING

Utility of Marginal Costing

1. It is used as one of the techniques of costing, *i.e.*,to ascertain the cost of production.
2. It aids management for making useful decision by providing useful information.
3. It serves as a tool of cost control.

Advantages

1. It avoids the complications of over or under absorption of fixed cost by excluding it from cost of production.
2. The technique provides useful data for managerial decision-making.
3. By not carrying forward fixed cost from period to period, it facilitates cost comparison.
4. The impact of profit on sales fluctuations are clearly shown under marginal costing.
5. The technique is flexible in the sense it can be used along with other techniques such as budgetary control and standard costing.
6. It establishes a clear relationships between cost, sales and volume of output and break-even analysis which shows the effect of increasing and decreasing production activity on the profitability of the company.

7. It provides useful data for the management in determination of policies regarding future production and sales.
8. Stock of work-in-progress and finished goods are valued at marginal cost, which is uniform.

Limitations

1. The segregation of semi-variable costs often poses a problem.
2. Closing stock of work-in-progress and finished goods are understated which is not acceptable to tax authorities.
3. With the change of technology and owing to automation of industries, it results in more fixed cost. Marginal costing fails to reflect the exact change because of adoption of new technology.
4. It does not provide any yardstick to exercise control. So an effective means of control cannot be exercised.
5. The technique is not suitable under cost plus contract because the technique ignores fixed cost in calculating total cost.
6. Variable cost per unit remains constant only in the short run but not in the long run.
7. Cost comparison of two jobs will be difficult. Though marginal costing may be same for both the jobs, yet a job which takes more time to complete involves more of overhead.
8. When sales are based on marginal cost or marginal cost with some contribution, it may result in losses or low profit.

PRECAUTIONS TO BE TAKEN WHILE ADOPTING MARGINAL COSTING

1. As fixed cost is not included in the cost of production, care must be taken to see that while fixing the selling price, profit is realised after full recovery of fixed cost.
2. It is necessary to check composition of marginal cost from time to time. There can be a change of variable cost into fixed and *vice versa* when production method is changed.
3. Based on contribution margin, orders are accepted with a view to increase profits. This may lead to uneconomic expansion in the business involving more capital. However, during the period of falling business this proves to be burden.

APPLICATIONS OF MARGINAL COSTING

The technique of marginal costing is largely used in the managerial decision-making process. The application of marginal costing in the day-to-day decision-making process are as follows:

1. Make or Buy Decision

Very often management is confronted with the problem of deciding whether to buy a component or product from an outside source or to manufacture the same if it is economical as compared to the price quoted by a supplier. In deciding this, absorption costing (total cost analysis) would mislead. If the decision is to buy from an external source the price quoted by the supplier should be less than marginal cost. If the decision is to make within the organisation, the cost of production should include all additional cost such as depreciation on new plant, interest on capital, etc. If this cost of production is less than the quotation price, it should be decided in making the product rather than procure it from an external source.

Problem 12. *(Make or buy decision):* 'A' Company Ltd. is producing a spare part no. 007, for its product. The cost of manufacturing 5,000 units of 0.07 is as under

	Rs.
Direct material	11,750
Direct wages	94,000
Variable overheads	47,000
Fixed overheads	58,750

Another manufacturer is offering to sell the same spare part for Rs. 41. It is estimated that by avoiding the production of this spare part, the company has to incur Rs. 35,200 as fixed overheads.

Should the company make or buy this spare part ?

Solution:

(1) Calculation of variable cost per unit

Direct material	11,750
Direct wages	94,000
Variable overheads	47,000
	1,52,750

$$= \frac{1,,52,750}{5,000} = \text{Rs. } 30.55$$

(2) Calculation of avodiable fixed cost

Present fixed cost	58,750
Less: Future fixed cost	35,200
Avodiable fixed cost	23,550

(3) Calculation of future fixed cost per unit

$$= \frac{35,200}{5,000} = 7.04 \text{ per unit}$$

(4) Calculation of total cost per unit:

Variable cost	30.55	per unit
Future fixed cost	7.04	
Total cost to make	37.59	
Purchase price	41.00	

The cost analysis shows that producing the component is preferable.

Problem 13. *(Make or buy decisions):* **A radio manufacturing company finds that the cost of a component, when manufactured comes to Rs. 6.25. But the same is available in the market at Rs. 5.75 with an assurance of continued supply.**

The cost details of the component per unit:

Materials	–	2.75
Labour	–	1.75
Other variable overhead	–	0.50
Depreciation and other fixed cost	–	1.25
		6.25

(a) Should the company make the component or buy from outside?

(b) What would be your answer if the outside price is Rs. 4.85?

CHAPTER 10

Solution:

STATEMENT OF MAKE OR BUY DECISION

	Cost to make	Cost of buying	
		at Rs. 5.75	at Rs. 4.85
Variable cost:			
Material	2.75		
Labour	1.75		
Variable overhead	0.50		
Total variable cost	5.00	5.75	4.85
Fixed cost	1.25	1.25	1.25
Total cost	6.25	7.00	6.10

Note: Fixed cost will be incurred for making the component as well as for buying the component.

Comment

As per above statement, the total variable cost of making components Rs. 5, is lower than cost of buying a component at Rs. 5.75. Because total cost including fixed overhead in buying component is higher, *i.e.,* Rs. 7.

In case component is available at Rs. 4.85 it is better to buy, because total cost of buying a component including fixed overhead is lower at Rs. 6.10.

Problem 14. *(Make or buy decision):* Ashoka Ltd., has an annual production of 90,000 units for a motor component. The cost structure is as below:

Materials	Rs. 270 per unit
Labour (25% fixed)	Rs. 180 per unit
Expenses:	
– Variable	Rs. 90 per unit
– Fixed	Rs. 135 per unit
Total	Rs. 675 per unit

(*i*) The Purchase Manager has an offer from a supplier who is willing to supply the component at Rs. 540. Should the component be purchased and production stopped ?

(*ii*) Assume the resources now used for this components manufacture are to be used to produce another new product for which the selling price is Rs. 485. In the latter case the material price will be Rs. 200 per unit. 90,000 units of this product can be produced at the same cost basis as above, for labour and expenses. Discuss whether it would be advisable to divert the resources to manufacture that new product on the footing that the component presently being produced would instead of being produced, be purchased from the market ?

(University of Delhi, B.Com. (Hons.), 2004)

Solution:

TOTAL VARIABLE COST

Material	270
Labour $\left(180 \times \frac{75}{100}\right)$	135
Variable expenses	90
Total	495

Supplier is willing to supply the component part at Rs. 540 which is Rs. 45 more than the variable cost of Rs. 495. As the supplier's price is more than the variable cost, the component should be manufactured and should not be purchased.

(*ii*) Selling price of the new product	=	485
Less: Variable cost (200 + 135 + 90)	=	425
Contribution		60

If the component is purchased, then the loss will be = Rs. (540 – 495) = Rs. 45 per unit

If the resources are diverted to produce the new product, profit will be = Rs. (60 – 45) = 15 per unit.

When 90,000 units are manufactured, saving will be = 90,000 × 15 = Rs. 13,50,000. Thus the company should divert the resources for manufacturing of new product and the present component should be bought from the supplier.

Problem 15. *(Make or buy decision):* Expansion Ltd. manufactures, automobiles accessories and parts. The following are the total costs of processing 1,00,000 units:

Direct materials cost	Rs. 5 lakhs
Direct labour cost	Rs. 8 lakhs
Variable factory overhead	Rs. 6 lakhs
Fixed factory overhead	Rs. 5 lakhs

The purchase price of the component is Rs. 22. The fixed overhead would continue to be incurred even when the component is bought from outside, although there would have been reduction to the extent of Rs. 2,00,000.

Required:

(*a*) Should the part be made or bought considering that the present facility when released following a buying decision would remain idle ?

(*b*) In case the released capacity can be rented out to another manufacturer for Rs. 1,50,000 having good demand. What should be the decision ? *(C.S. Inter, June 1990)*

Solution:

COST ANALYSIS

Particulars	*1. Whether to make*	*2. Whether to buy at Rs. 22 each present facility remaining idle*	*3. Whether to buy at Rs. 22 each when released capacity is rented out*
1. Variable cost of production	19	–	–
2. Cost of buying from outside	–	22	22
3. Saving in fixed cost when bought	–	(2)	(2)
4. Rent receipt when idle capacity is rented out	–	–	(1.5)
	19	20	18.5

Recommendations:

1. For decision to situation (*a*) — make the component.
2. For decision to situation (*b*) — buy the component

Problem 16. *(Make or buy decision):* Stoner Company use three different components (materials) in manufacturing its primary product. Stoner manufactures two of the components and purchases one (designated as component 1) from outside suppliers. The company is currently developing the annual

CHAPTER 10

profit plan. Sales are highly seasonal. Component 2 cannot be acquired from outsiders. However, component 3 can be purchased. The three components have critical specifications. The annual profit plan provided data for the following components:

	Components 3 (Unit cost at 12,000 units)
Material (direct)	Rs. 1.40
Labour (direct)	Rs. 2.20
Factory overhead (apportioned)	Re. 0.40
Annual machine rental (special machine used only for component 3)	Re. 0.50
Variable factory overhead	Re. 1.00
Average storage cost per year (fixed)	Re. 0.40
Total	Rs. 5.90

Average inventory level 500 units

The purchase manager searching outside suppliers found one that would sign a one year contract to deliver "12,000 top quality units as needed during the year at Rs. 5.20 per unit". Serious consideration is being to this alternative. Should Stoner make or buy component 3 ? Explain the relevant factors influencing your decision. *(ICWA, Inter, Dec, 1988)*

Solution:

Total cost of component 3 can be bifurcated into fixed and variable portion as follows:

Variable cost:	
Material (direct)	Rs. 1.40
Labour (direct)	Rs. 2.20
Annual machine rental (special machine used only for component 3)	Re. 0.50
Variable factory overhead	Re. 1.00
	Rs. 5.10
Fixed cost:	
Fixed overhead apportioned	Re. 0.40
Average storage cost per year	Re. 0.40
	Re. 0.80
Total	Rs. 5.90

On comparison of the price offered by outside supplier with the variable cost, if made in-house, it is observed that variable cost of in-house production is lower (fixed cost remains same whether made in house or purchased). This is also on the assumption that labour (direct) is variable if it is possible to get rid of the workers when required. Hence, labour cost is treated as fixed. If labour is treated as fixed, variable cost of in-house production is Rs. 2.90 (5.10 – 2.20) per unit which is less than the purchase price per unit (5.20 – 2.90 = 2.30).

Hence, it is advisable to make component no. 3.

2. Comparing Mannual with Machine Labour: Whenever a new product is launched management may have to decide whether the new product is to be manufactured mannually or mechanically. Some of the factors which are to be considered in this regard are the (*i*) savings arising in the labour cost, (*ii*) additional investment on machinery. (*iii*) additional capacity of machine, (*iv*) the possible displacement of employees and (*v*) the danger of obsolescence of the machine. However, cost of production is the significant factor which determines the method of production. Under marginal costing that method which gives the largest contribution, *i.e.*, lowest marginal cost would be preferred. Under absorption costing that method of manufacture which gives lowest cost will be selected.

Problem 17. *(Manual Vs. mechanical production):* The Navabharth Enterprise has a plan to introduce a new product *A* numbering 10,000 units for the year 2010. The following details are available:

	Machine (*Rs.*)	*Manual labour (Rs.)*
Purchase price of machine	20,000	–
Direct material	5,000 PA	5,000 PA
Direct labour	500 PA	3,000 PA
Variable overheads	2,000 PA	1,000 PA
Fixed overheads (not including depreciation)	1,500 PA	1,000 PA

The selling price of the products has been fixed at Rs. 3 each. If the machines is purchased it will have an estimated life of 10 years with little or no residuals value.

You are required to show the decision to be taken regarding the method of production under (*i*) marginal costing and (*ii*) absorption costing including interest @ 5%.

Solution:

STATEMENT OF MARGINAL COST

			Machine (*Rs.*)	*Manual labour* (*Rs.*)
Sales			30,000	30,000
Less: Variable cost:				
Direct material	5,000	5,000		
Direct labour	500	3,000		
Variable overheads	2,000	1,000	7,500	9,000
Contribution			22,500	2,000

On this basis (the lower marginal cost and large contribution) the machine is preferable.

STATEMENT OF TOTAL COST

		Machine (*Rs.*)	*Manual labour (Rs.)*
Direct material		5,000	5,000
Direct labour		500	3,000
Overheads:	Variable	2,000	1,000
	Fixed	1,500	1,000
Depreciation $\frac{1}{10} \times 20{,}000$		2,000	–
		11,000	10,000
Interest @ 5%		550	500
		11,550	10,500

From the above statement, manual method of production is cheaper. This problem illustrates a weakness of the marginal costing. Obviously, fixed costs have to be covered as they are incurred on the specific products.

3. Replacing Existing Machinery with New Machinery: Sometimes with a view to derive maximum efficiency an existing plant may have to be replaced by a new one. Again the guiding factors mentioned earlier will help in such decision-making process.

Problem 18. A factory engaged in the manufacture of electronic good has a ten-year old equipment depreciated on straight-line basis. The useful life of the equipment was estimated to be 20 years with a residual value of Rs. 3 lakhs (original cost of the equipment being Rs. 23 lakhs). The output of the equipment is 1,200 units per hour.

The management now proposes to install a new equipment worth Rs. 50 lakhs which has an estimated life of 15 years and a residual value of Rs. 5 lakhs. The payment terms for the new equipment include a part exchange provision of Rs. 6 lakhs in respect of the existing equipment. The output of the new equipment is 3,000 units per hour.

Other comparative annual cost data relating to the two equipments are as under:

	Existing equipment (Rs.)	*New equipment (Rs.)*
Wages	1,00,000	1,20,000
Repairs and maintenance	20,000	52,000
Consumables	3,20,000	4,80,000
Power	1,20,000	1,50,000
Allocation of fixed cost	60,000	80,000
Total hours run per year	2,400	2,400

You are required to prepare a comparative schedule showing total conversion cost as well as cost per 1,000 units after considering interest @ 10% on net cash outflow for procuring the new equipment and also for providing for the yearly recovery of the loss suffered in the transaction.

(ICWA, Inter, Dec. 1991)

Solution:

COMPARATIVE STATEMENT OF COST OF EQUIPMENT

	Existing equipment (Rs.)	*New equipment (Rs.)*
Capital cost of equipment including cost of installation	23,00,000	50,00,000
Less: Residual value	3,00,000	5,00,000
	20,00,000	45,00,000
Less: Depreciation written off	10,00,000	
	10,00,000	45,00,000

COMPARATIVE STATEMENT OF OPERATION COST OF EQUIPMENT

		Rs.	*Rs.*
Annual depreciation	— New	–	3,00,000
(See Note 1)	— Old	1,00,000	70,000
Interest on capital (new cash outflow) *(See Note 2)*		–	4,40,000
Wages		1,00,000	1,20,000
Repairs and maintenance		20,000	52,000
Consumables		3,20,000	4,80,000
Power		1,20,000	1,50,000
Allocation of fixed expense		60,000	80,000
		7,20,000	16,92,000
Total hours run per annum		2,400	2,400
Operating cost per hour		Rs. 300	Rs. 705
Output per hour (units)		1,200	3,000
Operating cost per 1,000 unit		Rs. 250	Rs. 235

Therefore, there is a net saving in cost of Rs. 15 per 1,000 units.

Working Note 1: Depreciation on new equipment

$$= \frac{\text{Original cost} - \text{Residual value}}{\text{Estimated life}}$$

$$= \frac{45,00,000}{15} = 3,00,000$$

Depreciation (yearly recovery) on loss on sale of existing equipment

$$= \frac{\text{Book value of existing equipment} - \text{Part exchange value}}{\text{Residual life of existing equipment}}$$

$$= \frac{13,00,000 - 6,00,000}{10 \text{ years}} = \text{Rs. } 70,000$$

Working Note 2: Interest is calculated on net cash outflow in procuring the new equipment is Rs. 44,00,000 (Cost of old equipment Rs. 50 lakhs less part exchange value Rs. 6 lakhs).

4. Alternative Use of Plant or Productive Facility: To take advantage of alternative use of production facility or alternative use of plant it is necessary to know the contribution margin. That alternative which yields highest contribution margin shall be selected.

Problem 19. The management of Alpha Co. Ltd. are considering the following two alternative proposals:

(*a*) A semi-automatic lathe costing Rs. 58,000. The annual cost for its operation being estimated at Rs. 42,000 and

(*b*) A fully automatic lathe costing Rs. 95,000 the annual costs for its operation being estimated at Rs. 38,000

The life of each is estimated as 12 years with zero salvage value. The capacity of the automatic plant is assessed as 30% more than that of semi-automatic plant, but this extra capacity is likely to remain unutilised for two years. Using straight line depreciation with an interest charge of 6% on the capital investment make your recommendations as to which alternative is preferable.

Solution:

	Semi-automatic lathe (Rs.)	*Full automatic lathe (Rs.)*
Purchase cost	58,000	95,000
Expenses on machines:		
Operating expenses	42,000	38,000
Depreciating $\frac{1}{12}$ of cost	4,833	7,917
Interest 6%	3,480	5,700
	50,313	51,617
Percentage increase		2.5%

In the first two years, the semi-automatic lathe is more profitable because extra 30% capacity of fully automatic lathe is not utilised. But in the remaining 10 years the fully automatic lathe will be much more profitable because the extra cost incurred by that plant is only about 2.5%, whereas the extra output capacity is 30%. So the fully automatic lathe is preferable.

However, it is assumed that (*i*) additional 30% output will have market and (*ii*) revenue from the same is more than the additional expenses incurred above.

Problem 20. Given below are the details of cost of production of two products X and Y manufactured by Adam's Company Ltd. Both the products use the same raw materials:

	X	*Y*
% of total capacity utilised	60%	40%
Units produced	1,200	2,000
Selling price per unit	Rs. 2,000	Rs. 1,200
Cost per unit:		
Raw materials	Rs. 1,400	700
Direct labour	Rs. 200	150
Variable expense	Rs. 180	230
Contribution per unit	Rs. 220	10
Fixed overhead	Rs. 3,00,000	–

Would you recommend increasing the capacity utilised on product X?

Solution:

	1		2		3	
	X	Y	X	Y	X	Y
Percentage of capacities	60%	40%	50%	50%	70%	30%
Units produced	1,200	2,000	1,000	2,500	1,400	1,500
Contribution per unit	220	120	220	120	220	120
Contribution	2,64,000	2,40,000	2,20,000	3,00,000	3,08,000	1,80,000
Total contribution		5,04,000		5,20,000		4,88,000
Fixed overhead		3,00,000		3,00,000		3,00,000
Net profit		2,04,000		2,20,000		1,88,000

Thus by increasing capacity usage on product Y profit is increased whereas capacity usage on product X, profit is reduced.

This is so because contribution per unit percentage of capacity for product *X* is $\frac{2,64,000}{60\%}$ = Rs. 4,400 and for product Y is $\frac{2,40,000}{40\%}$ = Rs. 6,000. Hence it is better to increase 40% of the capacity utilised on product Y and not on product X.

5. Product-mix, profit planning and profit maximisation: Companies manufacturing varieties of products often have to decide which product-mix is more profitable. The product-mix which gives maximum contribution is to be considered as best product-mix. Similarly, profit planning is often considered so as to earn reasonable profit if not maximum profit. The profit planning is affected by factors such as (*i*) volume of output, (*ii*) product mix, (*iii*) costs to be incurred, (*iv*) prices to be charged and so on. Marginal costing techniques guides the management in this regard.

Problem 21. *(Product mix):* A limited company manufactures three different products and the following information has been collected from the books of account.

	Products		
	X	*Y*	*Z*
Sales mix	35%	35%	30%
Selling price	Rs. 30	Rs. 40	Rs. 20
Variable cost	Rs. 15	Rs. 20	Rs. 12
Total fixed cost			Rs. 1,80,000
Total sales			Rs. 6,00,000

The company has a proposal to discontinue the product 'Z' and replace it with 'A' and the following results are anticipated.

	Products		
	X	*Y*	*Z*
Sales mix	50%	25%	25%
Selling price	Rs. 30	Rs. 40	Rs. 30
Variable cost	Rs. 15	Rs. 20	Rs. 15
Total fixed cost			Rs. 1,80,000
Total sales			Rs. 6,40,000

Will you advise the company to change over to product A ?

Solution:

PRESENT POSITION

(1) Calculation of sales and units sold for each product

$$X = 6,00,000 \times \frac{35}{100} = \text{Rs. } 2,10,000$$

$$= \frac{2,10,000}{30} = 7,000 \text{ units}$$

$$Y = 6,00,000 \times \frac{35}{100} = \text{Rs. } 2,10,000$$

$$= \frac{2,10,000}{40} = 5,250 \text{ units}$$

$$Z = 6,00,000 \times \frac{30}{100} = \text{Rs, } 1,80,000$$

$$= \frac{1,80,000}{20} = 9,000 \text{ units}$$

(2) Calculation of variable cost

$X = 7,000 \times 15 = 1,05,000$

$Y = 5,250 \times 20 = 1,05,000$

$Z = 9,000 \times 12 = 1,08,000$

STATEMENT OF CONTRIBUTION AND PROFIT FOR THE PRESENT POSITION

		X	*Y*	*Z*	*Total*
	Sales	2,10,000	2,10,000	1,80,000	6,00,000
Less:	Variable cost	1,05,000	1,05,000	1,08,000	3,18,000
	Contribution	1,05,000	1,05,000	72,000	2,82,000
Less:	Fixed cost				1,80,000
	Profit				1,02,000

CHAPTER 10

FOR THE FUTURE PERIOD

(1) Calculation of sales and units sold for each product

X — $6{,}40{,}000 \times \frac{50}{100}$ = Rs. 3,20,000

— $\frac{3{,}20{,}000}{30}$ = 10,667 units

Y — $6{,}40{,}000 \times \frac{25}{100}$ = Rs. 1,60,000

— $\frac{1{,}60{,}000}{40}$ = 4,000 units

Z — $6{,}40{,}000 \times \frac{25}{100}$ = Rs. 1,60,000

— $\frac{1{,}60{,}000}{30}$ = 5,333 units

(2) Calculating the variable cost

X = 10,667 × 15 = 1,60,005

Y = 4,000 × 20 = 80,000

Z = 5,333 × 15 = 79,995

STATEMENT OF CONTRIBUTION AND PROFIT FOR FUTURE PERIOD

		X	*Y*	*Z*	*Total*
Sales		3,20,000	1,60,000	1,60,000	6,40,000
Less:	Variable cost	1,60,005	80,000	79,995	3,20,000
	Contribution	1,59,995	80,000	80,005	3,20,000
Less:	Fixed cost	—	—	—	1,80,000
					1,40,000

It can be seen from the above statement that the change in the product mix has resulted in an increase in profit by Rs. 38,000. Hence it is suggested to the company for a change in the product mix.

Problem 22. The technical director has submitted the following three proposals of sales mix:

(a) 100 units of X and 300 units of Y

(b) 300 units of X and 100 units of Y

(c) 200 units of X and 200 units of Y

The cost structure is as follows:

	Product X	*Product Y*
Direct materials	25	22
Direct wages	7	5
Variable expenses	7	5
Selling price	50	40

Fixed expenses Rs. 1,000

As a management accountant, recommend which of the sales mixture is to be adopted and why ?

Solution:

(1) Calculation of variable cost per unit

$X = 25 + 7 + 7 = 39$

$Y = 22 + 5 + 5 = 32$

(2) Calculation of sales value

Proposal A = (100 × 50) + (300 × 40) = 17,000

Proposal B = (300 × 50) + (100 × 40) = 19,000

Proposal C = (200 × 50) + (200 × 40) = 18,000

(3) Calculating variable cost for each proposal

Proposal A = (100 × 39) + (300 × 32) = 13,500

Proposal B = (300 × 39) + (100 × 32) = 14,900

Proposal C = (200 × 39) + (200 × 32) = 14,200

STATEMENT SHOWING CONTRIBUTION AND PROFIT

		Proposal A	*Proposal B*	*Proposal C*
Sales		17,000	19,000	18,000
Less:	Variable cost	13,500	14,900	14,200
	Contribution	3,500	4,100	3,800
Less:	Fixed cost	1,000	1,000	1,000
	Profit	2,500	3,100	2,800

Proposal 'B' is recommended as it gives higher contribution and more profit.

Problem 23. *(Profit planning and profit maximisation):* BEST Company Ltd. is presently working at 75% capacity by producing 12,000 units. The unit cost of the product is as under:

Direct material	10
Direct labour	5
Variable overhead	5
Selling price	32
Fixed cost	Rs. 10,000

The expected selling price at 25% level is Rs. 40, at 50% level Rs. 35 and at 100% level Rs. 28. The fixed cost will increase by 10% when the production reaches 100%. Prepare a marginal cost statement and offer your comments.

Solution:

AT 25% LEVEL

(*a*) Output: For 75% capacity the output is 12,000 units

For 50% capacity the output is $\frac{25}{75} \times 12{,}000$ = 4,000 units

(*b*) Sales value = 4,000 × 40 = 1,60,000

(*c*) Variable cost per unit = 10 + 5 + 5 = 20

(*d*) Total variable cost = 4,000 × 20 = Rs. 8,000

AT 50% LEVEL

(*a*) Output: For 75% capacity the output is 12,000 units

For 50% capacity the output is $\frac{50}{75} \times 12{,}000$ = 8,000 units

CHAPTER 10

(*b*) Sales value = 8,000 × 35 = 2,80,000
(c) Variable cost = 8,000 × 12= 96,000

AT 100% LEVEL

(*a*) Output: For 75% capacity the output is 12,000 units

For 100% capacity $\frac{100}{75}$ × 12,000 = 16,000 units

(*b*) Sales value = 16,000 × 28 = 4,48,000
(c) Variable cost = 16,000 × 20 = 3,20,000
(*d*) Fixed cost = 10,000 + 10% = 11,000

STATEMENT SHOWING CONTRIBUTION AND PROFIT

	25% capacity		*50% capacity*		*75% capacity*		*100% capacity*	
	Unit cost	*Total (4,000) Units*	*Unit cost*	*Total (8,000) Units*	*Unit cost*	*Total (12,000) Units*	*Unit*	*Total (16,000) Units*
Sales	40	1,60,000	35	2,80,000	32	3,84,000	28	4,48,8000
Less: Variable cost	20	80,000	12	96,000	20	2,40,000	20	3,20,000
Contributions	20	80,000	23	1,84,000	12	1,44,000	8	1,28,000
Less: Fixed cost	2.5	10,000	1.25	10,000	0.83	10,000	0.68	11,000
Profit	17.5	70,000	21.75	1,74,000	11.17	1,34,000	7.32	1,17,000

Comments:

(1) The contribution per unit is decreasing
(2) At 75% level of activity there is highest profit
(3) When the activity is at 100% level, the contribution and profit per unit is less due to decrease in selling price and increase in fixed cost.

Problem 24. *(Sales mix / Product mix):* The following three alternative plans are being considered for the next accounting year

Plant A: Sell 1,000 units of product X and 500 units of product Y
Plant B : Sell 800 units of product X and 700 units of product Y
Plant C : Sell 760 units of each product

The budgeted figures are:

	X (*Rs.*)	*Y* (*Rs.*)
Selling price	8	7
Direct material cost	3	3
Direct wages	2	2
Variable overhead	0.50	0.25

Calculate the budgeted profit that would result from each of the three alternatives and suggest the most profitable alternative. The budgeted fixed cost are Rs. 1,500.

Solution:

PROFITABILITY STATEMENT

	Product X		Product Y	
Selling price		8		7
Less: Variable cost:				
Direct material cost	3.00		3.00	
Direct wages	2.00		2.00	
Variable overhead	0.50	5.50	0.25	5.25
Contribution per unit		2.50		1.75

(1) Plan A

Contribution of X – 1,000 units × 2.50		= 2,500
Contribution of Y – 500 units × 1.75		= 8,75
		3,375
Less: Fixed overhead		1,500
	Profit	1,875

(2) Plan B

Contribution of X – 800 units × 2.50		= 2,000
Contribution of Y – 700 units × 1.75		= 1,225
		3,225
Less: Total fixed overhead		1,500
	Profit	1,725

(3) Plan C

Contribution of X – 760 units × 2.50		= 1,900
Contribution of Y – 760 units × 1.75		= 1,330
		3,230
Less: Fixed cost		1,500
	Profit	1,730

Comment:

Plan 'A' *i.e., X* 1000 units and *Y* 500 units gives more profit. Therefore, it is more profitable.

Problem 25. ***(Product mix):*** **A company manufactures 3 products A, B and C. There are no common processes and the sale of one product does not affect prices or sales of any other.**

The budget estimate of 2008 is given below:

	Total	*A*	*B*	*C*
Sales	12,00,000	1,80,000	9,00,000	1,20,000
Prdouction cost:				
– Variable cost	7,20,000	96,000	5,76,000	48,000
– Fixed cost	2,40,000	12,000	1,92,000	36,000

Factory cost	9,60,000	1,08,000	7,68,000	84,000
Selling cost:				
– Variable cost	96,000	32,400	32,400	31,200
– Fixed cost	24,000	8,400	7,200	8,400
Total	10,80,000	1,48,800	8,07,600	1,23,600
Profit	1,20,000	31,200	92,400	– 3,600

On the basis of above the Bound has almost decided to eliminate product C on which a loss is budgeted. As the company's cost a accountant what would you advise ?

Solution:

PROFIT STATEMENT

	A		*B*		*C*		*Total*
Sales		1,80,000		9,00,000		1,20,000	12,00,000
Less: Variable cost:							
– Production cost	96,000		5,76,000		48,000		
– Selling cost	32,400	12,8400	32,400	6,08,400	31,200	79,200	8,16,000
Contribution		51,600		2,91,600		4,08,00	3,84,000
Less: Fixed cost		–		–		–	2,64,000
(2,40,000 + 24,000)							
Profit							1,20,000

Comment:

If product 'C' is eliminated, the profit of the company is reduced to the extent of C's contribution of Rs. 40,800. Therefore, the net profit after C's elimination is Rs. 79,200 (Rs. 1,20,000 – Rs. 40,800). Therefore, it is not advisable to eliminate 'C'.

Problem 25. The directors of A Ltd. are considering the sales budget for the budget period 2010. You are required to present to the board a statement showing the marginal cost of each product and to recommend which of the following product-mix should be adopted:

(*a*) 450 units of A and 300 units of B

(*b*) 900 unit of A only

(*c*) 600 units of B only

(*d*) 600 units of A and 200 units of B

The following additional information is furnished:

	Product A	*Product B*
Fixed overhead Rs. 10,000 p.a.		
Direct labour @ Re. 1 per hour	10 hrs.	15 hrs.
Variable overheads — 100% of labour		
Direct material	Rs. 20	Rs. 25
Selling price	Rs. 60	Rs. 100

Solution:

MARGINAL COST STATEMENT

	Product A	Product B
Direct materials	20	25
Direct labour	10	15
Variable overhead	10	15
Marginal cost	40	55
Contribution	20	45
Sales	60	100

Product mix

	A	*B*	*Total*
(*i*) 450 units of *A* and 300 units of *B*			
Contribution	9,000	13,500	22,500
Less: Fixed overheads			10,000
Profit			12,500
(*ii*) 900 units of *A*			
Contribution	18,000	—	18,000
Less: Fixed overheads			10,000
Profit			8,000
(*iii*) 600 units of *B*			
Contribution	10,000	17,000	27,000
Less: Fixed overhead			10,000
Profit			17,000
(*iv*) 600 units of *A* and 200 units *B*			
Contribution	12,000	9,000	21,000
Less: Fixed overhead			10,000
Profit			11,000

Thus alternative (*iii*) is the one recommended.

Problem 26. From the following particulars find the most profitable product-mix and prepare a statement of profitability of that product-mix.

	Product A	*Product B*	*Product C*
Units budgeted to produced and sold	1,800	3,000	1,200
Selling price per unit (Rs.)	60	55	50
Requirement per unit:			
Direct materials	5 kg.	3 kg.	4 kg.
Direct labour	4 hrs.	3 hrs.	2 hrs.
Variable overhead	Rs. 7	Rs. 13	Rs. 8

Fixed overhead	Rs. 10	Rs. 10	Rs. 10
Cost of direct material per kg.	Rs. 4	Rs. 4	Rs. 4
Direct labour hour rate	Rs. 2	Rs. 2	Rs. 2
Maximum possible units of sale	4,000	5,000	1,500

All the three products are from the same direct material using the same type of machines and labour, direct labour which is the key factor, is limited to 18,600 hours. *(C.A. Inter, May 1990)*

Solution:

STATEMENT SHOWING MOST PROFITABLE PRODUCT-MIX

				A	*B*	*C*
Selling price per unit				60	55	50
Less: Variable cost:						
Direct material	20	12	16			
Direct labour	8	6	4			
Variable overhead	7	13	8	35	31	28
Contribution				25	24	22
Contribution per hour = $\frac{\text{Contribution per unit}}{\text{Direct labour hour per unit}}$				$\frac{25}{4}$	$\frac{24}{3}$	$\frac{22}{2}$
				= 6.25	= 8	= 11
Ranking of most profitable product-mix				III	II	I

STATEMENT OF PROFITABILITY OF THE MOST PROFITABLE PRODUCT-MIX.

	A	*B*	*C*	*Total*
Total contribution on unit sold	3,750	1,20,000	33,000	1,56,750
Less: Fixed cost	18,000	30,000	12,000	60,000
			Profit	96,750

Note: Here direct labour hour is the key factor and only 18,600 hours are available to produce the three products. The available 18,600 hours are utilised in the order of the ranking assigned, *i.e.,* product *C*, *B* and *A*.

The number of units of each product to be produced will depend upon the maximum possible sale of each product. The details of the products to be produced keeping in view of available hours and their ranking are as below:

		Total hrs. utilised
Product *C*	1,500 × 2 Hrs.	3,000
Product *B*	5,000 × 3 Hrs.	15,000
Product *A*	150 × 4 Hrs.	600
		18,600

Problem 27. A farmer wants your recommendations for optimal mix of production for the coming year.

The current data is given below:

	Items produced			
	A	*B*	*C*	*D*
Area occupied (acres)	25	20	30	25
Yield per acre (tonnes)	10	8	9	12
Sale price per tonne (Rs.)	1,000	1,250	1,500	1,350
Variable cost per acre:				
Material (Rs.)	700	600	950	900
Labour (Rs.)	2,000	2,500	3,000	3,700
Variable overhead (Rs.)	2,000	2,000	2,000	2,000

Fixed overhead:	(Rs.)
Cultivation and growing	1,00,000
Harvesting and transport	2,40,000
Land revenue	90,000
Administration	1,10,000
	5,40,000

The land which is being used for producing items *A* and *B* can be used for either items but not items *C* and *D*. The land which is being used for producing items *C* and *D* can be used for either items but not for items *A* and *B*.

In order to provide adequate market service, the farmer must produce each year at least 40 tonnes each of *A* and *B* and 36 tons each of *C* and *D*.

You are required to calculate the following:

(*a*) The profit for the current year and

(*b*) The profit for the production-mix which you would recommend.

(C.S., Inter, Dec., 1989)

Solution:

STATEMENT SHOWING PROFIT FOR THE CURRENT YEAR

	Products			
	A	*B*	*C*	*D*
Sale value (per acre)	10,000	10,000	13,500	16,200
Variable cost (per acre)	4,700	5,100	5,950	6,600
Contribution per acre	5,300	4,900	7,550	9,600
Total acres	25	20	30	25
Total contribution	1,32,500	98,000	2,26,500	2,40,000
Total contribution (*A* + *B* + *C* + *D*)		6,97,000		
Less: Fixed cost		5,40,000		
Profit		1,57,000		

STATEMENT OF PROFIT FOR THE PROFITABLE PRODUCT-MIX

	A	*B*	*C*	*D*
Minimum required production (tonnes)	40	40	36	36
Yield per acre (tonnes)	10	8	9	12
Acre required for minimum production	4	5	4	3
Surplus area allocated to profitable production (*i.e.*, *A* and *B*)	36	–	–	48
Total acres	40	5	4	51
Contribution per acre	5,300	4,900	7,550	9,600
Total contribution	2,12,000	24,500	30,200	4,89,600
Total contribution (*A* + *B* + *C* + *D*) =	7,56,300			
Less: Fixed overhead	5,40,000			
Profit	2,16,300			

Note: The distribution of area to different products has been made in the manner given because the area is not interchangeable between all products but only partly interchangeble, *i.e.*, between *A*, *B*, *C*, *D* only. Therefore, the total area denoted to *A* and *B* is 45 acres (25 + 20) so that, on the basis of the per acre contribution, maximum possible land to the production of product *A* and similarly to product *D*.

6. Profitability of the department or products: The preparation of a departmental profit and loss account under marginal costing is useful in determining which department is making profit and which department is incurring a loss. This enables the management to decide whether a particular department must continue operation or it should be eliminated. The decision is taken by referring to the contribution made or loss incurred by the department of product.

Problem 28. A company produces three products. The cost data are as under:
The current data is given below:

		A	*B*	*C*
Direct materials		Rs. 64	152	117
Direct labour				
Department	Rate per hour (Rs.)	Hrs.	Hrs.	Hrs.
1	5	18	10	20
2	6	5	4	7
3	4	10	5	20
Variable overhead		Rs. 16	9	21

Fixed overheads Rs. 4,00,000 per annum

The budget was prepared at a time, when the market was sluggish. The budgeted quantities and selling prices are as under:

Product	*Budgeted qty.*	*Selling price.*
A	9,750	270
B	7,800	280
C	7,800	400

Later the market improved and the sales quantities could be increased by 20 for product '*A*' and 25% for product '*B*' and '*C*'. The sales manager confirmed that the increased quantities could be achieved at the prices originally budgeted. The production manager stated that the output cannot be increased beyond the budgeted level due to limitation of direct labour hours in department *Z*.

Required:

(*a*) Present a statement of budgeted profitability.

(*b*) Statement of optimal product mix and calculate the optimal profits.

(C.S. Inter, May 1988)

Solution:

STATEMENT OF BUDGETED PROFITABILITY

	A	*B*	*C*	*Total*
Selling price (p.u)	270	280	400	
Variable cost (p.u.):				
Direct materials	64	152	117	
Direct labour	160	94	222	
Variable overheads	16	9	21	
Total variable cost	240	255	360	
Contribution per unit	30	25	40	
Budgeted qty. (units)	9,750	7,800	7,800	
Total contribution	2,92,500	1,95,000	3,12,000	7,99,500
Less: fixed cost				4,00,000
			Profit	3,99,500

STATEMENT OF OPTIMAL PRODUCT MIX AND PROFIT

	A	*B*	*C*	*Total*
Contribution (*A*)	30	25	40	
Direct labour hors. in dept. 2 (*B*)	5	4	7	
Contribution per hr. $\frac{A}{B}$	6	6.25	5.71	
Ranking	II	I	III	
Optimal product-mix				
Units (C) (*See Notes*)	11,700	9,750	5,292	
Total contribution (*A* × *C*)	3,51,000	2,43,750	2,11,680	8,06,430
Less: Fixed cost				4,00,000
			Optimal profit	4,06,430

CHAPTER 10

Working Note:

Total hours avuilable in dept. 2

Products	*Units*	*Hours per unit*	*Total hours*
A	9,750	5	48,750
B	7,800	4	31,200
C	7,800	7	54,600
			1,34,550

Maximum sales quantities of products (under improved market conditions)

Products	*Units*	*Increase in percentage*	*Total number of units*
A	9,750	20	11,700
B	7,800	25	9,750
C	7,800	25	9,750

Problem 29. A Ltd. Company has three departments. The following data relates to the period ending 31st Dec. 2009:

	Department		
	A	*B*	*C*
Sales (Rs.)	80,000	40,000	60,000
Marginal cost:			
Direct material	10,000	5,000	10,000
Direct labour	4,000	5,000	16,000
Variable overhead	10,000	5,000	20,000
Fixed overheads:	20,000	10,000	20,000
Total cost	44,000	25,000	66,000

The manager in charge of department is disappointed with the results because of higher marginal cost and there is no hope of being reduced further. You are required to present the information in the most suitable manner indicating whether or not department *C* should be closed down.

Solution:

MARGINAL PROFIT AND LOSS OF 'A' LTD. COMPANY

Dept.	*Sales value (Rs.)*	*Marginal cost (Rs.)*	*Contribution (Rs.)*
A	80,000	24,000	56,000
B	40,000	15,000	25,000
C	60,000	46,000	14,000
	1,80,000	85,000	95,000
Less: Fixed cost			50,000
Net profit			45,000

Department *C* makes a contribution of Rs. 14,000 towards fixed cost. If this department is eliminated altogether, the following would be the position:

Dept.	*Sale value*	*Marginal cost*	*Contribution*
A	80,000	24,000	56,000
B	40,000	15,000	25,000
	1,20,000	39,000	81,000
Less: Fixed cost			50,000
Net profit			31,000

From the above it is clear that the net profit has fallen down to Rs. 31,000 from 45,000, if Department *C* is eliminated.

Problem 30. *XYZ* Ltd. is selling three brands of its products in the brand names *X*, *Y* and *Z*. The details regarding unit costs and selling prices are as under:

	X (Rs.)	*Y (Rs.)*	*Z (Rs.)*
Direct materials	6	12	16
Direct labour	8	8	20
Variable overheads	6	20	14
Selling price	36	50	96

The monthly fixed expenditure is Rs. 5,40,000.

Sales volume for the months of July and August of 1989 are as follows:

	X	*Y*	*Z*
July	20,000	20,000	20,000
August	40,000	26,000	10,000

Find out the monthly profits and if your computation brings out that higher profit was earned in the month having lower sales volume, kindly justify the finding with reasons.

(ICWA, Inter, Dec. 1989)

Solution:

STATEMENT OF COST PER UNIT

	X (Rs.)	*Y (Rs.)*	*Z (Rs.)*
Direct material	6	12	16
Direct labour	8	8	20
Variable overhead	6	20	14
Marginal cost	20	40	50
Selling price	36	50	96
Contribution	16	10	46

For July	*X*	*Y*	*Z*	*Total*
Sales (units)	20,000	20,000	20,000	60,000
Contribution per unit	16	10	46	
Total contribution	3,20,000	2,00,000	9,20,000	1,44,000
Less: Fixed cost				5,40,000
			Profit	Rs. 9,00,000

CHAPTER 10

For August				
Sales (units)	40,000	26,000	10,000	76,000
Contribution per unit	16	10	46	
Total contribution	6,40,000	2,60,000	4,60,000	13,60,000
Less: Fixed cost				5,40,000
			Profit	Rs. 8,20,000

Here it can be seen that although the sale quantity has been more in August at 76,000 than 60,000 in July, both contribution and profit were lower in August. The reason for this apparent mismatch is a drop in sales of product *Z* having the highest contribution of Rs. 46 per unit when compared to figure of Rs. 16 and 10 for products *X* and *Y* respectively. This means one unit of *Z* earns Rs. 20 more compared to one unit of *X* and *Y* taken together. Thus the increase in sales of *X* and *Y* though significant in quantitative measurement could not compensate the loss in contribution due to reduction of sales of *Z* in the month of August.

7. Selling at or below marginal cost: Some time it may become necessary to sell the goods at a price below the marginal cost. Some such situations are as follows:

(*a*) Where materials are of perishable nature.

(*b*) Where large quantities of stock is accumulated and whose market prices have fallen. This will save the carrying cost of stocks.

(*c*) In order to popularise a new product.

(*d*) In order to increase sales of those products which have higher margin of profit.

If the selling price is below the total cost but above the marginal cost, the contribution will leave an under-recovering of fixed cost. If the selling price fixed is equal to marginal cost, there will be a loss which is equal to fixed cost. However, where the selling price fixed is lesser than the marginal cost, the loss will be greater than fixed cost.

Problem 31. *(Selling at or below marginal cost):* The following details have been furnished to you regarding two proposals which are for consideration before a firm.

(*a*) Improvement in the quality of the product, which will result in an additional sale of 5,000 units at the existing price. However, this improvement in quality will result in increase in the variable cost by 10 paise per unit.

(*b*) Reduction in the selling price of the product by 12 paise per unit. This will push up sales by 5,000 units.

In both cases the fixed expenses will increase by Rs. 1,000. The present sales of the firm are 10,000 units at the rate of Rs. 2.10 per unit. The variable cost is Rs. 1.60 per unit and the total fixed costs are Rs. 3,000.

You are required to state whether it will be appropriate for the firm to select any of the new proposals or should it continue with the existing scheme. *(University of Madras, M.Com., 2002)*

Solution:

SELECTION OF ALTERNATIVES

I. Alternatives- present position

Present sales	10,000	units
Selling price per unit	2.10	
Less: Variable cost per unit	1.60	
Contribution per unit	0.50	

Total present contribution		=	5,000
10,000 × units × 0.50			
Less: Total present fixed cost		=	3,000
Present profit			2,000

II. Alternative-proposal (*a*)

Present sales		10,000	units
Add: Additional sales due to improvement in quality		5,000	units
		15,000	units
Present selling price per unit			2.10
Less: Existing variable cost per unit	1.60		
Increase in variable cost due to improvement in quality	0.10		1.70
Revised contribution per unit			0.40
Total revised contribution			6,000
15000 × 0.40			
Less: Total present fixed cost	=	3,000	
Increase in fixed cost	=	1000	4,000
Profit under proposal (a)			2,000

III. Alternative-proposal (*b*)

Present sales		10,000	units
Add: Sales push up due to Reduction in selling price		5,000	units
Revised sales unit		15,000	units
Present selling price per unit			Rs. 2.10
Less: Reduction in selling price			Re. 0.12
Revised selling price per unit			1.98
Less: Variable cost per unit			1.60
Revised contribution per unit			0.38
Total revised contribution	=		5,700
15,000 × 0.38			
Less: Total fixed cost	=	3,000	
Add: Increase in fixed cost	=	1,000	4,000
Profit under proposal (*b*)			1,700

Comment

Present position of 10,000 units gives profit of Rs. 2,000. However, in future, if any proposal to increase sale units it is better to consider proposal (*a*). It gives same present profit of 2,000 compared to proposal (*b*) which gives profit of Rs. 1,700.

Problem 32. Garden Product Limited manufactures the 'Rainpour' garden spray. The accounts of the company for the year 1981 are expected to reveal profit of Rs. 14,00,000 from the manufacturer of 'Rainpour' after charging fixed costs of Rs. 10,00,000. The rainpour is sold for Rs. 50 per unit and has a variable unit cost of Rs. 20.

Market sensitivity tests suggest the following response to price changes.

Alternatives	*Selling price reduced by*	*Quantity sold increased by*
A	5%	10%
B	7%	20%
C	10%	25%

Evaluate these alternatives and state which, on profitability consideration, should be adopted for the forthcoming year assuming cost structure unchanged from 1981.

(University of Delhi, B.Com. (Hons.), 1988)

Solution:

	Present	*Proposed*		
		A	*B*	*C*
Selling price	50	47.50	46.50	45
Less: Variable cost	20	20	20	20
Contribution	30	27.50	26.50	25
Quantity (units)	80,000	88,000	96,000	1,00,000
Total contribution	24,00,000	24,20,000	25,44,000	25,00,000
Less: Fixed cost	10,00,000	10,00,000	10,00,000	10,00,000
	14,00,000	14,20,000	15,44,000	15,00,000

Contribution and profit of proposal *B* is the largest and should, therefore, be adopted.

Note: Units produced and sold is calculated by employing the formula of:

$$= \frac{\text{Total contribution}}{\text{Contribution per unit}}$$

8. Determination of selling price and volume of output: The determination of selling price and volume of output is based on 'differential costing'. The difference in total cost due to increase in sales volume is known as differential costing. The increase in sales value due to increase in sales volume is known as 'incremental revenue'. The analysis of differential cost and incremental revenue helps in determining selling price which will yield the optimum profit. So long as incremental revenue is more than the differential cost it is advantageous to increase the output. But as soon as incremental revenue equals the differential cost further increase in output is not advantageous. Differential cost analysis thus helps to determine the selling price and the level of activity which are expected to yield the highest profit.

9. Determination of sales volume for a desired profit: Sometimes one of the decisions that is to be taken by the management is to determine the volume of sales in order to earn a desired profit. The cost-volume-profit analysis which is again based on marginal costing helps management is taking such decisions.

Problem 33. Frazer Ltd. manufactures and sells a product the selling price and raw material cost which have remained unchanged during the past two years. The followings are the relevant data:

	Year 1	Year 2
Quantity sold (kgs.)	100	150
	Rs.	Rs.
Sales value	20,000	?
Raw materials	10,000	?
Direct wages	3,000	?
Factory overheads	5,000	5,700
Profit	2,000	2,550

During year 2 direct wages rates increased by 150% but there was a saving of Rs. 300 in fixed factory overhead.

Required: What quantity (in kgs) the company should have produced and sold in year 2 in order maintain the same amount of net profit per kg. as it earned during the year.

(C.A. Inter, Nov., 1988)

Solution:

STATEMENT OF QUANTITY TO BE PRODUCED AND SOLD DURING YEAR 2

		Rs.
Selling price per kg. (*See Working Note 1*)		200
Variable costs per kg:		
Raw material	100	
Direct vages (*See Working Note 2*)	45	
Variable factory overhead (*See Working Note 3*)	20	165
		35
Less: Profit required per kg. (at 100 kgs. of sales)		20
Balance contribution per kg. for meeting fixed cost		15
Total fixed cost (*See Working Note 4*)		2,700

Quantity of kgs. produced and sold $\frac{2,700}{15}$ = 180 kgs.

Working Notes:

1. Selling price per kg $= \frac{\text{Rs. } 20,000}{100 \text{ kg.}} = \text{Rs. } 200$ per kg.

2. Direct wages $= \text{Rs. } \frac{\text{Rs. } 30,000}{100 \text{ kg.}} \times \text{Rs. } 150\% = \text{Rs. } 45$

3. **Variable factory overhead per kg:**

	Rs.
Total factory overhead in year 2	5,700
Add: Savings in fixed factory overhead	300
	6,000
Less: Total factory overhead in year 1	5,000
Increase in factory overhead	1,000

Increase in quantity (150 kg. – 100 kg.) = 50 kg.

∴ Variable factory overhead per kg. $\frac{\text{Rs. } 1,000}{100 \text{ kg.}}$ = Rs. 20

CHAPTER 10

4. **Fixed factory overhead:**	*Year 1*	*Year 2*
Total factory overhead	5,000	5,700
Less: Variable factory overhead	2,000	3,000
	3,000	2,700

Problem 34. X Ltd. manufactures and markets a single product. The following information is available:

	Rs. per unit
Materials	8.00
Conversion costs (variable)	6.00
Dealers margin	2.00
Selling price	20.00

Fixed cost Rs. 2,50,000

Present sales 80,000 units

Capacity utilisation 60 per cent

There is acute competition. Extra efforts are necessary to sell. Suggestions have been made for increasing sales:

1. By reducing sales price by 5%
2. By increasing dealers margin by 25% over the existing rate.

Which of the two suggestions you would recommend if the company desires to maintain the profit? Give reasons:

Solution:

PRESENT MARGINAL COST PER UNIT

Materials	8.00
Conversion costs	6.00
Dealer's margin	2.00
Total	16.00

Contribution per unit = Selling price – Marginal cost

= 20 – 16 = 4

Total contribution = 90,000 × 4 = 3,60,000

Profit = Contribution – Fixed cost

= 3,60,000 – 2,50,000 = 1,10,000

Since in both suggestions, fixed costs remain unchanged, the present profit can be maintained by keeping the total contribution at the present level, *i.e.,* Rs. 3,60,000.

1. Reducing sales price by 5%

New sales price = (20 – 1) = 19

New dealers margin = 10% of Rs. 19

= 1.90

∴ Varable cost = 8 + 6 + 1.90 = Rs. 15.90

Contribution per unit = 19 – 15.90 = Rs. 3.10

Sales (units) required to maintain the present level of profit

$$= \frac{\text{Total contribution}}{\text{Contribution per unit}} = \frac{3{,}60{,}000}{3.10}$$

= 1,16,111 units

2. Increasing dealers margin by 25%

New dealer's margin $= 8 + 2 \times \frac{25}{100} = 2.50$

New variable cost $= 8 + 6 + 2.50 = 16.50$

Contribution $= 20 - 16.50 =$ Rs. 3.50

Sales (units) $= \frac{3,60,000}{3.50} = 1.02$

The second proposal is recommended because the contribution per unit is higher and the sales (in units) are lower. Lower sales efforts and less finance would be required in implementing the proposal (*ii*).

10. Acceptance or rejection of a special order within the country: Sometimes a decision regarding acceptance of a special order is to be made. This, however, depends on the availability of spare capacity. The contribution available from such an order helps in making such decision.

Problem 35. *(Accepting or Rejecting an offer):* Narvous company is presently working at 75% capacity. The following are the financial results

	Cost per units
Direct materials	2.00
Direct labour	1.50
Variable overhead	2.50
Fixed overhead	3.00
Total cost	9.00
Production for the month	30,000 Units
Selling price per unit	8.00

A special order is received for 6,000 units @ Rs. 7. Advise the company, whether to accept or reject the offer.

Solution:

STATEMENT SHOWING THE PRESENT AND PROPOSED PROFIT

	Existing position 30,000 units	*Special order 6,000 units*	*Total 36,000 units*
Sales	2,40,000	42,000	2,82,000
Less: Marginal cost	1,80,000	36,000	2,16,000
Contribution	60,000	6,000	66,000
Less: Fixed cost	90,000	–	90,000
Loss	30,000		24,000

Suggested decision:

The company can accept the offer. On account of this the loss can be reduced by Rs. 6,000

Working Note:

(1) Sales = 30,000 × 8 = 2,40,000

6,000 × 7 = 42,000

2,82,000

(2) Marginal cost per unit

2 + 1.50 + 2.50 = Rs. 6		
30,000 × Rs. 6	=	1,80,000
6,000 × Rs. 6	=	36,000
		2,16,000
(3) Fixed cost = 30,000 × Rs. 3	=	90,000

Problem 36. Z Ltd. manufactures and sells a drink at 0.20 paise per cup. Current output is 4,00,000 cups per month which represents 80% of capacity. They have the opportunity to utilise their surplus capacity by selling their product at 0.13 paise per cup to a supermarket chain who will sell it as 'own lable' product.

Total costs for the previous month were Rs. 56,000 out of which Rs. 16,000 were fixed costs. This represented a total cost of 14 paise per cup.

Based on the above data should Z accept the super market order. What other factors should be considered?

Solution:

STATEMENT SHOWING THE PRESENT SITUATION

Sales 4,00,000 @ 0.20 p		80,000
Less:	Marginal cost 0.10 per cup	40,000
	Contribution	40,000
Less:	Fixed cost	16,000
	Net profit	24,000

On the assumption that fixed costs are unchanged, the special order will produce the following contribution:

Sales 1,00,000 × 0.13		13,000
Less:	Marginal cost 1,00,000 × 0.10	10,000
	Contribution	3,000

However, there are several other factors which would need to be considered before a final decision is taken:

(*a*) Will the acceptance of one order at a lower price lead other customers to demand lower prices as well ?

(*b*) Is this special order the most profitable way of using the spare capacity ?

(*c*) Will the special order lock up capacity which could be used for future full price business ?

Problem 37. A mechanical toy factory presents the following information for the year 1990:

Material cost (Rs.)	1,20,000
Labour cost (Rs.)	2,40,000
Fixed overhead (Rs.)	1,20,000
Variable overhead (Rs.)	60,000
Units produced	12,000
Selling price per unit (Rs.)	50

The available capacity is a production of 20,000 units per year. The firm has an offer for the purchase of 5,000 additional units at a price of Rs. 40 per units. It is expected that by accepting this offer there will be saving of rupee one per unit in the material cost on all units manufactured, the fixed overhead will increase by Rs. 35,000 and the overall efficiency will drop by 2% on all production.

State whether offer is acceptable or not. *(C.S. Inter, June 1993)*

Solution:

Available capacity — 20,000 units: 100%

Capacity utilised — 12,000 units: 60%

STATEMENT OF PRODUCTION AND SALES OF 12,000 UNITS

	Total	*Per unit*
(A) Sales	6,00,000	50.00
Variable cost of sales		
(*a*) Materials	1,20,000	10.00
(*b*) Labour cost	2,40,000	20.00
(*c*) Variable overheads	60,000	5.00
(B) Total variable cost of sales	4,20,000	35.00
(C) Contribution (A – B)	1,80,000	15.00
(D) Fixed overheads	1,20,000	10.00
(E) Profit	60,000	5.00

Profit prospect of the offer for 5,000 units is accepted

Existing sales revenue		6,00,000	
Additional sales revenue (5,000 × Rs. 40)		2,00,000	80,00,000
Less: Variable cost of sales:			
Materials 17,000 units × Rs. 9	=	1,53,000	
Labour 17,000 units × Rs. 20.40	=	3,46,800	
Variable overhead 17,000 units × Rs. 5	=	85,000	
			5,84,800
Total contribution from sales			2,15,200
Less: Original fixed overhead		1,20,000	
Additional fixed overhead		35,000	1,55,000
		Total profit	60,200
Total profit			60,200
Original profit before acceptance of offer			60,000
Additional profit			200

The offer is not acceptable in view of insignificant incremental profit. However, if the offer is likely to result in a wider market for the future, it can be accepted.

Working Note:

New marginal cost Rs. 10 – 10% of 10 = Rs. 9

New labour cost Rs. 20 + 2% of 20 = Rs. 22

Note: Efficiency is related to labour cost.

Problem 38. PQR Ltd. manufactures medals for winners of athletic events and other contests. Its manufacturing plant has the capacity to produce 10,000 medals each month. The company has current production and sales level of 7,500 medals per month. The current market price of the medal is Rs. 150.

The cost data for the month of October 1999 is as under:

Variable costs: (That vary with units produced)

	Rs.
Direct materials	2,62,500
Direct labour cost	3,00,000

CHAPTER 10

Variable costs (That vary with number of batches)	
Setups: Materials handling:	
Quality control	
150 batches × Rs. 500 per batch	57,000
Fixed manufacturing costs	2,75,000
Fixed marketing costs	1,75,000
	10,87,500

PQR Ltd. has received a special one-time-only order for 2,500 medals at Rs. 100 per medal.

PQR Ltd. makes medals for its existing customers in batch sizes of 50 medals (150 batches × 50 medals per batch = 7,500).

The special order for 2,500 medals required *PQR* Ltd. to manufacture the medals in 25 batches of 100 each:

Required:

(*a*) Should *PQR* Ltd. accept the special order ? Why ? Explain briefly.

(*b*) Suppose the plant capacity was 9,000 medals instead of 10,000 medals each month. The special order must be taken either in full or rejected totally. Should *PQR* Ltd. accept the special order ? Why ? *(C.A. Inter, November 1999)*

Solution:

STATEMENT OF CONTRIBUTION MARGIN ON ACCEPTING THE SPECIAL ORDER OF 2,500 MEDALS

Sales revenue (2,500 medals × Rs. 100 per medal)		2,50,000
Less: Variable costs:		
Direct materials (2,500 × 35)	87,500	
Direct wages (2,500 × 40)	1,00,000	
Setups, materials handling, quality control (25 batches × 500)	12,500	2,00,000
	Contribution	50,000

Decision:

PQR Ltd. should accept the sepcial order, because its acceptance would increase the operating profit of the concern by Rs. 50,000.

STATEMENT SHOWING PRESENT CONTRIBUTION ON 7,500 MEDALS

Sales revenue (7,500 × 150)		11,25,000
Less: Variable cost:		
Direct materials (7,500 × 35)	2,62,500	
Direct wages (7,500 × 40)	3,00,000	
Setups: Material handling and quality control	75,000	6,37,500
	Contribution	4,87,500

STATEMENT SHOWING CONTRIBUTION MARGIN ON 6,500 MEDALS

Sales revenue (6,500 × 150)			9,75,000
Less:	Variable cost		
	Direct materials (6,500 × 35)	2,27,500	
	Direct labour cost (6,500 × 40)	2,06,000	
	Setups: Material handling and quality control	65,000	
			5,52,000
		Contribution	4,22,500

STATEMENT SHOWING ACCEPTANCE OF SPECIAL ORDER WHEN THE PLANT CAPACITY WAS 9,000 MEDALS

Gain in contribution because of special order of 2,500 medals	50,000
Contribution on reduction of 1,000 medal sales in the internal market (4,87,500 – 4,22,500)	(–65,000)
Loss of contribution	(–15,000)

Decision:

The special order of 25,000 medals (when the plant capacity was reduced to 9,000 medals) should not be accepted, as its acceptance will result in a loss of margin by Rs. 15,000.

11. Acceptance or rejection of an offer from foreign market: The acceptance or rejection of an offer from a foreign market depends upon the incremental revenue.

Problem 39. *(Acceptance or rejection of an offer from foreign market):* A Company annually manufactures 10,000 units of a product at a cost of Rs. 4 per unit and there is home market for consuming the entire volume of production at the sale price of Rs. 4.25 per unit. In the year 1997, there is a fall in the demand for home market which can consume 10,000 units at a sale price of Rs. 3.72 per unit. The analysis of the cost per 10,000 units is as follows:

	Rs.
Materials	15,000
Wages	11,000
Fixed overhead	8,000
Variable overheads	6,000

The foreign market is explored and it is found that this market can consume 20,000 units of the product if offered at a sale price of Rs. 3.55 per unit. It is also discovered that for additional 10,000 units of the product (over initial 10,000 units). The fixed overheads will increase by 10%.

Is it worthwhile to try to capture the foreign market?

(University of Madras, M. Com., April 2002)

CHAPTER 10

Solution:

PROFITABILITY STATEMENT FOR CURRENT HOME MARKET

	10,000 Units	
	Total	*Per unit*
Sales value	42,500	4.25
Less: Variable cost:		
Material	15,000	1.50
Wages	11,000	1.10
Variable overheads	6,000	0.60
Total variable cost	32,000	3.20
Contribution (Sales – V.C.)	10,500	1.05
Less: Fixed overheads	8,000	0.80
Profit	2,500	0.25

PROFITABILITY STATEMENT AT REVISED SELLING PRICE IN HOME MARKET SALES AND FOREIGN MARKET SALES

	Home market		*Foreign market*		*Total*
	10,000 Units		*20,000 Units*		*30,000 Units*
	Total	*Per unit*	*Total*	*Per unit*	*Total*
Sales (revised)	37,200	3.72	71,000	3.55	1,08,200
Less: Variable cost:					
Material	15,000	1.50	30,000	1.50	45,000
Wages	11,000	1.10	22,000	1.10	33,000
Variable overheads	6,000	0.60	12,000	0.60	18,000
Total variable cost	32,000	3.20	64,000	3.20	96,000
Contribution (Sales – V.C.)	5,200	0.52	7,000	0.35	12,200
Less: Fixed overheads	8,000		1,600		9,600
Profit/Loss (–)	(–) 2,800		5,400		2,600

Comment:

It is better to accept foreign market order because after accepting foreign market order the company earns a profit of Rs. 2,600.

Working Note:

Upto 10,000 units fixed overhead = Rs. 8,000

For additional 10,000 units, 10% is increase in fixed cost

∴ For 20,000 units 20% is increase in Fixed cost

$$= \text{Rs. } 80{,}000 \times \frac{20}{100} = \text{Rs. } 1{,}600$$

Problem 40. *(Export Order):* A company now operating at 70% capacity gives the following data relating to the year 2009.

Number of units produced and sold –	3,500
Cost per unit:	
	Rs.
Material	15
Labour	10
Variable overheads	5
Fixed overheads	5
Selling price	50

An export order has been received which requires the rise of the balance of capacity. What minimum price should be quoted per unit for the export order if the company wants to make an overall total profit of Rs. 75,000 in the next year? Assume that the price to be quoted for export order will not affect home market.

Solution:

STATEMENT OF MARGINAL COST

	Per unit	*Existing 70% = 3,500 Units*	*Export order Balance 30% = 1,500 Units*	*Total for 5,000 Units*
Material	15	52,500	22,500	75,000
Labour	10	35,000	15,000	50,000
Variable overhead	5	17,500	7,500	25,000
Total V.C	30	1,05,000	45,000	1,50,000
Fixed cost	5	17,500	–	17,500
Total cost	35	1,22,500	45,000	1,67,500
Profit	15	52,500	22,500 See Note (3)	75,000
Sales	50	1,75,000	67,500	2,42,500

As per above statement, the export order selling price = $\frac{67,500}{1,500 \text{ units}}$ = Rs. 45 per unit.

Working Notes:

(1) Company produced 3500 units at 70% capacity. The balance 30% is export order

i.e., $\frac{3,500}{70} \times 30 = 1,500$ units

(2) The fixed cost in additional production for export order is taken as "NIL", because total fixed cost is already covered in existing 3,500 units.

CHAPTER 10

(3) Total profit required in next year	=	75,000
Less: Profit earned in existing production of 3,500 units	=	52,500
Balance profit to be required in export order	Rs.	22,500

Problem 41. A manufacturer has planned his level of operation at 50% of his plant capacity of 30,000 units. His expenses are estimated as follows, if 50% of the plant capacity is utilised:

(*i*)	Direct materials	Rs. 8,280
(*ii*)	Direct wages	Rs. 11,160
(*iii*)	Variable and other manufacturing expenses	Rs. 3,960
(*iv*)	Total expenses irrespective of capacity utilisation	Rs. 6,000

The expected selling price in the domestic market is Rs. 2 per unit. Recently, the manufacturer has received a trade enquiry from an overseas organisation interested in purchasing 6,000 units at a price of Rs. 1.45 per units.

As a professional management what would be your suggestion regarding acceptance or rejection of the offer ? Support your suggestion with suitable quantitative information.

(ICWA, Inter, Dec. 1991)

Solution:

STATEMENT OF COSTS AND PROFITS

Particular	*15,000 units (Expected output for domestic market)*		*6,000 units (Additional output for the foreign market)*		*Total*
	Amount	*Unit*	*Amount*	*Unit*	
Direct materials	8,280	0.552	3,312	0.552	11,592
Direct wages	11,160	0.744	4,464	0.744	15,624
Variable and other mfg. expenses	3,960	0.264	1,584	0.264	5,544
Variable cost	23,400	1.560	9,360	1.560	32,760
Sales	30,000	2.000	8,700	1.450	38,700
Contribution	6,600	0.440	(660)	0.11	5,940
Less: Fixed cost	6,000				6,000
Profit (Loss)	600		(660)		(60)

The price that can be obtained from the foreign market is Rs. 1.45 per unit of additional product which is less than the estimated variable cost of production, *i.e.,* Rs. 1.56 per unit. This will result in an estimated negative contribution of Re. 0.11 per unit and thereby will generate an expected loss of Rs. 660 on 6,000 additional costs.

Therefore, if the offer from the overseas organisation is accepted, the profit of Rs. 6,000 on 15,000 units (based on the 50% plant capacity utilisation) will be wiped out and expected net loss of Rs. 60 will arise. Therefore, it is suggested not to accept the offer.

Problem 42. A company manufacturing electric motors at a price of Rs. 6,900 each made up as under:

Direct material	Rs. 3,200
Direct labour	400
Variable overheads	1,000
Fixed overheads	200
Depreciation	200
Variable selling overheads	100
Royalty	200
Profit	1,000
	6,300
Central excise duty	600
	6,900

1. A foreign buyer has offered to buy 200 such motors at Rs. 5,000 each. As a cost accountant of the company, would you advise acceptance of the offer ?
2. What should the company quote for a motor to be purchased by a company under the same management if it should be at cost? *(CS, Inter, June 1990)*

Solution:

STATEMENT SHOWING CONTRIBUTION

Price offered			5,000
Less:	Variable cost		
(i)	Direct material	3,200	
(ii)	Direct labour	400	
(iii)	Variable overhead	1,000	
(iv)	Variable selling overheads	100	
(v)	Royalty (presumed to be production based)	200	4,900
		Contribution	100

From the above it appears that it would be worthwhile to accept the offer since the price offered covers marginal cost in full and gives Rs. 100 contribution towards the recovery of fixed cost. Secondly there will not be the incidence of central excise duty because this is an export order. Further there will also be government incentive for export order. This will help the company to recover further the fixed cost and balance if any the depreciation.

Problem 43. *ABC* Ltd. has been operating at 70% capacity and the following particulars relate to the present level of activity:

Units	3,500
Direct materials	Rs. 10,500
Direct labour	Rs. 5,250
Variable overhead (20% of direct labour)	
Fixed overhead	14,000
Sales	47,250

ABC Ltd. received an export for 1,000 units of its product @ Rs. 10 per unit. The export order will entail on extra expenditure in the form of special packing at Re. 0.50 per unit.

Is the export order worth trying? *(University of Madras, B.Com., March 95)*

CHAPTER 10

Solution:

STATEMENT SHOWING PROFIT

	Present 70% capacity 3,500 units @ Rs. 13.50		*Extra order 1,000 units @ Rs. 10*		*Total*	
Sales:	47,270		1,000			57,250
Less: Variable cost:						
Direct materials @ Rs. 3	10,500		3,000		13,500	
Direct labour @ Rs. 1.50	5,250		1,500		6,750	
Variable cost @ Rs. 3	10,500		3,000		13,500	
Packing @ Rs. 0.50	–		500		500	
		26,250		8,000		34,250
Contribution		21,000		2,000		23,000
Less: Fixed overhead		14,000		–		14,000
Profit		7,000		2,000		9,000

Comments:

The extra order is worth and it will earn extra profit of Rs. 2,000 subjected to reduction of selling price for Rs. 13.50 to Rs. 10 in export will not affect the local market. Therefore, the export order may be worth trying.

12. Differential pricing in the different markets/pricing in depression or recession: Under normal circumstances, the selling price is fixed by adding a margin of profit to the total cost. However, selling price is not always determined by the total cost of production. The market conditions play an important factor and hence, selling price is sometimes fixed considering market condition. In the long-run they tend to equal the cost of production of marginal firm. Occasionally, a factory may have to sell below the total cost. Such pricing has the following advantages:

(*a*) The services of talented employees can be utilised without discharging them.

(*b*) The idle capacity of plant may be prevented.

(*c*) The factory will be in a position to take the benefit arising out of a favourable condition at a later stage.

(*d*) The business can compete successfully.

Problem 44. A Company is at present working at 90% of its capacity and producing 13,500 units per annum. It operates a flexible budgetary control system. The following figures (excluding material and labour cost) are obtained from its budget.

		90%	*100%*
(*a*)	Sales	Rs. 15,00,000	16,00,000
(*b*)	Fixed expenses	Rs. 3,00,500	3,00,500
(*c*)	Semi-fixed expenses	Rs. 97,500	1,00,500
(*d*)	Variable expenses	Rs. 1,42,000	1,49,500

Material and labour cost per unit are constant under present conditions. Profit margin is 10% at 90% capacity.

(*a*) You are required to determine the cost of producing an additional 1,500 units.

(*b*) What would you recommend for an export price for these 1,500 units taking into account that overseas are much lower than indigenous prices? *(ICWA, Inter, June 1990)*

Solution:

(*i*) **Calculation of material and labour cost for 13,500 units:**

		Amount (Rs.)	Per Unit (Rs.)
Sales of 13,500 units		15,00,000	111.11
Less:	10% profit	1,50,000	11.11
		13,50,000	100.00
Less:	Overhead-Fixed, semi-fixed and variable	5,40,000	40.00
Material and labour cost		8,10,000	60.00

STATEMENT OF COST OF PRODUCEING 1,500 UNITS

	Total	Per Unit
Material & labour cost	90,000	60.00
Semi-fixed expenses (Differential cost 1,00,500 – 97,500)	3,000	2.00
Variable expenses (Differential cost 1,49,500 – 1,42,000)	7,500	5.00
Marginal cost	1,00,500	67.00

(*ii*) Indegenous price is Rs. 111.11. Though overseas prices are much lower than indigenous prices, an export price of Rs. 67 per unit plus any direct costs incurred for export is recommended in view of the foreign exchange earnings.

Problem 45. Banu Company Ltd. produce three products *A, B* and *C* through two departments, *viz.,* machining department and finishing department. The production costs are as follows:

	A (Rs.)	*B (Rs.)*	*C (Rs.)*
Material cost per unit	46	70	100
Labour Hours:			
Machine department at Rs. 5 per hr.	3	4	2
Finishing department at Rs. 10 per hr.	2	2	6
Selling price per unit	140	160	275

Overhead rates based on normal production budget are:

Machine department. Rs. 15 per hour (40% variable)

Finishing department at Rs. 10 per hour (50% variable)

Present information to the management to enable relative profitability of the three products to be assessed in:

(*a*) a time of depression

(*b*) normal times

(*c*) when there is a shortage of labour in finishing department. *(ICWA, Inter, June 1990)*

Solution:

		A	*B*	*C*
Selling price unit		140	160	274
Less: Material		46	70	100
Added value		94	90	174
Variable cost:				
Machine dept:	Labour	15	20	10
	Overhead	18	24	12
Finishing dept:	Labour	20	20	60
	Overhead	8	8	24
	Total variable cost	61	72	106

Contribution		33	18	69
Fixed cost:	Machinery	27	36	18
	Finishing	8	8	24
	Total fixed cost	35	39	42
Profit/Loss		– 2	– 26	27

(*a*) During depression all the products are to be manufactured.

(*b*) During normal times, *Z is* the only profitable product to be manufactured.

(*c*)

	A	B	C
Finishing hours per unit	2	2	6
Contribution per hour	Rs. 16.5	Rs. 9	Rs. 11.5
Profitability ranking	A	C	B

13. Decision-making and profit maximisation based on key factor: It is a situation under which a company has a choice between various types of product which it can manufacture but subjected to limiting factor. A limiting factor or a key factor or a factor involved in the productive activity of company, which at a point of time will limit the production capacity. It is also, therefore, termed as scarce factor, principal budget factor and so on. The limiting factor is affected by both internal and external environment. The contribution per unit of key factor is ascertained and it can be maximised according to the priority needs. Some of the examples of key factors are scarce raw materials, shortage of labour, plant capacity, inefficient management, shortage of capital and shortage of demand.

Problem 46. *(Key factor):* A manufacturing company has submitted the following particulars for the year 2009. The limiting factor is the availability of labour. The unit cost details are as under

		Product X	*Product* Y
No of units produced		1,000	1,100
Materials		24	26
Labour 8 Hrs @ Rs. 2		16	
3 Hrs. @ Rs. 2			6
Variable overheads		2	1
Fixed overheads		5	2
	Total cost	47	35
Selling price		50	40
	Profit	3	5

Find out which product is more profitable.

Solution:

STATEMENT SHOWING CONTRIBUTION AND PROFIT

	Product X Unit cost	Product X Total (1000 units)	Product Y Unit cost	Product Y Total (1100 units)
Sales	50	50,000	40	44,000
Less: Variable cost	42	42,000	33	36,300
Contribution	8	8,000	7	7,700
Less: Fixed cost	5	5,000	2	2,200
Profit	3	3,000	5	5,500

Contribution per unit of labour which is limiting factor

$$\text{Profitability} = \frac{\text{Contribution}}{\text{Key factor}}$$

$$\text{Product } X = \frac{\text{Rs. } 8}{8 \text{ Hrs,}} = \text{Re. } 1.00$$

$$\text{Product } Y = \frac{\text{Rs. } 7}{3 \text{ Hrs.}} = \text{Rs. } 2.33$$

Comment

(1) On the basis of contribution product 'X' is profitable.

(2) On the basis of net profit product 'Y' is profitable.

(3) On the basis of labour limiting key factor product 'Y' is profitable.

Problem 47. *(Key factor):* The following are the particulars submitted to you. Find out the profitability if the key factor is (*a*) Material (*b*) Labour

	Product P	*Product Q*
Materials:		
5 kg. @ 20	100	
6 kg @ 20		120
Wages:		
10 hrs. @ Rs. 4	40	
6 hrs. @ Rs. 4		24
Variable overheads	10	6
Fixed overheads	2	2
Total cost	152	152
Selling price	175	175

Solution:

STATEMENT OF CONTRIBUTION AND PROFIT

		Product P		*Product* Q
Selling price		175		175
Less: Variable expenses:				
Materials	100		120	
Wages	40		24	
Variable overheads	10		6	
		150		150
Contribution		25		25
Less: Fixed overheads		2		2
Profit		23		23

(1) Calculation of profitability when material is the key factor

$$\text{Profitability} = \frac{\text{Contribution}}{\text{Material consumption}}$$

$$\text{Product } P = \frac{\text{Rs. } 25}{5 \text{ kg.}} = \text{Rs. } 5$$

$$\text{Product } Q = \frac{\text{Rs. } 25}{6 \text{ kg.}} = \text{Rs. } 4.167$$

(2) Calculation of profitability when labour is the key factor

$$\text{Profitability} = \frac{\text{Contribution}}{\text{Labour hours}}$$

$$\text{Product } P = \frac{\text{Rs. } 25}{10 \text{ Hrs.}} = \text{Rs. } 2.50$$

$$\text{Product } Q = \frac{\text{Rs. } 25}{6 \text{ Hrs.}} = \text{Rs. } 4.167$$

Comments:

(1) The contribution and net profit for both the products are same

(2) If material is the key factor, product '*P*' is profitable.

(3) If labour is the key factor, product *Q* is profitable.

Problem 48. *(Key factor):* From the following particulars find out the profitable product mix and prepare a statement of profitability:

	Product A	*Product* B	*Product* C
Maximum units produced and sold	1,500	2,000	1,000
Selling price per unit	Rs. 60	Rs. 55	Rs. 50
Requirement per unit:			
Direct material	5 kg.	3 kg.	4 kg.
Cost of direct material per kg.	Rs. 4	Rs. 4	Rs. 4
Direct labour	4 Hrs.	3 Hrs.	2 Hrs.
Direct wages per hour	Rs. 2	Rs. 2	Rs. 2
Variable overhead	Rs. 9	Rs. 14	Rs. 6
Fixed overhead	Rs. 5	Rs. 5	Rs. 5

Total availability of direct materials – 12,000 kg.

Total availability of direct labour – 10,000 Hrs.

All the products *A*, *B* and *C* are produced from the same direct material. Consider both material and labour as the key factors.

Solution:

STATEMENT OF PROFITABILITY

	A		*B*		*C*	
Selling price		60		55		50
Less: Variable cost:						
– Direct material (Rs. 4 per kg)	20		12		16	
– Direct wages Rs. 2 per Hr	8		6		4	
– Variable overheads	9	37	14	32	6	26
Contribution		23		23		24
Less: Fixed cost		5		5		5
Profit		18		18		19

Key Factors:

$$\frac{\text{Contribution per unit}}{\text{Key factor}}$$

(1) Material as a key factor $= \dfrac{\text{Rs. } 23}{5 \text{ kg.}} = \dfrac{\text{Rs. } 23}{3 \text{ kg.}} = \dfrac{\text{Rs. } 24}{4 \text{ kg.}}$

$= 4.6 = 7.67 = 6$

(2) Labour as a key factor $= \dfrac{\text{Rs. } 23}{4 \text{ Hrs.}} \quad \dfrac{\text{Rs. } 23}{3 \text{ Hrs.}} \quad \dfrac{\text{Rs. } 24}{2 \text{ Hrs.}}$

$= 5.75 = 7.67 = 12$

Comment on profitable product mix

(1) If material is considered as key factor. Product '*B*' is first product '*C*' is second product '*A*' is third.

Product mix on availability of 12,000 kgs are:

First: Product '*B*' maximum production is 2,000 units × 3 kg. per unit = 6,000 kg.

Second: Product '*C*' maximum production is 1,000 units × 4 kg per unit = 4,000 kg.

Third: Product '*A*' produced only 400 units × 5 kg = 2,000 kg

i.e. Balance material is 12,000 kg – 6,000 kg – 4,000 kg = 2000 kg

$\dfrac{2{,}000 \text{ kg}}{5} = 400$ units

12,000 kgs.

(2) If labour is considered as key factor product 'C' is first product 'B' is second, product 'A' is third:

Product mix on availability of 10,000 hrs.

First:	Product 'C' maximum production is is 1,000 units × 2 hrs. per unit	=	2,000 hrs.
Second:	Product 'B' maximum production is 2,000 units × 3 hrs. per unit	=	6,000 hrs.
Third:	Product 'A' is produced only 500 units × 4 hrs per unit	=	2,000 hrs.
	Balance hours is 10,000 Hrs – 2,000 – 6,000		
	$= \frac{2{,}000 \text{ hrs}}{4} = 500$ units		
			10,000 hrs.

Problem 49. You are given the following information in respect of products X and Y of AB Company Ltd.

	Product X	*Product* Y
Selling price	Rs. 42	Rs. 33
Direct materials	Rs. 15	Rs. 15
Labour hours	18 hrs.	9 hrs.

(50 paise per hours)

Variable overheads 50% of direct wages Rs. 6,750

Show which product is more profitable during labour shortage.

Solution:

STATEMENT SHOWING CONTRIBUTION

	Product X	*Product* Y
Selling price	42	33.00
Less: Variable cost	28.50	21.75
Contribution	13.50	11.25

$$\text{Profitability} = \frac{\text{Contribution}}{\text{Key factor}}$$

$$\text{Product } X = \frac{13.5}{18} = 0.75$$

$$\text{Product } Y = \frac{11.25}{9} = 1.25$$

Thus, product 'Y' is more profitable than X during labour shortage.

14. Dropping out a product: When a company produces several range of products, a decision that is to be often taken is whether all the products are to be produced in spite of the fact that one or two products are responsible for incurring losses. Marginal costing helps in taking decision of such nature by calculating the margin of contribution.

Problem 50. *(Dropping out a product):* The cost per unit of the three products *A*, *B* and *C* of a company are given below:

	A	*B*	*C*
Direct materials	20	16	18
Direct labour	12	14	12
Variable overhead	8	10	6
Fixed expenses	6	6	4
Total	46	46	40
Profit	18	14	12
Selling price	64	60	52
No. of units produced	10,000	5,000	8,000

Production arrangements are such that if one product is given up the production of the others can be raised by 50%. The directors propose that product '*C*' should be given up because the contribution from that product is the lowest. Do you agree ? *(University of Madras, M.Com., May 2005)*

Solution:

STATEMENT OF MARGINAL COST

Particulars	*A*		*B*		*C*	
Sales		64		60		52
Less: Variable cost						
– Material	20		16		18	
– Labour	12		14		12	
– Variable overhead	8	40	10	40	6	36
Contribution		24		20		16
Less: Fixed cost		6		6		4
Profit		18		14		12
Production		10,000 Units		5,000 Units		8,000 Units

Total Fixed Cost:

A	=	10,000 units × Rs. 6	=	60,000
B	=	5,000 units × Rs. 6	=	30,000
C	=	8,000 × Rs. 4	=	32,000
		Total fixed cost	=	1,22,000

(1) If Product 'A' is given up

	A	*B*	*C*
Existing production	NIL	5,000	8,000
Add: 50 % increase	NIL	2,500	4,000
	NIL	7,500	12,000

Profit

Contribution of '*B*' (7,500 units × 20)	=	1,50,000
Contribution of '*C*' (12,000 units × 16)	=	1,92,000
		3,42,000
Less: Total fixed cost		1,22,000
Profit		2,20,000

CHAPTER 10

(2) If Product 'B' is given up

	A	B	C
Existing production	10,000	NIL	8,000
Add: Increase 50%	5,000	NIL	4,000
	15,000	NIL	12,000

Profit

Contribution of 'A' (15,000 units × 24)	=	3,60,000
Contribution of 'C' (12,000 units × 16)	=	1,92,000
		5,52,000
Less: Total fixed cost		1,22,000
Profit		4,30,000

(3) If Product 'C' is given up

	A	B	C
Existing production	10,000	5,000	NIL
Add: Increase 50%	5,000	2,500	NIL
	15,000	7,500	NIL

Profit

Contribution of 'A' (15,000 × 24)	=	3,60,000
Contribution of 'B' (7,500 × 20)	=	1,50,000
		5,10,000
Less: Total fixed cost		1,22,000
		3,88,000

Comments:

As per above when product 'B' given up, we get maximum profit. Therefore product 'C' given up is not advisable.

Problem 51. Synthetic Rubber Company Ltd. manufactures and sell three varieties of shoes. The income statement for the year ending 31.12.2004 is as follows:

		Varieties of shoes (Rs. in lakh)		
		A	B	C
Sales		1.00	4.00	2.00
Works cost:	Variable	50	1.80	70
	Fixed	30	1.00	40
Selling cost:	Variable	12	50	20
	Fixed	10	40	20
Profit/Loss		(2)	30	50

The income statement reveals loss in the product *A*, previous years also. The management is considering a proposal to discontinue *A* and intensify production of *C*. It is anticipated that an advertisement of Rs. 20,000 annually would yield an increase of 30% in its sales volume and this would exactly fit in the plant facilities released by discontinuance of the product *A*.

The variable overheads both works and selling, etc. are directly proportional to sales. The total fixed overheads will remain the same irrespective of the change in production policy. The present allocations are based on capacity utilisation on works side and sales effort on selling side.

The management desires to know comparative positions whether (*a*) the present production/sales policy shoud be continued, (*b*) the product *A* to be discontinued without any further change or (*c*) the proposal of increasing the product *C* in lieu of the product *A* should be accepted. Present the information assuming sales and costs will remain the same as in last year.

Solution:

(*a*) Present policy

				Products (Rs. in lakh)			
				A	*B*	*C*	*Total*
Sales				1.00	4.00	2.00	7.00
Variable cost:				0.62	2.30	0.90	3.82
Works	0.50	1.80	0.70				
Selling	0.12	0.50	0.20				
	0.62	2.30	0.90				
Contribution				0.38	1.70	1.10	3.18
Fixed cost:							
Works	0.30	1.00	0.40				
Selling	0.10	0.40	0.20	0.40	1.40	0.60	2.40
	0.40	1.40	0.60				
Profit or loss				0.02	0.30	0.50	0.78

(*b*) If 'A' is discontinued

	Products		
	B	*C*	*Total*
Sales	4.00	2.00	6.00
Variable cost	2.30	0.90	3.20
Contribution	1.70	1.10	2.80
Less: Fixed cost			2.40
Profit			0.40

(*c*) If 'C' is expanded

	Products		
	B	*C*	*Total*
Sales	4.00	2.60	6.60
Variable cost	2.30	1.17	3.47
Contribution	1.70	1.43	3.13
Fixed cost & Advertisement			2.60
Profit			0.53

From the above statement it is clear that the profit is highest under the present policy. Hence, the present policy should continue.

Problem 52. *(Henry demand/low demand situation):* Estimated profit and loss for 2008 for two companies are given

	A Company	*B Company*
Sales	3,00,000	3,00,000
Fixed cost	50,000	25,000
Variable cost	2,00,000	2,25,000
Estimated profit	50,000	50,000

Which company can earn more profit under conditions of:

(*i*) Heavy demand for the product?

(*ii*) Low demand for the product?

Solution:

PROFITABILITY STATEMENT

	A Company	*B Company*
Sales	3,00,000	3,00,000
Less: Variable cost	2,00,000	2,25,000
Contribution	1,00,000	75,000
Less: Fixed cost	50,000	25,000
Profit	50,000	50,000
P/V Ratio $= \frac{\text{Contribution}}{\text{Sales}} \times 100$		
'A' Company $= \frac{1,00,000}{3,00,000} \times 100$	$= 33\frac{1}{3}\%$	
'B' Company $= \frac{75,000}{3,00,000} \times 100$		$= 25\%$
BEP $= \frac{\text{Fixed expenses}}{\text{P / V Ratio}}$		
'A' Company $= \frac{50,000}{33\frac{1}{3}\%}$	$= 1,50,000$	
'B' Company $= \frac{25,000}{25\%}$		$= 1,00,000$

Profit position:

(*a*) In case of heavy demand for product:

'*A*' company will do better than '*B*' company, since additional sales will contribute a profit of $33\frac{1}{3}\%$ over break-even point.

(*b*) In case of low demand for product:

'*B*' company will do better profit than '*A*' company, since even in low additional sales will contribute a profit of 25% over break-even point. Because break-even point of '*B*' company is at lower level of Rs. 1,00,000 compared to '*A*' company.

Problem 53. *(Selection of best proposal or product mix):* A plant is capable of operating 5,000 machine hours during a month, interchangeable for any one or more products, Alpha, Beta or Gamma. The following details are ascertained:

Product	*No. of units per 100 machine hrs.*	*Material cost per unit (Rs.)*
Alpha	100	2
Beta	50	3
Gamma	25	5

(*a*) Labour cost per machine hour is Rs. 2 variable overhead per machine hour is Rs. 3.

(*b*) The fixed overhead of the department per month is Rs. 10,000.

(*c*) The share capital is Rs. 5,40,000 and the directors expect 20% profit before taxation.

(*d*) All financial expenses are included in the cost. The following two proposals are to be considered:

(*i*) To fix the selling price by adding 20% on cost.

(*ii*) To fix prices arbitrarily Rs. 11, Rs. 20, Rs. 40 for Alpha, Beta, and Gamma respectively.

Evaluate the aforesaid proposals and select the best proposal.

(Osmania University, M.Com., May 2001)

Solution:

Working Note

(1) Calculation of labour hours per unit

$$\text{Alpha} = \frac{100}{100} = 1$$

$$\text{Beta} = \frac{100}{50} = 2$$

$$\text{Gamma} = \frac{100}{25} = 4$$

(2) Calculation of monthly production

$$\text{Alpha} = \frac{5,000}{1} = 5,000$$

$$\text{Beta} = \frac{5,000}{2} = 2,500$$

$$\text{Gamma} = \frac{5,000}{25} = 1,250$$

(3) Calculation of selling price per unit

$$\text{Alpha} = \frac{54,000}{5,000} = 10.8$$

$$\text{Beta} = \frac{51,000}{2,500} = 20.4$$

$$\text{Gamma} = \frac{49,500}{1,250} = 39.6$$

CHAPTER 10

COST SHEET

Particulars	Alpha	Beta	Gamma
Units produced in a month	5,000	2,500	1,250
Materials	2	3	5
Labour	2	4	8
Overheads	3	6	12
Variable cost per unit	7	13	25
Total variable cost	35,000	32,500	31,250
Fixed cost	10,000	10,000	10,000
Total cost	45,000	42,500	41,250
Profit (20% on cost)	9,000	8,500	8,250
Sales	54,000	51,000	49,500

	Alpha	Beta	Gamma
Proposal 'A'			
Selling price per unit	10.8	20.4	39.6
Less: Variable cost	7.0	13.0	25
Contribution	3.8	7.4	14.6
Total contribution	19,000	18,500	18,250
Proposal 'B'			
Selling price	11	20	40
Less: Variable cost	7	13	25
Contribution per unit	4	7	15
Total contribution	20,000	17,500	18,750

Decision making:

After making calculations, it is suggested to the management, that they should take up the second proposal because it rewards more contribution against the first proposal.

Problem 54. *(Product mix):* A producer installed a machine which can produce product *'A'* as well as product *'B'*. Annual maximum machine running capacity is 4,000 hours. Cost details about the product are as follows:

	Product 'A'	*Product 'B'*
Selling price per unit	Rs. 50	Rs. 20
Variable cost per unit	Rs. 30	Rs. 12
Machine hours required per unit of product	10 hrs	2 hrs
Annual demand	300 units	1600 units
Annual fixed cost Rs. 10,000		

Calculate optimum product mix showing annual contribution and profit. Give necessary explanation. Also show that a product-mix other than that suggested by you will affect the profit.

(University of Delhi, B.Com., (Hons) 2005)

Solution:

STATEMENT OF MARGINAL COST

	Product 'A'	*Product 'B'*
Selling price	50	20
Less: Variable cost	30	12
Contribution	20	8
Machine hours per unit	10	2
Contribution per hour	2	4
Ranking	II	I

Suggested sales mix

Product *A* 80 units × 10 hours	=	800 hrs.
Product *B* 1600 units × 2 hours	=	3,200 hrs
		4,000 hrs

Profit at the suggested sales mix:

Product *'A'* contribution:		
80 units × Rs. 20	=	1,600
Product *'B'* contribution:		
1600 units × Rs. 8	=	12,800
Total		14,400
Less: Fixed cost		10,000
Profit		4,400

In order to earn maximum profit, we have to choose a product-mix where contribution per unit of key factor is maximum. If we choose any other product-mix, it will show lower profit. For example, if we have the following product mix.

Product *'A'* 100 units × 10 hrs	=	1,000 hrs
Product *'B'* 1,500 units × 2 hrs	=	3,000 hrs
		4,000 hrs

Profit will be:

A's contribution 100 × Rs. 20	=	2,000
B's contribution 1500 × Rs. 8	=	12,000
Total		14,000
Less: Fixed cost		10,000
Profit		4,000

Questions

I. Choose the correct answer

1. Production cost under marginal costing include:
   (*a*) Prime cost only
   (*b*) Prime cost and variable overhead
   (*c*) Prime cost and fixed overhead
   (*d*) Prime cost, variable overhead and fixed overhead [ ]
2. One of the primary difference between marginal costing and absorption costing is regarding the treatment of:
   (*a*) Direct material (*b*) Variable overhead
   (*c*) Fixed overhead (*d*) Prime cost [ ]
3. Period costs are:
   (*a*) Variable costs (*b*) Fixed cost
   (*c*) Prime cost (*d*) Overhead costs [ ]
4. Absorption costing differs from marginal costing in the:
   (*a*) Fact that standard costs can be used with absorption costing but not with marginal costing.
   (*b*) Amount to fixed costs that will be incurred
   (*c*) Kind of activities for which each can be used.
   (*d*) Amount of costs assigned to individual units of products. [ ]
5. Choose the true statement from the following:
   (*a*) Variable costing is a method of costing in which only direct materials and direct labour are charged to work-in-process.
   (*b*) Under variable costing, all costs which vary with sales volume are charged to work-in-progress.
   (*c*) Because certain manufacturing costs are treated as period costs, net earnings will always be lower under variable costing than under full costing.
   (*d*) Variable costing clearly points out the fact that if additional units of product can be sold for more than variable costs of producing and selling them, net earnings will increase. [ ]

[***Answers:*** 1. (*b*), 2. (*c*), 3. (*c*), 4. (*d*), 5. (*d*)]

II. Mark true or false

1. In marginal costing, fixed costs are excluded in the valuation of work-in-progress and finished goods stocks. *T/F*
2. Marginal costing may be used in conjunction with standard costing or budgetary control. *T/F*
3. In marginal costing, fixed costs are apportioned on some arbitrary basis. *T/F*
4. In absorption costing, the valuation of stock is higher than in marginal costing. *T/F*
5. In marginal costing, managerial decisions are guided by contribution margin than by profit. *T/F*
6. Absorption costing is generally accepted approach to product cost determination. *T/F*
7. Under marginal costing, variable overhead is treated as part of the cost of an assets, the goods produced, while fixed overhead is viewed as an expense. *T/F*
8. Marginal costing is sometimes referred to as variable costing. *T/F*

9. Under marginal costing, variable overhead is a period cost. *T/F*

10. Under variable costing only the costs of materials and labour are reviewed as product cost. *T/F*

11. Fixed factory overhead is not treated as a product cost under variable costing while it is so treated under full costing. *T/F*

12. Variable selling expenses are deducted from revenue under the variable costing technique in arriving at net earnings. *T/F*

13. In general, if production volume exceeds sales volume, net earnings under full costing will be greater than under marginal costing. *T/F*

14. In the earning statement under marginal costing, marginal earnings are directly related to the volume of units sold. *T/F*

15. An unresolved issue in the variable full costing controversy is whether fixed overhead costs are period cost or products costs. *T/F*

16. The effect upon net earnings of the sale of some additional units of product is readily determinable under full costing. *T/F*

[***Answers:*** *True*: 1, 2, 4, 5, 6, 7, 8, 11, 13, 14, 15, *False*: 3, 9, 10, 12, 16]

III. Short answer questions

1. What is 'analysis of margin of contribution'? Discuss the need for it.

(Calicut University, B.Com., April 1992)

2. Define marginal costing. What are the features of marginal costing?

(Bangalore University, B.Com., April 1992)

Exercises

Income Statement under Absorption Costing and Marginal Costing

1. "LMN" Limited sells its products at Rs. 3 per unit. The company uses a first-in-first-out actual costing system. A new fixed manufacturing overhead allocation rate is computed each year by dividing the actual fixed manufacturing overhead cost by the actual production costs. The following simplifed data are related to its first two years of operation.

	Year I	Year II
Unit data:		
Sales	1,000	1,200
Production	1,400	1,000
Cost:	Rs.	Rs.
Variable manufacturing	700	500
Fixed manufacturing	700	700
Variable marketing and administration	1,000	1,200
Fixed marketing and administration	400	400

Required:

(1) Prepare income statement based on

(*a*) absorption costing

(*b*) marginal costing for each year. *(University of Delhi, B.Com. (Hons.) 2001)*

[**Ans.**

	Absorption costing	
	I year	II year
Profit	600	640

	Marginal costing	
	I Year	II Year
Profit	400	700

2. Your company has a production capacity of 12,500 units and normal capacity utilisation is 80%. Opening inventory of finished goods on 1-1-2002 was 1,000 units. During the year ending 31-12-2002, it produced 11,000 units while it sold only 10,000 units.

Standard variable cost per unit is Rs. 6.50 and standard fixed factory cost per unit is Rs. 1.50. Total fixed selling and administration overhead amounted to Rs. 10,000. The company sells its product at Rs. 10 per unit.

Prepare income statements under absorption costing and marginal costing.

(University of Delhi, B.Com. (Hons.), 2000)

[**Ans.** Profit under Absorption costing Rs. 11,500 and under Marginal costing Rs. 10,000]

3. From the following information, recommend the best sales mix

Sales mixtures:

(*a*) 100 units of A and 200 of B

(*b*) 150 units of A and 150 of B

(c) 200 units of A and 100 of B

Other information:

Direct materials	A Rs. 10 per unit
	B Rs. 9 per unit
Direct wages	A Rs. 3 per unit
	B Rs. 2 per unit
Selling price	A Rs. 20 per unit
	B Rs. 15 per unit

Fixed expenses are allotted to the products at 100% of direct wages.

(University of Madras, B.Com., May 2004)

[**Ans.** Plan 'C' *i.e.,* A = 200 units, B = 100 units gives more profit. Therefore, it is most profitable mix]

Export Order

4. The Novelties Ltd. due to industrial depression is operating its plant at present at 50% of its capacity. The following details are available:

	Rs.
Cost of production per unit	2
Direct labour	1
Variable overhead	3
Fixed overhead	2
Total	Rs. 8
Production per month	20,000 Units
Total cost of production	Rs. 1,60,000
Selling price	Rs. 1,40,000
Loss	Rs. 20,000

An exporter offers to buy 5,000 units per month at the rate of Rs. 6.50 per unit and the company hesitates to accept the offer for fear of increasing its already large operating losses.

Advise whether the company should accept or decline this offer

(University of Madras, B.Com., October 2003)

[**Ans.** The export of 5,000 units contribute Rs. 2,500 as profit. Therefore, it helps to reduce existing loss of Rs. 20,000 to 17,500. It is advisable to accept the export order @ Rs. 6.50]

Accepting Order in Domestic Market

5. A firm can purchase a spare part from an outside source at Rs. 11 per unit. There is a proposal that the spare part be produced in the factory itself. For this purpose a machine costing Rs. 1,00,000 with annual capacity of 20,000 units and a life of 10 years will be required. A foreman with a monthly salary of Rs. 500 will have to be engaged. Materials required will be Rs. 4 per unit and wages Rs. 2 per unit. Variable overheads are 150% of direct labour. The firm can easily raise funds for buying machine at 10% interest per annum. The firm's annual requirement of the said spare part is 15,000 units. Advise the firm whether the proposal should be accepted or not. *(Osmania University, M.Com., June 2002)*

[**Ans.** As the cost production comes to Rs. 10.60 per unit as compared to purchase price of Rs. 11, it is better for the firm to manufacture the product]

Key Factor

6. From the following particulars, find the most profitable product mix and prepare a statement of profitability of that product-mix

	Product A	*Product B*	*Product C*
Units budgeted to be produced and sold	1,800	3,000	1,200
Selling price per unit (Rs)	60	55	50
Direct materials required per unit (kg)	5	3	4
Direct labour per unit (Rs.)	4	3	2
Variable overhead (Rs.)	7	13	8
Fixed overhead (Rs.)	10	10	10
Cost of direct material per kg (Rs.)	4	4	4
Direct labour hour rate (Rs.)	2	2	2
Maximum possible units of sale	4,000	5,000	1,500

All the three products are produced from the same direct material using the same type of machine and labour. Direct labour which is the key factor is limited to 18,600 hours.

(University of Delhi, B.Com. (Hons.), 2004)

[**Ans.** Profit Rs. 96,750]

Product Mix

7. A company manufactures three products. The budgeted quantity, selling price and unit costs are as under

	A Rs.	*B* Rs.	*C* Rs.
Raw materials @ Rs. 20 per kg	80	40	20
Direct wages @ Rs. 5 per hour	5	15	10
Variable overheads	10	30	20

CHAPTER 10

Fixed overheads	9	22	18
Budgeted production (in units)	6,400	3,200	2,400
Selling price per unit: (in Rs.)	140	120	90

Required:

(*i*) Present a statement of budgeted profit

(*ii*) Set optimal product-mix and determine the profit, if the supply of raw material is restricted to 18,400 kg. *(University of Delhi, B.Com. (Hons), 2002)*

[**Ans.** (*i*) Budgeted profit of A 2,30,400, of B 41,600, of C 52,5800 (*ii*) Total profit: Rs. 1,44,800]

Profit Planning

8. The accounts of 'X' Ltd are expected to reveal a profit of Rs. 28,00,000 after charging fixed cost of Rs. 20,00,000 for the year ended 31st March 2002. The selling price of the product is Rs. 100 per unit and variable cost per unit is Rs. 40.

Market investigation suggest the following responses to the price changes:

Alternatives	*Selling price reduced by*	*Quantity sold increased by*
I	5%	10%
II	7%	20%
III	10%	25%

Evaluate these alternatives and state which of the alternatives on profitability consideration, should be adopted for the forthcoming year. *(University of Delhi, B.Com. (Hons), 2001)*

[**Ans.** Profit of proposal II (*i.e.*, Rs. 30,88,000) is higher than the other proposals. So this is to be chosen]

Make or Buy Decision

9. The management of a company finds that while the cost of making a component part is Rs. 10. The same is available in the market at Rs. 9 with an assurance of continuous supply. Give a suggestion to make or buy this part. The cost information is as follows.

	Rs.
Material	3.50
Direct labour	4.50
Other variable expenses	1.00
Fixed expenses	1.50
Total	10.00

(University of Madras, M.Com., May 2005)

[**Ans.** Cost to make is Rs. 10 and cost to buy is 10.50. So the decision should be to make]

Export Order

10. A company is functioning at 50% of its capacity. The following details are available

Production per month	20,000 Units
Direct material per unit	Rs. 2
Direct labour per unit	Rs. 1
Variable overhead per unit	Rs. 3
Fixed overhead per unit	Rs. 2
Selling price per unit	Rs. 7

An exporter offers to buy 5,000 units per month at the rate of Rs. 6.50 per unit. Advise whether the company should accept or reject the offer.

(University of Madras, M.Com., May 2005)

[Ans. An export of 5,000 units contribute Rs. 2,500 as profit. Therefore, it helps to reduce existing loss of Rs. 20,000 to Rs. 17,500. It is advisable to accept the export order @ Rs. 6.50]

Dropping Out a Product

11. The Excel Ltd. has three products. The budgeted data for the next year are as follows:

	A	*B*	*C*
	Rs.	*Rs.*	*Rs.*
Sales	1,20,000	56,000	84,000
Costs:			
Direct material	14,000	7,000	14,000
Direct labour	56,000	7,000	22,400
Variable overhead	14,000	7,000	28,000
Fixed cost	28,000	14,000	28,000

The management is considering to close down product C. There is no possibility of reducing fixed costs. Advise whether or not product 'C' should be closed down.

(University of Madras, M.Com., May 2005)

[Ans. Proposal to close down product 'C' is not advisable as it results in a loss of Rs. 7,000]

Sales Mix

12. From the following data you are required to present to the management:

(*a*) The marginal cost of products X, Y and the contribution per unit.

(*b*) The total contribution and profit resulting from each of the suggested sales mixture.

	Rate Per unit
Direct materials	
Product X	10.50
Product Y	8.50
Direct wages	
Product X	3.00
Product Y	2.00
Fixed expenses (Total) 800	
Variable expenses	
100 % of direct wages per product	
Selling price:	
Product X	20.00
Product Y	14.50

Suggested Sales mix	*No. of Units*	
	Product X	*Product Y*
(*a*)	100	120
(*b*)	150	150
(c)	200	100

(University of Madras, M.Com., April 2002)

CHAPTER 10

[**Ans.** Sales mix of 200 units of X and 100 units of Y gives more profit of Rs. 100. Therefore it is recommended]

Determining of the Volume of Output

13. A company produces and markets industrial containers and packing cases. Due to competition the company proposes to reduce the selling price. If the present profit is to be maintained indicate the number of units to be sold if the proposed reduction in selling price is (*a*) 5% and (*b*) 10%

The following additional information is available:

Present sales turnover (30,000 Units)		3,00,000
Variable cost (30,000 Units)	1,80,000	
Fixed cost	70,000	
		2,50,000
	Net profit	50,000.

(University of Madras, M.Com., May 2001)

[**Ans.** At 5% reduction in selling price, Number of units sold is 34,286 and at 10% reduction in selling price, number of units sold is 40,000 units]

Make or Buy Decision

14. An automobile manufacturing company finds that while the cost of making its own workshop part no 0028 is Rs. 6 each the same is available in the market at Rs. 5.60 with an assurance of continuous supply. The cost data is as follows:

		Rs.
Materials		2.00
Direct labour		2.50
Other variable costs		0.50
Depreciation and other fixed cost		1.00
	Total	6.00

Write a report to the Managing Director giving your views whether to make or buy this part. Give also your views in case the suppliers reduce the prices from 5.60 to Rs. 4.60.

(University of Madras, M.Com., May 2001)

[**Ans.** (1) The company should manufacture the part, if the part is available in the market at Rs. 5.60. Because the production of each part gives 0.60 paise contribution (Rs. – 5.60 – Rs. 5); (2) The company should not manufacture the part, if the part is available in the market at Rs. 4.60. Because addition cost of 0.40 paise (Rs. 5 – Rs. 4.60) is saved in manufacture.]

High/Low method of segregating semi-variable cost

15. From the following data segregate semi-variable cost into fixed and variable cost:

Month	*Units*	*Semi-variable cost*
January	15,000	12,000
February	12,000	11,400
March	14,000	11,800
April	16,000	12,200
May	18,000	12,600
June	17,000	12,400
	92,000	72,400

[**Ans.** For the month of January variable cost is Rs. 3,000 and fixed cost is Rs. 9,000. For the month of February the variable cost is Rs. 2,400 and fixed cost is Rs. 9,000 and so on]

High/Low method and method of least square

16. From the following semi-variable cost calculate the fixed and variable cost under (*i*) High and low method, (*ii*) Method of least squares:

Month	*Units*	*Semi-variable cost*
January	500	2,500
February	600	2,800
March	700	3,100
April	940	3,820
May	740	3,220
June	880	3,640
July	900	3,700
August	840	3,520
September	800	3,400
October	860	3,580
November	760	3,280
December	540	2,620

[**Ans.** For the month of December the fixed cost under high-low method is Rs. 1,000 and variable cost is Rs. 2,820. Under method of least squares the fixed cost is Rs. 1,000 and variable cost per unit is Rs. 3]

17. The following figures have been extracted form the books of a manufacturing company for the first half year ending 30th June, 1989:

Month	*Volume (Units)*	*Semi-variable cost overhead cost (Rs.)*
January	150	2,700
February	300	3,900
March	450	5,100
April	750	7,500
May	1,050	9,900
June	1,350	12,300

Find out the fixed and variable cost. *(University of Kerala, B.Com., May 1989)*

[**Ans.** Adopting indirect method the variable cost per unit is Rs. 8. The fixed cost is Rs. 1,500]

Make or Buy Decision

18. X Ltd. manufactures component *A*-100 and the costs for the year 2010 when 50,000 units are produced is as follows:

Materials	Rs. 2.50
Labour	Rs. 1.25
Variable overheads	Rs. 1.75
Fixed overhead	Rs. 3.50
Total cost	Rs. 9.00

Component *A*-100 could be bought in for Rs. 7.75 and if so the production capacity as utilised in 2010 would be unused. Assuming that there is no technical consideration. Should component *A* be bought in or manufactured?

CHAPTER 10

[**Ans.** The component should be manufactured and not bought because marginal cost of manufacture is Rs. 5.50 and the buying price is Rs. 7.75. The reason for this is that the fixed cost of Rs. 1,75,000 (50,000 units × Rs. 3,50) would continue and as the capacity is not used, the fixed overhead would not be absorbed into production. If component *A*-100 is bought, overall profit will fall by Rs. 1,12,500 being the difference between buying price and marginal cost of manufacture *i.e.*, (7.75 – 5.50 × 50,000)]

Make or Buy Decision based On Cost Savings

19. A firm is considering whether to manufacture or purchase a particular component. This would be in batches of 10,000 and the buying price is Rs. 6.50. The marginal cost of manufacturing this component is Rs. 4.75 per unit and the component will have to be made on a machine which was currently working at full capacity. If the component is manufactured, it is estimated that the sales of the finished product would be reduced by 1,000 units. The finished product has a marginal cost of Rs. 60 per unit and is sold at Rs. 80 per unit.

Should the firm manufacture or purchase the component?

[**Ans.** Marginal cost of manufacture is Rs. 47,500. Add lost contribution of the finished product 100 × Rs. 20 = 20,000/67,500. Buying price is Rs. 65,000 (10,000 × Rs. 6.50). Thus there is a saving of Rs. 1,500 per 10,000 batch by buying rather than manufacture.]

20. A Radios manufacturing company finds that while it costs Rs. 6.75 each to make component of 376 radios, the same is available in the market at Rs. 5.75 each with the assurance of continued supply. The breakdown of cost is as follows:

Materials	Rs. 2.75 each
Labour	Rs. 1.75 each
Other variable	Rs. 0.50 each
Fixed cost	Rs. 1.25 each
	Rs. 6.25

(*a*) Should you make or buy.

(*b*) What would be your decision if the supplier offered the component at Rs. 4.55 each?

(Bangalore University, M.Com., May 1990)

[**Ans.** The marginal cost of production is Rs. 5 whereas market price is Rs. 5.75. Since market price is more, the component is to be manufactured. If the offer is 4.85, the proposal to buy may be examined in relation to other factors, *viz.*, more profitable use of available resources, the fixed cost to be borne by other products, the ability to sell at the same rate in the long run by the supplier etc.]

Product Mix

21. Tourus Ltd. produces three products: *A*, *B* and *C* from the same manufacturing facilities. The cost and other details of the three products are as follows:

	A	*B*	*C*
Selling price per unit (Rs.)	200	160	100
Variable cost per unit (Rs.)	120	120	40
Fixed expenses per unit (Rs.)			2,76,000
Maximum production per month (units)	5,000	8,000	6,000
Total hours available for the month			
Maximum demand per month (unit)	2,000	4,000	2,400

The processing hours cannot be increased beyond 200 hours per month.

You are required to compute the most profitable product mix. *(C.A. Inter, May 1988)*

[**Ans.** Contribution of A = 1,60,000, B = 64,000. C = 1,44,000. Total contribution is 3,68,000 and profit is Rs. 92,000]

Product Mix

22. A company engaged in plantation activities has 200 hectares of virgin land which can be used for growing jointly or individually tea, coffee and cardamom. The yield per hectare of the difference crops and their selling prices per kg. are as under:

	Yield kg.	*Selling price Rs. per kg.*
Tea	2,000	20
Coffee	500	40
Cardamom	100	25

The relevant cost data are given below:

(*a*) Variable cost per kg.

	Tea (Rs.)	*Coffee (Rs.)*	*Cardamom (Rs.)*
Labour charges	8	10	120
Packing materials	2	2	10
Other costs	4	1	20
Total cost	14	13	150

(*a*) Fixed cost per annum

Cultivation and growing cost	10,00,000
Administrative cost	2,00,000
Land revenue	50,000
Repairs and maintenance	2,50,000
Other costs	3,00,000
Total cost	18,00,000

The policy of the company is to produce and sell all the three kinds of products and maximum and minimum area to be cultivated per products is as follows:

	Maximum area (Hectares)	*Minimum area (Hectares)*
Tea	160	120
Coffee	50	30
Cardamom	30	10

Calculate the priority of production, the most profitable product mix and the maximum profit which can be achieved. (*CA Inter, June 1988*)

[**Ans.** Contribution per hectare for tea is Rs. 12,000, for coffee Rs. 13,500 and for cardamom Rs. 10,000. Profit is Rs. 6,55,000]

Acceptance or Rejection of an Offer)

23. A factory produces 5,000 articles for home consumption at the following costs:

		Rs.
Materials		50,000
Wages		30,000
Factory overheads		
Fixed	30,000	
Variable	10,000	40,000
Administrative overheads (fixed)		28,000

Selling and distributed overheads		
Fixed	15,000	
Variable	10,000	25,000
Total		1,73,000

The home market can consume only 5,000 articles at a price of Rs. 40 each and no more. The foreign market for this product can, however, consume 3,000 additional pieces at a price of Rs. 22 each C.I.F. If this export order is extended, the following additional costs will be incurred:

1. Special packing, forwarding charge, etc., Rs. 1.00 per unit.
2. Freight, insurance etc. Rs. 2.00 per unit.

The following 'export benefits' should also be considered:

1. Duty drawback on direct materials cost at 10%.
2. Cash subsidy at 10% on F.O.B. value if foreign market worth trying ?

(Bangalore University, M.Com., May 1990)

[**Ans.** The acceptance of foreign market result in an additional profit of Rs. 5,000. Hence the foreign market is worth trying]

❑❑❑

CHAPTER 11

COST-VOLUME-PROFIT ANALYSIS

INTRODUCTION

Earning of maximum profit is the ultimate objective of all business establishments. The profit in its turn is determined by a number of factors, both internal and external. One such factor is the sales revenue of the business. Increase in the sales revenue will lead to increase in the profits. But the sales itself is based on other factors such as demand for the product, the competition, the selling price fixed, marketing strategies adopted by the management. The other important factor that determines the amount of profit is the cost of production. A reduced cost of production will result in an increased profit, keeping the selling price constant. But cost of production itself is affected by many factors such as volume of production, product mix, capacity utilisation, efficiency in production and so on. Though all these factors affect the profit earning capacity of a business, a special mention about the volume of output deserves to be mentioned in this regard. This is because volume of output changes more frequently and rapidly, and are not susceptible to management control. More than this the profit is more closely related to volume rather than cost. This is because costs seldom vary in direct proportion to volume. Hence, small change in the volume has a significant effect on profit. Whereas a change in other factors such as an order size or lot size will have insignificant effect on profit. However, it is the duty of management to consider the cost and volume while planning the profit earning capacity of the business. One such technique which is used by the management is known as cost-volume-profit analysis. As the name suggests, cost-volume-profit analysis examines the relationship of costs and profit to the volume of business with a view to maximise profit.

Kohler in his **Dictionary for Accountants** defines cost-volume-profit relationship as "the area of interest within an organisation, management and accountants in observing and controlling the relations between prospective and actual manufacturing costs—both fixed and variable—rates of production and gross profit.

OBJECTIVES OF COST-VOLUME-PROFIT

According to NAA **Research Bulletin,** Vol. 31, "the cost-volume-profit analysis appears to be useful principally as a technique for the study of problems encountered in business planning. As such it is a tool used largely by those executives responsible for strategic planning and policy-making". More specifically the objectives of cost-volume-profit analysis are as follows:

(*a*) To forecast the profit accurately.

(*b*) To facilitate in the preparation of flexible budgets.

(*c*) To evaluate the performance of the business. For evaluating the profit earned and cost incurred, it is necessary to know the impact of cost on the changed volume of output.

(*d*) To enable management in determining the pricing policies.

(*e*) To enable the charging of overheads to cost of production at different levels of operation.

ASSUMPTIONS OF COST-VOLUME-PROFIT ANALYSIS

1. The analysis is valid for a limited range of values, *i.e.*, 'the relevant ranges' and for a limited period of time.
2. Costs can be classified as fixed and variable. The latter type of cost changes proportionately with the volume within the relevant volume range. Fixed costs are constant within the relevant volume of range.
3. Revenues change proportionately with volumes.
4. There exists a constant product mix.
5. There is no significant change in the inventories in terms of physical units. In other words, the units produced are assumed to be sold.
6. Changes in volume alone are responsible for changes in costs and revenues.
7. The analysis is deterministic in nature. It ignores uncertainty and probabilistic approach.

LIMITATIONS OF COST-VOLUME-PROFIT ANALYSIS

1. It is presumed that the anticipated capacity of production remains same. But it may be increased depending upon the need.
2. The analysis of cost-volume-profit gives satisfactory result only if elements of costs remain stable. But in actual practice it varies.
3. It is again presumed that plant capacity remains same. However, the cost-volume-profit relationship does not hold good if manual labour is replaced by machines or high cost of materials are substituted by low cost materials.
4. In a business with many varieties of products, it becomes difficult to forecast the profits more accurately.

PRESENTATION OF COST-VOLUME-PROFIT ANALYSIS OR BREAK-EVEN ANALYSIS

One method of presenting cost-volume-profit analysis is by the method of break-even-analysis. It is a technique which is designed to help management in planning and decision-making functions involving the effect of change in volume on the profitability. This method helps management by establishing relationship between cost, volume and profit at different levels of activity. The management of every business desires to know the impact of changes in sales volume on profit. It is interesting in knowing the level at which the cost of production is equal to the sales value. This point where no profit or no loss is incurred is known as break-even point. Below this point any production will result only loss and beyond this point it brings profit to the business. Hence, break-even analysis and its study beyond the break-even point only relates to cost-volume-profit analysis.

There are two methods of presenting break-even-analysis. They are: (*i*) Algebraic method and (*ii*) Graphic method. The method to be used depends upon the choice of management and the data available.

ALGEBRAIC METHOD OF PRESENTING CVP ANALYSIS

This method of analysing cost-volume-profit relationship is based on the fundamental equation of marginal costing, *i.e.,* Sales – Variable cost = Fixed cost + Profit. Given any of the three variables, it is possible to find out the fourth unknown variable. This formula can be modified in different ways according to the need. For example, at the break-even point, the profit is nil. So the equation can be written as S – V = F + 0 (zero). Similarly, in order to calculate the sales, at the break-even point, it is possible to arrive at the following formula by multiplying it with sales:

$$S\,(S - V) = F \times S$$

$$\text{or S (Sales at break-even point)} = \frac{F \times S}{S - V}$$

Because sale minus variable cost constitute contribution, the above formula may also be written as follows:

$$S = \frac{F \times S}{C}$$

Alternative, fixed cost and profit must be equal to contribution. So the formula can be written as follows:

$$S = \frac{F \times S}{F + P}$$

In the same way for determining different requirements different formula are available. They are indicated below:

1. To determine the break-even sales in terms of units, the formula is

$$\text{BEP (units of Sales)} = \frac{\text{Fixed cost}}{\text{Contribution per unit}}$$

$$\text{or} \quad = \frac{\text{Fixed cost}}{\text{Unit selling price} - \text{Unit variable cost}}$$

2. To determine break-even sales in terms of rupees

$$\text{BEP (Sales in Rs.)} = \frac{\text{Fixed cost} \times \text{Selling price}}{\text{Contribution per unit}}$$

$$\text{or BEP (Sales in Rs.)} = \frac{\text{Fixed cost} \times \text{Sales}}{\text{Fixed cost} + \text{Profit}}$$

$$\text{or BEP (Sales in Rs.)} = \frac{\text{Fixed cost}}{1 - \dfrac{\text{Variable cost}}{\text{Sales}}}$$

3. Level of sales in units to earn a desired amount of profit

$$\text{BEP (Sales in units)} = \frac{\text{Fixed cost} + \text{Desired profit}}{\text{Contribution per unit}} \times \text{SP}$$

4. To determine the value of sales to earn a desired profit before tax

$$\text{BEP (Sales in Rs.)} = \frac{\text{Fixed cost} + \text{Desired profit}}{1 - \dfrac{\text{Variable cost}}{\text{Sales}}}$$

5. To determine the level of sales (in units) to result a desired profit after tax.

$$\text{BEP (Sales in units)} = \frac{\text{Fixed cost} + \dfrac{\text{After tax profit}}{1 - \text{Tax rate}}}{\text{Contribution rate per unit}}$$

6. To determine BEP after additional fixed cost owing to plant expansion.

$$\text{BEP} = \frac{\text{Present fixed cost} + \text{Additional fixed cost}}{1 - \dfrac{\text{Variable cost}}{\text{Sales}}}$$

7. To determine sales volume required to earn existing profit

$$\text{BEP (Sales in units)} = \frac{\text{Present fixed cost} + \text{Additional fixed cost} + \text{Existing profit}}{1 - \dfrac{\text{Variable cost}}{\text{Sales}}}$$

8. To determine shut down point

$$= \frac{\text{Fixed cost} - \text{Shut down cost}}{1 - \dfrac{\text{Variable cost}}{\text{Sales}}}.$$

Problem 1. From the following particulars find out break-even point:

Variable cost per unit	Rs. 10
Selling price per unit	Rs. 15
Fixed expenses	Rs. 40,000

What will be the selling price per unit if break-even point is brought down to 5,000 units ?

(SV University, B.Com., October 1999)

Solution:

$$\text{BEP} = \frac{\text{Fixed cost}}{\text{Sales} - \text{Variable cost}} = \frac{40,000}{15 - 10} = \frac{40,000}{5} = 8,000 \text{ units}$$

SELLING PRICE PER UNIT IF BEP IS BROUGHT DOWN TO 5,000 UNITS

Variable cost	5,000 × 10	= 50,000
Fixed cost		= 40,000
	Total cost	90,000
	Profit/loss	NIL
	Sales value	90,000

$$\text{Selling price per unit} = \frac{90,000}{5,000} = 18$$

Note: There is no profit or loss as is clear from the following:

Sales (8,000 × 15)		1,20,000
Less: Variable cost	8,000 × 10	80,000
	Contribution	40,000
	Less: Fixed cost	40,000
	Profit	–

Problem 2. You are given the following particulars:

Selling price	Rs. 200 per unit
Variable cost	Rs. 100 per unit
Total fixed cost	Rs. 96,000

Calculate:

1. Break-even units and value
2. Sales to earn a profit of Rs. 20 per unit *(University of Delhi, B.Com. (Pass), April 1993)*

Solution:

Contribution = Sales – Variable cost
= 200 – 100 = 100 per unit

(*i*) BEP $= \dfrac{\text{Fixed cost}}{\text{Contribution per unit}}$

$= \dfrac{96{,}000}{100} = 960$ units

BEP in value = 960 × 200 = Rs. 1,92,000

(*ii*) Sales to earn a profit of Rs. 20 per unit

Let Sales = x units

Then total sales = $200x$

Sales = Fixed cost + Variable cost + Profit

$200x = 96{,}000 + 100x + 20x$

$200x - 120x = 96{,}000$

$80x = 96{,}000$

$x = \dfrac{96{,}000}{80} = 1{,}200$ units.

Problem 3. From the following data calculate:

(*i*) Break-even point expressed in amount of sales in rupees

(*ii*) How many units must be sold to earn a net profit of 10% of sales ?

Selling price	Rs. 20 per unit
Variable cost	Rs. 12 per unit
Fixed cost	Rs. 2,40,000

(University of Delhi, B.Com. (Pass), April 1991)

Solution:

BEP $= \dfrac{\text{Fixed cost}}{\text{Contribution per unit}}$

$= \dfrac{2{,}40{,}000}{20-12} = \dfrac{2{,}40{,}000}{8}$

= 30,000 units.

(*i*) BEP in rupees = 30,000 units × 20

= Rs. 6,00,000

Units to be sold to earn a net profit of 10% on sales

Let Sales units = x

Selling price = Rs. 20

Total sales = $20x$

Profit 10% of sales $= \dfrac{10}{100} \times 20x = 2x$

$$\text{Sales value} = \text{Fixed cost} + \text{Variable cost} + \text{Profit}$$
$$20x = 2{,}40{,}000 + 12x + 2x$$
$$20x - 12x - 2x = 2{,}40{,}000$$
$$6x = 2{,}40{,}000$$
$$x = \frac{2{,}40{,}000}{6} = 40{,}000 \text{ units.}$$

Problem 4. Gifts Land Ltd., is considering hiring a machine at an annual charge of Rs. 12,000 to increase the output of a product from its present level of 6,000 units. It is anticipated that with the introduction of the machine, the variable cost per unit will be reduced by Re. 1 due to saving in labour cost. The new machine will not affect fixed cost in total apart from the hiring charges. The selling price of the product is Rs. 12 per unit. The present cost structure of the products is variable cost Rs. 9 per unit and fixed cost per unit Re. 1 per unit.

(University of Madras, B.Com., September 1995)

Solution:

$$\text{Break-even point for new machine} = \frac{\text{Revised total fixed cost}}{\text{Revised contribution per unit}}$$

Calculation of revised contribution per unit:

Selling price per unit			12
Less:	Revised variable cost:		
	Variable cost	9	
Less:	Saving in labour cost	1	8
Revised contribution per unit			4

Calculation of revised total fixed cost:

Present fixed cost	6,000 × 1	=	6,000
Add: Additional hire charges		=	12,000
Revised total fixed cost		=	18,000

$$\text{New BEP} = \frac{18{,}000}{4} = 4{,}500 \text{ units}$$

$$\text{Existing BEP} = \frac{\text{Total Fixed cost}}{\text{Contribution per unit}}$$

$$\text{Existing fixed cost} = 6{,}000 \times 1 = 6{,}000$$

$$\text{Contribution per unit} = S - V$$
$$= 12 - 9 = 3$$

$$\text{Existing BEP} = \frac{6{,}000}{3} = 2{,}000 \text{ units}$$

∴ Extra units produced in justifying hiring machine is (4,500 units – 2,000 units) = 2,500 units for break-even.

Problem 5. Indian Plastics make plastic buckets. An analysis of their accounting reveal

Variable cost per bucket	Rs. 20
Fixed cost	Rs. 50,000 for the year
Capacity	2,000 buckets per year
Selling price per bucket	Rs. 70

Required:

(*i*) Find the break-even point.

(*ii*) Find the number of buckets to be sold to get a profit of Rs. 30,000.

(*iii*) If the company can manufacture 600 buckets more per year with an additional fixed cost of Rs. 2,000 what should be the selling price to maintain the profit per bucket as at (*ii*) above?

(University of Delhi, B.Com. (Hons.), April 1991)

Solution:

(*i*) BEP $= \dfrac{\text{Fixed cost}}{\text{SP} - \text{VC per unit}}$

$= \dfrac{50{,}000}{70-20} = \dfrac{50{,}000}{50} = 1{,}000$ units

Verification:

Total cost (at 1,000 units) = FC + VC = 50,000 + 20,000

= Rs. 70,000

Sales = 1,000 buckets @ Rs. 70 = Rs. 70,000

Total cost of Rs. 70,000 is equal to total sales of Rs. 70,000, *i.e.*, no profit-no loss or BEP.

(*ii*) Number of buckets to be sold to earn profit of Rs. 30,000

$= \dfrac{\text{Fixed cost + Desired profit}}{\text{Contribution}}$

$= \dfrac{50{,}000 + 30{,}000}{50}$ or $\dfrac{80{,}000}{50} = 1{,}600$ buckets

Verification:

Total cost (at 1,600 buckets) = FC + VC

= 50,000 + 32,000 = 82,000

Sales = 1,600 buckets @ Rs. 70 = Rs. 1,12,000

Profit = Sales – Total cost

= 1,12,000 – 82,000 = Rs. 30,000

(*iii*) Selling price to maintain the profit per bucket as at (*ii*):

Further profit per bucket under (*ii*) will be

$\dfrac{\text{Rs. } 30{,}000}{1{,}600 \text{ buckets}} = \text{Rs. } 18.75$ per bucket

And fixed cost will be Rs. 52,000 (50,000 + 2,000)

Desired sales (units) $= \dfrac{\text{FC + Desired profit}}{\text{Contribution}}$

$2{,}600 = \dfrac{52{,}000 + (2{,}600 \times 18.75)}{\text{SP} - \text{VC}}$

$2{,}600 = \dfrac{52{,}000 + 48{,}750}{\text{SP} - 20}$

$2{,}600\ (\text{SP} - 20) = 1{,}00{,}750$

$\text{SP} - 20 = \dfrac{1{,}00{,}750}{2{,}600}$

$\text{SP} - 20 = 38.75$

$\text{SP} = 38.75 + 20 = \text{Rs. } 58.75$

Verification:

At 2,600 buckets, capacity utilisation :

Fixed cost = Rs. 52,000

Variable cost = 20 × 2,600 = 52,000

Total cost	=	52,000 + 52,000 = 1,04,000
Sales	=	2,600 buckets @ Rs. 58.75 = Rs. 1,52,750
Profit	=	Sales – Total cost
	=	1,52,750 – 1,04,000 = Rs. 48,750

Problem 6. Quick well Ltd., manufactures pressure cookers, the selling price of which is Rs. 300 per unit. Currently, the capacity utilisation is 60% with a sales turnover of Rs. 18 lakhs. The company proposes to reduce the selling price by 20% but desires to maintain the same profit position by increasing the output. Assuming that the increased output could be made and sold, determine the level at which the company should operate to achieve the desired objective.

The following further data are available:

(*i*) Variable cost per unit = Rs. 60.

(*ii*) Semi-variable cost (including a variable element of Rs. 10 per unit) = Rs. 1,80,000

(*iii*) Fixed cost Rs. 3,00,000 will remain constant up to 80% level. Beyond this an additional amount of Rs. 60,000 will be incurred. *(ICWA, Inter, December 1998)*

Solution:

STATEMENT OF PRESENT PROFITABILITY AT 60% CAPACITY

		Amount	*Cost*
(*a*)	Sales	18,00,000	300
	Variable cost	3,60,000	60
	Variable portion of semi-variable cost	60,000	10
(*b*)	Total variable cost	4,20,000	70
(*c*)	Contribution (*a* – *b*)	13,80,000	230
(*d*)	Fixed cost	3,00,000	50
	Fixed portion of semi-variable cost	1,20,000	20
(*e*)	Total fixed cost	4,20,000	70
	Profits (*c* – *e*)	9,60,000	160
	Revised selling price per unit Rs. 300 – 20% of Rs. 300		240
	Total variable cost per unit		70
	Revised contribution per unit		170

$$\text{BEP (Sales in units)} = \frac{\text{FC} + \text{Profit}}{\text{Contribution per unit}}$$

$$= \frac{4{,}20{,}000 + 9{,}60{,}000}{170}$$

$$= \frac{13{,}80{,}000}{170} = 8{,}118 \text{ units } i.e., \text{ 81.18\% of capacity}$$

But if the level of capacity is beyond 80%, an additional amount of Rs. 60,000 will be increased towards fixed cost. Then in such case the position will be as follows:

$$\text{Sales in unit} = \frac{(4{,}20{,}000 + 60{,}000) + 9{,}60{,}000}{170}$$

$$= \frac{14{,}40{,}000}{170} = 8{,}470.58$$

8,471 units are to be sold to maintain the same profit. Therefore, the company should operate at 84.71% level of capacity.

Problem 7. You are given the following data:

Sales price	Rs. 350 per unit
Variable cost	Rs. 200 per unit
Fixed expenses	Rs. 16,50,000

Ascertain:

(*a*) Break-even point.

(*b*) Sales per unit if break-even point is brought up to 15,000 units.

(*c*) Sales per unit if break-even point is brought down to 10,000 units.

(Bangalore University, B.Com., April 1992)

Solution:

Contribution = Sales price – Variable cost
= 350 – 200 = Rs. 150

(*a*) $\text{BEP} = \dfrac{\text{Fixed expenses}}{\text{Contribution}} = \dfrac{16,50,000}{150} = 11,000 \text{ units}$

(*b*) Selling price if BEP is brought up to 15,000 units.

Variable cost 15,000 × 200	30,00,000
Add: Fixed cost	16,50,000
Total cost	46,50,000

$\text{Selling price} = \dfrac{46,50,000}{15,000} = \text{Rs. } 310$

(*c*) Selling price if break-even point is brought down to 10,000 units:

Variable cost	10,000 × 200	20,00,000
Add: Fixed cost		16,50,000
Total cost		36,50,000

$\text{Selling price} = \dfrac{36,50,000}{10,000} = \text{Rs. } 365$

Problem 8. A manufacturer provides you the following data regarding his operations for the year 1991:

	Rs.
Break-even sales	6,66,667
Direct materials	2,20,000
Gross profit	2,50,000
Contribution margin	3,00,000
Direct labour	3,00,000
Sales	10,00,000
Variable manufacturing overhead	5,000

Calculate the following using the above data:

(*a*) Fixed manufacturing overhead.

(*b*) Variable selling and administrative overhead.

(*c*) Fixed selling and administrative overhead.

(Calicut University, B.Com., April 1992)

Solution:

Cost of goods sold = Sales – Gross profit
= 10,00,000 – 2,50,000 = 7,50,000

(*a*) Fixed factory overhead

Let fixed factory overhead = x

Then cost of goods sold will be equal to

$$2,20,000 + 3,00,000 + 5,000 + x = 7,50,000$$
$$5,25,000 + x = 7,50,000$$
$$x = 2,25,000$$

(*b*) Calculation of fixed selling and administrative overhead

Sales at BEP	6,66,667
Less: Contribution	3,00,000
Total fixed cost	3,66,667
Less: Fixed factory overhead	2,25,000
Fixed selling and administrative overhead	1,41,667

(*c*) Calculation of variable selling and administrative overhead

$$\text{Sales} - \text{Variable cost} = \text{Contribution}$$
$$10,00,000 - (2,20,000 + 3,00,000 + 5,000 + x) = 3,00,000$$
$$10,00,000 - 5,25,000 + x = 3,00,000$$
$$4,75,000 + x = 3,00,000$$
$$x = -4,75,000 + 3,00,000$$
$$x = 1,75,000$$

Problem 9. Sharda Painters Ltd., manufactures and sells four types of paints under the brand names of P, Q, R and S. The sales mix in value comprises of $33\frac{1}{3}\%$, $41\frac{2}{3}\%$, $16\frac{2}{3}\%$ and $8\frac{1}{3}\%$ respectively.

The total budgeted sales are Rs. 60,000 per month. Operating costs are:

Variable cost of P	60% of selling price
Variable cost of Q	68% of selling price
Variable cost of R	80% of selling price
Variable cost of S	40% of selling price

Fixed cost amount to Rs. 14,700 per month. Calculate BEP for all paints on an overall basis.

(Bangalore University, M.Com., May 1989)

Solution:

	Products				*Total*
	P	*Q*	*R*	*S*	
Sales mix	$33\frac{1}{3}\%$	$41\frac{2}{3}\%$	$16\frac{2}{3}\%$	$8\frac{1}{3}\%$	100%
Sales (Rs.)	20,000	25,000	10,000	5,000	60,000
Variable cost	12,000	17,000	8,000	2,000	39,000
Contribution					21,000
Less: Fixed cost					14,700
				Profit	6,300

$$\text{BEP} = \frac{\text{Fixed cost}}{\text{Contribution}} \times \text{Sales}$$

$$= \frac{14{,}700}{21{,}000} \times 60{,}000 = \text{Rs. } 42{,}000$$

PROFIT-VOLUME RATIO OR CONTRIBUTION—SALES PERCENTAGE

The profit volume (*P/V*) ratio establishes the relation between contribution to sales. The ratio is expressed as a percentage and it furnishes the details of profitability of various products, processes or departments. A high *P/V* ratio shows that even a slight rise in the volume without a corresponding increase in fixed cost would result in high profit. Therefore, it is advisable for management to increase sales by taking suitable measures such as advertising and other sales promotional measures. The *P/V* ratio can be increased by maximising contribution, which is possible by increasing the selling price, reducing the variable cost and by improving the product mix. The formula to calculate *P/V* ratio is as follows:

$$P/V \text{ Ratio} = \frac{\text{Contribution}}{\text{Sales}} \times 100$$

Uses of P/V Ratio

1. It helps in the determination of BEP. The formula is:

$$\text{BEP} = \frac{\text{FC}}{\text{P/V Ratio}}$$

2. It helps in the determination of profit at any volume of sales. The formula is:

$$\text{Volume} \times P/V \text{ Ratio} = \text{Contribution} - \text{FC} = \text{Profit}$$

3. It helps in the determination of sales to earn a desired amount of profit. The formula is:

$$\text{Sales} = \frac{\text{F} + \text{P}}{\text{P/V Ratio}}$$

4. It helps in determining the required selling price per unit. The formula is:

$$\text{Selling price} = \frac{\text{Variable cost}}{(100 - \text{P/V Ratio})}$$

5. It helps in ascertaining the variable cost for any volume of sales by reverse method, *i.e.,* by deducting P/V ratio from sales considering it as 100%.

Margin of Safety

The excess of actual sales over the break-even sales is known as margin of safety. The formula is as follows:

$$\text{MS} = \text{Sales Volume} - \text{break-even sales volume}$$

The soundness of the business can be known by looking into the margin of safety. If the distance between sales revenue and break-even point on the graph is long, it shows the soundness, of the business. On the other hand, a small margin of safety involving a fall in sales revenue results in a loss. The margin of safety can be improved by taking the following measures:

(*a*) Increasing the selling price.

(*b*) Increasing the sales volume by increasing the capacity.

(*c*) By improving the contribution margin through reducing the variable cost.

(*d*) By lowering BEP through reduction of fixed cost.

(*e*) By adopting a better profitable product mix.

Margin of safety can be calculated by using the following formula:

$$\text{MS (in rupees)} = \frac{\text{Profit}}{\text{P/V Ratio}} \text{ or Profit} \times \frac{\text{Contribution}}{\text{Sales}}$$

$$\text{MS (in units)} = \frac{\text{Profit}}{\text{Contribution per unit}}$$

It can also be expressed as a percentage using the following formula:

$$\text{MS Ratio} = \frac{\text{MS}}{\text{Sales}} \times 100$$

$$\text{MS Ratio} = \frac{\text{Actual Sales} - \text{BEP Sales}}{\text{Sales}} \times 100$$

The other formulae used to calculate P/V ratio are as follows:

$$\text{P/V Ratio} = \frac{\text{Contribution}}{\text{Sales}}$$

$$\text{P/V Ratio} = \frac{\text{FC} + \text{Profit}}{\text{Sales}}$$

$$\text{P/V Ratio} = \frac{\text{FC} - \text{Loss}}{\text{Sales}}$$

$$\text{P/V Ratio} = \frac{\text{Change in contribution}}{\text{Change in sales}}$$

$$\text{P/V Ratio} = \frac{\text{Change in profit (or loss)}}{\text{Change in sales}}$$

Angle of Incidence

The angle at which sales line cuts the total cost line is known as angle of incidence. In other words, it represents the angle between sales line and total cost line. The angle of incidence shows the profit earning capacity of the business. Hence, higher the angle of incidence, higher the profit and *vice versa*. A high margin of safety with a broader angle of incidence reveals a highly favourable position of the business.

Problem 10. Given the following information:

Fixed cost	=	Rs. 4,000
Break-even sales	=	Rs. 20,000
Profit	=	Rs. 1,000
Selling price per unit	=	Rs. 20

You are required to calculate:

(*i*) Sales and marginal cost of sales, and
(*ii*) New break-even point if selling price is reduced by 10%.

(University of Delhi, B.Com. (Pass), April 1999)

Solution:

Contribution at BEP = Fixed cost is 4,000

$$\text{P/V ratio} = \frac{\text{Contribution}}{\text{Sales}} \times 100$$

$$= \frac{4{,}000}{20{,}000} \times 100 = 20\%$$

(*i*) Sales to earn a profit of Rs. 1,000

$$= \frac{\text{Fixed cost + Profit}}{\text{P/V Ratio}}$$

$$= \frac{4{,}000 + 1{,}000}{20\%} = \frac{5{,}000}{20\%} = 25{,}000$$

Marginal cost at sales of Rs. 25,000 is found as follows:

Contribution = Sales × P/V ratio

$$= 25{,}000 \times \frac{20}{100} = 5{,}000$$

Marginal cost = Sales – Contribution

= 25,000 – 5,000 = Rs. 20,000

(*ii*) New BEP if selling price is reduced by 10%

New sales = 25,000 – 10%

= 22,500

New contribution = 22,500 – 20,000 = Rs. 2,500

$$\text{New P/V Ratio} = \frac{\text{New Contribution}}{\text{New Sales}}$$

$$= \frac{2{,}500}{22{,}500} = \frac{1}{9}$$

$$\text{New BEP} = \frac{\text{Fixed cost}}{\text{P/V Ratio}}$$

$$= \frac{4{,}000}{\frac{1}{9}} \text{ or } = 36{,}000$$

Problem 11. Ascertain profit when

Sales = Rs. 2,00,000
Fixed cost = Rs. 40,000
BEP = Rs. 1,60,000

(CA Inter, May 1999)

Solution:

$$\text{P/V Ratio} = \frac{\text{Fixed cost}}{\text{BEP}} \times 100$$

$$= \frac{40{,}000}{1{,}60{,}000} \times 100 = 25\%$$

Contribution $= \text{Sales} \times \text{P/V Ratio} = \text{FC} + \text{Profit}$

$= 2{,}00{,}000 \times 25\% = \text{FC} + \text{profit}$

$50{,}000 = 40{,}000 + \text{Profit}$

or Profit $= 10{,}000$

Problem 12. Ascertain sales, when

Fixed cost = Rs. 20,000
Profit = Rs. 10,000
BEP = Rs. 40,000

(CA Inter, May 1999)

Solution:

Contribution $= \text{Fixed cost} + \text{Profit}$

$= 20{,}000 + 10{,}000 = 30{,}000$

P/V Ratio $= \frac{\text{Fixed cost}}{\text{BEP}} \times 100$

$= \frac{20{,}000}{40{,}000} \times 100 = 50\%$

Also P/V Ratio $= \frac{\text{Contribution}}{\text{Sales}} \times 100$

or Sales $= \frac{\text{Contribution}}{\text{P / V Ratio}} \times 100$

$= \frac{30{,}000}{50\%}$

$= \frac{30{,}000 \times 100}{50} = 60{,}000$

Problem 13. Raj Ltd., manufactures three products *X, Y* and *Z*. The unit selling prices of these products are Rs. 100, Rs. 160 and Rs. 75 respectively. The corresponding unit variable costs are Rs. 50, Rs. 80 and Rs. 30. The proportions (quantity wise) in which these products are manufactured and sold are 20%, 30% and 50% respectively. The total fixed costs are Rs. 14,80,000.

Calculate break-even quantity and the product wise break up of such quantity.

(CA. Inter, May 1999)

Solution:

OVERALL BREAK-EVEN QUANTITY

Products	*X*	*Y*	*Z*
Selling price per unit	100	160	75
Less: Variable cost per unit	50	80	30
Contribution per unit	50	80	45

Contribution at break-even point	$0.20x \times 50$	$0.30x \times 80$	$0.50x \times 45$
(see working note)	$10x$	$24x$	$22.5x$

At break-even point, contribution = Fixed cost

Hence, $10x + 24x + 22.5x$ = Rs. 14,80,000

or $x = \dfrac{14,80,000}{56.5} = 26,195$ units.

Product wise break-up of overall break-even quantity

Product X = 26,195 units × 0.20 = 5.239 units

Product Y = 26,195 units × 0.30 = 7,858 units

Product Z = 26,195 units × 0.50 = 13,098 units

Working note:

Let x be the overall break-even quantity of three products X, Y and Z. At break-even X has 20%, 30% and 50% units of X, Y and Z. The product wise production and sale of three given products in terms of overall break-even quantity are $0.20x$, $0.30x$ and $0.50x$ units respectively.

Problem 14. When volume is 3,000 units, average cost is Rs. 4 per unit. When volume is 4,000 units, average cost is Rs. 3.50. The break-even point is 5,000 units. Find the P/V ratio.

(ICWA, Inter, Dec. 1999)

Solution:

	Output	*Average Cost*	*Total Cost*
	3,000	4	12,000
	4,000	3.50	14,000
Change	1,000		2,000

Variable cost per unit $= \dfrac{2,000}{1,000} =$ Rs. 2

Total fixed cost = Total cost – Variable cost

= 12,000 – 3,000 units × Rs. 2

= 12,000 – 6,000

= 6,000

or = 14,000 – 4,000 units × Rs. 2

= 14,000 – 8,000

= 6,000

At BEP, BEP Sales = Variable cost + Fixed cost

5,000 units = 5,000 × 2 + 6,000

= 10,000 + 6,000

= 16,000

$$\text{BEP} = \frac{\text{Fixed Cost}}{\text{P/V Ratio}}$$

$$\text{or P/V Ratio} = \frac{\text{Fixed cost}}{\text{BEP}}$$

$$= \frac{6,000}{16,000} = \frac{3}{8} \text{ or } 37.5\%$$

Problem 15. ABC Ltd., fixed costs of Rs. 2,00,000. It has two products, that it can sell; *A* and *B*. The company sells these products at a rate of 2 units of *A* to 1 unit of *B*. The unit contribution is Re. 1 per unit for A and 2 per unit for B. How many units of *A* and *B* would be sold at the break-even point.

(ICWA, Inter, Dec. 1999)

Solution:

Products	*A*	*B*	*Total*
Sales unit be (in the ratio of 2 : 1)	2	1	3 units
Contribution per unit	Re. 1	Rs. 2	
Total contribution	Rs. 2	Rs. 2	Rs. 4

Hence, contribution per unit (composite) $= \frac{4}{3}$

Total fixed cost = Rs. 2,00,000

BEP (composite) $= \frac{\text{Fixed cost}}{\text{Contribution per unit}}$

$$= \frac{2,00,000}{\frac{4}{3}}$$

or BEP in units $= \frac{2,00,000 \times 3}{4}$

= 1,50,000 units

$\frac{2}{3}$ A = 1,00,000 units

$\frac{1}{3}$ B = 50,000 units

Problem 16. A company had incurred fixed expenses of Rs. 2,25,000 with sales of Rs. 7,50,000 and earned a profit of Rs. 1,50,000 during the first half year. In the second half year, it suffered a loss of Rs. 75,000 calculate:

(*i*) The P/V ratio, break-even point and margin of safety.

(*ii*) Expected sales-volume for the second half year assuming that selling price and fixed expenses remained unchanged during second half year.

(CS Inter, June 1998)

Solution: CALCULATIONS FOR THE FIRST HALF-YEAR

(*i*) P/V Ratio $= \frac{\text{Fixed cost + Profit}}{\text{Sales}}$

$$= \frac{2,25,000 + 1,50,000}{7,50,000} = 50\%$$

BEP $= \frac{\text{Fixed Cost}}{\text{P/V Ratio}}$

$$= \frac{2,25,000}{50\%} = 4,50,000$$

Margin of safety = Actual sales – Sales at BEP

= 7,50,000 – 4,50,000 = Rs. 3,00,000

(*ii*) Expected sales volume for second half year

$$= \frac{\text{Fixed Cost} - \text{Loss}}{\text{P/V Ratio}}$$

$$= \frac{2,25,000 - 75,000}{50\%}$$

$$= \frac{1,50,000}{50} \times 100 = 3,00,000$$

Problem 17. The profit volume ratio of X Ltd., is 50% and the margin of safety is 40%. You are required to calculate the net profit if the sales value is Rs. 1,00,000 *(CA Inter, November 1998)*

Solution:

$$\text{Margin of safety ratio} = \frac{\text{Margin of safety in absolute terms}}{\text{Actual sales}} \times 100$$

$$40 = \frac{\text{Margin of safety in absolute terms}}{1,00,000} \times 100$$

$$\text{or Margin of safety in absolute terms} = \frac{1,00,000 \times 40}{100}$$

$$= 40,000$$

$$\text{Also Margin of safety in absolute terms} = \frac{\text{Profit}}{\text{P/V Ratio}}$$

$$\text{or} \quad 40,000 = \frac{\text{Profit}}{50\%}$$

$$\text{or} \quad \text{Profit} = 40,000 \times \frac{50}{100} = \text{Rs. } 20,000$$

Problem 18. A single product company sells its products at Rs. 60 per unit. In 1996, the company operated at a margin of safety of 40%. The fixed costs amounted to Rs. 3,60,000 and the variable cost ratio to sales was 80%.

In 1997, it is estimated that the variable costs will go up by 10% and the fixed costs will increase by 5%. Find the selling price required to be fixed in 1997 to earn, the same P/V ratio as in 1996.

Assuming the same selling price of Rs. 60 per unit in 1997, find the number of units required to be produced and sold to earn the same profit as in 1996. *(CA Inter, May 1998)*

Solution: CALCULATION OF P/V RATIO (IN 1996)

$$\text{P/V Ratio} = \frac{\text{Selling price} - \text{Variable Cost per unit}}{\text{Selling price}} \times 100$$

$$= \frac{60 - 48}{60} \times 100$$

$$= \frac{12}{60} \times 100 = 20\%$$

CALCULATION OF UNITS SOLD (IN 1996)

$$\text{BEP} = \frac{\text{Fixed cost}}{\text{Contribution per unit}}$$

$$= \frac{3,60,000}{12} = 30,000 \text{ units}$$

Since, Margin of safety is 40%, therefore BEP is 60% of units sold or No. of units sold $= \dfrac{\text{BEP}}{60\%}$

$= \dfrac{30{,}000 \text{ units}}{60\%} = 50{,}000$

CALCULATION OF PROFIT EARNED IN 1996

Profit = Total contribution – Fixed cost
= 50,000 units × Rs. 12 per unit – 3,60,000
= 6,00,000 – 3,60,000 = 2,40,000

SELLING PRICE TO BE FIXED IN 1997

Variable cost per unit in 1997 = Rs. 52.80 (Rs. 48 + 4.80)
Fixed cost in 1997: = Rs. 3,78,000 (3,60,000 + 18,000)
P/V ratio in 1996 = 20%
Since, P/V ratio is 20% variable cost is 80%

Hence, the required SP $= \dfrac{52.80}{80\%}$

or $= \dfrac{52.80}{80} \times 100$
= Rs. 66

NUMBER OF UNITS TO BE PRODUCED AND SOLD IN 1997
TO EARN THE SAME PROFIT AS IN 1996

Profit in 1996 = Rs. 2,40,000
Fixed cost in 1997 = Rs. 3,78,000
Desired contribution in 1997 = (2,40,000 + 3,78,000)
= Rs. 6,18,000
Contribution per unit in 1997 = SP – Variable cost per unit
= 60 – 52.80
= 7.20

No. of units to be produced and sold in 1997 $= \dfrac{\text{Fixed cost in 1997}}{\text{Contribution per unit in 1997}}$

$= \dfrac{3{,}78{,}000}{7.20}$
= 52,500 units.

Problem 19. A Ltd. maintains, a margin of safety of 37.5% with an overall contribution to sales ratio of 40%. Its fixed costs amount to Rs. 5 lakhs.

Calculate the following:

(*i*) Break-even sales.
(*ii*) Total sales.
(*iii*) Total variable costs.
(*iv*) Current profit
(*v*) New margin of safety if the sales value is increased by $7\frac{1}{2}$ %.

(ICWA, Inter, December 1998)

Solution:

1\. Break-even sales $= \dfrac{\text{Fixed Cost}}{\text{Contribution/Sales}}$

$= \dfrac{5 \text{ lakhs}}{40\%} = 12.50 \text{ lakhs}$

2\. Total sales = Break-even sales + Margin of safety

$= \text{Break-even sales} + \dfrac{37.5}{100} \times \text{Sales}$ [Given MS = 37.5% of sales]

or Break-even sales $= \text{Sales} - \dfrac{37.5}{100} \text{ sales}$

or 12.50 lakhs $= \dfrac{62.5}{100} \text{ sales}$

or, Sales $= \dfrac{12.50 \text{ lakhs} \times 100}{62.5} = 20 \text{ lakhs}$

3\. Total variable cost = 60% of Rs. 20 lakhs

= 12 lakhs

Because, Sales – Variable cost = Contribution

4\. Current profit = Sales – (Variable cost + Fixed cost)

= 20 lakhs – (12 lakhs + 5 lakhs)

= 20 lakhs – 17 lakhs = 3 lakhs

5\. New margin of safety if sales value is increased by $7\frac{1}{2}\%$

New sales value = Rs. 20 lakhs + $7\frac{1}{2}\%$

= 21.50 lakhs

Hence, new margin of safety = New sales – Break-even sales

= 21.50 lakhs – 12.50 lakhs

= Rs. 9 lakhs

Problem 20.

(*i*) Find out BEP sales if budgeted output is 80,000 units. Fixed cost is Rs. 4,00,000, Selling price per unit is Rs. 20 and variable cost per unit is Rs. 10.

(*ii*) Calculate selling price, if marginal cost is Rs. 2,400 and P/V Ratio is 20%.

(*iii*) Find out margin of safety if profit is Rs. 20,000 and P/V Ratio is 40%.

(University of Delhi, B.Com. (Pass), April 1997)

Solution:

(*i*) $\text{BEP} = \dfrac{\text{Fixed Cost}}{\text{Contribution}} = \dfrac{4{,}00{,}000}{20 - 10} = \dfrac{4{,}00{,}000}{10} = 40{,}000 \text{ units}$

BEP Sales = 40,000 units × Rs. 20 = Rs. 8,00,000

(*ii*) When P/V Ratio is 20%, then variable cost is 80% of sales

Thus, Selling price $= 2{,}400 \times \dfrac{100}{80} = \text{Rs. } 3{,}000$

(*iii*) Margin of safety $= \dfrac{\text{Profit}}{\text{P/V Ratio}} = \dfrac{20{,}000}{40\%} = \text{Rs. } 50{,}000$

Problem 21. From the following data calculate:

(*a*) P/V ratio

(*b*) Profit when sales are Rs. 20,000 and

(*c*) New break-even point if selling price is reduced by 20%

Fixed expenses	Rs. 40,000
Break-even point	Rs. 10,000

(University of Madras, B.Com., April 2005)

Solution:

(*a*) P/V Ratio

$$\text{Break-even point} = \frac{\text{Fixed expenses}}{\text{P/V Ratio}}$$

$$\frac{10{,}000}{1} = \frac{\text{Rs. } 4{,}000}{\text{P/V Ratio}}$$

$$10{,}000 \text{ P/V Ratio} = \text{Rs. } 4{,}000 \text{ (on cross multiplication)}$$

$$\text{P/V Ratio} = \frac{4{,}000}{10{,}000} = 0.40$$

$$\text{P/V Ratio (in percentage)} = 0.40 \times 100 = 40\%$$

(*b*) Profit when sales are 20,000

Sales = Rs. 20,000

P/V Ratio = 40%

Contribution = Sales × P/V ratio

$$= 20{,}000 \times \frac{40}{100} = 8{,}000$$

Less: Fixed expenses	=	4,000
Profit	Rs.	4,000

(*c*) New BEP if selling price is reduced by 20%

Existing selling price (assumed)	= Rs.	100
Less: Reduction $\frac{20}{100} \times 100$	= Rs.	20
Revised selling price	Rs.	80

Calculation of Variable Cost

Existing selling price	Rs.	100
Less: Contribution 40%	Rs.	40
Variable cost		60

Revised P/V Ratio at Rs. 80 selling price:

$$= \frac{\text{Revised selling price} - \text{Variable cost}}{\text{Revised selling price}} \times 100$$

$$= \frac{\text{Rs. } 80 - \text{Rs. } 60}{\text{Rs. } 80} \times 100$$

$$= \frac{20}{80} \times 100 = 25\%$$

$$\text{New BEP} = \frac{\text{Fixed expenses}}{\text{New P/V Ratio}}$$

$$= \frac{4{,}000}{25\%}$$

$$= \frac{4{,}000}{25} \times 100 = \text{Rs. } 16{,}000$$

Problem 22. The P/V Ratio of a firm is 50% and the margin of safety is 40%. You are required to work out BEP and net profit, when sales is Rs. 50,00,000

(University of Madras, B.Com., May 2005)

Solution:

(a) Calculation of BEP

Sales		= 50,00,000
Less: Margin of safety – (40% of sales)		
$\frac{40}{100} \times 50{,}00{,}000$		= 20,00,000
	BEP	30,00,000

Calculation of Fixed Cost

$$\text{BEP} = \frac{\text{Fixed cost}}{\text{P/V Ratio}}$$

$$\frac{30{,}00{,}000}{1} = \frac{\text{Fixed cost}}{50\%}$$

Fixed cost = 15,00,000 (on cross multiplication)

(b) Net Profit if sales volume is Rs. 50,00,000

Contribution $\frac{50}{100} \times 50{,}00{,}000$		= 25,00,000
Less: Fixed cost		= 15,00,000
	Profit	10,00,000

Problem 23. From the following particulars calculate the following:

(*a*) P/V Ratio

(*b*) BEP

(*c*) Profit when sales volume is Rs. 2,00,000

(*d*) Sales required to get a profit of Rs. 30,000

Year	*Sales*	*Profit*
I	1,00,000	12,000
II	1,40,000	22,000

(University of Madras, B.Com., April 2004)

Solution:

CALCULATION OF CHANGE IN SALES AND CHANGE IN PROFIT

Year	*Sales*	*Profit*
I	1,00,000	12,000
II	1,40,000	22,000
Difference	40,000	10,000

(a) P/V Ratio

$$= \frac{\text{Change in profit}}{\text{Change in sales}} \times 100$$

$$= \frac{10,000}{40,000} \times 100 = 25\%$$

(b) BEP

$$= \frac{\text{Total fixed cost}}{\text{P/V Ratio}}$$

CALCULATION OF TOTAL FIXED COST

Taking Ist year's sales			1,00,000
Contribution	= Sales × P/V Ratio		
	$= 1,00,000 \times \frac{25}{100}$		= 25,000
Less:	Profit of I year		= 12,000
		Fixed cost	13,000

$$\text{BEP} = \frac{\text{Fixed cost}}{\text{P/V Ratio}}$$

$$= \frac{13,000}{25\%} = \frac{13,000}{25} \times 100 = \text{Rs. } 52,000$$

(c) Profit required when sale volume is Rs. 2,00,000

Contribution	= Sales × P/V Ratio		
	$= 2,00,000 \times \frac{25}{100}$		= 50,000
Less:	Total fixed cost		= 13,000
		Profit	37,000

(d) Sales required to get a profit of Rs. 30,000

$$= \frac{\text{Total fixed cost} + \text{Profit required}}{\text{P/V Ratio}}$$

$$= \frac{13,000 + 30,000}{25\%} = \frac{43,000}{25} \times 100 = \text{Rs. } 1,72,000$$

Problem 24. Hastinapur Engineering Limited manufactures and sells four products *A, B, C* and *D*. The sales mix in value comprises $33\frac{1}{3}\%$, $41\frac{2}{3}\%$, $16\frac{2}{3}\%$ and $8\frac{1}{3}\%$ of *A, B, C* and *D* respectively. The total budgeted sales (100 %) are Rs. 60,000 per month.

Operating costs are as follows:

Product A – 60 % of selling price
Product B – 68% of selling price
Product C – 80 % of selling price
Product D – 40 % of selling price

Fixed expenses Rs. 14,700

Calculate the break-even point for the enterprise as a whole.

(University of Madras, B.Com., May 2003)

Solution:

CALCULATION OF BEP

	Sales	*Variable cost*	*Contribution*
A	20,000	12,000	8,000
B	25,000	17,000	8,000
C	10,000	8,000	2,000
D	5,000	2,000	3,000
Total	60,000	39,000	21,000

$$\text{BEP} = \frac{\text{Total fixed cost}}{\text{Total contribution}} \times \text{Total sales}$$

$$= \frac{14{,}700}{21{,}000} \times 60{,}000 = \text{Rs. } 42{,}000$$

Problem 25. The sales turnover and profit / loss during two years were as follows

Year	*Sales*	*Profit/Loss*
1995	7,00,000	(Loss) 10,000
1996	9,00,000	(Profit) 10,000

The selling price per unit is Rs. 100.

Calculate:

(*a*) The amount of fixed cost
(*b*) The number of units to break-even
(*c*) The number of units to earn a profit of Rs. 50,000

(University of Madras, B.Com., October 2002)

Solution:

Year	*Sales*	*Profit / Loss*
1995	7,00,000	– 10,000 (Loss)
1996	9,00,000	+ 10,000 (Profit)
Difference	2,00,000	+ 20,000 [10,000 – (– 10,000)]

$$\text{P/V Ratio} = \frac{\text{Difference in profit}}{\text{Difference in sales}} \times 100$$

$$= \frac{20{,}000}{2{,}00{,}000} \times 100 = 10\%$$

(a) The amount of fixed cost:

Taken any sales (1996 sales)		9,00,000
Contribution	= Sales × P/V Ratio	
	$= 9{,}00{,}000 \times \frac{10}{100}$	= 90,000
Less: Fixed cost		= 10,000
	Total fixed cost	80,000

(b) Break-even point:

$$\text{BEP (in rupees)} = \frac{\text{Total fixed cost}}{\text{P/V Ratio}}$$

$$= \frac{80{,}000}{10\%} = \text{Rs. } 8{,}00{,}000$$

$$\text{BEP (in units)} = \frac{8{,}00{,}000}{\text{Rs. 100 per unit}}$$

$$= 8{,}000 \text{ units}$$

(c) The number of units to earn a profit of Rs. 50,000

$$= \frac{\text{Total fixed cost} + \text{profit}}{\text{P/V Ratio}}$$

$$= \frac{80{,}000 + 50{,}000}{10\%} = \frac{1{,}30{,}000}{10} \times 100 = \text{Rs. } 13{,}00{,}000$$

$$\text{In units} = \frac{\text{Rs. } 13{,}00{,}000}{\text{Rs. 100 per unit}} = 13{,}000 \text{ units.}$$

Problem 26. Arun Ltd. has prepared the following budget estimates for the year 2002–2003

Sales units	15,000
Fixed expenses	Rs. 34,000
Sales	Rs. 1,50,000
Variable cost	Rs. 6 per unit

You are required to:

(*a*) Find the P/V Ratio, Break-even point and margin of safety.

(*b*) Calculate the revised P/V Ratio, break-even point and margin of safety in each of the following cases.

(*i*) Decrease of 10% in selling price

(*ii*) Increase of 10% in variable cost

(*iii*) Increase of sales volume by 2,000 units

(*iv*) Increase of Rs. 6,000 in fixed costs. *(University of Madras, B.Com., May 2004)*

Solution:

(1) Calculation of P/V Ratio:

$$= \frac{\text{Sales} - \text{Variable cost}}{\text{sales}} \times 100$$

$$= \frac{\text{Rs. } 1{,}50{,}000 - 90{,}000}{\text{Rs. } 1{,}50{,}000} \times 100$$

$$= \frac{\text{Rs. } 60{,}000}{\text{Rs. } 1{,}50{,}000} \times 100 = 40\%$$

Variable cost = Rs. 6 per unit × 15,000 units
= Rs. 90,000

(2) Break-even point

$$\frac{\text{Fixed cost}}{\text{P/V Ratio}} = \frac{34{,}000}{40\ \%}$$

$$= \frac{34{,}000}{40} \times 100$$

= Rs. 85,000

(3) Margin of safety

= Actual sales – Break-even sales

= Rs. 1,50,000 – Rs. 85,000 = Rs. 65,000

(*b*) (*i*) Decrease in 10% of selling price:

Existing selling price $= \frac{1{,}50{,}000}{15{,}000 \text{ units}} =$ Rs. 10

Less: Decrease 10% $= \left(10 \times \frac{10}{100}\right) =$ Re. 1

Rs. 9

(1) P/V Ratio

$$= \frac{S - V}{S} \times 100 = \frac{\text{Rs. } 9 - \text{Rs. } 6}{\text{Rs. } 9} \times 100$$

$$= \frac{3}{9} \times 100 = 33\frac{1}{3}\%$$

(2) Break-even point

$$= \frac{\text{Fixed Expenses}}{33\frac{1}{3}\%} = \frac{34{,}000}{\frac{1}{3}}$$

$$= \frac{34{,}000}{1} \times 3$$

= Rs. 1,02,000

(3) Margin of safety

= Actual sales – Break-even sales

Actual sales at revised selling price

= 15,000 units × 9 = Rs. 1,35,000

= 1,35,000 – 1,02,000 = Rs. 33,000

(b)(ii) Increase of 10% in variable cost

Assuming no change in selling price of Rs. 10

i.e., $\frac{\text{Rs. 1,50,000}}{\text{15,000 units}}$ = Rs. 10

Existing variable cost = Rs. 6

Add: Increase of 10% of Rs. 6 = Rs. 0.60

Revised variable cost 6.60

(1) P/V Ratio

$$\frac{S - V}{S} \times 100 = \frac{\text{Rs. } 10 - \text{Rs. } 6.60}{\text{Rs. } 10} \times 100$$

$$= \frac{\text{Rs. } 3.40}{10} \times 100$$

$$= 34\%$$

(2) Break-even point

$$= \frac{\text{Fixed Expenses}}{\text{P/V Ratio}} = \frac{34{,}000}{34\%}$$

$$= \frac{34{,}000}{34} \times 100 = \text{Rs. } 1{,}00{,}000$$

(3) Margin of safety

= Actual sales – Break-even sales

= Rs. 1,50,000 – Rs. 1,00,000 = Rs. 50,000

Note: **Selling price unchanged, therefore existing sales Rs. 1,50,000 is taken.**

(b)(iii) Increase of sales volume by 2,000 units

Existing sales units = 15,000

Add: Increase in sales units = 2,000

17,000

Revised sales value 17,000 × Rs. 10 = 1,70,000

Revised variable cost 17,000 × Rs. 6 = 1,02,000

(1) P/V Ratio

$$= \frac{S - V}{S} \times 100 = \frac{1{,}70{,}000 - 1{,}02{,}000}{1{,}70{,}000} \times 100$$

$$= \frac{68{,}000}{1{,}70{,}000} \times 100$$

$$= 40\%$$

(2) Break-even point

$$= \frac{\text{Fixed Expenses}}{\text{P/V Ratio}} = \frac{34{,}000}{40\%}$$

$$= \frac{34{,}000}{40} \times 100$$

$$= \text{Rs. } 85{,}000$$

(3) Margin of safety

= Actual sales – Break-even sales

= 1,70,000 – 85,000 = Rs. 85,000

***(b)(iv)* Increase of Rs. 6,000 in fixed expenses**

Existing fixed expenses	=	34,000
Add: Increase in Fixed expenses	=	6,000
		40,000

(1) P/V Ratio

$$= \frac{S - V}{S} \times 100 = \frac{10 - 6}{10} \times 100$$

$$= \frac{4}{10} \times 100 = 40\%$$

(2) Break-even point

$$= \frac{\text{Fixed Expenses}}{\text{P/V Ratio}} = \frac{40,000}{40\%}$$

$$= \frac{40,000}{40} \times 100$$

$$= \text{Rs. } 1,00,000$$

(3) Margin of safety

= Actual sales – Break-even sales

= 1,50,000 – 1,00,000 = Rs. 50,000

Problem 27. From the following data calculate.

(*a*) P/V Ratio

(*b*) Profit when sales are Rs. 20,000 and

(*c*) New Break-even point if selling price is reduced by 20%

Fixed expenses	Rs. 4,000
Break-even point	Rs. 10,000

(University of Madras, B.Com., May 2003)

Solution:

(*a*) Calculation of P/V Ratio:

$$\text{Break-even point} = \frac{\text{Fixed expenses}}{\text{P/V Ratio}}$$

$$10,000 = \frac{\text{Rs. } 4,000}{\text{P/V Ratio}}$$

10,000 P/V Ratio = Rs. 4,000 (on cross multiplication)

$$\text{P/V Ratio} = \frac{4,000}{10,000} = 0.40$$

P/V Ratio (in percentage) = 0.40 × 100

= 40%

(*b*) Profit when sales are 20,000

Sales	=	Rs. 20,000
P/V Ratio	=	40%
Contribution	=	Sales × P/V ratio
	=	20,000 × 40%

	=	8,000
Less: Fixed expenses	=	4,000
Profit		4,000

(c) New BEP if selling price is reduced by 20%

Existing selling price (assumed)	=	Rs.	100
Less: Reduction 20%	=	Rs.	20
Revised selling price	=	Rs.	80
Existing selling price	=	Rs.	100
Less: Contribution 40%	=	Rs.	40
Remaining representing variable cost	=	Rs.	60

Revised P/V Ratio at Rs. 80 selling price is:

$$= \frac{\text{Revised selling} - \text{Variable cost}}{\text{Revised selling price}} \times 100$$

$$= \frac{80-60}{80} \times 100$$

$$= \frac{20}{80} \times 100 = 25\%$$

New BEP is, $= \dfrac{\text{Fixed expenses}}{\text{New P/V Ratio}}$

$$= \frac{4{,}000}{25\%} = \frac{4{,}000}{25} \times 100$$

$$= \text{Rs. } 16{,}000$$

Problem 28. The Fixed costs amount to Rs. 1,50,000. The percentage of variable costs to sales is given to be $66\frac{2}{3}$. At 100% capacity level, normal sales are Rs. 9,00,000. Calculate (*a*) P/V Ratio (*b*) Break-even point sales and the percentage of sales at which it occurs. (*c*) Determine profit at 80% capacity sales.
(University of Madras, B.Com., October 2002)

Solution:

(*a*) Calculation of P/V Ratio:

Let sales be taken is	100%
Less: Variable cost percentage on sales	$66\frac{2}{3}\%$
Balance is P/V Ratio	$33\frac{1}{3}\%$

(*b*) (*i*) Break-even sales

$$= \frac{\text{Fixed cost}}{\text{P/V Ratio}} = \frac{1{,}50{,}000}{33\frac{1}{3}\%}$$

$$= \frac{1{,}50{,}000}{\frac{100}{3}} \times 100$$

$$= \frac{1,50,000}{1} \times \frac{3}{100} \times \frac{100}{1}$$

$$= 4,50,000$$

(ii) Percentage of Break-even point on sales

$$= \frac{4,50,000}{9,00,000} \times 100 = 50\%$$

(a) Profit at 80% sales

Sales = Rs. 9,00,000

80% $= 9,00,000 \times \frac{80}{100}$ = Rs. 7,20,000

Profit on a sale of Rs. 7,20,000

Contribution = Sales × P/V Ratio

$= 7,20,000 \times 33\frac{1}{3}\%$

Contribution	=	2,40,000
Less: Fixed cost	=	1,50,000
Profit	=	90,000

Problem 29. The information about a firm is given below:

(*a*) Profit volume ratio 20%

(*b*) Fixed cost Rs. 36,000

(*c*) Selling price per unit Rs. 150

Calculate:

(*a*) Break-even point (in Rs.)

(*b*) Break-even point (in units)

(*c*) Variable cost per unit

(*d*) Profit on sale of Rs. 4,00,000 *(University of Madras, B.Com., May 2002)*

Solution:

(a) Calculation of BEP (in Rs.)

$$= \frac{\text{Fixed cost}}{\text{P/V Ratio}} = \frac{\text{Rs. } 36,000}{20\%}$$

= Rs. 1,80,000

(b) Calculation of BEP (in units)

$$= \frac{\text{Fixed cost}}{\text{Contribution per unit}}$$

Contribution Per unit = Selling price per unit × P/V Ratio

$= 150 \times \frac{20}{100}$ = Rs. 30

BEP units $= \frac{\text{Rs. } 36,000}{\text{Rs. } 30}$ = 1,200 units

(c) Variable cost per unit

Selling price per unit	Rs.	150
Less: Contribution per unit $\left(150 \times \frac{20}{100}\right)$	Rs.	30
i.e. Fixed cost + profit per unit		
Balance is variable cost		120

(d) Profit on sale of Rs. 4,00,000

Total sales		Rs. 4,00,000		
Total contribution	=	Sales × P/V Ratio		
	=	4,00,000 × 20%	=	80,000
Less: Fixed cost			=	36,000
		Profit		44,000

Problem 30. From the following information, find out (*a*) Contribution (*b*) BEP in units (*c*) Margin of safety (*d*) Profit (*e*) Sales to earn a profit of Rs. 6,000

Total fixed cost	Rs.	4,500
Total variable cost	Rs.	7,500
Total sales	Rs.	15,000
Number of units sold		5,000

(University of Madras, B.Com., April 2002)

Solution:

(a) Calculation of contribution

Total sales	15,000
Less: Variable cost	7,500
Contribution	7,500

$$\text{Contribution per unit} = \frac{\text{Rs. } 7{,}500}{5{,}000 \text{ units}} = 1.50 \text{ per unit}$$

(b) BEP in units

$$= \frac{\text{Total fixed cost}}{\text{Contribution per unit}} = \frac{\text{Rs. } 4{,}500}{\text{Rs. } 1.50 \text{ per unit}}$$

$$= 3{,}000 \text{ units}$$

(c) Margin of safety (in units)

= Actual sales – BEP sales

= 5,000 units – 3,000 units

= 2000 units

(d) Profit

Total sales		15,000
Less: Fixed cost	4,500	
Variable cost	7,500	12,000
Profit		3,000

(e) Sales to earn a profit of Rs. 6,000

$$= \frac{\text{Total fixed cost + Profit required}}{\text{Contribution per unit}}$$

$$= \frac{4,500 + 6,000}{1.50} = \frac{\text{Rs. } 10,500}{\text{Rs. } 1.50} = 7,000 \text{ units}$$

Problem 31. Estimate the impact of Break-even point on the following changes.

(*a*) Variable cost increase by 10%

(*b*) Variable cost decrease by 50% and fixed cost increase by 30%

Given particulars are below:

(*i*) Fixed cost Rs. 15,000, (*ii*) Selling price per unit Rs. 15, (*iii*) Variable cost per unit Rs. 9 *(University of Madras, B.Com., May 2002)*

Solution:

(*a*) Break-even point

$$\text{Present BEP} = \frac{\text{Fixed cost}}{\text{Contribution per unit}}$$

Contribution = Selling price per unit – Variable cost per unit

= Rs. 15 – Rs. 9

= Rs. 6

$$\text{Present BEP} = \frac{\text{Rs. } 15,000}{\text{Rs. } 6} = 2500 \text{ units}$$

(*a*) In case variable cost increases by 10%

Variable cost per unit	=	9.00
Add: 10% increase	=	0.90
		9.90

$$\text{Revised BEP} = \frac{\text{Fixed cost}}{\text{Revised contribution per unit}}$$

Revised contribution per unit

Selling price per unit – Revised variable cost per unit

= Rs. 15 – Rs. 9.90

= Rs. 5.10

$$\text{Revised BEP} = \frac{15,000}{5.10} = 2,941 \text{ units}$$

(*b*) In case variable cost decrease by 50% and fixed cost increases by 30%

Variable cost per unit	9.00
Less: Decrease 50%	4.50
Revised variable cost	4.50
Fixed cost	15,000
Add: Increase 30%	4,500
	19,500

$$\text{Revised BEP} = \frac{\text{Revised toal fixed cost}}{\text{Revised contribution per unit}}$$

Revised contribution = S.P. per unit – Revised V.C. per unit

= 15 – 4.50 = 10.50

Revised BEP = $\frac{19,500}{10.50}$ = 1,857 units.

Comment. If variable cost decreases by 50% and fixed cost increases by 30% then, break-even point comes down to 1,857 units.

Problem 32. A company is considering expansion. Fixed costs account to Rs. 4,20,000 and are expected to increase by Rs. 1,25,000 when plant expansion is completed. The present plant capacity is 80,000 units a year. The capacity will increase by 50% with the expansion. The current selling price is Rs. 16 per unit and is expected to remain same under both alternatives. What are the Break-even points under each alternative ? Which alternative is better and why ?

(University of Madras, M.Com., April 2005)

Solution:

STATEMENT SHOWING DIFFERENTIAL COST

Particulars	*Present 80,000 units*	*Proposed 50% increase 80,000 + 40,000 = 1,20,000*	*Differential cost*
Sales	12,80,000 (@ Rs. 16)	19,20,000	6,40,000
Less: Variable cost	5,44,000 (@ Rs. 6.80)	7,68,000 (@ Rs. 6.80 – 0.40 = 6.40)	2,24,000
Contribution	7,36,000 (@ Rs. 9.20)	11,52,000 @ Rs. 9.60	4,16,000
Less: Fixed cost	4,20,000	5,45,000 (4,20,000 + 1,25,000)	1,25,000
Profit	3,16,000	6,07,000	2,91,000
BEP = $\frac{\text{Fixed cost}}{\text{Contribution per unit}}$	$\frac{4,20,000}{9.20}$ = 45652 units	$\frac{5,45,000}{9.60}$ = 56.771 units	

Comment. The proposed expansion is better and it gives additional profitability of Rs. 2,91,000.

Problem 33. S. Ltd. furnishes you the following information relating to the half year ending 30th June, 2002.

Fixed expenses	Rs. 50,000
Sales value	Rs. 2,00,000
Profit	Rs. 50,000

During the second half of the same year the company has suffered a loss of Rs. 10,000. Calculate.

(*a*) The P/V Ratio, break-even point and margin of safety for six months ending 30th June, 2002.

(*b*) Expected sales volume for second half of the year assuming that selling price and fixed expenses remain unchanged in the second half of the year.

(University of Madras, M.Com., April 2004)

Solution:

(*a*) Present half year ending 30th June 2002

Contribution = Sales – Variable cost

or

= Fixed cost + Profit

= 50,000 + 50,000

= 1,00,000

(*i*) P/V Ratio $= \dfrac{S - V\,(\text{or})\,F + P}{\text{Sales}} \times 100$

$= \dfrac{1,00,000}{2,00,000} \times 100 = 50\%$

(*ii*) Break-even point $= \dfrac{\text{Fixed cost}}{\text{P/V Ratio}}$

$= \dfrac{\text{Rs. } 50,000}{50\%} = 1,00,000$

(*iii*) Margin of safety = Actual sales – Break-even sales

= 2,00,000 – 1,00,000

= 1,00,000

(*b*) Next half-year ending 31st December, 2002 with a loss of Rs. 10,000

Sales required for the next half year ending with a loss of Rs. 10,000

$= \dfrac{\text{Fixed cost} - \text{Loss (or)} + \text{Profit}}{\text{P/V Ratio}}$

$= \dfrac{50,000 - 10,000}{50\%}$

$= \dfrac{40,000}{50\%} = \text{Rs. } 80,000$

Problem 34. Two competing companies Ram Ltd. and Mohan Ltd. produce and sell same type of product in the same market. For the year ended 31st March 2001, their estimated profit and loss account is as follows:

	Ram Ltd.		*Mohan Ltd.*	
Sales		2,50,000		2,50,000
Less: Variable cost	2,00,000		1,50,000	
Fixed cost	25,000	2,25,000	75,000	2,25,000
Estimated profit before Tax		25,000		25,000

You are required to compute:

(a) P/V Ratio
(b) Break-even sales volume
(c) Margin of safety

You are also required to state which company is likely to earn greater profit in conditions of:

(i) low demand
(ii) high demand

(University of Madras, M.Com., May 2003)

Solution:

	Ram Ltd.	*Mohan Ltd.*
Sales	2,50,000	2,50,000
Less: Variable cost	2,00,000	1,50,000
Contribution	50,000	1,00,000
Less: Fixed cost	25,000	75,000
Profit	25,000	25,000
(a) P/V ratio $= \frac{S - V}{S} \times 100$	$\frac{50,000}{2,50,000} \times 100$ = 20%	$\frac{1,00,000}{2,50,000} \times 100$ = 40%
(b) Break-even sales $= \frac{\text{Fixed cost}}{\text{P/V Ratio}}$	$\frac{25,000}{20\%}$ = 1,25,000	$\frac{75,000}{40\%}$ = 1,87,500
(c) Margin of safety = Actual sales – Break-even sales	2,50,000 – 1,25,000 = 1,25,000	2,50,000 – 1,87,500 = 62,500

(a) In case of Low Demand:

Ram Ltd. company will do better profit than Mohan Ltd. since additional sales will contribute a profit of 20% over break-even point. Because the break-even point of Ram Ltd. is at lower level of Rs. 1,25,000 compared to Mohan Ltd.

(b) In case of High Demand:

Mohan Ltd. company will do better profit than Ram Ltd. since additional sales will contribute a profit of 40% over break point.

Problem 35. Company *X* and Company *Y* sell the same line of product at a unit price of Rs. 40. The variable costs per unit for each company are Rs. 24 and the fixed costs are Rs. 4,00,000. Company *X* reduces its price by 10% and company *Y* increases its variable cost by 10%.

(a) Compute the break-even point before the change in price and costs.
(b) Compute the break-even point for each company after the changes.
(c) Which company must sell a greater volume to break-even ?

(University of Madras, M.Com., April 2001)

Solution:

(a) Break-even point before change in price and costs (for both X and Y companies)

$$\text{BEP in value} = \frac{\text{Total fixed cost}}{\text{SP per unit} - \text{VC per unit}} \times \text{SP per unit}$$

$$= \frac{\text{Rs. } 4{,}00{,}000}{40 - 24} \times 40$$

$$= 25{,}000 \text{ units} \times 40$$

$$= \text{Rs. } 10{,}00{,}000$$

(b) BEP after the changes in selling price of *X* and changes in variable cost of *Y*

***(i)* Selling price of 'X' reduced by 10%**

Existing selling price per unit	Rs.	40
'X' reduces SP by 10%	Rs.	4
Revised S.P		36

Variable cost per unit remains same = Rs. 24

Total fixed cost remains same = 4,00,000

$$\text{Revised BEP} = \frac{\text{Total FC}}{\text{Revised SP} - \text{VC}} \times \text{Revised SP per unit}$$

$$= \frac{\text{Rs. } 4{,}00{,}000}{36 - 24} \times 36$$

$$= 33333.33 \text{ units} \times 36$$

$$= \text{Rs. } 12{,}00{,}000$$

***(ii)* Variable cost of 'Y' increases by 10%**

Existing variable cost of *Y*	= 24.00
Increase 10%	= 2.40
	26.40

Selling price remains same = Rs. 40

Total fixed cost remains same = Rs. 4,00,000

Revised BEP:

$$= \frac{\text{Total FC}}{\text{SP per unit} - \text{Revised VC per unit}} \times \text{SP per unit}$$

$$= \frac{\text{Rs. } 4{,}00{,}000}{40 - 26.40} \times 40$$

$$= 29411.76 \text{ units} \times \text{Rs. } 40$$

$$= \text{Rs. } 11{,}76{,}470$$

(*c*) Company 'X' sells greater volume to break-even.

Problem 36. The XL Ltd., furnish the following information:

	Ist Period	*IInd Period*
Sales	20,00,000	30,00,000
Profit	2,00,000	4,00,000

From the above, calculate the following:

(1) P/V Ratio

(2) Fixed expenses

(3) BEP

(4) Sales to earn profit of Rs. 5,00,000

(5) Profit when sales are Rs. 15,00,000 *(University of Bombay, B.Com., October 2006)*

Solution:

(1) P/V Ratio $= \frac{\text{Change in profit}}{\text{Change in sales}} \times 100$

$= \frac{2,00,000}{10,00,000} \times 100 = 20\%$

(2) Fixed expenses = Contribution – Profit

= 4,00,000 – 2,00,000

= Rs. 2,00,000

Contribution = Sales × P/V Ratio

= 20,00,000 × 20%

= 4,00,000

(3) BEP (in Rs.) $= \frac{\text{Fixed cost}}{\text{P/V Ratio}} \times 100$

$= \frac{2,00,000}{20} \times 100 = \text{Rs. } 10,00,000$

(4) Sales to earn profit of Rs. 5,00,000

$= \frac{\text{Fixed cost + Desired profit}}{\text{P/V Ratio}} \times 100$

$= \frac{2,00,000 + 5,00,000}{20} \times 100$

= Rs. 35,00,000

(5) Profit when sales are Rs. 15,00,000

Desired contribution = Sales × P/V Ratio

= 15,00,000 × 20%

= 3,00,000

Desired profit = Desired contribution – Fixed cost

= 3,00,000 – 2,00,000

= Rs. 1,00,000

Problem 37. From the following data compute:

(*a*) P/V Ratio

(*b*) BEP in rupees and unit

(*c*) Number of units to be sold to earn a profit of Rs. 7,50,000

Sales price	Rs. 20 per unit
Direct material	Rs. 5 per unit
Direct wages	Rs. 6 per unit
Variable administrative overheads	Rs. 3 per unit
Fixed factory overheads	Rs. 6,40,000 per year
Fixed administrative overhead	Rs. 1,52,000 per year

(University of Bombay, B.Com., April 2006)

Solution:

Working notes:

Sale price per unit = Rs. 20

Variable cost per unit = Direct expenses + Variable expenses

= 5 + 6 + 3 = Rs. 14

Contribution per unit = S.P. – VC per unit

= 20 – 14 = Rs. 6

Fixed cost = Factory overhead + Administrative overhead

= 6,40,000 + 1,52,000

= Rs. 7,92,000

(1) P/V Ratio $= \frac{\text{Contribution}}{\text{Sales}} \times 100$

$= \frac{6}{20} \times 100 = 30\%$

(2) BEP in Rs. $= \frac{\text{Fixed cost}}{\text{P/V Ratio}} = \frac{7,92,200}{30\%}$

= Rs. 26,40,000

BEP in units $= \frac{\text{BEP}}{\text{S.P.}} = \frac{26,40,000}{20}$

= 1,32,000 units

(3) Number of units to be sold to earn a profit of Rs. 7,50,000

$= \frac{\text{Fixed cost + Profit}}{\text{Contribution per unit}}$

$= \frac{7,92,000 + 7,50,000}{6} = \text{Rs. } 2,57,000$

Problem 38. The following is the cost structure of a product. Selling price is Rs. 100 per unit.

Variable cost per unit:

Material Rs. 38

Labour Rs. 14

Direct expenses Rs. 8

Fixed overheads for the year:

Factory overhead Rs. 2,80,000

Office overhead Rs. 2,20,000

No. of units produced and sold 40,000

Calculate:

(1) P/V Ratio
(2) BEP in units
(3) Margin of safety
(4) BEP if fixed overhead increased by 20%
(5) Revised P/V Ratio when selling price increased by 20%.

Solution:

(1) P/V Ratio $= \frac{\text{Contribution}}{\text{Sales}} = \frac{40}{100} = \frac{2}{5}$

(2) BEP in Units $= \frac{\text{Fixed cost}}{\text{Contribution per unit}}$

$= \frac{5,00,000}{40} = 12,500$ units

(3) Margin of safety

Units = 40,000 – 12,500 = 27,500 units

Amount =40,00,000 – 12,50,000 = 27,50,000

(4) Break-even point if fixed overheads increased by 20%

Contribution	=	16,00,000
New fixed overheads	=	6,00,000
Profit		10,00,000

BEP in units $= \frac{\text{Fixed cost}}{\text{Contribution per unit}}$

$= \frac{6,00,000}{40} = 15,000$ units

BEP in Rs. = Rs. 15,00,000

(5) Selling price increased by 20%

PV Ratio $= \frac{SP - VC}{SP} = \frac{120 - 60}{120} = 50\%$

Problem 39. A product is sold at Rs. 80 per unit its variable cost is Rs. 60. Fixed cost is Rs. 6,00,000. Compute the following:

(1) P/V Ratio
(2) BEP
(3) Margin of safety at a sale of 50,000 units
(4) At what sale the product will earn profit at 15% on sale

(University of Bombay, B.Com., April 2005)

Solution:

Sales		Rs.	80.00
Less:	Variable cost	Rs.	60.00
	Contribution		20.00

(1) P/V Ratio $= \dfrac{\text{Contribution}}{\text{Sales}} \times 100$

$= \dfrac{20}{80} \times 100 = 25\%$

(2) BEP $= \dfrac{\text{Fixed cost}}{\text{P/V Ratio}} = \dfrac{6,00,000}{25\%} = \text{Rs. } 24,00,000$

(3) Margin of safety at a sale of 50,000 units

BEP units $= \dfrac{\text{BEP (Rs.)}}{\text{SP}} = \dfrac{24,00,000}{80} = 30,000$

Margin of Safety = Sales units – BEP units

= 50,000 – 30,000 = 20,000 units

(4) Sale to earn a profit of 15% on sale

Sales = Rs. 80 per unit

VC = Rs. 60 per unit

Profit $= \dfrac{15 \times 80}{100} = 12$ per unit

Fixed cost = Rs. 6,00,000

Let units sold be x

Sales = $80 \times x$, since sales = VC + FC + Profit

$\therefore$ $80x = 60x + 6,00,000 + 12x$

$8x = 6,00,000$

$x = \dfrac{6,00,000}{8} = 75,000$ units

Cross verification

S = VC + FC + P

(75,000 × 80) = (75,000 × 60) + 6,00,000 + (75,000 × 12)

60,00,000 = 45,00,000 + 6,00,000 + 9,00,000

Profit 9,00,000 = 15% of sales 60,00,000

Problem 40. Z Ltd. produces and sells a single article at Rs. 10 each. The marginal cost of production is Rs. 6 each and fixed cost is Rs. 400 per month.

Calculate:

(*i*) P/V ratio

(*ii*) The break-even sales (in Rs. and Numbers)

(*iii*) The sales to earn a profit of Rs. 500

(*iv*) Profit at sales of Rs. 3000

(*v*) New break-even point if sales price is reduced by 10%

(*vi*) Margin of safety at sales of Rs. 1,500

(*vii*) Selling price per unit if the break-even point is reduced to 80 units.

(University of Madras, B.Com., October 2004)

Solution:

(1) P/V Ratio $= \dfrac{\text{Contribution}}{\text{Sales}} \times 100 = \dfrac{4}{10} \times 100$

$= 40\%$

(2) BEP

(a) in Rs. $= \dfrac{\text{Fixed cost}}{\text{P/V Ratio}} = \dfrac{400}{40\%}$

$= \dfrac{400}{40} \times 100 = \text{Rs. } 1{,}000$

(a) in Numbers $= \dfrac{\text{Fixed cost}}{\text{Contribution per unit}} = \dfrac{400}{4} = 100$ units

(3) The sales to earn a profit of Rs. 500

$= \dfrac{\text{Fixed cost + Desired profit}}{\text{P/V Ratio}}$

$= \dfrac{400 + 500}{40\%} = \dfrac{900}{40} \times 100 = \text{Rs. } 2{,}250$

(4) Profit on sale of Rs. 3,000

Profit = Contribution – Fixed cost

= 1,200 – 400 = Rs. 800

Contribution = 40% of Rs. 3,000 = Rs. 1,200

(5) New Break-even point if sales price is reduced by 10%

Old sales price Rs. 10

New sales price = 10 – 10% of 10

= 10 – 1 = Rs. 9

New contribution = SP – VC

= 9 – 6 = Rs. 3

New P/V Ratio $= \dfrac{3}{9} \times 100 = 33\dfrac{1}{3}\%$

New BEP in Rs. $= \dfrac{\text{FC}}{\text{P/V Ratio}}$

$= \dfrac{400}{33\frac{1}{3}\%}$

$= 400 \times \dfrac{1}{100} \times \dfrac{3}{100} = \text{Rs. } 1{,}200$

(6) Margin of safety at sales of Rs. 1,500

Old margin of safety = Actual sales – BEP sales

= 1,500 – 1,000 = Rs. 500

New margin of safety = 1,500 – 1,200 = Rs. 300

(7) Selling price per unit if BEP is reduced to 80 units

BEP in units $= \dfrac{\text{Fixed cost}}{\text{Contribution per unit}}$

$80 = \dfrac{400}{\text{Contribution per unit}}$

Contribution per unit $= \dfrac{400}{80} = \text{Rs. } 5$

Selling price per unit = Contribution per unit + VC per unit

= 5 + 6 = Rs. 11 per unit

Problem 41. K.T. and Company has prepared the following budget estimates for the year 2002–03:

Sales 15,000 units
Sales value Rs. 1,50,000
Fixed expenses Rs. 34,000
Variable cost per unit Rs. 6
You are required to find:
(1) P/V ratio (2) BEP (3) Margin of safety

Also calculate revised P/V Ratio, BEP and margin of safety, if selling price per unit is reduced by 10% *(University of Bombay, B.Com., October 2003)*

Solution:

Present sales	15,000 units × Rs. 10	=	Rs. 1,50,000
Present V.C	15,000 units × Rs. 6	=	Rs. 90,000
Present contribution	= Sales – Variable cost		
	= 1,50,000 – 90,000	=	60,000
Proposed sales	= 15,000 × Rs. 9	=	1,35,000
Proposed V.C.	= 15,000 × Rs. 6	=	90,000
Proposed contribution	= 1,35,000 – 90,000	=	45,000

(1) P/V Ratio $= \dfrac{\text{Contribution}}{\text{Sales}} \times 100$

Present $= \dfrac{60,000}{1,50,000} \times 100 = 40\%$

Proposed $= \dfrac{45,000}{1,35,000} \times 100 = 33.33\%$

(2) BEP $= \dfrac{\text{Fixed cost}}{\text{P/V Ratio}} \times 100$

Present $= \dfrac{34,000}{40} \times 100 = 85,000$

Proposed $= \dfrac{34,000 \times 3}{1} = 1,02,000$

(3) Margin of safety = Actual sales – Break-even sales

Present = 1,50,000 – 85,000 = 65,000

Proposed = 1,35,000 – 1,02,000 = 33,000

(4) Contribution per unit $= \dfrac{\text{Contribution}}{\text{Units}}$

Present $= \dfrac{60,000}{15,000} = 4$

Proposed $= \dfrac{45,000}{15,000} = 3$

(5) BEP (Units) $= \dfrac{\text{BEP}}{\text{SP}}$

Present $= \dfrac{85{,}000}{10} = 8{,}500$

Proposed $= \dfrac{1{,}02{,}000}{9} = 11{,}333$

Problem 42. M/S M. Ltd. made sales of Rs. 2,50,000 during a certain period. The net profit for the same period was Rs. 24,000 and the fixed overheads were Rs. 38,000.

Find out (1) BEP (Sales) and (2) Volume of sales to earn a profit of Rs. 40,000.

(Calcutta University, B.Com. (Hons.), 2005)

Solution:

Contribution = Profit + Fixed cost

= 24,000 + 38,000 = 62,000

P/V Ratio $= \dfrac{\text{Contribution}}{\text{Sales}} = \dfrac{62{,}000}{2{,}50{,}000} = \dfrac{31}{125}$

(1) **BEP Sales** $= \dfrac{\text{Fixed cost}}{\text{P/V Ratio}} = \dfrac{38{,}000}{\frac{31}{125}}$

$= 38{,}000 \times \dfrac{125}{31}$

= 1,53,226

(2) **Volume of sales to earn a profit of Rs. 40,000**

$= \dfrac{\text{Fixed cost} + \text{Desired profit}}{\text{P/V Ratio}}$

$= \dfrac{38{,}000 + 40{,}000}{\frac{31}{125}}$

$= 78{,}000 \times \dfrac{125}{31} = 3{,}14{,}516$

Problem 43. Alpha Ltd. furnished the following information:

Selling price – Rs. 10 per unit

Variable cost – Rs. 6 per unit

It is expected that variable cost will increase by 20%. What will be the selling price per unit if the company wishes to maintain the same P/V Ratio ? *(Calcutta University, B.Com. (Hons.), 2004)*

Solution: Contribution per unit is 10 – 6 = 4

P/V Ratio $= \dfrac{\text{Contribution per unit}}{\text{Selling price per unit}} \times 100$

$= \dfrac{4}{10} \times 100 = 40\%$

If variable cost is increased by 20%, *i.e.*, it becomes Rs. 7.20 per unit then contribution per unit will be $x - 7.20$ (assuming new selling price is Rs. x per unit).

$$\text{New P/V Ratio} = \frac{\text{Contribution per unit}}{\text{Selling price}}$$

$$40\% = \frac{x - 7.20}{x}$$

$$x = 12$$

New selling price will be Rs. 12 per unit.

Problem 44. From the following information calculate profit

Sales	Rs. 8,00,000
Variable cost	Rs. 6,00,000
Break-even sales	Rs. 6,00,000

(Calcutta University, B.Com. (Hons.), 2002)

Solution:

$$\text{Contribution} = \text{Sales} - \text{Variable cost}$$

$$= 8{,}00{,}000 - 6{,}00{,}000 = \text{Rs. } 2{,}00{,}000$$

$$\text{P/V Ratio} = \frac{\text{Contribution}}{\text{Sales}} \times 100$$

$$= \frac{2{,}00{,}000}{8{,}00{,}000} \times 100 = 25\%$$

Profit is earned on sales beyond BEP, *i.e.*, margin of safety

$$\text{So, Profit} = (8{,}00{,}000 - 6{,}00{,}000) \times \frac{25}{100}$$

$$= \text{Rs. } 50{,}000$$

Problem 45. On the basis of the following information you are required to ascertain: (*a*) Sales to break-even and (*b*) Sales to earn a profit of Rs. 60,000 if selling price per unit is reduced by Re. 1.

Sales (10,000 Units)	Rs. 1,60,000
Variable cost	Rs. 96,000
Fixed cost	Rs. 48,000

(Calcutta University, B.Com. (Hons.), 2001)

Solution:

(*a*) $$\text{Selling price per unit} = \frac{1{,}60{,}000}{10{,}000} = \text{Rs. } 16$$

$$\text{Contribution} = \text{Sales} - \text{Variable cost}$$

$$= 1{,}60{,}000 - 96{,}000 = 64{,}000$$

$$\text{Contribution per unit} = \frac{64{,}000}{10{,}000} = 6.40$$

$$\text{BEP in units} = \frac{\text{Fixed cost}}{\text{Contribution per unit}}$$

$$= \frac{48{,}000}{6.40} = 7{,}500 \text{ units}$$

BEP in (Rs.) = 7,500 units × Rs. 16
= Rs. 1,20,000

(*b*) If selling price is reduced by Re. 1, then contribution per unit will be 6.40 – 1 = Rs. 5.40.

Sales required to earn a profit of Rs. 60,000 will be

$$= \frac{\text{Fixed cost + Profit}}{\text{Contribution per unit}}$$

$$= \frac{48{,}000 + 60{,}000}{5.40} = 20{,}000 \text{ units}$$

Sales will be 20,000 × Rs. 15 = Rs. 3,00,000

Problem 46. Rainbow Ltd. sold goods for Rs. 30,00,000 in a year. In that year the variable costs were Rs. 6,00,000 and fixed costs were Rs. 8,00,000. Find out:

(*i*) P/V Ratio
(*ii*) Break-even sales
(*iii*) Break-even sales if selling price was reduced by 10% and fixed costs were increased by Rs. 1,00,000. *(Calcutta University, B.Com. (Hons.), 2000)*

Solution:

Sales Rs. 30,00,000, variable costs Rs. 6,00,000

∴ Contribution = 30,00,000 – 6,00,000
= 24,00,000

(1) So P/V Ratio $= \frac{\text{Contribution}}{\text{Sales}} \times 100$

$$= \frac{24{,}00{,}000}{30{,}00{,}000} \times 100 = 80\%$$

(2) Break-even sales $= \frac{\text{Fixed cost}}{\text{P/V Ratio}}$

$$= \frac{8{,}00{,}000}{80\%} = 10{,}00{,}000$$

(3) If selling price is reduced by 10% then new selling price will be 30,00,000 – 3,00,000 = 27,00,000 and fixed cost will be Rs. 8,00,000 + 1,00,000 = Rs. 9,00,000. Therefore contribution will be 27,00,000 – 6,00,000 = 21,00,000.

So new P/V Ratio will be

$$= \frac{21{,}00{,}000}{27{,}00{,}000} \times 100 = 77.77\%$$

$$\text{BEP} = \frac{9{,}00{,}000}{77.77\%} = \text{Rs. } 11{,}57{,}143$$

Problem 47. A business produces 200 units of a product by making following expenditure

(*i*) Materials	Rs. 30,000
(*ii*) Labour	Rs. 20,000
(*iii*) Factory overheads	Rs. 4,000
(*iv*) Administrative overheads	Rs. 5,754
(*v*) Selling and distribution overhead	Rs. 1,500

The products are sold at a price of Rs. 400 per unit.

The above expenditure are classified into fixed and variable as follows:

	Expenditure	*Fixed*	*Variable*
(*i*)	Materials	NIL	100%
(*ii*)	Labour	50%	50%
(*iii*)	Factory overhead	25%	75%
(*iv*)	Administrative overhead	100%	NIL
(*v*)	Selling and distribution overhead	60%	40%

From the above information determine (*a*) Total variable costs and fixed cost (*b*) Contribution (*c*) P/V ratio (*d*) Break-even point in units and sales value.

(Calcutta University, B.Com. (Hons.), 1999)

Solution:

(A) TOTAL VARIABLE COSTS AND FIXED COSTS

	Total	*Variable*	*Fixed*
Materials	30,000	30,000	
Labour	20,000	10,000	10,000
Factory overheads	4,000	3,000	1,000
Administrative overheads	5,754	–	5,754
Selling and distribution overhead	1,500	600	900
Total	61,254	43,600	17,654

(*b*) Contribution = Selling price – Variable cost

= 200 × 400 – 43,600

= 36,400

$$\text{Contribution per unit} = \frac{36,400}{200} = \text{Rs. } 182$$

$$\textbf{(c) P/V Ratio} = \frac{\text{Contribution per unit}}{\text{Selling price per unit}} \times 100$$

$$= \frac{182}{400} \times 100 = 45.5\%$$

$$\textbf{(d) BEP in Units} = \frac{\text{Total fixed cost}}{\text{Contribution per unit}}$$

$$= \frac{17,654}{182} = 97 \text{ units}$$

BEP in sales value = BEP in units × SP per unit

= 97 × 400 = Rs. 38,800

Problem 48. X Ltd. made sales during a certain period for Rs. 1,00,000. The net profit for the same period was Rs. 10,000 and the fixed overheads were Rs. 15,000. Find out

(1) P/V Ratio

(2) BEP sales

(3) Number of units to be sold to earn a profit of Rs. 15,000 and

(4) Net profit from the sales of Rs. 1,50,000. *(Calcutta University, B.Com. (Hons.), 1995)*

Solution:

Variable cost = Sales – (Fixed overhead + Profit)

= 1,00,000 – (15,000 + 10,000)

= 75,000

Contribution = Sales – Variable cost

= 1,00,000 – 75,000 = 25,000

Contribution = Fixed overhead + Profit

= 15,000 + 10,000

= 25,000

(1) P/V Ratio $= \frac{\text{Profit}}{\text{Sales}} \times 100$

$= \frac{25,000}{1,00,000} \times 100 = 25\%$

(2) BEP Sales $= \frac{\text{Fixed overheads}}{\text{P/V Ratio}}$

$= \frac{15,000}{25\%} = \frac{15,000}{25} \times 100 = \text{Rs. } 60,000$

(3) Sales volume to earn a profit of Rs. 15,000

$= \frac{\text{Fixed overhead + Desired profit}}{\text{P/V Ratio}}$

$= \frac{15,000 + 15,000}{25\%} = \frac{30,000}{25} \times 100 = \text{Rs. } 1,20,000$

(4) Net profit from the sales of Rs. 1,50,000

= Sales × P/V Ratio – Fixed overheads

$= \left(1,50,000 \times \frac{25}{100}\right) - 15,000$

= 37,500 – 15,000 = Rs. 22,500

Problem 49. M/s Natraj Stationers manufacturer of plastic files for office use. The break-up of its cost and sales is as follows:

Variable cost per file	:	Rs. 40
Fixed cost	:	Rs. 60,000 per year
Production capacity	:	3,000 files per year
Selling price	:	Rs. 100 per file

You are required to compute the following.

(*i*) Break-even point

(*ii*) Number of files to be sold to earn a net profit of Rs. 30,000

(*iii*) If the firm manufactures and sells 500 files more per year with an additional fixed cost of Rs. 2,000, what should be the selling price to earn the same amount of profit per file as in (*ii*) above. *(University of Delhi, B.Com. (Hons.), 2005)*

Solution:

(1) BEP $= \dfrac{\text{Fixed cost}}{\text{Contribution per unit}}$

$= \dfrac{60{,}000}{100 - 40} = \dfrac{60{,}000}{60} = 1{,}000$ files

(2) No. of files to be sold $= \dfrac{\text{Fixed cost + Profit}}{\text{Contribution per unit}}$

$= \dfrac{60{,}000 + 30{,}000}{60} = 1{,}500$ files

(3) Selling price to earn the same amount of profit:

Let x be the selling price

$$\text{Sales} = \frac{\text{Fixed cost + Profit}}{\text{Contribution}}$$

$$1500 + 500 = \frac{(60{,}000 + 2{,}000) + 30{,}000}{x - 40}$$

$$2{,}000\,(x - 40) = 92{,}000$$

$$2000x - 80{,}000 = 92{,}000$$

$$2000x = 92{,}000 + 80{,}000$$

$$x = \frac{1{,}72{,}000}{2{,}000} = \text{Rs. } 86$$

Problem 50. ABC Ltd. manufactures three products *P, Q* and *R*. The unit selling prices of these products are Rs. 100, Rs. 80 and Rs. 50 respectively. The corresponding unit variable costs are Rs. 50, Rs. 40 and Rs. 20. The proportions (quantity-wise) in which these products are manufactured and sold are 20%, 30% and 50% respectively. The total fixed cost is Rs. 14,80,000.

Given the above information, you are required to work out the overall break-even quantity and the product-wise break-up of such quantity. *(University of Delhi, B.Com. (Hons.), 2004)*

Solution:

Contribution = SP per unit – VC per unit

P = 100 – 50 = Rs. 50

Q = 80 – 40 = Rs. 40

R = 50 – 20 = Rs. 30

Contribution per composite unit

P = 50 × 20% = 10

Q = 40 × 30% = 12

R = 30 × 50% = 15

Rs. 37

Overall BEP $= \dfrac{\text{Fixed cost}}{\text{Contribution}}$

$= \dfrac{14{,}80{,}000}{37} = 40{,}000$ units

BEP of P = 40,000 × 20% = 8,000 units

BEP of Q = 40,000 × 30% = 12,000 units

BEP of R = 40,000 × 50% = 20,000 units

Problem 51. The trading results of P J Ltd. for the two years have been:

Year	*Sales*	*Profit*
2002	Rs. 5,40,000	Rs. 12,000
2003	Rs. 6,00,000	Rs. 30,000

Compute the following:

(*i*) P/V Ratio

(*ii*) Fixed cost

(*iii*) Break-even sales

(*iv*) Margin of safety at a profit of Rs. 48,000

(*v*) Variable cost during the two years *(University of Delhi, B.Com. (Hons.), 2004)*

Solution:

(1) P/V Ratio $= \dfrac{\text{Difference in profit}}{\text{Difference in sales}} \times 100$

$= \dfrac{30{,}000 - 12{,}000}{6{,}00{,}000 - 5{,}40{,}000} \times 100$

$= \dfrac{18{,}000}{60{,}000} \times 100 = 30\%$

(2) Fixed costs = (Sales × P/V Ratio) – Profit

$= \left(6{,}00{,}000 \times \dfrac{30}{100}\right) - 30{,}000$

= 1,50,000

(3) Break-even sales $= \dfrac{\text{Fixed cost}}{\text{P/V Ratio}} = \dfrac{1{,}50{,}000}{30\%}$ = Rs. 5,00,000

(4) Margin of safety at a profit of Rs. 48,000

Sales $= \dfrac{\text{Fixed cost + Profit}}{\text{P/V Ratio}}$

$= \dfrac{1{,}50{,}000 + 48{,}000}{30\%}$ = Rs. 6,60,000

Margin of safety = Sales – Break-even sales

= 6,60,000 – 5,00,000

= Rs. 1,60,000

(5) Variable cost ratio = 100 – P/V Ratio

= 100 – 30% = 70%

Variable cost = Sales × Variable cost ratio

2002 = 5,40,000 × 70% = Rs. 3,78,000

2003 = 6,00,000 × 70% = Rs. 4,20,000

Problem 52. The ratio of variable cost to sales is 70%. The break-even point occurs at 60% of the capacity sales. Find the capacity sales when fixed costs are Rs. 90,000. Also compute profit at 75% of the capacity sales. *(University of Delhi, B.Com. (Hons.), 2004)*

Solution:

P/V Ratio = 100 – 70% = 30%

$$\text{BEP} = \frac{\text{Fixed cost}}{\text{P/V Ratio}} = \frac{90{,}000}{30\%} = \text{Rs. } 3{,}00{,}000$$

$$\text{Sales at 100\% capacity} = \frac{\text{Rs. } 3{,}00{,}000}{60\%} = 5{,}00{,}000$$

$$\text{Sales at 75\% capacity} = \text{Rs. } 5{,}00{,}000 \times \frac{75}{100} = 3{,}75{,}000$$

Less:	Break-even point =	3,00,000
	Margin of safety	Rs. 75,000

∴ Profit = M/s × P/V Ratio
= 75,000 × 30% = Rs. 22,500

Problem 53. The Manager of ABC Ltd., provides you with the following information

		Rs.
Sales		4,00,000
Costs:		
Variable (60% of sales)	2,40,000	
Fixed	80,000	3,20,000
Profit before tax		80,000
Income-tax (60%)		48,000
Net profit		32,000

The ABC Ltd., is thinking of expanding the plant. The fixed costs with plant expansion will be increased by Rs. 40,000. The company also wants to earn additional income after tax of Rs. 3,200 on investment.

On the basis of computations, give your opinion on plant expansion.

(University of Delhi, B.Com. (Hons.), 2004)

Solution:

Contribution = Rs. 4,00,000 – Rs. 2,40,000
= Rs. 1,60,000

$$\text{BEP} = \frac{\text{Fixed cost}}{\text{Contribution}} \times \text{Sales}$$

$$= \frac{80{,}000}{1{,}60{,}000} \times 4{,}00{,}000 = \text{Rs. } 2{,}00{,}000$$

$$\text{P/V Ratio} = \frac{1{,}60{,}000}{4{,}00{,}000} \times 100 = 40\%$$

$$\text{New Break-even point} = \frac{\text{Fixed cost} + \text{Additional fixed cost}}{\text{P/V Ratio}}$$

$$= \frac{1{,}20{,}000}{40\%} = \text{Rs. } 3{,}00{,}000$$

$$\text{Sales to earn existing profit} = \frac{\text{Fixed cost + Profit}}{\text{P/V Ratio}}$$

$$= \frac{1,20,000 + 80,000}{40\%}$$

$$= \text{Rs. } 5,00,000$$

$$\text{Income before tax} = \frac{3,200}{40\%} = \text{Rs. } 8,000$$

$$\text{Sales to earn additional income} = \frac{2,00,000 + 8,000}{40\%} = \text{Rs. } 5,20,000.$$

In order to earn additional income after tax of Rs. 3,200, the sales should be Rs. 5,20,000.

Opinion:

If the company can achieve sales of Rs. 5,20,000 the idea of plant expansion should be acceptable.

Problem 54. Suman Ltd., manufactured and sold 1,000 Electric Irons last year at a price of Rs. 800 each. The cost structure of Electric Irons:

Materials	200
Labour	100
Variable cost	50
Marginal cost	350
Factory overheads (Fixed)	200
Total cost	550
Profit	250
Sales price	800

Due to heavy competition, price had to be reduced to Rs. 750 for the coming year. Assuming no change in costs, state the number of Electric Irons that would have to be sold at the new price to ensure the same amount of total profit as that of the last year.

(University of Delhi, B.Com. (Hons.), 2004)

Solution:

Total fixed cost = 1,000 × 200 = 2,00,000

Contribution = 800 – 350 = Rs. 450 per unit

New contribution = 750 – 350 = 400 per unit

Existing profit = 1,000 × 250 = Rs. 2,50,000

$$\text{Sales required} = \frac{\text{Fixed cost + Existing profit}}{\text{New contribution per unit}}$$

$$= \frac{2,00,000 + 2,50,000}{400} = 1,125 \text{ Irons}$$

Problem 55. A company has a fixed cost of Rs. 20,000. It sells two products A and B in the ratio of 2 units of A and 1 unit of B. Contribution is Re. 1 per unit of A and Rs. 2 per unit of B. How many units of A and B would be sold at break-even point ? *(University of Delhi, B.Com. (Hons.), 2003)*

Solution:

	A	*B*	*Total*
Sale units	2	1	3
Contribution per unit Rs.	1	2	–
Total contribution Rs.	2	2	4

Contribution per composite unit = Rs. $\frac{4}{3}$

Total fixed cost = Rs. 20,000

$$\text{Break-even point} = \frac{\text{Fixed cost}}{\text{Contribution per unit}}$$

$$= \frac{20{,}000}{\frac{4}{3}} = 15{,}000 \text{ units}$$

$$A = 15{,}000 \times \frac{2}{3} = 10{,}000 \text{ units}$$

$$B = 15{,}000 \times \frac{1}{3} = 5{,}000 \text{ units}$$

Problem 56. A Company having annual sales of Rs. 10 crores is earning 12% profit before charging interest and depreciation. Interest and depreciation amount to Rs. 60 lakhs and Rs. 100 lakhs respectively. If the $\frac{\text{Contribution}}{\text{Sales}}$ ratio of the company is 0.4, calculate its break-even sales.

(University of Delhi, B.Com. (Hons.), 2002)

Solution:

Sales = Rs. 10 crores

Profit = Rs. 10 crores × 12%

= Rs. 1.2 crores

Interest + Depreciation = (60 + 100) lakhs

= 1.6 crores

Loss = (Rs. 1.6 – 1.2) crores = Rs. 40 lakhs

Contribution = Sales × P/V Ratio

= 10 crores × 40%

= Rs. 4 crores

Fixed cost = Contribution + loss

= Rs. 4 crores + Rs. 40 lakhs

= Rs. 4.4 crores

$$\text{BEP} = \frac{\text{Fixed cost}}{\text{P/V Ratio}}$$

$$= \frac{\text{Rs. 4.4 crores}}{40\%} = \text{Rs. 11 crores}$$

Problem 57. Profit is product of P/V ratio and margin of safety ratio. Explain giving a suitable example.

(University of Delhi, B.Com. (Hons.), 2002)

Solution:

Profit = Margin of Safety Ratio × P/V Ratio

Example: Let sales = Rs. 10 lakhs

Let P/V Ratio = 30%

Let M/S Ratio = 20%

Profit = 30% × 20% of sales

= 6% of sales

= Rs. 10 lakhs × 6%

= Rs. 60,000

Contribution = Sales × P/V Ratio

= Rs. 10 lakhs × 30%

= Rs. 3 lakhs

Margin of safety Ratio = Rs. 10 lakhs × 20%

= Rs. 2 lakhs

Profit = Contribution × MS Ratio

= Rs. 3 lakhs × 20%

= Rs. 60,000

Profit = M/S × P/V Ratio

= Rs. 2 lakhs × 30%

= Rs. 60,000

$$\text{M/S Ratio} = \frac{\text{Profit}}{\text{P/V Ratio}} = \frac{60{,}000}{30\%}$$

= Rs. 2 lakhs

Problem 58. When sales of a company decline from Rs. 9,00,000 to Rs. 7,00,000 its profit of Rs. 50,000 is converted into a loss of Rs. 50,000. Determine contribution margin ratio.

(University of Delhi, B.Com. (Hons.), 2002)

Solution:

$$\text{P/V ratio} = \frac{\text{Difference in profit}}{\text{Difference in sales}}$$

$$= \frac{50{,}000-(-50{,}000)}{7{,}00{,}000-9{,}00{,}000}$$

$$= \frac{1{,}00{,}000}{2{,}00{,}000} = 50\%$$

Problem 59. A company sells its products at Rs. 15 per unit. In a period, if it produces and sells 8,000 units, it incurs a loss of Rs. 5 per unit. If the volume is raised to 20,000 units, it earns a profit of Rs. 4 per unit.

Calculate break-even point in terms of rupees as well as in units.

(University of Delhi, B.Com. (Hons.), 2001)

Solution:

When sales = 8,000 × Rs. 15 = Rs. 1,20,000

Loss = 8,000 × Rs. 5 = Rs. 40,000

When sales = 20,000 × Rs. 15 = Rs. 3,00,000

Profit = 20,000 × Rs. 4 = Rs. 80,000

$$\text{P/V Ratio} = \frac{\text{Difference in profit}}{\text{Difference in sales}}$$

$$= \frac{80,000 - (-40,000)}{3,00,000 - 1,20,000}$$

$$= \frac{1,20,000}{1,80,000} = 66\frac{2}{3}\%$$

Fixed cost = Contribution – Profit

Contribution = Sales × P/V Ratio

$$= 3,00,000 \times 66\frac{2}{3}\% = 2,00,000$$

Less: Profit = 80,000

Fixed cost 1,20,000

$$\text{BEP} = \frac{\text{Fixed cost}}{\text{P/V Ratio}} = \frac{1,20,000}{66\frac{2}{3}\%} = 1,80,000$$

$$\text{BEP in units} = \frac{1,80,000}{15} = 12,000 \text{ units}$$

Problem 60. P/V Ratio of a company is 50%, while its margin of safety is 40%. If sales volume of the company is Rs. 50 lakhs, find out its break-even point and net profit.

(University of Delhi, B.Com. (Hons.), 1999)

Solution:

P/V Ratio × M/S Ratio × Sales = Profit

50% × 40% × Rs. 50 lakhs = Rs. 10 lakhs

BEP = Sales – M/S Ratio

= Rs. 50 lakhs – 40% = Rs. 30 lakhs

Problem 61. For the coming year, a manufacturing company has budgeted as under:

$$\frac{\text{Contribution}}{\text{Sales}} \text{ Ratio} = 45\%$$

$$\text{Margin of safety Ratio} = 33\frac{1}{3}\%$$

Fixed costs = Rs. 5,85,000

Required: Determine total sales-volume for the coming year and profit thereon.

(University of Delhi, B.Com. (Pass), 2005)

Solution:

$$\text{BEP} = \frac{\text{Fixed cost}}{\text{P/V Ratio or C/S Ratio}}$$

$$= \frac{5,85,000}{45\%} = \frac{5,85,000}{45} \times 100$$

= Rs. 13,00,000

% Profit = P/V Ratio × M.S. Ratio × 100

$$= \frac{45}{100} \times \frac{100}{300} \times 100 = 15\%$$

$$\text{Sales volume} = \frac{\text{Fixed cost + Profit \% on sales}}{\text{P/V Ratio}}$$

$$\text{Sales} = \frac{5,85,000 + 15\% \text{ Sales}}{45\%}$$

45% Sales = Rs. 5,85,000 + 15% sales

30% Sales = 5,85,000

$$\text{Sales} = 5,85,000 \times \frac{100}{30} = \text{Rs. } 19,50,000$$

Sales = Rs. 19,50,000

Profit = 15% of Rs. 19,50,000 = Rs. 2,92,500

Problem 62. A factory engaged in manufacturing plastic buckets is working at 40% capacity and produces 10,000 buckets per annum.

The present cost break-up for one bucket is as under:

Material	Rs. 10
Labour cost	Rs. 3
Overheads	Rs. 5 (60% Fixed)

The selling price is Rs. 20 per bucket.

If it is decided to work the factory at 50% capacity, the selling price falls by 3%. At 90% capacity, the selling price falls by 5% accompanied by a similar fall in the prices of material.

You are required to calculate profit at 50% and 90% capacity and also calculate break-even point for the same capacity. *(University of Delhi, B.Com. (Pass), 2004)*

Solution:

Production at 40% capacity = 10,000 buckets

Production at 50% capacity

$$= \frac{10,000}{40} \times 50 = 12,500 \text{ buckets}$$

Production at 90% capacity

$$= \frac{10,000}{40} \times 90 = 22,500 \text{ buckets}$$

When production is 10,000 buckets, fixed cost per unit = Rs. 5 × 60% = Rs. 3

Total fixed cost = 10,000 × 3 = Rs. 30,000

Profit and Break-even point at 50% capacity:

Production = 12,500 buckets

Selling price per unit = Rs. 20 – 3% of Rs. 20

= 20 – 0.60

= Rs. 19.40

Sales = 12,500 × 19.40 = Rs. 2,42,500

Variable cost per unit:

Materials	Rs. 10
Labour	Rs. 3
Overheads $\left(\frac{40}{100} \times 5\right)$	Rs. 2
	Rs. 15

Contribution per unit = SP per unit – VC per unit
= 19.40 – 15 = Rs. 4.40

Total Contribution = Output × Contribution per unit
= 12,500 × Rs. 4.40
= Rs. 55,000

BEP in Units $= \frac{\text{Fixed cost}}{\text{Contribution per unit}}$

$= \frac{30,000}{4.40} = 6,818$ units

BEP in Rs. = BEP in units × SP per unit
= 6,818 × 19.40
= Rs. 1,32,269

Profit and BEP at 90% capacity

Production = 22,500 buckets

Selling price per unit = Rs. 20 – 5% of Rs. 20
= Rs. 20 – Re. 1
= Rs. 19

Sales = 22,500 × 19 = Rs. 4,27,500

Materials cost per unit = Rs. 10 – 5% of Rs. 10
= 10 – Re. 0.50
= 9.50

Variable cost per unit

Materials	Rs. 9.50
Labour	Rs. 3.00
Variable overheads	Rs. 2.00
	14.50

Contribution per unit = SP per unit – VC per unit
= 19 – 14.50
= Rs. 4.50

Total Contribution = Output × Contribution per unit
= 22,500 × 4.50 = 1,01,250

Profit = Contribution – Fixed cost
= 1,01,250 – 30,000
= Rs. 71,250

BEP in Units $= \frac{\text{Fixed cost}}{\text{Contribution per unit}}$

$= \frac{30,000}{4.50} = 6,667$ units

BEP in Rs. = BEP in units × SP per unit
= 6,667 × Rs. 19 = 1,26,673

Problem 63. A company annually manufactures and sells 20,000 units of a product. The selling price of which is Rs. 50 and profit earned is Rs. 10 per unit

The analysis of cost of 20,000 units is:

Materials cost	Rs. 3,00,000
Labour cost	Rs. 1,00,000
Overheads	Rs. 4,00,000 (50% variable)

You are required to compute:

(1) Break-even sales in units and in Rupees

(2) Sales to earn a profit of Rs. 3,00,000.

(3) Profit when 15,000 units are sold. *(University of Delhi, B.Com. (Pass), 2002)*

Solution:

MARGINAL COST OR VARIABLE COST

	Total	*Per unit*
Materials cost	3,00,000	15
Labour cost	1,00,000	5
Overheads (50% of 4,00,000)	2,00,000	10
Total		30

Contribution per unit = SP per unit – VC per unit
= Rs. 50 – Rs. 30 = Rs. 20

(1) BEP in units $= \dfrac{\text{Fixed cost}}{\text{Contribution per unit}}$

$= \dfrac{2,00,000}{20} = 10,000$ units

BEP in Rs. = BEP in units × SP per unit
= 10,000 × 50 = Rs. 5,00,000

(2) Sales to earn desired profit

$= \dfrac{\text{Fixed cost + Desired profit}}{\text{Contribution per unit}} \times \text{SP per unit}$

$= \dfrac{2,00,000 + 3,00,000}{20} \times 50$

$= \dfrac{5,00,000}{20} \times 50$

= 25,000 × 50 = Rs. 12,50,000

(3) Profit when 15,000 units are sold

Contribution = 15000 × 20	=	3,00,000
Less: Fixed cost	=	2,00,000
Profit		1,00,000

Problem 64. XYZ Company Ltd. produced and sold 1,000 toys last year at a price of Rs. 500 each. The cost structure per toy is as follows:

Materials	Rs.100
Labour	50
Variable overheads	25
	175
Fixed overheads	200
	375
Profit	125
Selling price	500

Due to heavy competition, the price has to be reduced to Rs. 425 for the coming year. Assuming that there will be no change in costs, find out how many toys shall be sold to ensure the same amount of total profit as last year. *(University of Delhi, B.Com. (Pass), 2001)*

Solution:

Total profit = 1,000 units × Rs. 125 = Rs. 1,25,000
Fixed cost = 1,000 units × Rs. 200 = Rs. 2,00,000
New selling price = Rs. 425 per unit
Variable cost = Rs. 175 per unit
New contribution = Rs. 425 – Rs. 175 = Rs. 250 per unit

Desired sales to earn the same total profit of Rs. 1,25,000

$$= \frac{\text{Fixed cost + Desired profit}}{\text{Contribution per unit}}$$

$$= \frac{2{,}00{,}000 + 1{,}25{,}000}{250} = 1{,}300 \text{ units}$$

Problem 65. The sales turnover and profit of M/s A Ltd. during the two years 1998 and 1999 were as follows:

Year	*Sales (Rs.)*	*Profit (Rs.)*
1998	4,50,000	60,000
1999	5,10,000	75,000

You are required to calculate:

(*i*) Profit-volume ratio.
(*ii*) Sales at which company will neither lose nor gain anything.
(*iii*) Sales required to earn a profit of Rs. 1,20,000.
(*iv*) The profit made when sales are Rs. 7,50,000. *(University of Delhi, B.Com. (Pass), 2000)*

Solution:

(1) P/V Ratio

$$= \frac{\text{Change in profit}}{\text{Change in sales}} \times 100$$

$$= \frac{75{,}000 - 60{,}000}{5{,}10{,}000 - 4{,}50{,}000} \times 100$$

$$= \frac{15{,}000}{60{,}000} \times 100 = 25\%$$

(2) P/V Ratio $= \dfrac{\text{Contribution}}{\text{Sales}} \times 100$

$25 = \dfrac{\text{Contribution}}{4,50,000} \times 100$

$\text{Contribution} = \dfrac{4,50,000 \times 25}{100}$

$= \text{Rs. } 1,12,500$

$\text{Fixed cost} = \text{Contribution} - \text{Profit}$

$= 1,12,500 - 60,000$

$= \text{Rs. } 52,500$

BEP $= \dfrac{\text{Fixed cost}}{\text{P/V Ratio}} = \dfrac{52,500}{25\%}$

$= \text{Rs. } 2,10,000$

(3) Sales to earn a profit of Rs. 1,20,000

$= \dfrac{\text{Fixed cost + Desired profit}}{\text{P/V Ratio}}$

$= \dfrac{52,500 + 1,20,000}{25\%} = \dfrac{1,72,500}{25\%}$

$= \text{Rs. } 6,90,000$

(4) P/V Ratio $= \dfrac{\text{Contribution}}{\text{Sales}} \times 100$

$25 = \dfrac{\text{Contribution}}{7,50,000} \times 100$

$\text{Contribution} = \dfrac{7,50,000}{100} \times 25 = \text{Rs. } 1,87,500$

∴ Profit made when sales are Rs. 7,50,000

$= \text{Contribution} - \text{Fixed cost}$

$= 1,87,500 - 52,500$

$= \text{Rs. } 1,35,000$

GRAPHIC METHOD OF PRESENTING CVP ANALYSIS

This method is preferred under two conditions viz. (*a*) where a simple overview is sufficient and (*b*) to avoid detailed calculations involved under equation method. The following procedure is involved under the graphic method:

(*a*) **Vertical line:** Draw a vertical line on the left side of the graph paper to show cost and sales revenue.

(*b*) **Horizontal line:** Draw a horizontal line to show the number of units in such a way it should intersect the vertical line at the point of zero.

(*c*) **Fixed cost line:** A line parallel to X-axis (horizontal line) is to be drawn to represent the fixed cost considering the units manufactured and fixed cost incurred.

(*d*) **Total sales revenue line:** This line is drawn starting at the zero point on the left hand corner and ending on the right hand side considering thc sales revenue.

(*e*) **Total cost line:** This is the total of fixed and variable cost. This line is drawn starting at the Y-axis fixed cost point and moving to the right considering the total cost.

(*f*) **BEP:** This is the point where total cost line intersects the sales line.

A typical graph depicting the BEP is shown in the Fig. 11.1.

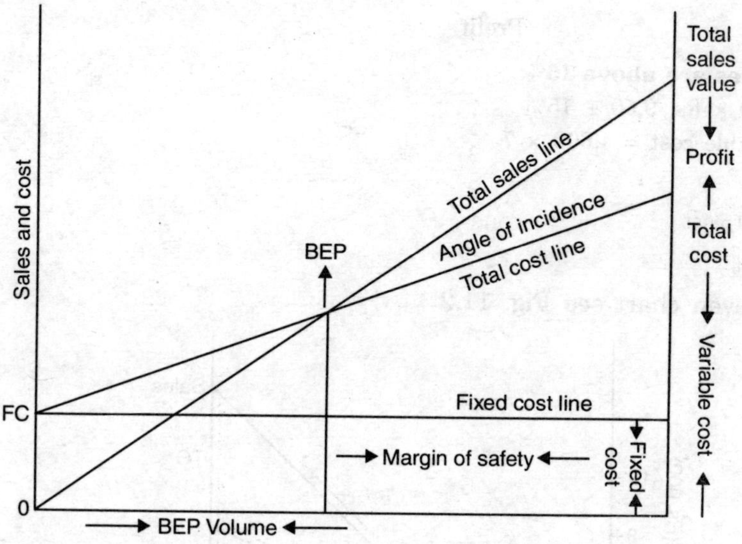

Fig. 11.1 *Break-even chart*

Problem 66. From the following data, draw a single break-even chart:

Selling price per unit	Rs. 10.00
Trade discount	5%
Direct material cost per unit	Rs. 3.00
Direct labour cost per unit	Rs. 2.00
Fixed overhead	Rs. 10.000

Variable overhead 100% of direct labour cost. If sales are 10% and 15% above the break even-volume, determine the net profit. *(Bangalore University, M.Com., April 1992)*

Solution:

Contribution = Selling price – Variable cost

			Rs.
Sales			10
Less:	5% trade discount (10 – 0.50)		9.50
Less:	Variable costs:		
	Direct materials	3.00	
	Direct labour	2.00	
	Variable overhead:		
	100% of labour cost	2.00	7.00
	Contribution		Rs. 2.50

$$\text{BEP} = \frac{\text{Fixed cost}}{\text{Contribution per unit}} = \frac{10{,}000}{2.50} = 4{,}000 \text{ units}$$

Profit if sale is 10% above BEP:

Sales = 4,000 × Rs. 9.50 + 10%		41.800
Less:	Variable cost 4,400 × 7	30,800
		11,000
Less:	Fixed cost	10,000
	Profit	1,000

Profit if sales are above 15%:

	4,000 × Rs. 9.50 + 15%	43,700
Less:	Variable cost = 4,600 × 7	32,200
		11,500
Less:	Fixed cost	10,000
	Profit	1,500

For break-even chart see Fig. 11.2

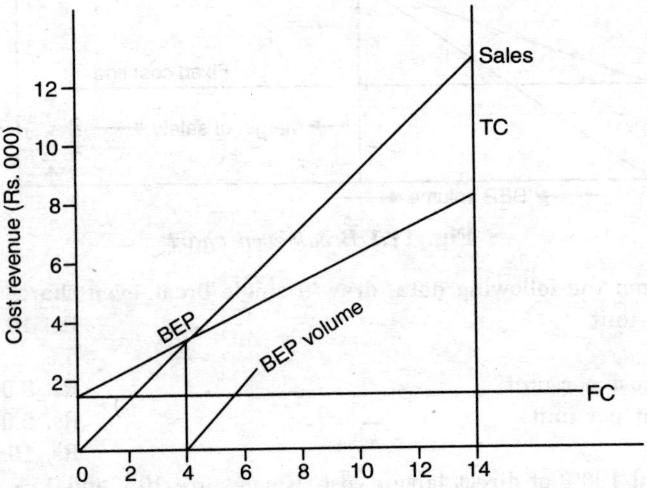

Fig. 11.2

Problem 67. You are given the following data for a costing year for a factory:

Budgeted output	1,00,000 units
Fixed expenses	Rs. 5,00,000
Variable expense per unit	Rs. 10
Selling price per unit	Rs. 20

Draw a break-even chart showing the break-even point. If the selling price is reduced to Rs. 18 per unit, what will be the new break-even point ? *(Bangalore University, M.Com., April 1991)*

Solution:

See Figs. 11.3 and 11.4.

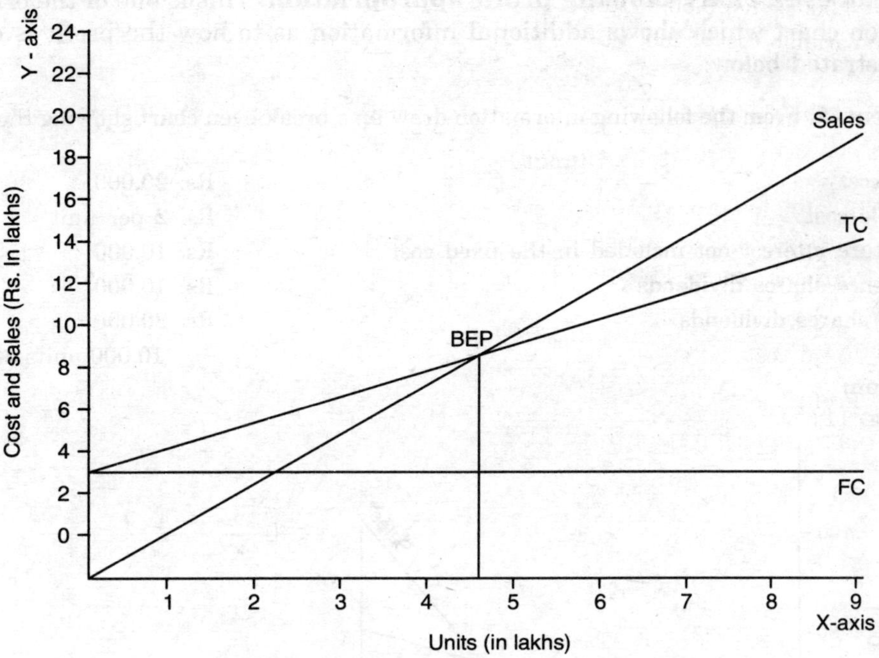

Fig. 11.3 *BEP if selling price is Rs. 20*

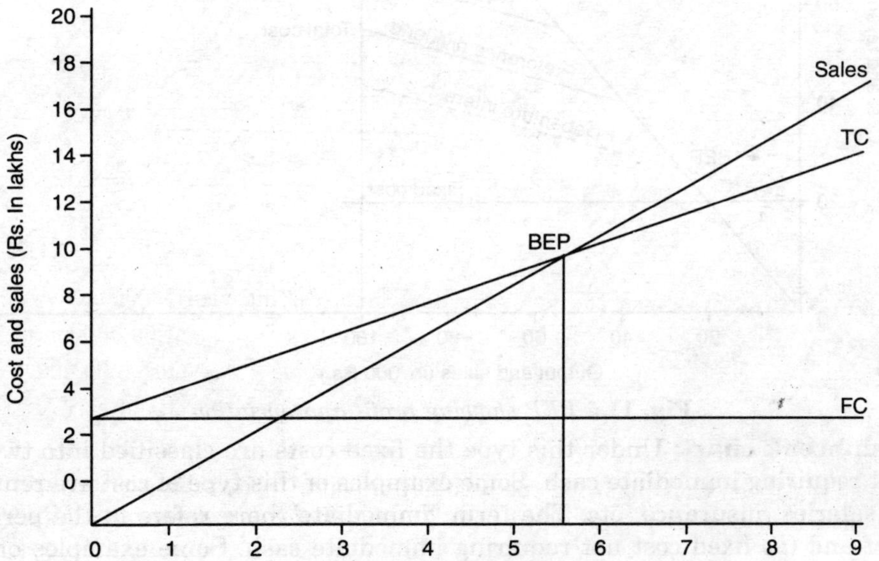

Fig. 11.4 *BEP if selling price is reduced to Rs. 18*

TYPES OF BREAK-EVEN CHARTS

1. Break-even chart showing profit appropriation: This is one of the orthodox type of break-even chart which shows additional information as to how the profit is distributed. This is illustrated below:

Problem 68. From the following information draw up a break-even chart showing the distribution of profit:

Fixed cost	Rs. 20,000
Variable cost	Rs. 2 per unit
Debenture interest not included in the fixed cost	Rs. 10,000
Preference shares dividends	Rs. 10,000
Equity shares dividends	Rs. 20,000
Sales	10,000 units at Rs. 10 each

Solution:

See Fig. 11.5

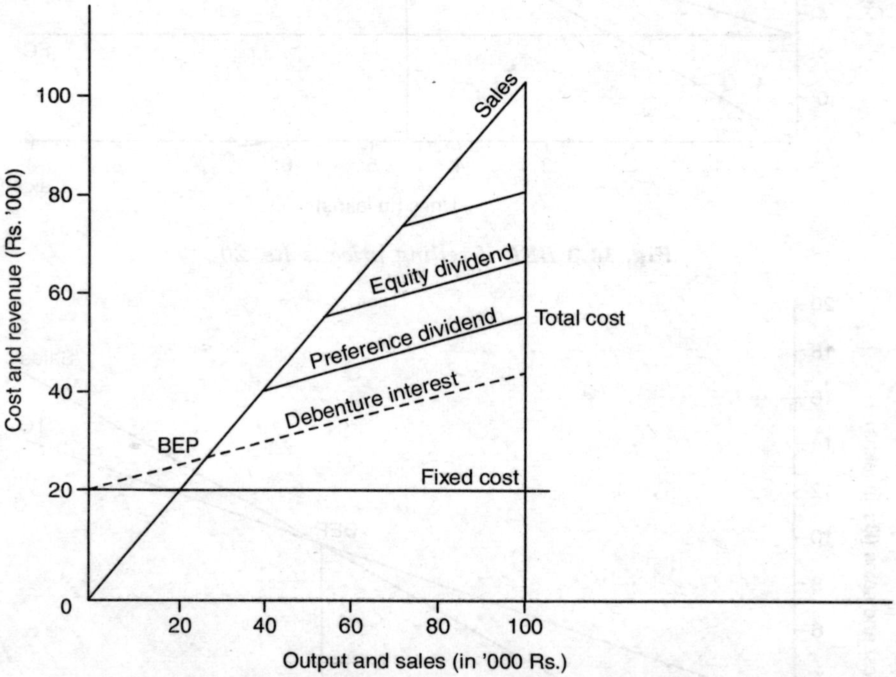

Fig. 11.5 *BEP showing profit appropriation*

2. Cash break chart: Under this type the fixed costs are classified into two types viz. (*i*) fixed cost requiring immediate cash. Some examples of this type of cost are rent and rates, wages and salaries, insurance, etc. The term "immediate" here refers to the period covered by the chart and (*ii*) fixed cost not requiring immediate cash. Some examples of such costs are depreciation, advertising, research and development cost. The presentation of these two types of fixed cost in the chart deserve special mention. The fixed cost requiring immediate cash is shown at the base line (horizontal line) of the chart. Whereas the fixed cost not requiring immediate cash is shown last. The variable cost is assumed to be payable in cash. Where credit transactions are involved their impact on cash available is measured and cash payments are adjusted. This is shown in the following problem:

Problem 69. From the following particulars prepare a cash break-even chart:

Fixed cost	Cash	Rs. 1,00,000
Already paid		Rs. 10,000
Variable cost		Rs. 2 per unit
Preference dividend and debenture interest		Rs. 10,000 each
Sales		10,000 units of Rs. 10 each

Solution:

See Fig. 11.6

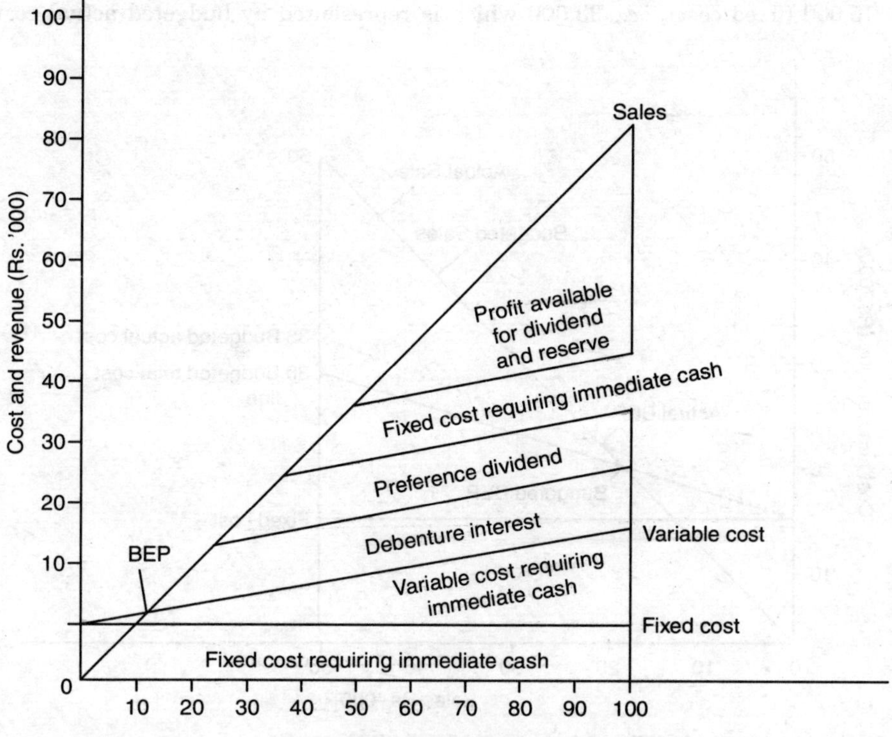

Fig. 11.6 *Cash break-even chart*

3. Control break-even chart: This chart is extremely useful for comparing budgeted and actual profits, break-even points and sales. This is shown in the following problem:

Problem 70. From the following particulars prepare a break-even chart:

Budgeted fixed costs	Rs. 15,000
Budgeted variable costs	Rs. 12,000 for budgeted sales
Budgeted sales	Rs. 40,000
Actual fixed costs	Rs. 15,000
Actual variable costs	Rs. 16,200
Actual sales	Rs. 45,000

Solution:

See Fig. 11.7

Notes: Profit for budgeted sales = Rs. 13,000

	Rs.
Profit for actual sales:	16,500
Budgeted profit — Actual profit	13,800
Profit variance	2,700

Note: The budgeted total cost line and actual cost line have been drawn on the variable costs for sales amount of Rs. 50,000. For 40,000 amount of sales, the variable costs are Rs. 12,000. Therefore, for Rs. 50,000 amount of sales, the variable cost should be Rs. 15,000, *i.e.,* at the rate of Rs. 3,000 for every Rs. 10,000 amount of sales. Hence, the total budgeted costs for Rs. 50,000 amount of sales should be Rs. 15,000 variable + Rs. 15,000 fixed costs, *i.e.,* total Rs. 30,000 as shown by budgeted total cost line. For actual sales amount of Rs. 45,000, variable costs are Rs. 16,200 considering Rs. 50,000 amount of sales, the actual respective costs will be Rs. 18,000 and the actual total cost will be Rs. 18,000 + 15,000 (fixed cost), *i.e.,* 33,000 which is represented by budgeted actual cost.

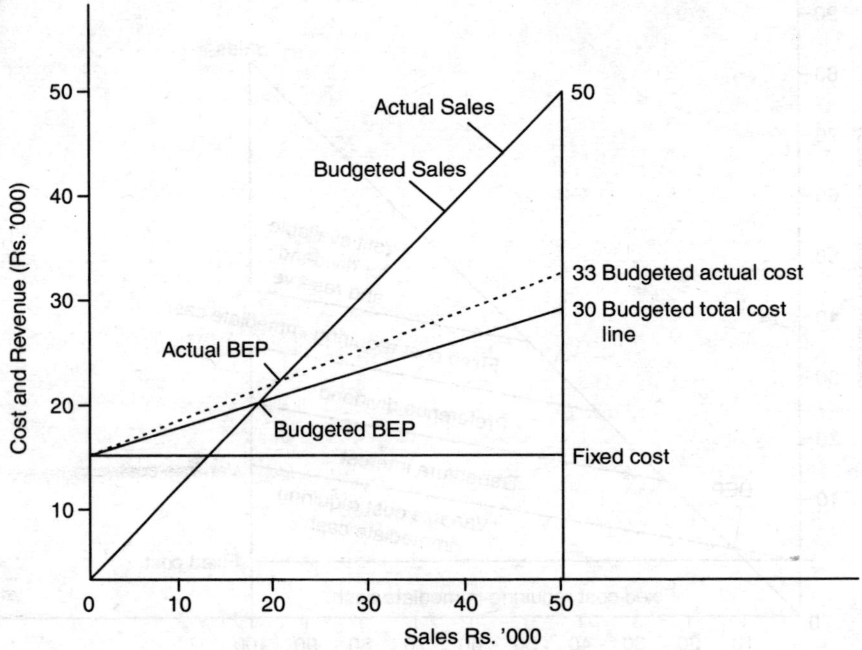

Fig. 11.7. *Control break-even chart.*

4. Profit chart: It is a variety of break-even chart. It shows the profit or loss at different levels of sales. As usual, volume of sales is shown under horizontal line. The vertical lines show the profit or loss position. The line shown above the horizontal line shows profit, whereas the line below shows loss.

Problem 71. From the following particulars prepare a profit chart:

Fixed cost	Rs. 40,000
Variable cost	Rs. 2 per unit
Sales	10,000 units @ Rs. 10 per unit

Solution:

$$\text{BEP} = \frac{\text{Fixed cost}}{\text{Contribution per unit}} = \frac{40,000}{8} = \text{Rs. } 5,000$$

For chart see Fig. 11.8

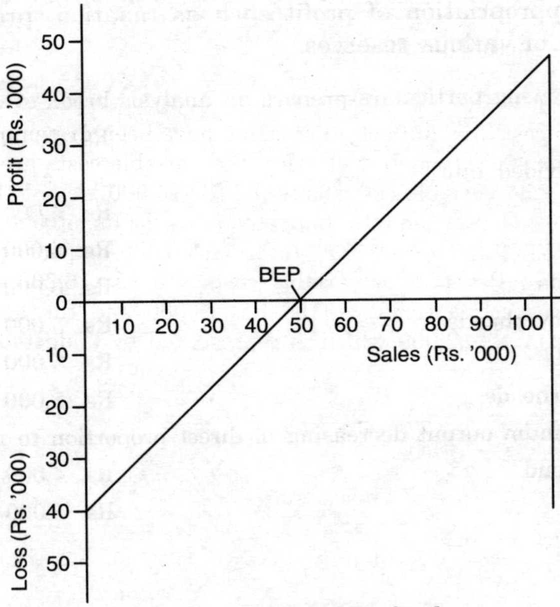

Fig. 11.8 *Profit chart.*

5. Profit chart for different prices: This chart shows the effect on profit of charging different prices.

Problem 72. From the following particulars prepare a profit break-even chart considering selling prices to be Rs. 11, Rs. 10 and Rs. 8

Fixed cost	Rs. 40,000
Variable cost per unit	Rs. 2
Sales units	10,000

Solution: See Fig. 11.9

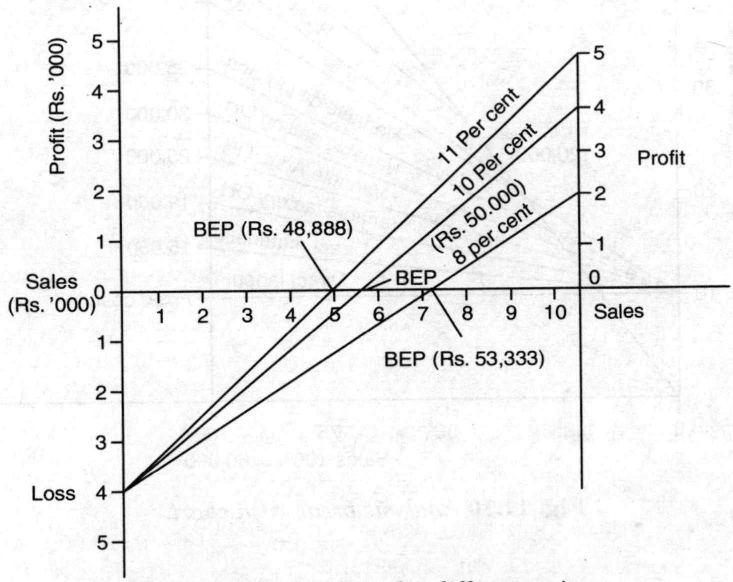

Fig. 11.9 *Profit chart for different prices.*

6. Analysis break-even chart: This chart portrays the components of variable cost such as direct material, direct labour and variable factory, administrative and selling and distribution overheads, the appropriation of profit such as taxation, preference dividend, equity dividends and creation of various reserves.

Problem 73. From the following particulars prepare an analysis break-even chart:

Fixed cost	Rs. 10,000
Variable cost Rs. 30,000 divided into	
Direct labour	Rs. 5,000
Direct materials	Rs. 4,000
Factory overheads	Rs. 6,000
Administration overheads	Rs. 5,000
Selling overheads	Rs. 5,000
Distribution overheads	Rs. 5,000

Taxation Rs. 2,000 at maximum output decreasing in direct proportion to reduction in profit

Preference dividend	Rs. 4,000
Equity dividend	Rs. 5,000

Solution:

See Fig. 11.10

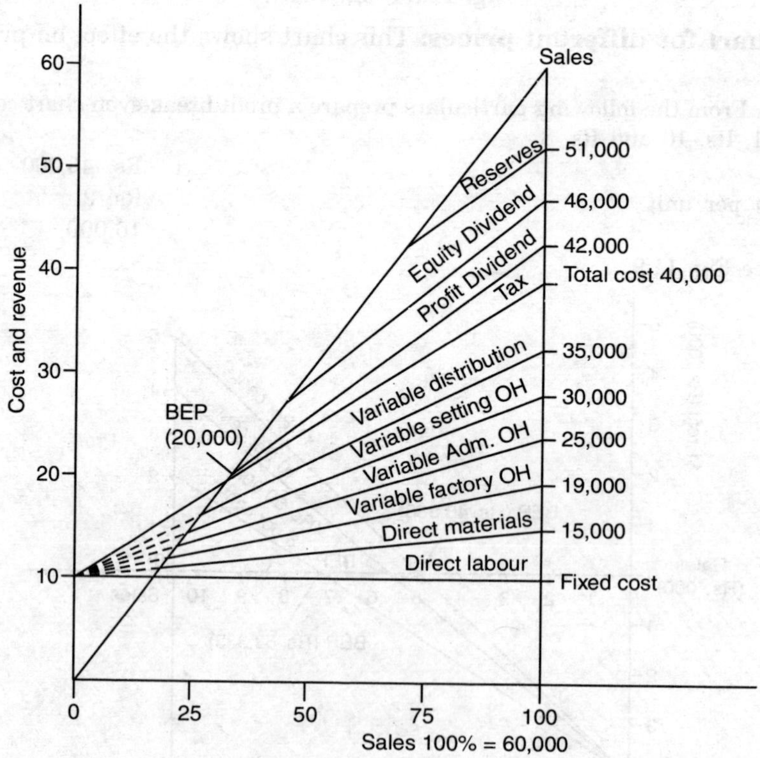

Fig. 11.10 *Analysis break-even chart.*

LIMITATIONS OF BREAK-EVEN CHARTS

1. The preparation of BEP involves separation of semi-variable cost into fixed and variable, which cannot be carried out accurately.
2. It is presumed that the revenue and cost can be represented by straight lines which may not always be true.
3. It assumes that a selling price is constant at different levels of sales which may not be true. Selling price either increases or decreases due to decrease or increase in sales units.
4. Apportionment of fixed cost over a variety of products poses a problem.
5. It assumes that the business conditions may not change which is not true.
6. Under a break-even chart only one product is studied. It fails to provide the effects of various product mixes on profit.
7. Capital employed which is one of the important guiding factors determining profitability is ignored by the break-even chart.

PROFIT-VOLUME CHART OR P/V CHART

Vance in his *Theory and Techniques of Cost Accounting* defines a profit graph as a graph showing the amounts of fixed and variable costs and sales revenue at different volumes of operation. The name P/V chart has arisen from the fact that the difference between the total of fixed and variable costs and the sales revenue at any volume gives the profit at that volume. Because of the importance of the point at which the total costs and revenue are equal the graph is often called a break-even chart. It shows the volume at which the firm covers all costs with revenue or "break-even".

The following points are to be considered while constructing a P/V graph:

(*a*) The graph is divided into two areas, viz., a loss area and a profit area. It is the sales line which divides the graph into these two areas. The horizontal line represents sales line.

(*b*) Total fixed costs are marked below the sales line on the left hand vertical line.

(*c*) The profit is computed and is marked on the right hand vertical line above the sales line.

(*d*) The two points, viz., fixed cost and profit point are joined by a diagonal line. The point of intersection is the break-even point.

Problem 74. From the data given below you are required to present on graph paper a profit volume (P/V) graph to show the expected company performance based on the budget for one year:

	('000 Rs.)
Sales	600
Marginal cost	350
Fixed cost	150

Determine the break-even point and the margin of safety. *(ICWA, Inter, June 1990)*

Solution:

Sales		600
Less: Marginal cost		350
	Contribution	250
Less: Fixed cost		150
	Profit	100

Calculation of BEP (for verification)

C/S Ratio $= \dfrac{\text{Contribution}}{\text{Sales}} = \dfrac{250}{600} = 41.66\%$

BEP $= \dfrac{\text{Fixed cost}}{\text{C/S ratio}} = \dfrac{1,50,000}{41.66} \times 100$

$= 3,60,000$

Margin of safety = Sales − BEP

6,00,000 − 3,60,000 = Rs. 2,40,000

For P/V graph see Fig. 11.11.

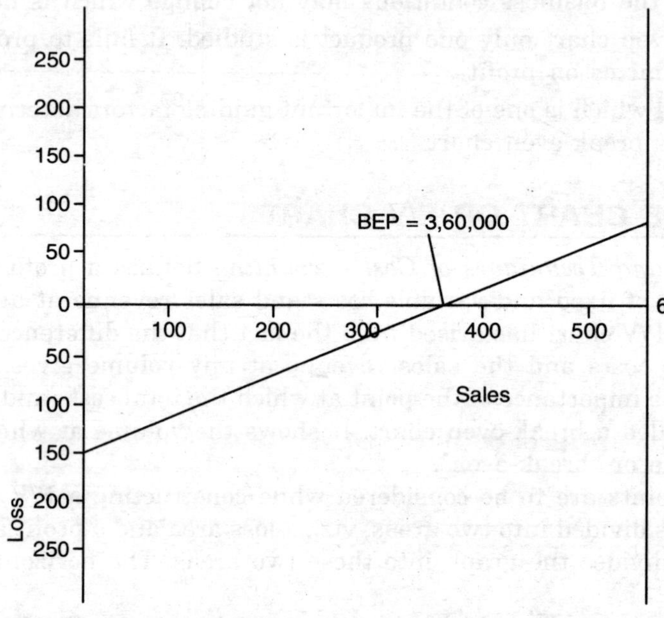

Fig. 11.11 *P/V graph.*

Problem 75. From the particulars given below relating to Avinash (Pvt.) Company Ltd. find out (*a*) P/V ratio. (*b*) BEP, (*c*) Margin of safety and (*d*) Plot P/V graph:

	Rs.
Sales	1,00,000
Total cost	80,000
Fixed cost	20,000
Net profit	20,000

Solution:

P/V Ratio $= \dfrac{\text{Sales} - \text{Variable expenses}}{\text{Sales}} \times 100$

$= \dfrac{1,00,000 - 60,000}{1,00,000} \times 100 = 40\%$

BEP $= \dfrac{\text{Fixed cost}}{\text{P/V ratio}} = \dfrac{20,000}{40\%} = \text{Rs. } 50,000$

$$\text{Margin of safety} = \frac{\text{Profit}}{\text{P/V ratio}} = \frac{20{,}000}{40\%} = \text{Rs. } 50{,}000$$

For P/V graph see Fig. 11.12.

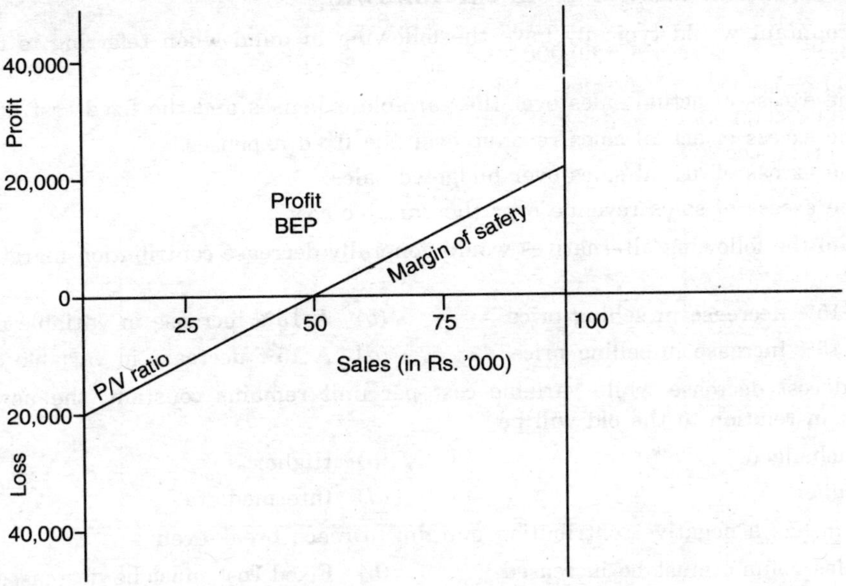

Fig. 11.12 *P/V Graph*

Difference Between Be Chart and P/V Chart

P/V Chart	*BE Chart*
1. Only one line, viz., profit line is shown.	1. Profit line is not shown instead sales and cost lines are shown.
2. Profit line starts below the horizontal line. Fixed cost incurred at zero point of production indicates loss. This loss is indicated below the horizontal line.	2. All lines are drawn above the horizontal line. The area up to the point of break-even shows loss and the area beyond the point shows profit.
3. The break-even point is found on the horizontal axis.	3. The break-even point is at the intersection of cost and sales lines.
4. Fixed cost is always considered as fixed and the ratio of sales and variable cost is again assumed as constant at different levels of sales.	4. The break-even chart can even measure break-even point when sales line is a curve because of price changes at different volumes of sales.
5. It is possible to measure the impact of individual products in a sales mix.	5. This chart can show the effect of a sales mix only in aggragate. For different items of sales, separate charts are to be prepared.
6. The chart shows profit and loss at different sales.	6. It can be drawn even to show the various elements of cost and profit appropriation.

Questions

I. Choose the correct answer from the following:

1. An accountant would typically have the following in mind when referring to the "margin of safety":

(*a*) The excess of actual sales over the variable expenses and the fixed cost at break-even.
(*b*) The excess of actual sales revenue over the fixed expenses.
(*c*) The excess of actual sales over budgeted sales.
(*d*) The excess of sales revenue over the variable cost. [ ]

2. Which of the following alternatives would generally decrease contribution margin per unit the most ?

(*a*) A 15% decrease in selling price. (*b*) A 15% increase in variable cost.
(*c*) A 15% increase in selling price. (*d*) A 15% decrease in variable cost. [ ]

3. If fixed cost decrease while variable cost per unit remains constant, the new contribution margin in relation to the old will be

(*a*) Unchanged (*b*) Higher
(*c*) Higher (*d*) Intermediate [ ]

4. If a firm has a negative contribution margin, to reach break-even:

(*a*) Sales volume must be increased. (*b*) Fixed cost must be decreased.
(*c*) Sales volume must be decreased. (*d*) Fixed cost must be increased. [ ]

5. If total contribution margin is decreased by a given amount, operating profit would

(*a*) Decrease by the same amount.
(*b*) Decrease by more than the given amount.
(*c*) Increase by the same amount.
(*d*) Remain unchanged. [ ]

6. The break-even point would be increased by ?

(*a*) A decrease in fixed cost.
(*b*) An increase in contribution margin ratio.
(*c*) An increase in variable cost.
(*d*) A decrease in variable cost. [ ]

7. Given the following formula, which one represents the break-even sales level in units. P = Selling price per unit, F = Fixed cost, V = Variable cost per unit.

(*a*) $\frac{F}{P - V}$ (*b*) $\frac{F}{F \div V}$

(*c*) $\frac{F}{V \div P}$ (*d*) $\frac{F}{SP - V}$ [ ]

8. Which one of the following assumptions is not made in break-even analysis ?

(*a*) Volume is the only factor affecting cost.
(*b*) No change between beginning and closing stock.
(*c*) The sales mix is maintained as volume changes.
(*d*) All of the above are assumptions sometimes required in break-even analysis. [ ]

9. The cost-volume-profit analysis underlying the conventional break-even chart does not assume that

(*a*) Prices will remain fixed.

(*b*) Production will equal sales.

(*c*) Some costs vary inversely with volume.

(*d*) Costs are linear and continuous over the relevant range. [ ]

10. The most useful information derived from a break-even chart is the

(*a*) Amount of sales revenue needed to cover enterprise variable cost.

(*b*) Amount of sales revenue needed to cover enterprise variable costs.

(*c*) Relationship among revenues, variable cost and fixed costs at various levels of activity.

(*d*) Volume or output level at which the enterprise break-even. [ ]

11. The major assumption as to cost and revenue behaviour underlying conventional cost-volume-profit calculation is the

(*a*) Constancy of fixed cost.

(*b*) Variability of unit prices and efficiency.

(*c*) Curvilinearity or relationships.

(*d*) Linearity of relationships. [ ]

12. Given the following notations, what is the break-even sales level in units ? SP = Selling price, FC = Fixed cost, VC = Variable cost per unit:

(*a*) $\dfrac{\text{SP}}{\text{FC} \div \text{VC}}$ (*b*) $\dfrac{\text{VC}}{\text{SP} - \text{FC}}$

(*c*) $\dfrac{\text{FC}}{\text{VC} \div \text{SP}}$ (*d*) $\dfrac{\text{FC}}{\text{SP} - \text{FC}}$ [ ]

13. P/V ratio is an indicator of:

(*a*) The volume of sales.

(*b*) The volume of profit.

(*c*) The rate at which goods are sold.

(*d*) All of the above. [ ]

14. When P/V ratio is 40% and sales value Rs. 10,000, the variable cost will be

(*a*) Rs. 4,000

(*b*) Rs. 6,000

(*c*) Rs. 10,000

(*d*) Variable cost cannot be calculated from the data given. [ ]

15. Cost-volume-profit analysis is most important for the determination of the

(*a*) Volume of operations necessary to break-even.

(*b*) Variable revenues necessary to equal fixed cost.

(*c*) Relationship between revenues and costs at various levels of operations.

(*d*) Sales revenue necessary to equal fixed cost. [ ]

II. Mark True or False in the space provided

1. At break-even point, contribution is equal to fixed cost. (*T*/*F*)

2. Margin of safety shows the excess of actual sales over budgeted sales. (*T*/*F*)

3. When P/V ratio is 50% and margin of safety is 40%, the net profit is 20% of sales. (*T/F*)
4. Margin of safety may come before or after the break-even point. (*T/F*)
5. Marginal costing and cost-volume-profit analysis is useful in profit planning. (*T/F*)
6. If the selling price per unit is assumed to be constant, total revenue will be proportionate to number of units sold. (*T/F*)
7. Decreasing the selling price decreases the break-even point. (*T/F*)
8. The contribution margin per unit is proportional to volume. (*T/F*)
9. It is impossible to obtain a break-even point with a negative contribution margin. (*T/F*)
10. To obtain the break-even point in units, the total fixed costs should be divided by the contribution margin ratio. (*T/F*)
11. The margin of safety is found by subtracting fixed cost from sales. (*T/F*)
12. If the variable cost per unit increase, the contribution margin ratio will increase. (*T/F*)
13. As production volume increases, the contribution margin rate increases. (*T/F*)
14. Accountants usually assume that variable costs are linear. (*T/F*)

III. Simple Questions

1. Define cost-volume-profit analysis.
2. State any four objectives of cost-volume-profit analysis.
3. State any four assumptions of cost-volume-profit analysis.
4. State any four limitations of cost-volume-profit analysis.
5. Define break-even analysis.
6. Mention any four assumptions underlying break-even analysis.
7. What do you mean by margin of safety ? What purpose does it ?
8. What do you mean by angle of incidence ?
9. Define profit-volume ratio.
10. State any four usefulness of profit-volume ratio.
11. Distinguish between break-even chart and profit-volume chart.

IV. Short Answer Questions

1. What do you mean by break-even point ? Discuss in detail. *(ICWA, Inter., Dec. 1990)*
2. What is meant by cost-volume-profit analysis ? What are its limitations ? Discuss the utility of determining cost-volume-profit relationship ? *(Kakatiya University, M.Com., August 1991)*
3. What do you understand by the term break-even analysis ? Enumerate its assumptions and uses. *(Calicut University, B.Com., October 1989)*
4. What is break-even point and state its usefulness to management ? *(University of Kerala, B.Com., April 1990)*
5. What are the uses of cost-volume-profit analysis ? Discuss the various ways of presenting the CVP relationship ? *(Bangalore University, M.Com., May 1990)*

Exercises

1. The Reliable Battery Ltd. furnishes you the following income information for the current year divided in two sub-parts

	First half	Second half
Sales	8,10,000	10,26,000
Profit earned	21,600	64,800

From the above, you are required to compute the following, assuming that the fixed cost remains the same in both the period

(*a*) Profit/volume ratio

(*b*) Fixed cost

(*c*) Amount of profit or loss when sales are Rs. 6,48,000

(*d*) Amount of sales required to earn a profit of Rs. 1,08,000

(University of Madras, B.Com., October 2003)

[**Ans.** (*a*) P/V Ratio 20%, (*b*) Fixed cost Rs. 1,40,400, (*c*) Loss of Rs. 10,800, when sales are Rs. 6,48,000, (*d*) Sales required is Rs. 12,42,000 when profit is Rs. 1,08,000]

2. From the information given below, determine break-even point in units

	Rs.
Selling price per unit	16.50
Trade discount per unit	1.50
Fixed expenses (aggregate)	7,50,000
Variable expenses per unit	11.00
Units of production	3,00,000 units

(University of Madras, B.Com., April 2003)

[**Ans.** BEP in units are 1,87,500 units]

3. You are given the following information

Fixed cost	Rs. 4,000
Break-even sales	Rs. 20,000
Profit	Rs. 1,000
Selling price	Rs. 20 per unit

Calculate: (*a*) Sales and marginal cost of sales

(*b*) New B.E. point if selling price is reduced by 10%.

(University of Madras, B.Com., October 2002)

[**Ans.** (*a*) Sales = Rs. 25,000 marginal cost of sales *i.e.*, variable cost Rs. 20,000, (*b*) New BEP if selling price is reduced by 10% is Rs. 36,000]

4. From the following particulars calculate the break-even point

	Rs.
Fixed cost	75,000
Sales	3,00,000
Direct materials	1,00,000
Direct labour	60,000
Direct expenses	40,000

(University of Madras, B.Com., April 2002)

[**Ans.** BEP Rs. 2,25,000, P/V Ratio $33\frac{1}{3}\%$]

5. On the basis of the following data find out the amount of profit if

(*i*) P/V Ratio is 30% and margin of safety is Rs. 3,30,000

(*ii*) P/V Ratio is 30%, margin of safety ratio is $33\frac{1}{3}\%$ and sales are Rs. 9,90,000

(University of Madras, B.Com., May 2002)

[**Ans.** (*a*) Profit Rs. 99,000 (*b*) Profit Rs. 99,000]

6. The sales turnover and profit during two periods were as follows:

Period	Sales (Rs.)	Profit (Rs.)
1991	2,00,000	20,000
1992	3,00,000	40,000

(*a*) What would be the probable trading results with sales of Rs. 1,80,000?

(*b*) What amount of sales will yield a profit of Rs. 50,000?

(University of Madras, B.Com., May 2002)

[**Ans.** (*a*) Profit Rs. 16,000, (*b*) Sales required to earn a profit of Rs. 50,000 is Rs. 3,50,000]

7. Margin of safety is 20%, P/V Ratio is 60%, Fixed cost = Rs. 30,000

Find out: (*a*) BEP sales, (*b*) Actual sales for the year, (*c*) Variable cost for the year, (*d*) Profit for the year. *(University of Madras, B.Com., April 2001)*

[**Ans.** (*a*) BEP sales Rs. 50,000, (*b*) Actual sales for the year Rs. 62,500, (*c*) Variable cost for the year 25,000, (*d*) Profit for the year Rs. 7,500]

8. The sales turnover and profit during two years were as follows:

Year	Sales	Profit
1999	1,50,000	20,000
2000	1,70,000	25,000

You are required to calculate:

(*a*) P/V Ratio, (*b*) Break-even point, (c) Sales required to earn a profit of Rs. 40,000, (*d*) Profit when sales are Rs. 2,50,000, (*e*) Margin of safety at a profit of Rs. 50,000, (*f*) Variable cost of the two periods. *(University of Madras, B.Com., April 2001)*

[**Ans.** (*a*) P/V Ratio 25%, (*b*) BEP Rs. 70,000, (c) Sales at a profit of Rs. 40,000 is Rs. 2,30,000, (*d*) Profit for a sales of Rs. 2,50,000 is Rs. 45,000, (*e*) Margin of safety at a profit of Rs. 50,000 is Rs. 2,00,000, (*f*) Variable cost of two period are Rs. 1,12,500 and Rs. 1,27,500]

9. Given:

	Rs.
Fixed overhead	1,20,000
Variable overhead	2,00,000
Direct wages	1,50,000
Direct material	4,10,000
Sales	10,00,000

Calculate BEP and P/V Ratio

(University of Madras, B.Com., April 2005)

[**Ans.** P/V Ratio 24%, BEP Rs. 5,00,000]

10. From the following particulars, calculate BEP: Sales Rs. 1,00,000, Variable cost Rs. 60,000, Fixed overhead Rs. 15,000. Also calculate new BEP if variable cost increases by 10%.

(University of Madras, B.Com., May 2005)

[**Ans.** (*a*) BEP Rs. 37,500, (*b*) New BEP 44,118]

11. The following information relating to a company is given to you

Sales	Rs. 4,00,000
Fixed cost	1,80,000
Variable cost	2,50,000

Ascertain how much the value of sales must be increased for the company to break-even

(University of Madras, B.Com., May 2004)

[**Ans.** Extra sales to be increased for the company to break-even is Rs. 80,000 (4,80,000 – 4,00,000)]

12. From the following particulars calculate the sales required to earn a profit of Rs. 1,20,000

Sales	6,00,000
Variable cost	3,75,000
Fixed cost	1,80,000

(University of Madras, B.Com., April 2003)

[**Ans.** Sales required Rs. 8,00,000]

13. From the following particulars, find BEP in units

Variable cost per unit	Rs.	15
Selling price per unit	Rs.	24
Fixed cost	Rs.	54,000

What should be the selling price per unit if the company wants to bring down BEP to 9,000 units.

(University of Madras, B.Com., April 2002)

[**Ans.** The revised selling price at BEP of 9,000 units is Rs. 21]

14. From the following data calculate:

(*a*) P/V ratio

(*b*) Profit when sales are Rs. 20,000 and

(*c*) New break-even point if selling price is reduced by 20%

Fixed expenses	Rs.	4,000
Break-even point	Rs.	10,000

(University of Madras, M.Com., April 2005)

[**Ans.** (*a*) P/V Ratio 40%, (*b*) Profit when sales are Rs. 20,000 is Rs. 4,000, (*c*) New BEP is Rs. 16,000]

15. Assuming that the cost structure and selling prices remain the same in periods I and II, find out:

(*a*) P/V ratio

(*b*) Break-even sales

(*c*) Profit when sales are Rs. 1,00,000

(*d*) Sales to earn a profit of Rs. 10,000

Period	*Sales* *Rs.*	*Profit* *Rs.*
I	1,20,000	9,000
II	1,40,000	13,000

(University of Madras, M.Com., April 2005)

[**Ans.** (*a*) P/V Ratio 20% (*b*) BEP sales Rs. 75,000 (*c*) Profit when sales are Rs. 1,00,000 is Rs. 5,000 (*d*) Sales to earn a profit of Rs. 10,000 is Rs. 1,25,000]

16. Calculate break-even point in units and in rupees from the following particulars

Fixed expenses	Rs. 1,50,000
Variable cost per unit	Rs. 10
Selling price per unit	Rs. 15

(University of Madras, M.Com., April 2004)

[**Ans.** BEP in units: 30,000 units, BEP in Rs.: 4,50,000 Rs.]

17. From the following data, calculate:

(*a*) P/V ratio

(*b*) Profit when sales are Rs. 20,000

(*c*) New break-even point if selling price is reduced by 20%

Fixed expenses	Rs. 4,000
Break-even point	Rs. 10,000

(University of Madras, M.Com., April 2003)

[**Ans.** (*a*) P/V Ratio 40% (*b*) Profit when sales are Rs. 20,000 is Rs. 4,000 (*c*) New BEP, if selling price is reduced by 20% is Rs. 16,000]

18. A company produces and sells 100 units of A per month at Rs. 20. Marginal cost per unit is Rs. 12 and fixed costs are Rs. 300 per month. It is proposed to reduce the selling price by 20%. Find the additional sales required to earn the same profit as before.

(University of Madras, M.Com., April 2003)

[**Ans.** Additional sales required is 100 units per month to earn same profit]

19. The position of Y Ltd. was as follows:

Sales		1,20,000
Variable overheads		96,000
	Gross profit	24,000
Fixed overhead		16,000
	Net profit	8,000

Find out:

(*a*) P/V Ratio

(*b*) BEP

(*c*) Net profit from the sales of Rs. 1,30,000

(*d*) Required sales for a net profit Rs. 10,000 *(University of Madras, M.Com., April 2003)*

[**Ans.** (*a*) P/V Ratio 20% (*b*) BEP Rs. 80,000 (*c*) Net profit required when sales is Rs. 1,30,000 is Rs. 10,000 (*d*) Sales required when profit is Rs. 10,000 is Rs. 1,30,000]

20. The P/V ratio of a firm is 50% and the margin of safety is 40%. You are required to work out BEP and net profit if the sales volume is Rs. 50,00,000.

(University of Madras, M.Com., May 2003)

[**Ans.** (*a*) BEP = 30,00,000 (*b*) Net profit if sale volume is Rs. 50,00,000 is Rs. 10,00,000]

21. Company A and Company B, both under the same management, make and sell the same type of product. Their budgeted profit and loss accounts for January-June 1996 are as under

		Company A		Company B
	Rs.	*Rs.*	*Rs.*	*Rs.*
Sales		3,00,000		3,00,000
Less: Variable cost	2,40,000		2,00,000	
Fixed cost	30,000	2,70,000	70,000	2,70,000
		30,000		30,000

You are required to:

(*a*) Calculate the break-even point for each company.

(*b*) Calculate the sales volume at which each of the two companies with make a profit of Rs. 10,000.

(*c*) Assess how their profitability will change with decrease or increase in volume.

(University of Madras, M.Com., April 2002)

[**Ans.** (*i*) P/V Ratio = 20% and $33\frac{1}{3}\%$; (*ii*) BEP = 1,50,000 and 2,10,000, (*iii*) Sales volume required to earn a profit of Rs. 10,000 = 2,00,000 and 2,40,000 (*iv*) In case of heavy demand company B will do better in case of low demand company A will do better]

22. From the following data, calculate:

(*a*) P/V ratio

(*b*) Profit when sales are Rs. 20,000

(*c*) New break-even point if selling price is reduced by 20%

Fixed cost	Rs. 4,000
Break-even point	Rs. 10,000

(University of Madras, M.Com., April 2002)

[**Ans.** (*a*) P/V Ratio 40%, (*b*) Profit is Rs. 4,000 when sales are 20,000 (*c*) New BEP, if selling price is reduced by 20% is Rs. 16,000]

23. You are given the following data for the year 1996 for a factory

Output	40,000 units
Fixed expenses	Rs. 2,00,000
Variable cost per unit	Rs. 10
Selling price per unit	Rs. 20

How many units must be produced and sold in the year 1997, if it is anticipated that selling price would be reduced by 10%, variable cost per unit would be Rs. 12 per unit and fixed cost will increase by 10%. The factory would like to make a profit in 1997 equal to that of the profit in 1996.

(University of Madras, M.Com., April 2001)

[**Ans.** (1) Profit for the year Rs. 2,00,000 (2) Production units in 1997 is 12,60,000 units]

24. The following data have been extracted from the books of Alfa Ltd.

Year	*Sales Rs.*	*Profit Rs.*
2002	5,00,000	50,000
2003	7,50,000	1,00,000

You are required to calculate:

(1) P/V Ratio

(2) Fixed cost

(3) Break-even sales

(4) Profit on sales of Rs. 4,00,000

(5) Sales to earn profit of Rs. 1,25,000 *(University of Mumbai, B.Com., April 2004)*

[**Ans.** (1) P/V Ratio 20% (2) Fixed cost Rs. 50,000 (3) Break-even sales Rs. 2,50,000 (4) Profit on sales of Rs. 4,00,000 is Rs. 30,000 (5) Sales to earn profit of Rs. 1,25,000 is Rs. 8,75,000]

25. From the following particulars you are required to calculate:

(1) P/V ratio (2) BEP (3) Profit when sale is Rs. 2,00,000 (4) Sales required to earn a profit of Rs. 40,000 (5) Margin of safety in the 2nd year.

Year	*Sales (Rs.)*	*Profit (Rs.)*
I	2,40,000	18,000
II	2,80,000	26,000

You may assume that the cost structure and selling prices remain constant in the two years. *(University of Mumbai, B.Com., April 2003)*

[**Ans.** (*a*) P/V Ratio 20% (*b*) BEP Rs. 1,50,000 (*c*) Profit when sale of Rs. 2,00,000 is Rs. 10,000 (*d*) Sales required to earn profit of Rs. 40,000 is Rs. 3,50,000 (*e*) Margin of safety in IInd year is Rs. 1,30,000].

❑❑❑

CHAPTER 12

DIFFERENTIAL COSTING

INTRODUCTION

Differential costing is a broader and more fundamental concept than marginal costing and hence has a much wider application. It helps in making appropriate decision by examining all the revenues and cost differences between alternatives. While marginal costing analyses the changes in only the variable costs, differential costing examines the differences in both variable and fixed costs. Hence this technique is more useful for those situations where fixed costs also differ and thus more appropriate for both short run and long run decisions.

Differential costs are costs that change in respect of an alternate course of action. The A.A.A. Committee on cost concepts and standards (Accounting Reviews, Vol. 27) defines it as "the increase or decrease in total cost or the change in specific elements of cost that result from any variation in operation". The alternative actions may be owing to change in sales, volume, price, product mix, or methods of production or change in decision. When two levels of activity are under review, differential cost is obtained by deducting the cost at one level from another level.

The technique used for analysing differential costs is known as differential costing. The ICMA terminology defines differential costing as "a technique used in the preparation of *ad hoc* information in which only cost and income differences between alternative courses of action are taken into consideration".

It is the process of determining how costs in particular and profit in general will be affected if one alternative is chosen over another.

The term "incremental cost" is used to denote differential cost when the increase in cost is due to increase in the level of production. Similarly the term "decremental cost" is used when the difference in cost is due to decrease in the level of production.

FEATURES AND IMPLICATIONS OF DIFFERENTIAL COSTING

The features of differential costing can be enumerated under the following points:

(*a*) It is expressed in total but not cost per unit.

(*b*) The data considered for such analysis are (*i*) cost, (*ii*) revenue, (*iii*) investment which are relevant to the problem under consideration.

(*c*) It is not a part of accounting though it is a form of budgeting. It is used only by the management to make decisions.

(*d*) The constant cost at different levels are ignored and only the differentials are considered. Absolute cost is not much important in this analysis.

(*e*) Differentials are measured from a common position or level of activity.

(*f*) Where the difference between revenue and cost is highest, that course of action is adopted.

The differential costing is based on the implication that only the relevant cost, which will change as a result of the decision is useful for decision-making. Any costs which are not expected to alter, then they are irrelevant for decision-making. The following are the examples of cost that are irrelevant:

(*a*) Sunk cost, *i.e.,* those costs which have already been incurred are irrelevant.
(*b*) Book values of assets.
(*c*) Cost of fully utilised resources, *i.e.,* where a limiting factor exists it will be used to the full extent and will, therefore, cost the same whatever alternative is considered.
(*d*) *Fixed cost:* Any item which remains constant whichever alternative is chosen is not a differential cost and can be ignored in choosing alternatives.

MARGINAL VS. DIFFERENTIAL COSTING

The differential costing is similar to marginal costing under the following respects:

(*a*) Both the techniques are based on cost analysis for presenting the information to management.
(*b*) Costs are analysed on the basis of their behaviour under both the techniques.
(*c*) Both the techniques serve the same purpose of planning and decision-making by the management.
(*d*) Both the techniques will be same if the fixed costs remain same for alternative proposals.
(*e*) Differential cost resembles to economist's concept of marginal cost.

However, differential costing differs from marginal costing in the following respects:

(*a*) The scope of marginal costing is narrow, while that of differential costing is broader and hence it has wider applications.
(*b*) Marginal costing technique is useful in making decisions in the short run, whereas differential costing is useful in taking both short-run and long-run decisions where fixed costs do alter.
(*c*) Marginal cost is defined as the sum of all variable costs. Such a concrete definition is not possible under differential cost. It can at best be defined as an increase or decrease of total cost owing to rise or fall in production.
(*d*) Marginal costing presents cost information under contribution margin approach. Whereas differential costing can be presented both under absorption costing and marginal costing techniques.
(*e*) Under marginal costing technique the performance is judged by the "contribution" and P/V ratio. Whereas under differential costing, the performance is evaluated by comparing differential cost with incremental revenue.
(*f*) Marginal costing is a part of accounting system whereas differential cost analysis is not a part of accounting system. It may be a part of budgeting.
(*g*) Marginal cost is expressed in terms of cost per unit whereas differential cost is expressed in total.

APPLICATIONS OF DIFFERENTIAL COSTING

Differential costing techniques is used to solve the problems relating to the following:

(*a*) Make or buy decisions.
(*b*) Expanding the marketability of the products.
(*c*) Changing the product mix.
(*d*) Further processing of products.
(*e*) Changing the method of production.
(*f*) Introducing a new product line.
(*g*) Replacing manual labour with mechanical labour.
(*h*) Fixation of selling price below the competitive price.
(*i*) Accepting or rejecting a new order.
(*j*) Dealing about the most profitable level of production.
(*k*) Shut down or continue operation.

Problem 1. *(Determination of level of production):* A company's flexible budget reveals the following market conditions and costs:

Output in units	*Selling price per unit (Rs.)*	*Total semi-fixed cost (Rs.)*	*Total variable cost (Rs.)*	*Total fixed cost (Rs.)*
60,000	12	1,50,000	4,18,000	1,42,000
1,20,000	11	1,50,000	8,18,000	1,42,000
1,80,000	10	1,70,000	12,78,000	1,42,000
2,40,000	9	1,70,000	15,78,000	1,42,000
3,00,000	8	2,00,000	17,78,000	1,42,000
3,60,000	7	2,00,000	19,02,000	1,42,000

(*a*) Prepare a schedule showing the total differential costs and incremental revenue.
(*b*) At what level should the company set its level of production ?
(*c*) What selling price should be established in order to obtain the most profitable operations for the year ?

Solution:

(A) STATEMENT SHOWING PROJECTED COST AND REVENUE

	Output (Units)					
	60,000	*1,20,000*	*1,80,000*	*2,40,000*	*3,00,000*	*3,60,000*
Fixed cost	1,42,000	1,42,000	1,42,000	1,42,000	1,42,000	1,42,000
Semi-fixed cost	1,50,000	1,50,000	1,70,000	1,70,000	2,00,000	2,00,000
Variable cost	4,18,000	8,18,000	12,78,000	15,78,000	17,78,000	19,02,000
Total cost	7,10,000	11,10,000	15,90,000	18,90,000	21,20,000	22,44,000
Differential cost		40,00,000 (Col 2– Col 1)	4,80,000 (Col 3 – Col 2)	3,00,000 (Col 4 – Col 3)	2,30,000 (Col 5 – Col 4)	1,24,000 (Col 6 – Col 5)
Sales value	7,20,000	13,20,000	18,00,000	21,60,000	24,00,000	25,20,000
Marginal revenue	–	6,00,000	4,80,000	3,60,000	2,40,000	1,20,000

(b) As long as the increment in total revenue exceeds the differential costs, it pays to increase output, but as soon as the incremental revenue equals differential costs, it is no longer profitable to read the volume of production. In the above case the incremental revenue is about equal to differential costs at 3,00,000 output level and beyond that it is decreasing. So the production should be fixed at this level (correctly between 3,00,000 and 3,60,000 levels).

(c) The sale price at this level is Rs. 8 per unit which should be adopted for the year, which would bring in maximum profit.

Problem 2. *(Acceptance or rejection of an offer):* The following particulars are extracted from the books of Adarsh Company Ltd. for the period ending 31st Dec., 2010:

Capacity	*Unit cost (Rs.)*	*Unit selling price (Rs.)*
6,000	80	100
7,000	75	97
8,000	74	95
9,000	72	
10,000	71	

The firm is operating at 8,000 units capacity and has received an order for 2,000 units from an export market at a price of Rs. 70 per unit. Advise the company as to whether the export order should be accepted or not ?

Solution:

STATEMENT SHOWING INCREMENTAL COST AND REVENUE

Capacity	*Unit cost*	*Total cost*	*Incremental cost*	*Unit price*	*Total sales*	*Incremental revenue*
6,000	80	4,80,000		100	6,00,000	
7,000	75	5,25,000	45,000	97	6,79,000	79,000
8,000	74	5,92,000	67,000	95	7,60,000	8,000
9,000	72	6,48,000	56,000			
10,000	71	7,10,000	62,000			

At 8,000 level of output, the total sales revenue is Rs. 7,60,000 and the total cost is Rs. 5,92,000 leaving a profit of Rs. 1,68,000. The fact that this level of output leaves a profit which means that the fixed costs have been recovered already. Hence we have to take only the incremental cost for further level of output. For an additional sales of 2,000 units the incremental cost is Rs. 7,10,000 – Rs. 5,92,000 = Rs. 1,18,000. The cost per unit, therefore, is $\frac{1,18,000}{2,000}$ = Rs. 59 for which the price quoted is Rs. 70 per unit. Hence the offer is to be accepted.

Problem 3. *(Increasing production capacity):* A company is at present working at 90% of its capacity and producing 13,500 units per annum. It operates at flexible budgetary control system. The following figures are obtained from its budget:

	Capacity 90% (Rs.)	*Utilisation 100% (Rs.)*
Sales	15,00,000	16,00,000
Fixed	3,00,500	3,00,600
Semi-fixed expenses	97,500	1,00,500
Variable expenses	1,45,000	1,49,500
Units manufactured	13,500	15,000

Labour and material costs per unit are constant under the present conditions. Profit margin is 10%.

(a) You are required to determine the differential cost of producing 1,500 units by increasing capacity utilisation to 100 per cent.

(b) What would you recommend as an export price for these 1,500 units after considering that overseas prices are much lower than in land prices ? *(ICWA, Inter., June 1992)*

Solution:

Sales at 90% capacity utilisation	15,00,000
Less: Profit (10%)	1,50,000
Cost of goods sold	13,50,000
Less: Expenses (fixed, semi-fixed and variable)	5,43,000
Cost of materials and labour	8,07,000

∴ Cost of materials and labour at 100% capacity utilisation

$$8,07,000 \times \frac{100}{90} = \text{Rs. } 8,96,667$$

Differential cost analysis will be as follows:

	90% Capacity utilisation	*100% capacity utilisation*
Production (Units)	13,500	15,000
Materials and Labour (Rs.)	8,07,000	8,96,667
Variable expenses (Rs.)	1,45,000	1,49,500
Semi-fixed expenses (Rs.)	97,500	1,00,500
Fixed expenses (Rs.)	3,00,500	3,00,600
	13,50,000	14,47,267

(a) Differential cost = 14,47,267 – 13,50,000 = Rs. 97,267

(b) Minimum price per export = $\frac{97,267}{1,500}$ = Rs. 64.84 per unit.

At this price, there is no addition to revenue. Any price above Rs. 64.84 per unit may be accepted. A price below this may be considered, if other benefits, *i.e.*, other than mere sales and revenue are likely to accrue. It is assumed that no capital investment is necessary and no export charges have to be incurred and that the export price will have no effect on the home market where the product will continue to be sold at the old price.

Problem 4. *(Equipment replacement):* A factory engaged in the manufacture of electronic goods has a ten-year old equipment depreciated on straight-line basis. The useful life of the equipment was estimated to be 20 years with a residual value of Rs. 3 lakhs (original cost of the equipment being Rs. 23 lakhs). The output of the equipment is 1,200 units per hour.

The management now proposes to instal a new equipment worth Rs. 50 lakhs which has an estimated life of 15 years and a residual value of Rs. 5 lakhs. The payment terms for the new equipment include a part exchange provision of Rs. 6 lakhs in respect of the existing equipment. The output of the new equipment is 3,000 units per hour.

Other comparative annual cost data relating to the two equipments are as under:

CHAPTER 12

	Existing equipment (Rs.)	*New equipment (Rs.)*
Wages	1,00,000	1,20,000
Repairs and maintenance	20,000	52,000
Consumables	3,20,000	4,80,000
Power	1,20,000	1,50,000
Allocation of fixed cost	60,000	80,000
Total hours run per year	2,400	2,400

You are required to prepare a comparative schedule showing total conversion cost as well as cost per 1,000 units after considering interest @ 10% on net cash outflow for procuring the new equipment and also for providing for the yearly recovery of the loss suffered in the transaction.

(ICWA, Inter, Dec. 1991)

Solution:

COMPARATIVE STATEMENT OF COST OF EQUIPMENT

	Existing equipment	*New equipment*
Capital cost of equipment including cost of installation	23,00,000	50,00,000
Less: Residual value	3,00,000	5,00,000
	20,00,000	45,00,000
Less: Depreciation written off	10,00,000	–
	10,00,000	45,00,000

COMPARATIVE STATEMENT OF OPERATION COST OF EQUIPMENT

	Existing	New
Annual depreciation		
New equipment	–	3,00,000
Old equipment	1,00,000	70,000
Interest on capital (net cash outflow)	–	4,40,000
Wages	1,00,000	1,20,000
Repairs and maintenance	20,000	52,000
Consumables	3,20,000	4,80,000
Power	1,20,000	1,50,000
Allocation of fixed expenses	60,000	80,000
	7,20,000	16,92,000
Total hours run per annum	2,400	2,400
Operating cost per hour (Rs.)	300	705
Output per hour (Units)	1,200	3,000
Operating cost per 1,000 units (Rs.)	250	235

Working Note 1: Depreciation on new equipment

$$\text{Depreciation} = \frac{\text{Original cost} - \text{Residual value}}{\text{Estimated life}}$$

$$= \frac{45{,}00{,}000}{15} = 3{,}00{,}000$$

Working Note 2: Depreciation (Yearly recovery) on loss on sale of existing equipment

$$= \frac{\text{Book value of existing equipment} - \text{Part exchange value}}{\text{Residual life of existing equipment}}$$

$$= \frac{\text{Rs. } 13{,}00{,}000 - \text{Rs. } 6{,}00{,}000}{10 \text{ years}} = \text{Rs. } 70{,}000$$

Working Note 3: Interest is calculated on net cash outflow in procuring the new equipment is Rs. 44,00,000 (cost of old equipment Rs. 50 lakhs less part exchange value of Rs. 6 lakhs).

Problem 5. *(Evaluating profitability of different products):* G. Ltd. is a process industry which manufactures three joint products, viz. A, B and C. The joint costs of production are Rs. 4,80,000 and sales are:

A = 30,000 units @ Rs. 6 per unit
B = 10,000 units @ Rs. 20 per unit
C = 20,000 units @ Rs. 12 per unit

It is possible to adopt two alternative production programmes after completion of the joint products. First, product C can be further processed, which will create a new product D. This additional processing will cost Rs. 1,60,000 p.a., but the 20,000 units of product D can be sold for Rs. 24 per unit. Secondly, processing of product C need not be continued further, but the resources which could have been used in converting product C to product D can be used to produce an additional product E. The cost of this programme would be Rs. 60,000 per annum, but sales of E would be obtained of 5,000 units at Rs. 16 per unit. Evaluate the two alternative production programmes.

Solution:

STATEMENT SHOWING INCREMENTAL COST

Details	*Proposal I*	*Total*	*Proposal II*	*Total*	*Incremental cost*
Sales					
A	1,80,000		1,80,000		
B	2,00,000		2,00,000		
C	–		2,40,000		
D	4,80,000		–		
E	–	8,60,000	80,000	7,00,000	1,60,000
Costs					
A, B, C	4,80,000		4,80,000		
D	1,60,000		–		
E	–	6,40,000	60,000	5,40,000	1,00,000
Profit		2,20,000		1,60,000	60,000

STATEMENT OF PROFIT FOR THE OPPORTUNITY FOREGONE

Sales of D			4,80,000
Less: Costs of D			1,60,000
			3,20,000
Sales foregone if D is produced	C =	2,40,000	
	D =	80,000	
	=	3,20,000	
Costs foregone if D is produced	E =	60,000	2,60,000
		Profit	60,000

Problem 6. *(Dropping or continuing an area for effecting sales):* X Company Ltd. produces and sells three products: A, B and C. These products are sold in a local market and in a regional market. At the end of the first quarter of the year 1933, the following income statement has been prepared:

	Total	*Local*	*Regional*
Sales Revenue	13,00,000	10,00,000	3,00,000
Cost of goods sold	10,10,000	7,75,000	2,35,000
Gross margin	2,90,000	2,25,000	65,000
Marketing cost	1,05,000	60,000	45,000
Administrative cost	52,000	40,000	12,000
Cost production and sales	1,57,000	1,00,000	57,000
Operating profit	1,33,000	1,25,000	8,000

The management has expressed special concern with the regional market because of the extremely poor return on sales. This market was entered a year ago because of excess capacity. It was originally believed that the return on sales would improve with time, but after a year, no noticeable improvement can be seen from the results as reported in the above quarterly statement.

In attempting to decide whether to eliminate the regional market, the following information has been gathered:

	Product A	*Product B*	*Product C*
Sales revenue (Rs.)	5,00,000	4,00,000	4,00,000
Variable cost as a percentage to sales	60%	70%	60%
Variable marketing cost as a percentage to sales	3%	2%	2%

	Sales by market	
	Locai	*Regional*
A	Rs. 4,00,000	Rs. 1,00,000
B	Rs. 3,00,000	Rs. 1,00,000
C	Rs. 3,00,000	Rs. 1,00,000

All administrative cost and fixed manufacturing costs are common to the three products and the two markets and are fixed for the period. Remaining marketing costs are fixed for the period and separable by market. All fixed costs are based upon a prorated yearly amount.

Required:

(*a*) Assuming there are no alternative used for the X Company Ltd.'s present capacity, would you recommend dropping the regional market.

(*b*) Prepare the quarterly income statement showing contribution margin by products.

(*c*) It is believed that a new product can be ready for sale next year. The new product can be produced by simply converting equipment presently used in producing product C. This conversion will increase fixed cost by Rs. 10,000 per quarter. What must be the minimum contribution margin per quarter for the new product to make the changeover financially feasible.

Solution:

(*a*) The regional market should not be dropped as this market not only covers all the variable costs and separable fixed costs but also gives net market contribution of Rs. 65,000 towards the common fixed cost.

Sales = Rs. 3,00,000

Variable manufacturing cost = (6 × 1,00,000) + (7 × 1,00,000) + (6 × 1,00,000)

= 1,90,000

Marketing costs = Rs. 45,000

(b) Statement showing quarterly income statement

In thousands

		A	*B*	*C*	*Total*
Sales (Rs.)		500	400	400	1,300
Less:	Variable cost				
	Manufacturing	300	280	240	820
	Marketing	15	8	8	31
	Total variable cost	315	288	248	851
	Contribution margin	185	112	152	449
Less:	Fixed costs:				
	Manufacturing				190
	Marketing				74
	Administrative				52
	Total fixed cost				316
	Operating profit (449 – 316)				133

(c) The new product must contribute at least Rs. 1,62,000 (Rs. 1,52,000 + Rs. 10,000) per quarter so as not to leave the company worse off when product C is replaced.

Problem 7. *(Acceptance or rejection of an export order):* MX Ltd. having an installed capacity of 1,00,000 units of a product is currently operating at 70% utilisation. At current levels of input prices the FOB unit cost (after credit for applicable export incentives) works out as follows:

Capacity utilisation %	*FOB unit cost (Rs.)*
70	97
80	92
90	87
100	82

The company has received three foreign offers from different sources as under:

Source A	5,000 units @ Rs. 54 per unit FOB
Source B	10,000 units @ Rs. 52 per unit FOB
Source C	10,000 units @ Rs. 51 per unit FOB

Advise the company as to whether any or all the export orders should be accepted or not.

Solution:

STATEMENT SHOWING DIFFERENTIAL COST

Capacity	*Production (Units)*	*FOB unit cost (Rs.)*	*Installed total cost*	*Differential cost*	*Per unit differential cost*
70	70,000	97	67,90,000	–	–
80	80,000	92	73,60,000	5,70,000	57
90	90,000	87	78,30,000	4,70,000	47
100	1,00,000	82	82,00,000	3,70,000	37

CHAPTER 12

STATEMENT SHOWING GAIN OR LOSS ON ACCEPTING VARIOUS EXPORT ORDERS

Source of export order	*Export, order in units*	*Capacity utilisation*	*Differential cost per unit*	*Differential cost Total*	*FOB price per unit*	*Sales revenue from the export*	*Gain or loss*
A	5,000	75	57	2,85,000	55	2,75,000	(10,000)
B	10,000	85	5,000 @ 57 5,000 @ 47	5,20,000	52	5,20,000	NIL
C	10,000	95	5,000 @ 47 5,000 @ 37	4,20,000	51	5,10,000	90,000
	25,000	95%		12,25,000		13,05,000	80,000

From the above analysis it can be said that when all the three export orders are accepted, the company will make a profit of Rs. 80,000. But accepting only source A and B orders will give a loss of Rs. 10,000.

Problem 8. *(Accepting an order at a lower selling price):* The overhead expenses of a factory, producing a single article at different operating levels are as follows:

Operating level capacity	Works overhead (Rs.)
80%	36,000
100%	40,000
120%	50,000
60%	33,000

The factory is currently working at 60% operating level and its annual sales amount is Rs. 1,44,000.

Selling prices have been based on 100% capacity and have the following relationship with costs at this level:

Factory cost	66.67% of sales value
Prime cost	75% of factory cost
Administrative and selling expenses (of which 75% is variable)	20% of sales value.

The management receives an offer for carrying out some work for another company valued at Rs. 33,000 per annum which will take up 40% of capacity. The prime cost for the work is estimated at Rs. 20,000. There will be an addition to administration expenses of Rs. 15,00 per annum.

The Sales Manager estimates that the sales of the company's own product will increase to 80% capacity by the time new order materialises.

Calculate the profit on current production. Give your views, supported by figure, on the advisability of taking on the new work. *(Sri Venkateshwara University, M.Com., September, 1988)*

Solution: STATEMENT SHOWING PROFITABILITY

			100%	60%
Sales			2,40,000	1,44,000
Less: Cost sales:				
Prime cost	1,20,000	72,000		
Works overhead	40,000	33,000		
Factory cost	1,60,000	1,05,600		
Administrative and selling	48,000	33,600	2,08,000	1,38,600
		Profit	32,000	5,400

At 100% capacity:

Factory cost 66.67% of sales value or $\frac{2}{3}$ of sales

$\frac{2}{3} \times 2,40,000 \quad = \text{Rs. } 1,60,000$

Prime cost = 75% of factory cost

$\frac{75}{100} \times 1,60,000 \quad = \text{Rs. } 1,20,000$

Factory overhead at 100% level given in the problem is 40,000
Administration and selling overhead

$\frac{20}{100} \times 2,40,000 \quad = \text{Rs. } 48,000$

of 48,000, 75% is variable, *i.e.*, Rs. 36,000 and Rs. 12,000 is fixed.
At 60% capacity:
Prime cost is 60% of Rs. 1,20,000 = Rs. 72,000
Works overhead is Rs. 33,000 as is given
Administration and selling overhead
Variable 75% and fixed Rs. 12,000

STATEMENT SHOWING PROFITABILITY FOR NEW WORK

	Capacity	80%	40% Additional	Total 100%
	Sales	1,92,000	33,000	2,25,000
Less:	Cost of sales			
	Prime cost	96,000	20,000	1,16,000
	Works overhead	36,000	14,000	50,000
	Factory overhead	1,32,000	34,000	1,66,000
Administrative and selling		40,800	6,450	47,250
		1,72,800	40,450	2,13,250
	Profit or Loss	19,200	7,450	11,750

As profit is reduced by Rs. 7,450 additional order is not profitable.

Problem 9. *(Expanding the marketability of the product):* The following extracts are taken from sales budget of a company for the current year

		(Rs. '000)
Sales: 40,000 units @ Rs. 25 per unit		1,000
Selling costs:		
Advertising	100	
Salesmen salaries	80	
Travelling expenses	50	
Rent of sales office	10	
Others	10	
		250

The management is considering a proposal to establish a new market in the eastern region in the next year. It is proposed to increase the advertising expenditure by 25% and appoint an additional sales supervisor at a salary of Rs. 30,000 per year to establish a market. This will involve additional travelling, and the travelling expenses shall increase by 10%.

Target annual sales volume at the existing selling price for the new market is 10,000 units. The estimated variable cost of production is Rs. 12 per unit. Should the company try to establish the new market ? *(C.S., Inter, December 1997)*

Solution:

STATEMENT SHOWING PROFIT

	Current cost and Revenue		*Budget cost and Revenue*		*Incremental cost and Revenue*	
Sales (units)		40,000		50,000		10,000
			(Rs. in '000)			
Sales value		1,000		1,250		250
Less: Advertising	100		125		25	
Sales salaries	80		110		30	
Travelling expenses	50		55		5	
Rent	10		10		–	
Others	10		10		–	
	250		310		60	
Variable cost of production	480		600		120	
		730		910		180
Profit		270		340		70

On the incremental sale, return is 28% of additional sales value and is higher than the current return. Hence the proposal is acceptable.

Questions

Simple Questions

1. Define "differential cost."
2. Define differential costing.
3. State any four characteristics of differential costing.
4. State any four implications of differential costing.
5. State any four similarities between marginal costing and differential costing.
6. State any four dissimilarities between marginal costing and differential costing.
7. Mention any eight uses of differential costing technique to management of an industry.

Exercises

Acceptance or Rejection of an Offer

1. A company currently operating at full capacity, manufactures and sells, sauce pans at Rs. 2 each. Current volume is 1,00,000 pans per annum with the following cost structure.

Operating statement for the year 2010

Sales 1,00,000 @ Rs. 2		Rs.	2,00,000
Less: Materials cost	50,000		
Labour cost	80,000	Rs.	1,30,000
	Contribution		70,000
Less: Fixed cost			30,000
	Profit	Rs.	40,000

An opportunity has arisen to supply 30,000 pans per annum at Rs. 1.80.

Acceptance of this order would incur extra materials cost of Rs. 15,000 per annum and fixed cost of Rs. 8,000 per annum for the hire for additional machinery and the payment of an overtime premium of 20% for the extra direct labour required.

Should this order be accepted ?

[**Ans.** The incremental revenue generated by accepting order is Rs. 2,200. So the order is to be accepted.]

Equipment Replacement

2. P. Ltd. is considering the replacement of machine A by a new machine B. Comparative data are as follows:

	Existing machine A	Proposed machine B
Purchase price	Rs. 2,000	6,000
Disposal value now	600	
Disposal value in one year's time	450	
Rate of depreciation	10%	10%
Estimated scrap value (Rs.)		500
Indirect labour and materials	1,000	1,000
Scrap losses	200	100
Direct wages	2,000	1,000

[**Ans.** Saving is Rs. 700 per annum. So new machine is to be bought.]

Acceptance or Rejection of Order

3. A company has a capacity of producing 1,00,000 units of a certain product in a month. The sales department reports that the following schedule of sale prices is possible:

Volume of production	Selling price per unit (Re.)
60%	0.90
70%	0.80
80%	0.75

CHAPTER 12

90%	0.67
100%	0.61

The variable cost of manufacture between these levels is Re. 0.50 per unit and fixed cost Rs. 40,000.

(*a*) Prepare a statement showing incremental revenue and differential cost of each stage. At which volume of production will the profit be maximum ?

(*b*) If there is a bulk offer at Re. 0.50 per unit of the balance capacity over the maximum profit volume for export and the price quoted will not affect the internal sale, will you advise this bid ?

[**Ans.** As the incremental revenue is higher than differential cost up to the level of 80%, the level of production should be set out 80%.

The bulk offer for export for the balance capacity of 2,000 units (over and above 80%) at 50 paise per unit will add Rs. 7,000 to profit. Hence the offer for export is to be accepted]

❑❑❑

CHAPTER 13

BUDGETARY CONTROL

INTRODUCTION

A budget is a quantitative expression of a plan of action prepared in advance for the period to which it relates. It may be prepared for the entire organisation or for various departments or for various functions involved in that organisation. Budget is a means of translating the overall objectives of the business into detailed feasible plan of action.

The concept of "budget" is being made use of by every individual who undertake a work involving expenditure. While some express it orally others put it in written form. For example, a person who wants to go on a holiday tour will prepare a budget involving expenditure on fares, boarding, lodging, purchasing, etc. After returning, he will compare the actual expenses incurred with budgeted expenses to know whether he spent more or less as compared to budgets and if so what factors were responsible for it. This enables him to increase or decrease his budget for the next year.

In the same way, every business undertakes to budget its expenditure for utilising the available funds more judiciously. Similarly, to ensure proper utilisation of scarce raw materials and other factors of production, the management of every business will prepare a budget relating to material, labour, production and various expenditure. This enables in proper planning of all activities, their co-ordination and finally controlling such activities. In this process it enables the management to know the performance of business for a given period of time.

BUDGET AND BUDGETARY CONTROL

Definition of a Budget

The ICMA terminology defines a budget as "a plan quantified in monetary items, prepared and approved prior to a defined period of time, usually showing planned income to be generated and/or expenditure to be incurred during that period and the capital to be employed to attain a given objective".

George R. Terry defines budget as "an estimate of future needs arranged according to an orderly basis, covering some or all the activities of an enterprise for a definite period of time".

Definition of Budgetary Control

According to F.H. Rowland and W.H Barr, budgetary control is a "Tool of management used to plan, carry out and control the operations of the business".

In the words of C.L. Van Sickle, "a budgetary control system is a carefully worked out financial plan, including the procedure involved in its operations for conducting the various divisions of a business for the ultimate purpose of earning a profit".

The ICMA terminology defines budgetary control as, "the establishment of budgets relating the responsibilities of executives to the requirements of a policy and the continuous comparison of actual with budgeted result either to secure by individual action the objective of that policy or to provide a basis for its revision".

An analysis of this definition reveal the following important points:

(*a*) **Executive responsibility:** It means that every manager has to do his job directed towards the overall objects of the business, *i.e.,* the policy. In other words, every manager is responsible for attaining the task delegated to him and thereby the ultimate object of the business.

(*b*) **The requirements of a policy:** The budget is a statement of policy relating to the position of business and plans to attain it. The business has to establish certain goals and necessary actions must be taken to achieve those aims. This is essential for proper growth and development of the business and for efficient utilisation of labour force and other assets.

(*c*) **Comparison of actual with budgeted results:** This facilitates control over the planned activities. The actual result is measured and compared with budgets. When there exists any unfavourable difference, management can take remedial steps.

(*d*) **The revision of policy:** The policy may have to be revised in order to make the best use of prevailing situation after considering unfavourable factors.

The budgetary control is exercised by applying the following steps:

1. **Preparation of the budget:** The process of preparing the budget will be explained under the heading "budget organisation".
2. **Publishing the budget:** This implies informing each executive what is expected of him. The publishing of the budget is most important, as no system is able to work unless the people concerned with it understand and are prepared to make it work.
3. **Measuring the results:** This implies the measurement of actual results achieved in order to know whether they are as per plan or otherwise.
4. **Comparing the results with the budget:** The comparison of actual results with the budget enables to know the efficiency or otherwise of the activity.
5. **Reporting the results of the above activity:** This is done by means of budgetary control statements, which enable the budget officer to know whether the objectives set out in the budget are being fulfilled and if not, in which areas attention should be concentrated.
6. **Correcting the unfavourable variances:** Necessary steps are then taken to avoid the occurrence of unfavourable variances. This is the function of the departmental manager or the supervisor. The budgetary control statements will assist as tools to the manager in correcting the unfavourable differences.

OBJECTIVES OF BUDGETARY CONTROL

1. Planning and Coordination

Budgeting involves preparation of a detailed operational plans to achieve the objectives of a business. The success of every business depends upon the planning of activities and budgeting forces planning to become more effective. As budgets are prepared covering all the activities and departments of a business, it also serves as a means of coordinating the efforts of all concerned in such a way that every department contributes towards the overall plan. The summary of all such plan is known as master plan. The budget forces every departmental manager to establish relationship with other departments and contribute to the achievement of objectives.

2. Clarification of Authority and Responsibility

Budgeting enables the superior to delegate the authority to his subordinates. This also clarifies the responsibility of each manager who can be held accountable if the targets are not reached. Thus budgeting facilitates management by exception.

3. Communication

Since all levels of management are involved in the preparation of budget, it facilitates communication process to become more effective. The objective of the business, problems involved in achieving them, and finalisation of budgets are all promptly communicated. It also coordinates the various functions such as sales, purchases and production more efficiently.

4. Motivation

The preparation of budgets by middle and lower management against which performance can be judged serves as a good motivation for them.

5. Control

The control over various activities is achieved by comparing the achievements with the targets. If there are any deviations, the causes for the same is investigated and remedial action taken.

ESSENTIAL REQUIREMENTS OF BUDGETARY CONTROL

The essential requirements of a budgetary control system as quoted by Robert I Dickey are as follows:

1. Budget must have the complete cooperation of the chief executive.
2. The ultimate realisation of the maximum amount of profit should always be kept uppermost.
3. Responsibility for the preparation of the estimate should rest on those individuals responsible for performance.
4. The budget must be realistic and the goals attainable.
5. A budget committee should be established consisting of the budget director, the chief executive officer, and the executives of the various divisions of the organisation.
6. The budget should cover all phases of operations.
7. Budgeting should be continuous.
8. Periodic reports should be prepared promptly, comparing budget and actual results.
9. The accounting system must be adequate.
10. A good organisation must be developed.

ADVANTAGES AND LIMITATIONS OF BUDGETARY CONTROL

Advantages

The important advantages of budgetary control are as follows:

1. It helps the process of planning by reducing it to concrete numerical goals. Through the budgets, the executives know what they are to produce or sell, how much they can spend, how much income to expect and so on.
2. It provides an effective means by which top management can delegate authority and responsibility without sacrificing its overall control. Limits for each department or division are laid down in the budget.
3. It keeps expenditure under check and constantly reminds employees and management of the targets and goals to be achieved. It helps in the cautious utilisation of resources and promotion of efficiency.
4. As a control devise, it supplies the means of checking results and comparing performance, of revealing weaknesses and making corrections.
5. The budget is not merely an instrument of planning but also a tool of coordination. It brings together the activities of various sections, departments and divisions in an overall perspective.
6. It helps in determining the policies of the factory.
7. It gives complete information in advance regarding the amount of capital needed for the budget period.
8. It gives the idea of where executive action is required.
9. It aids in measuring performance of each department of the factory.
10. It promotes cooperation among the different executives for determining future plans.
11. It acts as a control tool for administration.
12. It centralises management control.

Limitations of Budgetary Control

1. The budget is always based on estimates. The success or failure of a budget, to a large extent, depends upon the accuracy of estimates. The estimates cannot be accurately made in this dynamic world, although many statistical techniques are available.
2. To evolve a budgetary control system, normally, it takes several years as it has to be tried, improved, and discarded, depending upon the changing circumstances.
3. The success of budgetary control depends upon the enthusiastic participation of all levels of management. But it is difficult to secure the wholehearted cooperation of all in a factory.
4. Budgeting is only a tool of management but it cannot replace management.
5. It may be difficult to install a system of budgetary control in small factories owing to expenditure involved.

ORGANISATION FOR BUDGETARY CONTROL

A sound organisation is essential for an effective budgeting programme. It ensures cooperation from all levels of management. The organisation should facilitate control of all elements of cost. The essential ingredients of a sound organisation are as follows:

1. Organisation Chart

An organisation chart depicts the functional responsibilities of every manager. It shows the relative position of every manager and his relationship with others. A typical organisation chart is shown below:

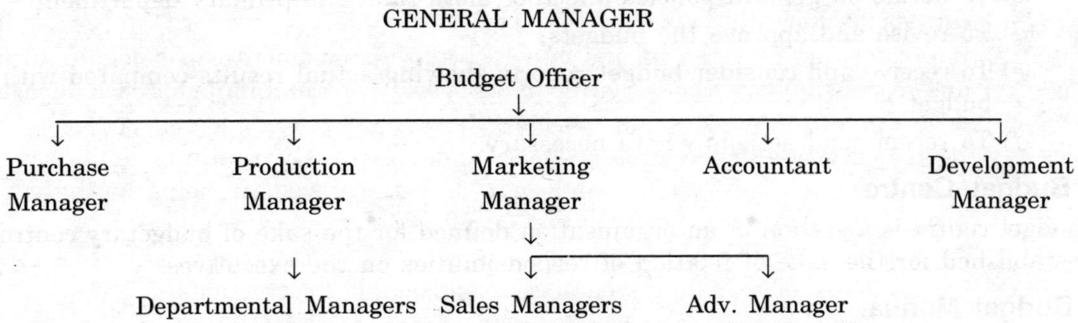

Fig. 13.1. *Budgetary control organisation chart.*

From the chart it is clear that, the entire responsibility of budgetary control process lies with general manager. He entrusts the authority to implement the budget programme to the budget officer. He shall coordinate the budgets prepared by all the departmental managers.

2. Budget Officer

A budget officer is a person who is nominated by the general manager to look after the functions of budgeting. He is responsible to general manager for completing the budgeting process. He acts as coordinator and adviser integrating the budgets prepared by the departmental managers so as to get a master budget. His functions are as follows:

(*a*) Issuing instructions to departments regarding requirements, dates of submission of data, etc.

(*b*) Providing historical information to departmental managers to help them in their forecasting.

(*c*) Receiving and checking budget estimates.

(*d*) Suggesting possible revisions.

(*e*) Discussing difficulties with managers.

(*f*) Ensuring that budgets are received in agreed time.

(*g*) Preparing budget summaries.

(*h*) Charting the departmental estimates on a master plan.

(*i*) Submitting budgets to the committee and furnishing explanations on particular points.

(*j*) Informing the departmental managers of any revisions made in their budgets by the committee.

(*k*) Preparing the final master plan approved by the committee.

(*l*) Coordinating all budget work.

3. Budget Committee

It serves as a coordinating authority of the budgeting process. It is concerned with resolving difficulties or disputes which may arise among functional heads and to take decisions so as to alter production, price, etc. The budget officer will submit the draft master budget to the committee for consideration after which it is finally approved by board of directors. The budget committee will function in an advisory capacity and it performs the following functions:

(*a*) To receive and review individual budget estimates.

(b) To suggest revisions.

(c) To decide on general policies affecting more than one primary department.

(d) To revise and approve the budgets.

(e) To receive and consider budget reports showing actual results compared with the budget.

(f) To recommend actions where necessary.

4. Budget Centre

A budget centre is a section of an organisation defined for the sake of budgetary control. It is established for the sake of fixation of responsibilities on the executives.

5. Budget Manual

The ICMA terminology defines a budget manual as "a document which sets out standing instructions governing the responsibilities of persons and the procedures, forms and records relating to the preparation and use of budgets". It serves as a basis of guidance and information about budgeting process. It is more like an instructional manual about the way budget operates. Lucey, in his book *Costing* sets out the contents of a budget manual as under:

Foreword:

Preferably by managing director

Objectives/explanation of the budgetary process.

- Explanation of budgetary control.
- Objectives of each stage of the budgetary process.
- Relationship to long-term planning.

Organisation structure and responsibilities:

- Structure of the organisation showing titles, responsibilities and relationships.
- Titles and names of current budget holders.

Main budgets and relationship:

- Outline of all main budgets and their accounting relationships.
- Explanation of key budgets (master budget, cash budget).

Budget development:

- Budget committee, membership and terms of reference.
- Sequence of budget preparation.
- Time-table for budget preparation and publication.

Accounting procedure:

- Name and terms of reference of the budget officer.
- Coding lists.
- Sample forms.
- Time-table for accounting procedures, production of reports, closing date.

6. Chart of Accounts

It refers to maintaining a systematic set of accounting books which should be able to record and analyse the information required. A chart of accounts may be maintained separately for every budget centre. In simple words, a chart of accounts denote accounting structure useful for budgetary control.

7. Budget Period

It is a period relating to which a budget is prepared. A budget period may relate to a short-term period or a long-term period. Sometime these period may relate even to a month for monitoring and control of budgets. It is always desirable to have a short-term budget period excepting those functions which may prolong for a longer period as for example, research and development. The duration of the budget period depends upon type of business, the length of manufacturing cycle, the ease or difficulty of forecasting future market conditions and so on. The budget period may be accordingly short-term budget period or long-term budget period.

The short-term budget period may cover a period of 3 months to 1 year. According to Matz-Curry and Frank (*Cost Accounting*) the following factors should be considered in determining the length of budget period:

(*a*) The budget period should be long enough to complete production of the various products.

(*b*) For a business of a seasonal nature, the budget period should cover at least one entire seasonal cycle.

(*c*) The budget period should be long enough to allow for the financing of production well in advance of actual needs.

(*d*) Major operations and drastic changes in plant layout or manufacturing methods must be planned far in advance to determine financial requirements.

(*e*) The budget period should coincide with the financial accounting period to compare actual results with budget estimates.

A long-term budget may be taken to mean a budget prepared for a period of more than a year. Such a budget is best suited for research and development activities, long-term capital investment, financial and profit planning. It is mostly based on probability of events. It is very much affected by risk factor.

8. Key Factor

The factor which governs the quantity which is manufactured or sold is known as key factor. The ICMA terminology defines a key factor as "a factor which at any time or over a period may limit the activity of an entity often one where there is a shortage or difficulty of supply". This will have an adverse effect over the achievement of budget. Therefore it must be identified and its effect on each of the budget is to be carefully considered during the preparation of budget. Key factor can be changed by the management action. For example, when production process is hampered by plant capacity, the purchase of additional plant will allow increase in production. The following are considered as key factors:

(*a*) **Materials:**

(*i*) Availability of supply.

(*ii*) Restrictions imposed by licences, quotas, etc.

(*b*) **Labour:**

(*i*) General shortage.

(*ii*) Shortage in certain key processes.

(*c*) **Plant:**

(*i*) Insufficient capacity due to lack of capital.

(*ii*) Insufficient or lack of space.

CHAPTER 13

(*iii*) Insufficient or lack of market.

(*iv*) Bottlenecks in certain key processes.

(*D*) **Sales:**

(*i*) Low market demand.

(*ii*) Shortage of experienced salesmen.

(*iii*) Inadequate advertisement for want of money.

(*E*) **Management:**

(*i*) Lack of "know-how".

(*ii*) Inefficient executives.

(*iii*) Insufficient research into product design and methods.

(*iv*) Lack of capital thereby restricting policy.

ROLLING BUDGET

The concept of rolling budget or continuous budgeting is an attempt to update budgets. The ICMA terminology defines it as "the continuous updating of a short-term budget by adding, say, a further month or quarter and deducting the earliest month or quarter, so that the budget can reflect current condition". Rolling budget is used to indicate continuity in the budgeting process. Suppose a budget is prepared for a period of 1 year, then at the end of each month, the results of the previous month's operations together with any new information or changes in business conditions are used to revise the budget for the next 11 months and to prepare a budget for one additional month beyond so that the budget always covers the next 12 months. Each month the budget "rolls" forward one month by adding one month to replace the month just past. The advantage of the rolling budget is that it makes planning a continuous activity.

TYPES OF BUDGET

The various types of budgets are shown in the following chart:

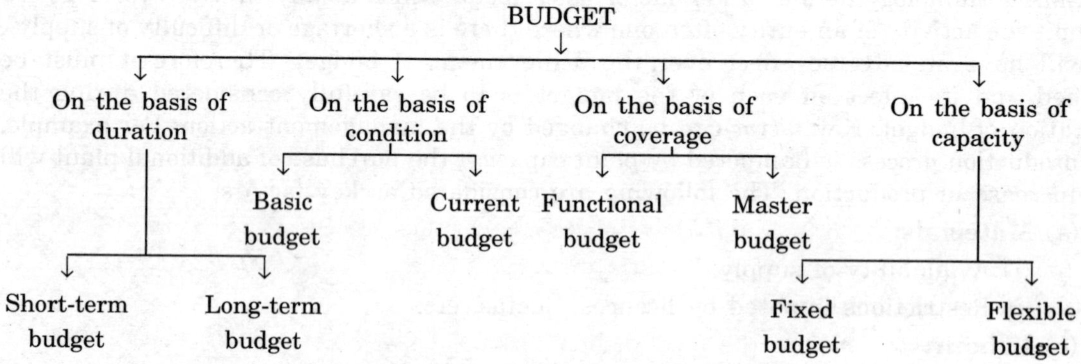

Fig. 13.2. *Chart showing types of budgets*

1. On the Basis of Duration

On the basis of duration, budget can be classified into two types, *viz.*, (*a*) short-term budget and (*b*) long-term budget.

(*a*) **Short-term budget:** A short-term budget is prepared to cover a duration of less than 1 year. Sometimes they are prepared for a month also. Cash budget is an example of short-term budget. Short-term budget is followed in many cases as it is difficult to forecast events on a long-term basis.

(*b*) **Long-term budget:** A budget which is prepared to cover a period of more than a year is known as long-term budget. The examples of a long-term budget are research and development budget, capital expenditure budget.

2. On the Basis of Condition

On the basis of conditions prevailing, a budget can be classified into two types: (*a*) basic budget and (*b*) current budget.

(*a*) **Basic budget:** The ICMA terminology defines a basic budget as "a budget which is established for use unaltered over a long period time". The budget does not take into account changes occurring from external environment which are beyond the control of management. This budget is more useful to top level management for formulating policies.

(*b*) **Current budget:** The ICMA terminology defines a current budget as "a budget which is established for use over a short period and is related to the current conditions". The budget under review will be adjusted to the current conditions prevailing in the business. This budget motivates the people preparing the budgets as they are sure of attaining the budget.

3. On the Basis of Coverage

On the basis of coverage of information and activities of a business, budget can be classified into two types, *viz.*, (*a*) Functional budget and (*b*) Master budget.

(*a*) **Functional budget:** As the name suggests functional budgets are prepared in respect of various functions performed in a business. Accordingly we have the following functional budgets.

1. Sales budget: A sales budget is a statement of planned sales in terms of quantity and value and analysed by products. It is the most important budget as it is most difficult to forecast and attain. It is prepared by the sales manager based on sales forecast. The factors to be considered in forecasting sales are (*a*) Past year sales, (*b*) Forecast of business conditions, (*c*) Market analysis and (*d*) Assessment by the sales department on the basis of (*i*) Products or group of similar products, (*ii*) Areas, (*iii*) Salesmen, (*iv*) Types of customers, (*v*) Period, such as a month or a quarter. While assessing the probable sales, the following are to be considered:

(*i*) The firms own sales and those of its competitors, areawise.

(*ii*) Whether its share of business is increasing or decreasing.

(*iii*) Seasonal fluctuations.

(*iv*) The effect of rise in the level of population.

(*v*) The changes in consumer's tastes and habits.

(*vi*) The effect of introducing new products.

(*vii*) Advertising.

(*viii*) The effect of sales promotion.

(*ix*) The possibility of extending the market.

Problem 1. Dutta Enterprises sells two products, A and B. During the year 2009, it plans to sell the following quantities of each product.

Sales budget (in units)

	Total	*I Quarter*	*II Quarter*	*III Quarter*	*IV Quarter*
A	7,00,000	90,000	2,30,000	3,00,000	80,000
B	3,00,000	85,000	75,000	55,000	85,000

Each of these two products is sold on a seasonal basis. Product A tends to sell better in summer months, while product B sells better during the winter. Dutta Enterprises plan to sell product A throughout the year at a price of Rs. 10 a unit and product B at a price of Rs. 20 a unit.

A study of the past experience reveals that Dutta Enterprises has lost about 3% of its invoice each year because of returns (constituting 2% of loss of revenue) allowances and bad debts (1% of loss).

Prepare a sales budget incorporating the above information.

Solution:

SALES BUDGET

	I Quarter	*II Quarter*	*III Quarter*	*IV Quarter*	*Total*
Product *A*	9,00,000	23,00,000	30,00,000	8,00,000	70,00,000
Product *B*	17,00,000	15,00,000	11,00,000	17,00,000	60,00,000
(*A*) Total sales	26,00,000	38,00,000	41,00,000	25,00,000	1,30,00,000
Loss of returns @ 2% on total sales	52,000	76,000	82,000	50,000	2,60,000
Loss for bad debts and allowance @ 1% of total sales	26,000	38,000	41,000	25,000	1,30,000
(*B*) Total deductions	78,000	1,14,000	1,23,000	75,000	3,90,000
Net sales (*A* – *B*)	25,22,000	36,86,000	39,77,000	24,25,000	1,26,10,000

Problem 2. Ambitions Company Ltd., has three sales divisions at Mumbai, Chennai and Kolkata. It sells two products – product X and product Y. The budgeted sales for the year ending 31st December, 2009 at each place are given below:

Mumbai:	Product *X*	1,00,000 units @ Rs. 8 each
	Product *Y*	70,000 units @ Rs. 5 each
Chennai:	Product *Y*	1,10,000 units @ Rs. 5 each
Kolkata:	Product *X*	1,50,000 units @ Rs. 8 each

The actual sales during the same period were as follows:

Mumbai:	Product *X*	1,25,000 units @ Rs. 8 each
	Product *Y*	75,000 units @ Rs. 5 each
Chennai:	Product *Y*	1,25,000 units @ Rs. 5 each
Kolkata:	Product *X*	1,55,000 units @ Rs. 8 each

From the reports of the sales personal, it was considered that the sales budget for the year ending 31st December, 2010 would be higher than 2009 budget in the following aspects

Mumbai:	Product *X*	8,000 units
	Product *Y*	5,000 units
Chennai:	Product *Y*	13,000 units
Kolkata:	Product *X*	10,000 units

Intensive sales campaign in Chennai and Kolkata is expected to result in additional sales of 25,000 units in product *X* in Chennai and 18,000 units of product *Y* in Kolkata.

You are required to prepare a sales budget for the period ending 31st December, 2010.

Solution:

ASCERTAINMENT OF QUANTITY OF BUDGETED SALES

(for the year 2010)

Mumbai Division

Product X:

		Units
	Budgeted sales for the year 2009	1,00,000
Add:	Expected increase in the sales for the year 2010	8,000
	Budgeted sales for 2010	1,08,000

Product Y:

	Budgeted sales for the year 2009	70,000
Add:	Expected increase in the sales for the year 2010	5,000
	Budgeted sales for 2010	75,000

Chennai Division

Product X:

		Units
	Budgeted sales for the year 2009	NIL
Add:	Expected sales for the year 2010	25,000
	Budgeted sales for the year 2010	25,000

Product Y:

	Budgeted sales for the year 2009	1,10,000
Add:	Expected increase in the sales for the year 2010	13,000
	Budgeted sales for the year 2010	1,23,000

Kolkata Division:

Product 'X'

		Units
	Budgeted sales for the year 2009	1,50,000
Add:	Expected increase in the sales for the year 2010	10,000
		1,60,000

Product Y:

	Budgeted sales for the year 2009	NIL
Add:	Expected sales for the year 2010	18,000
	Budgeted sales for the year 2010	18,000

SALES BUDGET

Sales Division	*Product*	*Budget for the year 2009*			*Actuals for the year 2009*			*Budget for the year 2010*		
		Qty.	*Rate*	*Amt.*	*Qty.*	*Rate*	*Amt.*	*Qty.*	*Rate*	*Amt.*
Mumbai	*X*	1,00,000	8	8,00,000	1,25,000	8	10,00,000	1,08,000	8	8,64,000
	Y	70,000	5	3,50,000	75,000	5	3,75,000	75,000	5	3,75,000
		1,70,000		11,50,000	2,00,000		13,75,000	1,83,000		12,39,000
Chennai	*X*	—	—	—	—	—	—	25,000	8	2,00,000
	Y	1,10,000	5	5,50,000	1,25,000	5	6,25,000	1,23,000	5	6,15,000
		1,10,000		5,50,000	1,25,000		6,25,000	1,48,000		8,15,000

CHAPTER 13

Kolkata	X	1,50,000	8	12,00,000	1,55,000	8	12,40,000	1,60,000	8	12,80,000
	Y	—	—	—	—	—	—	18,000	5	90,000
		1,50,000		12,00,000	1,55,000		12,40,000	1,78,000		13,70,000

Problem 3. Adventurous Company Ltd., manufactures two products *X* and *Y* and sells them through two divisions Bangalore and Cochin. For the purpose of submission of sales budget to the budget committee, the following information has been made available.

Budgeted sales for the current year were:

Product	*Bangalore*	*Cochin*
X	2,000 units @ Rs. 9	3,000 units @ Rs. 9
Y	1,500 units @ Rs. 21	2,500 units @ Rs. 21

Actual sales for the current year were:

Product	*Bangalore*	*Cochin*
X	2,500 units @ Rs. 9	3,500 units @ Rs. 9
Y	1,000 units @ Rs. 21	2,000 units @ Rs. 21

Market survey reveal that product *'Y'* is popular but under-priced. It is observed that if the price of product *'X'* is increased by Re. 1, it will still find a ready market. On the other hand, product *Y* is over-priced to the customers and the market could absorb more if the sales price of product *Y* is reduced by Re. 1. The management has agreed to give effect to the above price changes.

From the information relating to these price changes and reports from salesmen, the following estimates have been prepared by divisional managers.

Percentage increase in sales over current budget is

Product	*Bangalore*	*Cochin*
X	+ 10%	+ 5
Y	+ 20%	+ 10%

With the help of an intensive advertisement campaign, the following additional sales above the estimated sales of divisional managers are possible

Product	*Bangalore*	*Cochin*
X	300 units	350 units
Y	200 units	250 units

You are required to prepare a budget for sales incorporating the above estimates and also show the budgeted and actual sales of the current year.

Solution:

Working Notes:

Ascertainment of Quantity of budgeted sales for future period

Bangalore Division:

Product X:

		Units
	Budgeted sales for the current year	2,000
Add:	Expected increase in sales	
	10/100 × 2,000	200
Add:	Further increase in sales with intensive advertising	300
	Budgeted sales for future period	2,500

Product Y:

	Budgeted sales for the current year	1,500
Add:	Expected increase in sales	
	$\frac{20}{100} \times 1{,}500$	300

Add:	Further increase in sales with intensive advertising	200
	Budgeted sales for future period	2,000
Cochin Division:		
Product X:		
	Budgeted sales for the current year	3,000
Add:	Expected increase in sales – 5%	150
Add:	Further expected increase	350
	Budgeted sales for future period	3,500
Product Y:		
	Budgeted sales for the current year	2,500
Add:	Expected increase in sales – 10%	250
Add:	Further expected sales	250
		3,000

SALES BUDGET

Sales Division	*Product*	*Budget for the current year*			*Actuals for the current year*			*Budget for the future period*		
		Qty.	*Rate*	*Amt.*	*Qty.*	*Rate*	*Amt.*	*Qty.*	*Rate*	*Amt.*
Bangalore	X	2,000	9	18,000	2,500	9	22,500	2,500	10	25,000
	Y	1,500	21	31,500	1,000	21	21,000	2,000	20	40,000
		3,500		49,500	3,500		43,500	4,500		65,000
Cochin	X	3,000	9	27,000	3,500	9	31,500	3,500	10	35,000
	Y	2,500	21	52,500	2,000	21	42,000	3,000	20	60,000
		5,500		79,500	5,500		73,500	6,500		95,000

CHAPTER 13

2. Selling and distribution cost budget: The success of sales depends upon selling and distribution expenses incurred. Selling expenses are incurred to increase the sales volume. While preparing a selling budget, a classification is made according to the variability of cost. This budget is prepared by the sales managers. Common examples of selling expenses included in a selling cost budget are salesmen's salaries, commissions and travelling expenses. Sometimes a separate distribution cost budget is prepared apart from selling cost budget. The purpose is to know the expenditure involved in making available finished goods safely to customers. Two important costs go to make up distribution cost budget, *viz.*, (*a*) storage and warehousing cost and (*b*) transporation cost. As in the case of selling cost budget, distribution costs are also classified into fixed and variable cost. A specimen of selling and distribution cost budget is shown below:

SELLING & DISTRIBUTION COST BUDGET
(for the period ending 31st Dec. 2009)

		Previous years' budget	*Previous year's actual*	*Current budget*
A.	Personnel cost:			
	Salaries			
	Commission			
	Travelling			

B. Sale's office cost
Office supplies
Salaries
Postage
Telephone
Rent & rates
C. Publicity:
Salaries
Press
Journals
Television
Samples
D. Warehousing, packing & despatch
Salaries
Packing wages
Driver's wages
Sundries

Fig. 13.3 *Specimen of selling and distribution cost budget.*

Problem 4. Prepare a sales overhead budget for the months of January, February and March from the estimates given below:

	Rs.
Advertisement	2,500
Salaries of sales department	5,000
Expenses of the sales department	1,500
Counter salesman's salaries & D.A.	6,000

Commission to counter salesmen at 1% of their sales
Travelling salesmen's commission at 10% on their sales and expenses at 5% on their sales.
The sales during the period were estimated as follows:

Month	*Counter sales (Rs.)*	*Travelling salesmen's sales (Rs.)*
January	80,000	10,000
February	1,20,000	15,000
March	1,40,000	20,000

(University of Delhi, B.Com., (Hons.) April 1989)

Solution:

SALES OVERHEAD BUDGET

	January	*February*	*March*
Estimated sales	90,000	1,35,000	1,60,000
Fixed overhead:			
Advertisement	2,500	2,500	2,500
Salaries of sales department	5,000	5,000	5,000
Expenses of sales department	1,500	1,500	1,500
Salaries of counter salesmen & their DA	6,000	6,000	6,000
	15,000	15,000	15,000
Variable overheads:			
Commission of counter salesmen—10% of counter sales	800	1,200	1,400
Commission of travelling salesmen—10% of their sales	1,000	1,500	2,000

Expenses of travelling salesmen (5% of their sales)	500	700	1,000
	2,300	3,450	4,400
Total sales overhead	17,300	18,450	19,400
Total sales	90,000	1,35,000	1,60,000
% of sales overhead to sales	19.22%	13.67%	12.13%

3. Advertising budget: In large factories, to undertake an intensive sales programme, a separate advertising department is established. It will be looked after by an advertising manager. The advertising manager will have to deal with the following while budgeting advertising cost:

(*i*) Determine the best method or methods of advertising for the business concerned.

(*ii*) Fix the total amount to be spent on advertising in the budget period. This amount is known as the advertising appropriation.

(*iii*) Coordinate the sales function and advertising.

(*iv*) Control the expenditure within the limits laid down and also attempt to measure the effectiveness of advertising.

The best method of advertisement to be selected depends upon reference to the product, channel of distribution, method of selling, types of consumers served. Care must be taken to see that the advertisement cost does not increase cost of production drastically. In other words, advertising cost should be incurred as long as there is an increase in sales and net profit. The matching of advertising cost and benefits derived therefrom is a vital part of advertising budget.

A specimen of advertising budget is shown below:

Advertising budget for the year ended..........

Expenses	*Media of Advertisement*						
	Total Rs.	*News-paper*	*Magazines & journals*	*T.V.*	*Cinema*	*Catalogue*	*Samples*
Salaries Rent Travelling expenses Heating & lighting Payment to agencies							
Total							

Fig. 13.4 *Specimen of advertising cost budget.*

4. Production budget: This budget is prepared by production manager based upon (*a*) sales budget, (*b*) the production capacity and (*c*) the budgeted finished goods stock requirements. According to Blocker and Weltmer, a production budget deals with:

(*i*) The determination of the total estimated volume of production.

(*ii*) The division of the estimated output into different types of products.

(*iii*) The scheduling of operations by days, weeks and months.

(*iv*) The establishment of finished goods inventory requirements.

(*v*) The storage of finished products until delivery can be made in accordance with sales orders.

Advantages: The preparation of a production budget has the following advantages:

(*i*) Plans can be made to keep inventories at reasonable levels consistent with production and sales requirements.

(*ii*) The requirements of raw materials and the sources of their supply can be selected for deriving best terms of purchases along with quality.

(*iii*) By maintaining production schedule, the promised delivery dates can be maintained. This increases reputation of the business.

Procedure: According to Heckert and Wilson (*Business Budgeting and Control*) the preparation of a production budget involves the following steps:

(*i*) Determine the period of time to be used as a basis for the production budget.

(*ii*) Ascertain what physical quantities should be produced to meet the sales budget and to provide properly balanced inventories.

(*iii*) Determine when the goods should be produced.

(*iv*) Determine where the goods should be produced.

(*v*) Determine the manufacturing operations required by the production.

(*vi*) Establish standards of production performance for use in measuring production efficiency.

(*vii*) Develop a programme of materials, labour and equipment requirements.

(*viii*) Use the production budget for purpose of cost control.

(*ix*) Make necessary revisions of the production budget.

5. Production cost budget: This budget shows the cost of production taking into account the elements of costs viz., direct materials cost, direct labour cost and production overheads.

Problem 5. *(Production Cost Budget):* From the following particulars, prepare production cost budget for the month of December 2000.

Particulars	*Opening stock* 1-1-2000	*Closing stock* 31-12-2000
Finished goods	1200 units	1600 units
Raw material A	5000 kg	4800 kg
Raw material B	2000 kg	3100 kg

(*a*) Budgeted sales for the month–7000 units.

(*b*) Raw materials required to produce one unit: A–4 kg at Rs. 8 per kg, B–2 kg at Rs. 25 per kg.

(University of Madras, B.Com., October 2002)

Solution:

PRODUCTION UNIT BUDGET

Budgeted sales units	7,000 units
Less: Opening finished goods	1200 units
	5800 units
Add: Closing finished goods	1600 units
Production units	7400 units

RAW MATERIAL CONSUMPTION BUDGET FOR 7400 UNITS PRODUCTION

		Raw Material 'A'	*Raw Material 'B'*
(*i*)	Raw material required for one unit of production	4 kg	2 kg
(*ii*)	Raw material required for 7400 units of production	7400 × 4 = 29,600 kg	7400 × 2 = 14,800 kg

RAW MATERIAL PURCHASE BUDGET FOR 7400 UNITS FINISHED UNITS

	Raw Material 'A'	*Raw Material 'B'*
Consumption of raw material	29,600 kg	14,800 kg
Less: Opening stock of raw materials	5,000 kg	2,000 kg
	24,600 kg	12,800 kg
Add: Closing stock of raw material	4,800 kg	3,100 kg
Purchase of raw materials	29,400 kg	15,900 kg

PURCHASE COST BUDGET

	Raw Material 'A'	*Raw Material 'B'*
Cost per unit of raw material	Rs. 8 per kg	Rs. 25 per kg
Purchase cost of raw material	29,400 × Rs. 8 = Rs. 2,35,200	15,900 × 25 = 3,97,500

PRODUCTION COST BUDGET FOR 74,000 UNITS OF PRODUCTION

(1)	Material cost		
	Raw material *A*	= 29,600 kg × Rs. 8	= 2,36,800
	Raw material *B*	= 14,800 kg × Rs. 25	= 3,70,000
			6,06,800
(2)	Labour cost		—
(3)	Overheads		—
		Total production cost	6,06,800

Problem 6. *(Production budget and production cost budget):* X Company Ltd., manufactures two products *A* and *B*. Forecast of the number of units to be sold in the first seven months of the year is given below:

	Product A	*Product B*
January	1,000	2,800
February	1,200	2,800
March	1,600	2,400

April	2,000	2,000
May	2,400	1,600
June	2,400	1,600
July	2,000	1,800

It is anticipated that:

(*i*) There will be no work-in-progress at the end of every month.

(*ii*) Finished units equal to half of the sales for the next month will be in stock at the end of each month (including previous December).

Budgeted production and costs for the whole year are as follows:

	Product A	*Product B*
Products (units)	22,000	24,000
Direct materials cost per unit	12.50	19
Direct labour cost per unit	4.50	7
Total factory overhead apportioned	66,000	96,000

Prepare for the six months period ending 30th June, 2003 (*a*) production budget for each month and (*b*) summarised production cost budget. *(University of Madras, B.Com., May 2004)*

Solution:

PRODUCTION BUDGET (IN UNITS)

Product A

	Jan.	*Feb.*	*Mar.*	*April*	*May*	*June*	*July*	*Total*
Sales of current month	1,000	1,200	1,600	2,000	2,400	2,400	2,000	
Add: Closing stock	600	800	1,000	1,200	1,200	1,000	—	
	1,600	2,000	2,600	3,200	3,600	3,400		
Less: Opening balance	500	600	800	1,000	1,200	1,200		
	1,100	1,400	1,800	2,200	2,400	2,200		11,100
Product B								
Sales of current month	2,800	2,800	2,400	2,000	1,600	1,600	1,800	
Add: Closing stock	1,400	1,200	1,000	800	800	900		
	4,200	4,000	3,400	2,800	2,400	2,500		
Less: Opening balance	1,400	1,400	1,200	1,000	800	800		
Total units to be produced	2,800	2,600	2,200	1,800	1,600	1,700		12,700

PRODUCTION COST BUDGET

	Product A	*Product B*
Direct material cost per unit	12.50	19.00
Direct labour cost per unit	4.50	7.00
Factory overhead	3.00	4.00
	20.00	30.00
Total budget cost	11,100 × Rs. 20 = Rs. 2,20,000	12,700 × Rs. 30 = Rs. 3,81,000

Problem 7. *(Production cost budget):* The production cost of Modern Machine Company Ltd., for the year 2009 are as under:

Direct material cost		Rs. 1,20,000
Direct wages		75,000
Overheads: Variable	70,000	
Fixed	45,000	1,15,000

The following changes are anticipated in 2010:

(*a*) The purchase price per unit of direct materials, and of other materials included in overheads will remain unchanged.

(*b*) The average rate for direct labour will fall from Rs. 4 per hour to Rs. 3 per hour, and direct labour hours will increase by 10%.

(*c*) Production efficiency will decrease by 4%.

The overheads are absorbed on a direct wages basis. Draw up a production cost budget for 2010.

Solution:

PRODUCTION COST BUDGET FOR 2010

		Original budget in 2009	*Budget for 2010*
Direct materials cost		1,20,000	1,32,000
Direct labour cost		75,000	64,350
	Prime cost	1,95,000	1,96,350
Overheads — Fixed		45,000	45,000
— Variable		70,000	77,000
	Total cost	3,10,000	3,18,350

Notes:

1. Labour hours will increase by 10%. Therefore it is presumed that production will increase by 10% and hence 10% increase in the cost of materials.

2. Direct labour is affected by production efficiency of 4% and increase in labour hours of 10%.

$$\text{Labour hours} = \frac{\text{Rs. } 75{,}000}{\text{Rs. 4 per hour}} = 18{,}750 \text{ hrs.}$$

Add: 10% increase in labour hrs. = 1,875 hrs.

= 20,625 hrs.

Add: 4% decrease in production efficiency

$$\frac{4}{100} \times 20{,}625 = 825 \text{ hrs.}$$

= 21,450 hrs.

Total labour cost = 21,450 × Rs. 3 = 64,350

3. Variable overheads are related to production, 10% rise in production will result 10% rise in variable overhead.

6. Plant utilisation budget: This budget indicates the plant capacity required to meet the production budget. While preparing this budget allowance must be made for the time lost in repairs and maintenance, setting up time, etc. For a smooth flow of production the capacity of plant must be balanced. For this purpose it is necessary to decide whether to (*a*) reduce the production volume, (*b*) purchase new machinery, (*c*) work extra shift, (*d*) use of subcontractors and so on. A specimen of plant utilisation budget is shown below:

CHAPTER 13

Plant utilisation budget for the period.......

Department	*Machines*	*Number of hours available during the period (2,000)*	*Normal lost time (200)*	*Standard capacity in hrs. (1,800)*	*Output per std. hr. (5)*	*Std. qty. (units) (9,000)*
A	1					
	2					
	3					
B	4					
	5					
	6					
	Total					

Fig. 13.5 *Specimen showing plant utilisation budget.*

7. Material purchase budget: This budget is prepared by the purchase manager for the purchase of raw materials and component parts. While preparing this budget priority must be given for the purchase of materials which are in short supply. This budget is prepared considering both quantities and value of materials. The following factors are to be considered before materials purchase budget is prepared:

(*a*) Planned increase or decrease in the stock of raw materials during the budget period.

(*b*) The date on which materials are required.

(*c*) The fluctuation in the prices of material.

(*d*) The duration of credit allowed.

Problem 8. *(Purchase budget):*From the following particulars, prepare a purchase budget for the year 2001 when the estimated price per kg of material is: A – Re. 1, B – Rs. 2 and C – Rs. 3

Materials	*Estimated consumption of materials (in kg)*
A	2,00,000
B	3,00,000
C	4,00,000

	Estimated stock (in kg)	
	On 1-1-2001	*On 31-12-2001*
A	20,000	25,000
B	40,000	30,000
C	50,000	40,000

(Calcutta University, B.Com. (Hons), 2002)

Solution:

CALCULATION OF PURCHASES

Purchases	=	Raw materials consumed	+ Closing stock	– Opening stock
A	=	2,00,000	+ 25,000	– 20,000
	=	2,05,000		
B	=	3,00,000	+ 30,000	– 40,000
	=	2,90,000		
C	=	4,00,000	+ 40,000	– 50,000

	=	3,90,000		
Purchase cost (budget)	=	Purchase (units) × Cost per unit		
A	=	2,05,000 × 1	=	2,05,000
B	=	2,90,000 × 2	=	5,80,000
C	=	3,90,000 × 3	=	11,70,000
				19,55,000

Problem 9. *(Material purchase budget):* The Sales Director of a manufacturing company reports that next year he expects to sell 50,000 units of a particular product.

Two kinds of raw materials *A* and *B* are required for manufacturing the finished product. Each unit of finished product requires 2 units of *A* and 3 units of *B*. The estimated opening balances at the commencement of next year are

Finished goods	—	10,000 units
Raw material *A*	—	12,000 units
Raw material *B*	—	15,000 units

The desirable closing balances at the end of next year:

Finished Product	—	14,000 units
A	—	13,000 units
B	—	16,000 units

Draw up a materials purchase budget. *(University of Madras, B.Com., April 2001)*

Solution:

PRODUCTION BUDGET

Sales units	50,000 units
Less: Opening finished goods	10,000 units
	40,000 units
Add: Closing finished goods	14,000 units
Production units	54,000 units

MATERIAL CONSUMPTION BUDGET

	Material 'A'	*Material* 'B'
Material required per unit of production	2 units	3 units
For production of 54,000 units	54,000 × 2 = 1,08,000 units	54,000 × 3 = 1,62,000 units

MATERIAL PURCHASE BUDGET

	A (Units)	*B* (Units)
Material consumption	1,08,000	1,62,000
Less: Opening stock of raw materials	12,000	15,000
	96,000	1,47,000
Add: Closing stock of materials	13,000	16,000
	1,09,000	1,63,000

Problem 10. A single product company estimated its sales for the next year quarterwise as under.

Quarter	*Sales units*
I	30,000 units
II	37,500 units
III	41,250 units
IV	45,000 units

The opening stock of finished goods is 10,000 units and the company expects to maintain the closing stock of finished goods at 16,250 units at the end of the year. The production pattern in each quarter is based on 80% of the sales of the current quarter and 20% of the sales of the next quarter. The opening stock of raw materials in the beginning of the year is 10,000 kg and the closing stock at the end of the year is required to be maintained at 5,000 kg. Each unit of finished output requires 2 kg of raw materials.

The company proposes to purchase the entire annual requirement of raw materials in the first three quarters in the proportion and that the prices given below:

Quarter	*Purchase of raw material % to total annual requirement in quantity*	*Price per kg (Rs.)*
I	30%	2
II	50%	3
III	20%	4

The value of the opening stock of raw materials in the beginning of the year is Rs. 20,000.

Required: Present the following for the next year, quarterwise.

(*i*) Production budget in units

(*ii*) Raw material consumption budget in quantity.

(*iii*) Raw material purchase budget in quantity and value.

(University of Delhi, B.Com. (Hons.), 2005)

Solution:

Working Note

TOTAL BUDGETED PRODUCTION

	Sale in Quarter	I	30,000
		II	37,500
		III	41,250
		IV	45,000
			1,53,750 units
Add:	Closing stock		16,250
			1,70,000
Less:	Opening stock		10,000
	Budgeted production		1,60,000 units

(1) Production budget in units

	Q_1	Q_2	Q_3	Q_4	*Total*
Budgeted sales units	30,000	37,500	41,250	45,000	
80% of sales of current quarter	24,000	30,000	33,000	36,000	
20% of sales of next quarter	7,500	8,250	9,000	12,250	
	31,500	38,250	42,000	48,250	1,60,000

(2) Raw material consumption budget in quantity

	Q_1	Q_2	Q_3	Q_4	*Total*
Budget production	31,500	38,250	42,000	48,250	1,60,000
Consumption per unit (kg)	2	2	2	2	2
Budgeted consumption (kg)	63,000	76,500	84,000	96,500	3,20,000

(3) Raw material purchase budget in quantity

Budgeted raw material required for production	kg.	3,20,000
Add: Closing stock		5,000
		3,25,000
Less: Opening stock		10,000
Material quantity to be purchased		3,15,000

RAW MATERIAL PURCHASE BUDGET IN VALUE

Quarter	*Qty to be purchased*	*Rate (Rs.)*	*Amount*
I	3,15,000 × 30% = 94,500 kg	2	1,89,000
II	3,15,000 × 50% = 1,57,500 kg	3	4,72,500
III	3,15,000 × 20% = 63,000 kg	4	2,52,000
	Total purchase value		9,13,500

Problem 11. The following are the estimated sales of a company for eight months ending 30.8.98

Months	*Estimated sales (units)*
April 1998	12,000
May 1998	13,000
June 1998	9,000
July 1998	8,000
August 1998	10,000
September 1988	12,000
October 1998	14,000
November 1998	12,000

As a matter of policy, the company maintains the closing balance of finished goods and raw materials as follows:

Stock Item	*Closing balance of a month*
Finished goods	50% of the estimated sales for the next month
Raw materials	Estimated consumption for the next month

Every unit of production requires 2 kg of raw material costing Rs. 5 per kg.

Prepare production budget (in units) and raw materials purchase budget (in units and cost) of the company for the half year ending 30th September 1998. *(ICWA, Inter, June 1999)*

Solution:

PRODUCTION BUDGET

Months	*Sales*		*Closing Balance 50% of the estimated sales for next month*		*Operating Balance*	*Production*
1998						
April	12,000	+	6,500	–	6,000	12,500
May	13,000	+	4,500	–	6,500	11,000
June	9,000	+	4,000	–	4,500	8,500
July	8,000	+	5,000	–	4,000	9,000
August	10,000	+	6,000	–	5,000	11,000
September	12,000	+	7,000	–	6,000	13,000
	64,000					65,000

PURCHASE BUDGET

Consumption 2 kg per unit	*Closing Balance*	*Opening Balance*	*Purchase in kg*	*Rate*	*Amt.*
25,000	22,000	25,000	22,000	5	1,10,000
22,000	17,000	22,000	17,000	5	85,000
17,000	18,000	17,000	18,000	5	90,000
18,000	22,000	18,000	22,000	5	1,10,000
22,000	26,000	22,000	26,000	5	1,30,000
26,000	26,000	26,000	26,000	5	1,30,000
1,30,000					6,55,000

Problem 12. A company is drawing its production plan for the year 1997–98 in respect of two of its products "Gamma" and "Delta". The company's policy is not to carry any closing work-in-progress at the end of any month. However, its policy is to hold a closing stock of finished goods at 50% of the anticipated quantity of sales of the succeeding month. For the year 1997–98 the company's budgeted production is 20,000 units of "Gamma" and 25,000 units of "Delta". The following is the estimated cost data.

	Gamma	*Delta*
	Rs.	*Rs.*
Direct materials per unit	50	80
Direct labour cost per unit	20	30
Other manufacturing expenses apportionable to each type of product based on production	2,00,000	3,75,000

The estimated units to be sold in the first 7 months of the year 1997–98 are as under:

	Apr.	*May*	*June*	*July*	*Aug.*	*Sept.*	*Oct.*
Gamma	900	1,100	1,400	1,800	2,200	2,200	1,800
Delta	2,900	2,900	2,500	2,100	1,700	1,700	1,900

You are required to:

(*i*) Prepare a production budget showing month-wise number of units to be manufactured.

(*ii*) Present a summarised production cost budget for the half year ending 30-9-97.

(ICWA, Inter, Dec. 1996)

Solution:

PRODUCTION BUDGET FOR HALF YEAR
(ending 30th September 1997)

	April	May	June	July	August	Sept.	Total
Product–Gamma							
Budgeted sales	900	1,100	1,400	1,800	2,200	2,200	9,600
Add: Closing stock to be built up (*See note 1*)	550	700	900	1,100	1,100	900	900
	1,450	1,800	2,300	2,900	3,300	3,100	10,500
Less: Carry-over stock -opening (*See note 2*)	450	550	700	900	1,100	1,100	4,50
Budgeted production	1,000	1,250	1,600	2,000	2,200	2,000	10,050
Product–Delta							
Budgeted sales	2,900	2,900	2,500	2,100	1,700	1,700	13,800
Add: Closing stock to be built up	1,450	1,250	1,050	850	850	950	950
	4,350	4,150	3,550	2,950	2,550	2,650	14,750
Less: Carry-over stock -opening (*See note 2*)	1,450	1,450	1,250	1,050	850	850	1,450
Budgeted production	2,900	2,700	2,300	1,900	1,700	1,800	13,300

Note 1:

Closing stock of finished goods at the end of each month is to be ascertained as per company's policy, *i.e.,* 50% of the anticipated quantity of sales of the succeeding month. Closing stock under total column is 50% of October month, *i.e.,* $\frac{50}{100} \times 1{,}800 = 900$

Note 2:

Opening stock of each month is the closing stock of preceeding month. The opening stock under total column is 50% of September closing stock, *i.e.,* $\frac{50}{100} \times 900 = 450.$

Summarised production cost budget for the half-year ending 30th September 1997

Production (units)	*Gamma* 10,050 Cost		*Delta* 13,300 Cost	
	Per unit	Amount	Per unit	Amount
Direct materials	50	5,02,500	80	10,64,000
Direct labour	20	2,01,000	30	3,99,000
Other manufacturing expenses	10	1,00,500	15	1,99,500
	80	8,04,000	125	16,62,500

Note 3:

Other manufacturing expenses are apportioned on the following basis:

$$\frac{\text{Other manufacturing expenses}}{\text{No. of units to be produced in 1997–98}}$$

$$\frac{2{,}00{,}000}{20{,}000} = 10, \quad \frac{3{,}75{,}000}{25{,}000} = 15$$

CHAPTER 13

Problem 13. *(Materials purchase budget):* Draw a material procurement budget (quantitative) from the following information:

Estimated sales of a product 40,000 units. Each unit of the product requires 3 units of material *A* and 5 units of material *B*.

Estimated opening stock at the commencement of the next year:	
Finished product	5,000 units
Material *A*	12,000 units
Material *B*	20,000 units
Materials on order:	
Material *A*	7,000 units
Material *B*	11,000 units
The desirable closing stock at the end of the next year:	
Finished product	7,000 units
Material *A*	15,000 units
Material *B*	25,000 units
Materials on order:	
Material *A*	8,000 units
Material *B*	10,000 units

Solution:

PRODUCTION BUDGET

Estimated sales	40,000 units
Add: Closing stock	7,000 units
	47,000 units
Less: Opening stock	5,000 units
Production during the year	42,000 units

MATERIALS PROCUREMENT BUDGET

		A (units)		*B (units)*
Materials required 3 units of *A* and 5 units of *B* for 42,000 units		1,26,000		2,10,000
Add: Closing stock required	15,000		25,000	
Material on order	8,000	23,000	10,000	35,000
		1,49,000		2,45,000
Less: Opening stock	12,000		20,000	
Material on order	7,000	19,000	11,000	31,000
Units to be procured		1,30,000		2,14,000

Problem 14. The XY Company expects the following sales by month in units for the first six months of next year:

January	–	5,400	April	–	5,700
February	–	5,700	May	–	6,000
March	–	7,500	June	–	4,500

The company has a policy of maintaining an inventory equal to budgeted sales for the following two months. The beginning inventory reflects this policy. Each unit cost Rs. 10.

You are required to prepare purchase budget for as many months as you can both in units and rupees. Explain also why you had to stop where you did. *(University of Kerala, M.Com., May 1992)*

Solution:

PRODUCTION BUDGET FOR 3 MONTHS

	Jan.	Feb.	March
Sales	5,400	5,700	7,500
Add: Closing stock	5,700	6,000	4,500
	11,100	11,700	12,000
Less: Opening stock	7,500	5,700	6,000
Units to be produced	3,600	6,000	6,000

PURCHASE BUDGET IN QUANTITY

	Jan.	Feb.	March
Units of materials required (*see note 1*)	3,600	6,000	6,000
Add: Desired stock at the end of the year	5,700	6,000	4,500
	9,300	12,000	10,500
Less: Expected stock at the beginning	7,500	5,700	6,000
Quantity of materials to be purchased	1,800	6,300	4,500

Note: In the problem the policy of opening stock is expected to be the sale units equal to the following two months. As the opening stock of every month is the closing stock of the previous month, it has been treated for every month in the same manner. The budget could be prepared only for three months because of the policy of the management in considering opening stock.

Regarding the purchase budget as separate unit of materials required for production, it is presumed that for every unit of finished goods to be produced, one unit of raw materials is required.

PURCHASE COST BUDGET

Month	*Qty.*	*Rate*		*Amount*
Jan.	1,800	10		18,000
Feb.	6,300	10		63,000
March	4,500	10		45,000
		Total	Rs.	1,26,000

8. Labour cost budget: This budget represents the direct labour required to meet the production demand of the factory during the budget period. The labour hours required for budgeted production will be determined on the basis of output which must conform standard. It shows the information relating to the number of employees along with the grade required to achieve the output and the probable labour cost for budget period. The advantages of preparing a labour cost budget are as follows:

(*a*) It determines the labour force required.

(*b*) It facilitates the personal department to plan ahead in recruitment and training of labourer.

(*c*) It helps in the preparation of production cost budget and cash budget. A specimen of labour cost budget is shown below:

Labour cost budget for the period........

Grade	*Number of employees*	*Hours employed*	*Rate per hour*	*Amount*
(1) Skilled (2) Semi-skilled (3) Unskilled				
			Total	

Fig. 13.6 *Specimen showing labour cost budget.*

Problem 15. *(Labour cost budget):* The direct labour hour requirements of three of the products manufactured in a factory each involving more than one labour operation are estimated as follows:

Direct labour hours per unit (in minutes)

		roducts		
		1	2	*3*
Operation:	1	18	42	30
	2	–	12	24
	3	9	6	–

The factory works 8 hours per day, 6 days in a week. The budget quarter is taken as 13 weeks and during a quarter, lost hours due to leave and holidays and other causes are estimated to be 124 hours.

The budgeted sales of the products during the quarter are:

Product	1	9,000 units
	2	15,000 units
	3	12,000 units

There is a opening balance of 5,000 units of product 2 and 4,000 units of product 3 and it is proposed to build up a stock at the end of the budget quarter as follows:

Product	1	1,000 units
	3	2,000 units

Prepare a manpower budget for the quarter showing for each operation (*i*) Direct labour hours, (*ii*) Direct labour cost and (*iii*) The number of workers.

Solution:

PRODUCTION BUDGET

	Products		
	1 (units)	*2 (units)*	*3 (units)*
Sales	9,000	15,000	12,000
Add: Closing stock	1,000	–	2,000
	10,000	15,000	14,000
Less: Opening stock	–	5,000	4,000
	10,000	10,000	10,000

Total available hours in a quarter per man

Total hrs. = 8 × 6 × 13	=	624 hrs.
Less: Hours lost	=	124 hrs.
Total available hours per man		500 hrs.

Total time duration and manpower requirement

Products →	1		2		3		Total
Units →	10,000		10,000		10,000		hrs.
Operation	Time in min.	Time in hrs.	Time in min.	Time in hrs.	Time in min.	Time in hrs.	
1	1,80,000	3,000	4,20,000	7,000	3,00,000	5,000	15,000
2	–	–	1,20,000	2,000	2,40,000	4,000	6,000
3	90,000	1,500	60,000	1,000	–	–	2,500

Number of workers required for each operation

Operation	*Total hours*	*No. of workers required*
1	15,000	30
2	6,000	12
3	2,500	5

QUARTERLY MANPOWER BUDGET

Operation	*Hourly rate (Rs.)*	*Product 1*		*Product 2*		*Product 3*		*Total*		*No. of workers*
		Direct labour hours	*Cost (Rs.)*	*Direct labour hours*	*Cost (Rs.)*	*Direct labour hours*	*Cost (Rs.)*	*Direct labour hours*	*Cost (Rs.)*	
1	2.00	3,000	6,000	7,000	14,000	5,000	10,000	15,000	30,000	30
2	2.50	–	–	2,000	5,000	4,000	10,000	6,000	15,000	12
3	3.00	1,500	4,500	1,000	3,000	–	–	2,500	7,500	5
		4,500	10,500	10,000	22,000	9,000	20,000	23,500	52,000	47

CHAPTER 13

9. Factory overhead budget: This budget is prepared by production manager or persons in charge of various production departments. To facilitate control it is necessary to classify the overheads into fixed and variable and controllable and uncontrollable overhead. The budget officer may assist the production manager in supplying the past year's data to prepare this budget more comprehensively.

Problem 16. *(Factory overhead budget):* The following particulars are extracted from the books of Jai Engineering Company. The production overhead apportionment has been prepared to show overheads for each of the production and service departments. Service departments overheads are apportioned as shown below:

Service department	*Production departments* A	B	C
1	30%	20%	50%
2	25%	30%	45%

Overhead	*Basis of apportionment*	*Total*	*A*	*B*	*C*	*1*	*2*
Salaries	Actual	14,700	4,000	3,000	5,000	1,500	1,200
Indirect wages	Actual	31,600	9,000	7,000	10,000	3,000	2,600
Consumable stores	No. of employees	1,650	400	350	500	300	100
Depreciation	Value of asset	22,200	6,000	5,000	10,000	700	500
Insurance	Value of asset	2,480	750	650	900	100	80
Rent	Area occupied	3,790	1,200	1,000	1,500	50	40
Power	Meter	2,170	600	700	800	40	30
Light	Area occupied	395	100	110	140	20	25

Maintenance	Value of asset	2,525	800	700	1,000	15	10
Employee's insurance	No. of workers	3,150	900	800	1,000	250	200
Scrap	Actual	775	225	205	345	–	–
Sundries	Actual	1,105	375	235	455	25	15
	Total	86,540	24,350	19,750	31,640	6,000	4,800
Service department 1		–	1,800	1,200	3,000		
Service department 2		–	1,200	1,440	2,160		
		86,540	27,350	22,390	36,800		

Solution:

FACTORY OVERHEAD BUDGET FOR THE PERIOD

	Total	*A*	*B*	*C*
Overhead				
Variable:				
Consumable stores	1,250	400	350	500
Scrap	775	225	205	345
Fixed:				
Salaries	12,000	4,000	3,000	5,000
Depreciation	21,000	6,000	5,000	10,000
Insurance	2,300	750	650	900
Rent	3,700	1,200	1,000	1,500
Sundries	1,065	375	235	455
Service department 1	6,000	1,800	1,200	3,000
Service department 2	4,800	1,200	1,440	2,160
Semi-variable:				
Indirect wages	26,000	9,000	7,000	10,000
Power	2,100	600	700	800
Light	350	100	110	140
Maintenance	2,500	800	700	1,000
Insurance	2,700	900	800	1,000
	86,540	27,350	22,390	36,800

10. Administration cost budget: This budget is prepared by chief accountant and approved by the managing director. This budget is likely to pose some problems since most of the administration cost is fixed in nature. The main budget is divided into separate budget covering separate administrative activity such as accounting, secretarial work and so on. The factors to be considered in preparing this budget are (*a*) the existing number of staff, (*b*) their rates of pay, (*c*) change in volume of work and (*d*) legal and other administrative overheads. A specimen of administrative cost budget is shown below:

Specimen of administration cost budget.

For the period ....

Items of expenses	Previous year's budget	Previous year's actual	Current budget
Materials:			
Stationery			
Supplies			
Salaries & wages:			
Manager's salary			
Director's salary			
Clerical salary			
Expenses:			
Rent & rates			
Telephone			
Postage			
Travelling			
Insurance			
Bank charges			
Audit fee			
Subscriptions			
Total			

11. Research and development budget: The preparation of this budget is governed by the policy of the Board of Directors and head of production departments. This budget may be prepared by elements of cost or by specific project. Similarly this budget can be prepared from the point of view of short-range or long-range activities. While long-range research programme is carried out in tune with future market trends and demands, short-range programme are carried out to achieve a satisfactory margin of profit. The research and development budget may relate to (*a*) Basic research, (*b*) New product development, (c) Product improvement, (*d*) Cost and capacity improvement and (*e*) Safety, health and welfare. A specimen of research and development budget is shown below:

Research and development budget for the period ....

	Projects				
	1	*2*	*3*	*4*	*Total*
Projects costs:					
Materials:					
Raw materials					
Small tools					
Oil & lubricants					
Sundries					
Wages & salaries:					
Chemists					
Technicians					
Others					

Expenses:				
Professional fee				
Total				
General costs:				
Supervisory salary				
Laboratory cost				
Technical publications				
Office costs				
Subscriptions				
			Total	

12. Capital expenditure budget: This budget shows the estimated expenditure on fixed assets during the period concerned. This budget is subject to strict management control as it involves heavy amount of investment requiring the approval of top management. This budget is prepared based on the requirements of various assets for the following departments:

(*i*) Production department—For the purchase of machineries and plant either for replacing the existing machinery or new machinery for producing new products.

(*ii*) Transport department—For the purchase of new vehicles.

(*iii*) Office department—For the purchase of new office equipments.

(*iv*) Service departments—For purchasing equipments peculiar to such departments.

(*v*) Equipments to control pollution such as liquid, gaseous and noise pollution.

The capital expenditure budget offers the following advantages:

(*i*) It estimates the capital expenditure requirements and accordingly provides or arranges for it.

(*ii*) The priority of procuring assets can be determined. Those assets which are very important and unavoidable is given first preference and others are postponed to a later period.

(*iii*) It serves as a tool of controlling capital expenditure.

13. Cash budget: This budget is prepared to know the estimated cash balance each month along with the estimated receipts and payments during the month. It performs two important functions, viz., (*i*) to ensure sufficient availability of cash and (*ii*) to take necessary action when there is shortage of cash. If there is any surplus, which is not immediately required, the same may be invested for a short period. According to Matz-Curry-Frank, a cash budget serves the following purposes:

(*i*) It indicates the effect on the cash position of seasonal requirements, large inventories, unusual receipts and slowness in collecting bills receiveable.

(*ii*) It indicates cash requirements for a plant or equipments expansion programme.

(*iii*) It points to the need for additional funds from external sources such as bank loans, issue of securities.

(*iv*) It indicates the availability of cash for taking advantage of discounts offered.

(*v*) It helps in planning redemption of preference shares or redeemable debentures, payments of pension, etc.

(*vi*) It shows the availability of excess funds for short- or long-term investments.

A cash budget can be prepared under any of the following three methods.

(i) **The receipts and payments method:** Under this method all the cash receipts and payments expected during the budget period is considered. However, care must be taken to ensure that cash adjustments and accruals are not shown in the cash budget. The cash transactions for preparing this budget is obtained from various functional budgets discussed above. For example, materials cost from material budget, labour cost from labour budget, overheads from overhead budget, sales revenue from sales budget and so on.

Problem 18. *(Cash budgets):* Murugan Company is expected to have Rs. 25,000 in its bank account on 1-4-1996. Prepare a cash budget for April, May, June 1996 from the following estimates.

Months	*Sales*	*Purchases*	*Salary*	*Adm. expense*	*Selling expense*
February	50,000	30,000	6,000	9,000	3,000
March	56,000	32,000	6,500	9,500	3,000
April	60,000	35,000	7,000	10,000	3,500
May	80,000	40,000	9,000	11,500	4,500
June	90,000	40,000	9,500	12,500	4,500

Other information:

(a) 20% sales on cash. Balance on credit and the amount to be collected in the next month.

(b) Suppliers are paid second month following the purchases.

(c) Workers salary paid in the same month.

(d) Administrative and selling expenses are paid in the next month.

(e) Dividend of Rs. 10,000 and Bonus to workers of Rs. 15,000 are to be paid in May.

(f) Income tax of Rs. 25,000 to be paid in June. *(University of Madras, B.Com., October 2003)*

Solution:

Working Notes

(1) Calculation of sales:

20% of sales is for cash and balance 80% is collected next month, *i.e.*, period of credit allowed to customers is one month

(a) Sales collection of April

$$= 80\% \text{ of March} = 56{,}000 \times \frac{80}{100} = 44{,}800$$

$$= 20\% \text{ of April} = 60{,}000 \times \frac{20}{100} = 12{,}000$$

56,800

(b) Sales collection of May

$$= 80\% \text{ of April} = 60{,}000 \times \frac{80}{100} = 48{,}000$$

$$= 20\% \text{ of May} = 80{,}000 \times \frac{20}{100} = 16{,}000$$

64,000

(c) Sales collection of June

$$= 80\% \text{ of May} = 80{,}000 \times \frac{80}{100} = 64{,}000$$

CHAPTER 13

= 20% of June	= $90{,}000 \times \frac{20}{100}$	=	18,000
			82,000

(2) Calculation of purchases:

Period of credit allowed by suppliers is second month following the purchases. That is two months credit

(*a*) Payment in April = Purchase of February for Rs. 30,000 is given in April

(*b*) Payment in May = Purchase of March for Rs. 32,000 is given in May

(*c*) Payment in June = Purchase of April for Rs. 35,000 is given in June

(3) All other expenses

Paid one month delay. Therefore Administrative expenses, selling expenses are paid one month delay, *i.e.,* March paid in April, April paid in May, May paid in June.

(4) Closing balance of April is opening balance of May and so on.

CASH BUDGET

		April	*May*	*June*
Cash on hand		25,000	32,300	16,800
Cash receipts:				
Sales collection	56,800	64,000	82,000	
	Total	81,800	96,300	98,800
Cash payment:				
Purchases		30,000	32,000	35,000
Salary		7,000	9,000	9,500
Administrative expenses		9,500	10,000	11,500
Selling expenses		3,000	3,500	4,500
Dividend		—	10,000	—
Bonus		—	15,000	—
Income tax		—	—	25,000
	Total	49,500	79,500	85,500
Closing balance		32,300	16,800	13,300

Problem 19. Bajaj Company, wishes to arrange overdraft facilities with its Bankers during the period from April to June 1996 when it will be manufacturing mostly for stock. Prepare a cash budget for the above period from the following data, indicating the extent of the bank overdraft facility the company will require at the end of each month.

(*a*)

1996	Sales	Purchases	Wages
February	90,000	62,400	6,000
March	96,000	72,000	7,000
April	54,000	1,21,500	5,500
May	87,000	1,23,000	5,000
June	63,000	1,34,000	7,500

(b) 50% of credit sales are realised in the month following the sales and the remaining 50% in the second month following: Creditors are paid in the month following the month of purchase and lag in payment of wages—1 month.

(c) Cash at bank on 1-4-1996 is estimated at Rs. 12,500.

(University of Madras, B.Com., April 2003)

Solution:

Working Notes

Debtors collection

April	=	50% of February + 50% of March	
	=	45,000 + 48,000	= 93,000
May	=	50% of March + 50% of April	
	=	48,000 + 27,000	= 75,000
June	=	50% of April + 50% of May	
	=	27,000 + 43,500	= 70,500

CASH BUDGET

		April	*May*	*June*
Cash at bank on 1-4-1996		12,500	26,500	(25,500)
Receipts				
Debtors collection		93,000	75,000	70,500
	Total	1,05,500	1,01,500	45,000
Payments				
Creditors payment		72,000	1,21,500	1,23,000
Wages		7,000	5,500	5,000
	Total	79,000	1,27,000	1,28,000
Closing balance (Bank OD)		26,500	(25,500)	(83,000)

Problem 20. From the following information, prepare a cash budget for the quarter ending 30-6-2000.

Months	*Sales* Rs.	*Purchases* Rs.	*Wages* Rs.	*Miscellaneous expenses.* Rs.
February	1,20,000	84,000	10,000	7,000
March	1,30,000	1,00,000	12,000	8,000
April	80,000	1,04,000	8,000	6,000
May	1,16,000	1,06,000	10,000	12,000
June	88,000	80,000	8,000	6,000

Additional information:

(1) Cash on hand on 1-4-2000 Rs. 5,000.

(2) Sales—20% realised in the month of sale. Discount allowed 2%, balance realised after two months?

(3) Purchases are paid one month after.

(4) Wages—25% in arrears paid in the following month.

(5) Other expenses are paid at a lag of one month.

(6) Income tax Rs. 25,000 due on or before 30-6-2000.

(Calcutta University, B.Com. (Pass), 2000)

CHAPTER 13

Solution:

Working Notes

(1) Cash sales

April $80,000 \times \frac{20}{100} \times \frac{98}{100} = 15,680$

May $1,16,000 \times \frac{20}{100} \times \frac{98}{100} = 22,736$

June $88,000 \times \frac{20}{100} \times \frac{98}{100} = 17,248$

(2) Collection from debtors : from sales of two months ago

April $1,20,000 \times \frac{80}{100} = 96,000$

May $1,30,000 \times \frac{80}{100} = 1,04,000$

June $80,000 \times \frac{80}{100} = 64,000$

CASH BUDGET

	April Rs.	*May* Rs.	*June* Rs.
Cash receipts:			
Opening cash balance	5,000	(–) 320	6,916
Cash sales (net of discount)	15,680	22,736	17,248
Collection from debtors	96,000	1,04,000	64,000
Total receipts	1,16,680	1,26,416	88,164
Cash payments:			
Payment to creditors	1,00,000	1,04,000	1,06,000
Wages of last month (25%)	3,000	2,000	2,500
Wages of current month	6,000	7,500	6.000
Other expenses (75%)	8,000	6,000	12,000
Income tax (assumed paid)	—	—	25,000
Total payments	1,17,000	1,19,500	1,51,500
Closing balance of cash	(–) 320	6,916	(– 1) 63,336

Problem 21. From the following information, compute collections from debtors for the months of June, July and August.

	April	*May*	*June*	*July*	*August*
Sales (Rs.)	1,00,000	1,20,000	90,000	1,05,000	1,25,000

20% of sales are made in cash. Debtors are allowed 2 months credit but will receive 5% cash discount if they pay off their dues within the month next to the month of sale. $\frac{4}{5}$th of the debtors normally clear their dues to avail the cash discount. Remaining debtors pay on the due date.

(Calcutta University, B.Com. (Hons.), 2004)

Solution:

STATEMENT SHOWING CALCULATION OF COLLECTION FROM DEBTORS

		June	*July*	*August*
I	From credit sales of one month ago	72,960	54,720	63,840
II	From credit sales of two months ago	16,000	19,200	14,400
	Collection from debtors	88,960	73,920	78,240

Working Notes

(*i*) Last months sale $\times \frac{80}{100} \times \frac{4}{5} \times \frac{95}{100}$

June $= 1,20,000 \times \frac{80}{100} \times \frac{4}{5} \times \frac{95}{100} = 72,960$

July $= 90,000 \times \frac{80}{100} \times \frac{4}{5} \times \frac{95}{100} = 54,720$

Aug. $= 1,05,000 \times \frac{80}{100} \times \frac{4}{5} \times \frac{95}{100} = 63,840$

(*ii*) Sales of two months ago $\times \frac{80}{100} \times \frac{1}{5}$

June $= 1,00,000 \times \frac{80}{100} \times \frac{1}{5} = 16,000$

July $= 1,20,000 \times \frac{80}{100} \times \frac{1}{5} = 19,200$

Aug. $= 90,000 \times \frac{80}{100} \times \frac{1}{5} = 14,400$

Problem 22. Prepare a cash budget for three months ending on 30th September 2000 from the following information.

(A)

Month	*Sales*	*Materials*	*Wages*	*Overhead*
May	56,000	19,200	6,000	3,400
June	60,000	18,000	6,000	3,800
July	64,000	18,400	6,400	4,000
Aug.	68,000	20,000	7,200	4,400
Sept.	72,000	20,800	8,000	4,600

(B) Credit terms are:

Sales/debtors – 10% sales are on cash. 50% of the credit sales are collected in the next month and the balance in the following month. Creditors – Raw material 1 month, wages $\frac{1}{2}$ month, overhead $\frac{1}{4}$ month.

(C) Other relevant information:

(*i*) Rent from a godown let out of 1-1-1999 at yearly contract of Rs. 2,25,000 with a condition of 12% increase in each year is receivable in equal instalment in each month.

(*ii*) Interest @ 15% p.a. on an amount of Rs. 2,00,000 borrowed on 1-1-1998 is to be paid in each month starting from 1-3-2000. Interest for the period from 1-1-1998 to 29-2-2000 is to be paid in 10 equal monthly instalments starting from 1-7-2000. Interest

CHAPTER 13

for the month of April 2000 remaining unpaid. This unpaid amount is to be paid in the month of September 2000 with a penal interest of Rs. 100.

(*iii*) An advance of Rs. 50,000 will be paid to a contractor for the construction of a building in September 2000.

(*iv*) Adequate overdraft facilities may be arranged for, if needed.

(Calcutta University, B.Com. (Hons.), 2001)

Solution:

Working Notes:

(1) Interest payable from 1-1-98 to 29-2-2000

$$= 2,00,000 \times \frac{15}{100} \times \frac{26}{12} = 65,000$$

Arrear interest to be paid in monthly instalments $= \frac{65,000}{10} = 6,500$

Interest for current period

$$= 2,00,000 \times \frac{15}{100} \times \frac{1}{12} = 2,500$$

Interest to be paid in September

$$= 6,500 + 2,500 + (2,500 + 100)$$

$$= 11,600$$

(2) Rent receivable:

$$= 2,25,000 \times \frac{112}{110} \times \frac{1}{12} = \text{Rs. } 21,000$$

(3) Payment of wages:

$$\text{July} = \frac{1}{2} \text{ of June} + \frac{1}{2} \text{ of July}$$

$$= 3,000 + 3,200 = 6,200$$

$$\text{Aug.} = \frac{1}{2} \text{ of July} + \frac{1}{2} \text{ of August}$$

$$= 3,200 + 3,600 = 6,800$$

$$\text{Sep.} = \frac{1}{2} \text{ of Aug.} + \frac{1}{2} \text{ of Sept.}$$

$$= 3,600 + 4,000 = 7,600$$

(4) Payment of overhead:

$$\text{July} = \frac{1}{4} \text{ of June} + \frac{3}{4} \text{ of July}$$

$$= 9,50 + 3,000 = 3,950$$

$$\text{Aug.} = \frac{1}{4} \text{ of July} + \frac{3}{4} \text{ of August}$$

$$= 1,000 + 3,300 = 4,300$$

$$\text{Sep.} = \frac{1}{4} \text{ of Aug.} + \frac{3}{4} \text{ of Sept.}$$

$$= 1,100 + 3,450 = 4,550$$

(5) Cash sales

$$\text{July} = 64{,}000 \times \frac{10}{100} = 6{,}400$$

$$\text{Aug.} = 68{,}000 \times \frac{10}{100} = 6{,}800$$

$$\text{Sept.} = 72{,}000 \times \frac{10}{100} = 7{,}200$$

(6) Collection from debtors

$$\text{July} = \frac{1}{2} \text{ of credit sales of May} + \frac{1}{2} \text{ of credit sales of June}$$

$$= \left(56{,}000 \times \frac{90}{100} \times \frac{1}{2}\right) + \left(60{,}000 \times \frac{90}{100} \times \frac{1}{2}\right)$$

$$= 52{,}200$$

$$\text{Aug.} = \frac{1}{2} \text{ of credit sales of June} + \frac{1}{2} \text{ of credit sales of July}$$

$$= \left(60{,}000 \times \frac{90}{100} \times \frac{1}{2}\right) + \left(64{,}000 \times \frac{90}{100} \times \frac{1}{2}\right)$$

$$= 55{,}800$$

$$\text{Sept.} = \frac{1}{2} \text{ of credit sales of July} + \frac{1}{2} \text{ of credit sales of August}$$

$$= \left(64{,}000 \times \frac{90}{100} \times \frac{1}{2}\right) + \left(68{,}000 \times \frac{90}{100} \times \frac{1}{2}\right)$$

$$= 59{,}400$$

CASH BUDGET FOR THE THREE MONTHS

(ending on 30-9-2000)

	July	*Aug.*	*Sept.*
	Rs.	*Rs.*	*Rs.*
Receipts:			
Opening cash balance	NIL	42,450	47,550
Cash received from cash sales	6,400	6,800	7,200
Cash collection from debtors	52,200	55,800	59,400
Rent receivable	21,000	21,000	21,000
Total receipts	79,600	1,26,050	1,75,150
Payments:			
Payment of wages	6,200	6,800	7,600
Payment of overhead	3,950	4,300	4,550
Payment to creditors	18,000	18,400	20,000
Payment of interest on loan	9,000	9,000	11,600
Payment to contractor (advance)	—	—	50,000
Total payments	37,150	38,500	93,750
Closing balance	42,450	87,550	81,400

Problem 23. *(Cash budget under receipts and payments method):* The following particulars are extracted from the books of Ajanta Company Ltd. for the quarter ending 31.3.2010:

Opening cash balance as on 1.1.2010 Rs. 30,000

Sales budget were as follows:

	Rs.
November 2009	80,000
December 2009	90,000
January 2010	75,000
February 2010	75,000
March 2010	80,000

Analysis of records shows that debtors settle according to the following pattern:

60% within the month of sale.

25% the month following.

15% the month following.

Extracts from the purchase budget were as follows:

	Rs.
December 2009	60,000
January 2010	55,000
February 2010	45,000
March 2010	55,000

All purchase are on credit and past experience show that 90% are settled in the month of purchase and the balance settled the month after.

Wages are Rs. 15,000 per month and overheads Rs. 20,000 per month (including Rs. 5,000 depreciation) are settled monthly.

Taxation of Rs. 8,000 has to be settled in February and the company will receive settlement of an insurance claim of Rs. 25,000 in March.

Prepare a cash budget for January, February and March 2010.

Solution:

CALCULATION OF SALES REVENUE

	January *Rs.*
November 2009 (15% of 80,000)	12,000
December 2009 (25% of 90,000)	22,500
January 2010 (60% of 75,000)	45,000
	79,500
	February
December 2009 (15% of 90,000)	13,500
January 2010 (25% of 75,000)	18,750
February 2010 (60% of 75,000)	45,000
	77,250
	March *Rs.*
January 2010 (15% of 75,000)	11,250
February 2010 (25% of 75,000)	18,750
March 2010 (60% of 80,000)	48,000
	78,000

PAYMENTS FOR PURCHASES

	January Rs.
December 2009 (10% of 60,000)	6,000
January 2010 (90% of 55,000)	49,500
	55,000

	February Rs.
January 2010 (10% of 55,000)	5,500
February 2010 (90% of 45,000)	40,500
	46,000

	March
February 2010 (10% of 45,000)	4,500
March 2010 (90% of 55,000)	49,500
	54,000

CASH BUDGET FOR THE QUARTER ENDING MARCH 2010

	January	*Feb.*	*March*
Opening balance	30,000	24,000	17,250
Receipts from sales	79,500	77,250	78,000
Insurance claim			25,000
Total cash available	1,09,500	1,01,250	1,20,250
Purchases	55,500	46,000	54,000
Wages	15,000	15,000	15,000
Overhead (less depreciation)	15,000	15,000	15,000
Taxation		8,000	
Total payment	85,500	84,000	84,000
Closing balance c/o	24,000	17,250	36,250

Problem 24. *(Cash budget)*: The Swadeshi Manufacturing Company has a cash balance of Rs. 27,000 at the beginning of March 2010, you are required to prepare a cash budget for March, April and May 2010 having regard to the following information:

Creditors give 1 month credit.

Salaries are paid in the current month.

Fixed costs are paid one month in arrears and include a charge for depreciation of Rs. 5,000 per month.

Credit sales are settled as follows:

40% in month of sale, 45% in next month and 12% in the following month. The balance represents bad debts.

Months	*Cash sales (Rs.)*	*Credit sales (Rs.)*	*Purchases (Rs.)*	*Salaries (Rs.)*	*Fixed overhead (Rs.)*
January	—	74,000	55,200	9,000	30,000
February	—	82,000	61,200	9,000	30,000
March	20,000	80,000	60,000	9,500	30,000
April	22,000	90,000	69,000	9,500	32,000
May	25,000	1,00,000	75,000	10,000	32,000

Solution:

STATEMENT SHOWING RECEIPTS FROM DEBTORS

	March		*April*		*May*
40% of March sales	32,000	40% of April sales	36,000	40% of May sales	40,000
45% of February sales	36,900	45% of March sales	36,000	45% of April sales	40,500
12% of January sales	8,880	12% of February sales	9,840	12% of March sales	9,600
	77,780		81,840		90,100

CASH BUDGET FOR THE MONTHS OF MARCH, APRIL & MAY 2010

	March	*April*	*May*
Opening balance	27,000	29,080	38,420
Receipts from debtors	77,780	81,840	90,100
Cash sales	20,000	22,000	25,000
Total cash available	1,24,780	1,32,920	1,53,520
Salaries	9,500	9,500	10,000
Fixed overheads	25,000	25,000	27,000
Purchases	61,200	60,000	69,000
Total disbursement	95,700	94,500	1,06,000
Closing balance c/o	29,080	38,420	47,520

Problem 25. *(Cash budget):* From the following budgeted figures prepare a cash budget in respect of the three months to June 30, 2010:

Months	*Sales*	*Materials*	*Wages*	*Overhead*
January	30,000	20,000	5,500	3,100
February	28,000	24,000	5,800	3,300
March	32,000	25,000	6,000	3,400
April	40,000	28,000	6,200	3,600
May	42,000	31,000	6,500	4,300
June	38,000	25,000	7,000	4,000

Estimated cash balance on April 1st 2010 Rs. 10,000.

Materials and overheads are paid during the month following the month of supply. Wages are paid during the month in which they are earned.

Credit items of sale are payment by the end of the month following the month of sale. It is estimated that one-half of sales are paid when due, the other half being paid during the next month.

A sales commission of 5% on sales is to be paid within the month following actual sales.

Preference share dividend of 10% on capital of Rs. 3,00,000 is to be paid on May 1, 2010.

Plant and machinery to be installed in May at a cost of Rs. 10,000 will be payable on 1st June, 2010.

10% calls on equity shares capital of Rs. 2,50,000 are due on April 1st and June 1st 2010.

Solution:

STATEMENT SHOWING SALES

	April		*May*		*June*
1/2 of March	16,000	1/2 of April	20,000	1/2 of May	21,000
1/2 of February	14,000	1/2 of March	16,000	1/2 of April	20,000
	30,000		36,000		41,000

CASH BUDGET FOR PERIOD ENDING 30TH JUNE, 2010

	April	*May*	*June*
Opening balance	10,000	28,800	(5,300)
Sales	30,000	36,000	41,000
Capital	25,000	–	25,000
Total receipts	65,000	64,800	60,700
Materials	25,000	28,000	31,000
Wages	6,200	6,500	7,000
Overheads	3,400	3,600	4,300
Sales commission	1,600	2,000	2,100
Pref. share dividend	—	30,000	—
Plant & machinery	—	—	10,000
Total payment	Rs. 36,200	70,100	54,400
Balance c/o	Rs. 28,800	Rs. (5,300)	Rs. 6,300

Problem 26. From the following data prepare a cash budget for the 3 months commencing 1st June, 1996, when the bank balance was Rs. 1,00,000.

Months	*Sales*	*Purchases*	*Wages*	*Production expenses*	*Adm. expenses*
	(Rs.)	*(Rs.)*	*(Rs.)*	*(Rs.)*	*(Rs.)*
April	80,000	41,000	5,600	3,900	10,000
May	76,500	40,500	5,400	4,200	14,000
June	78,500	38,500	5,400	5,100	15,000
July	90,000	37,000	4,800	5,100	17,000
August	95,000	35,000	4,700	6,000	13,000

There is two month credit period allowed to customers and received from suppliers. Wages, production expenses and administration expenses are payable in the following month.

(University of Madras, B.Com., September 1997)

Solution:

CASH BUDGET FROM JUNE TO AUGUST 1996

	June	*July*	*August*
Opening balance	1,00,000	1,15,400	1,25,900
Add: Receipts:			
Collection from debtors	80,000	76,500	78,500
	1,80,000	1,91,900	2,04,400

Less: Payment:			
paid to creditors	41,000	40,500	38,500
Wages	5,400	5,400	4,800
Production expenses	4,200	5,100	5,100
Administration expenses	14,000	15,000	17,000
	64,600	66,000	65,400
Closing balance	1,15,400	1,25,900	1,39,000

Note 1:

Delay in payment allowed to customers from suppliers is 2 months *i.e.,* April payment paid in June.

Note 2:

Delay in payment allowed to wages, production expenses, administration expenses are one month *i.e.,* May payment paid in June.

Problem 27. Infotech Ltd., commences business on 1st April, 2000 and deposits Rs. 1,00,000 in the Global Trust Bank. The sum deposited would not be sufficient to finance its operations over a period of four months. As a company secretary, you are asked to prepare a cash budget from 1st April, 2000 to 31st July, 2000 to ascertain the monthly overhead limits to seek from the company's bankers.

Requisite data is as under:

(i) Sales are made to one distributor only in 30 days terms 2% discount and cheques are received on the first date of the following due date.

(ii) Furniture purchases for Rs. 10,000 preferred to be made in April, 2000

(iii) Budget figures are:

	April	*May*	*June*	*July*
Purchases	50,000	40,000	30,000	40,000
Wages	40,000	50,000	40,000	40,000
Cash expenses	4,000	5,000	4,000	4,000
Sales	60,000	70,000	80,000	80,000

All purchases are made on net 30 days terms and cheques are posted to creditors on the last day of the month due. *(CS Inter, June 2000)*

Solution:

CASH BUDGET FOR APRIL TO JULY

		April	*May*	*June*	*July*
(A)	Cash receipts:				
	Cash receipts from distributor (Sales – 2% discount with one month time lag)	–	58,800	68,600	78,400
(B)	Cash payments:				
	Payment to creditors (one month time lag)	–	50,000	40,000	30,000
	Wages	40,000	50,000	40,000	40,000
	Cash expenses	4,000	5,000	4,000	4,000
	Furniture purchases	10,000	–	–	–
	Total cash payments	54,000	1,05,000	84,000	74,000

CALCULATION OF COLLECTION FROM DEBTORS (CREDIT SALES)

(C)	Net cash receipts/ Deficit (*A* – *B*)	(54,000)	(46,200)	(15,400)	4,400
	Balance (overdraft) at the start of the month (cumulative)	1,00,000	46,000	(200)	(15,600)
	(Overdraft) required monthwise	–	(200)	(15,400)	–

Problem 28. Prepare a cash budget for the three months ended 30th September 1998 based on the following information:

	Rs.
Cash at bank on 1st July 1998	25,000
Monthly salaries and wages (estimated)	10,000
Interest payable in August 1998	5,000

Estimated	*June* Rs.	*July* Rs.	*August* Rs.	*September* Rs.
Cash sales (actual)	1,20,000	1,40,000	1,52,000	1,21,000
Credit sales	1,00,000	80,000	1,40,000	1,20,000
Purchases	1,60,000	1,70,000	2,40,000	1,80,000
Other expenses	18,000	20,000	22,000	21,000

Credit sales are collected 50% in the month of sale and 50% in the month following. Collections from credit sales are subject to 10% discount if received in the month of sale and to 5% if received in the month following:

10% of the purchases are in cash and balance is paid in next month. *(CS, Inter, June 1999)*

Solution:

CALCULATION OF COLLECTION FROM DEBTORS (CREDIT SALES)

		July	*August*	*September*
July	50% of 1,00,000	50,000		
	50% of 80,000	40,000		
Aug.	50% of 80,000		40,000	
	50% of 1,40,000		70,000	
Sept.	50% of 1,40,000			70,000
	50% of 1,20,000			60,000
	Total collections	90,000	1,10,000	1,30,000
Less: Discount allowed				
(*a*)	5% on 50,000 + 10% on 40,000	6,500		
(*b*)	5% on 40,000 + 10% on 70,000		9,000	
(c)	5% on 70,000 + 10% on 60,000			9,500
	Net collections after discount	83,500	1,01,000	1,20,500

CALCULATION OF PAYMENTS TO CREDITORS

(*a*)	90% of 1,60,000 + 10% of 1,70,000	1,61,000		
(*b*)	90% of 1,70,000 + 10% of 2,40,000		1,77,000	
(c)	90% of 2,40,000 + 10% of 1,80,000			2,34,000
		1,61,000	1,77,000	2,34,000

CHAPTER 13

CASH BUDGET FOR THREE MONTHS JULY TO SEPTEMBER

		July	*Aug.*	*Sept.*
	Opening balance	25,000	57,500	96,500
Add:	Receipts;			
	Sales: Cash	1,40,000	1,52,000	1,21,000
	Credit	83,500	1,01,000	1,20,500
	Total cash	2,48,500	3,10,500	3,38,000
Payments:				
	Purchases	1,61,000	1,77,000	2,34,000
	Other expenses	20,000	22,000	21,000
	Interest	–	5,000	–
	Salaries & wages	10,000	10,000	10,000
	Total payment	1,91,000	2,14,000	2,65,000
	Closing cash balance	57,500	96,500	73,000

Problem 29. On 30th September, 1996, the balance sheet of M. Ltd. (retailer) was as under:

Equity shares of		Equipment	20,000	
Rs. 10 each fully paid	20,000	*Less:* Dep.	5,000	15,000
Reserves	10,000	Stock		20,000
Trade creditors	40,000	Trade debtors		15,000
Proposed dividend	15,000	Bank		35,000
	85,000			85,000

The company is developing a system of forward planning and on 1st October 1996 it supplies the following information:

	Sales		*Purchases*
	Credit	*Cash*	*Credit*
	Rs.	*Rs.*	*Rs.*
September 1996 (actual)	15,000	14,000	40,000
October 1996 (budget)	18,000	5,000	23,000
November 1996 (budget)	20,000	6,000	27,000
December 1996 (budget)	25,000	8,000	26,000

All trade debtors are allowed one months credit and are expected to settle promptly. All trade creditors are paid in the month following delivery:

On 1st October, 1996, all equipment were replaced at a cost of Rs. 30,000, Rs. 14,000 was allowed in exchange for the old equipment and a net payment of Rs. 16,000 was made.

The proposed dividend will be paid in December 1996.

The following expenses will be paid:

Wages Rs. 3,000 per month.

Administration Rs. 1,500 per month.

Rent Rs. 3,600 for the year upto 30th September 1997 (to be paid in October 1996).

You are required to prepare a cash budget for the months of October, November and December 1996. *(CS Inter, June 1997)*

Solution:

CASH BUDGET FOR THREE MONTHS ENDING 1996

		October	November	December
Opening balance of bank/Bank overdraft		35,000	(9,100)	(12,600)
Add:	Cash inflows:			
	Sales Cash sales of current month	5,000	6,000	8,000
	Credit sales of previous month	15,000	18,000	2,000
	Total receipts	55,000	14,900	15,400
Less:	Cash outflow			
	Credit purchases of previous month	40,000	23,000	27,000
	Equipment	16,000	–	–
	Wages	3,000	3,000	3,000
	Administration	1,500	11,500	1,500
	Rent	3,600	–	–
	Dividend	–	–	15,000
	Total payment	64,100	27,500	46,500
Closing balance/overdraft (A – B)		(9,100)	(12,600)	(31,100)

Problem 30. Prepare cash budget of fashion fabrics for the months April 1999 to July (four months) from the details given below:

(*i*) Estimated sales:

	Rs.
February 1999	12,00,000
March 1999	12,00,000
April 1999	16,00,000
May 1999	20,00,000
June 1999	18,00,000
July 1999	16,00,000
August 1999	14,00,000

(*ii*) On an average 20% of sales are cash sales. The credit sales are realised in the third month (January sales in March).

(*iii*) Purchases amount to 60% of sales. Purchases made in a month are generally sold in the third month and payment for purchases is also made in the third month.

(*iv*) Variable expenses (other than sales commission) constitute 10% of sales and there is a time lag of half a month in these payment.

(*v*) Commission on sales is paid at 5% of sales value and payment is made in the third month.

(*vi*) Fixed expenses per month amount to Rs. 75,000 approximately.

(*vii*) Other items anticipated:

	Due
Interest payable on deposits	Rs. 1,60,000 (April 99)
Sales of old assets	Rs. 1,25,000 (May 99)
Payment of Tax	Rs. 80,000 (June 99)
Purchase of fixed assets	Rs. 6,50,000

(*viii*) Opening cash balance Rs. 1,50,000

(ICWA, Inter, December 1999)

CHAPTER 13

Solution:

CASH BUDGET FOR APRIL TO JULY 1999

	April	*May*	*June*	*July*
Opening balance	1,50,000	35,000	5,000	1,40,000
Add: Receipts:				
Cash sales	3,20,000	4,00,000	3,60,000	3,20,000
Collection from debtors	9,60,000	9,60,000	12,80,000	16,00,000
Sales of old assets	–	1,25,000	–	–
Total	14,30,000	15,20,000	16,40,000	19,20,000
Payments:				
Payments for purchases	9,60,000	12,00,000	10,80,000	9,60,000
Variable expenses	1,40,000	1,80,000	1,90,000	1,70,000
Commission on sales	60,000	60,000	80,000	1,00,000
Fixed expenses	75,000	75,000	75,000	75,000
Interest on deposits	1,60,000	–	–	–
Tax	–	–	80,000	–
Purchase of fixed asset	–	–	–	6,50,000
Total	13,95,000	15,15,000	15,05,000	19,55,000
Closing balance	35,000	5,000	1,40,000	1,05,000

Working Notes:

(1) *Calculation of sales*

April

Cash sales 20% of April sales

$\frac{20}{100} \times 16,00,000 = 3,20,000$

Credit sales 80% of February sales

$\frac{80}{100} \times 12,00,000 = 9,60,000$

May

Cash sales 20% of May sales

$\frac{20}{100} \times 20,00,000 = 4,00,000$

Credit sales $\frac{80}{100}$ of March sales

$\frac{80}{100} \times 12,00,000 = 9,60,000$

June

Cash sales 20% of June sales

$\frac{20}{100} \times 18,00,000 = 3,60,000$

Credit sales 80% of April sales

$\frac{80}{100} \times 16,00,000 = 12,80,000$

July

Cash sales 20% of July sales

$\frac{20}{100} \times 16,00,000$ = 3,20,000

Credit sales 80% of May sales

$\frac{80}{100} \times 20,00,000$ = 16,00,000

(2) *Calculation of Purchases*

April	$16,00,000 \times \frac{60}{100}$	= 9,60,000
May	$20,00,000 \times \frac{60}{100}$	= 12,00,000
June	$18,00,000 \times \frac{60}{100}$	= 10,80,000
July	$16,00,000 \times \frac{60}{100}$	= 9,60,000

(3) *Calculation of variable expenses*

April

March time lag	$\frac{10}{100} \times 12,00,000 \times \frac{1}{2}$	=	60,000
April sales	$\frac{10}{100} \times 16,00,000 \times \frac{1}{2}$	=	80,000
		1,40,000	

May

April sales	$\frac{10}{100} \times 16,00,000 \times \frac{1}{2}$	=	80,000
May sales	$\frac{10}{100} \times 20,00,000 \times \frac{1}{2}$	=	1,00,000
		1,80,000	

June

May sales	$\frac{10}{100} \times 20,00,000 \times \frac{1}{2}$	=	1,00,000
June sales	$\frac{10}{100} \times 18,00,000 \times \frac{1}{2}$	=	90,000
		1,90,000	

July

June sales	$\frac{10}{100} \times 18,00,000 \times \frac{1}{2}$	= 90,000
July sales	$\frac{10}{100} \times 16,00,000 \times \frac{1}{2}$	= 80,000
		1,70,000

(4) Calculation of commission on sales

April, Feb. sales 12,00,000 × $\frac{5}{100}$ = 60,000

May, Mar. sales 12,00,000 × $\frac{5}{100}$ = 60,000

June, April sales 16,00,000 × $\frac{5}{100}$ = 80,000

July, May sales 20,00,000 × $\frac{5}{100}$ = 1,00,000

(ii) Adjusted profit and loss method: This method presents the cash budget in the form of cash flow statement. When compared to receipts and payments method, this method is less detailed but more useful for long-term budgeting. In preparing a long-term budget management is concerned with overall position of cash rather than cash receipts and payments. The preparation of a cash flow statement involves the use of the following information:

(*a*) Cash balance in the beginning.

(*b*) Net profit forecast for the period (before charging depreciation and other provisions)

(*c*) Changes in working capital.

(*d*) Capital expenditure and sale of plant and machinery.

(*e*) Capital receipts

(*f*) Dividends.

This method differs from the previous method in one respect, *i.e.,* it considers particularly non-cash transactions. That is to say, this method is based on the assumption that profit = cash. In other words, if there were no credit transactions, capital transactions, accrual, provisions, or appropriation of profit, the amount of profit shown in the profit and loss account will be equal to the balance of cash shown by the cash book. But in practice such a situation rarely arises and hence requires adjustments. Hence, the name adjusted profit and loss method.

(iii) Balance sheet method: This method resembles the adjusted profit and loss method explained above. In addition to a cash flow statement a budgeted balance sheet is prepared for the next period under this method. In preparing the balance sheet all assets and liabilities are taken into account except cash. The two sides of balance sheet is then balanced and the balancing figure represents cash. When the asset side is heavier than liabilities side, it denotes bank overdraft and if the liability side is heavier than asset side it denotes cash on hand or at bank.

Problem 31. *(Preparation of different budgets):* Ratna Enterprises manufactures three products *A, B* and *C*. You are required to prepare (*a*) Sales budget, (*b*) Production budget, (*c*) Material budget for the month of January 2010.

Sales of products	*Quantity*	*Rate*
A	1,000	100
B	2,000	120
C	1,500	140

Materials used in the company's products are:

Material	M_1	M_2	M_3
Rate	*Rs. 4*	*Rs. 6*	*Rs. 9*

Quantities used in	*Units*	*Units*	*Units*
A	4	2	–
B	3	3	2
C	2	1	1
Finished stock as on:	*A (units)*	*B (units)*	*C (units)*
1.1.2010	1,000	1,500	500
31.1.2010	1,100	1,650	550
Material stocks:	M_1	M_2	M_3
1.1.2010	(units) 26,000	20,000	12,000
31.1.2010	(units) 31,200	24,000	14,400

Solution:

SALES BUDGET

	A	*B*	*C*
Sales quantities	1,000	2,000	1,500
Selling price (per unit)	Rs. 100	120	140
Sales	Rs. 1,00,000	2,40,000	2,10,000

STATEMENT SHOWING PRODUCTION BUDGET

	A	*B*	*C*
Sales units	1,000	2,000	1,500
Add: Closing stock	1,100	1,650	550
	2,100	3,650	2,050
Less: Opening stock	1,000	1,500	500
Production	1,100	2,150	1,550

STATEMENT SHOWING MATERIAL USAGE BUDGET

	M_1		M_2		M_3	
Production Budget	*Unit per product*	*Total*	*Unit per product*	*Total*	*Unit per product*	*Total*
A—1,100	4	4,400	2	2,200	–	–
B—2,150	3	6,450	3	6,450	2	4,300
C—1,550	2	3,100	1	1,550	1	1,550
Material usage		13,950		10,200		5,850

MATERIAL PURCHASE BUDGET (QTY, AND VALUE)

	M_1	M_2	M_3
Material usage (units)	13,950	10,200	5,850
Add: Closing stock (units)	31,200	24,000	14,400
	45,150	34,200	20,250
Less: Opening stock	26,000	20,000	12,000
Required purchases (units)	19,150	14,200	8,250
Unit rate	Rs. 4	6	9
Value of materials	76,600	85,200	74,250

CHAPTER 13

(*b*) **Master budget:** It is a summary of all the functional budgets discussed above. According to the ICMA terminology, "A master budget is the summary budget incorporating its component functional budgets and which is finally approved, adopted and employed. A master budget shows the operating profit of the business for the budget period and budgeted balance sheet at its close.

Problem 32. *(Master budget):* A glass manufacturing company requires you to calculate and present the budget for the year from the following information:

Toughened glass	–	Rs. 3,60,000
Bent toughened glass	–	Rs. 5,40,000
Direct material cost	–	60% of sales
Direct wages of 20 workers at the rate of	–	Rs. 200 per month
Factory overheads:		
Works manager	–	Rs. 600 per month
Foreman	–	Rs. 500 per month
Stores and spares	–	$2\frac{1}{2}$% on sales
Depreciation on machinery	–	Rs. 12,500
Light and power	–	Rs. 5,000
Repairs and maintenance	–	Rs. 8,000
Other sundries	–	10% on direct wages
Administration, selling and distribution expenses	–	Rs. 16,00 per year.

(University of Kerala, M.Com., May 1992)

Solution:

MASTER BUDGET FOR THE PERIOD ENDING

Sales budget		
Toughened glass		3,60,000
Bent toughened glass		5,40,000
		9,00,000
Less: Administration and selling and distribution overheads		16,000
Net sales revenue		8,84,000
Product cost budget:		
Direct materials—60% of sales		5,30,400
Direct wages—20 × 200 × 12		48,000
Prime cost		5,78,400
Variable factory overheads:		
Stores and spares $2\frac{1}{2}$% on sales	22,500	
Light	5,000	
Repairs	5,000	37,500
Fixed factory overhead:		
Works manager's salary	7,200	
Foreman's salary	6,000	
Depreciation	12,500	
Sundries 10% of wages	4,800	30,500
Works cost		6,46,400

Expected proift = Sales – Works cost

= 8,84,000 – 6,46,400 = 2,37,600

4. On the Basis of Capacity

On the basis of capacity operated budgets are classified into two types. They are as follows:

(*a*) **Fixed budget:** The ICMA terminology defines fixed budget as "a budget which is designed to remain unchanged irrespective of the volume of output or turnover attained. All the budgets explained above are fixed budgets, *i.e.,* they are based on a fixed level of operation. In other words, it is assumed that the various departments of a factory work at a stated level of activity and that a specified production level is going to be achieved and that sales budget will be attained. A fixed budget does not help as a controlling tool. Hence, flexible budgets are made use of.

(*b*) **Flexible budget or variable budget:** It is defined by the ICMA terminology as "a budget which by recognising the difference in behaviour between fixed and variable costs in relation to fluctuations in output, turnover or other variable factors such as number of employees, is designed to change appropriately with such fluctuation". A flexible budget is not rigid as in the case of fixed budget. Instead it adapts itself to any level of activity. The budget varies according to a change in the level of output. Hence, it is also known as variable budget. It serves as a useful tool of controlling cost.

Problem 33. *(Flexible budget):* The expenses for the budgeted production of 10,000 units in a factory are furnished below

	Per unit Rs.
Material	70
Labour	25
Variable overheads	20
Fixed overhead (Rs. 1,00,000)	10
Variable expenses (direct)	5
Selling expenses (10% fixed)	13
Distribution expenses (20% fixed)	7
Administration expenses (Rs. 50,000)	5
Total	155

Prepare a flexible budget for production of

(*a*) 8000 units and

(*b*) 6000 units *(University of Madras, B.Com., April 2005)*

Solution:

FLEXIBLE BUDGET

	10,000 Units		*8,000 Units*		*6,000 Units*	
	Total	*Per unit*	*Total*	*Per unit*	*Total*	*Per unit*
Material	7,00,000	70	5,60,000	70	4,20,000	70
Labour	2,50,000	25	2,00,000	25	1,50,000	25
Variable overhead	2,00,000	20	1,60,000	20	1,20,000	20
Fixed overhead	1,00,000	10	1,00,000	12.5	1,00,000	16.67
Variable expenses	50,000	5	40,000	5	30,000	5.00
Selling expenses						
10% fixed	13,000	1.3	13,000	1.63	13,000	2.17
90% variable	1,17,000	11.7	93,600	11.7	70,200	11.70
Distribution overhead						
20% fixed	14,000	1.4	14,000	1.75	14,000	2.34
80% variable	56,000	5.6	44,800	5.60	33,600	5.60
Administration expenses						
Fixed	50,000	5	50,000	6.25	50,000	8.34
Total cost	15,50,000	155	12,75,400	159.43	10,00,800	166.800

Problem 34. *(Flexible budget):* The cost of an article at a capacity level of 5,000 units is given under 'A' below. For a variation of 25% in capacity above or below this level, the individual expenses as indicated under 'B' below

	A	*B*
	Rs.	
Material cost	25,000	(100% varying)
Labour cost	15,000	(100% varying)
Power	1,250	(80% varying)
Repairs and maintenance	2,000	(75% varying)
Stores	1,000	(100% varying)
Inspection	500	(20% varying)
Depreciation	10,000	(100% varying)
Administrative overheads	5,000	(25% varying)
Selling overhead	3,000	(25% varying)
	62,750	

Cost per unit Rs. 12.55

Find out the cost per unit and total cost for production levels of 4,000 units and 6000 units. Also show the total cost and unit cost of 5,000 units *(University of Madras, B.Com., October 2002)*

Solution:

FLEXIBLE BUDGET

	5,000 Units		4,000 Units		6,000 Units	
	Total	*Per unit*	*Total*	*Per unit*	*Total*	*Per unit*
Material (100% variable)	25,000	5	20,000	5	30,000	5
Labour (100% variable)	15,000	3	12,000	3	18,000	3
Prime cost	40,000	8.00	32,000	8	48,000	8
Power – 80% variable	1,000	0.20	800	0.20	1,200	0.20
– 20% fixed	250	0.05	250	0.06	250	0.04
Repairs – 75% variable	1,500	0.30	1,200	0.30	1,800	0.30
– 25% fixed	500	0.10	500	0.12	500	0.08
Stores – 100% variable	1,000	0.20	800	0.20	1,200	0.20
Inspection – 20% variable	100	0.02	80	0.02	120	0.02
– 80% fixed	400	0.08	400	0.10	400	0.07
Depreciation – 100% variable	10,000	2.00	8,000	2.00	12,000	2.00
Factory cost	54,750	10.95	44,030	11.00	65,470	10.91
Administration overhead						
– 25% variable	1,250	0.25	1,000	0.25	1,500	0.25
– 75% fixed	3,750	0.75	3,750	0.94	3,750	0.62
Selling overhead – 25% variable	750	0.15	600	0.15	900	0.15
– 75% fixed	2,250	0.45	2,250	0.56	2,250	0.37
Total	62,750	12.55	51,630	12.98	73,870	12.31

Problem 35. The expenses for the budgeted production of 20,000 units in a factory are furnished below

	Per unit Rs.
Materials	140
Labour	50
Variable overheads	40
Fixed overhead	20
Variable expenses (direct)	10
Selling expenses (10% fixed)	26
Distribution expenses (20% fixed)	14
Administrative expenses	10

Prepare a flexible budget for the production of (*a*) 16,000 units (*b*) 12,000 units (*c*) Indicate cost per unit at both the levels. Assume that the Administrative expenses are fixed for all levels of production.

(University of Madras, B.Com., May 2002)

CHAPTER 13

Solution:

FLEXIBLE BUDGET

	20000 units		*16000 units*		*12000 units*	
	Total	*Per unit*	*Total*	*Per unit*	*Total*	*Per unit*
Materials	28,00,000	140	22,40,000	140	16,80,000	140
Labour	10,00,000	50	8,00,000	50	6,00,000	50
Overhead variable	8,00,000	40	6,40,000	40	4,80,000	40
– Fixed	4,00,000	20	4,00,000	25	4,00,000	33.33
Administration expenses	2,00,000	10	2,00,000	12.50	2,00,000	16.67
Selling expenses						
– 10% fixed	52,000	2.60	52,000	3.25	52,000	4.33
– 90% variable	4,68,000	23.40	3,74,400	23.40	2,80,800	23.40
Distribution expenses						
– 20% fixed	56,000	2.80	56,000	3.50	56,000	4.67
– 80% variable	2,24,000	11.20	1,79,200	11.20	1,34,400	11.20
Total	62,00,000	310	51,01,600	318.85	40,03,200	333.60

Problem 36. *(Flexible budget):* AB Ltd. has the production capacity of 40,000 units p.a. Currently the company is producing 20,000 units. The cost data for the production of 20,000 units are given below:

Materials	Rs. 80 per unit
Wages (50% fixed)	Rs. 40 per unit
Manufacturing expenses (60% variable)	Rs. 20 per unit
Administrative expenses (100% fixed)	Rs. 10 per unit
Selling & Distribution expenses (60% fixed)	Rs. 10 per unit
Total cost	160 per unit
Profit	40 per unit
Selling price	200 per unit

Prepare a flexible budget and calculate the cost per unit of the company at a capacity level of 75% *(Calcutta University, B.Com., (Pass) 2005)*

Solution:

Flexible budget for 30,000 units

$$40{,}000 \times \frac{75}{100}$$

Materials 30,000 × 80 — 24,00,000

Wages – Variable $\left(40 \times \frac{50}{100}\right) \times 30{,}000$ — 6,00,000

– Fixed $\left(40 \times \frac{50}{100}\right) \times 20{,}000$ — 4,00,000

Manufacturing expenses

– Variable	$\left(20 \times \frac{60}{100}\right) \times 30{,}000$	3,60,000
– Fixed	$\left(20 \times \frac{40}{100}\right) \times 20{,}000$	1,60,000
Administrative expenses		
– Fixed	20,000 × 10	2,00,000
Selling & distribution expenses		
– Variable	$\left(10 \times \frac{40}{100}\right) \times 30{,}000$	1,20,000
– Fixed	$\left(10 \times \frac{60}{100}\right) \times 20{,}000$	1,20,000
	Cost of sales	43,60,000

Cost per unit $= \frac{43{,}60{,}000}{30{,}000}$

= 145.33

Problem 37. A factory is currently running at 50% capacity and produces 5,000 units at a cost of Rs. 90 per unit as per details given below

	Rs.
Material	50
Labour	15
Factory overheads	15 (Rs. 6 fixed)
Administrative overheads	10 (Rs. 5 fixed)

The current selling price is Rs. 100 per unit. At 60% working material cost per unit increases by 2% and selling price per unit falls by 2%.

At 80% working, material cost per unit increases by 5% and selling price per unit falls by 2.5%

Prepare a flexible budget showing profits of the factory at 60% and 80% working and offer your comments. *(University of Delhi, B.Com. (Hons), 2004)*

Solution:

FLEXIBLE BUDGET

	60% (6,000 units)		*80% (8,000 units)*	
	Per unit	*Total*	*Per unit*	*Total*
Materials	51	3,06,000	52.50	4,20,000
Labour	15	90,000	15	1,20,000
Overheads factory				
– Variable	9	54,000	9	72,000
– Fixed	5	30,000	3.75	30,000
Administrative overhead				
– Variable	5	30,000	5	40,000
– Fixed	4.17	25,000	3.125	25,000
Total cost	89.17	5,35,000	88.375	7,07,000
Profit	8.83	53,000	9.125	73,000
Sales	98	5,88,000	97.50	7,80,000

Comments. When capacity is increased from 60% to 80%, the profit has increased from Rs. 53,000 to Rs. 73,000. This is because, fixed cost has not increased.

Problem 38. G.S. Ltd., manufactures a single product for which market demand exists for additional quantity. Present sales of Rs. 60,000 per month utilises only 60% capacity of the plant. Marketing manager assures that with the reduction of 10% in the price he would be in a position to increase the sales by about 25% to 30%.

The following data are available.

(i)	Selling price	Rs. 10 per unit
(ii)	Variable cost	Rs. 3 per unit
(iii)	Semi-variable cost	Rs. 6,000 fixed + 50 paise per unit
(iv)	Fixed cost	Rs. 20,000 at present level estimated to be Rs. 24,000 at 80% output

You are required to prepare the following statements:

(1) The operating profit at 60%, 70% and 80% levels at current selling price and

(2) The operating profits at proposed selling price at the above levels.

(University of Delhi, B.Com. (Hons), 2004)

Solution:

FLEXIBLE BUDGET

	60%	*70%*	*80%*
Output (units)	6,000	7,000	8,000
	Rs.	Rs.	Rs.
Selling price	10	10	10
New selling price	9	9	9
Less: Variable cost	3	3	3
(i) Sales @ Rs. 10 (A)	60,000	70,000	80,000
Fixed cost	20,000	20,000	24,000
Variable cost	18,000	21,000	24,000
Semi-variable cost	9,000	9,500	10,000
Total cost (B)	47,000	50,500	58,000
Profit (A – B)	13,000	19,500	22,000
	Rs.	*Rs.*	*Rs.*
(ii) Sales at new S.P.	54,000	63,000	72,000
Less: Total cost	47,000	50,500	58,000
Profit	7,000	12,500	14,000

Problem 39. Figures regarding sales, cost and profit at 50% capacity are given below:

	Rs.
Sales	20,00,000
Direct cost	8,00,000
Factory overheads	4,00,000
Office overheads	2,00,000
Selling overheads	3,00,000
Profit	3,00,000

Every 10% increase in sales beyond 50% capacity is possible only after reducing the price by 1% on the base level of 50% capacity. Direct material cost is 25% of the total direct cost at 50% capacity. With every 10% increase in capacity above this level, the price of direct material comes down by 2%. 50% of the factory overheads are fixed and the rest are fully variable. Office overheads are of step character. Every 10% increase in output results in 2% increase in office overheads over 50% capacity. Selling overheads increase in proportion of sales value.

Prepare a flexible budget at 80% capacity level. *(University of Delhi, B.Com. (Hons.), 2003)*

Solution:

FLEXIBLE BUDGET (80% CAPACITY)

		Rs.
Direct materials		
$(2,00,000 - 6\%) \times \frac{80\%}{50\%}$		3,00,800
Other direct cost $\left(6,00,000 \times \frac{80\%}{50\%}\right)$		9,60,000
	Total direct cost	12,60,800
Variable factory overheads		3,20,000
$\left(2,00,000 \times \frac{80\%}{50\%}\right)$		
Variable selling overheads		4,65,000
$\left(3,00,000 \times \frac{31,04,000}{20,00,000}\right)$		
	Total variable cost	20,46,400
Fixed cost:		
– Factory overhead		2,00,000
– Office overhead (2,00,000 + 6%)		2,12,000
	Total cost	24,58,400
	Profit	6,45,600
	Sales	31,04,000

Working Note:

Calculation of sales

$$(20,00,000 - 3\%) \times \frac{80\%}{50\%} = \text{Rs. } 31,04,000$$

Problem 40. G. Ltd. is currently operating at 75% of its capacity. In the past two years the level of operations were 55% and 65% respectively. Presently the production is 75,000 units. The company is planning for 85% capacity level during 2002–03. The cost details are as follows:

Particulars	55% Rs.	65% Rs.	75% Rs.
Direct materials	11,00,000	13,00,000	15,00,000
Direct labour	5,50,000	6,50,000	7,50,000
Factory overheads	3,10,000	3,30,000	3,50,000
Selling overheads	3,20,000	3,60,000	4,00,000
Administrative overheads	1,60,000	1,60,000	1,60,000
	24,40,000	28,00,000	31,60,000

CHAPTER 13

Profit is estimated @ 20% on sales. The following increases in costs are expected during the year

	In %
Direct materials	8
Direct labour	5
Variable factory overheads	5
Variable selling overheads	8
Fixed factory overheads	10
Fixed selling overheads	15
Administrative overhead	10

Prepare a flexible budget for the year 2002–03 at 85% level of capacity and ascertain the profit on sales. *(University of Delhi, B.Com. (Hons.), 2002)*

Solution:

Working Notes:

(1) Direct material cost per unit

$$\frac{\text{Rs. }15{,}00{,}000}{75{,}000 \text{ units}} = \text{Rs. } 20$$

$= (20 + 8\%) \times 85{,}000$ units $=$ Rs. 18,36,000

(2) Direct labour per unit

$$= \frac{\text{Rs. }7{,}50{,}000}{75{,}000 \text{ units}} = \text{Rs. } 10$$

$= (10 + 5\%) \times 85{,}000$ units $=$ Rs. 8,92,500

(3) Fixed factory overheads

$= 2{,}00{,}000 + 10\% = 2{,}20{,}000$

(4) Variable factory overheads

$=$ (Rs. $2 + 5\%) \times 85{,}000$ units $=$ Rs. 1,78,500

(5) Administrative overheads

$= (1{,}60{,}000 + 10\%) =$ Rs. 1,76,000

(6) Fixed selling overheads

$= (1{,}00{,}000 + 15\%) =$ Rs. 1,15,000

(7) Selling overhead variable

$= (4 + 8\%) \times 85{,}000$ units $=$ Rs. 3,67,200

Factory overheads and selling overheads are semi-variable items and this are segregated as follows

$$\text{Variable factory overheads} = \frac{\text{Difference in overhead}}{\text{Difference in output}}$$

$$= \frac{3{,}30{,}000 - 3{,}10{,}000}{10{,}000}$$

$$= \frac{20{,}000}{10{,}000} = \text{Rs. } 2$$

Fixed factory overhead $=$ Total overheads $-$ (output $\times$ variable overhead per unit)

$= 3{,}10{,}000 - (55{,}000 \times 2)$

$=$ Rs. 2,00,000

Similarly, selling overheads are segregated.

FLEXIBLE BUDGET 85% CAPACITY 85,000 UNITS

		Rs.
Direct materials		18,36,000
Direct labour		8,92,500
Factory overheads – Fixed		2,20,000
– Variable		1,78,500
Administrative overheads		1,76,000
Selling overhead – Fixed		1,15,000
– Variable		3,67,200
	Total cost	37,85,200
	Profit	9,46,300
	Sales	47,31,500

Problem 41. *(Flexible budget):* Prepare a flexible budget from the following data:

Capacity	50%
Volume	10,000 units
Selling price per unit	Rs. 200
Material	Rs. 100
Labour	Rs. 30
Factory overhead	Rs. 30 (Rs. 12 fixed)
Administration overhead	Rs. 20 (Rs. 10 fixed)

At 60% working, material cost per unit increased by 2% and selling price per unit falls by 2%. At 80% working, material cost per unit increases by 5% and selling price per unit falls by 5%. Estimate the profit at 60% and 80% working. *(Nagarjuna University, M.Com., March 1992)*

Solution:

FLEXIBLE BUDGET

	Level of activity					
	50%		*60%*		*80%*	
	Per unit	*Amount*	*Per unit*	*Amount*	*Per unit*	*Amount*
Units	10,000		12,000		16,000	
Raw materials	100	10,00,000	102	12,24,000	105	16,80,000
Labour	30	3,00,000	30	3,60,000	30	4,80,000
Factory overhead – Fixed	12	1,20,000	10	1,20,000	7.5	1,20,000
– Variable	18	1,80,000	18	2,16,000	18.0	2,88,000
Administration overhead						
– Fixed	10	1,00,000	8.33	1,00,000	6.25	1,00,000
– Variable	10	1,00,000	10	1,20,000	10	1,60,000
Total cost	180	18,00,000	178.33	2,14,000	176.75	28,28,000
Profit	20	2,00,000	17.67	2,12,000	13.25	2,12,000
Sales	200	20,00,000	196.00	23,52,000	190	30,40,000

Problem 42. Draw up a flexible budget for overhead expenses on the basis of the following data and determine the overhead rates at 70%, 80% and 90% plant capacity.

	Capacity level		
Variable overheads:	70%	80%	90%
Indirect labour	–	12,000	–
Indirect materials	–	4,000	–
Semi-variable overheads:			
Power (30% fixed)	–	20,000	–
Repairs and maintenance 60% fixed	–	2,000	–
Fixed overhead:			
Depreciation	–	11,000	–
Insurance	–	3,000	–
Salaries	–	10,000	–
Total overheads	–	62,000	–
Estimated direct labour hours	–	1,20,000	–

(Madurai Kamaraj University, B.Com., April 1992)
(University of Madras, B.Com., April 2001)

Solution:

FLEXIBLE BUDGET

		Level of activity		
		70%	*80%*	*90%*
Variable overheads:				
Indirect labour		10,500	12,000	13,500
Stores		3,500	4,000	4,500
Semi-variable overhead:				
Power	—Fixed	6,000	6,000	6,000
	—Variable	12,250	14,000	15,750
Repairs	—Fixed	1,200	1,200	1,200
	—Variable	700	800	900
Fixed Overhead				
Depreciation		11,000	11,000	11,000
Insurance		3,000	3,000	3,000
Salaries		10,000	10,000	10,000
Total overheads		58,150	62,000	65,850
Estimated labour hours		1,08,500	1,24,000	1,39,500
Overhead rate		0.53	0.50	0.47

Problem 43. For production of 10,000 articles the following are budgeted expenses per unit:

	Rs.
Direct materials	60.00
Direct labour	30.00
Variable overhead	20.00
Fixed overhead (Rs. 1,60,000)	16.00
Variable expenses (direct)	5.00
Selling expenses (20% fixed)	15.00
Administration expenses (Rs. 50,000 fixed for all level of production)	5.00
Distribution expenses (20% fixed)	5.00
	156.00

Prepare a flexible budget for production of 6,000, 7,000 and 8,000 units of articles, showing clearly variable cost, fixed cost and total cost. *(University of Madras, B.Com., April 2004)*

Solution:

FLEXIBLE BUDGET

	6,000 units		7,000 units		8,000 units		10,000 units	
	Per unit	*Amount*	*Per unit*	*Amount*	*Per unit*	*Amount*	*Per unit*	*Amount*
Material	60	3,60,000	60	4,20,000	60	4,80,000	60	6,00,000
Labour	30	1,80,000	30	2,10,000	30	2,40,000	30	3,00,000
Variable overhead	20	1,20,000	20	1,40,000	20	1,60,000	20	2,00,000
Fixed overhead	26.66	1,60,000	22.85	1,60,000	20	1,60,000	16	1,60,000
Variable expense	5	30,000	5	35,000	5	40,000	5	50,000
Selling expense:								
—Fixed	5	30,000	4.28	30,000	3.75	30,000	3	30,000
—Variable	12	72,000	12	84,000	12.00	96,000	12	1,20,000
Administration expense	8.34	50,000	7.14	50,000	6.25	50,000	5	50,000
Distribution expense								
—Fixed	1.66	10,000	1.42	10,000	1.25	10,000	1	10,000
—Variable	4.00	24,000	4.00	28,000	4.00	32,000	4	40,000
Total cost	172.66	10,36,000	166.69	11,67,000	162.25	12,98,000	156	15,60,000

Problem 44. The following figures are available from sales and costs forecast of M/s. Asiad and Company for the year ended 31st December, 1984 at 50% (5,000 units) capacity utilisation:

(i) Fixed expenses remain constant for all levels of production and sales.

(ii) Selling price between 50% and 75% capacity is Rs. 25 per unit.

(iii) Semi-variable expenses will remain unchanged at 50% to 65% capacity but will increase by 10% between 65% and 80% capacity and 30% between 80% and 100% capacity.

(iv) At 90% level, material cost increases by 5% and selling price is reduced by 5%.

(v) At 100% level both material and labour costs increase by 10% and selling price is reduced by 8%.

(vi) Semi-variable expenses are Rs. 50,000.

(vii) Fixed expenses are Rs. 58,000.

(viii) Variable expenses are, material Rs. 5 per unit, labour Rs. 2 per unit and direct expenses, Re. 1 per unit.

Prepare a profit forecast statement through flexible budget at 60%, 75%, 90% and 100% capacity. *(Bangalore University, M.Com., May 1989)*

Solution:

FLEXIBLE BUDGET

	60% (6,000 units)	75% (7,500 units)	90% (9,000 units)	100% (10,000 units)
Materials	30,000	37,500	47,250	55,000
Labour	12,000	15,000	18,000	22,000
Expenses	6,000	7,500	9,000	10,000
Semi-variable cost	50,000	55,000	65,000	65,000
Fixed cost	58,000	58,000	58,000	58,000
Total	1,56,000	1,73,000	1,97,250	2,10,000
Profit/Loss	(6,000)	14,500	16,500	20,000
Sales	1,50,000	1,87,500	2,13,750	2,30,000

Working Note: Sales value for 9,000 units and 10,000 units

9,000 × 25	=	2,25,000	10,000 × 25	=	2,50,000
Less: 5%		11,250	*Less:* 8%		20,000
		2,13,750			2,30,000

Problem 45. The following data are available for a manufacturing company for a yearly period:

	Rs. (in lakhs)
Fixed expenses:	
Wages and salaries	9.5
Rent, rates and taxes	6.6
Depreciation	7.4
Sundry administration expenses	6.5
Semi-variable expenses (at 50% capacity)	
Maintenance and repairs	3.5
Indirect labour	7.9
Sales department salaries	3.8
Sundry administrative expenses	2.8
Variable expenses (at 50% capacity)	
Materials	21.7
Labour	20.4
Other expenses	7.9
Total cost	98.0

Assume that the fixed expenses remain constant at all levels of production, semi-variable expenses remain constant between 45% and 65% capacity increasing by 10% between 65% and 80% capacity and by 20% between 80% and 100% capacity.

Sales at various levels are:

	Rs. (in lakhs)
50% Capacity	100
60% Capacity	120
75% Capacity	150
90% Capacity	180
100% Capacity	200

Prepare a flexible budget for the year at 60% and 90% capacities and estimate the profit at these levels of budget. *(C.S., Inter., Dec. 1991)*

Solution:

FLEXIBLE BUDGET

	60%	90%
	Rs. (in lakhs)	
Variable cost:		
Materials	26.04	39.06
Labour	24.48	36.72
Other expenses	9.48	14.22
Semi-variable cost:		
Maintenance	3.5	4.20
Indirect labour	7.9	9.48
Sales department salaries	3.8	4.56
Sundry administration expenses	2.8	3.36
Fixed cost:		
Wages and salaries	9.50	9.50

Rent, rates and taxes	6.60	6.60
Depreciation	7.40	7.40
Sundry administration expenses	6.50	6.50
Total cost	108	141.60
Profit	12	38.40
Sales	120	180

Problem 46. The monthly budgets for manufacturing overhead of a concern for two levels of activity were as follows:

Capacity	*60%*	*100%*
Budgeted production (units)	600	1,000
	Rs.	*Rs.*
Wages	1,200	2,000
Consumable stores	900	1,500
Maintenance	1,100	1,500
Power and fuel	1,600	2,000
Depreciation	4,000	4,000
Insurance	1,000	1,000
	9,600	12,000

You are required to:

(*i*) Indicate which of the items are fixed, variable and semi-variable,

(*ii*) Prepare a budget for 80% capacity and

(*iii*) Find the total cost both fixed and variable per unit of output at 60%, 80% and 100% capacity. *(C.S., Inter., December 1990)*

Solution:

Fixed costs are: (*i*) Depreciation
(*ii*) Insurance

Variable costs are: (*i*) Wages at Rs. 2 per unit
(*ii*) Consumable stores at 1.50 p.u.

Semi-variable cost:

$$\text{Maintenance} = \left(\frac{1{,}500 - 1{,}100}{1{,}000 - 6{,}000}\right) = \frac{400}{400} = \text{Re. 1 p.u variable}$$

$$\text{Rs. } 1{,}100 - 600 \times \text{Re. } 1 = 500 \text{ fixed}$$

$$\text{Power and fuel} = \left(\frac{2{,}000 - 1{,}600}{1{,}000 - 600}\right) = \frac{400}{400} = \text{Re. 1 p.u variable}$$

$$\text{Rs. } 1{,}600 - 600 \times \text{Re. } 1 = 1{,}000 \text{ fixed}$$

BUDGET FOR 80% CAPACITY

	Output 800 units
Wages @ Rs. 2 per unit	1,600
Consumable stores @ Rs. 1.50 per unit	1,200
Maintenance Rs. 500 + Re. 1.00 per unit	1,300
Power and fuel Rs. 1,000 + Re. 1.00 per unit	1,800
Depreciation	4,000
Insurance	1,000
Total cost	10,900

STATEMENT SHOWING TOTAL COST

Capacity	*60%*	*80%*	*100%*
Units	600	800	1,000
	Rs.	*Rs.*	*Rs.*
Fixed cost:			
Depreciation	4,000	4,000	4,000
Insurance	1,000	1,000	1,000
Maintenance	500	500	500
Power and fuel	1,000	1,000	1,000
Total fixed cost	6,500	6,500	6,500
Variable cost:			
Wages @ Rs. 2 per unit	1,200	1,600	2,000
Consumable stores @ Rs. 1.50 p.u.	900	1,200	1,500
Maintenance cost @ Re. 1 p.u	600	800	1,000
Power and fuel at Re. 1 p.u.	600	800	1,000
Total variable cost	3,300	4,400	5,500
Fixed cost per unit	10.83	8.125	6.50
Variable cost per unit	5.50	5.50	5.50
Total cost per unit	16.33	13.625	12.00

ZERO BASED BUDGETING (ZBB)

Zero based budgeting is a new technique of budgeting introduced first in USA in the year 1969. This system of budgeting was developed by Peter Pyhrr of Texas Instruments of USA. This technique of budgeting is more useful in government budgeting but can also be used in factories for non-manufacturing activities, such as administration and selling activities.

The ICMA terminology defines ZBB as "a method of budgeting whereby all activities are re-evaluated each time a budget is formulated. Each functional budget starts with the assumption that the function does not exist and is a zero cost. Increments of cost are compared with increments of benefits culminating in the planned maximum benefits for a given budgeted cost".

The technique of budgeting is considered to be an improvement over traditional method of budgeting, which is also known as 'incremental budgeting'. Under incremental budgeting every departmental manager would prepare a budget for his department based upon the previous experience and allow for a certain increase in amount in the budget for meeting contingency. However, in spite of a best forecast, sometimes the targets may not be achieved owing to inefficiency. Thus, whenever previous budget is taken as a basis to prepare a current budget, the current budget involves an element of inefficiency that is carried forward from the previous year. Hence, under ZBB, the budget is prepared by considering the base for the current year as zero and this eliminates the accrual of inefficiency for preparing future years budget.

The basic steps in implementing ZBB are as follows:

(*i*) Identify each function and activity of the organisation—this is referred to as a 'decision packing'.

(*ii*) Evaluate each decision package so as to ensure that is cost effective.

(*iii*) Compare each activity with possible alternatives.

(*iv*) Rank each activity—in some cases decision packages can be evaluated in terms of profitability or in any subjective terms using cost-benefit analysis.

(*v*) Allocate resources in accordance with the ranking of activities and with resources available to the organisation.

BUDGET REPORT

The work of a budget officer does not end with the preparation and approval of budgets. He has to prepare reports on a continuous basis so as to facilitate comparison of actuals with budgets. The budget reports are sent to various departmental managers showing favourable or adverse variance from the budget. Based on this the departmental managers will prepare a report to be submitted to the managing director pointing out the reasons for the variances. This enable remedial actions to be taken to set right unfavourable variance. The reports so furnished will also help as a guide for future planning.

According to W.W. Bigg, while preparing reports on budgets it is necessary to follow the undermentioned principles so as to make reports more effective:

1. The report should be clearly headed and the period covered shown. The unit, viz., cash, tonnes, litre, etc., should be indicated.
2. Like must be compared with like and there must be no ambiguity of description. For example, it must be clear whether a sales report refers to 'deliveries made and invoiced' or to 'orders received'.
3. Information not relevant to the purpose for which the report is prepared should be omitted so that conclusions from the report can be drawn quickly and with certainty.
4. The report should not attempt to portray so much information that clarity is lost. If the information to be conveyed is complicated, more than one statement may be desirable. For example, to show actual sales compared with budget, analysed over both 'areas' and 'commodities', a separate statement for each analysis would improve clarity.
5. The information included should be limited to the sphere of the person to whom it is furnished. The data to be given to a foreman would normally be confined to that affecting his particular shop, but the factory manager would require broader information covering all departments for which he is responsible.
6. Promptness is to be preferred to absolute accuracy, the purpose is not merely to convey information but to convey it promptly and to the person who has the authority to take action.
7. All reports should be reviewed periodically to ensure that they are still useful and to ascertain whether they should be expanded, contracted or discontinued.

Questions

I. Choose the correct answer from the following:

1. A budget that gives a summary of all the functional budgets and projected profit and loss account is known as

(*a*) Capital budget (*b*) Flexible budget

(*c*) Master budget (*b*) Discretionary budget [ ]

2. The fixed variable cost classification has a special significance in the preparation of

(*a*) Flexible budget (*b*) Master budget

(*c*) Cash budget (*d*) Capital budget [ ]

3. The basis difference a fixed budget and a flexible budget is that a fixed budget

(*a*) includes only fixed costs and a flexible budget only variable cost.

(*b*) is a budget for a single level of some measures of activity, what a flexible budget consists of several budgets based on different activity levels.

(*c*) is concerned with future acquisition of fixed assets, while a flexible budget is concerned with expenses that vary with sales.

(*d*) cannot be changed after a fiscal period begins, while a flexibe budget can be changed after a fiscal period begins. [ ]

4. Which of the following is usually a long-term budget ?

(*a*) Sales budget (*b*) Cash budget

(*c*) Capital expenditure budget (*d*) Fixed budget [ ]

5. Under flexible budgeting

(*a*) statements included in the budget report vary from period to period.

(*b*) budget standards may be adjusted at will.

(*c*) reporting dates vary according to the activity level reported upon.

(*d*) planned activity level is adjusted to the actual activity level before the budget comparison report is prepared. [ ]

[***Answers:*** 1. (*c*), 2. (*a*), 3. (*b*), 4. (*c*), 5. (*d*)]

II. Mark 'T' or 'F' in the space provided

1. A budget is nothing but an estimate. (*T*/*F*)
2. Budgets are drawn up by the chief accountant. (*T*/*F*)
3. Budgets are blue prints for action. (*T*/*F*)
4. On the basis of budgets, next year's balance sheet and P & L a/c can be drawn up. (*T*/*F*)
5. Raw material supply is always the key factor. (*T*/*F*)
6. Diverse production is reduced to common basis on the basis of standard hours involved. (*T*/*F*)
7. Standing costing can operate without budgetary control. (*T*/*F*)
8. A system of budgetary control cannot be operated when standard costing system is in use. (*T*/*F*)
9. All functional budgets should be coordinated with sales budget which is always prepared first. (*T*/*F*)
10. A budget manual is a summary of all the functional budget. (*T*/*F*)
11. Budgetary control system does not suit small business concern. (*T*/*F*)
12. Purchase budget and material budget is the same thing. (*T*/*F*)
13. A flexible budget is one which changes from year to year. (*T*/*F*)
14. A flexible budget system will recast quickly for changes in the volume of activity. (*T*/*F*)
15. A flexible budget carefully differentiates between fixed and variable cost. (*T*/*F*)

[***Answers:*** True—3, 4, 5, 6, 7, 14. Rest are all false]

III. Short and Long answer questions

1. Define budgetary control and state its advantages. *(Kakatiya University, B.Com., October 1989)*
2. Write notes on: (*a*) Limiting factor, (*b*) Budget mannual.
(Calicut University, M.Com., April 1992)
3. What are 'budget' and 'budgetary control' ? Discuss various advantages and essentials for success of budgetary control. *(Kakatiya University, M.Com., August 1991)*
4. What is budgetary control and how it is exercised ? Discuss various advantages and essentials for the success of budgetary control. *(Mangalore University, B.Com., October 1991)*
5. Briefly explain the steps in the installation of a system of budgetary control.
(University of Kerala, B.Com., April 1989)
6. What is meant by 'budgetary control' ? State the essentials of good budgetary control system. What are the advantages and limitations of a budgetary control system ?
(Calicut University, B.Com., October 1989)
7. What is a flexible budget ? How does it differ from a fixed budget ?
(University of Kerala, B.Com., March 1990)
8. What are the facts to be considered in setting up an annual sales budget ?
(ICWA, Inter, Dec. 1989)
9. What is zero base budgeting ? What are the advantages of zero based approach over the traditional approach ? *(ICWA, Inter, Dec. 1989)*
10. (*a*) Define the terms 'budget' and 'budgetary control'.
(*b*) List down any five objectives of a budgetary control system. *(CA, Inter, Nov. 1990)*

CHAPTER 13

Exercises

Sales Budget

1. The Sunshine Company Ltd., has four sales divisions each consisting of four regions ; North, South, East and West. The company sells two products *X* and *Y*. Budgeted sales for the six months ended 30th June, 1993, in each area of division 1 were as follows:

North	—*X*	10,000 units @ Rs. 10 each
	—*Y*	6,000 units @ Rs. 5 each
South	—*Y*	12,000 units @ Rs. 5 each
East	—*X*	15,000 units @ Rs. 10 each
West	—*X*	8,000 units @ Rs. 10 each
	—*Y*	5,000 units @ Rs. 5 each

Actual sales for the same period in division 1 were as follows:

North	—*X*	11,500 units @ Rs. 10 each
	—*Y*	7,000 units @ Rs. 5 each
South	—*Y*	12,500 units @ Rs. 5 each
East	—*X*	16,500 units @ Rs. 10 each
West	—*X*	9,500 units @ Rs. 10 each
	—*Y*	5,250 units @ Rs. 5 each

From the salesmen's report, it is thought that sales could be budgeted for the six months ended June 30, 1994 as follows:

North	—*X*	Budgeted increase of 2,000	units on 30.6.93
	—*Y*	Budgeted increase of 500	units on 30.6.93
South	—*Y*	Budgeted increase of 1,000	units on 30.6.93
East	—*X*	Budgeted increase of 2,000	units on 30.6.93
West	—*X*	Budgeted increase of 1,000	units on 30.6.93
	—*Y*	Budgeted increase of 500	units on 30.6.93

At a meeting of area sales manager with the divisional sales manager it is decided that sales campaign will be undertaken in areas South and East. It is anticipated that these campaigns will result in additional sales of 3,000 units of *X* in the South area and 5,000 units of *Y* in the East area.

Prepare a sales budget for the period ending 30th June, 1994. Showing also the budgeted and actual sales for June 30th, 1993.

[**Ans.** Total sales for all areas and for products *X* and *Y* for the budgeted period 30.6.94 is Rs. 5,60,000 for the budgeted period 30.6.93 is Rs. 4,45,000 and actual total sales for 30.6.93 is 4,98,750]

Selling and Distribution Cost Budget

2. Hardcore Ltd., incurred the following selling costs in its last budget year.

	Sales area			*Head office*	*Total*
	A	*B*	*C*		
Personnel cost:					
Salesmen's salary	27,000	33,000	5,000	—	65,000
Commission	5,000	7,000	500	—	12,500
Travelling expenses	8,000	11,000	3,000	—	22,000
Sales management:					
Salaries	—	—	—	11,000	11,000
Building services	—	—	—	7,000	7,000
Other costs	—	—	—	1,500	1,500
Publicity:					
	30,000	45,000	8,000	—	83,000
	70,000	96,000	16,500	19,500	2,02,00

From the above particulars and the following additional information prepare a selling cost budget for the year 1993:

1. A new sales area *D* is to be formed to cover what is mostly virgin territory, but to take in certain areas previously covered by *B* sales are. Salemen's salaries of the new area are estimated to be Rs. 8,000, which includes Rs. 2,000 to be paid to *B* salesmen transferred to the *D* area. Travelling expenses of *B* representative in the *D* area last year was Rs. 900. For the whole *D* area they are expected to be Rs. 2,400 next year.
2. An additional clerk will be needed in the sales manager's office at a salary of Rs. 900 per annum and the building service budget shows an increase of 5%.
3. The publicity budget allows for expenditure of *A*—Rs. 25,000, *B*—Rs. 40,000, *C*—Rs. 10,000, *D*—Rs. 25,000. Rs. 15,000 is budgeted as the cost of an international exhibition.
4. Commission is paid at the rate of 1/2% of all sales. The sales budget specifies sales of *A*—Rs. 12,00,000, *B*—Rs. 15,00,000, *C*—Rs. 5,00,000, *D*—Rs. 4,00,000.

[**Ans.** Selling budget cost for A—Rs. 66,000, B—Rs. 88,600 C—Rs. 20,500, D—Rs. 37,400, H.O. Rs. 37,750, Total Rs. 2,48250.]

Production Budget and Production Cost Budget

3. Gaira Engineering Company Ltd. manufactures two products *X* and *Y*. An estimate of number of units expected to be sold in the first seven months of 1985 is given below

	Product X	*Product Y*
Jan.	500	1,400
Feb.	600	1,400
Mar.	800	1,200
Apr.	1,000	1,000
May	1,200	800
June	1,200	800
July	1,000	900

It is anticipated that:

(*a*) There will be no work-in-progress at the end of any month.

(*b*) Finished units equal to half the anticipated sales for the next month will be in stock at the end of each month (including December 1984). The budgeted production and production costs for the year ending 31st December 1985 are as follows:

	Product X	*Product Y*
Production units	11,000	12,000
Direct material cost per unit	Rs. 12	Rs. 19
Direct wages per unit	5	7
Other manufacturing charges apportioned	33,000	48,000

You are required to prepare:

(*a*) Production budget showing the number of units to be manufactured for each month.

(*b*) A summarised production cost budget for the 6 months period to June 1985.

[**Ans.** Production of *X* – 5550 units and that of *Y* – 6350 units. Total cost of *X* Rs. 1,11,000 and *Y* Rs. 1,90,500]

Flexible Budget

4. Draw up a flexible budget for overhead expenses on the basis of following data and determine the overhead rates at 70%, 80% and 90% plant capacity.

Expenses	*At 80% Capacity*
	Rs.
Indirect labour	12,000
Stores including spares	4000
Power (30% fixed)	20,000
Repairs & maintenance (60% fixed)	2,000
Depreciation	11,000
Insurance	3,000
Salaries	10,000
Total overhead	62,000
Estimated direct labour hours	1,24,000 hrs.

(*Calcutta University, B.Com. (Pass), 2003*)

[**Ans.** Overhead rates for 70% – 0.54, for 80% – 0.50, for 90% – 0.49]

CHAPTER 13

Cash Budget

5. From the following information prepare a cash budget for the three months ending 31st March 2004

Months	*Sales*	*Purchases*	*Other expenses*
December	Rs. 1,00,000	Rs. 60,000	10,000
January	Rs. 1,20,000	80,000	12,000
February	Rs. 1,50,000	90,000	15,000
March	Rs. 1,40,000	1,00,000	18,000

Other Informations:

(1) 10% of the sales are for cash and the remaining customers are allowed one month's credit.

(2) Lag in payment of other expenses is one month.

(3) Supplier's allow credit for one month

(4) Income tax of Rs. 20,000 is to be paid in March 2003

(5) Expected cash balance on 1-1-2004 will be Rs. 20,000

(Calcutta University, B.Com. (Pass), 2004)

[**Ans.** Closing cash balance for January is Rs. 52,000, February Rs. 83,000, March Rs. 1,07,000]

Cash Budget

6. From the following information prepare a cash budget for the quarter ending 30.6.2005

Months	*Sales*	*Purchases*	*Wages*	*Other expenses*
Feb.	2,40,000	1,68,000	20,000	14,000
Mar.	2,60,000	2,00,000	24,000	16,000
Apr.	1,60,000	2,08,000	16,000	12,000
May	2,32,000	2,12,000	20,000	24,000
June	1,76,000	1,60,000	16,000	12,000

Additional informations:

(1) Cash-in-hand on 1-4-2005 Rs. 10,000

(2) Sales 20% realised in the month of sale and discount allowed 2%: the balance realised after 2 months of sales.

(3) $\frac{4}{5}$th of the credit purchase is paid after one month of that purchase and next to that month the balance $\frac{1}{5}$th is paid.

(4) Wages are paid $\frac{3}{4}$th on the due date while $\frac{1}{4}$th during the next month.

(5) Other expenses are paid at a lag of one month.

(6) Income tax Rs. 25,000 due on or before 30.6.2005

(Calcutta University, B.Com. (Hons.), 2005)

[**Ans.** Closing cash balance on April Rs. 5,760, May Rs. (–) 11,832 and June Rs. (–) 1,26,536]

Flexible Budget

7. A Company increases the following expenses to produce 1,000 units of an article

	Rs.
Raw materials	30,000
Direct wages	15,000
Power (20% fixed)	10,000
Repairs & maintenance (15% fixed)	8,000
Depreciation (40% variable)	6,000
Administrative expenses (100% fixed)	12,000

Prepare a flexible budget showing individual expenses of production at 1500 units and 2000 units *(Calcutta University B.Com. (Hons.), 1998)*

[**Ans.** For 1500 units production the budgeted cost of production is Rs. 1,12,100, for 2,000 units production the budgeted cost of production is Rs. 1,43,200]

Flexible Budget

8. From the following data, prepare a flexible budget for production of 40,000 units and 75,000 units distinctly showing variable cost and fixed cost as well as total cost. Also indicate element-wise cost per unit. Budgeted output is 1,00,000 units and budgeted cost per unit is as follows:

	Rs.
Direct material	95
Direct labour	50
Production overheads (variable)	40
Production overhead (fixed)	5
Administration overhead (fixed)	5
Selling overhead (10% fixed)	10
Distribution overhead (20% fixed)	15

(University of Delhi, B.Com. (Hons.), 2000)

[**Ans.** At 75,000 units cost per unit is 224.67 and total cost is Rs. 1,68,50,000. At 40,000 units cost per unit is 241 and total cost is Rs. 96,40,000]

❑❑❑

CHAPTER 14

STANDARD COSTING

HISTORICAL COSTING

Historical costing is one of the technique of ascertaining cost of production. It is based on accumulation of actual or historical cost. The National Association of Accountants has defined historical cost as "the cost which is accumulated during the process of production by the usual historical costing technique as opposed to the cost which has been determined in advance of the production process. The term 'actual' is not intended to convey any implication as to the accuracy with which costs are measured. "Historical costing is not a very popular technique as it suffers from the following limitations:

(*a*) It provides cost information only after the completion of production. So cost of production cannot be ascertained until production is completed. The fixation of selling price becomes difficult under such situation.

(*b*) It is not possible to exercise control over the costs which are actually incurred.

(*c*) It becomes difficult to take decisions based on actual costs.

(*d*) Actual costs may vary from period to period and fixation of selling price on the basis of actual cost will lead to differences in selling price which may not be accepted by customers.

(*e*) It is an expensive and time-consuming technique. For example, in a larger sized business, to ascertain the cost of production of say 1 lakh units involving different materials and various expenses will involve more time and clerical work.

In order to overcome the above limitations, standard costing was developed.

STANDARD COST AND STANDARD COSTING

A standard cost has been defined by Lucey (*Costing*) as a "predetermined calculation of how much costs should be under specified working conditions".

Brown and Howard in their book *Principles and Practice of Management Accountancy* define a standard cost as a predetermined cost which determines what each product or service should cost under given circumstances.

The ICMA terminology defines a standard cost as "a predetermined cost calculated in relation to a prescribed set of working conditions, correlating technical specifications and scientific measurements of materials and labour to the prices and wage rates expected to apply during the period to which the standard cost is intended to relate, with an addition of an appropriate share of budgeted overhead".

The objects of standard costs are as follows:

1. Promoting and measuring efficiencies
2. Controlling and reducing costs

3. Simplifying costing procedure
4. Valuing inventories
5. Fixing selling price

Standard costing has been defined by Brown and Howard as "a technique of cost accounting which compares the 'standard cost' of each product or service with the actual cost to determine the efficiency of the operation, so that any remedial action may be taken immediately".

The ICMA terminology defines standard costing as "the preparation and use of standard costs, their comparison with actual cost and the analysis of variances to their causes and points of incidence".

According to the above two definitions, standard costing involves the following steps:

(*a*) Setting of standards.
(*b*) Measurement of results.
(*c*) Comparison of actuals with standards to determine the variance.
(*d*) Investigation of variance.
(*e*) Taking remedial action to set right adverse variance.

DIFFERENCE BETWEEN A BUDGET AND STANDARD

1. A standard is a unit concept, *i.e.,* that apply to a particular product, process or operation, whereas budget is concerned with totals, *i.e.,* for the firms as a whole.
2. Standards are revised only when they are inappropriate for current operating conditions, whereas budget is revised on a periodic basis generally once a year.
3. Standards and the resulting variances are part of the double entry system of accounting, whereas budgets are memorandum figures and do not form part of double-entry system of accounting.

Differences between A Budgetary Control and Standard Costing

Budgetary control	*Standard costing*
1. It is prepared to cover various functions of a business such as purchases, sales, production, finance, etc. In other words it has a macro-approach.	1. It is prepared in respect of a cost unit. In other words it has a micro-approach.
2. It is more extensive as it covers all the operations of the business.	2. It is more intensive technique of controlling costs.
3. It is a projection of financial accounts.	3. It is a projection of cost accounts.
4. It can be implemented even in parts, *i.e.,* to cover one or more than one area of business.	4. It covers all items of expenses without leaving any item. So it cannot be operated in part.
5. It can be operated without standards.	5. It cannot exist without budget.
6. It is more management oriented.	6. It is more engineering oriented.
7. It can be implemented in all industries.	7. It is not possible in certain industries.
8. It does not involve any accounting after computing variances.	8. Variances are accounted for under standard costing.
9. Its objects are formulation of policy, coordination of activities and delegation of authority.	9. Its aim is to enable management in making decisions, in price-fixing and valuation of closing stock.

CHAPTER 14

STANDARD COSTING AND STANDARDISED COSTING

Standard costing should not be confused with standardised costing as both are completely different. Standard costing is a tool of cost control whereas standardised costing refers to a uniform application of costing and cost accounting principles and procedures to a firm having number of plants. The object of standardised costing is to facilitate inter-firm comparison of results.

Secondly, standard costing technique makes use of only standard cost. Whereas standardised costing can be operated with the help of historical cost also.

Thirdly, standard costing is operated on the principles of exception, whereas, standardised costing is based on the principle of example.

Lastly, the objective of standard costing is to set and maintain the standard. Whereas, the objective of standardised costing is to improve upon the standard in addition to maintaining the standard.

ADVANTAGES AND LIMITATIONS OF STANDARD COSTING

The advantages of standard costing are as follows:

1. It provides a yardstick against which actuals are measured.
2. Standard costing helps in standardising the various activities of a business. Thus, it enables use of standard materials and better method of production.
3. It increases the cost consciousness among all concerned by fixing targets and to achieve them.
4. This technique follow the principles of management by exception.
5. It simplifies the accounting procedure which involves less clerical work and time.
6. It helps in valuing the closing stock.
7. It pinpoints the responsibility of everybody concerned in the organisation.
8. It facilitates cost control by analysing the causes of inefficiency and by taking remedial measures. This will result in reduced cost of production.
9. It enables in periodic preparation of profit and loss account which helps management in knowing the trends of the business.
10. It helps in introducing incentives to employees and provides a basis for motivating them.
11. It helps in fixing selling price in advance of production. Thus quotations can be sent and orders secured.

The limitations of standard costing are as follows:

1. It is difficult to establish standards in practice.
2. Standards established may have to be revised owing to changing conditions. But frequent revision of standards is costly and may create problems.
3. Standards which are inaccurate, unreliable and outdated may do more harm than good.
4. If the standards set are not attained it will lead to psychological effects resulting in frustration.
5. It may not be suitable for small concerns.

6. It cannot be applied to non-standardised industries and in industries where goods are to be manufactured according to specification of customers.

PRELIMINARIES IN ESTABLISHING SYSTEM OF STANDARD COSTING

Before establishing a system of standard costing the following preliminary factors are to be considered.

1. **Establishment of Cost Centre.** In order to identify the area of efficiency or otherwise of it and to fix responsibilities for the persons who are concerned with attaining the standards, it is necessary to establish cost centre. A main cost centre can again be sub-divided into sub-cost centre for exercising better control. Having established cost centres, the next requirement is to fix up responsibility on the persons who are incharge of such cost centres.
2. **Classification of Accounts.** It refers to classifying the costs under a suitable accounting heading. This facilitates quick identification and collection of expenses, their analysis and in prompt reporting. As was explained under chapter 2, costs can be classified on the basis of number of characteristics. Having classified, it is desirable to give a code number for each heading of cost. For example, raw materials may be given a code number from 0 to 10, direct labour cost 11 to 20 factory overheads 21 to 30 and so on. This results in convenience in dealing with costs and ensures secrecy.
3. **Determination of Duration.** This refers to the period for which standard is used. It may be a short or long duration.
4. **Determination of Capacity.** It refers to fixation of standard output for the sake of fixed and variable overheads.
5. **Types of Standard.** It refers to the level of attainment accepted by management as the basis on which standard costs are determined. The following are the important types of standards.

   (*a*) **Basic or fixed or static standard:** It is defined by the ICMA terminology as "a standard established for use over a long period from which a current standard can be developed. These standards remain constant for a long period of time and hence they are also termed as long-term standards. They show the trend over a period of time relating to material prices, labour rates, effects of changing methods. They serve at the most as the statistical data and is not used to evaluate current efficiency or inefficiency.

   (*b*) **Ideal standard:** It is defined by the ICMA terminology as "a standard which can be attained under the most favourable conditions". Such standards are based on the best possible operating conditions. In other words the productive activity is carried out assuming that there are no breakdown in machineries, no wastage in materials, no idle time. This standard, while using is revised periodically to reflect improvement in methods, materials and technology. It is difficult to attain ideal standard in practice and hence rarely used. Therefore, such standards are used for the purpose of investigation and development but not for controlling day-to-day activities.

   (*c*) **Attainable or expected or practical standard:** It is defined by the ICMA terminology as "a standard which can be attained if a standard unit of work is carried out efficiently, a machine properly operated or materials properly used.

This is most widely used in practice. This standard is based on efficient operating condition though not a perfect condition. This standards covers allowances for normal material wastages, for fatigue, machine breakdown, etc. This standard is more realistic in nature and hence used for cost control, valuing stock and a basis for budgeting.

(*d*) **Current standard:** It is defined by the ICMA terminology as "a standard established for use over a short period of time related to current conditions". This standard is used to reflect the current condition. It can be equated to attainable standard during the period of stable prices and normal conditions.

(*e*) **Normal standard:** It is defined by the ICMA terminology as "the average standard which it is anticipated can be attained over a future period of time, preferably, long enough to cover one trade cycle". Accordingly, this standard may be prepared once in 10 years.

STANDARD HOUR

It is a measure used in standard costing by means of which products in different forms such as solid, liquid and gases are measured. The standard hour is the quantity of output or amount of work which should be produced in one hour. In other words, a standard hour is the expression of the actual output in terms of standard time instead of units. The ICMA terminology defines it as "a hypothetical unit pre-established to represent the amount of work which should be performed in one hour under standard performance". The concept of standard hour has a practical advantage in the measurement of efficiency ratio and activity ratio of an organisation.

Problem 1. A factory produces three products X Y and Z. The standard time allowed are 10 hours, 15 hours and 12 hours respectively. The actual production in January 2010 and February 2010 are as under:

		Jan. 2010 (units)	*Feb. 2010 (units)*
Product	X	20,000	12,000
	Y	10,000	12,000
	Z	15,000	18,000

By what percentage has production changed in February over January. If the actual hours paid for each in January and February were 6,00,000 what is the rate of labour efficiency ?

Solution:

STATEMENT SHOWING PRODUCTION IN JANUARY & FEBRUARY

	January		*February*	
	Units	*Std. hrs.*	*Units*	*Std. hrs.*
X	20,000	2,00,000	12,000	1,20,000
Y	10,000	1,50,000	12,000	1,80,000
Z	15,000	1,80,000	18,000	2,16,000
Total	45,000	5,30,000	42,000	5,16,000

Output in February has declined by 14,000 hrs. (5,30,000 – 5,16,000), *i.e.*, by 2.6% over January.

Labour efficiency in:

January $\frac{5,30,000}{6,00,000} \times 100 = 88.33\%$

February $\frac{5,16,000}{6,00,000} \times 100 = 86\%$

STANDARD COST CARD

A standard cost card is used for recording the standard cost. It is prepared for each product or service. It contains particulars relating to quantity and price of materials consumed, time and rate of labour required, overheads and the total cost. A specimen of standard cost card is shown below:

Standard Cost Card

Part No. .... *Description* .... *Batch Qty.* ....

Elements of cost	Standard rate	Dept. 1	Dept. 2	Dept. 3	Total
Direct material					
Direct labour					
Production overhead					
Administration overhead					
Selling and distribution					

Fig. 14.1. *Specimen showing standard cost card.*

PROCEDURE FOR INTRODUCING STANDARD COSTING

Introduction of standard costing system involves a number of steps. Setting up of standard cost for each element of cost—direct material, direct labour and overheads is a complex task. Standards must be set up carefully since a low standard which everyone can achieve does not bring out the best performance. Too high a standard on the other hand, is being impossible to achieve, is treated with disdain and indifference. Setting up of standards is not merely an accounting job, it involves the cooperation of various functional departments. The following steps are to be followed while introducing a standard costing in a factory.

1. To study the technical aspects of the factory. This involves a study of the various methods of manufacture and the process involved. The input-output relationship of each process is to be worked out. For this purpose, the total production process has to be subdivided into various sub-process of manufacture. The pattern of losses — normal and abnormal in each sub-process over a considerable period of time should be examined. By going through the previous production records, the number of defectives and their cost of rectification should also be looked into. Previous records should be studied in order to work out the normal efficiency of labour in each production process. This study of the technical aspect of the factory is essential since a system of standard cost must be based on the actual situation in the factory.
2. The existing cost system should be reviewed with special reference to the existing records and forms. This is important since the standard costing system has to be established as an extension to the existing system of maintaining cost.
3. The organisation chart showing the various lines of authority and responsibilities should be studied so that the responsibility of supplying basic data regarding the various operations of the undertaking can be established.

4. One of the important steps involved in the implementation of standard costing is to secure the cooperation of all executive in the factory in order to fix the quality and efficient standard. The standard costing will be successful only if it receives full-fledged support from various line managers. If line managers view it merely as an imposition by the accounting department they will never cooperate with it. It is essential to classify that the system is for the benefit and would be run only if they find it useful. A standard cost committee may be formed comprising important line managers to discuss the various problems regarding standard costing system.
5. The existing cost policies specially with regard to the methods of allocation and apportionment of overheads should be studied so that what should constitute as the standard cost can be clearly defined.
6. The budgetary control and internal control procedures should also be reviewed since the standard costing system is to be built-up to the existing system of control and budgeting.
7. A detailed manual should be prepared for the guidance of the staff. This manual should briefly describe the system about to be enforced and should list benefit arising therefrom. It should then clearly demarcate responsibilities of generating the activity and cost data. The various definitions must be clearly defined and the following procedure of the system outlined:
   (*a*) Establishment of standard cost.
   (*b*) Target dates for fixing the standards.
   (*c*) Target dates for fixing standard cost.
   (*d*) Various reports of actual performances to be sent to accounts department for ascertaining the variances.
   (*e*) Variance analysis reports to be sent by the accounts department.
   (*f*) Any other relevant information.
8. The office-staff required to work the system should be properly trained. When the system goes into force, a number of alterations would be required in the scheme of making entries and in document flow.
9. To fix up the material quantity standard. The following is the procedure involved in setting material quantity standars.
   (*a*) *Standardisation of products:* This involves drawing detailed specifications, blue prints, norms for normal wastage.
   (*b*) *Classification of products:* This involves preparation of detailed lists of products to be manufactured.
   (*c*) *Standardisation of materials:* This involves determination of quality, specifications, etc. to be used in the standard products.
   (*d*) *Bill of materials:* This involves preparation of a bill of materials for each product showing the quantity of materials needed.
   (*e*) *Test runs:* This helps in setting standards when production is to be undertaken under regulated conditions.

The material price standards are then fixed based on the following factors:

(*a*) Stock of materials on hand and its value.
(*b*) The prices at which orders for future deliveries of materials have been placed.
(*c*) Possible price fluctuations.

10. To fix-up labour quantity standards by applying the following steps:
    (*a*) Standardisation and classification of products.
    (*b*) Training facilities required.
    (*c*) Time and motion studies.
    (*d*) Production planning techniques.
    (*e*) Selection of right machines, deciding the method of operation.

The labour cost standard can be set up by studying data regarding the wage rates paid.

11. *Determination of standard overhead rate:* This involves the following steps:
    (*a*) Determination of standard overhead cost.
    (*b*) Estimation of the level of activity.

By dividing standard overhead for the budget period by the estimated level of activity, the standard overhead rate can be compiled.

12. *Determination of standard administration cost:* The purpose of setting standard administration cost is to secure maximum service at minimum cost. This objective can be achieved by a study conducted by organisation and method department. Setting up of standards also depends upon standardisation and simplification of office procedure. The standard quantity of work to be performed can be set on the following basis:
    (*a*) Past performance.
    (*b*) On the advise of O & M studies.
    (*c*) Time and motion studies.
13. *Determination of standard selling and distribution cost:* Selling and distribution expenses bears a close relationship with sales volume. Hence a sales forecast is necessary before setting standard cost for selling and distribution. The classification into fixed and variable will help in setting standard cost.

VARIANCE ANALYSIS

The ICMA terminology defines a variance as "the difference between a standard cost and the comparable actual cost incurred during a given period. The purpose of knowing the variance is to enable the management to exercise control over cost. It enables to know whether the standards set is achieved or not.

Variance analysis has been defined by the terminology as "the process of computing the amount of and isolating the cause of variances between actual costs and standard cost". Thus variance analysis involves: (*a*) computation of individual variances and, (*b*) determination of the cause of each variance. The purpose of variance analysis is to enable management to improve operations, increase efficiency, utilize resources more effectively and reduce cost. To serve these purpose variance analysis must be easy to understand, they must be calculated immediately and the causes must be immediately enquired and remedial measures quickly taken.

A variance is said to be favourable when the actual results are better than the standards and an adverse or unfavourable variance when actuals are not up to the standard. Similarly, a variance is said to be controllable when it is amenable to control by an executive action and an uncontrollable variance when it is not amendable to control by executive action. An unfavourable variance is caused by external environment such as market conditions, fluctuations in demand, supply etc. In other words, no individual in the organisation can be held responsible for it.

Before calculating various cost variances it would be appropriate to know the meaning of the following concepts:

(*a*) *Actual production:* It means actual quantity produced during the actual hours worked.

(*b*) *Budgeted production:* It means the budgeted quantity to be produced during the budgeted hours to be worked.

(*c*) *Standard production:* It means the quantity which should have been produced during the actual hours worked.

(*d*) *Actual cost:* It means the actual quantity produced at the actual cost per unit.

(*e*) *Budgeted cost:* It means the budgeted quantity to be produced at the standard cost per unit.

(*f*) *Standard cost:* It means the actual quantity produced at the standard cost per unit.

The various cost variances are as follows:

1. Material Variance

The various material cost variances can be shown under the following chart:

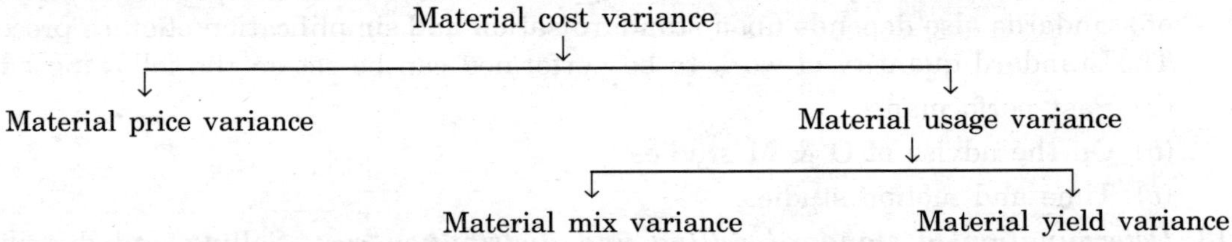

Fig. 14.2 *Chart showing material cost variance*

(*a*) **Material cost variance:** It refers to the difference between the standard cost of materials specified and the actual cost of materials used. It arises owing to variation in the price of the material or in its usage. The ICMA terminology defines it as, "the difference between the standard direct material cost of the actual production volume and the actual cost of direct material. "The following formula is used to calculate this variance.

Material cost variance = (Standard units × Standard price) – (Actual units × Actual price)

Material cost variance is the aggregate of material price and usage variance.

(*b*) **Material price variance:** It refers to that portion of material cost variance which is due to the difference between the standard price and the actual price paid. The ICMA terminology defines it as, "that portion of the direct material cost variance which is the difference between the standard price specified and the actual price paid for the direct material used". This variance arises on account of extra price paid on the units purchased. It can be calculated at the time of purchase or at the time of usage. Generally the former is preferable. The formula to calculate material price variance is as follows:

Direct material variance = (Standard price – Actual price) × Actual quantity

The direct materials price variance arises on account of the following causes:

(*a*) Higher or lower prices paid than planned.

(*b*) Discounts received or foregone depending upon quantity bought.

(*c*) Purchase of superior or inferior quality of materials than planned.

(*d*) Buying substitute materials.

(*e*) Increase in charges such as transport cost, etc.

(*c*) **Material usage variances:** It refers to that portion of material cost variance owing to the difference between the standard quantity of materials specified and the actual quantity used. The ICMA terminology defines it as "the difference between the standard quantity specified for the actual production and the actual quantity used at standard purchase price". The following formula is used to calculate this variance:

Material usage variance = (Standard quantity – Actual quantity) × Standard price

The following causes give rise to materials usage variance:

(*a*) Low or high yield from material expected.

(*b*) Gain or loss arising out of substitute materials.

(*c*) Difference in the quality of materials than planned.

(*d*) Increased or decreased quantity of scrap than expected.

(*e*) Use of sub-standard or defective materials.

(*f*) Carelessness in the use of materials.

(*g*) Pilferage.

(*h*) Defective method of production.

(*i*) Wrong mixture of raw materials.

The materials usage variance can be sub-divided into material mix and material yield variance.

(*d*) **Materials mix variance:** The calculation of this variance arises in those industries where different types of raw materials are mixed to obtain required output. Examples of such industries are fertilisers, chemical, cement, food processing industries and so on. One distinct feature of such industries is, it involves losses by way of evaporation, breakage, shrinkage, etc. and are responsible for difference in the output. The ICMA terminology defines it as "the difference between total quantity in standard proportion, priced at the standard price and the actual quantity of materials used at the standard price. In simple words it is that portion of material usage variance which is due to the difference between the standard and the actual composition of mixture.

Material mix variance may arise under the following two situations:

1. Where the ratio of standard mix differ from the ratio of actual mix, but the total quantity of both the actual mix and standard mix remaining the same.

The following formula is used in this situation.

Material mix variance = (Revised std. qty. – Actual qty.) × Std. price

In this case the revised standard quantity is the same as the standard quantity.

2. Where both the ratio as well as total quantity differ between the standard and the actual mix.

The above formula is applicable in this situation also with a difference in calculating the revised standard quantity. The following formula is used to calculate the revised standard quantity.

$$\text{Revised std. qty.} = \frac{\text{Total weight of actual mix}}{\text{Total weight of std. mix}} \times \text{Std. qty. of material in question}$$

CHAPTER 14

This is because, the standard quantity of each material will be revised, when the total weight of actual mix varies from the total weight of standard mix.

(*e*) **Material yield variance:** It is the difference between the standard yield specified and the actual yield obtained. The ICMA terminology defines it as "the difference between the standard yield of the actual material input and the actual yield both valued at the standard material cost. There are two situations under which material yield variance can arise, viz.

(*i*) Where actual mix does not differ from standard mix. In this situation the following formula is used:

$$\text{Material yield variance} = (\text{Total actual yield} - \text{Total std. yield}) \times \text{Std. yield rate}$$

$$\text{Where, Std. yield rate} = \frac{\text{Std. cost of std. mix}}{\text{Net standard output}}$$

(*ii*) Where actual mix differ from standard mix. The following formula is used in this situation

$$\text{Material yield variance} = (\text{Actual yield} - \text{Revised std. yield}) \times \text{Std. yield rate}$$

$$\text{Where, Std. yield rate} = \frac{\text{Std. cost of revised std. mix}}{\text{Net std. output}}$$

Material yield variance will arise owing to the following causes:

1. Use of sub-standard quality of materials.
2. Losses arising out of chemical reaction.
3. Defective method of production.
4. Inadequate supervision.

Problem 1. *(Material cost variance):* Product 'A' requires 20 kgs of material at Rs. 4 per kg. The actual consumption of material for the manufacturing of product 'A' came to 24 kgs of materials at Rs. 4.50 per kg. Calculate

1. Material cost variance
2. Material price variance
3. Material usage variance [*University of Madras, (B.Com.), April 2005*]

Solution:

(1) Material cost variance:

$$= \left(\begin{matrix}\text{Std. qty.} \\ \text{for actual output}\end{matrix} \times \text{Std. price}\right) - \left(\begin{matrix}\text{Actual} \\ \text{qty.}\end{matrix} \times \begin{matrix}\text{Actual} \\ \text{price}\end{matrix}\right)$$

= (20 kg × Rs. 4) – (24 kg × 4.50)

= Rs. 80 – Rs. 108 = Rs. 28 (A)

(2) Material price variance:

= (Std. price – Actual price) × Actual qty.

= (Rs. 4 – Rs. 4.50) × 24 kg = 12 (A)

(3) Material usage variance:

$$= \left(\begin{matrix}\text{Std. qty. for} \\ \text{actual output}\end{matrix} - \begin{matrix}\text{Actual} \\ \text{qty.}\end{matrix}\right) \times \text{Std. rate}$$

= (20 kg – 24 kg) × Rs. 4 = 16 (A)

Verification

Material cost variance	=	Material price variance	+	Material usage variance
28 (A)	=	12 (A) + 16 (A)		
28 (A)	=	28 (A)		

Problem 2. The standard material and standard cost per kg of material required for the production of one unit of product A is as follows:

Material = 5 kgs
Standard price = Rs. 5 per kg

The actual production and related material date are as follows:
400 units of product A
Material used 2200 kgs.
Price of material Rs. 4.50 per kg

Calculate:
(1) Material cost variance
(2) Material usage variance
(3) Material price variance *(University of Madras, B.Com., April 2004)*

Solution:

CALCULATION OF STANDARD MATERIAL

Std. qty. of material for one unit	= 5 kg
Std. qty. of material for 400 units	= 400 × 5 = 2,000 kgs
Std. material price	= Rs. 5 per kg

MCV = MUV + MPV

100 (F) = 1000 (A) + 1100 (F)

100 F = 100 F

Problem 3. From the following particulars, calculate:
(*a*) Material usage variance
(*b*) Material price variance
(*c*) Material cost variance

Standard : Materials for 70 kg of finished products = 100 kg
Price of material Re. 1 per kg.

Actual: Output 2,10,000 kg
Materials used 2,80,000 kg
Cost of materials Rs. 2,52,000

(University of Madras, B.Com., October 2002)

Solution:

CALCULATION OF ACTUAL PRICE

$$\text{Actual price} = \frac{\text{Cost of materials}}{\text{Materials used}}$$

$$= \frac{\text{Rs. } 2,52,000}{2,80,000 \text{ kg}} = 0.90 \text{ P}$$

CHAPTER 14

(1) Material price variance:

(Std. price – Actual price) × Actual qty.

= (Re. 1 – Re 0.90) × 2,80,000

= Rs. 28,000 (F)

(2) Material usage variance:

= (Std. qty. – Actual qty.) × Std. price

= (3,00,000 kgs – 2,80,000 kgs) × Re 1

= Rs. 20,000 (F)

Working Note: **Calculation of Std. Qty.**

For 70 kgs of output = 100 kg of material

For 2,10,000 kgs output = $\frac{100}{70} \times 2,00,000$

= 3,00,000 kgs

(3) Materials cost variance:

= (Std. quantity × Std. price) – $\left(\text{Actual Qty} \times \text{Actual price}\right)$

= (3,00,000 kg × Re 1) – (2,80,000 × 0.90)

= Rs. 3,00,000 – Rs. 2,52,000 = Rs. 48,000 (F)

Verification

Material cost variance	=	Material price variance	+	Material usage variance
Rs. 48,000 (F)	=	Rs. 28,000 (F) + Rs. 20,000 (F)		
48,000 (F)	=	48,000 (F)		

Problem 4. From the following data calculate:

(*a*) Material cost variance

(*b*) Price variance

(*c*) Usage variance

Product	*Std. qty.*	*Std. price*	*Actual qty.*	*Actual price*
A	1,050	2.00	1,100	2.25
B	1,500	3.25	1,400	3.50
C	2,100	3.50	2,000	3.75

(University of Madras, B.Com., April 2001)

Solution:

(1) Material cost variance

= (Std. price × Std. qty.) – $\left(\text{Actual price} \times \text{Actual qty.}\right)$

A = (2 × 1,050) – (2.25 × 1,000)

= 2,100 – 2,475 = 375 (A)

B = (3.25 × 1,500) – (3.50 × 1,400) = 25 (A)

C = (3.50 × 2,100) – (3.75 × 2,000) = 150 (A)

550 (A)

(2) Material price variance

= (Std. price – Actual price) × Actual qty.

A	=	(2.00 – 2.25) × 1,100	=	275 (A)
B	=	(3.25 – 3.50) × 1,400	=	350 (A)
C	=	(3.50 – 3.75) × 2,000	=	500 (A)
				1,125 (A)

(3) Material usage variance:

= Std. qty. – Actual qty.) × Std. rate

A	=	(1,050 – 1,100) × 2	=	100 (A)
B	=	(1,500 – 1,400) × 3.25	=	325 (F)
C	=	(2,100 – 2,000) × 3.50	=	350 (F)
				575 (F)

Verification

Material cost variance	=	Material price variance	+	Material usage variance
550 (A)	=	1,125 (A)	+	575 (F)
550 (A)	=	550 (A)		

Problem 5. From the following data, calculate

(*a*) Material cost variance
(*b*) Material price variance
(*c*) Material usage variance
(*d*) Material mix variance

Name of Material	*Standard*		*Actual*	
	Kg.	*Rate*	*Kg.*	*Rate*
X	8,000	1.05	7,500	1.20
Y	3,000	2.15	3,300	2.30
Z	2,000	3.30	2,400	3.50

(*University of Madras, B.Com., May 2002*)

Solution:

(*a*) Material cost variance:

$$= \left(\text{Std. price} \times \begin{array}{c}\text{Std. qty.}\\ \text{for actual}\\ \text{output}\end{array}\right) - \left(\begin{array}{c}\text{Actual}\\ \text{price}\end{array} \times \begin{array}{c}\text{Actual}\\ \text{qty.}\end{array}\right)$$

X	=	(1.05 × 8,000) – (1.20 × 7,500)		
	=	8,400 – 9,000	=	600 (A)
Y	=	(2.15 × 3,000) – (2.30 × 3,300)	=	1,140 (A)
Z	=	(3.30 × 2,000) – (3.50 × 2,400)	=	1,800 (A)
				3,540 (A)

(*b*) Material price variance

= (Std. price – Actual price) × Actual qty.

X	=	(1.05 – 1.20) × 7,500	=	1,125 (A)
Y	=	(2.15 – 2.30) × 3,300	=	495 (A)
Z	=	(3.30 – 3.50) × 2,400	=	480 (A)
				2,100 (A)

(*c*) **Material usage variance:**

$$= \left(\begin{array}{c}\text{Std. qty. for}\\ \text{actual output}\end{array} - \begin{array}{c}\text{Actual}\\ \text{output}\end{array}\right) \times \text{Std. rate}$$

X	=	(8,000 – 7,500) × 1.05	=	525 (F)
Y	=	(3,000 – 3,300) × 2.15	=	645 (A)
Z	=	(2,000 – 2,400) × 3.30	=	1,320 (A)
				1,440 (A)

Verification

Material cost variance	=	Material price variance	+	Material usage variance
3,540 (A)	=	2,100 (A)	+	1,440 (A)
3,540 (A)	=	3,540 (A)		

(*d*) **Material mix variance:**

$$\text{Revised std. qty.} = \frac{\text{Total actual mix qty}}{\text{Total std. mix qty.}} \times \text{Individual std. qty.}$$

$$X = \frac{13{,}200}{13{,}000} \times 8{,}000 = 8{,}123$$

$$Y = \frac{13{,}200}{13{,}000} \times 3{,}000 = 3{,}046$$

$$Z = \frac{13{,}200}{13{,}000} \times 2{,}000 = 2{,}031$$

Total = 13,200

$$\text{Mix variance} = \left(\begin{array}{c}\text{Revised std.}\\ \text{qty.}\end{array} - \begin{array}{c}\text{Actual}\\ \text{qty.}\end{array}\right) \times \text{Std. price}$$

X	=	(8,123 – 7,500) × 1.05	=	654 (F)
Y	=	(3,046 – 3,300) × 2.15	=	546 (A)
Z	=	(2,031 – 2,400) × 3.30	=	1,218 (A)
				1,110 (A)

Problem 6. The standard cost of a chemical mixture is as under:

4 tons of material 'X' at Rs. 20 per ton
6 tons of material 'Y' at Rs. 30 per ton
Standard yield is 90% of input
4.5 tons of material 'X' at Rs. 15 per ton
5.5 tons of material 'Y' at Rs. 34 per ton
Actual yield is 9.1 tons

Compute:

(*a*) Material price variance
(*b*) Material usage variance
(*c*) Material mix variance
(*d*) Material cost variance

(Osmania University, M.Com., June 2003)

Solution:

Materials	Std. cost			Actual cost		
	Std. qty	*Std. price*	*Std. cost*	*Actual qty.*	*Actual price*	*Actual cost*
X	4	20	80	4.5	15	67.5
Y	6	30	180	5.5	34	18.7
	10		260	10		254.5

(*a*) Material price variance

(Std. price – Actual price) × Actual qty.

X = (20 – 15) × 4.5

= (5) × 4.5 = 22.5 (F)

Y = (30 – 34) × 5.5

= (4) × 5.5 = 22.0 (A)

0.5 (F)

(*b*) Material usage variance:

(Std. qty. – Actual qty.) × Std. price

X = (4 – 4.5) × 20

= (0.5) × 20 = 10 (A)

Y = (6 – 5.5) × 30

= (0.5) × 30 = 15 (F)

5 (F)

(*c*) Material mix variance:

Actual weight of the mix and the standard weight of mix do not differ. In this case material mix variance is calculated using the following formula

Material mix variance = (Std. price – Actual qty.) × Std. price

X = (4 – 4.5) × 20 = 10 (A)

Y = (6 – 5.5) × 30 = 15 (F)

5 (F)

(*d*) Material cost variance:

Std. cost – Actual cost = 260 – 254.5 = 5.5 (F)

Varification

Material cost variance = Material price variance + Material usage variance

5.5 = 0.5 + 5

5.5 = 5.5

Problem 7. The standard material cost for 200 units of output is

Material	*Kg*	*Rate per kg*
A	50	12
B	100	9
C	100	10

The actual cost for 8,000 units is as follows

Material	*Kg*	*Rate per kg*
A	2,100	28,350
B	3,750	30,750
C	4,150	46,480

Calculate material cost variance, material price variance and material usage variance.

(University of Mumbai, B.Com., October 2005)

Solution:

WORKING NOTE

	Standard (200 units)			*Std. for actual output*			*Actual output 8000 units*		
Product	*qty.*	*Price*	*Rs.*	*Std. qty.*	*Std. price*	*Std. cost*	*Actual qty.*	*Actual price*	*Actual cost*
A	50	12	600	2,000	12	24,000	2,100	13.50	28,350
B	100	9	900	4,000	9	36,000	3,750	8.20	30,750
C	100	10	1,000	4,000	10	40,000	4,150	11.20	46,480
			2,500			1,00,000			1,05,580

(1) Material cost variance:

= Std. cost – Actual cost

= 1,00,000 – 1,05,580 = 5,580 (A)

(2) Material price variance:

= (Std. price – Actual price) × Actual qty.

A = (12 – 13.5) × 2,100 = 3,150 (A)

B = (9 – 8.2) × 3,750 = 3,000 (F)

C = (10 – 11.2) × 4,150 = 4,980 (A)

5,130 (A)

(3) Material usage variance

= (Std. qty. – Actual qty.) × Std. price

A = (2,000 – 2,100) × 12 = 1,200 (A)

B = (4,000 – 3,750) × 9 = 2,250 (F)

C = (4,000 – 4,150) × 10 = 1,500 (A)

450 (A)

Verification

Material cost variance = Material price variance + Material usage variance

5,580 (A) = 5,130 (A) + 450 (A)

5,580 (A) = 5,580 (A)

Problem 8. For producing 80 units of a product 30 kg of material 'X' and 20 kg of material 'Y' is the requirement. Standard price is Rs. 6 per kg of 'X' and Rs. 10 per kg of 'Y'. 80 units were actually produced using 50 kg of materials 'X' purchased for Rs. 200 and 10 kg of materials 'Y' purchased at Rs. 8 per kg. Compute: (1) Material cost variance, (2) Material price variance and (3) Material usage variance.

(University of Bombay, B.Com., April 2005)

Solution:

WORKING NOTE

	Standard			Actual		
	qty.	*Rate*	*Amount*	*Qty.*	*Rate*	*Amount*
X	30	6	180	50	4	200
Y	20	10	200	10	8	80
			380			280

(1) Material cost variance:

= Std. cost – Actual cost

= 380 – 280 = 100 (F)

(2) Material price variance:

= (Std. price – Actual price) × Actual qty.

X = (6 – 4) × 50 = 100 (F)

Y = (10 – 8) × 10 = 20 (F)

120 F

(3) Material usage variance

= (Std. qty. – Actual qty.) × Std. price

X = (30 – 50) × 6 = 120 (A)

Y = (20 – 10) × 10 = 100 (F)

20 (A)

Verification

Material cost variance = Material price variance + Material usage variance

100 (F) = 120 (F) + 20 (A)

100 (F) = 100 (F)

Problem 9. A manufacturing concern which has adopted standard costing furnishes the following information:

(1) Standard materials for 70 kg finished products, 100 kg
(2) Standard price of material Re 1 per kg
(3) Actual output 2,10,000 kg
(4) Actual material used 2,80,000 kg
(5) Cost of material used Rs. 2,52,000

Calculate (1) Material usage variance, (2) Material price variance, (3) Material cost variance.

(University of Mumbai, B.Com., October 2003)

Solution:

Working Notes:

(1) Calculation of Std. output

For 100 kg materials : the output is 70 kg

CHAPTER 14

For 2,80,000 kg materials = $\frac{2,80,000 \times 70}{100}$ = 1,96,000 kgs

(2) Calculation of Std. materials

For 70 kg output the raw materials is 100 kg

For 2,10,000 kg output = $\frac{2,10,000 \times 100}{70}$ = 3,00,000 kgs

(3) Calculation of Actual rate

$= \frac{\text{Cost of materials}}{\text{Actual materials used}}$

$= \frac{\text{Rs. } 2,52,000}{2,80,000}$ = Rs. 0.90

(1) Material cost variance:

$= \left(\begin{matrix}\text{Std.} \\ \text{qty.}\end{matrix} \times \begin{matrix}\text{Std.} \\ \text{rate}\end{matrix}\right) - \left(\begin{matrix}\text{Actual} \\ \text{qty.}\end{matrix} \times \begin{matrix}\text{Actual} \\ \text{rate}\end{matrix}\right)$

= (3,00,000 kgs × Re 1) – (2,80,000 × 0.90)

= 3,00,000 – 2,52,000 = Rs. 48,000 (F)

(2) Material price variance:

= (Std. price – Actual price) × Actual qty.

= (1.00 – 0.90) × 2,80,000 kgs

= 28,000 (F)

(3) Material usage variance:

= (Std. qty. – Actual qty.) × Std. rate

= (3,00,000 kgs – 2,80,000 kgs) × Re 1

= Rs. 20,000 (F)

Verification

Material cost variance	=	Material usage variance	+	Material price variance
48,000 (F)	=	28,000 (F)	+	20,000 (F)
48,000 (F)	=	48,000 (F)		

Problem 10. 1 ton of material input yields a standard output of 1,00,000 units. The standard price of material is Rs. 20 per kg. The actual quantity of material used is 10 tons and the actual price paid is Rs. 21 per kg. Actual output obtained is 9,00,000 units. Compute material variances.

(Calcutta University, B.Com. (Hons.), 2005)

Solution: ***Working Notes:***

(1) 1 ton = 1,000 kgs

(2) Standard yield of 1 ton material = 1,00,000 units

So, standard input of 9,00,000 units

$= \frac{9,00,000}{1,00,000}$ = 9 tons or 9,000 kgs

(3) Standard cost of material per unit of output

$= \frac{20 \times 1,000}{1,00,000}$ = 0.20

(1) Material cost variance:

= (Std. qty. × Std. price) – $\left(\begin{matrix}\text{Actual} \\ \text{qty.}\end{matrix} \times \begin{matrix}\text{Actual} \\ \text{price}\end{matrix}\right)$

= (9,000 × 20) – (10,000 × 21)

= 1,80,000 – 2,10,000 = 30,000 (A)

(2) Material price variance:

= (Std. rate – Actual rate) × Actual qty.

= (20 – 21) × 10,000

= 1 × 10,000 = 10,000 (A)

(3) Material usage variance:

= (Std. qty. – Actual qty.) × Std. rate

= (9,000 – 10,000) × 20

= 1,000 × 20 = 20,000 (A)

(4) Material yield variance:

= (Std. yield for actual mix – Actual yield) × Std. rate

= (10,00,000 – 9,00,000) × 0.20

= 1,00,000 × 0.20 = 20,000 (A)

Problem 11. From the following data, calculate the following variances:

(1) Material cost variance
(2) Material price variance
(3) Material quantity variance
(4) Material mix variance
(5) Material yield variance

Material	*Standard*		*Actual*	
	Qty.	*Unit price*	*Qty.*	*Unit price*
A	60%	Rs. 20	88	Rs. 30
B	40%	Rs. 10	132	Rs. 10

Standard loss 10%
Actual output : 180 units *(University of Delhi, B.Com., (Hons.) 2005)*

Solution:

Material	Standard			Actual		
	Qty.	*Rate*	*Amount*	*Qty.*	*Rate*	*Amount*
A	120	20	2,400	88	30	2,640
B	80	10	800	132	10	1,320
	200		3,200	220		3,960
Normal loss	20					
Output	180			220		

(1) Material cost variance = Std. cost – Actual cost

= 3,200 – 3,960

= 760 (A)

(2) Material price variance:

= (Std. price – Actual price) × Actual qty.

A	=	(20 – 30) × 88	=	880 (A)
B	=	(10 – 10) × 132	=	NIL
				880 (A)

(3) Material quantity variance:

= (Std. qty. – Actual qty.) × Std. rate

A	=	(120 – 88) × 20	=	Rs. 640 (F)
B	=	(80 – 132) × 10	=	Rs. 520 (A)
				120 (F)

(4) Material mix variance:

$$= \left(\text{Revised std. qty.} - \text{Actual qty.}\right) \times \text{Std. rate}$$

$$A = \left(220 \times \frac{60}{100} - 88\right) \times 20 = \text{Rs. } 880 \text{ (F)}$$

$$B = \left(220 \times \frac{40}{100} - 132\right) \times 10 = \text{Rs. } 440 \text{ (A)}$$

440 (F)

(5) Material yield variance:

$$= (\text{Actual yield} - \text{Std. yield}) \times \frac{\text{Std. amount}}{\text{Std. output}}$$

$$= \left(180 - 220 \times \frac{90}{100}\right) \times \frac{3{,}200}{180}$$

$$= (180 - 198) \times 17.77$$

$$= 18 \times 17.77 = 320 \text{ (A)}$$

Problem 12. Bello Chemical Industries provide the following information from their records:

Material	*Qty. (kg.)*	*Rate per kg*
X	8	6
Y	4	4

During December 1990, 100 kgs of OMO were produced. The actual consumption of material is as under:

Material	*Qty. (kg)*	*Rate per kg*
X	75	7
Y	50	5

Calculate:

(i) Material cost variance.
(ii) Material price variance.
(iii) Material usage variance. *(University of Delhi, B.Com. (Hons.), 1991)*

Solution:

Material	*Standard*		*For 100 kgs of OMO Actual*	
	Qty.	*Rate*	*Qty.*	*Rate*
X	80*1	6	75	7
Y	40*2	4	50	5

(*i*) **Material cost variance = (Std. units × Std. price) – (Actual units × Actual price)**

X	=	(80 × 6) – (75 × 7)		
	=	480 – 525	=	45 (A)
Y	=	(40 × 4) – (50 × 5)		
	=	160 – 250	=	90 (A)
		Total		135 (A)

(*ii*) **Material price variance = (Std. price – Actual price) × Actual qty.**

X	=	(6 – 7) × 75	=	75 (A)
Y	=	(4 – 5) × 50	=	50 (A)
		Total		125 (A)

(*iii*) **Material usage variance = (Std. qty. – Actual qty.) × Std. price**

X	=	(80 – 75) × 6	=	30 F
Y	=	(49 – 50) × 6	=	40 (A)
		Total		10 (A)

Verification:

Material cost variance = Material price variance + Material usage variance

135 (A) = 125 A + 10 A

***1 For 10 kgs → 8 kg is required**

For 100 kgs → $\frac{8}{10} \times 100 = 80$

***2 For 10 kgs → 4 kg is required**

For 100 kgs → $\frac{4}{10} \times 100 = 40$

Problem 13. **From the given data, calculate:**

(*a*) Material price variance.

(*b*) Material usage variance.

(*c*) Material cost variance.

Standard:

1. **250 kg. of raw materials is required for producing 175 kgs. of finished products.**
2. **Price of material per kg. Rs. 4.**

Actuals:

1. **Production 52,500 kg.**
2. **Materials consumed 70,000 kg.**
3. **Cost of materials Rs. 2,73,000.**

(S.V. University, B.Com., April 1999)

Solution:

Calculation of actual price $= \dfrac{\text{Actual material cost}}{\text{Actual material consumed}}$

$= \dfrac{2,73,000}{70,000 \text{ kg}} = \text{Rs. } 3.90 \text{ per kg.}$

(1) Material price variance = (Standard price – Actual price) × Actual usage

= (4.00 – 3.90) × 70,000

= 0.10 × 70,000 = 7,000 (F)

CALCULATION OF STANDARD USAGE

For 175 kgs of production, 250 kgs of raw materials

For 52,500 kgs → ?

$\dfrac{52,500}{175} \times 250 = 75,000 \text{ kgs}$

(2) Material usage variance = (Std. qty. – Actual qty.) × Std. price

= (75,000 – 70,000) × 4

= 5,000 × 4 = 20,000 (F)

(3) Material cost variance = (Std. qty. × Std. price) – (Actual qty. – Actual price)

= (75,000 × 4) × (70,000 × 3.90)

= 3,00,000 – 2,73,000

= 27,000 (F)

Verification:

Material cost variance = Material price variance + Material usage variance

27,000 F = 7,000 (F) + 20,000 (F)

27,000 F = 27,000 (F)

Problem 14. From the given data below,

Calculate:

(*i*) The material price variance

(*ii*) Material usage variance

(*iii*) Material cost variance

Quantity of materials purchased	3,000 units
Value of materials purchased	Rs. 9,000
Standard quantity of materials required per tonne of output	30 units
Standard rate of materials	Rs. 2.50 per unit
Opening stock of materials	NIL
Closing stock of materials	500 units
Output during the period	80 tonnes

Solution:

CALCULATION OF QUANTITY OF MATERIALS USED

Qty. of opening stock of materials	Nil
Add: Qty. of materials purchased	3,000
	3,000
Less: Qty. of closing stock of materials	500
	2,500

CALCULATION OF ACTUAL PRICE OF MATERIALS PURCHASED

$$= \frac{\text{Value of materials purchased}}{\text{Qty. of materials purchased}}$$

$$= \frac{9,000}{3,000} = 3$$

CALCULATION OF STANDARD QTY. OF MATERIALS FOR ACTUAL OUTPUT

= Std. qty. of materials required per tonne of output × Actual output during the year

= 30 × 80 = 2,400 units

1. Materials price variance = (Std. price – Actual price) × Actual qty.
   = (2.50 – 3.00) × 2,500
   = 0.50 × 2,500 = 1,250 (A)
2. Material usage variance = (Std. qty. – Actual qty.) × Std. price
   = (2,400 – 2,500) × 2.50
   = 100 × 2.50 = 250
3. Materials cost variance = (Std. qty. × Std. rate) – (Actual qty. × Actual rate)
   = (2,400 × 2.50) – (2,500 × 3)
   = 6,000 – 7,500 = Rs. 15,000 (A)

Problem 15. From the following data extracted from the books Arvind Parimal Works, calculate the material mix variance:

	Standard	*Actual*
Materials A	70 units @ Rs. 6 per unit	60 units @ Rs. 10 per unit
Materials B	30 units @ Rs. 4 per unit	40 units @ Rs. 2 per unit
Output	100 units	100 units

Solution:

In this problem the total weight of the standard mix (70 + 30) = 100 units and the total weight of the actual mix (60 + 40) = 100 units are the same. So the revised standard quantity of each material will be the same as its standard quantity.

Materials mix variance = (Revised std. qty. – Actual qty.) × Std. price

A = (70 – 60) × 6
= 10 × 6 = 60 (F)

B = (30 – 40) × 4
= 10 × 4 = 40 (A)

Total material mix variance:

Material A = 60 (F)

Material B = 40 (A)

20 (F)

Problem 16. Calculate the material mix variance from the following data:

	Standard			Actual		
	Qty. *kg.*	*Rate* *Rs.*	*Amount* *Rs.*	*Qty.* *kg.*	*Rate* *Rs.*	*Amount* *Rs.*
Material *X*	6	1.50	9	5	2.40	12
Material *Y*	2	3.50	7	1	6.00	6
	8	2.00	16	6	3.60	18

Solution:

In the problem, the total standard mix is different from the total actual mix. So, we have to calculate the revised standard quantity in respect of each materials.

$$\frac{\text{Total weight of actual mix}}{\text{Total weight of standard mix}} \times \text{Std. qty. of the material in the mix}$$

Material A $= \frac{6}{8} \times 6 = 4.5$ kgs

Material B $= \frac{6}{8} \times 2 = 1.5$ kgs

Materials mix variance = (Revised std. qty. – Actual qty.) × Std. price

A = (4.5 – 5) × 1.50

= 0.5 × 1.50 = 0.75 (A)

B = (1.5 – 1) × 3.50

= 0.50 × 3.50 = 1.75 (F)

Net material mix variance:

Material A = 0.75 (A)

Material B = 1.75 (F)

1.00 (F)

Problem 17. From the given data below, calculate the material price variance, the material usage variance, material cost variance and material mix variance.

Consumption per 100 units of product

Raw material	*Standard*	*Actual*
A	40 units @ Rs 50 per unit	50 units @ Rs. 50 per unit
B	60 units @ Rs. 40 per unit	60 units @ Rs. 45 per unit

(S.V. University, B.Com., October 1999)

Solution:

CALCULATION OF ACTUAL QUANTITY USED

A = 100 units @ 50 units per unit = 5,000 units

B = 100 units @ 60 units per units = 6,000 units

1. Material price variance = (Std. price – Actual price) × Actual qty.

A = (50 – 50) × 5,000

= 0 × 5,000 = Nil

B = (40 – 45) × 6,000

= 5 × 6,000 = 30,000 (A)

CALCULATION OF STANDARD QUANTITY

A = 100 × 40 = 4,000 units

B = 100 × 60 = 6,000 units

2. Materials usage variance = (Std. qty. – Actual qty.) × Std. price

A = (4,000 – 5,000) × 50

= 1,000 × 50 = 50,000 (A)

B = (6,000 – 6,000) × 40

= 0 × 40 = NIL

3. Materials cost variance = (Std. qty. × Std. price) – (Actual qty. × Actual price)

A = (4,000 × 50) – (5,000 × 50)

= 2,00,000 – 2,50,000 = 50,000 (A)

B = (6,000 × 40) – (6,000 × 45)

= 2,40,000 – 2,70,000 = 30,000 (A)

Verification:

Material cost variance = Material price variance + Material usage variance

A 50,000 (A) = NIL + 50,000 (A)

B 30,000 (A) = 30,000 (A) + NIL

4. Calculation of Revised std. qty.

$$\text{Revised std. qty.} = \frac{\text{Total weight of actual mix}}{\text{Total weight of std. mix}} \times \text{Std. qty. of material in question}$$

$$A = \frac{11{,}000}{10{,}000} \times 4{,}000 = 4{,}400$$

$$B = \frac{11{,}000}{10{,}000} \times 6{,}000 = 6{,}600$$

Material mix variance = (Revised std. qty. – Actual qty.) × Std. price

A = (4,400 – 5,000) × 50

= 600 × 50 = 30,000 (A)

B = (6,600 – 6,000) × 40

= 600 × 40 = 24,000 (F)

Problem 18. Mixers Ltd. is engaged in producing a standard mix by using 60 kgs. of Chemical X and 40 kgs. of Chemical Y. The standard loss of production is 30%. The standard price of X is Rs. 5 per kg. and Y is Rs. 10 per kg.

The actual mix and yield were as follows:

X — 80 kgs. @ Rs. 4.5 per kg.

Y — 70 kgs. @ Rs. 8 per kg.

Actual yield is 115 kgs.

Calculate material variance. *(C.S., Inter, December 1998)*

CHAPTER 14

Solution:

	Standard cost Qty.	Rate	Amt.	Actual cost Qty.	Rate	Amt.	Standard cost per actual input qty. Qty.	Rate	Amt.
$X = \frac{60}{70} \times 115$	= 98.571	5	493	80	4.50	360	80	5	400
$Y = \frac{40}{70} \times 115$	= 65.714	10	657	70	8.00	560	70	10	700
	164.3		1,150	150		920	150		1,100
Less: 30%	49.3		–	35		–			
	115		1,150	115		920			

1. Material cost variance = (SQ × SR) – (AQ × AR)

   X = (98.571 × 5) – (80 × 4.50)
   = 492.86 – 360 = 132.56 (F)

   Y = (65.714 × 10) – (70 × 8)
   = 657.14 – 560 = 97.14 (F)

   230 (F)

2. Material usage variance = (SQ – AQ) × SP

   X = (98.571 – 80) × 5 = 92.86 (F)

   Y = (65.714 – 70) × 10 = 42.86 (A)

   50.00 (F)

3. Materials price variance = (SP – AP) × SQ

   X = (5 – 4.50) × 80 = 40 (F)

   Y = (10 – 8) × 70 = 140 (F)

   180 (F)

4. Calculation of Revised std. qty.

   $$\frac{\text{Total weight of actual mix}}{\text{Total weight of std. mix}} \times \text{Std. qty. of material in question}$$

   $$X = \frac{150}{164.3} \times 98.6 = 90$$

   $$Y = \frac{150}{164.3} \times 65.7 = 60$$

   Material mix variance = (Revised std. qty. – Actual qty.) × Std. price

   X = (90 – 80) × 5 = 50 (F)

   Y = (60 – 70) × 10 = 100 (A)

   50 (F)

5. Calculation of standard yield

   $$\frac{115}{164.3} \times 150 = 105 \text{ kgs}$$

   Material yield variance = (Actual yield – Std. yield) × Std. price
   = (115 – 105) × 10 = 100 (F)

Problem 19. The standard cost of a certain chemical mixture was:

40% Materials A at Rs. 200 per ton

60% Materials B at Rs. 300 per ton

A standard loss of 10% is expected in production. During the period the following materials were used:

90 tons. Materials A at the cost of Rs. 180 per ton.

110 tons. Materials B at the cost of Rs. 340 per ton.

The weight produced was 182 tons of good production.

Calculate:

(*i*) Material price variance

(*ii*) Material usage variance

(*iii*) Material mix variance

(*iv*) Material yield variance (*CS, Inter., December 1997*)

Solution:

CALCULATION OF VARIANCES

		Standard			*Actual*		
		Qty.	*Rate*	*Amt.*	*Qty.*	*Rate*	*Amt.*
A	40%	80	200	16,000	90	180	16,200
B	60%	120	300	36,000	110	340	37,400
		200		52,000	200		53,600
Less: Normal loss		20			18		
		180			182		

(*i*) Material price variance = (SP – AP) × AQ

A = (200 – 180) × 90

= 20 × 90 = 1,800 (F)

B = (300 – 340) × 110

= 40 × 110 = 4,400 (A)

2,600 (A)

(*ii*) Material usage variance:

Calculation of Std. Qty. for actual output

$$\frac{\text{Std. qty.}}{\text{Std. output}} \times \text{Actual output}$$

$$A = \frac{80}{100} \times 182 = 80$$

$$B = \frac{120}{100} \times 182 = 121$$

Material usage variance = (Std. qty. for actual output – Actual qty.) × Std. price

A = (80 – 90) × 200 = 2,000 (A)

B = (121 – 110) × 300 = 330 (F)

= 1,670 (A)

(*ii*) Material mix variance:

As the Std. mix (200) is equal to actual mix (200), Revised std. qty. for both A and B is the same as Std. qty.

CHAPTER 14

Material mix variance = (Revised std. qty. – Actual qty.) × Std. price

A = (80 – 90) × 200
= 10 × 200 = 2,000 (A)

B = (120 – 110) × 300
= 10 × 300 = 3,000 (F)

= 1,000 (F)

Calculation of std. cost per unit

$$= \frac{\text{Total cost of std. mix}}{\text{Std. output}}$$

$$= \frac{52{,}000}{180} = \text{Rs. } 288.88$$

(*iv*) Material yield variance = (Std. output for actual mix – Actual output) × Std. yield price
= (180 – 182) × 288.88
= 22 × 288.88 = 577

2. Labour Cost Variance

The various labour cost variances can be shown under the following chart:

Labour cost variance
↓
Labour rate or price variance | Labour efficiency or usage variance
↓ (Labour efficiency or usage variance)
Labour mix variance | Idle time variance

Fig. 14.3 *Chart showing material cost variance.*

A. Labour Cost Variance

This is also known as direct wages variance and represents the difference between standard direct wages specified for the actual production and the actual direct wages paid. The formula to calculate labour cost variance is as follows:

Labour cost variance = (Actual hours × Actual rate) – (Std. hours × Std. rate)

Labour cost variance is analysed into the following two variances.

(*a*) **Labour rate or price variance:** The ICMA terminology defines labour rate variance as "the difference between the standard and actual direct labour hour rate per hour for the total hours worked". In order words, it is that portion of labour cost variance owing to difference between the actual rate and standard rate of pay specified. The following formula is used for calculating this variance.

Labour rate variance = (Standard rate – Actual rate) × Actual hour

The following factors are responsible for direct labour rate variance:

(*i*) Payment of high wage rate than estimated.
(*ii*) Use of more skilled or unskilled labourers than planned.
(*iii*) Payment of excess bonus than estimated.
(*iv*) Payment of more overtime premium than estimated.

(*b*) **Labour Efficiency or usage variance:** It refers to that portion of the wage variance which is due to the difference between the standard labour hours specified and the actual labour hours expended. The following formula is used to calculate this variance.

Labour efficiency variance = (Standard hours – Actual hours) × Standard rate.

The following factors cause this variance:

(a) Use of inferior grade of labour.

(b) Lack of good supervision.

(c) Use of sub-standard quality of materials.

(d) Breakdown of machines during production process.

(e) Inadequate working conditions such as poor lighting, lack of ventilation, heating, etc.

(f) Lack of proper production organisation involving defective production planning, routing, scheduling, inspection, etc.

B. Idle Time Variance

Sometimes idle time may occur in spite of fixing standard carefully. Such idle time should not be included in efficiency variance but shown separately. Otherwise, employees are made responsible for inefficiency although idle time is beyond their control as for example, breakdown of machine or power supply. The following formula is used to calculate idle time variance:

Idle time variance = Idle hours × Standard rate variance

Idle time variance is always an adverse variance.

C. Labour Mix Variance or Gang Composition Variance

Sometimes employees of different grade may have to be used in place of specific grade because of shortage of labourers. In such a situation a labour mix variance is calculated to show management how much of labour cost variance is due to change in labour force. It is defined as the portion of the wage variance which is due to the difference between the standard labour grades specified and the actual grades utilised. It arises under two situations.

(a) Where standard composition of labour force is revised owing to shortage of a particular type of labour but the total labour time spent is equal to the total standard time.

The following formula is used:

Labour mix variance = (Revised std. mix on time – Actual mix or time) × Std. wage rate per hour

Note: The revised standard mix is the same as the standard mix or time.

(b) Where the standard composition of labour force is revised due to shortage of a particular type of labour and total actual time of labour differs from total standard time of labour, the following formula is used.

Labour mix variance = (Revised std. mix – Actual std. mix or time) × Std. wage rate per hour

Where,

Revised standard mix of time

$$= \frac{\text{Total time of actual mix of workers}}{\text{Total time of std. mix of workers}} \times \text{Std. time of the respective category of workers}$$

Problem 20. *(Labour cost variance):* The standard time and rate for unit component are given below:

	Rs.
Standard hours	20 hrs
Standard rate Rs. 5 per hour	
Actual production	1,000 units

CHAPTER 14

Actual hours	20,500 hrs.
Actual rate per hour	Rs. 4.80

Calculate:

(a) Labour cost variance

(b) Labour rate variance

(c) Labour efficiency variance *(University of Madras, B.Com., May 2005)*

Solution:

Standard hours = 20 hrs per unit

Actual production = 1,000 units

∴ Standard time for actual output

= 20 hrs. × 1,000 units

Standard rate = Rs. 5 per hour

Actual hours given = 20,500 hours

Actual rate = Rs. 4.80 per hour

(1) Labour cost variance:

$$\left(\begin{array}{l}\text{Std. time for}\\ \text{actual output}\end{array} \times \text{Std. rate}\right) - \left(\begin{array}{l}\text{Actual}\\ \text{time}\end{array} \times \begin{array}{l}\text{Actual}\\ \text{rate}\end{array}\right)$$

= (20,000 hrs. × 5) – (20,500 × 4.80)

= 1,00,000 – 98,400

= 1600 (A)

(2) Labour rate variance:

= (Std. rate – Actual rate) × Actual time

= (Rs. 5 – 4.80) × 20,500

= Rs. 4,100 (F)

(3) Labour efficiency variance:

$$= \left(\begin{array}{l}\text{Std. time for}\\ \text{actual output}\end{array} \times \begin{array}{l}\text{Actual}\\ \text{time}\end{array}\right) \times \text{Std. rate}$$

= (20,000 hrs. – 20,500 hrs.) × Rs. 5

= 2,500 (A)

Verification

Labour cost variance = Labour rate variance + Labour efficiency

1,600 A = 4,100 (F) + 2,500 (A)

1,600 (A) = 1,600 (A)

Problem 21. The budgeted labour force for producing 1,000 articles is

Particulars	*Total Std. hours*	*Total Std. cost Rs.*
30 men @ Rs. 40 per hour for 50 hours	1,500	60,000
20 women @ Rs. 30 per hour for 30 hours	600	18,000
10 boys @ Rs. 20 per hour for 20 hours	200	4,000
	2,300 hrs.	82,000

The actual data and related work force for 1,000 articles are as follows:

Particulars	*Total Actual hours*	*Total Actual cost*
25 men @ Rs. 45 per hour for 50 hours	1,250	56,250
30 women @ Rs. 30 per hour for 30 hours	900	27,000
10 boys @ Rs. 20 per hour for 15 hours	150	3,000
	2,300	86,250

Calculate:

(1) Labour cost variance
(2) Labour rate variance
(3) Labour efficiency variance
(4) Labour mix variance

(Osmania University, B.Com., October 2007)

Solution:

(1) Labour cost variance:

	=	Standard labour cost − Actual labour cost		
Men	=	(30 × 40 × 50) − (25 × 45 × 50)		
	=	60,000 − 56,250	=	3,750
Women	=	(20 × 30 × 20) − (30 × 30 × 30)		
	=	18,000 − 27,000	=	− 9,000
Boys	=	(10 × 20 × 20) − (10 × 20 × 15)		
	=	4,000 − 3,000	=	1,000
				4,250 F

(2) Labour rate variance:

	=	(Std. rate − Actual rate) × Actual time		
Men	=	(40 − 45) × 1,250	=	− 6,250 (A)
Women	=	(30 − 30) × 900	=	0
Boys	=	(20 − 20) × 150	=	0
				6,250 (A)

(3) Labour efficiency variance:

	=	(Std. time − Actual time) × Std. rate		
Men	=	(1,500 − 1,250) × 40	=	10,000
Women	=	(600 − 900) × 30	=	9,000 A
Boys		(200 − 150) × 20	=	1,000
				2,000 (F)

Verification

Labour cost variance	=	Labour rate variance	+	Labour efficiency variance
4,250 (F)	=	6,250 (A) + 2,000 (F)		
4,250 (F)	=	4,250 (F)		

(4) Labour mix variance:

The total actual and standard hours are the same. Labour mix variance is therefore equal to labour efficiency variance.

CHAPTER 14

Problem 22. The details regarding the composition and the weekly wage rates of labour force engaged on a job scheduled to be completed in 30 weeks are as follows:

	Standard		*Actual*	
Category of workers	*No. of labourers*	*Weekly wage rate per worker*	*No. of labourers*	*Weekly wages rate per worker*
Skilled	75	600	70	700
Semi-skilled	45	400	30	500
Un-skilled	60	300	80	200

The work is actually completed in 32 weeks

Calculate:
(1) Labour cost variance
(2) Labour rate variance
(3) Labour efficiency variance
(4) Labour mix variance *(Osmania University, B.Com., March 2007)*

Solution:

(1) Labour cost variance:

Standard labour cost – Actual labour cost

Standard labour cost:

Skilled	75 × 600 × 30	=	13,50,000
Semi-skilled	45 × 400 × 30	=	5,40,000
Unskilled	60 × 300 × 30	=	5,40,000
	Total		24,30,000

Actual labour cost

Skilled	70 × 700 × 32	=	15,68,000
Semi-skilled	30 × 500 × 32	=	4,80,000
Unskilled	80 × 200 × 32	=	5,12,000
			25,60,000

Total labour cost variance = 24,30,000 – 25,60,000 = 1,30,000 (A)

(2) Labour rate variance:

(Std. rate – Actual rate) × Actual time

Skilled	=	(600 – 700) × 2,240	=	2,24,000 (A)
Semi-skilled	=	(400 – 500) × 960	=	96,000 (A)
Unskilled	=	(300 – 200) × 2,560	=	2,56,000 (F)
				64,000 (A)

(3) Labour efficiency variance :

(Std. time – Actual time) × Std. rate

Skilled	=	(2,250 – 2,240) × 600	=	6,000 (F)
Semi-skilled	=	(1,350 – 960) × 400	=	1,56,000 (F)
Unskilled	=	(1,800 – 2,560) × 300	=	2,28,000 (A)
				66,000 (A)

(4) Labour mix variance:

Calculation of Revised standard hours:

	Std. Labour hrs.			*Actual labour hrs.*		
Skilled	75 × 30	=	2,250	70 × 32	=	2,240
Semi-skilled	45 × 30	=	1,350	30 × 32	=	960
Unskilled	60 × 30	=	1,800	80 × 32	=	2,560
			54,00			5,760

RSH of skilled $= \dfrac{2,250}{5,400} \times 5,760 = 2,400$

RSH of semi-skilled $= \dfrac{1,350}{5,400} \times 5,760 = 1,440$

RSH of unskilled $= \dfrac{1,800}{5,400} \times 5,760 = 1,920$

Labour mix variance = (Revised std. hrs. – Actual hrs.) × Std. rate

Skilled	=	(2,400 – 2,240) × 600	=	96,000 (F)
Semi-skilled	=	(1,440 – 960) × 400	=	1,92,000 (F)
Unskilled	=	(1,920 – 2,560) × 300	=	1,92,000 (A)
				96,000 (F)

(5) Labour sub-efficiency variance:

= (Std. hours – Revised Std. hrs) × Std. rate

Skilled	=	(2,250 – 2,400) × 600	=	90,000 (A)
Semi-skilled	=	(1,350 – 1,440) × 400	=	36,000 (A)
Unskilled	=	(1,800 – 1,920) × 300	=	36,000 (A)
				1,62,000 (A)

Problem 23. Calculate labour variances from the following data:

The standard labour force for producing product 'A' is:

20 semi-skilled workers @ Rs. 75 per hour for 50 hours

10 skilled workers @ Rs. 125 per hour for 50 hours

The actual labour force employed for producing product 'A' is:

22 semi-skilled workers @ Rs. 80 per hour for 50 hours

8 skilled workers @ Rs. 120 per hour for 50 hours. *(Osmania University, M.Com., June 2002)*

Solution:

(1) Labour cost variance:

= Standard labour cost – Actual labour cost

Semi-skilled = 20 × 75 × 50 – 22 × 80 × 50
= 75,000 – 88,000
= 13,000 (A)

Skilled = 10 × 125 × 50 – 8 × 120 × 50
= 62,500 – 48,000
= 14,500 (F)

(2) Labour rate variance:

= (Std. rate – Actual rate) × Actual time

Semi-skilled	=	(75 – 80) × 1,100	=	5,500 (A)
Skilled	=	(125 – 120) × 400	=	200 (F)
			=	5,300 (A)

CHAPTER 14

(3) Labour efficiency variance:

= (Std. time – Actual time) × Std. rate

Semi-skilled	(1,000 – 1,100) × 75	=	750 (A)
Skilled	(500 – 400) × 125	=	12,500 (F)
		=	11,750 (F)

(4) Labour mix variance:

Calculation of standard cost of standard mix:

20 semi-skilled workers @ Rs. 75 for 50 hours	=	Rs. 75,000
10 skilled workers @ Rs. 120 for 50 hours	=	Rs. 60,000
		1,35,000

Standard cost of Actual mix

22 semi-skilled workers @ Rs. 75 for 50 hours	=	Rs. 82,500
8 skilled workers @ Rs. 125 for 50 hours	=	Rs. 50,000
		Rs. 1,32,500

Labour mix variance = Standard cost of Std. mix – standard cost of Actual mix

= 1,35,000 – 1,32,500

= 2,500 (F)

Problem 24. A gang of workers usually consists of 10 men, 5 women, and 5 boys in a factory. They are paid at standard hourly rates of Rs. 1.25, Re 0.80 and Re. 0.70 respectively. In a normal working week of 40 hours the gang is expected to produce 1,000 units of output.

In a certain week, the gang consisted of 13 men, 4 women and 3 boys. Actual wages were paid at the rates of Rs. 1.20, Re 0.85 and Rs. 0.65 respectively. Two hours were lost due to abnormal idle time and 960 units of output were produced. Calculate different labour variances

(Osmania University, M.Com., May 2001)

Solution:

	Standard for 1,000 units				*Actual for 960 units*			
Category of workers	*No. of workers*	*Hours*	*Rate per hr.*	*Amt.*	*No. of workers*	*Hours*	*Rate per hr.*	*Amt.*
Men	10	400	1.25	500	13	520	1.20	624
Women	5	200	0.80	160	4	160	0.85	136
Boys	5	200	0.70	140	3	120	0.65	78
	20	800		800	20	800		838

Standard labour cost for 1,000 units = Rs. 800

Therefore, Std. labour cost per unit = $\frac{800}{1,000}$ = 0.80

(1) Labour cost variance:

= Std. labour cost – Actual labour cost

For Actual output

= 960 units × Re 0.80 – Rs. 838

= Rs. 768 – 838 = Rs. 70 (A)

(2) Labour rate variance:

= (Std. rate – Actual rate) × Actual time

Men	= (1.25 – 1.20) × 520	=	Rs.	26 (F)
Women	= (0.80 – 0.85) × 160	=	Rs.	8 (A)
Boys	= (0.70 – 0.65) × 120	=	Rs.	6 (F)
			Rs.	24 (A)

(3) Labour efficiency variance:

$$= \left(\begin{matrix}\text{Std. time for} \\ \text{actual output}\end{matrix} - \begin{matrix}\text{Actual time} \\ \text{worked}\end{matrix}\right) \times \text{Std. rate}$$

$$\text{Men} = \left(\frac{400 \text{ hrs.}}{1{,}000 \text{ units}} \times 960 \text{ units} - 12 \times 38 \text{ hrs.}\right) \times \text{Rs. } 1.25$$

= 137.50 (A)

$$\text{Women} = \left(\frac{200 \text{ hrs.}}{1{,}000 \text{ units}} \times 960 \text{ units} - 4 \times 38 \text{ hrs.}\right) \times \text{Re. } 0.80$$

= Rs. 32 (F)

$$\text{Boys} = \left(\frac{200 \text{ hrs.}}{1{,}000 \text{ units}} \times 960 \text{ units} - 3 \times 38 \text{ hrs.}\right) \times \text{Re. } 0.70$$

= Rs. 54.60 F

Total labour efficiency variance = Rs. 50.90 (F)

Note:

Actual time worked = No. of workers × hrs. worked in the week

Hours worked in the week = 40 hours – 2 hrs. idle time = 38 hours.

(4) Labour idle time variance:

= No. of workers × idle time per worker × Std. rate

Men	= 13 × 2 hrs. × Re. 1.25	=	Rs	32.50 (A)
Women	= 4 × 2 hrs. × Re 0.80	=	Rs.	6.40 (A)
Boys	= 3 × 2 hrs. × Re 0.70	=	Rs.	4.20 (A)
			Rs.	43.10 (A)

(5) Labour mix variance:

$$= \left(\frac{\text{Total time of actual labour mix (excluding idle time)}}{\text{Total time of std. labour mix}} \times \text{Std. cost of std. labour mix}\right) -$$

Std. cost of Actual labour mix (excluding idle time)

$$= \frac{760 \text{ hrs.}}{800 \text{ hrs.}} \times \text{Rs. } 800 - (494 \times \text{Rs. } 1.25 + 152 \times \text{Re. } 0.80 + 114 \times \text{Re. } 0.70)$$

= 760 – (617.50 + 121.60 + 79.80)

= Rs. 760 – Rs. 818.90 = Rs. 58.90 (A)

Problem 25. The standard and actual figures of Ashok Metal Works are as under:

Standard time for the job	1,000 hours
Standard rate per hour	Re. 0.50
Actual time taken	900 hours
Actual wages paid	Rs. 360

Compute (*i*) Labour rate variance, (*ii*) Labour efficiency variance and (*iii*) Labour cost variance.

Solution:

Calculation of standard cost = 1,000 hrs. × 0.50 = 500

Calculation of actual rate per hour = $\frac{\text{Rs. } 360}{900} = 0.40$

(*i*) Labour rate variance = (Standard rate – Actual rate) × Actual hrs.
= (0.50 – 0.40) × 900 = Rs. 90 (F)

(*ii*) Labour efficiency variance = (Standard hours – Actual hrs.) × Standard rate
= (1,000 – 900) × 0.50 = Rs. 50 (F)

(*iii*) Labour cost variance = Standard labour cost – Actual labour cost
= (500 – 360) = 140 (F)

Problem 26. From the following particulars calculate:

(1) Labour rate variance,
(2) Labour mix variance,
(3) Labour efficiency variance and
(4) Labour cost variance.

	Skilled	*Semi-skilled*	*Unskilled*
No. in Std. gang	16	6	3
Standard rate per hour	3	2	1
Actual number in the gang	14	9	2
Actual rate of pay	4	3	2

In a 40-hours week the gang as a whole produced 900 standard hours.

Solution:

Calculation of standard hours

In a 40 hours week, the standard gang should have produced 1,000 standard hours as shown below:

Skilled	16 × 40	=	640
Semi-skilled	6 × 40	=	240
Unskilled	3 × 40	=	120
			1,000

But the actual output is 900 standard hours. Hence to find out the total labour cost variance the standard cost or cost charged to production is to be computed with reference to 900 standard hours. This is shown in the following statement.

STATEMENT SHOWING THE STANDARD COST, ACTUAL COST AND STANDARD COST OF ACTUAL GANG FOR ACTUAL OUTPUT, *i.e.*, 900 STANDARD HOURS:

Gang	*Standard cost*			*Actual cost*			*Standard cost of actual gang*		
	Hrs.	*Rate*	*Amt.*	*Hrs.*	*Rate*	*Amt.*	*Hrs.*	*Rate*	*Amt.*
Skilled $\frac{640}{1,000} \times 900$	576	3	1,728	14 × 40 = 560	4	2,240	560	3	1,680
Semi-skilled $\frac{240}{1,000} \times 900$	216	2	432	9 × 40 = 360	3	1,080	360	2	720
Unskilled $\frac{120}{1,000} \times 900$	108	1	108	2 × 40 = 80	2	160	80	1	80
	900	2.52	2,268	1,000	3.84	3,480	1,000	2.48	2,480

1. Labour rate variance = (Standard rate – Actual rate) × Actual hours
   = (3 – 4) × 1,000 = 1,000 (A)
2. Labour mix variance = (Std. rate of std. gang or mix – Std. rate of actual gang or mix) × Actual hours
   = (2.52 – 2.48) × 1,000 = 40 (F)
3. Labour efficiency variance = (Standard hours – Actual hours) × Std. rate
   = (900 – 1,000) × 2.52 = 252 (A)
4. Labour cost variance = (Standard labour cost – Actual labour cost)
   = (2,268 – 3,480) = 1,212 (A)

Problem 27. From the following particulars relating to Karnataka Toys Co. Ltd.

Calculate:

(1) Wage rate variance,

(2) Labour efficiency variance,

(3) Abnormal idle time variance and

(4) Total labour cost variance.

Standard time per unit	2.5 hours
Actual hours worked	2,000 hours
Standard rate of pay	Rs. 2 per hour

25% of the actual hours has been lost as abnormal idle time.

Actual output	1,000 units
Actual wages	Rs. 4,500

Solution:

Calculation of standard cost charged to Production = 1,000 units × 2.5 hrs. × Rs. 2
= Rs. 5,000

Actual wages paid = Rs. 4,500

Actual wage rate per hour = $\frac{4,500}{2,000}$ = 2.25

Standard wage rate per hour = 2.00

CHAPTER 14

Abnormal idle time = $\frac{25}{100}$ × 2,000= 500 hrs.

1. Wage rate variance = (Standard rate – Actual rate) × Actual hours
   = (2 – 2.25) × 2,000 = 500 (A)
2. Labour efficiency variance = (Standard hours – Actual hours) × Standard rate
   = (2,500 – 1,500) × 2 = 2,00 (F)

   Standard hrs. = Std. time per unit × Actual ourput
   = (2.5 × 1,000) = hrs. 2,500
3. Abnormal idle time variance = Idle time × Standard rate
   = 500 × Rs. 2 = Rs. 1,000 (A)
4. Total labour cost variance = Standard labour cost – Actual labour cost
   = 5,000 – 4,500 = 500 (F)

Problem 28. Compute labour variance from the following data of a ship building company. The particulars of labour force engaged on job no. 777, scheduled to be completed in 30 weeks are :

Category	*Standard*		*Actual*	
	Number	*Wage rate (Rs.)*	*Number*	*Wage rate (Rs.)*
Skilled	75	60	70	70
Semi-skilled	45	40	30	50
Un-skilled	60	30	80	20

The actual time taken is 32 weeks. *(Nagarjuna University, M.Com., March 1992)*

Solution:

Category	*Standard*			*Actual*		
	*Weeks**	*Rate*	*Amt.*	*Weeks**	*Rate*	*Amt.*
Skilled	2,250	60	1,35,000	2,240	70	1,56,800
Semi-skilled	1,350	40	54,000	960	50	48,000
Unskilled	1,800	30	54,000	2,560	20	51,200
	5,400		2,43,000	5,760		2,56,000

* Total weeks = No. of workers × No. of weeks

1. Labour cost variance = Standard cost – Actual cost
   = 2,43,000 – 2,56,000
   = Rs. 13,000 (A)
2. Labour rate variance = (Standard rate – Actual rate) × Actual hours
   Skilled = (60 – 70) × 2,240 = 22,400 (A)
   Semi-skilled = (40 – 50) × 960 = 9,600 (A)
   Unskilled = (30 – 20) × 2,560 = 25,600 (F)
   6,400 (F)

3. Labour efficiency variance = (Standard hours – Actual hours) × Standard rate

Skilled	=	(2,250 – 2,240) × 60	=	600 (F)
Semi-skilled	=	(1,350 – 960) × 40	=	15,600 (F)
Unskilled	=	(1,800 – 2,560) × 30	=	22,800 (A)
Total labour efficiency variance				6,600 (A)

Verification:

Labour cost variance = Labour rate variance + Labour efficiency variance

13,000 (A) = 6,400 (A) + 6,600 (A)

4. Labour mix variance = (Revised std. time* – Actual time) × Std. rate

Revised std. time $= \dfrac{\text{Actual mix}}{\text{Std. mix}} \times \text{Std. time}$

Skilled $= \dfrac{5,760}{5,400} \times 2,250 = 2,400$

Semi-skilled $= \dfrac{5,760}{5,400} \times 1,350 = 1,440$

Unskilled $= \dfrac{5,760}{5,400} \times 1,800 = 1,920$

Skilled	=	(2,400 – 2,240) × 60	=	9,600 (F)
Semi-skilled	=	(1,440 – 960) × 40	=	19,200 (F)
Unskilled	=	(1,920 – 2,560) × 30	=	19,200 (A)
		Labour mix variance		9,600 (F)

Revised labour efficiency variance = (Std. time – Revised std. time) × Std. rate

Skilled	= (2,250 – 2,400) × 60	=	9,000 (A)
Semi-skilled	= (1,350 – 1,440) × 40	=	3,600 (A)
Unskilled	= (1,800 – 1,920) × 30	=	3,600 (A)
		=	16,200 (A)

Verification:

Labour efficiency variance = (Labour mix variance + Revised labour efficiency variance)

6,600 (A) = 9,600 (F) + 16,200 (A)

CHAPTER 14

Problem 29. *(Material and labour variances):* From the following data, calculate:

(1) Material cost variance
(2) Material price variance
(3) Material yield variance
(4) Labour cost variance
(5) Labour yield variance

	Standard				*Actual*	
Kg.			*Amt. Rs.*	*Kg.*		*Amt. Rs.*
Input	500	Material: @ 39 per kg	19,500	500	Material @ Rs. 42 per kg	21,000
		Labour: 4,000 hrs @ Rs. 1.50 per hour	6,000		Labour: 4,000 hrs @ Rs. 1.50 per hr.	6,000
	20	Normal loss		40	Actual loss	
Output	480		25,500	460		27,000

(Osmania University, M.Com., May 2000)

Solution:

(1) Material cost variance:

= (Std. qty. × Std. rate) – (Actual qty. × Actual rate)
= (500 × Rs. 39) – (500 × Rs. 42)
= Rs. 19,500 – Rs. 21,000
= Rs. 1,500 (A)

(2) Material price variance:

= (Std. rate – Actual rate) × Actual qty.
= (Rs. 39 – Rs. 42) × 500
= 1500 (A)

(3) Material yield variance:

= (Actual yield – Std. yield) × Std. rate

= Std. rate = $\frac{\text{Std. cost of std. mix}}{\text{Net std. output}}$

	Std. mix			*Actual mix*		
Material	500	@ Rs. 39 =	19,500	500	@ Rs. 42 =	2,100
Less: Net loss	20			40		
	480		19,500	460		2,100

Std. cost per unit = $\frac{19{,}500}{480}$ = Rs. 40.625

Yield variance = (Actual yield – Std. yield) Std. rate
= (460 – 480) × 40.625
= Rs. 812.50 (A)

(4) Labour cost variance:

= Std. hrs. × Std. rate – Actual hrs. × Actual rate

= 4,000 × Rs. 1.50 – 4,000 × Rs. 1.50

= 0

(5) Labour yield variance:

= (Actual yield – Std. yield) × Std. labour cost per unit

= (460 – 480) × 12

= Rs. 240 (A)

Problem 30. From the following particulars, calculate:

(1) Material cost variance

(2) Material price variance

(3) Material yield variance

(4) Labour cost variance

(5) Labour rate variance

Standard Data			*Actual Data*		
Kgs	*Material*	*Amt.*	*Kg*	*Material*	*Amt.*
450	of material 'A' @ Rs. 20 per kg	9,000	450	of material 'A' @ Rs. 19 per kg	8,550
360	of material 'B' @ Rs. 10 per kg	3,600	360	of material 'B' @ Rs. 11 per kg	3,960
810		12,600	810		12,510
90	Normal loss		50	Actual loss	
720			760		
Hrs.	Labour		Hrs.	Labour	
2,400	Skilled hrs. @ Rs. 2 per hour	4,800	2,400	Skilled hrs. @ Rs. 2.25 per hr.	5,400
1,200	Unskilled hrs. @ Rs. 1 per hour	1,200	1,200	Unskilled hrs. @ Rs. 1.25 per hour	1,500
		18,600			19,410

(Osmania University, M.Com., May 1999)

Solution:

(1) Material price variance:

= (Std. qty. × Std. rate) – (Actual qty. × Actual rate)

= Rs. 12,600 – Rs. 12,510 = Rs. 90 (F)

(2) Material price variance:

= (Std. rate – Actual rate) × Actual qty.

A = (20 – 19) × 450 = 450 (F)

B = (10 – 11) × 360 = 360 (A)

90 (F)

(3) Material yield variance:

= (Actual yield – Std. yield) × Std. rate

$$\text{Std. Rate} = \frac{\text{Std. cost of std. mix}}{\text{Net std. output}}$$

	Std. mix		Actual mix
	Units	*Amount*	*Units*
Material A	450	9,000	450
Material B	360	3,600	360
	810	12,600	810
Less: loss	90	–	50
Net output	720	12,600	760

$$\text{Std. cost per unit} = \frac{12{,}600}{720} = \text{Rs. } 17.5$$

Yield variance = (Actual yield – Std. yield) × Std. cost per unit
= (760 – 720) × 17.5
= 40 × 17.5 = Rs. 700 (F)

(4) Labour cost variance:

Std. labour cost – Actual labour cost

Skilled = 2,400 × 2 – 2,400 × 2.25
= 4,800 – 5,400 = 600 (A)

Skilled = 1,200 × 1 – 1,200 × 1.25
= 1,200 – 1,500 = 300 (A)

= 900 (A)

(5) Labour rate variance:

(Std. rate – Actual rate) × Actual time

Skilled workers = (2 – 2.25) × 2,400 = Rs. 600 (A)

Un-skilled workers = (1 – 1.25) × 1,200 = Rs. 300 (A)

900 (A)

Problem 31. Calculate material and labour variances from the following data:

For 5 units of product A, the standard data are:
Material 40 kg @ Rs. 25 per kg
Labour 100 hours @ Rs. 2.50 per hour

Actual data are:
Actual Production – 1000 units
Material 7,840 kg @ Rs. 27 per kg
Labour 19,800 hours @ Rs. 2.60 per hour. *(University of Bombay, B.Com., April 2006)*

Solution:

Working Notes:

(1) Standard quantity for actual output

$$1{,}000 \times \frac{40}{5} = 8{,}000$$

(2) Standard hours for actual output

$$1{,}000 \times \frac{100}{5} = 20{,}000$$

Material variances:

(1) Material cost variance:

(Std. qty. × Std. rate) – (Actual qty. × Actual rate)

= (8,000 × 25) – (7,840 × 27)

= 2,00,000 – 2,11,680 = 11,680 (A)

(2) Material usage variance

= (Std. qty. – Actual qty.) × Std. price

= (8,000 – 7,840) × 25

= 160 × 25 = 4,000 (F)

(3) Material price variance

= (Std. price – Actual price) × Actual qty.

= (25 – 27) × 7,840 = 15,680 (A)

Labour variances:

(1) Labour cost variance:

= (Std. hrs. × Std. rate) – (Actual hrs. × Actual rate)

= (20,000 × 2.50) – (19,800 × 2.60)

= 50,000 – 51,480 = 1,480 (A)

(2) Labour efficiency variance:

= (Std. hrs. – Actual hrs.) × Std. rate

= (20,000 – 19,800) × 2.50

= 200 × 2.50 = 500

(3) Labour rate variance:

= (Std. rate – Actual rate) × Actual hours

= (2.50 – 2.60) × 19,800

= 0.10 × 19,800 = 1,980 (A)

Verification

(1) Material cost variance	=	Material usage variance	+	Material price variance
11,800 (A)	=	4,000 (F) + 15,680 (A)		
11,680 (A)	=	11,680 (A)		
(2) Labour cost variance	=	Labour efficiency variance	+	Labour rate variance
1,480 (A)	=	500 (F) + 1,980 (A)		
1,480 (A)	=	1,480 (A)		

Problem 32. The standard cost on "Material" and "Labour" for the making of a unit of a certain product is estimated as under:

Material 80 kg at Rs. 1.50 per kg

Labour 18 hrs. at Rs. 1.25 per hr.

On completion of the production of a unit, it was found that 75 kg of material costing Rs. 1.75 per kg has been consumed and that the time taken was 16 hours, the wage rate being Rs. 1.50 per hour.

You are required to analyse material and labour variances.

(University of Delhi, B.Com. (Hons), 2004)

Solution:

(1) Material cost variance:

= Std. cost – Actual cost

= (80 kg × Rs. 1.50) – (75 kg × Rs. 1.75)

= 120 – 131.25 = Rs. 11.25 (A)

(2) Material price variance:

= (Std. rate – Actual rate) × Actual qty.

= (1.50 – 1.75) × 75 = Rs. 18.75 (A)

(3) Material quantity variance:

= (Std. qty. – Actual qty.) × Std. rate

= (80 – 75) × 1.50 = Rs. 7.50 (F)

(4) Labour cost variance:

= Std. cost – Actual cost

= (18 × 1.25) – (16 × 1.5) = Rs. 1.50 (A)

(5) Labour rate variance:

= (Std. rate – Actual rate) × Actual hrs.

= (1.25 – 1.50) × 16 = Rs. 4 (A)

(6) Labour efficiency variance:

= (Std. hrs. – Actual hrs.) × Std. rate

= (18 – 16) × 1.25 = Rs. 2.50 (F)

3. Overhead Variance

Overhead variances arise on account of difference between actual overhead and absorbed overheads. In order to calculate overhead variance it is necessary to know the actual and absorbed overheads. The absorbed overhead rates are also known as the standard overhead recovery rate or standard overhead absorption rate. It is calculated by applying the following steps:

(*a*) To make the estimate of probable overheads to be incurred for each department for the next year.

(*b*) This estimated overhead is to be classified into fixed and variable overheads.

(*c*) The capacity utilisation either in terms of machine hours or labour hours or production units is to be estimated.

(*d*) Apply the following formula for calculating the standard overhead absorption rate:

$$\text{Standard fixed overhead rate} = \frac{\text{Budgeted fixed overhead}}{\text{Normal volume}}$$

$$\text{Standard variable overhead rate} = \frac{\text{Budgeted variable overhead}}{\text{Normal volume}}$$

(*A*) **Variable overhead cost variance:** It is defined as the difference between the actual variable overheads incurred and the variable overheads absorbed. This variance arises on account of over or under absorption of overheads. Calculation of this variance can be based on either hours or output. The following formula is used:

Variable overhead cost variance = Std. overhead cost recovered – Actual overhead cost

Where Std. overhead cost when overhead rate is per hours

= (Std. hours for actual output × Std. overhead rate per hour)

Standard overhead cost when overhead rate is output

= (Actual output × Std. overhead rate per unit)

(*a*) **Variable overhead expenditure variance:** It is defined as the difference between the actual variable overheads incurred and the allowed variable overheads based on the actual hours worked. Calculation of this variance is based on hours or output. The following formula is used:

Based on rate per hour

Variable overhead expenditure variance = (Std. variable overhead absorption rate per hr. × Actual hrs. worked) – Actual variable overhead

Based on rate per unit

Variable overhead expenditure variance = (Std. variable overhead absorption rate per unit – Actual variable rate per unit) × Actual output

(*b*) **Variable overhead efficiency variance:** This is defined as the difference between the allowed variable overheads and the absorbed variable overhead. The following formula is used to calculate this variance.

Variable overhead efficiency variance = (Actual hours – Std. hours for actual output) × Standard variable overhead aborption rate

or, = (Std. qty. of output – Actual qty. of output) × Std. rate per unit.

(*B*) **Fixed overhead variance:** This is defined as the difference between the standard cost of fixed overhead absorbed in the production achieved, whether completed or not, and the fixed overhead attributed and charged to that period. The formula is:

(Std. hrs. for actual output × Std. fixed overhead rate) × Actual fixed overheads

(*a*) **Fixed overhead expenditure overhead:** This is defined as the difference between the budget cost allowance for production for a specified control period and the actual fixed expenditure attributed and charged to that period. In other words, it represents the difference between the fixed overhead as per budget and the actual fixed overhead incurred. The following formula is used to calculate this variance:

Fixed overhead expenditure variance = (Budgeted qty. × Std. fixed overhead rate per unit) – Actual fixed overhead

or, = (Budgeted fixed overhead – Actual fixed overhead)

(*b*) **Fixed overhead volume variance:** It is that portion of the fixed overhead variance which is the difference between the standard cost of the overhead absorbed in

CHAPTER 14

actual output and the standard allowance for that output. In this definition standard allowance means the budgeted overhead. The fixed overhead volume variance arises mainly because of the use of a pre-determined overhead recovery rate based on a normal volume of activity and of the actual level of activity being less or more than the normal volume so selected. The following formula is used to calculate this variance.

Fixed overhead variance = (Actual qty. – Budgeted qty.) × Standard rate.

or Fixed overhead volume variance = (Std. hours for actual production – Actual hours worked) × Std. fixed overhead rate

(*c*) **Fixed overhead efficiency variance:** It is that portion of the volume variance which reflects the increased or reduced output arising from efficiency above or below the standard which is expected. The following formula is used to calculate this variance:

Fixed overhead efficiency variance = (Actual qty. of production – Std. qty. of production for actual capacity) × Std. fixed overhead absorption rate per unit

or = (Std. hours for actual production – Actual hours worked) × Std. fixed overhead rate.

(*d*) **Fixed overhead capacity variance:** This is defined as that portion of the fixed overhead volume variance which is due to working at higher or lower capacity than standard capacity.

The following formula is used to calculate this variance:

Fixed overhead capacity or usage variance = (Std. qty. of production – Revised budgeted qty. of production) × Std. fixed overhead rate per unit

or = (Std. hours for actual production – Actual hours of the period) × Std. fixed overhead rate per hour

(*e*) **Calender variance:** It is that portion of overhead volume variance which is due to the difference between the number of working days in the period to which the budget is applied.

The following formula is used to calculate this variance.

Calender variance = (Standard units – Revised budgeted units) × Std. fixed overhead rate per unit

or = (Std. number of working days or hrs. – Possible number of working days or hours) × Std. fixed overhead rate per day or hour.

Calender variance = (Budgeted hrs. – Actual hrs.) × Std. rate (*f*) **Seasonal variance:** It is that portion of the overhead volume variance which is due to the difference between the seasonally budgeted output and the average output on which standards have been calculated.

The following are the main causes for the overhead volume variance.

Causes Controllable by Management

(*i*) Employees waiting for work.
(*ii*) Avoidable machine breakdown.
(*iii*) Lack of operator.
(*iv*) Lack of tools.
(*v*) Lack of instructions.

Causes Uncontrollable by Management

(*i*) Decrease in customer demand.
(*ii*) Excess plant capacity.

Problem 33. *(Overhead variance):* From the following information compute fixed overhead cost, expenditure and volume/capacity variances.

Normal capacity is 5,000 hours.

Budgeted fixed overhead rate is Rs. 10 per standard hours. Actual level of capacity utilised is 4,400 standard hours.

Actual fixed overheads Rs. 52,000. *(University of Delhi, B.Com., (Hons.), 1991)*

Solution:

1. Fixed overhead variance = Budgeted fixed overhead – Actual fixed overhead
= 44,000 – 52,000 = 8,000 (A)

Budgeted fixed overhead = Std. hrs. × Budgeted fixed overhead rate
= 4,400 × Rs. 10 = 44,000

Note: Budgeted fixed overheads are taken at the actual capacity utilised.

2. Expense variance = Budgeted total factory overhead – Actual fixed factory overhead
= 50,000 – 52,000 = 2,000 (A)

3. Volume/capacity variance= (Actual capacity – Budgeted capacity) × Std. rate
= (4,400 – 5,000) × 10 = 6,000 (A)

Verification:

Fixed overhead variance = Expense variance + Volume variance
8,000 (A) = 2,000 (A) + 6,000 (A).

Problem 34. From the following particulars calculate all variances relating to fixed overheads.

	Budgeted	*Actual*
Fixed overhead for July 2010	Rs. 10,000	Rs. 10,200
Units produced in July 2010	5,000	5,200
Standard for one unit	4 hours	
Actual hours worked		20,100 hours

Solution:

1. Total fixed overhead variance = Budgeted fixed overhead – Actual fixed overhead
= 10,400 – 10,200 = 200 (F)

CHAPTER 14

Absorbed fixed overhead = 5,200 × 2 = 10,400

$$= \frac{\text{Budged overhead}}{\text{Units}} = \frac{10,000}{5,000} = 2$$

2. Overhead expenditure variance = Actual overheads – Budgeted overheads
= 10,200 – 10,000 = 200 (A)

3. Overhead volume variance = (Actual qty. – Budgeted qty.) × Std. rate
= (5,200 – 5,000) × 2 = 400 (F)

4. Overhead capacity variance = (Actual hours – Budgeted hours) × Std. hourly rate
= (20,100 – 20,000) × 0.50 = 50 (F)

Std. hourly rate $= \frac{\text{Rs. 2}}{\text{4 hrs.}} = \text{Rs. 0.50}$

5. Overhead efficiency variance = (Actual hrs. × Standard hour) × Std. rate per hr.
= (21,000 – 20,800) × 0.50 = 350 (F)

Standard hours produced = 5,200 × 4 = 20,800

Problem 35. Determine the budget and capacity variance from the following data:

Estimated factory overhead	Rs. 25,000
Estimated direct labour hours	5,000
Actual overhead expenses	Rs. 26,500
Applied overhead expenses	Rs. 22,500

(ICWA, Inter, June 1990)

Solution:

Overhead budget variance = Budgeted overhead – Actual overhead
= 25,000 – 26,500 = 1,500 (A)

Overhead capacity variance = (Std. hrs. for actual production – Actual hrs.) × Std. overhead rate per hr.
= (4,500 – 5,000) × 5 = 2,500 (A)

Problem 36. The following information is available from the records of a factory

	Budgeted	*Actual*
Fixed overhead for May (Rs.)	5,000	6,000
Production in May (units)	1,000	1,050
Standard time per units (hrs.)	10	–
Actual hours worked in May	–	11,000

Compute:

(*i*) Fixed overhead cost variance
(*ii*) Expenditure variance
(*iii*) Volume variance
(*iv*) Capacity variance
(*v*) Efficiency variance

(CS, Inter, December 2,000)

Solution:

Calculation of Std. overhead rate $= \frac{\text{Budgeted Fixed overhead}}{\text{Budgeted production units}}$

$$= \frac{5,000}{1,000} = \text{Rs. } 5$$

Calculation of hourly rate:

$$= \frac{\text{Rs. 5}}{\text{10 hrs.}} = \text{Re. } 0.50$$

(*i*) Fixed overhead cost variance	=	Overhead recovered on actual output – Actual fixed overheads
	=	(1,050 units × Rs. 5) – 6,000
	=	5,250 – 6,000 = Rs. 750 (A)
(*ii*) Expenditure variance	=	Budgeted fixed overhead – Actual fixed overhead
	=	5,000 – 6,000 = Rs. 1,000 (A)
(*iii*) Volume variance	=	(Budgeted output – Actual output) × Std. rate per unit
	=	(1,000 – 1,050) × 5 = 250 (F)
(*iv*) Capacity variance	=	(Budgeted hours – Actual hours) × Std. rate per hr.
	=	(10,000 – 11,000) × 0.50 = 500 F
(*v*) Efficiency variance	=	(Std. hours for actual output – Actual hours) × Std. rate per unit
	=	(10,500 – 11,000) × 0.50 = 250 (A)

Problem 37. Adarsh Ltd. has furnished you the following data:

	Budgeted	*Actual for July 1998*
Number of working days	25	27
Production in units	20,000	22,000
Fixed overhead (Rs.)	30,000	31,000

Budged fixed overhead rate is Re. 1 per hour. In July 1998, the actual hours worked were 31,500. Calculate the following variances:

(*i*) Efficiency variance
(*ii*) Capacity variance
(*iii*) Calender variance
(*iv*) Volume variance
(*v*) Expenditure variance
(*vi*) Total overhead variance. *(CS, Inter, June 1999)*

Solution:

CALCULATION OF BUDGETED HOURS

Budgeted overheads Rs. 30,000

Budgeted overhead rate per hour Re. 1

$$\text{Budgeted hours} = \frac{\text{Rs. 30,000}}{\text{Re. 1 per hour}} = 30,000 \text{ hrs.}$$

CALCULATION OF STANDARD TIME PER UNIT OF OUTPUT

$$\frac{\text{Budgeted hours}}{\text{Budgeted output}} = \frac{30,000}{20,000} = 1.5 \text{ hrs.}$$

CALCULATION OF STD. HOURS FOR ACTUAL OUTPUT

Actual output × Std. time per unit

22,000 × 15 = 33,000 hrs.

CHAPTER 14

CALCULATION OF STD. RATE PER UNIT

Std. rate per unit of output = 1.5 hrs. × Re. 1 per hr.
= Rs. 1.50

1. Efficiency variance = (Std. hrs. for actual output – Actual hrs.) × Std. rate per hr.
= (33,000 – 31,500) × 1
= 1,500 × 1 = 1.500 (F)

OR

Efficiency variance = (Std. output for actual hrs. – Actual ourput hrs.) × Std. rate per unit
= (21,000 units – 22,000 units) × 1.50
= 1,000 × 1.50 = 1.500 (F)

Note: *Calculation of std. output for actual hours:*

$$= \frac{\text{Actual hours}}{\text{Std. time per unit of output}}$$

$$= \frac{31{,}500 \text{ hours}}{1.5 \text{ hours}} = 21{,}000 \text{ units}$$

2. Capacity variance = (Revised budgeted hours – Actual hours) × Std. rate per hour
= (32,400 – 31,500) × 1 = 900 (A)

Note: *Calculation of revised budgeted hours:*

Actual working days × Budgeted hours per day
27 × 1,200 = 32,400

3. Calender variance = (Budgeted no. of working days – Actual no. of working days) × Budgeted overheads/Budgeted working days

$$= (25 - 27) \times \frac{30{,}000}{25}$$

= 2 × 1,200 = 2.400 (F)

4. Volume variance = (Budgeted output – Actual output) × Std. rate per unit
= (20,000 – 22,000) × 1.50
= 2,000 × 1.50 = 3,000 (F)

5. Expenditure variance = Budgeted overheads – Actual overheads
= 30,000 – 31,000 = 1,000 (A)

6. Total overhead variance = (Actual output × Std. rate per unit – Actual overheads)
= (22,000 × 1.50 – 31,000)
= 33,000 – 31,000 = 2,000 (F).

Problem 38. *(Overhead variance):* The following data is available in connection with the fixed overheads of a factory

Budgeted fixed overhead for January	Rs. 1,00,000
Budgeted output for January	50,000 units
Standard time for one unit	5 hrs.
Actual hours worked	2,55,000
Actual fixed overhead for the month	Rs. 1,10,000
Units produced during the month	52,000

Calculate:

(1) Fixed overhead variance
(2) Fixed overhead expenditure variance
(3) Volume variance
(4) Capacity variance

(Osmania University, B.Com., March 2007)

Solution:

(1) Fixed overhead variance:

(Actual output × Std. overhead rate) – (Actual fixed overhead)

= (52,000 × 2) – 1,10,000 = Rs. 6,000 (A)

$$= \text{Std. overhead rate} = \frac{1,00,000}{50,000} = \text{Rs. 2 per unit}$$

(2) Fixed overhead expenditure variance:

Budgeted fixed overhead – Actual fixed overhead

= 1,00,000 – 1,10,000 = Rs. 10,000 (A)

(3) Volume variance:

(Actual output × Std. rate) – Budgeted fixed overheads

= (52,000 × 2) – 1,00,000 = 4000 (F)

(4) Capacity variance:

(Revised budgeted units – Budgeted units) × Std. rate

$$= \left(\frac{2,55,000}{5} - 50,000\right) \times 2$$

= (51,000 – 50,000) × 2

= Rs. 2,000 (F)

Problem 39. From the following particulars, calculate Variable overhead expenditure variance. Actual variable overhead expenses Rs. 28,500, Budgeted variable overhead expenses Rs. 34,000, Actual level of production 8,000 standard hours, Normal capacity 10,000 standard hours.

(Calcutta University, B.Com. (Hons.), 2003)

Solution:

Variable overhead expenditure variance:

= (Std. variable overhead rate × Actual hrs.) – Actual variable overhead

= (Rs. 3,40 × 8,000) – Rs. 28,500

= Rs. 27,200 – Rs. 28,500 = Rs. 1,300 (A)

Working Note:

$$\text{Std. variable overhead rate} = \frac{34,000}{10,000} = \text{Rs. 3.40}$$

Problem 40. A company has a normal capacity of 120 machines working 8 hours per day of 25 days in a month. The fixed overheads are budgeted at Rs. 1,44,000 per month. The standard time required to manufacture one unit of production is 4 hours.

In April 1998, the company worked 24 days of 840 machine hours per day and produced 5,305 units of output. The actual fixed overheads were Rs. 1,42,000.

Compute:

(1) Efficiency variance

(2) Capacity variance

(3) Calender variance

(University of Delhi, B.Com. (Hons.), 2005)

Solution:

Working Notes

Actual hours = 840 × 24 = 20,160 per month

Budgeted hours = 120 × 8 × 25 = 24,000 per month

$$\text{Budgeted output} = \frac{24,000}{4 \text{ hrs.}} = 6,000 \text{ units}$$

CHAPTER 14

Standard overhead rate $= \dfrac{1,44,000}{24,000}$ = Rs. 6 per hr.

1. Efficiency variance = Absorbed overhead – Standard overhead
   = (5,305 × 4 × Rs. 6) – (20,160 × Rs. 6)
   = 6,360 (F)

2. Capacity variance = Std. overhead – Revised budgeted overhead
   $= (20,160 \times \text{Rs. } 6) - \left(1,44,000 - \dfrac{14 \text{ days}}{25 \text{ days}}\right)$
   = 1,20,960 – 1,38,240 = Rs. 17,280 (A)

3. Calender variance
   = (Actual days – Standard days) × Std. rate per day
   $= (24 - 25) \times \dfrac{\text{Rs, } 1,44,000}{25 \text{ days}}$ = Rs. 5,760 (A)

Problem 41. From the following data, calculate:

(1) Fixed overhead expenditure variance
(2) Fixed overhead volume variance
(3) Fixed overhead cost variance

	Budgeted	*Actual*
Fixed overhead for July	Rs. 10,000	Rs. 10,200
Units of production in July	5,000	5,200
Standard time for one unit	4 hours	–
Actual hours worked	–	20,100 hrs.

(University of Delhi, B.Com. (Hons.), 2002)

Solution:

(1) Fixed overhead expenditure variance:
= Budgeted overhead – Actual overhead
= 10,000 – 10,200 = Rs. 200 (A)

(2) Fixed overhead volume variance:
= Absorbed overheads – Budgeted overheads
= (Std. hrs. for actual output × Std. rate per hour) – Budgeted overheads
$= \left[(5,200 \text{ units} \times 4) \times \dfrac{10,000}{20,000}\right] - 10,000$
= 10,400 – 10,000 = Rs. 400 (F)

Note: Std. hours for budgeted output
= 5,000 × 4 = 20,000

(3) Fixed overhead cost variance:
= Absorbed overheads – Actual overheads
= (20,800 × 0.50) – 10,200 = Rs. 200 (F)

Problem 42. From the following information, calculate variable overhead variances

	Budgeted	*Actual*
Production units	400	360
Labour hours	8,000	7,000
Variable overheads	10,000	9,150

Standard time to produce one unit of output is 20 hours.

(University of Delhi, B.Com. (Hons.), 2000)

Solution:

Std. overhead rate per hour = $\frac{\text{Rs. } 10{,}000}{8{,}000 \text{ hrs.}}$ = Rs. 1.25

(1) Cost variance:

= (Std. hrs. for actual output × Std. rate) – Actual cost

= (360 × 20 × 1.25) – 9,150 = Rs. 150 (A)

(2) Budgeted overhead

= (Actual hours × Std. rate) – Actual cost

= (7,000 × 1.25) – 9,150 = 400 (A)

(3) Efficiency variance

(Std. hours for actual output – Actual hours) × Std. rate

= (7,200 – 7,000) × 1.25 = Rs. 250 (F)

Problem 43. *(Material, labour and overhead variances):* The following information is available from the cost records of a company for December 2003:

	Rs.
Materials purchased 20,000 pieces	88,000
Materials consumed 19,000 pieces	
Actual wages paid for 4,950 hrs.	24,750
Fixed factory overheads budgeted	40,000
Fixed factory overhead incurred	44,000
Units produced—1,800	

Standard rates and prices are:

Direct material rate—Rs. 4 per piece

Standard input—10 pieces per unit

Direct labour rate—Rs. 4 per hour

Standard hours required to produce a unit—2.5 hours

Overheads—Rs. 8 per labour hour

Compute the following variances:

(*a*) Material price and usage

(*b*) Labour rate and efficiency

(*c*) Fixed overhead expenditure variance

(*d*) Fixed overhead volume variance.

(University of Delhi, B.Com. (Hons.), 2004)

Solution:

(1) Material variance:

Actual price = $\frac{88{,}000}{20{,}000}$ = Rs. 4.40

Std. price = Rs. 4.00

Actual quantity = 19,000 pieces

Std. qty. = 1,800 × 10 = 18,000 pieces

(2) Material price variance:

= (Std. rate – Actual rate) × Actual qty.

= (4 – 4.40) × 19,000 = Rs. 7,600 (A)

(3) Material usage variance:

= (Std. qty. – Actual qty.) × Std. rate

= (18,000 – 19,000) × 4 = Rs. 4,000 (A)

CHAPTER 14

(4) Labour variance:

Actual rate = $\frac{24{,}750}{4{,}950}$ = Rs. 5

Std. rate = Rs. 4

Actual hrs. = 4,950

Std. hrs. = 1,800 × 2.5 = Rs. 4,500

(5) Labour rate variance:

= (Std. rate – Actual rate) × Actual hours

= (4 – 5) × 4,950 = Rs. 4,950 (A)

(6) Labour efficiency variance:

= (Std. hours – Actual hours) × Std. rate

= (4,500 – 4,950) × 4 = Rs. 1,800 (A)

(7) Fixed overhead expenditure variance:

= Absorbed overheads – Budgeted overhead

= (4,950 × 8) – 40,000 = Rs. 400 (A)

4. Sales Variance

Sales variance is useful in controlling sales and thereby earn more profit. Sales variances can be calculated on the basis of (1) Sales margin or (2) Turnover.

(*a*) **Sales variances based on sales margin:** Before the various sales margin variances are discussed, it is better to know the concept of standard sales margin.

Standard sales margin: This refers to the difference between the standard selling price of a product and its standard cost and it is the same as the standard profit for the product. The term 'Standard cost' used above is to indicate 'total standard cost'. It includes both fixed and variable cost. When fixed cost is excluded it becomes standard marginal cost. The difference between standard selling price and standard marginal cost is known as standard sales contribution.

The various sales margin variances are as follows:

(*a*) **Total sales margin variance:** This is the difference between the budgeted margin from sales and the actual margin, when the cost of sales is valued at the standard cost of production. The following formula is used to calculate this variance:

Total sales margin variance = Standard margin or profit – Actual margin or profit.

(*b*) **Sales margin price variance:** This is that portion of total sales margin variance which is the difference between the standard margin per unit and the actual margin per unit for the number of units sold in the period. The following formula is used to calculate this variance:

Sales margin price variance = (Std. margin per unit – Actual margin per unit) × Actual quantity

(*c*) **Sales margin quantity variance:** This is that portion of total sales margin variance which is the difference between the budgeted number of units sold and the actual number sold valued at standard margin per unit. The following formula is used to calculate this variance:

Standard margin qty. variance = (Standard proportion for actual sales – Budgeted qty.) × Standard profit.

When more than one product is sold, sales margin quantity variance is sub-divided into (*a*) Mix variance and (*b*) Volume variance.

(*d*) **Sales margin mix variance:** This is that portion of sales margin quantity variance which is the difference between the actual total number of units at that actual mix and the actual total number of units at standard mix valued at the standard margin per unit. The following formula is used to calculate this variance:

Sales margin mix variance = (Actual qty. – Std. proportion for actual sales) × Std. profit

(*e*) **Sales margin volume variance:** This is that portion of the sales margin quantity variance which is the difference between the actual total quantity of units sold and the budgeted total number of units at the standard mix valued at the standard margin per unit. The following formula is used to calculate this variance:

Sales volume variance = (Budgeted units sold – Actual units sold) × Std. margin per unit.

(*B*) **Sales variances based on turnover or sales value:** Sales variances based on turnover is used, by those organisations which would like to exercise control on actual sales by comparing it with budgeted sales.

The following are the variances based on turnover.

(*a*) **Sales value variance:** This refers to the difference between the budgeted sales and the actual sales. The following formula is used to calculate this variance:

Sales value variance = Budgeted sales – Actual sales

(*b*) **Sales price variance:** This refers to that portion of the sales variance which is due to the difference between the standard price specified and the actual price charged. The following formula is used to calculate this variance:

Sales price variance = (Actual price – Std. price) × Actual qty.

(*c*) **Sales volume variance:** It refers to that portion of the sales value variance which is due to the difference between the budgeted sales and the standard value of actual mix of sales. The following formula is used to calculate this variance:

Sales volume variance = (Std. qty. of sales – Actual qty. of sales) × Std. price

(*d*) **Sales quantity variance:** This refers to that portion of sales volume variance which is due to the difference between standard value of actual sales at standard mix and the budgeted sales. The following formula is used to calculate this variance:

Sales quantity variance = Budgeted sales – Revised standard sales

(*e*) **Sales mix variance:** This refers to that portion of the sales volume variance which is due to the difference between the standard and the actual composition of the sales mix. It is the difference between the standard value of actual mix and the standard value of actual sales at standard mix.

The following formula is used to calculate this variance:

$$\text{Sales mix variance} = \left(\frac{\text{Std. mix}}{\text{Revised mix of actual qty. sold}} - \text{Actual mix}\right) \times \text{Std. price}$$

Problem 44. *(Sales variance):* Ultra Modern Cassette Ltd. had budgeted the following sales for February 1991.

Cassette A	1,100 units @ Rs. 50 per unit
Cassette B	950 units @ Rs. 100 per unit
Cassette C	1,250 units @ Rs. 80 per unit

As against this, the actual sales were:

Cassette A	1,300 units @ Rs. 55 per unit
Cassette B	1,000 units @ Rs. 96 per unit
Cassette C	1,200 units @ Rs. 78 per unit

The cost per unit of Casette A, B and C was Rs. 45, Rs. 85 and Rs. 70 respectively.

Compute the different variance to explain the difference between the budgeted and actual product.

(C.S., Inter, June 1992)

Solution:

	Cassette A	*Cassette B*	*Cassette C*
Standard selling price *(see assumption 1)*	50	100	80
Standard cost *(see assumption 2)*	45	85	70
Standard profit	5	15	10

Assumption 1: Standard selling price is assumed to be equal to the budgeted selling price.
Assumption 2: Standard cost is assumed to be equal to actual cost.

CALCULATION OF ACTUAL PROFIT

	A	B	C
Actual selling price	55	95	78
Actual cost	45	85	70
Actual profit	10	10	8

Actual profit:				
Cassette A	—	1,300 @ Rs. 10	=	13,000
Cassette B	—	1,000 @ Rs. 10	=	10,000
Cassette C	—	1,200 @ Rs. 8	=	9,600
		Total	=	32,600
Budgeted profit:				
Cassette A	—	1,100 @ Rs. 5	=	5,500
Cassette B	—	950 @ Rs. 15	=	14,250
Cassette C	—	1,250 @ Rs. 10	=	12,500
		Total	=	32,250

1. Total sales margin variance = (Std. profit or budgeted profit – Actual profit)
   (32,250 – 32,600) = 350 (F)
2. Sales margin price variance = (Std. margin per unit – Actual margin per unit) × Actual qty.

Cassette A	—	(50 – 55) × 1,300	=	6,500 (F)
Cassette B	—	(100 – 95) × 1,000	=	5,000 (A)
Cassette C	—	(80 – 78) × 1,200	=	2,400 (A)
			=	900 (A)

3. Sales margin volume variance = (Budgeted sales qty. – Actual qty. sold) × Std. margin per unit

Cassette A	—	(1,100 – 1,300) × 5	=	1,000 (F)
Cassette B	—	(950 – 1,000) × 15	=	750 (A)
Cassette C	—	(1,250 – 1,200) × 10	=	500 (F)
			=	1,250 (F)

4. Sales mix variance = (Actual qty. – Std. proportion for actual sales) × Std. profit

Cassette A	—	(1,300 – 1,67) × 5	=	665 (F)
Cassette B	—	(1,000 – 1,008) × 15	=	120 (A)
Cassette C	—	(1,250 – 1,325) × 10	=	1,250 (A)
			=	705 (A)

Calculation of standard proportion for actual sales

Cassette A $= \frac{3,500}{3,300} \times 1,100$ = 1,167 units

Cassette B $= \frac{3,500}{3,300} \times 950$ = 1,008 units

Cassette C $= \frac{3,500}{3,300} \times 1,250$ = 1,325 units

= 3,500 units

5. Sales margin qty. variance = (Std. proportion for actual sales – Budgeted qty.) × Std. profit

Cassette A	—	(1,167 – 1,100) × 5	=	335 (F)
Cassette B	—	(1,008 – 950) × 15	=	870 (F)
Cassette C	—	(1,325 – 1,250) × 10	=	750 (F)
			=	1,995 (F)

Volume variance = Mix variance + Quantity variance
1,250 F = 705 (A) + 1,955 (F)

PROFIT AND LOSS STATEMENT

	Cassette A	*Cassette B*	*Cassette C*
Budgeted Sales	55,000	95,000	1,00,000
Less: Budgeted cost	49,500	80,750	87,500
Budgeted profit	5,500	14,250	12,500
Variances:			
Price variance	6,500 (F)	5,000 (A)	2,400 (A)
Mix variance	665 (F)	120 (A)	1,250 (A)
Qty. variance	335 (F)	870 (F)	750 (F)
	13,000	10,000	9,600

Problem 45. Compute the following variances from the data given below:

1. Total sales margin variance.
2. Sales margin volume variance
3. Sales margin price variance.
4. Sales margin mix variance.
5. Sales margin quantity (sub-volume) variance.

Product	*Budgeted quantity (Units)*	*Actual quantity (Units)*	*Budgeted sales price per unit (Rs.)*	*Actual sales price per unit (Rs.)*	*Standard cost per unit (Rs.)*
X	240	400	50	45	30
Y	160	200	25	20	15

(CA, Inter, May 1992)

CHAPTER 14

Solution:

1. Total sales margin variance = Actual profit – Budgeted profit
= 7,000 – 6,400 = 600 (F)
(see working note 2)

2. Sales margin volume variance = (Actual qty. – Budgeted qty.) × Budgeted margin per unit

X	= (400 – 240) × 20	= 3,200 (F)
Y	= (200 – 160) × 10	= 400 (F)
		3,600 (F)

3. Sales margin price variance = (Actual margin per unit – Budgeted margin per unit) × Actual qty.

X	= (15 – 20) × 400	= 2,000 (A)
Y	= (5 – 10) × 200	= 1,000 (A)
		3,000 (A)

4. Sales margin mix variance = (Budgeted margin per unit on actual mix – Budgeted margin per unit as budgeted mix) × Total actual qty.
(Refer working note 3)
= (Rs. 16.666 – 16) × 600 = 400 (F)

5. Sales margin qty. (sub-volume) variance = (Total actual qty. – Total budgeted qty.) × Budgeted margin per unit on budgeted mix
= (600 – 400) × 16 = 3,200 (F)
(Refer working note 3)

Verification:

Sales margin variance = Sales margin price variance + Sales margin volume variance
600 (F) = 3,000 (A) + 3,600 (F)

Sales margin volume variance = Sales margin mix variance + Sales margin qty. variance
3,600 (F) = 400 (F) + 3,200 (F)

Working Notes:

1. (*a*) Actual margin per unit = (Actual sales prices per unit – Standard cost per unit)

X = (45 – 30) = Rs. 15
Y = (20 – 15) = Rs. 5

(*b*) Budgeted margin per unit = Budgeted sales price per unit – Std. cost per unit

X = (50 – 30) = Rs. 20
Y = (25 – 15) = Rs. 10

2. (*a*) Actual profit = Actual qty. of product unit sold – Actual margin per unit

X = 400 × 15	=	6,000
Y = 200 × 5	=	1,000
		7,000

(*b*) Budgeted profit = Budgeted qty. of units to be sold × Budgeted margin per unit.

X = 240 × 20	=	4,800
Y = 160 × 10	=	1,600
		6,400

3. (*a*) Budgeted margin per unit on actual mix

$$= \frac{20 \times 400 + 10 \times 200}{600}$$

$$= \frac{8{,}000 + 2{,}000}{600} = 16.666$$

(b) Budgeted margin per unit on budgeted mix

$$= \frac{20 \times 240 + 10 \times 160}{400}$$

$$= \frac{4{,}800 + 1{,}600}{400} = \frac{6{,}400}{400} = \text{Rs. } 16$$

Control Ratios

Ratios are often used as a control technique in addition to monetary value of variance. The following ratios are commonly used in industries.

Efficiency Ratio

This ratio measures the efficiency at which the factory operates. The ICMA terminology defines efficiency ratio as "the standard hours equivalent to the work produced, expressed as a percentage of the actual hours spent in producing that work. Expressed as a formula:

$$\text{Efficiency ratio} = \frac{\text{Actual production in terms of standard hrs.}}{\text{Actual hours worked}} \times 100$$

Activity Ratio

This ratio measures the level of activity at which the factory is operating. The ICMA terminology defines it as "the number of standard hours equivalent to the work produced, expressed as a percentage of the budgeted standard hours". Expressed as a formula:

$$\text{Actual ratio} = \frac{\text{Production in terms of standard hours}}{\text{Budgeted production in terms of standard hours}} \times 100$$

Capacity Ratio

The ICMA terminology defines it as "the relationship between the actual number of working hours and the budgeted number". Expressed as a formula:

$$\text{Capacity ratio} = \frac{\text{Actual hours worked}}{\text{Budgeted production in terms of standard hours}} \times 100$$

Problem 46. In a manufacturing shop, product X required 2.5 man hours and product Y requires 6 man hours. In a month of 25 working days of 8 hours a day 2,000 units of X and 1,000 units of Y were produced. The company employed 50 workers in the shop and the budgeted man-hours are 1,08,000 for the year. You are required to work out the capacity ratio, activity ratio, and efficiency ratio. *(ICWA, Inter, June 1990)*

Solution:

Standard man-hours produced

Product X:	2,000 units @ 2.5 man hours	=	5,000 man-hours
Product Y:	1,000 units @ 6 man-hours	=	6,000 man hours
Total			11,000 man-hours

Budgeted standard man-hours per month

$$= \frac{1{,}08{,}000}{12} = 9{,}000 \text{ man-hours}$$

Actual man-hours worked = 50 workers × 25 days × 8 hrs. = 10,000 man-hours.

1. Capacity ratio $= \frac{\text{Actual man-hours worked}}{\text{Std. man-hours budgeted}} \times 100$

CHAPTER 14

$$= \frac{10{,}000}{9{,}000} \times 100 = 111.11\%$$

2. Activity ratio $= \dfrac{\text{Standard man-hours produced}}{\text{Standard man-hours budgeted}} \times 100$

$$= \frac{11{,}000}{9{,}000} \times 100 = 122.22\%$$

3. Efficiency ratio $= \dfrac{\text{Standard man-Hours produced}}{\text{Actual man-hours worked}} \times 100$

$$= \frac{11{,}000}{10{,}000} \times 100 = 110\%$$

Problem 47. Calculate from the following figure:

1. Efficiency ratio
2. Activity ratio
3. Capacity ratio

Budgeted production	880 units
Standard hours per unit	10
Actual production	750 units
Actual working hours	6,000

(CA, Inter, May 1992)

Solution:

1. Efficiency $= \dfrac{\text{Std. hours for actual production}}{\text{Actual hours worked}} \times 100$

$$= \frac{750 \text{ units} \times 10 \text{ hours}}{6{,}000} \times 100$$

$$= 125\%$$

2. Activity ratio $= \dfrac{\text{Std. hours for actual production}}{\text{Budgeted hours}} \times 100$

$$= \frac{750 \times 10}{880 \text{ units} \times 10 \text{ hours}} \times 100$$

$$= 85.23\%$$

3. Capacity ratio $= \dfrac{\text{Actual hours worked}}{\text{Budgeted hours}} \times 100$

$$= \frac{6000}{880 \times 10} \times 100$$

$$= 68.18\%$$

Problem 48. If the activity ratio and capacity ratio of a company is 104% and 96% respectively find out its efficiency ratio. *(CA, Inter, May 1997)*

Solution:

Activity ratio $= \dfrac{\text{Std. hours required for actual production}}{\text{Budgeted hours}} \times 100$

$$\text{Capacity ratio} = \frac{\text{Actual hours worked}}{\text{Budgeted hours}} \times 100$$

$$\text{Efficiency ratio} = \frac{\text{Std. hours required for actual production}}{\text{Actual hours worked}} \times 100$$

From the above ratios it is clear that efficiency ratio can be obtained by dividing activity ratio by capacity ratio.

So,

$$\text{Efficiency ratio (in percentage)} = \frac{\text{Activity Ratio}}{\text{Capacity Ratio}} \times 100$$

$$= \frac{104}{96} \times 100 = 108.33\%$$

INVESTIGATION OF VARIANCE AND REPORTING TO MANAGEMENT

Investigation of variance enables to take prompt control over the variance. The following benefits are derived by the process of investigating the variance:

(*a*) The person responsible for the variance is revealed.

(*b*) It reveals the accuracy of accounting system.

(*c*) Whether variance is on account of abnormal conditions and if so how to take precautionary measures in such situation.

(*d*) To appraise the current standard and revise if need be.

Having investigated the variance, it is necessary to prepare variance analysis reports for the sake of submitting to management. In preparing and submitting such reports the following guidelines may be followed:

(*a*) The reports must be prepared and submitted promptly to the management. Otherwise its purpose is lost.

(*b*) Only those reports which are relevant to the persons jurisdiction to exercise control should be sent.

(*c*) Analysis of variance must be based on cost centre.

(*d*) Ratios, percentages, etc. can also be used while reporting in addition to monetary value of variances.

(*e*) The cost accountant should interact with various levels of management in addition to preparation and submission of reports.

The important reports to be sent to various levels of management are (1) Material price variance reports. (2) Material usage variance report, (3) Wage rate variances report, (4) Labour efficiency variance report, (5) Sales value variance report.

Based on the reports of variance analysis, the variances are disposed off. One method of treating variance is to transfer it to costing profit and loss account. The second method advocated by some accountants is to properly distribute over closing stock and cost of sales so that both these items can be shown at actual costs in financial statement.

REVISION AND METHODS VARIANCE

Sometimes standards are revised owing to reasons such as change in wage rate, fiscal policy, etc. If such changes are incorporated in the standards, it may affect the variances and controllable factors may escape the attention of executives. Hence any changes in the factors is included in the revised budget but the standard cost is retained as a matter of policy. This process of retaining the basic standards in spite of change in the factors and analysing the variances on the basis of basic standards is known as revision variance.

A method variance refers to the difference between the standard cost of manufacture performed by normal method and the cost of manufacturing by an alternate method.

Questions

I. Fill in the blanks

1. The system which determines what the cost 'should be' in advance of production is called .... technique of costing.
2. For labour, the difference between actual hours at actual rate and actual hours at standard rate is called the .... variance.
3. Gang composition variance is a sub-variance of .... variance.
4. When the actual cost is less than the standard cost, the difference is known as .... variance.
5. Overhead variance is the difference between the .... cost of overhead absorbed in the output and the .... overhead cost.
6. Material price variance = (Std. price – Actual price) × ....

[***Answers:*** (1) standard costing, (2) labour rate, (3) labour efficiency (4) favourable, (5) standard, actual, (6) actual quantity]

II. Choose the correct answer

1. The cost of a product as determined under standard costing is
   (*a*) Fixed cost (*b*) Historical cost
   (*c*) Direct cost (*b*) Predetermined cost [ ]
2. While evaluating deviations of actual cost from standard cost, the technique used is
   (*a*) Regression analysis (*b*) Variance analysis
   (*c*) Trend analysis (*d*) Linear progression [ ]
3. Which one of the following standards can be attained under the most favourable conditions possible ?
   (*a*) Theoretical standard (*b*) Expected standard
   (*c*) Normal standard (*d*) Basic standard [ ]
4. Standard costs are useful in all of the following except
   (*a*) Reducing cost
   (*b*) Speeding up preparation of operating reports
   (*c*) Establishing records
   (*d*) Costing inventories [ ]
5. Controllable variance are best disposed of by transferring to
   (*a*) Cost of goods sold
   (*b*) Inventories of work-in-progress and finished goods

(c) Cost of goods sold and inventories

(d) Costing profit and loss account [ ]

[**Answers:** 1. (d), 2. (b), 3. (a), 4. (c), 5. (d)]

III. Mark True or False

1. Standard costs are fixed for each industry by the trade association concerned. (T/F)
2. Standards do not allow for any idleness and wastage and are therefore idealistic. (T/F)
3. Standards are arrived at on the basis of past performance. (T/F)
4. Standards represent achievable targets but after a purposeful effort. (T/F)
5. Standards for materials labour and overheads are inter-connected. (T/F)
6. Variance means the difference between the budget and standard costs. (T/F)
7. Favourable variances are those that increase the amounts and unfavourable ones reduce the amounts. (T/F)
8. All variances are transferred to the cost of sales account. (T/F)
9. Standard costing help management by exception by showing up the total difference between actual profit and standard profit. (T/F)
10. Price variances are calculated at actual quantity multiplied by differences in prices. (T/F)
11. Efficiency variance are concerned with quantities used and are calculated as standard quantity multiplied by difference in price. (T/F)
12. Yield variance shows the efficiency of labour. (T/F)

[**Answers:** True—4, 5, 10, Rest are all false]

IV. Short and Long Questions

1. Explain the term 'variance' under standard costing and discuss its significance. *(University of Kerala, M.Com., May 1992)*
2. Discuss the utility of variance analysis in cost control. What are the major causes for efficiency, volume, capacity and calender variance ? *(Calicut University, M.Com., April 1992)*
3. Distinguish between standard costing and budgetary control. Discuss the advantages and disadvantages of standard costing. *(Kakatiya University, M.Com., May 1991)*
4. Distinguish between standard costing and budgetary control. Are both these systems inter-related ? *(Bangalore University, M.Com., August 1989)*
5. Define standard costing. Explain the advantages and limitations of standard costing. *(Bangalore Univeristy, M.Com., May 1990)*
6. What are the various circumstances under which material price and usage variances are likely to arise ? *(ICWA, Inter, June 1990)*

Exercises

Material variance

1. Product 'A' requires 10 kgs of materials at the rate of Rs. 4 per kg. The actual consumption of material for the manufacturing of product A came to 12 kgs of materials at the rate of Rs. 4.50 per kg.

CHAPTER 14

Calculate: (*a*) Material cost variance (*b*) Material price variance (*c*) Material usage variance

(University of Madras, B.Com., April 2002)

[**Ans.** (*a*) Material cost variance = Rs. 14 (A) (*b*) Material price variance = Rs. 6 (A), (*c*) Material usage variance = Rs. 8 (A)]

2. The standard material cost of producing each unit of product K is as follows:

4.8 kg of material @ Rs. 5 per kg.

Actual material cost of producing 200 units of product K is as follows:

1200 kg of material costing Rs. 4,800

Compute: (1) Material cost variance (2) Material price variance (3) Material usage variance

(Calcutta University, B.Com., 2004)

[**Ans.** (1) Material cost variance = NIL (2) Material price variance = 1,200 (F), (3) Material usage variance = 1,200 (A)]

3. The standard raw material costs of producing 300 units of product 'A' were as follows

600 units of raw materials @ Rs. 50 per unit = Rs. 30,000

But actual raw material costs of producing 300 units of product A were:

1,000 units of raw materials @ Rs. 40 per unit = Rs. 40,000

Determine: (1) Material cost variance (2) Material price variance (3) Material usage variance

(Calcutta University B.Com. (Hons.), 1998)

[**Ans.** (1) Material cost variance = Rs. 10,000 (A) (2) Material price variance = Rs. 10,000 (F) (3) Material usage variance = Rs. 20,000 (A)]

4. The standard material cost for 100 kg of chemical D is made up of:

Chemical A	30 kg @ Rs. 4.00 per kg
Chemical B	40 kg @ Rs. 5.00 per kg
Chemical C	80 kg @ Rs. 6.00 per kg

In a batch 500 kg of chemical D was produced from a mix of:

Chemical A	140 kg at a cost of Rs. 500
Chemical B	220 kg at a cost of Rs. 1,056
Chemical C	440 kg at a cost of Rs. 2,860

Compute material variances contributing in the actual cost of 100 kg of chemcial D over the standard cost. *(University of Delhi, B.Com. (Hons.), 2001)*

[**Ans.** Total material cost variance = 100.80 (A), Total material price variance = 40.80 (A), Total material usage variance = 60 (A), Total material mix variance = 6.67 (A), Material yield variance = 53.33 (A)]

Labour Variances

5. From the following information, calculate labour variances

Standard for 100 units

500 labour hours

Rate Rs. 24 per hour

Actual production

1,000 units were produced

Total wages paid Rs. 1,30,000 for 5,200 hours

(University of Mumbai, B.Com., 2006)

[**Ans.** Labour cost variance = 10,000 (A), Labour rate variance = 5,200 (A), Labour efficiency variance = 4,800 (A)]

6. The following information is gathered from the labour records of KT & Company.

(1) Payroll allocation for direct labour Rs. 20,000

(2) Time card analysis shows that 8,000 hours were worked on production lines.

(3) Production reports for the period shows that 4,000 units have been completed each having standard labour time for $1\frac{1}{2}$ hours and standard labour rate Rs. 2 per hour.

Calculate: (1) Labour rate variance (2) Labour cost variance (3) Labour efficiency variance

(Mumbai University, B.Com., 2003)

[**Ans.** (1) Labour rate variance = 4,000 (A), (2) Labour cost variance = 8,000 (A), (3) Labour efficiency variance = 4,000 (A)]

7. From the following data of orient manufacturing Company Ltd. determine (1) Labour cost variance (2) Labour efficiency variance (3) Labour rate variance

(1) Standard labour cost per unit of production is Rs. 15

(2) Time allowed per unit is 30 hours

(3) During the month of March 1996, 3,000 units are produced in 75,000 hours

(4) Actual payment of wages for the month is Rs. 45,000

(Calcutta University, B.Com.(Hons.), 1996)

[**Ans.** (1) Labour cost variance = NIL (2) Labour efficiency variance = 7,500 (F) (3) Labour rate variance = 7,500 (A)]

Material and labour variance

8. The following details are related to the product 'A' for the month of March 2,000. You are required to compute material and labour cost variances from the given details.

Actual production 100 units

Standard cost per unit:

Material 50 kg @ Rs. 40 per kg

Labour 400 hours @ Rs. 1 per hour

Actual cost for the month:

Materials 4,900 kg @ Rs. 42 per kg

Labour 39,600 hours @ Rs. 1.10 per hour.

(University Calcutta, B.Com. (Hons.), 2002)

[**Ans.** (1) Material cost variance = 5,800 (A), (2) Material price variance = 9,800 (A), (3) Material usage variance = 4,000 (F), (4) Labour cost variance = 3,560 (A), (5) Labour rate variance = 3,960 (A), (6) Labour efficiency variance = 400 (F)]

9. SV Ltd., manufactures BXE by mixing three raw materials. For every batch of 100 kgs. of BXE, 125 kgs. of raw materials are used. In February 1986, 60 batches were prepared to produce and output of 5,600 kgs. of BXE. The standard and actual particulars for February 1986 are as under:

Raw	*Standard*		*Actual*		*Quantity of raw materials*
Materials	*Mix %*	*Price per kg. (Rs.)*	*Mix %*	*Price per kg.*	*Purchased kg. (Rs.)*
A	50	20	60	21	5,000
B	30	10	20	8	2,000
C	20	5	20	6	1,200

Calculate:

(1) Material cost variance.

CHAPTER 14

(2) Material price variance
(3) Material mix variance.
(4) Material yield variance. *(University of Delhi, B.Com. (Hons.), 1986)*

[**Ans.** Material cost variance = Rs. 17,500 (A), Material price variance A = 4,500 (A), B = 3,000 (F), C = 1,500 (A), Material mix variance A = 15,000 A, B = 7,500 F C = 0, Material yield variance = Rs. 7,000 F]

10. Gemini Chemical Industries provide the following information from their records: For making 10 kgs. of Gemco the standard material requirement is:

Materials	*Qty. (kgs.)*	*Rate per kg. (Rs.)*
A	8	6.00
B	4	4.00

During April 1,000 kgs. of Gemco were produced. The actual consumption of material is as under:

Materials	*Qty. (kgs.)*	*Rate per kg. (Rs.)*
A	750	7
B	500	5

Calculate: (*a*) Material cost variance, (*b*) Material price variance, (*c*) Material usage variance. *(CA, Inter, May 1989)*

[**Ans.** Total material cost variance = 13.50 (A), Total material price variance = 1,250 (A), Total material usage variance = 100 (A)]

11. A company manufacturing 'distempers' operates a costing system. The standard cost for one of the products of the company shows the following materials standard:

Materials	*Quantity*	*Standard price per kg.*	*Total (Rs.)*
A	40 kg.	75	3,000
B	10 kg.	50	500
C	50 kg.	20	1,000
		Material cost per units	4,500

The standard input mix is 200 kg. and the standard output of the finished product is 90 kg. The actual results for a period are:
You are required to calculate the material price, mix, yield, usage and cost variance. *(ICWA, Inter, December 1989)*

[**Ans.** Material price variance = 15,00.00, Material mix variance = 19,00.00, Material yield variance = 15,00.000. Material usage variance = 34,00.000. Material cost variance = 4,90,000]

12. Using the following information calculate labour cost variance, labour rate variance, labour efficiency variance and idle time variance.

Standard hours	:	5,000
Standard wage rate	:	Rs. 4 per hour
Actual hours	:	6,000
Actual wage rate	:	Rs. 3.50 per hr.
Time cost on account of machine breakdown	:	3,000 hrs.

(University Calicuit, B. Com., April 1986)

[**Ans.** Labour cost variance = Rs. 1,000 (A), Labour rate variance = Rs. 3,000 (F), Labour efficiency variance = Rs. 2,800 (A), Idle time variance = Rs. 1,200 (A)]

13. (Overhead variance): Following information is available from the records of a factory:

	Budget	*Actual*
Fixed overhead for June	Rs. 10,000	Rs. 12,000
Production in June (units)	2,000	2,100

Standard time per unit (hrs.)	10
Actual hours worked in June	

Compute: (*i*) Fixed overhead cost variance, (*ii*) Expenditure variance, (*iii*) Volume variance, (*iv*) Capacity variance, (*v*) Efficiency variance. *(CA, Inter, November 1989)*

[**Ans.** Fixed overhead expenditure variance = 2,000 (A), Fixed overhead capacity variance = 1,000 (F), Fixed overhead efficiency variance = 500 (A), Fixed overhead volume variance = 500 (F), Fixed overhead cost variance = 1,500 (A)]

14. (Sales variance): X Ltd. operate a budgetary control and standard costing system. From the following data calculate: (*i*) Sales variance, (*ii*) Sales volume variance, (*iii*) Sales price variance.

Product	*Budgeted*		*Actual*	
	Unit to be sold	*Sales value (Rs.)*	*Units sold*	*Sales value (Rs.)*
A	100	1,200	100	1,100
B	50	600	50	600
C	100	900	200	1,700
D	75	450	50	300
	325	3,150	400	3,700

(CS. (Final) June 1987)

[**Ans.** Sales volume variance = Rs. 200 (A), Sales price variance = Rs. 750 (F), Sales variance = Rs. 550 (F)]

15. Budgeted and actual sale for the month of December 1984 of two products A and B of M/s. XY Ltd. were as follows:

Product	*Budgeted sales*		*Actual sales*	
	Units	*Price per unit*	*Units*	*Price per unit*
A	6,000	5	5,000	5.00
	1,5000	4.75		
B	10,000	2	7,500	2.00
	1,750	1.90		

Budgeted cost for products *A* and *B* were Rs. 4 and Rs. 1.50 per unit respectively. Work out from the data the following variances:

(*i*) Sales value variance.

(*ii*) Sales volume variance

(*iii*) Sales prices variance.

CHAPTER 14

CHAPTER 15

UNIFORM COSTING

INTRODUCTION

Uniform costing is a system of costing which is applied to a number of plants belonging to a firm or number of firms belonging to an industry with a view to standardise accounting system. Under this system a uniform set of principles are followed for ascertaining cost and for fixing the selling price. Such practice of applying uniform principles of costing facilitates comparing efficiency and as a means of controlling costs in different undertakings. Uniform costing is different from costing methods and techniques. The use of uniform costing system does not mean that costs in different businesses will be the same. However, it means that comparison of costs of different firms will be logical if they practice uniform costing. On cost comparison it facilitates investigations into specific cost problems of the industry.

DEFINITION AND FEATURES AND APPLICATIONS OF UNIFORM COSTING

The ICMA terminology defines uniform costing as "the use by several undertakings of the same costing principles and/or practices".

The important features of uniform costing are as follows:

1. The same costing principles are applied by all member units for ascertaining costs.
2. Cost statements and reports are prepared on a uniform basis.
3. The accounting period is common for all the member units.
4. All the member units adopt the same costing methods, techniques and systems for collection, ascertainment and control of cost.

Uniform costing can be applied under the following two situations:

(*a*) Where an industry has number of firms or a firm has number of plants under same ownership and control. Where the units manufactured are identical, then cost per unit of different firms or plants can be compared. Where the units manufactured are dissimilar the cost comparison of similar operation, process or service can be made.

(*b*) Where number of business enterprises are affiliated to a trade association in order to derive advise regarding costing problems. Under this situation the trade association undertakes to collect costing data from various member firms, exchanges the ideas with them and offers valuable advise as to how to improve upon under an adverse situation. Examples of undertakings which may adopt uniform costing are printing press, hotels, sugar mill, cement mill, textile mill, plastic industries, road transport undertakings of various states and so on.

OBJECTIVES OF UNIFORM COSTING

The following are the important objectives of uniform costing:

1. To facilitate cost comparison of several firms belonging to the same industry in order to know efficiency or otherwise of it.
2. To take suitable steps to eliminate inefficiency of firms.
3. To furnish information to the government to get subsidy and other concessions.
4. To provide cost structure to government to enable it to formulate policies relating to wage structure, pricing, import, export and so on.
5. To determine standardisation in the method of production and distribution.
6. To avoid competition among the member firms by fixing a common selling price acceptable by all.
7. To enable member firms to exchange their ideas and problems and to equip them with the current trends and technology.
8. To enable the trade association to have a general control over the member firms by way of regulating the production capacity, fixation of selling price and so on.

ADVANTAGES AND DISADVANTAGES OF UNIFORM COSTING

(A) Advantages to Member Units

1. It facilitates cost comparison between different firms.
2. The services of professional cost accountant and consultant can be engaged to devise a uniform costing for all firms. This would prove economical which otherwise would have become expensive if different costing methods and techniques are used.
3. It makes easier to computerise accounting system of the various firms.
4. The appointment of a full-fledged cost accountant can be dispensed with for every member unit. Instead, one competent cost accountant can be appointed at head office who can be incharge of cost accounting system of all firms.
5. It enhances efficiency and productivity.
6. It encourages standardisation of materials and labour.
7. It reveals profitable and unprofitable products or operations of all firms.

(B) Advantages to Workers

1. It ensures a uniform wage structure in all member units.
2. Employees will benefit in sharing bonus and other amenities because of their efficiency and productivity.

(C) Advantages to Trade Association

1. Trade association can exercise better control over industries in eliminating wasteful competition.
2. It regulates the output and prices of member units and improves the profitability of the organisations.
3. Trade association represents the problems of its member units to the government in terms of getting subsidy and other forms of protection.

(D) Advantages to the Government

1. It enables the government in formulating various policies based upon the costing information of several industries.
2. It enables the government to give priority to public or private sector industry, based upon their progress and prosperity.
3. It provides useful information regarding profitability of the industry. The government can formulate tax laws based on this information.

Disadvantages of Uniform Costing

1. It may not suit every firm or member unit belonging to trade association. This is particularly so when there is difference in size, structure, diversification of production, etc. in different firms.
2. It is not flexible. It can not be changed whenever the need arises.
3. In some cases the members of trade association may not furnish reliable information as they feel that confidential matters when revealed to others may lead to competitions.

PROCEDURE INVOLVED UNDER UNIFORM COSTING

The procedure that is involved in the installation of uniform costing can be outlined as follows:

1. The determination of costing technique to be adopted such as historical or marginal or standard costing, etc.
2. The determination of costing method to be adopted such as process, single or output, etc.
3. The determination of cost unit such as a kg, tonne, etc.
4. The basis for cost classification and codification.
5. The identification of production and service departments.
6. The method of pricing the material issues that is to be followed such as FIFO, LIFO. etc.
7. The method of valuing closing stock and work-in-progress.
8. The method of wage payment to be followed such as time rate, price rate, etc.
9. The method of treating O.T. and idle time wages.
10. The determination of basis of apportionment and method of absorption of overheads.
11. The method of depreciating assets and interest.
12. The method of treating notional cost.
13. The method of presentation of reports and use of standard forms.
14. Method of treating office and administration overhead and selling and distribution overheads.
15. The method of treating over and under absorption of overheads.
16. Cost accounting system to be followed such as integral, non-integral accounts, etc.

UNIFORM COSTING MANUAL

It is a booklet which contains a set of guidelines relating to uniform costing system. In other words it contains details of various principles and the method of operating the system of uniform costing. It serves as a reference book for all the firms. It contains the objectives, the procedure to be followed and the rules to be followed under this system. The advantages of uniform costing manual may be outlined as under:

(a) It facilitates every firm to ascertain the cost of production as per the prescribed rules. Thus, it simplifies the accounting procedure in all firms.

(b) It evokes interest with everybody concerned.

(c) It enables the firms to overcome problems in practising this system.

(d) It facilitates the preparation of reports under a definite form.

INTER-FIRM COMPARISON

It is one of the management technique which is used to control the operations of weaker firms belonging to an industry. Though it is one of the oldest technique, it is being used extensively in modern days. It is a technique by means of which the management of similar undertakings compare the relative performance with a view to know the efficiency of various firms. Uniform costing system facilitates inter-firm comparison.

Though the final accounts consisting of trading account, profit and loss account and balance sheet can be taken as a basis for inter-firm comparison, they only serve as a general basis. However, they enable to know the overall profitability and financial position of various firms. They do not help to know the efficiency, productivity, and how best the firms utilise various resources available with them. Therefore different firms are compared with one another to know their performance and to take remedial steps if their performance is poor.

Objectives of Inter-firm Comparison

The objectives of inter-firm comparison may be summarised as under:

(a) To spot out weakness of the firms so that they can be improved upon.

(b) To attain and maintain overall efficiency of the firms.

(c) To enable the firms to standardise various functions so that wastages can be minimised.

(d) To enable the member-firms to know their relative position compared to other firms.

Procedure Involved Under Inter-firm Comparison

The following steps are involved under inter-firm comparison as a management technique:

(a) Adoption of uniform costing: As inter-firm comparison is based upon the uniform costing, it is essential to instal uniform costing in all the firms. To enable the system function efficiently, a uniform costing manual is also to be prepared and distributed among all firms.

(b) Establishment of a central or head office: The central office which undertakes to compare the performance of firms is to be established. It is headed by a chief cost accountant whose duty is to coordinate the functions of all the firms belonging to the same industry. In UK, an agency known as centre for inter-firm comparison was established in 1959. This agency undertakes to study the data furnished by the member-firms, analyses and prepares the reports of all the firms for the sake of inter-firm comparison. The other activities performed by this agency are as follows:

(a) It conducts seminars on inter-firm comparison.

(b) It carries out research with a view to discover best method of inter-firm comparison.

(*c*) It provides opportunities for the British firms to compare their performances with similar units in other countries such as Germany, France, etc.

However, there is no specialised agency in India, though certain organisations like National Productivity Council, and Trade Association facilitate inter-firm comparison.

(*c*) **Collection of data:** The central office collects the data from member-firms by preparing and sending a questionnaire. All the member-firms will furnish the required information which will be kept in secret by the central office. The information so collected will be processed, analysed and interpreted. Usually ratios are made use of to know the solvency, profitability, efficiency, productivity and financial soundness of the firm.

(*d*) **Preparation of report:** Based on the interpretation of data, the central office will prepare comprehensive report indicating the comparative position of other firms. It also provides concrete suggestions for improving upon with reference to special problems faced by the weak firms. This enables the member firms to set right in those areas where they lack behind.

Advantages of Inter-firm Comparison

The technique of inter-firm comparison is not only useful to management of a firm in improving its activities but also to different category of people such as bankers, investors, creditors, customers, trade associations and so on. Some of the advantages derived from this technique are as follows:

(*a*) It reveals the weakness of the firm. Such weakness can be overcome by taking remedial action.

(*b*) It reveals the trend of sales, profit, costs of the firm over a period of time.

(*c*) It encourages the member firms to become more cost conscious which helps in cost reduction.

(*d*) It provides useful information to management of every firm to make proper decisions.

(*e*) It ensures standardisation of production process.

(*f*) The benefit of research is passed on to member-firms by the central office.

(*g*) External parties such as bankers, investors, creditors also benefit by the technique of inter-firm comparison.

Questions

I. Simple Questions

1. Define uniform costing.
2. State the two situations under which uniform costing can be adopted.
3. State the features of uniform costing.
4. What do you mean by inter-firm comparison.
5. State any four objectives of inter-firm comparison.

II. Short Answer Questions

1. Outline the objectives of uniform costing.
2. State the advantages and disadvantages of uniform costing.
3. Outline the procedure involved under uniform costing.
4. Briefly explain the procedure involved under inter-firm comparison.

CHAPTER 16

COST REDUCTION

INTRODUCTION

One of the main objectives of a business organisation is to earn maximum possible profit. Profit is the resultant of interaction of two factors, viz., sales revenue and cost. Therefore it is imperative to probe into the possibilities of increasing the revenue and reducing the cost in order to achieve the desired goal. The profit may be increased through some methods such as (*a*) Increasing the selling price of the products, (*b*) Reducing the selling price to penetrate the market extensively and increase the turnover, (*c*) Increasing the demand through sales promotion, and (*d*) Reducing the cost of production for a given output or increasing the production for the same cost. Though profits may be increased by one of these methods, yet the last method seems to be the best. This is because; (*a*) Increasing the selling price may, other things remaining the same, result in losing the customer; (*b*) If the firm reduces the prices, competitors may also match the price cut. Thus, cutthroat competition may set in. (*c*) Demand can be increased through sales promotion but other firms may also increase their sales promotion outlay.

The ICWA publication of London defines cost reduction as "cost reduction is to be understood as the achievement of real and permanent reductions in the units costs of the goods manufactured or services rendered without impassing their suitability for the use that is intended". From this definition it is clear that cost reduction must be real, permanent and should not be done sacrificing the quality of goods.

ADVANTAGES

(A) To the Particular Concern

(*a*) Cost reduction enables improving the profits. This leads to distributing more dividends to the shareholders, paying better remuneration and bonus to staff and permits ploughing back of profit to finance the desired growth of the firm.

(*b*) Improvement in the financial position provides the opportunities to invest the funds in research which eventually helps in providing the customers with products of improved quality.

(*c*) Cost reduction improves the competitive capabilities of the firm.

(*d*) The extent to which the costs are reduced becomes an index of efficiency attained by the organisations.

(B) To the Industry

Cost reduction causes benefits not only to the particular concern but to entire industry. If a particular concern is running efficiently, other concerns in the industry will have to follow and adopt the efficient practices.

(C) To the Nation

(*a*) If the cost reduction programme is introduced, the scarce resources can be used properly.

(*b*) Higher taxes will be available to the government because the firms can earn better profit through productivity.

(*c*) In international markets cost reduction and concomitant reduction in prices of exports help in retaining the markets and gaining new buyers.

(*d*) Cost reduction helps in combating inflation.

COST CONTROL AND COST REDUCTION

Cost control and cost reduction are efficient tools of management but are different concepts. The differences between these two concepts are summarised below:

(1) Cost control is the means to achieve the end, *i.e.,* cost reduction. Thus cost control is the first step to achieve end objective. Cost reduction begins where cost control ends.

(2) Cost control is exercised by fixing standards or budgets and taking remedial measures, in case actuals deviate adversely from standards. Cost reduction is achieved by improving upon standards and improving upon production methods.

(3) Cost control is management oriented whereas cost reduction is oriented towards economising all resources with a view to reduce cost of production.

(4) Cost control ends after reaching the target. But under cost reduction programme, "products", "process", "procedures", "organisation", "methods" and "personnel" are continuously scrutinised in order to improve efficiency and reduce cost. It is thus a process of continuous self-analysis and self-criticism.

AREAS COVERED BY COST REDUCTION

A cost reduction programme should be systematic investigation into the techniques and methods of production and other functions. It involves an analysis of every aspect of the business with the object of ascertaining the most efficient method of performing the various functions. The broad areas to be covered by a cost reduction programme are as follows:

1. Materials

In most manufacturing concerns, the materials form a substantial portion of the total cost. A small percentage of cost reduction in material cost will lead to considerable savings to the organisation. Therefore materials must be managed and controlled as carefully as the cash. The important areas on which material cost reduction can be attempted are as under:

(*a*) Design: The material cost reduction programme should always begin at the design stage. The quality and quantity of material depends upon the design of the product. Continuous research should be undertaken in selecting the design. The effect of introduction of any new design or improvement of existing design in relation to cost of production should be thoroughly investigated with the help of cost accountant.

(*b*) Purchasing: Purchasing of materials is extremely important particularly to a manufacturing concern because it has its bearing on every vital factor concerning manufacture, *i.e.,* quantity, quality, cost-efficiency, economy, prompt delivery, volume of production, etc. For efficient buying, purchasing departments must be equipped with the list of various sources of supply and the latest prices ruling in the market for different materials. Ordering costs must be reduced as far as possible by purchasing in economic lot size quantities.

(*c*) **Storage:** After receiving the materials from the seller, they must be stored properly. The carrying cost must be reduced by keeping the inventory at optimum level. Various levels of materials, viz., maximum, minimum and reordering level must be fixed properly and must be revised keeping in view the changes in production and market condition. Materials must be properly classified, coded and standardised. A, B, C analysis of material control is highly useful in cost reduction. Continuous check of materials must be undertaken to detect the misappropriation of or misuse of materials.

(*d*) **Transport:** External and internal transport costs must be reduced as far as possible. Costs must be reduced by eliminating or reducing empty or idle trips, the distance travelled by the vehicles and consumption of petrol and diesel.

(*e*) **Production:** Production can be improved upon by training the employees, replacing the defective machinery, using better quality of materials, providing adequate and proper tools, reducing the wastage, scrap and defectives. Cost of production can be minimised by taking prompt action in respect of above factors.

The following are some of the techniques of materials cost reduction:

(*a*) **Value analysis:** Value analysis is one important technique of material cost reduction. This technique was first developed by Lawrence D. Miles of General Electric Company, USA. In a broad sense, it is a study of characterwise cost of an item such as quality, design, methods of manufacture, etc. with a view to reduce the ultimate cost of the product. Value analysis also means the determination of the value of items proposed to be sought with a view to examine the worth of money spent on buying the materials. It is based upon the quality of materials in relation to their functional value and price. It helps in deciding what is actually required and what amount should be paid so as to derive the maximum benefit. This reduces the cost of purchase and enhances the profit of the factory. Value analysis is carried out for the following purposes:

(1) Elimination of an item altogether if possible.

(2) Examination of cheaper quality materials for their advantageous use in the factory.

(3) Examination of costlier quality in relation to its functional value and life and to give up purchase of specified cheaper quality materials.

(4) Examinations of alternative source and price in relation to the money to be spent and realisation of its worth in due course.

(*b*) **Simplification and standardisation:** Simplification means elimination of superfluous varieties, size, dimension and grades of materials and component parts. It is also known as variety, reduction, as it is essentially a reduction process, a cutting down of varieties of materials. The basic principle of simplification is that a single item shall serves as many different applications as possible.

Standardisation in a production sense means the reduction of a line of product to fixed types, sizes and characteristics. It has been defined by E.C. Peck thus;

"Standardisation primarily means setting up standards by which the extent, quantity, value, performance, or service may be gauged or determined, and is the crystallised best thought and practice of industry into definite forms for general use".

(*c*) **Linear programming:** Sometimes it is possible to reduce the cost without making any substantial change in the design of the product or price or number of items in inventory. This is possible through the use of the linear programming. This is a quantitative technique, which is used to solve problems such as inventory management, scheduling, make or buy problem transportation problem and so on.

2. Labour

According to the old theory the direct labour is a variable cost. But this does not hold good today. Once an employee is appointed it is not easy to retrench him. Therefore the selection of an employee is a very important task. Cost reduction is possible if the right worker having the right attitude is employed to do the right work. Efficiency of employees can be increased by motivating them suitably to do their best for the organisation. By increasing the productivity, the labour cost can be reduced. It is possible through introduction of incentive schemes, bonus schemes, profit sharing, accompanied by proper working conditions, welfare facilities and safety measures.

3. Overheads

Overhead can be reduced by utilising the resources to the full extent. The following are the methods which can be used by the management for reducing the overhead cost.

(*a*) **Increasing the production volume:** By undertaking the production on optimum scale the costs can be decreased as the incidence of overheads chargeable to each unit of output declines. The idle capacity can be reduced by proper planning and control.

(*b*) **Extension of market, and price differentials:** Overheads can be reduced by extending the market area, increasing sales by offering discounts and charging suitable prices to attract different types of buyers.

(*c*) **Size of business units—Expansion and contraction policies:** By expanding or contracting the size of business units the overheads can be reduced. As some overheads are fixed in nature, they can be reduced by increasing the scale of production. On the contrary, costs may be more because the optimum scale of operations has been exceeded. Therefore by reducing the size, costs can be brought down.

COST REDUCTION COMMITTEE

As cost reduction plan requires coordination and cooperation, a high level committee must be formed. It will be entrusted with the responsibility of cost reduction programme. It is constituted by involving the heads of various functional departments. It will perform functions like agreeing on a target of cost reduction over a period of time, identifying the areas to be covered for cost reduction programmes and constantly reviewing the programme from time to time.

Questions

I. Simple Questions

1. Define cost reduction.
2. Distinguish between cost control and cost reduction.
3. State the advantages of reducing cost in a factory.

II. Long Answer Questions

1. Explain the various areas covered by cost reduction programme.

❑❑❑

CHAPTER 17

REPORTING TO MANAGEMENT

INTRODUCTION

The term 'reporting' conveys different meanings under different circumstances. In a narrow sense it means supplying facts and figures. Where a committee is appointed to study a problem, a report is taken to mean review of certain matter with its pros and cons and offering suggestions. In case of dealing with routine matters, a report refers to supplying the information at regular intervals in standardised forms. In this chapter reporting is taken to mean furnishing of cost accounting information to management. A report is a means of communication which is in written form and is meant for use of management for the purpose of planning, decision-making and controlling. Simply stated it is a communication of result by a subordinate to superior. It serves as a feedback to the management. The contents of report, the details of the data reported and the method of presentation depend upon the size and type of the business enterprise, extent of power delegated to subordinates and the existence of various levels of management for whom information is meant.

IMPORTANCE

Cost accounting embraces three important divisions, viz., cost ascertainment, cost presentation and cost control. Cost presentation serves as a link between cost ascertainment and cost control. The management of every organisation is interested in maximisation of profit through minimisation of wastages, losses and ultimately cost. So management will have to be furnished with frequent reports on all functional areas of business to achieve these objectives. One of the important functions of cost accounting is to provide the required information to all levels of management at the appropriate time. The various aspects of reporting such as nature of reports to be prepared, the details of information to be included and mode of presentation are all decided at the time of installation of cost accounting system. In fact, cost ascertainment and cost control are designed in such a way that they suit the scheme of information to be presented so that they serve all levels of management but not the other way round.

REQUISITES OF A GOOD REPORT

A good report should satisfy the following requisities in order to enable the receiver of report to understand and get interested in the report.

(*a*) **Title:** This contains the subject-matter of the report. It should be brief but not vague. Where a lengthy report is to be prepared the subject-matter is to be presented in various paragraphs under different sub-titles.

(*b*) **Period:** It should mention the duration covered by the report.

(*c*) **Units of measurement:** In case of quantitative information is to be reported the units in which quantities are expressed should be clear. For example, production in tonnes, sales in lakh rupees, idle time in hours.

(*d*) **Date:** The date on which the report is presented is to be mentioned. This helps receiver of the report to know what changes must have occurred during the time lag of period covered under the report and date of presentation of report.

(*e*) **Name:** The report must contain the name of the person by whom a report is prepared, the name of person to whom it is meant and the names of those for whom copies are sent.

(*f*) **Standard:** The reports prepared must meet the standard expected by its receiver. Use of highly technical words may be readily understood by lower level management.

(*g*) **Use of diagrams:** Wherever possible the reports must be illustrated by diagrams and charts in addition to description of the report. This facilitates ready understanding.

(*h*) **Recommendations:** Recommendations are to be offered to facilitate the reader as to what course of action is to be taken to set right the defects.

(*i*) **Promptness:** The reports should be prepared periodically and submitted to all levels of management promptly. It is said that report delayed is report denied. If the time lag between the period of preparation and period of submission is more it may give rise to wrong decisions.

(*j*) **Accuracy:** The information furnished in the report must be accurate. It is important to avoid furnishing unnecessary details in the report.

(*k*) **Comparison:** A comparative study must be incorporated in the report so as to facilitate the receiver of the report to know the progress and prospects of the performance. Comparison can be based on past performance or predetermined performance.

(*l*) **Economy:** The expenses incurred in maintaining reporting system must be less than the benefits derived therefrom or loss sustained by not reporting.

(*m*) **Simplicity:** The report should be brief, clear and simple to understand. The form of report should be designed to suit different levels of management. Where it is inevitable to prepare a lengthy report, a brief synopsis should precede the report.

(*n*) **Controllability:** Where variances are incorporated it is essential to stress on controllable aspects and to drop out uncontrollable element. But this depends upon the circumstance under which the report is prepared.

(*o*) **Source of information:** The source of information must be included in the report.

TYPES OF REPORTS

Reports are classified into different types according to different bases. This is shown in the following chart:

TYPES OF REPORT

- On the basis of purpose
  - External report
  - Internal report
- On the basis of period of submission
  - Routine report
  - Special report
- On the basis of function
  - Operating report
  - Financial report

I. On the Basis of Purpose

On the basis of purpose, reports can be classified into two types, *viz.*, (*a*) External report and (*b*) Internal report.

(*a*) **External report:** External report is prepared for meeting the requirements of persons outside the business, such as shareholders, creditors, bankers, government, stock exchange and so on. An example of external report is the published accounts, viz., profit and loss account and balance sheet. External report is brief in size as compared to internal report and they are prepared as per the statutory requirements.

(*b*) **Internal report:** Internal report is meant for different levels of management. This can again be classified into three types: (*a*) Report meant for top level management, (*b*) Report meant for middle level management and (*c*) Report meant for lower level management. Report to top level management should be in summary form giving an overall view of the performance of the business. Whereas external reports are prepared annually, internal reports are prepared frequently to serve the needs of management. Internal report need not conform to any standard form as it is not statutorily required to be prepared.

II. On the Basis of Period of Submission

According to this basis, reports can be classified into two types, viz., (1) Routine reports and (2) Special reports.

(*a*) **Routine reports:** They are prepared periodically to cover normal activities of the business. They are submitted to different levels of management according to a time schedule fixed. While some reports are prepared and submitted at a very short intervals, some are prepared and submitted at a long interval of time. Some examples of routine reports relate to monthly profit and loss account, monthly balance sheets, monthly production, purchases, sales etc.

(*b*) **Special reports:** Special reports are prepared to cover specific or special matters concerning the business. Most of the special reports are prepared after investigation or survey. There is no standard form used for submitting this report. Some of the matters which are covered by special reports are : causes for production delays, labour disputes, effects of machine breakdown, problems involved in capital expenditure, make or buy problems, purchase or hire of fixed assets, price fixation problems, closing down or continuation of certain departments, cost reduction schemes, etc.

III. On the Basis of Function

According to the purpose served by the reports, it can be classified into two types, viz., (*a*) Operating report and (*b*) Financial report.

(*a*) **Operating report:** These reports are prepared to reveal the various functional results. These reports can again be classified into three types, viz., (*a*) Control reports, which are prepared to exercise control over various operation of business, (*b*) Information report which are prepared for facilitating planning and policy formulation in a business, (*c*) Venture measurement report which is prepared to show the result of a specific venture undertaken as for example a new product line introduced.

(*b*) **Financial report:** Such reports provide information about financial position of the undertaking. These reports may be prepared annually to show the financial position for the year as in the case of balance sheet or periodically to show the cash position for a given period as in the case of fund flow analysis and cash flow analysis.

FORMS OF REPORT

Reporting of information to management takes different forms. They are explained below:

(*A*) Oral Report

An oral report is not very popular as it does not serve any evidence and cannot be referred to in future. Oral report may take the form of a meeting with individuals or a conference.

(*B*) Descriptive Reports

These are written in narrative style. They are frequently supported by tables and charts to illustrate certain points covered in the report. One important point that must be considered in drafting this form of report is the language. The language used must be simple, easy to understand and lucid. Where the report is very long, it must be suitably divided into paragraphs with headings. They must cover all the principles of good report discussed earlier.

(*C*) Comparative Statement

This form of report is used for preparing the routine report. Under this method the particulars of information are shown in a comparative form, *i.e.,* the actual results are compared with planned results and the deviations between the two are indicated. The various tools used to prepare this form of report are comparative financial statements, ratio analysis, fund flow analysis and so on.

(*D*) Diagrammatic and Graphic Representation

This is more popular form of preparing reports. They occupy lesser space and gives at a glance the whole picture about a particular aspect of study. They also facilitate the comparative study and shows the trend over a period of time. This form of report can be used where a report contains presentation of statistical numbers and other facts and figures. It overcomes the language barrier and is very easily understood by everyone. Of course when large numbers are involved, it is to be reduced by selecting a convenient scale. Diagrammatic representation involves the following forms:

(*a*) **Bar diagram:** They make use of horizontal and vertical axes to show the magnitude of values, quantity and period. Bar diagrams are of the following types.

(*i*) ***Simple bar diagram:*** These are most popularly used in preparing reports. They consider only length but not the width to indicate the change. Information relating to volume of production, cost of production, sales, etc. for different years can be shown under this form.

(*ii*) ***Multiple bar diagram:*** This type of diagram is used to report related matters such as production and sales, sales and profit, advertisement and sales and so on.

(*iii*) ***Sub-divided bar diagram:*** This form of diagram is used to report matters which involve different component parts as for example, the components of total cost of production such as prime cost, factory cost, office cost, cost of sales.

(*iv*) ***Percentage bar diagrams:*** These diagrams depict the information on a percentage basis.

(*b*) **Pie-diagram:** They take the form of circles instead of bars. They facilitate comparison besides depicting the actual information under review.

(*E*) Break-even Chart

This type of chart is prepared to show the relationship between variable and fixed cost and sales. It shows the point of no-profit and no-loss or where total cost equals total revenue received.

(*F*) Gantt Chart

This chart was first introduced by Henry L. Gantt. It is a special type of bar diagram under which bars are drawn horizontally. This chart shows the bars of planned schedule and attained performance. They are largely used to denote utilisation of machine capacity.

REPORTS SUBMITTED TO VARIOUS LEVELS OF MANAGEMENT

1. Top Level Management

The top level management comprises of board of directors, managing director and other executives who are concerned with determination of objectives and formulation of policies. Top management is to be furnished with reports at regular intervals in order to enable them to exercise control over the activities of the business. The following are some of the matters to be reported to board of directors.

(*a*) Master budget which covers all functional budgets for taking remedial actions where there are significant deviations from budgeted figures.

(*b*) Various functional budgets prepared by various departmental managers for holding departmental managers responsible for any shortfall in their performance.

(*c*) Capital expenditure budget and cash budget to know the extent of variances for taking remedial measures.

(*d*) Reports relating to production and sales, which shows the trend of the performance of business.

(*e*) Report covering important ratios such as stock turnover ratio, fixed assets turnover ratio, liquidity ratio, solvency ratio, profitability ratios, etc. to know the improvement in business.

(*f*) Appraisal of various projects undertaken by the organisation.

2. Middle Level Management

It comprises of different department managers such as production manager, purchase manager, sales manager, chief accountant, etc. These managers require reports to improve the efficiency of their respective departments. The following are some of the matters reported to production manager:

(*a*) Report relating to actual capacity utilised as compared to budgeted capacity.

(*b*) Report relating to actual output as against standard output.

(*c*) Labour and machine capacity utilised.

(*d*) Idle time lost.

(*e*) Report on scraps, wastages and losses in production.

(*f*) Report relating to stock of raw materials, work-in-progress and finished goods.

(*g*) Report relating to cost of production, operation of different departments.

The following are some of the matters reported to sales manager:

(*a*) Report relating to number of orders executed, orders received and orders on hand.

(*b*) Report relating to actual sales and budgeted sales and actual selling and distribution expenses and budgeted selling and distribution expenses.

(*c*) Summary of selling expenses incurred in different territories and their corresponding sales.

(*d*) Gross profit earned on different products and in different areas.

(*e*) Market survey reports.

(*f*) Report relating to present and potential demand.

The following are some of the matters reported to financial manager:

(*a*) Report relating to cash position.

(*b*) Summary of receipts and payments.

(*c*) Report relating to outstanding debts on credits sales.

(*d*) Report on debts due on credit purchases.

(*e*) Monthly profit and loss account.

(*f*) Quarterly report on capital expenditure.

3. Lower Level Management

The lower level management include supervisors, forement and inspectors who are concerned with the operations of the factory. They are interested in increasing the efficiency of the production departments. The reports that are to be sent to them are variances relating to planned and actual performance. The report must also emphasise cost control aspects.

Illustration 1

The profit of Bhaskar Ltd. is declining year after year. You, as a Management Accountant of that company, draft a report to the management exploring the reasons for the declining profits and suggesting corrective measures:

Answer:

From

The Management Accountant

....

....

To

The Managing Director

....

....

Dear Sir,

In pursuance of your instructions for ascertaining the reasons for declining profits year after year I submit hereunder my observations and suggestions.

Observations

Profit is basically the difference between the selling price of the products and their costs. Thus, lower the sales or higher the costs, any of them or both of them can be the cause of lower profits. Lower sales may be due to lower selling price or lower volume of sales. Lower selling price may have to be kept on account of market competition or inadequate demand for the goods. Lower volume of sales may be due to tough market competition, insufficient sales promotion activities, short fall in output, customer dissatisfaction due to inferior quality of goods, non-fulfilment of delivery promises, etc. Cost of production may be higher because of under-utilisation of resources, inadequate control measures leading to more waste of materials, labour, etc., lack of coordination, motivation or incentive among employees, inflationary market conditions resulting in rise in the cost of material, labour and overheads.

Causes

In our company the prime causes of declining profits can be summarised as follows:

(1) **Lower sales:** The sales figure are constantly declining both in terms of quantity and value. Over a period of 5 years from 2005 to 2009, the sales have almost been reduced to one half. Fall in sales is partly due to more competition and partly due to poor quality of our products.

(2) **Higher costs:** Lower sales have resulted in considerable under utilisation of factory's capacity. As compared to that in 2005, in 2009 about 50% of the factory's capacity is lying idle. However the fixed costs have remained constant. In other words, the fixed cost per unit have gone up resulting in lower profits. Besides, the company has to pay penal interest on borrowings from banks on account of non-payment of loan instalment on time.

(3) **Faulty product mix :** At present, the company is manufacturing four products A, B, C and D. Products A, B and C are giving contribution of Rs. 4, Rs. 6 and Rs. 2 respectively. However, product D is giving a loss of Rs. 3 per unit. The company manufactured 10,000 units of each of these products in the year 2009. This means that manufacture of product 'D' alone has resulted in an avoidable loss of Rs. 30,000 in 2009.

(4) **Faulty dividend policy:** In spite of the fact that the profits of the company are declining every year, the company is paying dividend almost at uniform rate of 15% p.a. for the last five years to its equity shareholders. This has resulted in almost negligible ploughing back of profits in the company's business.

Suggestions

(1) **Improving sales:** The company has to push up its sales. This will require improving the quality of the company's products and intensive publicity campaign highlighting the quality, utility and reliability of the company's products.

(2) **Full utilisation of company's capacity:** Till the sales of the company picks up requiring full installed capacity of the company, the extra capacity of the company may be hired out. This will result in partial recovery of fixed costs and improving the profitability of the company.

(3) **Change in product mix:** The product mix requires a change. The production of product D should be discontinued. The company should use the capital rendered surplus for manufacturing other products in the priority order of B, A and C subject to the overall market demand. The company can also think of diversifying into new areas.

(4) **Review dividend policy:** The dividend policy of the company is to be reviewed. The dividends are to be paid only when there are profits. Moreover sound financial planning requires that company should not go for 100% distribution of its profits. A part of the profits of the company to the extent required and feasible should be

retained by the company with itself. This will considerably strengthen the financial resources of the company. It will also help the company in the payment of loan instalments in time, thus resulting in saving of penal interest charged by the bankers.

Illustration 2

The Production Manager of Raju Co. Ltd is experiencing difficultes like power shortage, high labour absenteeism and non-availability of raw-materials on time. These have caused decline in production. You as a management consultant of the company, make an indepth study of the problems and report to the management suggesting suitable solution:

Answer:

From

The Management Consultant

....

....

To

The Managing Director

Raju Co. Ltd.

....

....

Dear Sir,

Sub : Report suggesting solutions to problems owing to power shortage, high labour absenteeism and non-availability of raw-materials on time.

With reference to your letter no......dated........ assigning me the task of recommending solutions for the problems relating to shortage of power and raw-materials and labour absenteeism. I wish to report as under.

The reason for shortage of power supply is mainly due to failure of monsoon. Due to non-availability of sufficient water, the hydro-power generation and supply is insufficient to meet the ever increasing demands of the industry. Same is the case in the thermal power sector where coal is insufficient for generating the required units of electricity.

There is increasing rate of labour absenteeism which adds to the agony. The labourers who feel their work is not sufficiently rewarded feel it is same as staying at home. Due to power breakdown many labourers are sent home due to stoppage in work.

Due to the increased number of firms and failure of monsoon, the basic raw-material production is very low, again leading to stoppage in work. Even the materials that are procured are at a high cost.

Recommendations

1. Planning the production schedule in advance and preparation of material procurrent budget, labour requirement budget and proper projection of power requirements would go a long way in eliminating or at least in minimising the problem.
2. Construction of self-generating power supply plants like solar power plants, wind mill etc., would curb the problem. We could also try to store the power when available for future use. Hence if the company could prepare project feasibility

report on these matters it would be possible to ascertain whether these can solve our power problems or not. Acquiring power from private sector can also solve the problem.

3. As far as labour absenteeism is concerned, providing them adequate monetary and non-monetary benefits would reduce the turnover and absenteeism. Fringe benefits and retirement benefits provided would go a long way to reduce the rate of absenteeism.
4. Raw-materials supply can be properly regulated by inviting tenders and choosing the price that would be most favourable for us. Centralised purchases would eliminate unnecessary cost involved if any. Dividing the entire loss into most economic order quantities and placing them with different suppliers would give us the twin benefits of cost reduction and uninterrupted supply. Proper material control system should be formulated.

If we try these, I have no doubt, that we shall minimise the problems which we are experiencing at present.

Yours truly

Sd/- Production Manager

Illustration 3

The Directors of Anita Steels Ltd. are facing the problem of working capital. They are not in a position to coordinate the inflow and outflow of cash. Examine the existing management of working capital and submit a report to the management on your findings and recommendations to correct the situation.

Answer:

The Board of Directors
Anita Steel Ltd.
....
....

Sir,

Sub : Problem of working capital management.

With reference to your letter no. .... dated .... assigning me the task of studying the working capital management and suggesting ways and means to improve them, I wish to report as under:

1. The main reason for the inadequacy of working capital is the lack of coordination between cash inflows and outflows. The liquidity management should be sound. Then only coordination is possible in any organisation.
2. Only when the credit management is efficient the inflow of cash would be sound. In our organisation, the credit management is very weak.
3. In addition to credit management, management of inventory also is our important area which will boost the cash inflow. The inventory of finished goods are on the higher side which is due to the ineffective and poor selling line.
4. The cash inflow management is not sound enough in our organisation to monitor and to meet the regular payment to our suppliers, our organisation is not a prompt and regular payer to our suppliers and we never get the discounts and allowances.

Suggestions

1. The credit collection policy should be streamlined in order to collect promptly from our debtors and bills receivable.
2. Our selling force must be motivated and to provide more concessions and discounts to our distribution network and also to customers which will pave the way for minimum inventory of finished goods.
3. The improvement of credit collection and selling will boost our sales and minimise our inventory and both will reward us with sound cash inflow management.
4. The effective and efficient cash flow will provide us opportunity to coordinate with cash outflow and obtain concessions and discounts from our suppliers.
5. Proper coordinated cash inflow and outflow management will maintain sound and better working capital management. The above recommended line of action will provide us fruitful results in our working capital management.

Yours truly

Sd/- Finance Manager

Illustration 4

ABC Company is facing the problem of raw-materials shortage. As a production manager, you are required to draft a report suggesting reformulation of purchase policy to have uninterrupted supply of raw-materials.

Answer:

The Managing Director
ABC Co. Ltd.
....
....

Dear Sir,

Sub : Interrupted supply of raw-materials — Reformulation of purchase policy.

With reference to your letter no. .... dated .... assigning me the task of reformulation of purchase policy to have uninterrupted supply of raw-materials, I wish to report as under:

The main reason for delay in supply has been due to lack of alertness of the stores department which has led to a fall in the stock level below danger level interrupting production. This has occurred more than once.

There is a boom in demand for the goods continuously in the market and this particular product line where we specialise has given profits to producers which as usual has brought a number of new entrants as our competitors. Unfortunately, the basic raw-materials production has not increased proportionately. Thus dividing total available raw-materials among many has led to shortage in availability and has failed to cater to the needs of the industry.

Recommendations

1. Formulation of a policy to limit the disadvantages of shortage is to begin with proper coordination between the purchase and stores department.
2. The stores department should continuously evaluate the needs in accordance with the production schedule which is available with any department well in advance.

Timely requisitions given would lead to easy procurement at the most favourable prices.

3. Proper materials management should be devised by adopting ABC analysis and VED (Vital, Essential, Desirable) analysis that could lead to avoidance of unnecessary procurement and wastage.
4. As far as the general scarcity in the market is concerned, this is a situation which would fade away once the new firms leave the market in the long run. But as far as the present time is concerned, we could invite tenders from suppliers from other states and if found favourable, orders could be placed with them to be sure of supply which would however be difficult from our old suppliers due to the market conditions.

If we try these, I have no doubt that we shall be able to overcome the challenge and survive in this market.

Yours truly

Sd/- Production Manager

Questions

Long Answer Questions

1. What do you understand by the term 'reporting to management'? Discuss briefly the matters that you would deal with while reporting to the board of directors.
2. Discuss the general principles to be observed while preparing reports.
3. Describe the various forms of reporting to management.
4. Distinguish between routine and special reports. State the various matters which are sent to management under routine and special reports.
5. Explain different types of reports submitted to the management of an organisation.
6. Explain the information submitted to different levels of management.

Exercises

1. The profits of XC Company are declining year by year. As the management accountant of the company draft a report to the management exploring the possible reasons with recommendation.

❑❑❑

CHAPTER 18

MECHANISED COST ACCOUNTING

INTRODUCTION

Accounting machines constitute one of the modern method of performing accounting work. It is an improvement over manual method in the sense it performs the job more quickly and efficiently. However, use of machines need operation and supervision by accounting department's staff. Today a wide variety of machines are available to perform different types of accounting work. Though the entry of mechanisation is considered to be late entry in accounting department when compared to production process in factories, it is now increasing at a pace which seems likely to surpass its progress in the factory. The increased importance of accounting machines is felt by the management of most of the organisations and many sophisticated machines have been introduced to simplify the mental work of the organisation. This is because in most of the large sized business organisations, the accounting transactions are so numerous that the manual system of account keeping becomes inaccurate, monotonous, inefficient, costly and takes too much of time. The use of accounting machines to a considerable extent save time and labour, improves quality of work, maintains accuracy and provides information to management more quickly when compared to manual system.

POLICIES GUIDING MECHANISATION

The following aspects serve as guiding principles of mechanisation of accounting department:

(*a*) Volume of work.

(*b*) Savings in time.

(*c*) Bottlenecks in the mechanisation process.

(*d*) Training the operator.

(*e*) Resistance to change.

(*f*) Cost of the machine.

(*g*) Maintenance of service cost.

(*h*) Speed of maintenance and availability of spare parts.

(*i*) Operating costs.

(*j*) Special paperwork required

(*k*) Space available.

(*l*) Accuracy and quality required.

(*m*) Relief (or creation) of boredom and monotony.

(*n*) Less risk of fraud if checking built in.

(*o*) Obsolescence factor.

(*p*) Reaction from other staff and the staff union.

(*q*) Fear of redundancy and lower of morale.

(*r*) Lower fatigue factor.

(*s*) Degree of supervision.

(*t*) Information requirements of management.

FACTORS TO BE CONSIDERED BEFORE INTRODUCING MACHINES

Before introducing machines in the accounting department certain important factors ought to be taken into consideration. They are as follows:

(*a*) Need of the Machine

The need for machines in the accounting department may be felt for the reasons such as replacing manual labour with a view to reduce cost, to overcome shortage of staff or to improve management-staff communication.

(*b*) Determination of Appropriate Machine

Different machines serve different purposes and hence right choice must be made to serve a definite purpose. Suppliers can be relied upon to suggest the right equipment for performing a right job.

(*c*) Specification of the Machine

The specification of machines may relate to size, installation requirements and special structural requirements. Though the machines may satisfy most of the requirements, yet it is often discarded owing to inappropriate specification.

(*d*) Cost of the Machine

The investment on machines is an important factor to be considered as it has a profound impact on the financial policy of the business. Before the machines are bought, quotations are to be invited and the most competitive supplies is to be choosen.

(*e*) Flexibility of Machines

Certain machines can be used for performing different types of work when compared to other machines. Hence the extent of the flexibility of the machine should be thoroughly evaluated.

(*f*) Ease of Machine Operation

Some machines are easy to operate while others require complete knowledge about the functioning of machines.

(*g*) Cost of Machine Operation

The cost of operating the machine which include power and labour cost and stationery cost is to be taken into account before selecting the machine.

(*h*) Staff Consideration

Before installing machines, it is necessary to seek cooperation from the staff. The attitude of staff on mechanisation is to be considered to avoid antagonism toward's introduction of machines.

IMPORTANCE OF MECHANISATION

The importance of mechanisation can be summarised as follows:

(*a*) It saves labour.
(*b*) It saves time.
(*c*) It performs the work more accurately.
(*d*) It relieves monotony.
(*e*) It reduces the scope of fraud.
(*f*) It provides work of better appearance as for example a computerised invoice.
(*g*) It lessons the fatigue of accounting staff and thus improves the quality of work.
(*h*) It provides management with more information at a short notice of time.

PRELIMINARY ARRANGEMENTS TO BE MADE BEFORE INSTALLING MACHINES

In order to derive maximum benefits from the installation of accounting machines certain preliminary arrangements are essential. These are as follows:

(*a*) Preliminary Notice to the Staff

In order to seek cooperation from the staff, they must be convinced that machines are introduced only to increase their efficiency and it does not result in dismissal of staff.

(*b*) Gradual Introduction

There should be gradual but not sudden installation of machines.

(*c*) Flow of Work

There must be careful planning of office machines so as to ensure a smooth flow of work from one department to another instead of creating a bottleneck.

(*d*) Training

Wherever applicable it is necessary to train the existing staff or to recruit trained operator before the machines are installed.

(*e*) Allocation of Machines to Departments or Personnel

It is desireable to decide in advance what machines are to be used by different departments and which employee is to operate upon the machine.

(*f*) Accommodation

Sufficient space must be made available for the installation of machines as well as for the operators. Necessary arrangements must also be made to isolate noise producing machines so as to avoid distraction.

(*g*) Communication

The information relating to the following is to be communicated as and when it is applicable:

(*i*) Personnel entitled to use the machines.
(*ii*) Instructions regarding care and maintenance.
(*iii*) Procedural changes to be made in the respective departments resulting from the introduction of the new machines.

CHAPTER 18

After satisfying that all the above-mentioned factors are within the reach of the organisation, the actual installation of machines is to be considered. The actual installation of machines are in turn governed by the following:

(*a*) Size and weight of the machine for the sake of allocation of space.

(*b*) Availability of suitable power point.

(*c*) Vibration caused by the machine.

(*d*) Noise produced by the machine demanding a separate room to avoid distraction.

(*e*) Focusing the light so as to facilitate correct visibility of papers.

(*f*) Space for keeping service tools and storage facilities.

PUNCHED CARD EQUIPMENT

This is one of the data processing machine. The most important feature of a punched card machine lies in the common language provided by the punched card. This card can be used to record alphabetic information, numerical information and special symbols such as "Rs.", "%" etc.

Equipments used in Punch-Card System

(*a*) **Key punch machine:** A key punch machine is used to punch the holes in the punched cards. The data for punching the holes can be obtained from various documents such as purchase orders, invoices, payment vouchers, inventory reports, etc. This equipment has a key-board which resembles to a typewriter key-board. The operator after reading the information from the various documents mentioned above enters the same data (information) on the card.

The key punch machine has a feeding mechanism which permits the automatic feeding of a new card. This equipment is also capable of programming which means several operating functions can be performed automatically. It can also be used to interpret the information contained on the card which means that the letters, numbers, or symbols are printed at the top of the card. This can be used to verify the accuracy of the information punched in the cards.

(*b*) **Card verifier:** The purpose of the card verifier is to verify the accuracy of the information contained on the card so punched by the key punch machine. The operator using the key board, punches the same card for the second time based on the same information. If the punched hole this time is in a different position (not coinciding on the hole originally punched) then electrical contact is lost and the machine automatically stops. This indicates the error in punching. The operator then ascertains whether the fault lies in the first or in the repeat punching.

(*c*) **Interpreter:** The function of the interpreter is to print on the cards the data that are contained in the punches on the card. In other words it translates the punched data into alphabetic, numerical or special character and points it on the card. This is so because the key punch machines do not have a printing mechanism. Interpreting the punched data is necessary when the cards are to be used for filing purposes.

(*d*) **Reproducer:** This equipment is used to duplicate the data contained on one set of cards to another set of cards. This is useful when the original set is wornout.

(*e*) **Sorter:** The function of the sorter is to arrange the cards into alphabetic or numerical order. For example, sales invoice cards may have to be sorted out on the basis of territories.

(*f*) **Tabulating or accounting machine:** After sorting out the cards they are fed into a tabulator which senses the information punched in the cards, automatically adds up the various totals and prints the totals at the end of each section. The tabulator has a plugged programme.

Applications

Punched card equipments are put to the following uses:

(*a*) **Bank transactions:** The usual purposes served in the bank transactions are preparation of cheques, standing orders, reconciliation statements, payment of dividend, interest etc.

(*b*) **Budgetary control:** In preparing departmental budgets and master budget.

(*c*) **Cost accounting:** In ascertaining total cost of production.

(*d*) **General accounting:** In preparing invoice, sales and purchase ledger and depreciation accounting.

(*e*) **Sales analysis:** In preparing market and advertising research copies.

(*f*) **Stores accounting:** In preparing stock verification report.

(*g*) **Wage accounting:** In preparing wage sheets, wage slips etc.

(*h*) **Statistical records:** In analysing and tabulating the facts and figures.

Advantages

(*a*) **Speed of operation:** This is considered to be an outstanding advantage of punched card machine.

(*b*) **Accuracy:** The card verifier ensures 100% accuracy of the results.

(*c*) **Simplicity of operation:** Once the person is trained it is very simple to operate.

(*d*) **Economy of labour:** As most of the work is performed by the machine, labour cost is saved to a large extent.

(*e*) **Adaptability:** It is suitable to all types of business big or small.

(*f*) **Production of computer input:** The cards produced by punched card machine may be used as input for a computer.

(*g*) **Information to management:** It provides accurate and up-to-date information which is very essential to management.

Disadvantages

(*a*) **Cost:** The initial investment is heavy.

(*b*) **Operating cost:** The cost involved in punching and verifying the card increases the operating cost.

(*c*) **Noise:** The equipment produces excessive noise which disturbs people working in the same room.

(*d*) **Breakdown of equipment:** This is a serious disadvantage as the processing of punched card machine relies upon its continuous operation. If one component, say the sorter breakdown, the result may be chaotic.

COMPUTERS OR ELECTRONIC DATA PROCESSING

The introduction of computers has revolutionised the management function of most of the business enterprises. An electronic computer is a machine by which data is processed electronically at great speed which means the performance of calculations, analysis and printing the results. A computer differs from a calculator in respect of a "memory" which calculators do not have.

Characteristics of Computers

(*a*) A computer is capable of performing only such function which it is instructed. In other words, it will not undertake anything other than what it is instructed to do in its programme.

(*b*) It is capable of performing repetitive calculations rather than simple calculations. Its great capacity enables large volumes of data to be processed very quickly with greater accuracy.

(*c*) For programming the data in computers it is a prerequisite that the data is capable of representation in an alphabetic or numerical pattern recognised by the computer.

(*d*) Performing the work and presenting the results in a very short span of time is one of the distinctive features of computers.

(*e*) The magnetic storage devices which are part of computers enable a considerable degree of integration between systems to be achieved.

Functions

Mills and Standingford have summarised the functions of computers thus:

(*a*) To receive one or several programmes of instructions, stores these and obey them as may be required.

(*b*) To take in new information through one or more input channels and store it for reference as required by any of the programmes.

(*c*) To perform any arithmetical function as required by a programme.

(*d*) According to the state of information it produces, selects and carries out alternative courses of action.

(*e*) To apply checks to data it receives or produces and if the test indicates a failure produce a record or a signal to obtain human intervention and meanwhile proceed to next transaction or to another job.

(*f*) To store data produced for further reference.

(*g*) To select information from the stores, arrange it in any sequence and discharge it through one or more output channels to be printed for human use or to be recorded for subsequent computer use.

The Computer System

The computer system can be studied under two headings, viz., (*A*) hardware and (*B*) software.

(*A*) **Hardware:** The term hardware is used to denote the actual computer and its associated equipments. A computer consists of five components. They are as follows:

(*i*) ***Input unit:*** This component receives the data in an appropriate form such as punched cards, perforated paper tape and magnetic tape. In this equipment the data is converted into electrical pulses which represents the 'language' of the computer.

(*ii*) ***Storage unit:*** This component is also known as memory unit. The data which is in computer language is retained in this unit which may consist of magnetic tape, magentic discs, plated wire, etc. This equipment holds standing information, new information and the programmed instructions.

(*iii*) ***Control unit:*** This is also known as coordination unit which directs and brings into operation the various circuits required to carry out the programmed instructions.

(*iv*) ***Arithmetic unit:*** This is also known as function or process unit which performs all the arithmetical calculations at high speed.

(*v*) ***Output unit:*** This equipment converts the results from the language of computer to the ordinary language.

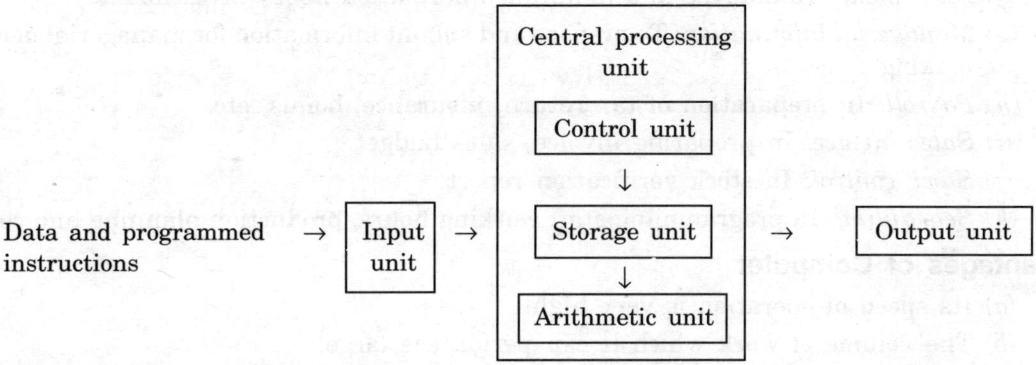

Working of a computer: The working of a computer can be discussed under the following steps:

(*a*) ***Programming:*** As a computer is capable of performing all the work relating to almost all the functions of a business such as purchase accounting, sales accounting, wages accounting, cost accounting, etc., it is necessary for the programmer at the first instance to prepare a programme for a particular task.

(*b*) ***Input:*** Having prepared a programme this is fed into the computer through any of the input media such as punched cards, paper tape, etc.

(*c*) ***Storage:*** The data compiled earlier and the information related to the current task is held on storage unit to facilitate further operation.

(*d*) ***Processing:*** This stage involves the following sub-stages:

(*i*) Transmission of data contained in the punched card to the central processing unit.

(*ii*) The existing data from the storage unit is fed simultaneously along with current data into central processing unit.

(*iii*) Acting upon the data supplied by the two sources, the process performs the necessary calculations and assembles information.

(*iv*) The processor then passes the result of its work to the output devices.

(*e*) ***Output:*** The result received by the output unit is finally presented in the required form, as for example on punched cards.

(*B*) **Software:** The term software is defined as a system of computer which includes the programming control instruments, training facilities, etc. provided to enable a computer to operate. In brief it is a means by which the computer can be put to work.

Applications

Perhaps there is no field where the utility of computers are not felt. The following are some of the areas in which the computers are used.

(a) *Analysis:* It is used to solve all problems involving analysis.

(b) *Apportionment:* To apportion common expenses incurred to various departments or products.

(c) *Budgetary control:* In preparing departmental and master budgets.

(d) *Conversion:* To convert the amount into various currencies.

(e) *Financial accounts:* In preparing debtors, creditors and nominal ledgers.

(f) *Investment:* To analyse and maintain information about investments.

(g) *Managerial information:* To prepare and submit information for managerial decision-making.

(h) *Payroll:* In preparation of tax return, insurance, bonus, etc.

(i) *Sales invoice:* In preparing invoice, sales budget.

(j) *Stock control:* In stock verification report.

(k) *Scheduling:* In programming staff working hours, production planning and control.

Advantages of Computer

(a) Its speed of operation is very high.

(b) The volume of work which it can perform is large.

(c) It reduces the labour cost.

(d) The results shown by it is accurate and reliable.

(e) It furnishes up-to-date information for management.

Disadvantages

(a) The initial investment on computers is very high.

(b) It may be subjected to obsolescence owing to increased technology of computers.

(c) It requires additional cost of maintenance such as air-conditioning.

(d) It results in waste of stationery in producing unnecessary records and increased labour cost in producing, handling and filing them.

(e) It is difficult to obtain experienced system analyst and programmers.

(f) Breakdown of machine leads to delay in product in results.

Questions

Short Answer Questions

1. Outline the policies which guide machanisation of cost accounting department.
2. State the various factors which are to be considered before introducing machines in the costing departments.
3. Explain the importance of machines in costing department.
4. State the features, advantages and disadvantages of punched card equipment.
5. State the features, advantages and disadvantages of computers.
6. Explain briefly the working of a computer.

❑❑❑

CHAPTER 19

ACTIVITY BASED COSTING

INTRODUCTION

Traditional approach to cost accounting is known by volume based cost accounting. It is a simple system of accounting whereby direct costs (*i.e.,* prime cost elements) are charged to products and overheads are allocated on the basis of some arbitrary selected criteria. The various bases that are commonly used relates to MHR, LHR, percentage methods and so on. These methods have their limitations and they tend to misinterpret the apportionment of overheads to the products.

Inadequacies of Traditional Methods of Overhead Absorption

The cost of product arrived in traditional cost accounting system is not so accurate owing to following reasons:

(1) The basis used in the traditional cost accounting are acceptable when valuing stock for financial reporting, but they are inappropriate when used for decision making.

(2) The traditional basis of classification of cost into fixed and variable is often unrealistic since, as business grows they often become more complex.

(3) Traditional accounting is confined merely to furnish information at product level. The new manufacturing technology demands feedback of performance while production is still in progress.

ABC is a recent development in cost accounting. It attempts to absorb overheads into product costs on a more realistic basis. Supporters of the technique argue that many costs are unnecessarily treated as common costs and are arbitrarily absorbed using a basis such as labour hours, machine hours etc. Many organisations have now adopted Advanced Manufacturing Technology (AMT) with the result that overheads are increasing and labour costs are becoming a small portion of total costs. Costing system which absorb overheads on a direct labour basis are therefore not relevant in an AMT environment. The basic idea of ABC is that costs are grouped according to what drives them or causes them to be incurred. The cost drivers are then used as an absorption basis.

Definition

The CIMA terminology defines ABC as a "Cost attribution to cost units on the basis of benefit received from indirect activities".

The ABC system was designed by Kaplan and Cooper of Harvard Business School. According to them total cost should be classified into long-term variable cost and short-term variable cost. Traditionally, short-term variable costs are known as *variable cost* and long-term variable costs are known as *fixed cost*. Short-term variable costs are volume related and change proportionately with the volume of production. Long-term variable cost vary in the long-term but not instantaneously. For example, production scheduling cost can be changed in the long-term by changing number of runs rather than changing number of units produced.

They further claim that this approach relates overhead costs to the forces behind them. The forces behind overheads are named "cost drivers". Cost drivers can be defined as those activities or transactions that are significant determinants of costs. A cost driver is an activity which generates cost. ABC system is based on the belief that activities cause costs and that a link should therefore be made between activities and products. This is done by assigning costs of activities to products based on an individual product's demand for each activity.

Characteristics of ABC

(1) Simple traditional distinction made between fixed and variable cost is not enough guide to provide quality of information to design a cost system.

(2) The more appropriate distinction between cost behaviour pattern are volume (scale) related, diversity (scope) related, events (decisions) related and time related.

(3) Cost drivers need to be identified. A cost driver is a structural determinant of cost related activity. The logic behind is that the cost behaviour pattern must be understood so that behaviour pattern is dictated by cost drivers.

ELEMENTS INVOLVED IN ACTIVITY BASED COSTING

The following elements are involved in Activity-Based Costing.

1. Activity

An aggregate of closely related tasks constitute an activity. Thus any activity can comprise of one or more of tasks which are associated with one another to accomplish an objective. For example customer order processing involves the following tasks:

(*a*) Receiving orders from customers

(*b*) Interacting with production regarding capacity to produce

(*c*) Giving commitment to the customer regarding delivery time etc.

2. Activity Cost Pool

A cost pool is a term used to indicate grouping of costs incurred on a particular activity which drives them. An example of cost pool is procurement to raw materials. In this transaction we see all costs associated with purchase of materials, viz, ordering cost, receiving and inspection cost, storage cost etc.

3. Cost Drivers

Cost drivers are used to trace costs to products by using a measure of resources consumed by each activity. For instance, in a consulting firm, for the activity consulting, consultant levels may be the cost driver.

4. Process

A process represents an aggregation of related function specific activities. For example, activities like inspection of materials, receiving material, recording and testing for its specifications or standards are all related and can be grouped under material handling process.

5. Cost Object

It is the final point to which costs are traced. The cost objects are linked to the objective of the organisation. For instance if the objective is to select a most profitable customer, individual customer is the cost object.

6. Non-Value Adding Activity

There are a large number of activities which continue to be carried out particularly in certain old factories, which do not contribute to the value of a product. The ideal example would be movement of material from one work station to another not adding any value to the product.

STEPS INVOLVED IN IMPLEMENTING ABC

1. Step 1

Identify the functional areas involved.

2. Step 2

Identify the relative activities involved in each functional areas.

3. Step 3

Allocate the common expenditure to various activities in each functional areas.

4. Step 4

Prepare the statement of expenditure functionwise/activitywise/account headwise.

5. Step 5

Identify the most suitable cost driver in each activity under functional areas.

6. Step 6

Identify the share of one cost driver to the product manufactured for absorption of overhead.

7. Step 7

Determine the rate or cost driver for a particular activity.

8. Step 8

Absorb overhead on the basis of rate or cost driver.

Differences between Conventional costing and ABC

Point of differences	*Conventional costing*	*ABC*
(1) Implied Assumptions	– Products incur cost. – Only short-term variable costs are recognised. – All the non-product costs are time cost (*i.e.*, fixed cost). They are overheads that cannot be traced with products.	– Products consume activities and activities incur cost. – Recognises long-term variable costs also. These costs do not vary with the variation in the volume of the product. – Long-term variable cost can be related with activities through which they can be logically linked with product and services.
(2) Cost centre	– Machine labour group, department etc. are identified as cost centre around which indirect costs are collected. – Types of cost centres are almost standard.	– Activities required for producing products are identified as cost centres. – Types of activity based cost centres depend on actual operation.

CHAPTER 19

(3) Cost pool	– The costs are generally pooled as per the accounts head like wages, salary, depreciation etc.	– Cost pools are based on the behaviour of cost item and its relation with a particular activity. Therefore salary is divided into many parts depending upon which part of salary varies with cost driver.
	– Cost pools are usually standard.	– Cost pools will depend upon activities causing them.
	– Cost pools are finally combined into one at operating cost centre level.	– Once cost pools are made they are never combined.
(4) Absorption methods	– They are volume related. Machine hours, Labour hours, are examples of absorption methods.	– Most logical methods are followed. They are activity related because activities consume resources.
(5) MIS	– Very little data required for being able to charge overheads. Data of actual machine and labour hours for each product would be sufficient.	– Strong MIS support is required for creating database.
(6) Decision Support Potential	– Because it does not facilitate accurate regrouping of cost required for various decision situations. For example, division of cost into fixed and variable is based on either intution or statistical techniques.	– It facilitates decision because regrouping of expenses is more accurate. Accuracy level is achieved because the behaviour of the cost is studied in the ABC.
(7) Applicability	– This system is more economical where operations are simple, few products are produced and proportion of direct costs is very high in total cost.	– The ABC is more economical where operation is complex. Product diversity is high and size of overhead is large in total cost.

Development of ABC System

The following steps are involved in ABC system:

(1) Identifty the main activities performed in the organisation, such as manufacturing, assembly etc. as well as support activities including purchasing, packing and dispatching.

(2) Identify the factors which influence the cost of each activity-cost driver.

(3) Collect accurate data on direct materials. Direct labour and overheads.

(4) Establish the demand made by particular products on activities using the cost drivers as a measure of demand.

(5) Trace the cost of activities to products according to a products demand for each activity.

Thus ABC is the process of tracing costs first from resources to activities and then from activities to specific products. The technique of ABC lays the importance of different costs for different purposes and the identification of just those costs which are relevant to a particular decision.

IMPLEMENTATION OF ABC SYSTEM

The implementation of ABC system involves the following steps:

(1) Planning for ABC System

In planning stage, management's problem is identified. The questions that are to be raised here are what are the company's critical issues ? What are the difficult decisions the management is faced with ? What type of culture prevails in the organisation ? The designer, by determining the critical decisions that management typically faces, can he able to identify the accounting information that the management would need to support decisions relating to those critical issues. Identification of the level of information (either by his analysis or interviewing the senior managers) would provide the framework for an actual cost system using ABC concepts.

(2) Organising ABC Teams

Designing and implementing ABC needs support from management at every level of the organisation. A committee of Senior Managers may be constituted first to set direction and make decisions regarding the strategic use of the ABC data. The committee may appoint persons as representatives of their respective functional organisations in order to serve on the ABC work group. The work group is to assist in identifying the data to be captured and to determine the operational use of ABC data. Accordingly 3 teams are constituted as indicated below:

Level	*Teams*	*Function*
(1) Top level	Steering Committee	To monitor strategic use of cost information and determine cost objects.
(2) Middle level	Project Manager and Work group	To assist in identify the data to be captured and to determine the operational use of ABC data.
(3) Operational	Project Teams Analyst	To identify the meaningful activities, cost drivers and the sources of data.

(3) Management Education

The provision of better cost information does not solve the organisational problem by itself. Management needs to be educated to the multiple uses of the ABC data and must be trained to play an interactive role in the ABC process. Thus a major task in the process of implementation is to educate the managers and the employees that in turn would realise their cooperation in successful implementation.

(4) Designing the Process

This involve:

(*a*) Defining cost objects by interviewing the top and middle level managers and perceiving their current problem such as individual product (if the problem is profitable product mix) distribution channel, customers etc.

CHAPTER 19

(*b*) Analysing and defining critical activities (both production and services) with a view to establishing them as a basis for determining cost and performance.

(*c*) Calculating the activity costs by selecting appropriate first stage cost drivers and applying them to distribute the natural expenses to various activities undertaken.

(*d*) Calculating the product costs by selecting appropriate second stage cost drivers and applying them to distribute the activity costs to various products manufactured.

(5) Maintaining the System

Implementing ABC system is not a single time job; it needs continuous maintenance and monitoring. The cost drivers may change along with the production dynamism. Hence periodical testing of the system is necessary particularly to ensure that the cost drivers both first and second stage are working well to indicate the resource consumption. It may so happen that the cost driver chosen for a particular resource type or activity type at the time of introducing the system subsequently may fail to justify the distribution of costs. This may not only cause distrust towards the system but also create other organisational problem.

Advantages or Benefits of ABC System

ABC system besides providing more accurate basis for calculating product costs has other benefits too, *i.e.*, it serves as a mechanism for managing cost. It is in this area of cost management and resource planning ABC has the greatest potential. Its benefits may be summarised as under:

(1) **Measurement of performance:** This is done by identifying responsibility or accountability for cost behaviour.

(2) **Price fixation:** This is done by identifying the true cost of product or services.

(3) **Product promotion:** This is done by analysing profit contribution and impact of inter-relationship of product or service.

(4) **Customer profitability:** It helps to know which customers yield more profit. This is done by identifying the cost or profitability of type of customer.

(5) **Choosing channel of distribution:** This is done by analysing the cost and performance of alternative distribution channels.

Usefulness of ABC to Management

(1) It supports management by furnishing data at operational level. Accurate product costing will help the management to compare the profits that various customers, product lines, brands or regions generate and to decide on pricing strategy, droppig unprofitable products or product lines.

(2) It helps management to redesign the products.

(3) It can be used to change technology, reduce set-up time, rationalisation of plant layout so as to reduce material handling cost, improve quality etc.

(4) It helps management in improving activities which helps in improving profits.

(5) It helps in preparing good budgets.

(6) The accurate information provided on product cost enables better decisions to be made on pricing, marketing, product design, product mix etc.

Limitations

(1) ABC is not appropriate for all companies particularly those where overheads are relatively small and which do not produce a wide range of products.

(2) This method cannot be used to prepare monthly or periodical profit statement. For this, traditional cost accounting system is more suitable.

(3) ABC is more suitable to support strategic decision making, profitability analysis and control of manufacturing.

When to use ABC

ABC system is useful under the following circumstances:

(1) Products which are difficult to produce and which involve more cost of production.

(2) Where profit earned is more than the accounting system suggest.

(3) Where competitors are not able to produce more in spite of increased profit which such products yield, because of increased cost of production.

(4) In quoting tender prices.

(5) When buy or make decision is involved *i.e.*, when suppliers are able to sell at lesser price as compared to high cost of production in the organisation.

(6) Where overheads incurred are more and apportionment becomes more complex.

(7) The cost of marketing and distribution of products varies considerably by distribution channel and yet cost accounting system effectively ignores marketing cost.

Problem 1. Sunlight company is manufacturing two products *A* and *B*, using the same equipment and similar processes. An extract of the production data for these products in one period is shown below:

Particulars	*A*	*B*
Quantity produced (units)	5,000	7,000
Direct labour hours per unit	1	2
Machine hours per unit	3	1
Set ups in the period	10	40
Orders handled in the period	15	60
Overhead costs	*Rs.*	
Relating to machine activity	2,20,000	
Relating to production run set-ups	20,000	
Relating to handling of orders	45,000	
	2,85,000	

You are required to calculate production overheads to be absorbed by one unit of each of the products using the following costing methods:

(1) Traditional costing approach using a direct labour hour rate to absorb overheads.

(2) An activity based costing approach using suitable cost drivers to trace overheads to products.

Solution:

(1) Traditional costing approach:

Calculation of Direct labour hours:

Product *A*	5,000 units × 1 hr	=	5,000 hrs
Product *B*	7,000 units × 2 hrs.	=	14,000 hrs
	Total		19,000 hrs

$$\text{Overhead Absorption Rate} = \frac{\text{Total Overheads}}{\text{Total Direct Labour Hours}}$$

$$= \frac{\text{Rs. } 2{,}85{,}000}{19{,}000}$$

$$= \text{Rs. } 15 \text{ per hour}$$

CHAPTER 19

Overheads to be absorbed are:

Product *A*	1 hr × Rs. 15	=	Rs. 15 per unit
Product *B*	2 hrs. × Rs. 15	=	Rs. 30 per unit

(2) ABC costing approach:

Calculation of Direct labour hours:

Product *A*	–	5,000 units × 3 hrs.	=	15,000 hrs.
Product *B*	–	7,000 units × 1 hrs.	=	7,000 hrs.
				22,000 hrs.

VERIFICATION

Product 'A'	=	5,000 units × 15 per unit	=	75,000
Product 'B'	=	7,000 units × 30 per unit	=	2,10,000
				2,85,000

(1) Machine–hour driver cost $= \dfrac{\text{Overhead relating to machine activity}}{\text{Machine hours}}$

$$= \frac{\text{Rs. } 2,20,000}{22,000 \text{ hrs.}} = \text{Rs. } 10 \text{ per machine hour}$$

(2) Set up driver cost $= \dfrac{\text{Overhead relating to production setup}}{\text{No. of setups}}$

$$= \frac{20,000}{50\,(10+40)} = \text{Rs. } 400 \text{ per set up}$$

(3) Order handling cost $= \dfrac{\text{Overhead relating to handling of orders}}{\text{No. of orders handled}}$

$$= \frac{45,000}{75\,(15+60)} = \text{Rs. } 600 \text{ per order}$$

OVERHEAD COST

		Product *A*			Product *B*		
Machine driver cost	=	15,000 × 10	=	1,50,000	7,000 × 10	=	70,000
Set-up cost		10 × 400	=	4,000	40 × 400	=	16,000
Order handling cost		15 × 600	=	9,000	60 × 600	=	36,000
				1,63,000			1,22,000
Overhead rate per unit			=	$\dfrac{1,63,000}{5,000}$		=	$\dfrac{1,22,000}{7,000}$
			=	32.60		=	17.43

Problem 2. The following particulars are extracted from the books of Mr. Colin. Calculate cost per unit under (*i*) Traditional volume based costing and (*ii*) Activity based costing.

	Machine hrs. per unit	*Direct labour hour per unit*	*Annual output (units)*	*Total Machine hours*	*Total direct labour hrs*	*No. of purchase order*	*No. of set ups*
Product *A*	2	4	1,000	2,000	4,000	80	40
Product *B*	2	4	10,000	20,000	40,000	160	60
				22,000	44,000	240	100

The cost of activities is as follows:

Volume related	1,10,000
Purchasing related	1,20,000
Set-up related	2,10,000
	4,40,000

Assume that under Traditional method product '*A*' absorbs overheads on Machine Hour Rate and product '*B*' on Labour Hour Rate method.

Solution:

TRADITIONAL VOLUME BASED COSTING SYSTEM

Total cost allocated to cost centre 4,40,000

$$\text{Machine hour rate} = \frac{\text{Total overheads}}{\text{Total machine hours}} = \frac{4,40,000}{22,000} = \text{Rs. } 20$$

$$\text{Labour hour rate} = \frac{\text{Total overheads}}{\text{Total labour hours}} = \frac{4,40,000}{44,000} = \text{Rs. } 10$$

Cost per unit = Machine hour rate × Machine hours per unit
= 20 × 2 = 40
= Labour hour rate × Labour hours per unit
= 10 × 4 = 40

Total cost charged to product = Annual output × Cost per unit
A = 1,000 × 40 = 40,000
B = 10,000 × 40 = 4,00,000

ABC SYSTEM

$$\text{Volume related cost} = \frac{1,10,000}{22,000} = \text{Rs. 5 per machine hour}$$

$$\text{Purchased related cost} = \frac{1,20,000}{240} = \text{Rs. 500 per order}$$

$$\text{Set-up related cost} = \frac{2,10,000}{100} = \text{Rs. 2,100 per set up}$$

Cost charged to products:

Product A:

Volume related cost	= 2,000 × 5	= 10,000
Purchase related cost	= 80 × 500	= 40,000
Set up related cost	= 40 × 2,100	= 84,000
		1,34,000

Product *B*:

Volume related cost	= 20,000 × 5	= 1,00,000
Purchase related cost	= 160 × 500	= 80,000
Set up related cost	= 60 × 2,100	= 1,26,000
		3,06,000

CHAPTER 19

Problem 3. The following particulars are given to you

Particulars	*Product* A	*Product* B
Units produced	20	20
Material handling per product unit	6	14
Direct labour hours per unit	870	870

Budgeted material handling costs are Rs. 1,74,000. You are required to

(*a*) Determine cost per unit of the products using traditional volume-based costing method under direct labour hour rate method.

(*b*) Determine cost per unit of the products under ABC method.

(c) Comment on the results. *(Sri Sathya Sai University, B.Com., (Hons) October 2008)*

Solution:

(*a*) Traditional volume-based costing:

$$\text{Direct labour hour rate} = \frac{\text{Budgeted overhead}}{\text{Budgeted direct labour hours}}$$

$$= \frac{1,74,000}{(20\times 870)+(20\times 870)}$$

$$= \text{Rs. } 5$$

	Product A	*Product* B
Total direct labour hours taken 870 × 20	17,400	17,400
Labour hour rate	Rs. 5	Rs. 5
Total material handling cost absorbed (17400 × Rs. 5)	87,000	87,000

$$\text{Cost per unit} = \frac{\text{Total material handling cost}}{\text{Units produced}}$$

	Product A	*Product* B
=	$\frac{87,000}{20}$	$\frac{87,000}{20}$
	= Rs. 4350	Rs. 4350

Comment: As per the above calculation each unit of product *A* and *B* absorbs the same amount of material handling costs even though product '*B*' consumes more material handling than product '*A*'. In other words, conventional volume based application rate fails to trace the large number of material handling for product '*B*'

(*b*) Under ABC Costing:

$$\text{Rate per material handling} = \frac{\text{Budgeted material handling cost}}{\text{Budgeted material handling}}$$

$$= \frac{1,74,000}{6+14} = \text{Rs. } 8,700$$

		Product A	Product B
Material handling per product unit		6	14
Material handling cost @ Rs. 8700 per material move	8700 × 6 8700 × 14	Rs. 52,200	Rs. 1,21,800
Material handling cost per unit	=	$\frac{52200}{20}$	$\frac{1,21,800}{20}$
	=	Rs. 2,610	Rs. 6,090

Comment: Product *'B'* that requires more material handling correctly receives more material handling costs under ABC.

Problem 4. 'X' Ltd. manufactures four products *A*, *B*, *C* and *D*. Output and cost data for January 2007 are as follows.

Product.	*Output (units)*	*No. of Production runs in the period*	*Material cost per unit*	*Direct Labour hours per unit*	*Machine hours per unit*
A	20	2	10	1	1
B	20	2	50	3	3
C	200	5	20	1	1
D	200	5	80	3	3

Direct labour cost is Rs. 10 per hour

Overhead costs are as follows:

Short run variable cost	Rs. 10,560
Set up cost	Rs. 14,000
Expediting and scheduling cost	Rs. 10,080
Material handling cost	Rs. 7,840
	Rs. 42,480

Calculate production cost under:

(*a*) Activity-based costing system using machine hour rate as cost driver for recovering short run variable cost and number of production runs as cost driver for recovering other overheads.

(*b*) Traditional volume based costing system

(Sri Sathya Sai University, B Com., (Hons) October 2007)

Solution: Working Notes

(1) Calculation of material cost:

A = 20 units × Rs. 10 per unit = 200

B = 20 units × Rs. 50 per unit = 1,000

C = 200 units × Rs. 20 per unit = 4,000

D = 200 units × Rs. 80 per unit = 16,000

CHAPTER 19

(2) Calculation of labour cost:

A = 20 units × 1 hr per unit × Rs. 10 = 200

B = 20 units × 3 hrs per unit × Rs. 10 = 600

C = 200 units × 1 hr per unit × Rs. 10 = 2,000

D = 200 units × 3 hrs per unit × Rs. 10 = 6,000

(3) Calculation of short-run variable cost:

$$\text{Machine hour rate} = \frac{\text{Short - run variable cost}}{\text{Total machine hours}}$$

$$= \frac{10{,}560}{(20 \times 1) + (20 \times 3) + (200 \times 1) + (200 \times 3)}$$

$$= \frac{10{,}560}{20 + 60 + 200 + 600}$$

$$= \frac{10{,}560}{880} = \text{Rs. 12 per machine hour}$$

A = 20 units × 1 hour per unit × Rs. 12 = 240

B = 20 units × 3 hours per unit × Rs. 12 = 720

C = 200 units × 1 hour per unit × Rs. 12 = 2,400

D = 200 units × 1 hour per unit × Rs. 12 = 7,200

(4) Calculation of set up cost:

$$= \frac{\text{Set up cost}}{\text{No. of production runs}}$$

$$= \frac{14{,}000}{14} = \text{Rs. 1,000 per production run}$$

A = 2 × 1,000 = 2,000

B = 2 × 1,000 = 2,000

C = 5 × 1,000 = 5,000

D = 5 × 1,000 = 5,000

(5) Calculation of expediting and scheduling cost:

$$= \frac{\text{Expediting and Scheduling cost}}{\text{No. of production order}}$$

$$= \frac{10{,}080}{14} = \text{Rs. 720 per production run}$$

A = 2 × 720 = 1,440

B = 2 × 720 = 1,440

C = 5 × 720 = 3,600

D = 5 × 720 = 3,600

(6) Calculation of material handling cost:

$$= \frac{\text{Material handing cost}}{\text{No. of production order}}$$

$$= \frac{7{,}840}{14} = \text{Rs. 560 per production run}$$

$A = 2 \times 560 = 1{,}120$
$B = 2 \times 560 = 1{,}120$
$C = 5 \times 560 = 2{,}800$
$D = 5 \times 560 = 2{,}800$

(1) **PRODUCTION COST UNDER ABC COSTING**

	A	B	C	D
Direct material cost	200	1,000	4,000	16,000
Direct labour cost	200	600	2,000	6,000
Short run variable cost	240	720	2,400	7,200
Set up cost	2,000	2,000	5,000	5,000
Expediting and Scheduling cost	1,440	1,440	3,600	3,600
Material handling cost	1,120	1,120	2,800	2,800
Cost of production	5,200	6,880	19,800	40,600

(2) **UNDER ABSORPTION COSTING**

	A	B	C	D
Materials	200	1,000	4,000	16,000
Direct labour	200	600	2,000	6,000
Overhead (see Note below)	965	2,896	9,655	28,964
Cost of production	1,365	4,496	15,655	50,964

Working Note

$$\text{Machine hour rate} = \frac{\text{Total overhead}}{\text{Total machine hours}}$$

$$= \frac{42{,}480}{880} = 48.2727$$

$A = 20 \text{ units} \times 1 \times 48.2727 = 965$
$B = 20 \text{ units} \times 3 \times 48.2727 = 2{,}896$
$C = 200 \text{ units} \times 1 \times 48.2727 = 9{,}655$
$D = 200 \text{ units} \times 3 \times 48.2727 = 28{,}964$

Problem 5. A manufacturing company produces a variety of products. You are given the following date for the year ended 31-3-2003

Factory overhead:

	Rs.	Rs.
Material related		
– Material handling	43,600	
– Materials receipts and processing	54,400	
– Freight inwards	24,000	1,22,000

Machine related:		
– Machine operating cost	1,00,800	
– Machine setup cost	32,040	1,32,840
Other overhead		3,60,000
Total overhead		6,14,840
Direct material cost		2,03,400
Direct labour hours	80,000	
Quantity of materials (kg)	30,000	
Set up hours	84	
Number of receipts	1,600	
Machine hours	7,200	

The company used volume-based cost drivers and allocated all materials related overheads based on direct material cost, all machine related overheads based on machine hours and all remaining overheads based on direct labour hours. The company, however, wants to switch over to ABC system for more accurate ascertainment of overhead costs.

You are required to ascertain:

(1) the three overhead rates currently used by the company.

(2) the overhead rates that the ABC system should use for the year ended 31-3-2005 based on new cost drivers. *(Sri Sathya Sai University, B Com., (Hons) October 2006)*

Solution:

(1) Overhead rates currently used by the company:

(*a*) Material related overhead cost rate $= \dfrac{\text{Material related overhead cost}}{\text{Direct material cost}} \times 100$

$= \dfrac{1,22,000}{2,03,400} \times 100 = 60\%$ of Direct material cost

(*b*) Machine hour rate $= \dfrac{\text{Machine related overhead cost}}{\text{Machine hours}}$

$= \dfrac{1,32,840}{7,200} = 18.44$ per hour

(*c*) Labour hour rate $= \dfrac{\text{Other overheads}}{\text{Labour hours}}$

$= \dfrac{3,60,000}{80,000} =$Rs. 4.50 per hour

(2) Using ABC:

(*a*) Material handling $= \dfrac{43,600}{30,000} = 1.45$ per kg. of materials

(*b*) Materials receipts and processing $= \dfrac{54,400}{1,600} = 34$ per number

(*c*) Freight inwards $= \dfrac{24,000}{30,000} = 0.80$ per kg of material received

(*d*) Machine operation $= \dfrac{1,00,800}{7,200} =$ Rs. 14 per machine hour

(e) Machine set up $= \frac{32,040}{84} = 381.43$ per set up hour

(f) Other overheads $= \frac{3,60,000}{80,000} = 4.50$ per hour

Problem 6. A company produces three products *A*, *B* and *C* for which the standard cost and quantities per unit are as follows:

	A	*B*	*C*
Quantity produced (units)	10,000	20,000	30,000
Direct materials per unit (Rs.)	50	40	30
Direct labour per unit (Rs.)	30	40	50
No. of purchase requisitions	1,200	1,800	2,000
No. of set ups	240	260	300

Production overhead split by activity:

– Receiving/Inspecting	Rs.	14,00,000
– Production scheduling /machine set up	Rs.	12,00,000
	Rs.	26,00,000

Number of batches received/inspected = 5,000

Number of batches for scheduling and set up = 800

You are required to prepare a cost statement under Activity Based costing method

(Sri Sathya Sai University, B.Com. (Hons.), October 2005)

Solution:

Working Notes

(1) Calculation of cost driver rates:

Receiving / Inspection $= \frac{14,00,000}{5,000 \text{ (No. of batches received)}}$

= 280 per requisition

Production scheduling/ machine set-up $= \frac{12,00,000}{800 \text{ (No. of batches for scheduling)}}$

= 1500 per set up

(2) Calculation of overhead per unit of receiving:

$A = \frac{280}{10,000} \times 1,200 = 33.6$ or 34

$B = \frac{280}{20,000} \times 1,800 = 25.2$ or 25

$C = \frac{280}{30,000} \times 2,000 = 18.66$ or 19

Calculation of overhead per unit of scheduling:

$A = \frac{1,500}{10,000} \times 240 = 36$

CHAPTER 19

$$B = \frac{1,500}{20,000} \times 260 = 19.5 \text{ or } 20$$

$$C = \frac{1,500}{30,000} \times 300 = 15$$

COST STATEMENT UNDER ABC

	A	*B*	*C*
Materials	50	40	30
Labour	30	40	50
Overhead:			
– Receiving	34	25	19
– Production Scheduling	36	20	15
Total cost per unit	150	125	114

Problem 7. From the following particulars calculate the cost of production and profit under (*a*) Traditional method and (*b*) ABC System

Particulars		*Products* *A*	*B*	*C*	*Total*
Selling price per unit		70	100	80	
Production & sales units		50,000	20,000	10,000	
Raw material usage (units)		5	5	12	
Direct material cost (Rs.)		20	25	10	16,00,000
Machine hours		1.0	1.5	2.5	1,05,000
Direct labour hours		1.0	2.5	1.5	1,15,000
No. of production runs		5	10	20	35
Direct labour cost (Rs.)		6	15	9	6,90,000
No. of Deliveries		5	2	20	27
No. of receipts on the basis of raw materials (units × production runs)		25	50	240	315
No. of production orders		10	5	25	40
Overhead costs:					
Set up activities	1,00,000				
Machine activities	8,00,000				
Receiving activities	5,00,000				
Packing activities	3,00,000				
Engineering activities	6,00,000				23,00,000

Solution:

TRADITIONAL COSTING APPROACH

		A per unit	B per unit	C per unit
Direct Materials		20	25	10
Direct labour		6	15	9

$$\text{Overhead rate} = \frac{\text{Total overhead}}{\text{Total D.L.H.}} = \frac{23,00,000}{1,15,000} = 20$$

		A	B	C
A =	20 × 1 D.L.H	20		
B =	20 × 2.5 D.L.H		50	
C =	20 × 1.5 D.L.H			30
	Total cost	46	90	49
	Profit	24	10	31
	Selling price	70	100	80

ABC COSTING APPROACH

	A	B	C
Direct Material cost	20	25	10
Direct labour cost	6	15	9
Overheads:			
(1) *Set up cost:*			

$$\text{Cost per set up} = \frac{\text{Total set up cost}}{\text{Total no. of production runs}}$$

$$= \frac{1,00,000}{35} = 2857.14$$

$$\text{Set up cost per unit} = \frac{\text{Cost per set up} \times \text{No. of production run}}{\text{No. of units produced}}$$

	A	B	C
$A = \frac{2,857.14 \times 5}{50,000}$	0.29		
$B = \frac{2,857 \times 10}{20,000}$		1.43	
$C = \frac{2,857 \times 10}{10,000}$	–	–	5.71

(2) *Machine cost:*

$$\text{Cost per machine hour} = \frac{\text{Machine cost}}{\text{Machine hrs.}}$$

CHAPTER 19

	A	B	C
$= \dfrac{8,00,000}{1,05,000}$			
$= 7.62$			
Machine cost per unit =			
Cost per machine hour × Machine hrs used per unit			
$A = 7.62 \times 1$	7.62		
$B = 7.62 \times 1.5$		11.43	
$C = 7.62 \times 2.5$			19.05
(3) *Receive cost:*			
Cost receiving order $= \dfrac{\text{Receiving cost}}{\text{No. of orders received}}$			
$= \dfrac{5,00,000}{315} = 1,587.30$			
Receiving cost per unit = $\dfrac{\text{Cost per receiving order} \times \text{No. of receipts}}{\text{No. of units produced}}$			
$A = \dfrac{1,587.30 \times 25}{50,000}$	0.79		
$B = \dfrac{1,587.30 \times 50}{20,000}$		3.97	
$C = \dfrac{1,587.30 \times 240}{10,000}$			38.10
(4) *Packing cost:*			
Cost per packing $= \dfrac{\text{Packing cost}}{\text{No. of Deliveries}}$			
$= \dfrac{3,00,000}{27} = 11,111.11$			
Packing cost per unit = Cost per packing order = No. of deliveries for the product/ No. of units produced			
$A = \dfrac{11,111 \times 5}{50,000}$	1.11		
$B = \dfrac{11,111 \times 2}{20,000}$		1.11	
$C = \dfrac{11,111 \times 20}{10,000}$			22.22
(5) *Engineering cost:*			
Cost per engineering transaction			

$= \dfrac{\text{Engineering cost}}{\text{No. of production order}}$			
$= \dfrac{6{,}00{,}000}{40} = 15{,}000$			
Engineering cost per unit $= \dfrac{\text{Cost per order} \times \text{No. of production order}}{\text{No. of units produced}}$			
$A = \dfrac{15{,}000 \times 10}{50{,}000}$	3		
$B = \dfrac{15{,}000 \times 5}{20{,}000}$		3.75	
$C = \dfrac{15{,}000 \times 25}{10{,}000}$			37.50
Total product cost	38.81	61.69	141.58
Profit/Loss	31.19	38.31	–61.52
Sales	70	100	80

❑❑❑

CHAPTER 19

CHAPTER 20

TARGET COSTING

TYPES OF COST ACCOUNTING SYSTEMS

The cost accounting systems can be grouped under two heads:

1. Conventional method

2. Modern method

1. The following conventional methods have practical applications:
   - Marginal Costing
   - Absorption Costing
   - Process Costing
   - Standard Costing
2. Under the modern methods of costing the following are the types:
   - Target Costing
   - Activity Based Costing

Target Costing—Meaning

Though target costing has been in use for several decades, its development as a key cost accounting approach has been rapid only since 1980s and it became more popular in Japan where it has become a major element in the cost management strategies of firms.

The target cost is maximum allowable cost at a particular point of time and is specified by reference to the successive stages in a product's life cycle. The important feature of the target costing is that it is used only after the product has reached its established production phase. However, it follows the conventional material, labour and overhead classification.

The important feature that is ascribed to target costing is its market orientation *i.e.*, it involves ascertaining of the product cost beforehand by working back from an expected future market price. Thus, it includes the following two fundamental steps:

- Keeping track of the external market information which would affect the organisation decision-making.
- Internal cost planning process that is based on the external market information.

Nature of Target Costing

The target costing makes the system to support the cost reduction process not only in the manufacturing phase of the existing product but also in the developing and designing phase of an entirely new product. It acts as a motivator of much of the creative and innovative activity in an organisation.

The following may be construed as the components in the product target costing:

- Type of product
- Technical specifications
- Technical requirements
- Customer
- Resource consumption (Acquisition price)
- Resource consumption (Cost)

Another important aspect of the target costing is the concept of Design to cost. This concept which originated in 1975, finds its use in the target costing by setting the target cost at level which can be achieved with the greatest possible efforts from planners and it aims at improving the internal capabilities of an organisation.

The approach of design to cost coupled with the market-based target cost provides an excellent basis for cost management.

Concept of target cost

The target cost which influence decisions pertaining to the product's nature and specification, is the cost that is to be achieved during the planning, design and preparation for manufacturing stages of a new product.

This, however, can be viewed from two ways:

1. The costs incurred by the manufacturer, such as planning and design costs, development, manufacturing and sales costs.
2. The costs incurred by the customer for the product, such as maintenance, operating and disposal costs.

It should be noted that various cost objects can be used for the target which in turn aid in the cost reduction process by pinpointing where efforts can be fruitful.

The cost target should be fixed taking into consideration the pros and cons of an organisation. It is because if the target is very high and unachievable it may act as a negative motivator: and if the target is easily achievable it would end up in underutilisation of efficient resources. And so setting up of target is something that comes only from previous experience and experiments.

The target can also be fixed with some element of flexibility or tolerance level, providing cushion for uncontrollable elements.

Methods of establishment of Target costs

Three basic methods are used for setting target costs:

First, there is the subtraction 'method which is based on the price of competitors' product, where the target cost is worked backwards from the market price. The result may represent a very rigorous target, and it may be impossible to achieve.

The second method of setting up the target cost is the addition method which is based on the existing technology and past cost data of the company. Normally it results in achievable targets because it is basically an extension of what has already been happening within the company. The greatest disadvantage of this method is that it is very inward looking and grossly ignores the market situation.

The third method is the integrated method, a mixture of the subtraction and addition methods. However, in practice this integrated method involves many difficult problems.

Subtraction Method

The key factor in establishing the target product cost by the subtraction method is the price charged by the competitors, *i.e.,*

Target Cost = Selling price – Required profit.

Hence the type of target cost varies with the type of expected profit. If the expected profit is only contribution, the target cost is only for the variable costs of the product and if the expected profit is gross profit, the target cost is for the full product cost. In practice, the target cost reflecting the full product cost are by far the most common.

The target cost for the full product cost can be calculated as follows:

$$\text{Target Cost} = \text{Price} - \left\{\text{price} \times \frac{\text{gross profit}}{\text{sales}}\right\}$$

The price will be determined after considering the market situation and the expected competition. Thus, it requires a great deal of market information and also much expertise especially where the product is new, the price differentiation is expected to exist.

While computing it should be noted that the future expected gross profit of that product is linked to the gross profit for that product and also to the corporate level of future expected gross profit. It should also be borne in mind that though the target is fixed based on market situation, it is important that the same should be subject to some modification by the internal cost estimating system.

The full product target cost includes costs which planners and designers cannot control, such as idle capacity costs and many general factory management costs. Therefore, for cost management purpose only the controllable costs are included.

Addition Method

As mentioned earlier, the addition method focuses on internal factors and capabilities.

Here, the target cost is defined as follows:

$$\text{Target Cost} = \text{Cost of main functions} \left\{1 + \frac{\text{number of extra functions}}{\text{number of existing functions}}\right\}$$

This method involves statistical techniques such as regression analysis and factor analysis. The simplest approach is using of the past data on similar products.

Integrated Method

Though the addition method does not take into consideration the market situation, it plays a vital role in the integrated method.

The integrated method involves a process of negotiation once the results of the subtraction and addition methods are available. The basic idea is that the integrated target cost should provide a reconciliation of the two methods and give a final target.

This method calls for an interactive approach whereby all the departments in an organisation reach consensus — thereby fixing individual targets.

Departmentalisation of Targets

For any system to be successful there should be a perfect co-ordination between the various departments. Hence the targets should be subdivided and assigned to the various function of the organisation.

The targets can be decentralised and assigned in the following manner:

- Based on the functions of the product
- Based on cost items
- Based on activities (for example operation and maintenance activity)

This kind of decentralisation of the targets enables easier comparison of the target costs with the actual costs and the assigning of the reasons for the variations, if any.

In a survey conducted in 9 leading corporates the following result was obtained in respect of assignment of Target Cost:

TABLE SHOWING ASSIGNMENT OF COST

	No of companies
Target cost not assigned	1
Target cost assigned	
– individual functions of product	4
– cost items	2
– activities	1
No response	1

Note on assignment based on functions

When the target cost is based on the function, the following points should be taken care of:

- **Defining the functions**

  The function that is to be covered under the target cost should be defined in an unambiguous manner, as, even a good system if installed in ambiguity would lead to dire circumstances.

- **Evaluation of importance of the function**

  Each function has its own uniqueness and hence there should be an extensive evaluation of how important is a function. Based on the importance of a function the critical factor can be fixed and analysed, which would espouse in the installation of the target cost.

- **Assigning target cost based on the functional priority**

  After evaluating the importance of function, the functions are to be arranged based on the order of importance and then the target cost must be assigned to each function area based on the relative importance of each function.

Tools used in the evaluation of function areas

The fundamental tools used in the evaluation of function areas are:

- Ratios
- Monetary values

In both the tools customer has a role to play.

Under the Ratio method, the customer considers all the functions of a product to represent 100% and then they divide this 100% between the different function areas based on a previously assigned factor to individual functions.

In the monetary value approach, the customer estimates the monetary amount which they would be willing to pay for each function. However, both the Ratio and Monetary value approaches overlooks certain factors such as technical aspects, law and regulations.

The best system would be that which has its locus in the customer's evaluation coupled with modification made by the manufacturer to suit his own requirement.

Note on assignment based on cost items

The nature of the target cost is directly related to the complexity of the product under consideration.

The cost items are classified into direct materials, direct labour and overhead costs. However, the classification of cost items will be determined by the degree of control required of actual costs at a later stage in the product life cycle.

It is in the best interest of the designers that there should be minimal subdivision of the target cost as greater the subdivisions, the greater are the restrictions placed on the planners and the less likely it is that new ideas will emerge.

Note on assignment of target cost based on activity

Assigning the target cost based on the activities though theoretically possible, its practical application is limited as it involves a complicated system. Also the activity based costings only feasible for the allocation of the overhead cost.

Conclusion

Target costing is the basis for effective cost management as it provides the goal towards which all cost control efforts are directed.

By interacting with the market situations it ensures that the management accounting system is generating information which will help the organisation to maintain and enhance its competitiveness. It also encourage innovation and creativeness.

❑❑❑